SLOW POSITRON BEAM TECHNIQUES FOR SOLIDS AND SURFACES

FIFTH INTERNATIONAL WORKSHOP

AIP CONFERENCE PROCEEDINGS 303

SLOW POSITRON BEAM TECHNIQUES FOR SOLIDS AND SURFACES

FIFTH INTERNATIONAL WORKSHOP
JACKSON HOLE, WY AUGUST 1992

EDITORS: **ERIC OTTEWITTE**
IDAHO NATIONAL ENGINEERING LABORATORY

ALEX H. WEISS
UNIVERSITY OF TEXAS AT ARLINGTON

American Institute of Physics   **New York**

Authorization to photocopy items for internal or personal use, beyond the free copying permitted under the 1978 U.S. Copyright Law (see statement below), is granted by the American Institute of Physics for users registered with the Copyright Clearance Center (CCC) Transactional Reporting Service, provided that the base fee of $2.00 per copy is paid directly to CCC, 27 Congress St., Salem, MA 01970. For those organizations that have been granted a photocopy license by CCC, a separate system of payment has been arranged. The fee code for users of the Transactional Reporting Service is: 0094-243X/87 $2.00.

© 1994 American Institute of Physics.

Individual readers of this volume and nonprofit libraries, acting for them, are permitted to make fair use of the material in it, such as copying an article for use in teaching or research. Permission is granted to quote from this volume in scientific work with the customary acknowledgment of the source. To reprint a figure, table, or other excerpt requires the consent of one of the original authors and notification to AIP. Republication or systematic or multiple reproduction of any material in this volume is permitted only under license from AIP. Address inquiries to Series Editor, AIP Conference Proceedings, AIP Press, American Institute of Physics, 500 Sunnyside Boulevard, Woodbury, NY 11797-2999.

L.C. Catalog Card No. 94-71036
ISBN 1-56396-267-5
DOE CONF-920829

Printed in the United States of America.

Contents

Preface and Acknowledgments xi
Conference Photograph xiii

**SUMMARY OF THE FIFTH INTERNATIONAL WORKSHOP ON
SLOW POSITRON BEAM TECHNIQUES FOR SOLIDS AND SURFACES** 1
    A. H. Weiss, E. H. Ottewitte, B. W. Augenstein, A. B. Denison, V. J. Ghosh,
    F. M. Jacobsen, K. Krištiaková, K. G. Lynn, R. M. Nieminen, K. A. Ritley,
    and P. J. Simpson

SECTION 1: DEFECTS—DEPTH PROFILING

Calculation of Positron Diffusion in Layered Systems 13
    G. C. Aers, K. O. Jensen, and A. B. Walker
Diffusion and Annihilation of Positrons in Solids 20
    D. T. Britton
Evidence of Deep Vacancy Formation in Fluorine Implanted Silicon 25
    N. B. Chilton and M. Fujinami
A Study of Implantation Induced Defects in SiO_2 31
    M. Fujinami and N. B. Chilton
**Monte Carlo Studies of Positron Implantation in Elemental Metallic
and Multilayer Systems** 37
    V. J. Ghosh, D. O. Welch, and K. G. Lynn
Characterization of Amorphous Silicon 48
    R. A. Hakvoort, A. van Veen, H. Schut, M. J. van den Boogaard,
    A. J. M. Berntsen, S. Roorda, P. A. Stolk, and A. H. Reader
Positron Stopping in Germanium 53
    A. Halec, P. Maguire, P. J. Simpson, P. J. Schultz, G. C. Aers, T. E. Jackman,
    and P. Marshall
**Slow Positron Beam Measurements on GaAs Grown by Molecular Beam
Epitaxy at Low Temperatures** 58
    N. Hozhabri, S. C. Sharma, R. N. Pathak, and K. Alavi
The Effects of Low-Energy Scattering on Positron Implantation 64
    K. A. Ritley, K. G. Lynn, V. Ghosh, and D. O. Welch
**The Effect of Channeling on the Defect Depth Distribution in 110 keV Rb
Implanted Poly-W** 73
    H. Schut, A. van Veen, R. A. Hakvoort, M. J. W. Greuter, and L. Niesen
Positron Experiments with Cadmium Mercury Telluride 78
    C. Smith, P. C. Rice-Evans, and N. Shaw
Characterization of Thin Films by a Pulsed Positron Beam 84
    R. Suzuki, T. Mikado, H. Ohgaki, M. Chiwaki, T. Yamazaki, K. Awazu, A. Matsuda,
    Y. Kobayashi, A. Uedono, and S. Tanigawa

Point Defects in As-Grown and Ion Implanted GaAs Probed by a Monoenergetic
Positron Beam.. 92
 A. Uedono, S. Fujii, L. Wei, and S. Tanigawa
Characterization of SiO_2 by Monoenergetic Positron Beams................... 101
 A. Uedono, S. Watauchi, Y. Ujihira, L. Wei, S. Tanigawa, R. Suzuki,
 H. Ohgaki, T. Mikado, H. Kametani, H. Akiyama, Y. Yamaguchi,
 and M. Koumaru
Defect in Amorphous Silicon Prepared by Ion Implantation.................... 107
 L. Wei, S. Tanigawa, Y. Hiroyama, T. Motooka, and T. Tokuyama
Investigation of Vacancy-Related Defects in Heavily Phosphorus-Doped
Si:P Grown by Plasma Chemical Vapor Deposition 113
 L. Wei, S. Tanigawa, Y. Jia, A. Yamada, and M. Konagai

SECTION 2: DEFECTS—BULK

Positron Lifetime Measurements in γ-Irradiated Polyethylene Under
Different Conditions.. 123
 G. Brauer, Th. Daniel, W. Faust, H. Schneider, and Z. Michno
Free Volume Model for Dielectric Constant of Polymer Films 128
 A. Eftekhari, A. St. Clair, D. M. Stockly, D. R. Sprinkle, and J. J. Singh
Preliminary Results of a Slow Positron Study on an Epoxy Polymer 129
 Y. C. Jean, G. H. Dai, H. Shi, R. Suzuki, and Y. Kobayashi
On the Interpretation of ACPAR Spectra with Respect to Electron Momentum
Density in Real Metals.. 140
 G. Kontrym-Sznajd and A. Rubaszek
The Correlation Between Lifetime and Momentum of e^+-e^- Pair............ 150
 K. Krištiaková, J. Krištiak, O. Šauša, M. Morháč, and P. Bandžuch
Creation and Evolution of Vacancy Clusters in Solids Under Irradiation. Theory
and Computer Simulation... 156
 A. I. Melker
Beam-Based Age-Momentum Correlation Studies of Positronium Spin
Conversion in Paramagnetic Solutions and of Positron Trapping at Defects
in Diamonds 179
 H. Stoll, M. Koch, U. Lauff, K. Maier, J. Major, A. Seeger, P. Wesolowski,
 I. Billard, J. Ch. Abbé, G. Duplâtre, S. H. Connell, J. P. F. Sellschop,
 E. Sideras-Haddad, K. Bharuth-Ram, and H. Haricharun

SECTION 3: SURFACE STUDIES

Positron Re-Emission Studies of the Growth and Annealing Properties of
Epitaxial Palladium Overlayers on Cu(100)................................. 193
 G. W. Anderson, K. O. Jensen, T. D. Pope, K. Griffiths, P. R. Norton,
 and P. J. Schultz

Verification of Focusing from a Hemispherically Shaped Surface 200
    B. L. Brown, T. S. Andrew, M. S. Clarkson, C. S. Sutton, P. Encarnación, A. Denison,
    H. Makowitz, and K. Bundy

A Model for the PAES Cu M_{23} VV Signal Versus Cs Coverage on the Cu(100) Surface 208
    N. G. Fazleev, J. L. Fry, J. H. Kaiser, A. R. Koymen, T. D. Niedzwiecki,
    and A. Weiss

Work-Function and Epithermal Positron Emission from Surfaces 218
    P. G. Coleman, A. Goodyear, and A. P. Knights

Positronium at a Nitric Oxide Monolayer on Graphite 223
    C. E. Haynes and P. C. Rice-Evans

Theory of Surface Adsorbate Analysis by Positronium Formation 227
    A. Ishii

The Temperature Dependence of the Atomic Composition of the Surface of Cu(100) after Deposition of Submonolayer Films of Au 234
    J. H. Kim, G. Yang, S. Yang, K. H. Lee, A. R. Koymen, and A. H. Weiss

Study of Submonolayer Films of Au on Cu(100) Using Positron Annihilation Induced Auger Electron Spectroscopy 239
    K. H. Lee, G. Yang, A. R. Koymen, and A. H. Weiss

Low Energy Electron and Positron Diffraction from Surfaces. What You Learn. How They Differ 246
    D. L. Lessor, K. F. Canter, and C. B. Duke

Annihilation Characteristics for Positrons Trapped at the Surfaces of Simple Metals 249
    A. Rubaszek, A. Kiejna, and S. Daniuk

Physisorbed Surfaces Studied by Positron Annihilation Spectroscopy 259
    S. J. Wang

Re-emitted Positron Spectroscopy of Cobalt and Nickel Silicide Films 264
    B. D. Wissman, W. E. Frieze, and D. W. Gidley

Study of the Structure of the Rh/Ag Surface Using Positron Annihilation Induced Auger Electron Spectroscopy (PAES) 274
    G. Yang, S. Yang, J. H. Kim, K. H. Lee, A. R. Koymen, and A. H. Weiss

Design of an Electrostatic Positron Beam for Background-Free High-Resolution Auger Lineshape Studies 279
    H. Q. Zhou, S. Yang, A. R. Koymen, and A. H. Weiss

SECTION 4: INTENSE POSITRON BEAM FACILITIES

A Proposed Intense Slow Positron Source Based on ^{58}Co 289
    B. L. Brown, A. Denison, H. Makowitz, D. Gidley, B. Frieze, H. Griffin,
    and P. Encarnación

Positrons at CEBAF 296
    W. J. Kossler, A. J. Greer, and L. D. Hulett, Jr.

The Intense Slow Positron Source Concept: A Theoretical Perspective on a Proposed INEL Facility 305
    H. Makowitz, J. D. Abrashoff, W. H. Landman, R. K. Albano, T. Tajima, and J. D. Larson

Report on Positron Spectroscopy for the BESAC Panel on Neutron Sources ... 335
    A. P. Mills, Jr.

The Design of a Nuclear-Reactor-Based Positron Beam for Materials Analysis 354
    A. van Veen, H. Schut, P. E. Mijnarends, L. Seijbel, and P. Kruit

Development of a High Intensity Low Energy Positron Beam 365
    W. B. Waeber, M. Shi, D. Taqqu, U. Zimmermann, D. Gerola, F. Hegedüs, and L. O. Roellig

SECTION 5: POSITRON MICROSCOPES

Brandeis Second Generation Positron Reemission Microscope 385
    K. F. Canter, V. Dharmavaram, A. G. Smirnov, S. A. Wesley, K. H. Wong, R. Xie, G. R. Brandes, and A. P. Mills, Jr.

An Overview of the Michigan Positron Microscope Program 391
    D. W. Gidley, W. E. Frieze, T. L. Dull, G. B. DeMaggio, E. Y. Yu, H. C. Griffin, M. Skalsey, R. S. Vallery, and B. D. Wissman

SECTION 6: FUTURE USES OF POSITRONS

Why Antihydrogen—and Not Just Bare Antiprotons? 401
    B. W. Augenstein

Formation of Electron-Positron Plasmas in the Laboratory 422
    H. Boehmer

SECTION 7: SLOW POSITRON BEAMS—ARTS AND TECHNIQUES

Extraction of Slow Positrons from the Magnetic Field 437
    T. Akahane

Electrostatic Lenses and Beam Optics Calculations 441
    G. Amarendra and B. Viswanathan

Slow Positron Beam Set Up at Kalpakkam—A Progress Report 452
    G. Amarendra, B. Viswanathan, G. Venugopal Rao, K. V. Thomas Kutty, B. Purniah, and K. P. Gopinathan

The Slow-Positron Beam Facility at the University of Hong Kong 462
    C. D. Beling, S. Fung, H. M. Weng, C. V. Reddy, S. W. Fan, Y. Y. Shan, and C. C. Ling

Heating of a Thin Tungsten Foil for Efficient Positron Moderation in a 4.2 K Environment 480
    B. L. Brown, T. S. Andrew, M. S. Clarkson, S. K. Makoski, S. Parikh, and S. Vemuri

Accumulation and Bunching of Positrons 487
    B. Ghaffari, R. S. Conti, and T. D. Steiger

Monte-Carlo Simulation of the Positron Production at a LINAC-Based Slow Positron Beam 496
    D. Segers, M. Dorikens, J. Paridaens, and L. Dorikens-Vanpraet

Optical Design of a Remoderation Section 502
    L. J. Seijbel, P. Kruit, J. E. Barth, A. van Veen, and H. Schut

Theoretical Simulation of Positron Premoderation 506
    M. Shi, W. B. Waeber, D. Taqqu, F. Foroughi, and J. Arkuszewski

A Low Energy Positron Flux Generator for Microstructural Characterization of Thin Polymer Films 516
    J. J. Singh, A. Eftekhari, and T. L. St. Clair

An Intense Pulsed Positron Beam and its Applications 526
    R. Suzuki, T. Mikado, H. Ohgaki, M. Chiwaki, T. Yamazaki, and Y. Kobayashi

Development of the Slow Positron Beam System as a Commercial Prototype 535
    H. Ueno, O. Azuma, T. Ogawa, T. Sato, T. Kawaratani, and O. Hara

Retarding Field and Timing Measurements of Positrons from a W[100] Foil 542
    P. Willutzki, J. Störmer, D. T. Britton, G. Kögel, P. Sperr, R. Steindl, and W. Trifshäuser

The ORNL Slow Positron Facility and Quadratic-Potenital Time-of-Flight Mass Spectrometer 550
    J. Xu, L. D. Hulett, Jr., T. A. Lewis

SECTION 8: FUNDAMENTAL PHYSICS STUDIES

Observation of Resonance-Like Structures in Positron-Krypton Elastic Scattering 559
    L. Dou, W. E. Kauppila, C. K. Kwan, and T. S. Stein

Measurements of Positronium Formation Cross Sections for Positron-Potassium Atom Scattering 562
    L. Jiang, W. E. Kauppila, C. K. Kwan, S. P. Parikh, T. S. Stein, and S. Zhou

Experimental and Monte-Carlo Studies of Electron and Positron Backscattering 564
    G. R. Massoumi, W. N. Lennard, P. J. Schultz, A. B. Walker, and K. O. Jensen

Search for Resonance-Like Structure in the Total Scattering of Positrons by Argon and Krypton 571
    D. Przybyla, C. K. Kwan, R. A. Lukaszew, W. E. Kauppila,
    and T. S. Stein

Experiments with Low Energy Positrons 574
    H. Schneider, I. Tobehn, M. Rückert, U. Brinkmann,
    and R. Hippler

Formation of Positronium Compounds in Slow Positron Beams 579
    D. M. Schrader

Energy Loss of Subionizing Charged Particles in Polar Media 587
    S. V. Stepanov

Measurements of Positronium Formation Cross Sections for Positron-Argon Scattering 599
    S. Zhou, L. Jiang, W. E. Kauppila, C. K. Kwan, S. P. Parikh,
    and T. S. Stein

List of Participants 603
Workshop Program 613
Author Index 621

PREFACE AND ACKNOWLEDGMENTS

The editors of this volume would first of all like to thank the participants of SLOPOS5 whose contributions made the workshop an outstanding success and made these proceedings possible. The international aspect of this workshop was heightened by an expanded spirit of camaraderie as manifested by the sharing of rooms, transportation to, from and during the meeting, and other resources, especially with those from countries with limited means. The organizers particularly sought to assist Central and East European scientists who had, for the first time, the legal freedom to attend, but not yet the financial freedom. These outreach efforts were also extended to some two dozen graduate students. To those who gave of themselves, this workshop surely had extra meaning.

The recipients of assistance (scientists and graduate students alike) physically contributed to the workshop in terms of papers and/or labor (copying, registration, chauffeuring, coffee, lunch couriers, etc.). This facilitated such amenities as low-cost wholesome meals, airport transportation, and hard copies of the transparencies preceding each talk. The latter service minimized the diversion of note-taking during talks, thereby enhancing concentration and understanding.

The *a priori* hard copies also complemented a key scientific feature of the workshop: introductory tutorials to each scientific session. One easily anticipates the value of these tutorials to graduate students and to scientists from non-first world countries. However, they actually provide "Continuing Education" to all. It is to the strong credit of these tutors that they provided excellent tutorial reviews to the positron community with virtually no remuneration. For most, their only incentive was that the talks did not have to be written up. We thank you.

Other somewhat novel aspects of the workshop included

1. use of horizontal (on-a-table) poster displays. This meant some bending over, but simultaneously allowed one to scan the entire panorama of two assembly rooms for people you wanted to catch for discussions or whatever. (The astute logic leading to this innovation was the absence of suitable wall space and an abundance of tables.)
2. the finding of a suitable convention center (The Virginian) with economical lodgings in a resort area at peak season: A double room cost $55 total.
3. the full-time commercial rental of a copying machine (should have rented two of them).
4. the import of speakers from outside the community to provide scientific and technical challenges (and possible sources of funding) to the positron community. This was not fully accomplished due to lack of available time.
5. a lunchtime session to address the needs of East European scientists facing diminishing support. The participants concluded on the usefulness of small-funded collaborations with Western institutions.

We mention all the above experiences here as a collection of ideas for possible consideration of others planning workshops.

In a time of decreasing resources, putting this meeting together became more difficult than anticipated. In this regard we would especially acknowledge the following:

1. financial support from the Physics Department, the College of Science and the Research Enhancement Fund of the University of Texas at Arlington; EG&G Idaho, Inc.; the Koshkiuczko Foundation; and the Idaho State University Physics Department.
2. guidance from the Idaho State University conference planning group in the selection of Jackson, Wyoming and The Virginian for the workshop location.
3. an early "seed money" grant from duPont New England Nuclear Corporation.

4. a scientific talk by Dr. Mark Boyce of the University of Wyoming on the ecology of Yellowstone National Park.
5. dedication of the academic positron leaders from the U. S. university communities to bring their graduate students to the meeting en masse.
6. the reflection of God's grace and glory in Yellowstone National Park.

Finally, we would like to recognize the exhaustive efforts of the Conference Secretary Elisabeth Leonard in helping to make this workshop and the proceedings so successful. We acknowledge the editorial contributions of Juli Hobdy who proofed the entire manuscript, catching some significant errors. However, only major errors were corrected: most papers are published as submitted. We must also thank our spouses and families for keeping the home fires burning while we had such a rewarding experience.

Eric Ottewitte, Idaho Falls, Idaho
Alex H. Weiss, Arlington, Texas

We close this Preface with the inspired thoughts of one participant:

> *Positrons are like people*
> *Who diffuse in all directions,*
> *But every person can be trapped*
> *So S-factor needs correction.*
>
> *What's the sense of any life trajectory*
> *Which is leading us through disturbances?*
> *Don't forget all the good and beautiful*
> *Which is experienced despite the disturbances!*
>
> *I'm trapped by Yellowstones's beauty,*
> *I'm trapped by my host's heart attraction.*
> *It will be my ever life duty*
> *To promote such science interaction.*

- Alexander Melker
St. Petersburg Technical University

SUMMARY OF THE FIFTH INTERNATIONAL WORKSHOP ON SLOW POSITRON BEAM TECHNIQUES FOR SOLIDS AND SURFACES

A. H. Weiss, E. H. Ottewitte, B. W. Augenstein, A. B. Denison,
V. J. Ghosh, F. M. Jacobsen, K. Krištiaková, K. G. Lynn,
R. M. Nieminen, K. A. Ritley, and P. J. Simpson

INTRODUCTION

The Fifth International Workshop on Slow Positron Beam Techniques for Solids and Surfaces (SLOPOS5) took place 5-10 August 1992 in Jackson Hole, Wyoming. One hundred thirty scientists, including two dozen students, from sixteen countries attended. We attempt here to briefly summarize the contributions to the workshop. The selection of topics and research inevitably reflect the interests and expertise of the authors. Much important research, presented at the workshop, has not been included in this short review for which we apologize to the reader.

As evidenced by the contributions to the workshop, the application of positron beams to studies of solids and surfaces has reached a certain level of maturity. For example we are now able to achieve not only qualitative but also quantitative understanding of the slowing down and implantation of positrons. Agreement between theoretical simulation and experiment, especially in the backscattering geometry, is very encouraging. Such detailed studies confirm that no implantation profile is universal. As accurate cross-section data become available, Monte Carlo codes can provide good implantation profiles for an increasing array of materials. Calculation of the transport of thermal positrons using the time-dependent diffusion equation is now tractable. Positron energetics have been elucidated for a wide variety of inorganic materials. Electronic structure calculations for elemental solids yield both the electron and positron chemical potentials to an accuracy of 0.1-0.2 eV for many metals, semiconductors and insulators. These chemical potentials provide the quantitative basis for work functions, affinities, deformation potential constants and Ps formation potentials. Understanding of scattering rates and diffusion mechanisms for thermal positrons has markedly improved. Similarly, studies of mobilities in semiconductors yield detailed and systematic information.

The organization of the rest of this paper corresponds to the major topics covered at the workshop. Section II summarizes depth profiling including developments in Monte Carlo simulations and semiconductor studies. Section III pertains to surface studies including the techniques of Low Energy Positron Diffraction, Re-emitted Positron Spectroscopy, and Positron-annihilation-induced Auger Electron Spectroscopy. Section IV analyzes progress in positron beam techniques including the development of intense positron beams, positron microscopes, and pulsed beam techniques. Section V covers experiments aimed at fundamental measurement of such quantities as scattering and ionization cross sections. Section VI contains concluding comments.

DEFECT PROFILING

A large fraction of the workshop papers were related to defect depth profiling. This reflects the fact that this technique, which has established itself as a powerful tool for characterizing near-surface materials, is one of the principal uses of low-energy positron beams. Defect profiling is accomplished by measuring changes in annihilation characteristics as a function of positron beam energy. This section will begin with a discussion of Monte Carlo techniques for determining positron stopping profiles. We will then discuss some examples of measurements of the Doppler broadening of annihilation (using Ge detectors) which account for the majority of defect profiling to date. There is a growing body of work on beam-lifetime measurements (see e.g. Suzuki et al., Willutzki et al., Stoll et al.). Since Doppler broadening can only provide averaged information on various defect species in a sample, beam-lifetime measurements, from which more detailed information can be extracted, hold great promise. Progress in the use of pulsed beams to perform defect profiling using positron lifetime measurements will be discussed in section IV.

Monte Carlo Calculations and Benchmarks

There has been considerable progress in the area of Monte Carlo routines to estimate implantation profiles for systems in which experimental data are not available, including heterostructures (Aers et al., Ghosh et al.). This effort is supported by related experimental work on implantation profiles (Halec et al), backscattering (Massoumi et al, Albrecht et al., Nieminen), transmission through thin foils (Schneider et al), and theoretical work on energy loss (Stepanov et al.), resonance trapping at defects (Dryzek), and trapping on metal surfaces (Rubaszek et al.). Monte Carlo simulations, combined with diffusion-based programs POSTRAP (G.C. Aers, SLOPOS4) and VEPFIT (A. van Veen et al, SLOPOS4) provide us with adequate theoretical understanding to model the experimental Doppler-broadening data. What is still lacking for truly quantitative analysis is accurate values for trapping rates and S-parameters for various defects.

Existing Monte-Carlo programs can track the energy of a positron from the initial keV implantation energies down to final endpoint energies of about 20 eV, at which point the positron is considered to be stopped. Ritley et al. pointed out that some error in the calculations may result from neglecting the final low-energy scattering (with conduction electrons and phonons). This especially pertains to positrons incident at low kinetic energies. Also discussed were the role of epithermal positrons (Britton) and the difficulty in developing time-dependent diffusion equations which can treat these effects. Calculation of epithermal positron transport requires a Boltzmann equation solution, which is much more demanding numerically. Even though the microscopic scattering processes and cross sections are well understood, the difficulty of the analysis presently limits the usage of epithermal positrons in surface and interface studies.

A number of experiments were reported (e.g. Chilton and Fujinami, Fujinami and Chilton, Uedono et al, Wei et al) on defects introduced by ion implantation, which satisfy dual goals: using positrons to provide new insights on a problem of considerable industrial significance, and using ion irradiation to provides us with a systematic way to introduce defects and thereby further our understanding of the defect profiling technique.

Compared with SLOPOS4, the present meeting saw greater use of correlations between positron results and those obtained from other longer-established techniques. Examples include the use of electron spin resonance (Fujinami and Chilton), Raman scattering (Hakvoort et al, Wei et al), Rutherford backscattering (Hakvoort et al), electron microscopy (Wei et al), and chemical etching to check for self-consistency in the fits to the positron data (Chilton and Fujinami). It is hoped that a growing database of such experiments in the literature can ultimately provide progress towards more quantitative analysis, as discussed above.

Discussion at SLOPOS5 revealed a further need for Monte-Carlo models: to help experimentalists extract accurate information from systems such as multilayers and buried interfaces. These are technologically important systems which can be probed with positrons. First steps toward understanding the behavior of positron stopping (Ghosh, Welch and Lynn) and diffusion (Aers, Jensen and Walker) in these systems were presented.

Semiconductor Experiments

Defect control in epitaxial layers on semiconductor substrates is critical in the fabrication of semiconductor devices. Defect profiling using positron beams is establishing itself as a useful adjunct in the characterization of defects in these systems. In particular, Lynn and Tanigawa et al. presented results indicating that Doppler-broadening measurements provide an extremely powerful tool for profiling defects in over-layers grown on semiconductor substrates using chemical vapor deposition (CVD) or molecular beam epitaxial (MBE) techniques.

Hakvoort et al. discussed how density, differences in structure, and presence of incorporated gas in various types of plasma-enhanced CVD amorphous-silicon affects the behavior of positrons. Fujinami used slow positron spectroscopy to study oxygen-related defects in Si implanted with high doses of H, B, O, F, P and As ions. Uedono et al. presented results of an investigation of vacancy-type defects in as-grown and ion-implanted GaAs by measuring S parameters. Wei et al. discussed studies of vacancy-related defects in heavily phosphorus-doped Si using variable energy positrons. Szeles measured dependence of the S parameter upon positron incident energy for the SiO_2/Si system to determine interface charge and trap densities. Tanigawa et al. presented results of studies of the formation of silicides via the thin-film deposition of transition metals on Si. Nielsen and Heald discussed the use of positrons to study defect character and density of both metals in bi-metallic layers of Al on Co, Cr, Cu, Mn, Ni, and Pd. Simpson et al. presented results from the study of the annealing of Si-implanted GaAs using the Doppler-broadening annihilation technique. Fujinami et al. compared S parameter measurements and ESR studies of ion implanted SiO_2.

SURFACE STUDIES

Most of the positron surface studies employed one of four methods: Low Energy Positron Diffraction (LEPD), Re-emitted Positron Spectroscopy (RPS), Positronium Spectroscopy, and Positron-annihilation-induced Auger Electron Spectroscopy (PAES). More details on research reported at the workshop in these areas is given below under the appropriate headings. In addition, positron-surface sticking has attracted recent interest. At low energies the behavior of the sticking (trapping) rate is somewhat controversial, and definitive experimental studies are still lacking. The results of Mills et al. for Ps desorption point to strong-coupling effects at low energies. Ishii presented a theoretical analysis of a new method of positron surface study: the use of Ps emission to study randomly adsorbed adatoms.

LEPD

4 SUMMARY OF THE WORKSHOP

Presentations of the work of Canter et al. and Lessor et al. on GaAs (110) showed that LEPD agrees better between calculations and experiment than does LEED. LEPD and LEED showed similar structure for GaAs(110) with interesting differences near the limits of experimental uncertainty. LEPD R-factors were lower and the visual quality of its fits was better. The differences between LEPD and LEED are attributed to the following: (1) positrons encounter a repulsive nucleus and attractive screening electrons; electrons encounter an attractive nucleus and repulsive screening electrons; (2) LEPD contains no exchange potential term; (3) inelastic processes differ in phase space availability: unlike electrons, the positrons do not have a filled Fermi sea of states, forbidden to inelastically scattered positrons. Hence, the attenuation length for positrons is, in general, shorter than that for electrons at low energies. As a consequence, LEPD exhibits weaker back-diffraction from a single layer at most energies and for most beams and multiple interlayer scattering is less important. As a result, LEPD computations are less sensitive to (1) the precision of computed phase shifts, (2) anisotropy of inelastic attenuation, and (3) anisotropy of scattering potentials. Differences in surface structures as determined by LEPD versus LEED are not yet understood. They present a challenge and an opportunity to examine questions which affect the ultimate accuracy of both LEED and LEPD.

RPS

Anderson et al. used RPS to study the growth and annealing properties of epitaxial palladium overlayers on Cu(100). They found that, in terms of the positron work function ϕ^+, a 1.5 ML- (monolayer) surface resembles pure Pd. A plateau in ϕ^+ near 0.5 ML indicates a change in the growth mode. In examining yield vs coverage, they observe two changes in the growth mode: (1) completion of the alloy layer at 0.5 ML, and (2) transition from interfacial layers to bulk Pd at 1 ML. They also observed that Pd dissolution facilitated annealing of the alloy surface. Positron re-emission yield has great potential as a structural probe for metal overlayer systems.

Wissman, et al. used RPS to study $CoSi_2$ and Ni/Si(100) films. This was the first systematic investigation of the positronic properties of Co and Ni silicide films grown in situ. The various silicide phases (M_2Si, MSi, MSi_2) all have negative ϕ^+ with S having surprisingly large variation. Thus each phase is easily distinguishable in an RPS spectrum. The growth of $NiSi_2$ from an Ni/Si film was investigated using a form of depth-profiling. Studies of $CoSi_2$ films using the thermal expansion technique yielded values of the deformation potential. The short e^+ diffusion length (~15 nm) in these films was attributed to defects.

PAES

The University of Texas at Arlington group presented a number of studies of the growth, inter-diffusion and alloy formation of ultrathin metal layers deposited on metal substrates. PAES along with Electron-induced Auger Electron Spectroscopy (EAES) and LEED were used to study the growth and inter-diffusion/alloying behavior of vapor-deposited Au and Pd films on Cu(100) and Rh on Ag(100). Plots of PAES intensities versus deposition time indicate that the PAES signal saturates at 1 ML, demonstrating that PAES is extremely surface selective. Strong non-linearities in the PAES intensity vs. deposition time plots indicate over-sampling of the adsorbate atoms by the positrons at submonolayer coverages. Inter-diffusion and/or alloying was clearly seen in all systems studied as the films deposited at -100°C were warmed to room temperature and slightly above.

The recent work of Koymen et al. indicated the power of the PAES technique to give detailed information about the otherwise elusive positron surface state. Progress was reported on a high-resolution PAES apparatus (being built at the University of Texas at Arlington) which will permit the exploitation of the ability of PAES to produce essentially background free spectra to obtain accurate Auger line-shapes.

PROGRESS IN POSITRON BEAM TECHNIQUES

Intense Positron Beams

There is world-wide interest in building next-generation beams with high intensity, good spatial resolution and timing capabilities. Groups in Japan, INEL, CEBAF, Delft, Munich and PSI are designing large-scale facilities for extremely intense positron beams. Research teams at Brandeis, Delft and Michigan are actively building positron microscopes and microprobes which will greatly benefit from the availability of high flux beams at these facilities. Such beams will also make possible and/or practical studies of electron-positron plasmas, antimatter production and antihydrogen experiments, ACAR and positron lifetime: microanalytical tools for surface and interface studies.

Makowitz et al. presented an analysis of the proposed INEL Intense Slow Positron Source (ISPS) concept. They showed the feasibility of producing a monoenergetic 5-keV positron beam of $\geq 10^{12}$ e$^+$/s on a <0.03 cm diameter target using existing INEL reactor facilities. They proposed using first a 6-cm and then a 30.0-cm diameter, large-area source dish, moderated with thin W films or solid Ne. The ^{58}Co source will be made by irradiating Ni with fast neutrons in the INEL Experimental Breeder Reactor. A. van Veen et al. presented the design of an intense (> 10^8 e$^+$/s) positron beam facility currently being developed at the Delft University of Technology. This beam will be extracted from a ^{64}Cu source inside the Delft University of Technology research reactor, produced by irradiation of an array of Cu cylinders mounted adjacent to the reactor core. W. Triftshäuser presented another novel approach to constructing a reactor-based high flux beam. Thermal neutron capture by ^{113}Cd would generate MeV gamma rays. An adjacent tungsten layer would then convert the gammas by pair production to e$^+$ and also moderate them to slow e$^+$ energies.

Kossler et al. presented Monte-Carlo calculations of low energy positron yields and problems of target design in connection with their concept for utilizing the beam dump of a proposed Free Electron Laser (FEL) to be built at CEBAF. They estimate that ~2.5 x 10^{10} slow positrons per second could be produced using 1.0 mA of 40-MeV electrons which will be incident on the FEL beam dump. The electron beam structure of 1 to 2 ps pulses repeats at the rate of 7.5 MHz. This high repetition rate would result in a quasi-DC positron beam even before any additional positron pulse stretching. That would make the CEBAF beam attractive for ACAR and other measurements limited by peak pulse counting rates.

Waeber et al. from PSI described development work on an intense positron beam that will make use of a scheme for obtaining an extremely high moderation efficiency, ~50%. The scheme uses a superconducting magnet to repeatedly pass the high energy β$^+$s through a thin foil until their energy decreases to ~10 keV. The premoderated positrons can then be further moderated to a few eV with very high efficiency using a cryogenic neon moderator. The PSI group estimates that this arrangement will be able to achieve ~10^{10} e$^+$/s fluxes from relatively modest (~1 Curie) sources.

K. G. Lynn reported on progress in achieving a high flux beam of ~10^8 e$^+$/s at Brookhaven National Laboratory. They will use a ^{64}Cu source produced at the BNL reactor and a solid neon moderator. R. Howell described operations of the high flux (10^8-10^9 e$^+$/s) linac-based positron beams at Livermore and the possibility of upgrading the system to achieve fluxes of 10^{10} e$^+$/s. T. Kurihara described progress in connection with plans for a linac-based Japanese positron factory.

Positron Microscopes

The development of the positron microscope continues to provide an exciting frontier in positron spectroscopy. The positron re-emission microscope (PRM) is based on a proposal of Hulett et al. to image thermalized positrons that are spontaneously re-emitted from materials having negative positron affinities. Because of its positive charge, the positron preferentially seeks out voids and defects. The possibility of non-destructively and efficiently detecting small features, particularly micro-defects, allows a unique new tool to investigate materials for both fundamental and applied studies. The parallel development of intense slow positron beams should enable the examination of ~10-nm features in a reasonable amount of time.

Several institutions are actively developing designs of positron microscopes. The present PRM at Brandeis University can operate in the transmission mode with a maximum magnification of 4450X and 80-nm resolution. Canter et al. project a magnification of 50,000X for their second generation microscope. The University of Michigan has been operating a PRM in the reflection mode achieving a maximum magnification of 56X and 2.3-μm resolution. A second generation version, currently under construction, will initially magnify 2200X. The University of Michigan group is currently collaborating with the Idaho National Engineering Laboratory (INEL) to build an intense positron beam using a reactor-produced ^{58}Co source. They plan to develop a third-generation positron microscope capable of 10-nm resolution. Seijbel et al. reported on progress in the construction of the scanning positron microprobe which will make use of a high flux reactor-based positron beam at Delft University of Technology. They anticipate achieving a 10^6 e$^+$/s beam with energy variable from 0.2-20 keV and a smallest diameter of 100 nm. It will be used in conjunction with a Ge detector to perform defect depth profiles which are spatially resolved parallel to the surface. The University of East Anglia also reported progress on their program of positron microscope development. Their first-generation microscope will use reflection mode geometry and two stages of brightness enhancement. The integrated incident and re-emitted beam optics will achieve a resolution of 1 μm and a magnification of 1700X.

Other Developments in Positron Beams

Suzuki and coworkers have developed an intense slow positron pulsing system which is able to measure positron lifetime of thin films. It can do this in the near-surface, interface or surface regions with high count rate, high peak-to-background ratio, wide time range and good time resolution. The system generates a short-pulsed positron beam of variable energy (0.4-25 keV) and variable pulse period from the ETL intense positron beam. Results of this variable-energy, positron lifetime spectroscopy include lifetime data of ~9 ns for an amorphous Si-H mixture, and ~ 1.3-8 ns for diamond-like carbon films. Also shown were data for polyethylene terephthalate films and SiO_2 layers on Si and MOS structures.

Jean et al. used pulsed beam techniques to obtain the positron lifetime spectra of an epoxy polymer as a function of temperature above and below the glass transition. They used positron energies of 0-5 keV at the ETL (Electrotechnical Laboratory at Tsukuba) slow positron facility. Willutzki et al. reported on progress in the development of a new system for prebunching and bunching the positrons from a radioisotope-based positron beam. The new system achieves higher efficiency, shorter pulse width and better resolution.

Ueno et al. reported on progress in the development of a prototype positron beam at Ishikawajima-Harima Heavy Industries with an eye toward possible commercial manufacture of "turn-key" variable-energy positron beams for defect studies.

A number of papers pertained directly to production and use of antihydrogen. The representatives of Novosibirsk research, who have been working for some time in this and related fields, brought welcome attention to their recent progress. Active cooperation between the positron and antiproton experimental teams is essential to progress in the sophisticated issues surrounding production of complex forms of antimatter.

FUNDAMENTAL PHYSICS

Many fundamental studies of scattering compare results of similar experiments in which positrons and electrons are used as projectiles. The availability of these two similar but distinct probes provides a unique opportunity in that it is possible to, in effect, experimentally turn on and off different interaction terms in the scattering cross section. At this workshop, such comparisons were presented for backscattering coefficients, K and L shell ionization and differential cross sections in atomic scattering. New results were also presented for Positronium (Ps) formation in e^+ collisions with gas atoms. The further possibility of the formation of Ps compounds was also discussed.

Massoumi et al. reported experimental and Monte-Carlo studies of e^+/e^- backscattering on different thick targets at 35 keV. In the range Z = 4-82, the total backscattering coefficients, η_{tot}, measured for e^- and e^+ show that the ratio η^-/η^+ equals approximately 1.3, in accord with earlier measurements. However, the new data may indicate a weak decrease of this ratio with increasing Z. It is encouraging to observe that the excellent agreement between parameter-free Monte Carlo calculations and experiment also holds for the examples of the angular and doubly-differential back scattering distribution presented in this work.

Schneider et al. showed new and preliminary results on the energy loss of keV e^+ passing through a 0.25-μm thick Al foil and compared them to the tabulated stopping power. In the same paper, previously published results on K and L shell ionization show that the e^- cross-section exceeds that of the e^+ at impact energies up to about four times the threshold energy. Double ionization by e^- and e^+ follow a similar trend with the e^- cross-section being larger than that of e^+. These observations are of interest in light of the fact that the stopping power for e^+ is greater than that for e^-.

The Detroit group presented several contributions on e^+-gas scattering. In one paper they show that data for elastic differential scattering cross-sections for e^+ in Kr suggest a resonance structure at 25 and 200 eV. The same group observed similar effects for e^+ in Ar, although at a different energy. At present, the origin of these structures is unknown. This group also presented results on the Ps formation cross-sections in e^+ collisions with K and Ar although only lower and upper limits were given. In view of the great theoretical interest in this collision channel, accurate data for Ps formation would be most welcome.

Schrader discussed the possibility of forming Ps compounds in rearrangement collisions of e^+ and molecules. A number of such compounds are known to be chemically stable. Examples are PsH, PsCl and the other Ps halides although only PsH has been produced in vacuum. Measurements of the binding energies of these Ps compounds would be important as a challenge to theory. Hulett et al. presented some very interesting data concerning the ionization of large organic molecules by slow positrons at energies above and below the energy required to form Ps. H. Boehmer discussed designs for apparatus for the creation of electron-positron plasmas. R. Howell et al. discussed apparatus for the creation of cold positron plasmas.

CONCLUSIONS

The conference was highly successful. It is clear from the papers that substantial progress has been made in the areas of defect profiling including improvements in theoretical modeling and the beginning of the realization of the power of pulsed variable-energy beam lifetime techniques. Contributions in the area of positron interactions with surfaces indicate the growth in the applications of techniques such as LEPD, RPS, and PAES which were at a more developmental stage at the time of SLOPOS4. Plans presented for the development of intense positron beams and the experiments that these beams will make practical, including positron microscopes, evoked great excitement: they indicate that great leaps of progress in the application of positron beams are possible in the near future. There has also been substantial incremental progress in the development of laboratory-based positron beam techniques. We look forward to SLOPOS6.

BIBLIOGRAPHY

All references in this paper are to talks presented at the workshop as well as to papers included in these proceedings (see Table of Contents). For the sake of completeness and continuity, the reader is also referred to the following related publications and events:

SLOPOS Series

(SLOPOS1)   Asko Vehanen (ed.), *Proceedings of the International Workshop "Slow Positrons in Surface Science,"* Pajulahti Sports Institute, Finland, June 25-29, 1984, Helsinki University of Technology Laboratory of Physics Report 135 (1984), ISSN 0359-6214

(SLOPOS2)   D. G. Reiche and W. B Yelon (eds.), *Proceedings of the MURR Slow Positron Beam Workshop*, October 2-4, 1985, (not officially published).

(SLOPOS3)   P. G. Coleman and A. B. Walker (eds.), *Proceedings of the International Workshop on Slow Positron Beams for Solids and Surfaces*, Norwich, England, 1986, (unpublished).

SLOPOS4   P. J. Schultz, G. R. Massoumi, P. J. Simpson (eds.), *Positron Beams for Solids and Surfaces, London, Ontario, Canada 1990*, 4th International

	Workshop on Slow-Positron Beam Techniques for Solids & Surfaces, July 3-6, 1990, New York: American Institute of Physics Conference Proceedings 218, 1990 (ISBN 0-88318-842-2).
SLOPOS5	E. H. Ottewitte, A. Weiss (eds.), *Proceedings of the Fifth International Workshop on Slow Positron Beam Techniques for Solids and Surfaces, Jackson Hole, Wyoming, 6-10 August 1992*, New York: American Institute of Physics Conference Proceedings, 1993 (this publication)
SLOPOS6 (future event)	6th International Workshop on Slow-Positron Beam Techniques for Solids & Surfaces, to be held in Makuhari, Japan, 18-22 May 1994 (immediately preceding the ICPA-10 meeting in Beijing, which starts 23 May 1994).

Intense Positron Beams

E. H. Ottewitte, W. P. Kells, (eds.), *Intense Positron Beams, Proceedings of a Workshop at INEL, 16-17 June 1987*, World Scientific Press, Singapore (ISBN 9971-50-593-2), 1988, 300p.

Other Recent Publications

1. L. Hulett, et al., *Report of 9-12 September 1992 Workshop at Palm Springs on the Application of Positron Spectroscopy to Materials Science,* to be published.
2. A. Ishii (Ed.), *Positrons at Metallic Surfaces*, Solid State Phenomena vol 28/29, Aedermannsdorf Switzerland: Trans Tech Publications Ltd, 1992, 388 pp. (ISBN 0-87849-648-3)
3. Zs. Kajcsos and Cs. Szeles, *Positron Annihilation*, Proceedings of the 9th Intl. Conf., Szombathely, Hungary, 1991, Materials Science Forum vols. 105-110, Aedermannsdorf Switzerland: Trans Tech Publications Ltd, 1991, 2146 pp., 3 vol. set, (ISBN 0-87849-636-x)

SECTION 1

DEFECTS—DEPTH PROFILING

CALCULATION OF POSITRON DIFFUSION IN LAYERED SYSTEMS

G.C. Aers
National Research Council of Canada, Montreal Rd, Ottawa, K1A0R6, Canada

Kjeld O. Jensen
Department of Physics, University of Essex, Colchester, CO43SQ, UK

and Alison B. Walker
School of Physics, University of East Anglia, Norwich, NR47TJ, UK

ABSTRACT

We describe calculations of the annihilating positron distribution necessary for the depth profiling of defects in layered systems. These calculations combine a Monte Carlo calculation of the stopping profile for variable energy positrons in a layered medium with a solution of the one-dimensional diffusion equation in the presence of defects, electric fields and variable diffusion coefficients. The stopping profile for arbitrary energies is obtained by interpolation of a conveniently parameterized fit to the Monte Carlo results at selected energies. The method will be illustrated for the case of a heavily defected gold layer on an aluminum substrate.

1. DEFECT PROFILING

The use of positrons for the depth profiling of defects has been described extensively in many earlier publications.[1,2,3] The one-dimensional diffusion equation is solved for an initial distribution, or stopping profile $P(E,z)$, of positrons slowed rapidly (~ps) from an incident energy E[1,2]:

$$D(z)n''(z) + n'(z)[D'(z) - v_d(z)] - n(z)[v'_d(z) + \lambda_{eff}(z)] + P(E,z) = 0 \qquad (1)$$

where $n(z)$ is the annihilating positron distribution with depth z, $D(z)$ is a depth dependent diffusion coefficient and $v_d(z)$ is an electric field term. The annihilation term $\lambda_{eff}(z) = \lambda_F + \upsilon C(z)$ consists of a 'free annihilation' component in undefected material, λ_F, and a defect component given by the product of the specific trapping rate υ and the distribution of defects $C(z)$ (for the case of a single defect type).

From the solution of the diffusion equation we can easily determine the fractions of positrons annihilating in undefected material - $F_F(E)$, in a given defect region j - $F_j(E)$, or at the surface - $F_S(E)$ and we can then calculate a total lineshape parameter $S(E)$ or $W(E)$ by a simple sum of the (hopefully known) specific lineshape parameters for these processes weighted by their respective fractions:

$$S(E) = S_S F_S(E) + S_F F_F(E) + \sum_j S_j F_j(E) \qquad (2)$$

We then minimise the difference between this calculated $S(E)$ or $W(E)$ and the experimental results to determine $C(z)$[2,3].

The stopping profile, P(E,z), is usually treated for convenience as a parameterized form

$$P(E,z) = \frac{mz^{m-1}}{z_0^m} \exp\left[-\left(\frac{z}{z_0}\right)^m\right] \qquad (3)$$

known as the Makhov function[1,2]. In much of the work done in recent years the shape parameter m has been taken to be 2 so that P(E,z) is the derivative of a gaussian function. The depth parameter z_0 is related to the mean depth $\bar{z}$ by $z_0 = \bar{z}/\Gamma(1+1/m)$ with $\bar{z}$ assumed to be of the form

$$\bar{z} = \left(\frac{A}{\rho}\right) E^n \qquad (4)$$

where E is in keV and ρ is the density in gm/cm^3. The constants A and n have empirically been set to 400 Å g cm^{-3} keV^{-n} and 1.6 repectively for much of the profiling work in the literature[1]. However, it is clear that this simple parameterizations of P(E,z) is not very accurate even for simple monatomic systems[4]. As the profiling method is applied to a larger number of systems, it becomes increasingly difficult to operate without going beyond the simple picture described above. An example of this was provided when a study of deep interfaces in silicon samples[5] at the University of Western Ontario (UWO) positron facility required the extension of the incident energy range from about 20keV to 60keV. Subsequent analysis indicated that a value of n = 1.65 was required to fit the data above 20keV. A more serious illustration will be given in the next section.

As we extend our interests to multilayer samples the problem of parameterizing P(E,z) becomes substantially more complicated and we have therefore investigated the use of sample specific Monte Carlo calculations in multilayers to obtain P(E,z) for the diffusion equation.

2. MONTE CARLO CALCULATIONS

In this section we will describe an extension of the Monte Carlo calculations of positron slowing down discussed in earlier work[6] to treat monatomic systems more efficiently and to calculate P(E,z) for multilayer systems.

In previous work the inelastic valence and core electron scattering processes were treated by a combination of random phase approximation (RPA)[7] and Gryzinski[8] or Salvat et. al.[9] cross sections. Many of the calculations are time-consuming and may not be very accurate for non nearly-free-electron systems. In the present work an alternative approach based on the Penn statistical model[10] has been included. In this method the dielectric loss function $\text{Im}[\varepsilon_0/\varepsilon(q,\omega)]$ is approximated by

$$\text{Im}\left(\frac{\varepsilon_0}{\varepsilon(q,\omega)}\right) \approx \text{Im}\left(\frac{\varepsilon_0}{\varepsilon_P(q,\omega)}\right) = \int_0^\infty d\omega' g(\omega') \text{Im}\left(\frac{\varepsilon_0}{\varepsilon_L(q,\omega,\rho(\omega'))}\right) \quad (5)$$

where ε_L is the Lindhard function[7] calculated at the density $\rho(\omega') = \varepsilon_0 m(\omega'/e)^2$ corresponding to a plasma frequency ω' and $g(\omega')$ is given by

$$g(\omega') = -\frac{2}{\pi\omega'} \text{Im}\left(\frac{\varepsilon_0}{\varepsilon_{opt}(\omega')}\right) \quad (6)$$

Here, $\varepsilon_{opt}(\omega) \equiv \varepsilon(0,\omega)$ is the optical dielectric function obtained by experiment[11]. This means that, given $\varepsilon_{opt}(\omega)$, the inelastic cross-sections required for the simulations can be generated much more efficiently than is possible with the Gryzinski or Salvat approximate calculations. The disadvantage is that $\varepsilon_{opt}(\omega)$ must be known over the entire energy loss range from eV to keV. In many cases the data are not available, e.g. for Be, and we must resort to the slower approximations.

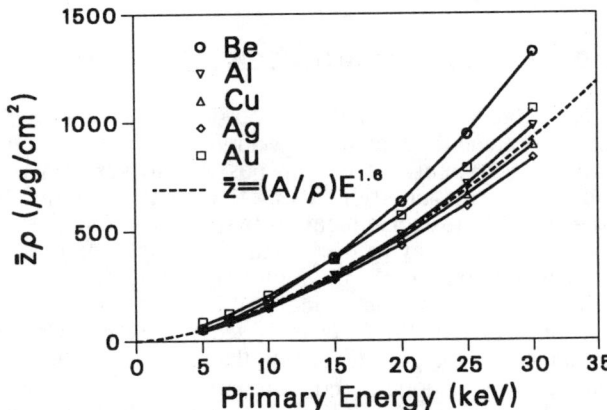

FIGURE 1.
Mean depth as a function of primary positron energy obtained from Monte Carlo calculations for a number of monatomic systems. Also shown is the empirical formula of Equation (4) with A=400 Å g cm^{-3} keV^{-n} and n=1.6.

Details of these calculations for monatomic systems are given elsewhere[12], including evidence that, for positrons at least, the Penn model leads to very good agreement with a number of experimental results from both Western Ontario and East Anglia. We will use just some of the results from that work to illustrate the non-universality of the constants in (3) and (4). In Figure 1 we show the mean depth $\bar{z}$ as a function of incident energy calculated for a number of monatomic systems, and compared to the expression (4) with A=400 Å g cm^{-3} keV$^{-1.6}$ and n=1.6. Clearly the mean depths diverge significantly from the simple result at energies above about 10 keV (the energy below which the constants were determined[13]). This point is further illustrated in Table 1 where we fit the mean depth data with A and n from (4). Also shown in this table is the value of the Makhov function shape parameter, m required to give the best fit to the implantation profile obtained from the Monte Carlo calculation.

TABLE 1.
Positron implantation profile parameters. A and n are fits of Equation (4) to the mean implantation depth obtained from Monte Carlo and m is the Makhov shape parameter obtained by fitting the Monte Carlo implantation profile.

Element	n	A (µg cm^{-2} keV^{-n})	m
Be	1.82	1.43	2.69
Al	1.74	2.64	1.97
Cu	1.61	3.78	1.78
Ag	1.57	3.98	1.76
Au	1.49	6.58	1.71

3. PROFILING MULTILAYER SYSTEMS

Having tested the validity of the Monte Carlo calculations for monatomic systems it is simple to extend the scheme to generate the depth distribution function for positrons in multilayer systems. It is then possible to perform profiling on such systems provided we can find an efficient way to convert such distribution functions into the simple functional forms convenient for the solution of the diffusion equation.

We illustrate how this may be done with an example for an 800Å layer of gold on a thick aluminum substrate, both of which are heavily defected. In Figure 2 we show the depth distribution function, calculated by Monte Carlo, for three incident positron energies (solid circles). We can clearly see that, as the energy is increased, the positrons penetrate through the gold layer into the aluminum where they can travel a much longer distance before stopping. For comparison we show the results for 3000Å of aluminum on gold in Figure 3. In this case the positrons tend to pile up just inside the gold layer as they encounter a larger stopping power.

Clearly we do not wish to perform calculations at all the energies required for the S (or W) parameter spectrum. We therefore parameterize the Monte Carlo results obtained at key energies and then interpolate the resulting coefficients between these energy points. The parameterization we have chosen for this initial study is a simple generalization of the Makhov function. In each layer of the system we fit the Monte Carlo results for the depth distribution to the form:

$$P_j(E,z) = A_j \frac{m_j z^{m_j-1}}{z_{0j}^{m_j}} \exp\left[-\left(\frac{z-z'_j}{z_{0j}}\right)^{m_j}\right] \tag{7}$$

where, in addition to the depth parameter z_{0j} and shape parameter m_j we have introduced a prefactor A_j and an origin z'_j (fixed at zero in the first layer). For a particular energy we interpolate between the coefficients obtained at the surrounding key energies and then renormalize the resulting profile to unity. Provided the key energies are sufficiently close together the renormalization is not significant. In Figures 2 and 3 we show the fits (solid lines) to the Monte Carlo results obtained using (7). The fits at 10keV in Figure 2 and at 6keV in Figure 3 are less than perfect indicating that for future work we will probably require a more flexible functional form.

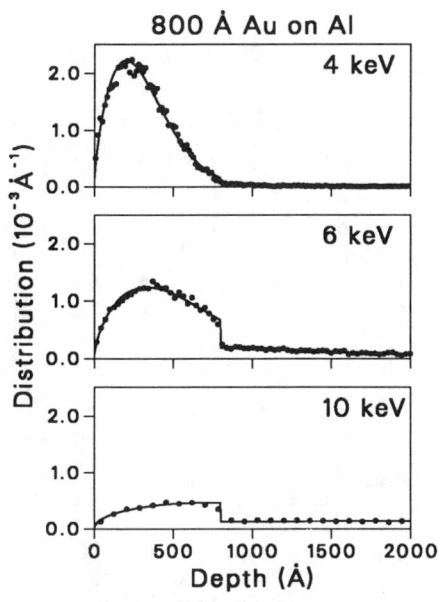

FIGURE 2.
Stopping profile generated by Monte Carlo for 800 Å of gold on aluminum. Solid lines are fits of the generalized Makhov function (equation (7)) in each region.

FIGURE 3.
Stopping profile generated by Monte Carlo for 3000 Å of aluminum on gold. Solid lines are fits of the generalized Makhov function (equation(7)) in each region.

We then solve the diffusion equation using the P(E,z) obtained in this way, together with values for the other parameters (D, λ_F, υ etc) taken from the literature, to obtain the fraction of defects F_j trapped in each heavily defected region (the free annihilation fractions are very small here). To do this we use a new program, POSTRAP5 which has been modified from the earlier POSTRAP4[2] to solve the diffusion equation for the case of multilayers consisting of different materials. In addition this program includes an option to construct the stopping profile from the fitted generalized Makhov function coefficients as described above. Figure 4 shows the trapped fractions calculated for this system, plotted as a function of both energy and mean depth. The dotted line shows the fraction assumed to be annihilating at the surface.

Experimental results were obtained from a number of samples of gold layers on aluminum as well as aluminum layers on gold grown by electron beam evaporation at the National Research Council of Canada. The lineshape parameters were measured at the UWO slow positron facility. A more detailed study will be published elsewhere but, for our present purpose, we will consider the results for the 800 Å gold layer on aluminum calculated above. In Figure 5 we compare the S and W parameters calculated from (2) with the experimental results, using the gold characteristic lineshape value as a free variable. The fit is obtained using $S_{Au}/S_{Al} = 0.916$ for the S-parameter data and $W_{Au}/W_{Al} = 1.19$ for the W-parameter data.

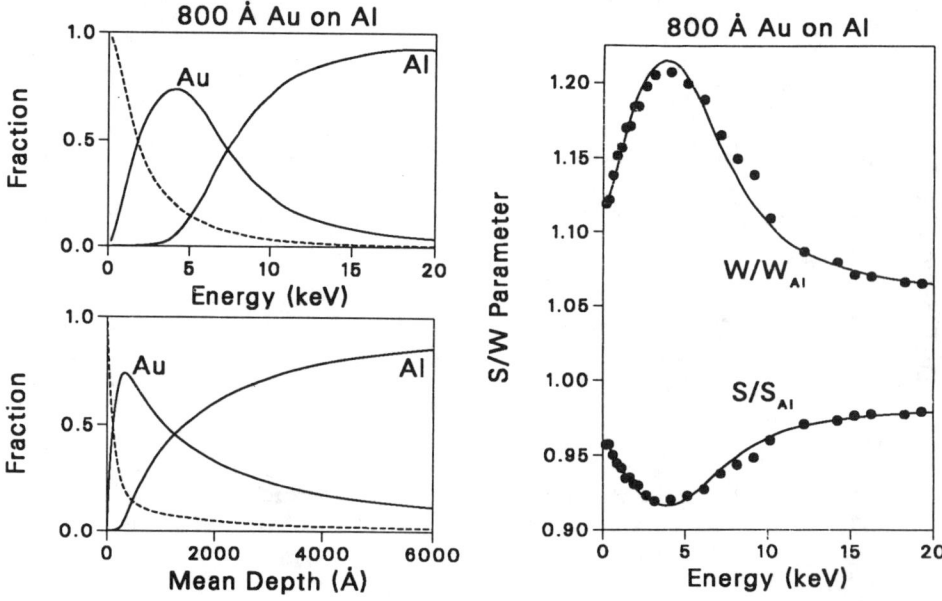

FIGURE 4.
Fraction of incident positrons which annihilate in defects or at surface as a function of incident energy (top panel) and mean implantation depth (bottom panel).

FIGURE 5.
Fit of model (solid line) to experimental S and W lineshape parameters using characteristic lineshape parameter of the gold layer as a variable.

In conclusion we suggest that the use of sample specific Monte Carlo calculations for the stopping profile in layered media is now a practical possibility. It would, of course, be preferable to have a scheme for deriving profiles based on the known properties of the constituent media and efforts are underway to address this aspect of the problem.

REFERENCES

1. P.J. Schultz and K.G. Lynn, Rev. Mod. Phys. 60, 701 (1988).
2. G.C. Aers, "Positron Beams for Solids and Surfaces", edited by P.J. Schultz, G. Massoumi and P.J. Simpson, AIP New York, 162 (1990).

3. A. van Veen, H. Schut, J. de Vries, R. A. Hakvoort and M.R. Ijpma, "Positron Beams for Solids and Surfaces", edited by P.J. Schultz, G. Massoumi and P.J. Simpson, AIP New York, 171 (1990).
4. J.A. Baker, N.B. Chilton, K.O. Jensen, A.B. Walker and P.G. Coleman, J. Phys. Condens. Matt. 3, 4109 (1991); Appl. Phys. Lett. 59, 2962 (1991).
5. P.J. Simpson, P.J. Schultz, T.E. Jackman, G.C. Aers, J.-P. Noël, D.C. Houghton, D.D. Perovic and G.C. Weatherly, "Positron Beams for Solids and Surfaces", edited by P.J. Schultz, G. Massoumi and P.J. Simpson, AIP New York, 125 (1990).
6. K.O. Jensen, A.B. Walker and N. Bouarissa, "Positron Beams for Solids and Surfaces", edited by P.J. Schultz, G. Massoumi and P.J. Simpson, AIP New York, 19 (1990).
7. C.J. Tung and R.H. Ritchie, Phys. Rev. B 16, 4302 (1977).
8. M. Gryzinski, Phys. Rev. A 138, 322 (1965).
9. F. Salvat, J.D. Martinez, R. Mayol and J. Parellada, J. Phys. D 18, 299 (1985).
10. D.R. Penn, Phys. Rev. B 35, 482 (1987).
11. e.g. D.E. Palik (ed.), "Handbook of Optical Constants of Solids", (Academic, Orlando, 1985).
12. K.O. Jensen and A.B. Walker, submitted to Surf. Sci.
13. A.P. Mills and R. Wilson, Phys. Rev. A 26, 490 (1982); A. Vehanen, K. Saarinen, P. Hautojärvi and H. Huomo, Phys. Rev. B 35, 4606 (1987); K.G. Lynn and H. Lutz, Phys. Rev. B 22, 4143 (1980).

DIFFUSION AND ANNIHILATION OF POSITRONS IN SOLIDS

D. T. Britton
Universität der Bundeswehr München
Institut für Nukleare Festkörperphysik
D–8014 Neubiberg
Germany

ABSTRACT

The diffusion–annihilation equations for positrons implanted both thermally and epithermally are presented. The epithermal case has been solved numerically assuming a two group model in which the epithermal distribution acts as a time varying source for the thermal positrons. Using experimental data from the metals Al, Ag and Au the model has been used to extract values for the positron diffusion and surface penetration coefficients. The two group model is shown to be inadequate in describing the transport of epithermal positrons as thermalisation must be treated as a continuous process. For deeper implantation, a single diffusion equation can be applied.

INTRODUCTION

Although it is usual to speak of positron lifetimes in solids, such an analysis is only strictly valid for homogeneous infinite media. In this case, both trapping and annihilation are purely statistical processes resulting in exponential components with lifetimes given by a simple trapping model[1]. In inhomogeneous finite samples the populations of positrons in different states, and hence the total annihilation rate, can deviate strongly from the exponential form[2,3]. Even when the spectra resemble a sum of exponential components the intensities cannot be simply related by a trapping model. These effects are most significant for timed positron beam experiments, where the positron can easily return to the vicinity of the surface or where the defects to be studied are non–uniformly distributed.

DIFFUSION MODELS

In this paper two different diffusion models are considered; a single group model in which positrons are assumed to be implanted thermally, and a two group model in which the positrons are implanted epithermally. In the latter case, the epithermal positron distribution is used as time dependent source of thermal positrons. Implantation is assumed to be instantaneous so that the initial distributions are given by the positron stopping profile $P(z)$.

In the thermal model, the evolution of the positron distribution $u_\theta(z,t)$ is determined by a single diffusion–annihilation equation[4]

$$D_\theta \partial^2 u_\theta/\partial z^2 - (\lambda + \mu_\theta(z))u_\theta = \partial u_\theta/\partial t \qquad (1)$$

with the initial condition $u_\theta(z,0) = P(z)$. Similarly for the epithermal two group model, the time development of the system is determined by two diffusion–annihilation equations

$$D_\epsilon \partial^2 u_\epsilon/\partial z^2 - (\lambda + \gamma + \mu_\epsilon(z))u_\epsilon = \partial u_\epsilon/dt$$

and
$$D_\theta \partial^2 u_\theta/\partial z^2 - (\lambda + \mu_\theta(z))u_\theta + \gamma u_\epsilon = \partial u_\theta/dt, \qquad (2)$$

with the initial conditions $u_\theta(z,0) = 0$ and $u_\epsilon(z,0) = P(z)$. The subscripts θ and ϵ refer to thermal and epithermal states respectively, λ is the bulk annihilation rate, μ the spatially dependent trapping rate to a distribution of defects and γ is the thermalisation rate.

In either case the surface plays an important rôle, acting as a sink for the positron flux, the effect of which is described by a general radiative boundary condition[4,5]

$$D \,\partial u/\partial z \big|_{z=0} = \nu u(0,t). \qquad (3)$$

Equation (3) expresses the condition that the flux through the surface is proportional to the probability of a positron being at the surface. The transition rate ν has the dimensions of velocity and is the sum of all partial transition rates to the different surface channels (positronium, free and surface trapped positrons). In the limiting cases of ν infinite or zero Dirichlet and Neumann boundary conditions respectively are obtained. Obviously the Neumann boundary condition, for which no positrons can leave the bulk, is of no interest here. The Dirichlet condition, for which the positron distribution at the surface is zero, is that usually applied in slow positron analysis[6].

The observed annihilation rate is given by the sum of the populations of the different states weighted with the intrinsic annihilation rates λ,

$$N(t) = \lambda_B n_B + \lambda_S n_S + \lambda_{Ps} n_{Ps} + \lambda_B n_\epsilon \qquad (4)$$

where the subscripts B, S and Ps refer to the bulk, surface state and parapositronium respectively. The intrinsic annihilation rates for orthopositronium and free flying positrons are assumed to be immeasurably small. The final term in Equation (4) describes the annihilation of epithermal positrons before thermalisation and only occurs in the two group model.

The populations of the different states are given by a set of rate equations, similar to those applied to normal bulk lifetime analysis, eg. for the surface state

$$dn_S/dt = -\lambda_S n_S(t) + (\nu_S/\nu)\kappa_\theta(t) + \epsilon_S \kappa_\epsilon(t). \qquad (5)$$

The principle difference between Equation (5) and the usual trapping model is that the sink rates are themselves time–dependent functions given by the boundary conditions (Eq. 3). For the epithermal flux, a branching ratio to the surface surface state ϵ_S can be written explicitly as (almost) infinite transition rates can be assumed.

For particular problems the differential equations (4) and (5) can be solved analytically[2,3], although for a general initial profile and boundary condition numerical solutions are necessary. For the diffusion equation with an initial distribution $P(z)$, the clearest and simplest method is to fold a suitable Green's function with the initial profile

$$u(z,t) = \int P(x)G(z|x,t)dx. \qquad (6)$$

A solution of particular interest is that for the usually applied Gaussian derivative implantation profile, $P(z) = -d/dz \exp(-z^2/z_0^2)$, and a perfectly absorbing surface:

$$u(z,t) = \frac{z_0 \exp(-\lambda t)}{(z_0^2 + 4Dt)^{1/2}} \frac{2z}{z_0^2 + 4Dt} \exp\left[\frac{-z^2}{z_0^2 + 4Dt}\right] \qquad (7)$$

Equation (7) has the same form as the initial profile but with mean depth which increases with time and a purely time dependent damping prefactor which describes both escape and annihilation of the positrons. For the two group model, this moving Makhovian profile is then the source for thermal positrons, which averaged over the thermalisation time has a minimum implantation depth corresponding to the thermalisation length. Integration of Eq (7) gives the time dependence of the bulk population,

$$n_B(t) = z_0/(z_0^2 + 4Dt)^{1/2} \exp(-\lambda t) \qquad (8)$$

which deviates very strongly from an exponential form for shallow implantation.

In general, however, there is no analytical solution for the diffusion–annihilation equations. Even with the Gaussian derivative initial profile, Eq (6) is only integrable for Dirichlet and Neumann boundary conditions. The simplest numerical method is to integrate the diffusion equation using a finite–difference method, and simultaneously for each successive time step calculate the change in

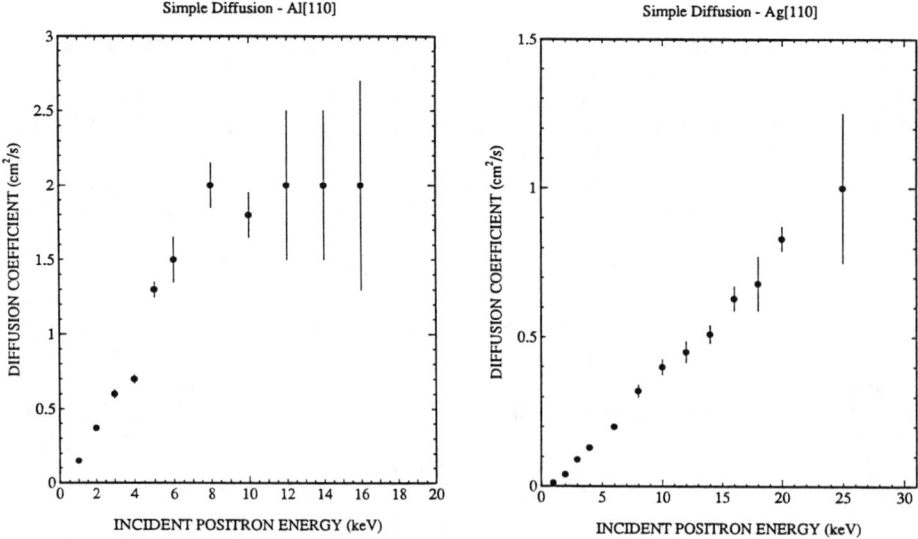

Fig 1 Apparent positron diffusion coefficient in aluminium (a) and silver (b) assuming positrons to be implanted thermally and a perfectly absorbing surface.

the respective populations and the total annihilation rate. Increased stability and accuracy can be achieved through a fully implicit (forward–time) calculation[7]. Solutions are the obtainable for any initial and boundary conditions and any defect profile. Where comparable the results are identical to those from the analytical solution to within the computation accuracy.

EXPERIMENTAL

Annihilation spectra for Al, Ag and Au (110) surfaces have been modelled and compared with data measured using the pulsed positron beam at Munich[8]. All samples were measured with a natural oxide layer to suppress positron and positronium emission, resulting in a single surface state for thermal positrons making the analysis much simpler. Initially the simplest model, single group diffusion and a perfectly absorbing surface, was applied. For the implantation profile, a Gaussian derivative was used.

The effective diffusion coefficient as a function of incident positron energy is shown in Fig.1 for aluminium and silver. Gold exhibits a similar behaviour to silver. The apparent, unphysical reduction for lower implantation energies for aluminium is predominantly an epithermal effect which will be discussed later. In silver and gold it is not.

By introducing a finite surface absorption coefficient, the effective diffusion coefficient for positrons implanted in silver takes on the same form as for aluminium (fig. 2a). The energy above which the diffusion coefficient takes on its physical value scales with the density for all three metals. An indication that the lower energy regime is dominated by epithermal positron transport is the dramatic increase in surface absorption coefficient (fig 2b). Similar behaviour is also seen in gold, but in this case the limiting energy is 20keV. The values for the thermal positron diffusion coefficients in aluminium and silver, 2.0(2) and 1.0(2) cm^2/s respectively are in good agreement with earlier results[9].

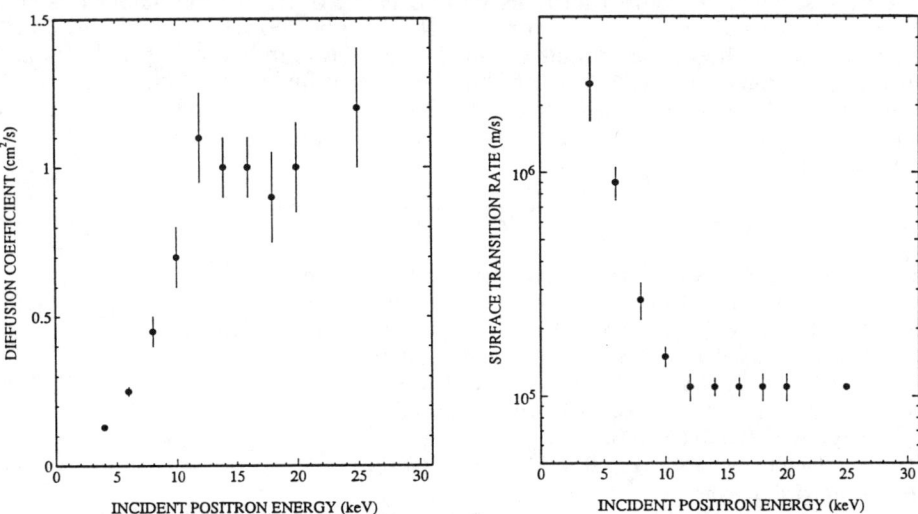

Fig 2 Apparent positron diffusion coefficient (a) and surface transmission coefficient (b) for positrons in silver assuming positrons to be implanted thermally and a general radiative boundary condition.

Qualitatively, the apparent reduction in diffusion coefficient at lower implantation energies can easily be explained with the epithermal diffusion model. Although the probability of a positron returning to the surface is enhanced by the back-diffusion of epithermals, not all epithermals are trapped in the surface state and the net effect is a depletion of the returned flux. This can be interpreted as an increase in implantation depth or a corresponding decrease in the apparent diffusion coefficient. If all epithermal positrons escape without being detected, the mean implantation depth will tend to a finite value at zero energy corresponding to the thermalisation length.

In quantitative analysis, using the two group diffusion model, several problems arise. Firstly, because of the short times involved and limited experimental resolution, it is not possible to independently determine the epithermal diffusion coefficient and thermalisation rate. However, a thermalisation length, analogous to the diffusion length in continuous beam measurements, can be derived from the product of the two. Secondly, the branching ratios to free positrons and positronium are non-zero for epithermal positrons, resulting in two many degrees of freedom to unambiguously model the spectra at intermediate energies. This is not, in itself, a problem if the assumption that the branching ratios are constant held. Unfortunately, it does not. At very low implantation energies the relative fractions of the surface state and parapositronium are in the range 3-4 : 1, but for deeper implantation more than 90% of epithermal positrons are trapped at the oxide layer. There is also some indication that, for aluminium, the thermalisation length increases with implantation depth.

CONCLUSION

Thermalisation cannot be modelled as a single stage process, those epithermal positrons reaching the surface from deeper inside the sample have a different energy distribution (more thermalised) and hence different characteristics than those implanted nearer the surface. To study positron dynamics fully thermalisation should be modelled continuously, preferably using a Boltzmann formalism. However, saturation of the effective diffusion coefficient for deeper implantation, where the positrons fully thermalise before reaching the surface, shows that a simple diffusion model can be used successfully for some applications.

REFERENCES

1. A.Vehanen and K.Rytsöla in *Positron Solid State Physics*
   (W.Brandt and A.Dupasquier Eds., North Holland 1983)
2. W.E.Frieze, K.G.Lynn and D.O.Welch Phys. Rev B31 15 (1985)
3. D.T.Britton J. Phys. Condens. Matter 3 681 (1991)
4. R.M.Nieminen and J.Oliva Phys. Rev. B22 2226 (1980)
5. P.A.Huttunen, J.Mäkinen, D.T.Britton, E.Soininen and A.Vehanen
   Phys. Rev. B42 1560 (1990)
6. P.J.Schultz and K.G.Lynn Rev. Mod. Phys. 60 701 (1989)
7. Numerical Recipes
8. D.Schödlbauer, P.Sperr, G.Kögel and W.Triftshäuser Nucl. Inst. Meth. B34 258 (1988)
9. E.Soininen, H.Huomo, P.A.Huttunen, J.Mäkinen, A.Vehanen and P.Hautojärvi
   Phys. Rev. B41 6227 (1990)

EVIDENCE OF DEEP VACANCY FORMATION IN FLUORINE IMPLANTED SILICON

N.B. Chilton and M. Fujinami
Advanced Materials & Technology Research Laboratories
Nippon Steel Corporation
1618 Ida, Nakahara-ku, Kawasaki,
Japan.

ABSTRACT

Variable energy positron defect profiling has revealed the sub-surface vacancy profile in Cz Si(100) implanted to a dose of 2×10^{14} cm^{-2} with 120keV fluorine ions. We demonstrate the presence of vacancies at ppm concentrations at depths exceeding 1.5µm, many times the depth at which implantation damage is expected to occur. The vacancy profile deduced was verified by removing layers of known thickness from the surface of the sample by chemical etching and repeating the defect profiling process. In the fitting procedure, the use of a Monte Carlo deduced positron implantation profile is *required*. The annealing behavior of the defects shows that the vacancies caused by the implantation process consist of two components. The deep vacancies are found to be relatively easily removed by a moderate temperature anneal, in contrast to those vacancies nearer to the surface. The former component is suggested to be the more mobile and thus may diffuse to the depths at which it is observed.

INTRODUCTION

Ion implantation is a useful tool for modification of device quality silicon wafers. Using ion implantation, one is able to produce a specific dopant profile to create a desired electrical characteristic. The study into the damage caused by ion implantation has a long history and at this time the nature of a substantial number of the defects present in implanted Si is known[1]. Techniques such as ESR have been highly successful in demonstrating that many of the defects present are correlated to vacancies or vacancy complexes[2]. In light of this, and the goal of ultimate miniaturization of electronic devices, it is thus of great importance that the sub-surface distribution of vacancies caused by ion implantation be known. RBS is able to indirectly map the sub-surface distribution of vacancies by mapping the ion displacements of the original lattice. However, the technique of defect profiling using a slow positron beam[3] is the only technique where the vacancy distribution can be obtained *directly* with a sensitivity of ppm order.

EXPERIMENTAL

A study of vacancy formation in Cz-Si(100) implanted with 120keV Fluorine ions to a dose of 2×10^{14} cm^{-2} is presented. The fluorine ion implantation angle was 7° from normal so ion channeling is not expected to occur. The sub-surface vacancy distribution obtained by fitting of the positron data is compared to that predicted by the TRIM[4] Monte Carlo ion implantation program. The effect of annealing, after implantation, on the vacancy profile was investigated.

For this series of experiments the Nippon Steel Corporation slow positron beam[5] was used. Using this beam, mono-energetic positrons in the energy range 0-30keV may be implanted into the sample under study. The Doppler broadening of the electron-positron annihilation gamma photopeak is measured with respect to positron implantation energy. The S parameter is used as the measure of Doppler broadening in this experiment. The reader unfamiliar with positron defect profiling is referred to reference 3 for definitions of the S parameter and related technical points.

In this system we assume that there are only three different and distinguishable sites at which the positron can annihilate. These are (i) the surface state, (ii) bulk Si with no defects and (iii) the vacancy. The electron momentum at these sites is characteristically different as, therefore, is the characteristic S parameter associated with that state. A high value of the S parameter is characteristic of vacancies as a vacancy contains only delocalised electrons.

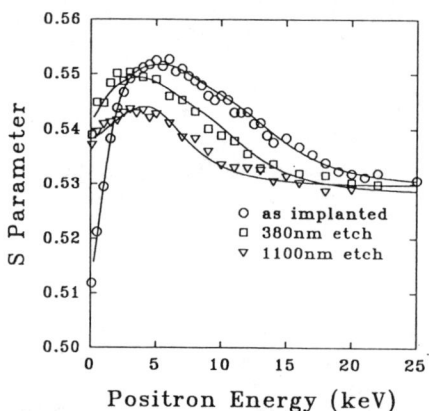

In figure 1 is shown the S parameter as a function of incident positron energy for the as implanted sample. At high positron energy, the value of the S parameter tends to that of defect free material (bulk Si) as the ion implantation causes damage only in the near surface region. On first inspection it is apparent that the higher value of S in the low energy region corresponds to the presence of vacancies near the surface. However, computer modeling is required in order to be able to deduce the true vacancy distribution.

Figure 1. A plot of the S parameter against the incident positron energy for the fluorine implanted sample in its as-implanted and etched states. The solid lines through the points are the computer generated fit to the data achieved using the vacancy profiles shown in figure 2.

RESULTS AND DISCUSSION

At any given positron implantation energy E, the S parameter measured is a linear combination (eq. 1) of S parameter signals from the 3 states mentioned above.

$$S(E) = F_{surface}S_{surface} + F_{bulk}S_{bulk} + F_{vacancy}S_{vacancy}$$
$$\text{where} \sum_i F_i = 1 \qquad (1)$$

The fraction of positrons which annihilate in each state (the F values) is found by solution of the positron diffusion equation (eq. 2) in the sample.

$$D \cdot \frac{\partial^2 n(z)}{\partial z^2} + P(z) - [\lambda_b + vC(z)]n(z) - \frac{\partial}{\partial z}[v_d n(z)] = 0 \qquad (2)$$

where n(z) is the positron density with respect to depth z, D_+ is the positron diffusion constant, C(z) is the defect distribution, P(z) is the positron implantation profile, v_d is the drift velocity, λ_b is the defect-free annihilation rate and the product $vC(z)$ is the defect annihilation rate.

For a trial defect distribution, C(z), the diffusion equation is solved[6] and the resultant S(E) values are fit to the experimental data. The quality of the fit can be seen in Fig. 1. In this way the vacancy distribution (histogram) shown in figure 2(a) is deduced. Figure 2(a) also shows the fluorine and vacancy profiles (scaled to fit the plot) predicted by the TRIM program. It is known that the TRIM code accurately predicts the Fluorine profile (by comparison to RBS and SIMS[7] data) but agreement between TRIM predictions for vacancy distribution and positron-deduced vacancy profiles has been shown to be limited[8,9]. In this case it is impossible to fit the data of figure 1 using a profile based on the TRIM result. We thus show a clear discrepancy between the two data sets - the TRIM prediction has a maximum depth of approximately 400nm whereas the positron deduced profile shows the presence of vacancies at much greater depths. In order to produce the good fit to the positron data seen in figure 1, the high depth tail on the vacancy distribution is *required*. The concentration of these, deeper, vacancies tails off to a few ppm - far below that at which detection by other means would allow.

As a check on the accuracy of the vacancy profile deduced and shown in figure 2(a) the samples were etched in a solution of HNO_3/HF to remove a surface layer of known depth. The S vs. E results of the implanted sample after removal of surface layers of 380nm and 1100nm are presented in figure 1. The change in surface S caused by the etching process is of no consequence to the modeling procedure. It is again apparent that there is still a high S component (vacancy-characteristic) in the near surface region of these etched samples. This result verifies that the vacancy profile originally deduced and shown in figure 2(a) does indeed extend to much greater depth than that predicted by the TRIM code.

Computer modeling of the S vs. E plots of the 380nm and 1100nm etched samples produced the vacancy profiles shown in figure 2(b) and (c) respectively. The vacancy profiles are shifted to larger depth by amounts equal to the etch depth to allow easy comparison between figure 2(a), (b) and (c). It is thus unequivocally shown that there is a small, but non-negligible, vacancy concentration present in the region of the sample below the maximum depth of 400nm predicted by TRIM. By use of a control sample, the etching process was proven to not introduce vacancies.

As can be seen from equation 2, the analytic form of the positron implantation profile P(z), strongly affects the solution of the diffusion equation and thus the deduced vacancy profile. The most commonly used form of P(z) is a derivative of a Gaussian profile. In our analysis, this form of P(z) was used initially and was able to fit any of the data in figure 1 individually. However, when we consider the data from the sequentially etched samples it is obvious that these vacancy profiles should be a sub-set of the initial profile as shown in figure 2. With this criterion applied, it is impossible to fit the data in figure 1 in a consistent manner if we use the Gaussian derivative profile.

Recently the Monte Carlo deduced positron implantation profile of Jensen *et al* [10] has been shown to be more accurate than the Gaussian derivative by two separate studies[11,12].

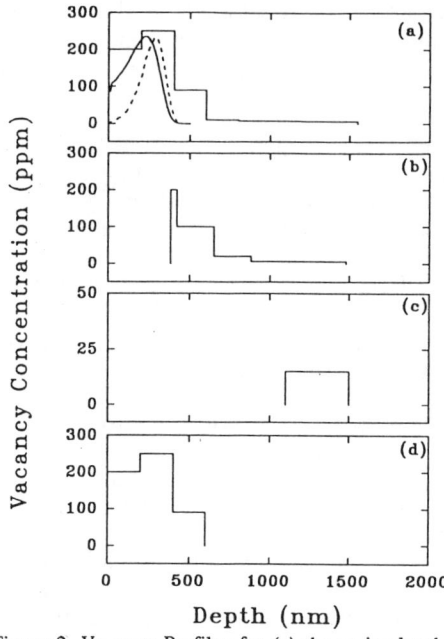

Figure 2. Vacancy Profiles for (a) the as implated sample, (b) the 380nm etched sample, (c) the 1100nm etched sample, and (d) the 300°C annealed sample. The solid line and dashed line shown in (a) are the TRIM predicted profiles for vacancies and implanted fluorine respectively.

When a parameterised form of this profile is used for P(z) in our analysis, we find that the vacancy profiles are able to be deduced as a consistent set - that is that the profiles shown in figures 2(b) and 2(c) are a sub-set of 2(a). In order to use this parameterised form of P(z), the program POSTRAP4 was slightly modified. This, the first defect profiling analysis to use the new Monte Carlo profile, adds weight to the argument that a properly derived material dependent P(z) is essential in fitting positron data. In this case it is important to note that the P(z) used was calculated for Al and not Si but the difference in this case is expected to be small as the densities are similar.

The production of vacancies is due to the energy loss of the incoming F ions. The F ions are known to be unable to extend down to the great depth at which vacancies have been found in this investigation. Thus the vacancy profile *caused directly* by implantation probably does not have the long tail that is present in our analysis. The vacancies present after ion implantation are expected to be predominantly of di-vacancy character (ref 11) and so should be immobile at room temperature. However, it may be possible that some of these divacancies break up and form mono-vacancies, which are mobile at room temperature. We suggest that the depth to which vacancies are present indicates that some vacancy component is free to diffuse in the sample from the production region (surface to ~400nm) to much greater depths.

In this analysis, the positron trapping rate is limited to one value throughout the sample - the di-vacancy[13] value was used. The possibility that another vacancy type (with different trapping rate) is present in the high depth region may be the cause of the discrepancy in the vacancy *concentrations* shown in the deep regions of figures 2(a)-(c).

To investigate the character of the vacancies present, the as implanted sample was subjected to annealing in vacuum for 30 minutes at 300°C and at 600°C - the results are shown in figure 3. The annealing was done in a separate vacuum chamber to that in which our measurements were made. The vacuum in the annealing chamber was rather poor (10^{-4} Torr) and resulted in the formation of a rough SiO_2 layer on the surface of the 600°C annealed sample. Thermally grown SiO_2 has an S parameter similar to that of undefected Si[14] - SiO_2 grown in a poor vacuum has many defects in its structure which result in an S parameter lower than that of undefected Si.

The S vs. E results for the sample annealed at 300°C can be fit by assuming the defect profile for the as implanted sample but with the long tail of defects removed - figure 2(d). The 600°C annealed sample shows no evidence of the presence of vacancies at all. The S vs. E curve shows only a transition between positron - electron annihilation at the surface (low energy) to annihilation in bulk undefected Si. The flatness of the S parameter in the very low energy region of the 600°C sample is a consequence of the SiO_2 surface growth only.

The anneal at 300°C appears to be insufficient to remove the vacancies in the surface to 600nm region but those in the deeper region are shown to be removed effectively by this anneal. This adds strength to the argument that there are at least two vacancy components present in the as implanted sample. The deep vacancies recover at a lower annealing temperature than those in the shallower region.

Figure 3. A plot of the S parameter against the incident positron energy for the annealed samples. The as implanted sample is also shown for purpose of comparison. The solid lines are the computer fit to the data.

CONCLUSIONS

In conclusion, we have presented data which shows that the vacancies present in F ion implanted Si are manifest to a much greater depth than that predicted by TRIM. The uniquely high sensitivity of the positron technique is demonstrated. The fitting of the data using the vacancy profiles shown in figure 2 is consistent in that the profiles used to fit the etched and annealed data are sub-sets of the profile for the as implanted sample. The method by which vacancies appear in the deep region is suggested to be that of diffusion of a component of the vacancies present in the 0-400nm region to large depths. This deep vacancy component is unstable to annealing at 300°C. An anneal at 600°C is sufficient to remove all the vacancies from the sample.

To model this data in a consistent manner it was not possible to use the Gaussian derivative positron implantation profile and the Monte Carlo deduced positron implantation profile of Jensen *et al* was required.

Acknowledgments

The authors would like to thank S. Hayashi and K. Tanaka for useful discussions. We are grateful to T. Sakon and K. Uemura for sample preparation. This research was financed in part by NEDO grant for the International Joint Research.

REFERENCES

1. *for example* W.R. Brown in *Beam Solid Interactions and Phase Transitions*, MRS Symposium proceedings edited by H. Kurz, G.L. Oesan and J.M. Poate, (Materials Research Society, Pittsburgh, 1985) Vol 51 p53.
2. *for example* D.F. Daly and K.A. Pickar, Appl. Phys. Lett. **15** 267 (1969).
   K.L. Brower and J.A. Borders, Appl. Phys. Lett. **16** 169 (1970).
   K.L. Brower and W. Beezhold, J. Appl. Phys. **43** 3499 (1972).
3. P.J. Schultz and K.G. Lynn, Rev. Mod. Phys. **60** 701 (1988).
4. J.P. Biersack and L.G. Haggmark, Nucl. Instrum. Methods **174** 257 (1980).
5. M. Fujinami, Houshasen *in press*.
6. G. C. Aers, in *Positron Beams for Solids and Surfaces*, edited by P.J. Schultz, G.R. Massoumi and P.J. Simpson (*American Institute of Physics* **218**, New York, 1990), p.162.
7. Samples such as this are used as calibration for the SIMS technique - the TRIM ion profile is assumed.
8. P.J. Simpson, M. Vos, I.V. Mitchell, C. Wu and P.J. Schultz, Phys. Rev. B. **44** 12180 (1991).
9. A. Uedono, S. Tanigawa, J. Sugiura and M. Ogasawara, Jpn. J. Appl. Phys. **29** 1867 (1990).
10. K.O. Jensen and A.B. Walker, in *Positron Beams for Solids and Surfaces*, edited by P.J. Schultz, G.R. Massoumi and P.J. Simpson (*American Institute of Physics* **218**, New York, 1990), p.44.
11. G.R. Massoumi, N. Hozhabri, K.O. Jensen, W.N. Lennard, M.S. Lorenzo, P.J. Schultz and A.B. Walker Phys. Rev. Lett. **68** 3873 (1992).
12. J.A. Baker, N.B. Chilton, K.O. Jensen, A.B. Walker and P.G. Coleman, J. Phys., Conden. Matter **3** 4109 (1991).
13. P. Mascher, S. Dannefaer and D. Kerr, Phys. Rev. B **40** 764 (1989).
14. N.B. Chilton and M. Fujinami - in these proceedings.

A STUDY OF IMPLANTATION INDUCED DEFECTS IN SiO$_2$

M. Fujinami and N.B. Chilton
Advanced Materials and Technology Research Laboratories
Nippon Steel Corporation
1618 Ida, Nakahara-ku, Kawasaki
JAPAN.

ABSTRACT

Boron ion implantation-induced defects in SiO$_2$ were investigated using slow positron annihilation spectroscopy and electron spin resonance (ESR). The defects caused by ion implantation are manifest as a particularly low S-parameter in the region of the SiO$_2$ layer in which B implantation damage occurs. The annealing behavior of the defect responsible for positron trapping was studied. The defect to which the positron is sensitive is found to be unobservable in ESR measurements. The defect is suggested to be dissolved O$_2$ or a charged Frenkel defect, such as the negative non bridging-oxygen hole center ($\equiv$S–O$^-$).

INTRODUCTION

Slow positron annihilation spectroscopy (the use of a positron beam for probing the sub-surface distribution of defects) is one of the most promising tools for the study of defects caused in the damaged region of ion implanted samples. The technique has been applied extensively and successfully to the study of defect behavior in ion implanted Si wafers[1]. There have been, however, very few studies into the nature of defects in SiO$_2$[2].

Defects in silicon dioxide, both intrinsic and extrinsic, cause absorption effects in the core of glass fiber waveguides and can degrade the electronic properties of MOS devices. The importance of further understanding of the defects present in SiO$_2$ is thus obvious. The enormous variety of defects present in SiO$_2$ after irradiation or implantation has been demonstrated by techniques such as electron spin resonance (ESR) and optical absorption spectroscopy, these are summarized in Table I[3].

Ion implantation causes the breaking of the network structure of the SiO$_2$. Before implantation, the normal bonding can be represented as $\equiv$Si–O–Si$\equiv$. The implantation process breaks the bonds in this structure resulting in Frenkel pairs; for example, if the above schematic is broken between the Si and O bond, the result is the formation of E' centers and nonbridging-oxygen hole centers (NBOHC). Some of these defects have a paramagnetic character which makes their detection by ESR possible[4]. By a process of *charge transfer*, whereby the electrons responsible for the original bond all go to one or other of the resultant defects, it is also possible for the defects to have charged states. These defects have the possible charge states shown in Table I. Importantly, these charged species are not detectable by ESR. In its as grown state, SiO$_2$ also contains non-stoichiometric defects (e.g. oxygen vacancy or peroxy linkage) as well as impurities (e.g. dissolved O$_2$ or OH).

Which, if any, of the defects shown in Table I may trap positrons is still largely unknown. For the positron technique to fulfill its potential as a defect probe it is necessary that the nature of positron trapping at defects is fully understood. In this, and future studies, we correlate positron data to that from ESR and other probes in order to understand the nature of positron trapping in SiO_2.

EXPERIMENTAL

A 1.1μm thick SiO_2 layer was thermally grown on a 10Ωcm p-type Cz Si(100) wafer. The growth took place in a wet O_2 environment at 1100°C for 4 hours. The thickness of the SiO_2 film was estimated by an optical interference method. 100keV boron ions were implanted into this sample to a dose of $1 \times 10^{14} cm^{-2}$. The B ion distribution is predicted by the TRIM Monte Carlo code[5] to be entirely within the SiO_2 layer. The sample was studied before implantation, in its as-implanted state and after annealing, in a N_2 atmosphere for 30 minutes, at various temperatures.

The Nippon Steel Corporation slow positron beam[6] was used to measure the Doppler broadening (S parameter) of the annihilation γ-ray with respect to incident positron energy, effectively measuring the positron trapping, with respect to depth, in the samples. In the positron beam, mono-energetic positrons of energy 0-30 keV impinge on the target. The experimental pressure was 10^{-9} Torr. All positron measurements were taken at room temperature. The ESR measurements were carried out at room temperature and at 5K with a JES-RE2X(JEOL) spectrometer at X-band frequency ($\nu \approx 9.1$ GHz). A microwave power of 100μW was applied.

The level of Doppler broadening of the γ-ray was quantified by the S parameter, which is defined as the ratio of the central area to the total area of the (511keV) annihilation photopeak[7]. The results of measurement of the S parameter with respect to positron energy for the various samples are shown in Fig. 1.

RESULTS AND DISCUSSION

The unimplanted sample shows an almost flat S parameter vs. Energy(S-E) curve indicative of an absence of positron diffusion in the sample. The sample also shows that the S parameter for this thermally grown SiO_2 layer is equal to that of the bulk Si. The S parameter in SiO_2 has previously been observed to have values which appear to be dependent on the growth conditions of the oxide (temperature, water content, atmosphere etc.)[8]. The ESR spectrum of this sample shows only the presence of the P_b centers ($Si_3 \equiv Si\bullet$), which are often observable at the SiO_2/Si interface[9].

Boron ion implantation causes the S parameter in the near surface region to drop enormously - the value of S in this layer is indicative of positron annihilation at a site where the momentum distribution of the electrons is very wide. The solid line in Fig. 1 is the computer generated fit to the data obtained by solution of the positron diffusion equation[10] in the sample.

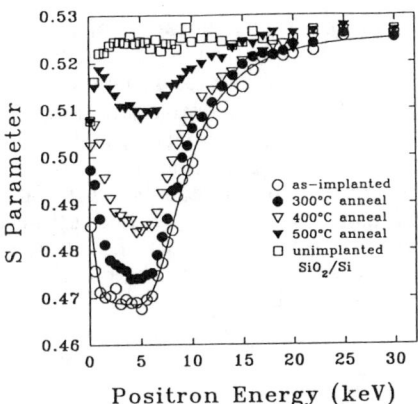

We model the system by assuming a 1.1μm layer of SiO_2 with zero positron diffusion on a substrate of crystalline Si. The TRIM code is used to predict the range of implantation-induced defects - the depth of such defects (see Fig. 2) extends to ≈5300Å. Of the 1.1μm layer we assume that the first 5300Å has an S parameter of 0.4685 (caused by implantation defects) and the remaining SiO_2 layer has an S parameter equal to that of bulk Si. This, rather simple, model is shown as the block in Fig. 2 - the model is, nevertheless, successful.

Figure 1. Plot of S parameter vs. incident positron energy for the sample described in the text. The effect of annealing, at various temperatures, can clearly be seen. The line through the data for the as-implanted sample is the computer generated fit to the data.

We therefore show that the defect(s) at which the positron traps are only present in the region of the SiO_2 where ion implantation causes damage to the structure. The implication here is that the defects present are immobile - in contrast to a previous study into *vacancy* formation in the ion implanted Si[11]. The as-implanted sample has the ESR spectrum shown in Fig. 3(a). It is confirmed that paramagnetic defects such as E' centers, NBOHC and peroxy radicals (PR) are induced by ion implantation[4,12], the presence of charged defects is implied.

name	schematic	charge transfer state
E' center	≡Si•	≡Si+
nonbridging-oxygen hole center (NBOHC)	≡Si-O•	≡Si-O-
peroxy radical (PR)	≡Si-O-O•	≡Si-O-O-
oxygen vacancy	≡Si-Si≡	
peroxy linkage (PL)	≡Si-O-O-Si≡	
dissolved oxygen	O_2	O_2^-
impurity	≡Si-OH	
impurity	≡Si-H	

Table I. Defects induced in silicon dioxide by ion implantation. The • mark indicates an unpaired electron (radical) and the ≡ mark indicates silicon bonding with 3 oxygen atoms.

Annealing of the sample at 300°C produces only a very small change in the value of the S parameter in the defected region (evidence of only a small reduction in positron trapping), as seen in Fig. 1. The ESR spectrum in Fig. 3(b) however, shows an absence of all paramagnetic defects.

The ESR measurement is not straightforward but we are confident that the E' centers, at least, disappear after a 300°C anneal as they have a distinct and strong ESR signal. Annealing at 400°C and 500°C further increases the S parameter (decreased positron trapping) in the damaged region, and an anneal at 600°C is sufficient to reproduce the curve of the unimplanted sample (positron trapping at implantation induced defects is no longer occurring).

The site at which the positron traps is the question to which this paper is addressed. It should be noted that, whatever the defect is, the positron technique is sensitive to it whereas ESR is not.

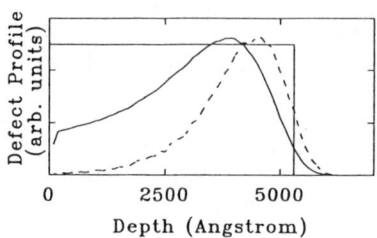

Figure 2. Plot of the TRIM prediction of the damage caused by 100keV B ion implantation into SiO2. The solid curve is the vacancy profile, the dashed curve is the B ion profile and the block figure is the defect model used in fitting the positron data.

We suggest that positron trapping is not dependent on the implantation species (a phosphorus implanted sample has a curve identical to the B implanted sample)[13], but on damage caused to the SiO_2 network by the implantation process. Starting from the defects listed in Table I and using a process of deduction we attempt to narrow down the choice of defects which could be responsible for positron trapping.

The criteria involved are that the defect be either neutral or, more likely, negatively charged (in order to trap the positron), that it gives rise to an S parameter which is much lower than that of SiO_2 or bulk Si and that the defect annealing behavior is consistent with that observed.

As the S parameter indicative of *vacancy trapping* is characteristically *high*, the defect here is clearly not the vacancy. It has been shown that a reduction in positronium (Ps) formation can cause a drop in the observed S parameter[14] by a *small* amount ($\approx 2.5\%$). Ion implantation can cause a drop in Ps formation by closing up free space in the SiO_2 structure. However, we do not favor this as an explanation of the behavior here, especially as the drop in S parameter is so large.

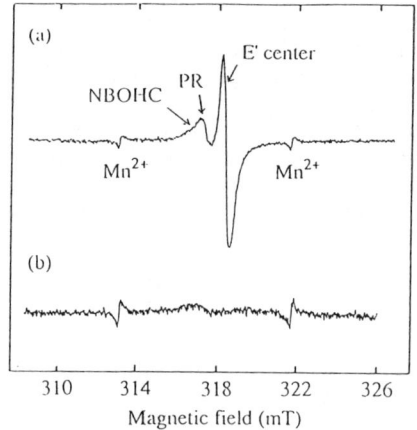

Figure 3. Electron Spin Resonance data showing the presence of the indicated paramagnetic defects in the as-implanted sample and the lack of any defect signal in the 300°C annealed sample.

The *impurity defects* (which are present before implantation) shown in Table I will, in fact, split up into the same defect species as SiO_2 when bombarded with B ions and hence are not distinguishable as separate defects. The implantation process probably releases atomic hydrogen and oxygen from these defects. Of the *non-paramagnetic defects* shown, there are a number which we can disregard.

As the positron data for the annealed samples show recovery at temperatures up to 600°C, much lower than the 900°C required for structural recovery of the SiO_2, we can discount any defect for which structural recovery (extensive re-ordering of the SiO_2 network) would be necessary in order for positron trapping to be reduced. Hence we can discount the intrinsic defects such as peroxy linkages and oxygen vacancies.

As mentioned above, no ESR signals are observed in the samples annealed above 300°C, but positron trapping patently remains. We therefore consider that the paramagnetic defects such as E' centers are not responsible for positron trapping. The E' center can probably also be discounted because its electronic structure is similar to Si and hence may not be able to give rise to the wide electron momentum that would be required to produce such a low S parameter.

The remaining options are dissolved O_2 and non-paramagnetic trapped electron centers (negatively charged NBOHC and PR). The dissolved oxygen molecule can also act as an electron trap thus becoming negatively charged. The annealing behavior probably gives the strongest clue as to the true identity of the positron trap. The recovery of S parameter level with temperature could be due to recombination or passivation of the NBOHC or PR defects or, it may be due to the out diffusion of interstitial molecular oxygen. As mentioned above, the implantation process could produce H and O atoms which could enhance the ability of existing impurities to passivate the NBOHC or PR defect. Further studies are planned to investigate in more detail these possibilities.

CONCLUSION

In conclusion we have shown that B ion implantation into SiO_2 produces a defect which strongly traps the positron giving rise to a particularly low S parameter value. The effect of annealing on the S-E data was investigated. The positron technique was shown to be highly applicable to the study of defects in SiO_2. The positron technique is found to be sensitive to a defect in SiO_2 which the ESR technique is not sensitive to. A first stage in the study into the defect(s) responsible for trapping of the positron has narrowed down the possible choice to dissolved molecular oxygen or charged Frenkel defects.

Acknowledgments

The authors would like to thank K. Watanabe and T. Sugiura for their help with the ESR measurement and M. Takiyama and S. Hayashi for stimulating discussions. We are grateful to M. Ishizaka for sample preparation. This work was financed in part by a NEDO grant for International Joint Research.

REFERENCES

1. *for example,* J. Keinonen, M. Hautala, E.Rauhala, V. Karttunen, A. Kuronen, J. Räisänen, J. Lautinen, A. Vehanen, E. Puukka, and P. Hautojärvi, Phys. Rev. B **37**, 8269 (1988).
   A. Uedono, L. Wei, C. Dosho, H. Kondo, S. Tanigawa, J. Sugiura and M. Ogasawara, Jpn. J. Appl. Phys. **30**, 201 (1991).
   A. Uedono, S. Tanigawa, J. Sugiura and M. Ogasawara, Jpn. J. Appl. Phys. **29**, 1867 (1990).
   P.J. Simpson, M. Vos, I.V. Mitchell, C. Wu and P.J. Schultz, Phys. Rev. B **44**, 12180 (1991).
2. B. Nielsen, K.G. Lynn, T.C. Leung, B.F. Cordts and S. Seraphin, Phys. Rev. B. **44**, 1812 (1991).
3. *for example,* D.L. Griscom and E.J. Friebele, Radiat. Effects **65**, 63 (1982).

E.J. Friebele and D.L. Griscom, *Treatise on Materials Science and Technology, vol. 17*, edited by M. Tomozawa and R.H. Doremus (Academic Press, New York, 1979) p.257.
D.L. Griscom, Nucl. Instrum. Methods Phys. Res. B **1**, 481 (1984).
H. Nishikawa, R. Nakamura, R. Tohmon, Y. Ohki, Y. Sakurai, K. Nagasawa and Y. Hama, Phys. Rev. B **41**, 7828 (1990).
E.H. Poindexter and P.J. Caplan, J. Vac. Sci. Technol. A **6**, 1352 (1988).
D.L. Griscom, J. Ceramic Soc. Jpn. **99**, 923 (1991).

4. H. Hosono and R.A. Weeks, Phys. Rev. B **40**, 10543 (1989).
   E.H. Poindexter and P.J. Caplan, Prog. Surf. Sci. **14**, 201 (1983).
5. J.P. Biersack and L.G. Haggmark, Nucl. Instrum. Methods **174**, 257 (1980).
6. M. Fujinami, Hoshasen *in press*.
7. P.J. Schultz and K.G. Lynn, Rev. Mod. Phys. **60**, 701 (1988).
8. B. Nielsen, K.G. Lynn, D.O. Welch and T.C. Leung, Phys. Rev. B. **40**, 1434 (1989).
   Yen-C. Chen, K.G. Lynn, and B. Nielsen, Phys. Rev. B **37**, 3105 (1988).
   B. Nielsen, K.G. Lynn, Yen-C. Chen and D.O. Welch, Appl. Phys. Lett. **51**, 1022 (1987).
9. E.H. Poindexter, P.J. Caplan, B.E. Deal and R.R. Razouk, J. Appl. Phys. **52**, 879 (1981).
10. G. C. Aers, in *Positron Beams for Solids and Surfaces*, edited by P.J. Schultz, G.R. Massoumi and P.J. Simpson (*American Institute of Physics* **218**, New York, 1990), p.162.
11. M. Fujinami and N.B. Chilton, Appl. Phys. Lett. submitted.
12. R. Tohmon, Ph. D thesis, Waseda University (1990).
13. M. Fujinami and N.B. Chilton unpublished.
14. Mbungu-Tsumbu, D. Segers, M. Dorikens, L. Dorikens-Vanpraet, C. Laermans and A. van den Bosch, J. Non-Cryst. Solids **65**, 131 (1984).

MONTE CARLO STUDIES OF POSITRON IMPLANTATION IN ELEMENTAL METALLIC AND MULTILAYER SYSTEMS

V. J. Ghosh, D. O. Welch, and K. G. Lynn
Brookhaven National Laboratory
Upton, NY 11973

ABSTRACT

We have used a Monte Carlo computer code developed at Brookhaven[1,2] to study the implantation profiles of 1-10 keV positrons incident on a wide range of semi-infinite metals and multilayer systems. Our Monte Carlo program accounts for elastic scattering as well as inelastic scattering from core and valence electrons, and includes the excitation of plasmons. The implantation profiles of positrons in many metals as well as Pd/Al, and Al/Co/Si multilayers are presented. Scaling relations and closed-form expressions representing the implantation profiles are also discussed.

INTRODUCTION

The motivation for the development of Monte Carlo methods to study positron slowing down in solids has been discussed before[1-4] (and references therein), and will not be discussed here, except to stress that the implantation profiles generated by Monte Carlo programs are needed in the analysis of data obtained by the positron beam study of solids and surfaces, in particular for defect profiling studies. Here, we will discuss, briefly, some results obtained by using a Monte Carlo program developed at BNL.[1,2]

A brief description of this Monte Carlo program is given in Section II and the results for simple metallic systems, the variation of mean depth with energy as well as the implantation profiles are discussed in Section III. A scaling relation which reduces profiles of different energies to the same functional form is proposed. Section IV contains the Monte Carlo results for multilayer systems composed of layers of metallic elements. Trends followed by the backscattered fraction, mean depth, and the implantation profiles are discussed and a scaling procedure to reduce the complicated implantation profiles for multilayers to a simple functional form is described. Section V contains a brief conclusion and suggestions for extensions of the present work which will lead to a better understanding of the positron implantation process.

DESCRIPTION OF MONTE CARLO PROGRAM

The Monte Carlo code developed at Brookhaven (BNLMC) is very similar to that code of Valkealahti and Nieminen[3,4] and is described in detail elsewhere.[1,2] The positron or electron implantation is modeled by following the trajectories of a large number of particles of a given starting energy E_0. The particle trajectory is changed by 'collisions' with the atoms of the host material. The collisions are either elastic, where the particle loses no energy, or inelastic, where there is a certain energy loss ΔE. During the collision the particle is also deflected through a scattering angle $\Omega(\theta,\phi)$. For each collision, the type of collision, the energy loss, the scattering angle, and the distance the particle travels between collisions are determined from a set of random numbers. The trajectory of the particle is followed until it either gets

© 1994 American Institute of Physics

backscattered out of the material or until its energy is reduced below a 'cutoff energy' when it is considered to be implanted in the material.

The physics of the positron-atom or electron-atom interactions is incorporated via the cross-sections which describe the scattering processes. The incident particles undergo elastic Coulomb scattering from a single atom potential which has been modified to account for the effects of the lattice on the electronic structure.[5] Such a potential, however, only partially accounts for crystalline order, and structural effects such as channeling are not included. The differential scattering cross-section is calculated using the method of partial waves[6] for scattering from a central potential, for which the phase shifts are obtained by solving the radial Schrodinger equation using Milne's method.[7]

Inelastic scattering cross-sections and energy-loss functions are determined separately for core and valence electrons. The total inelastic cross-section for relativistic scattering by K-shell ionization includes contributions from processes at both large and small impact parameters. For large impact parameters the cross-sections of Kolbensvedt[8] are used while for small impact parameters, Möller cross-sections[9] are used for electrons and Bhabha cross-sections[10] for positrons. For core electrons other than the K-shell electrons Gryzinski excitation cross-sections[11] are used. For all core inelastic scattering the method of Rohrlich and Carlson[12] is used to determine the energy loss functions and the Gryzinski formula relates the energy loss to the scattering angle.[11]

Inelastic scattering from valence electrons can proceed either by single particle processes or by the collective plasmon excitation of the electron gas. For the single particle processes, Penn mean free paths,[13,14] random-phase approximation energy loss functions,[15] and DeSalvo, Parisini and Rosa scattering angle formulae[16] are used. For the collective plasmon processes Quinn mean free paths,[17] Isaacson energy-loss tables[18] and the DeSalvo, Parisini and Rosa scattering angle[16] formulae are used.

Our Monte Carlo simulations are limited to electrons and positrons with initial energies of 1 to 10 keV. The particles are considered implanted when their energy reduces to 25 eV, which is well above the energy of thermalized positrons. Thus the BNLMC does not incorporate the final stages of thermalization. As the positron energy decreases from 25 eV to <1 eV, conduction electron scattering becomes less and less efficient and is replaced by phonon scattering as the major energy-loss mechanism. Phonon scattering is included in a low-energy Monte Carlo algorithm being developed by Ritley, Lynn, Ghosh and Welch.[19]

RESULTS FOR SEMI-INFINITE ELEMENTAL METALS

A. <u>Mean Depth</u>

Monte Carlo simulations were done for many semi-infinite metals: Al, Si (ignoring the effects of band gaps), Co, Ni, Cu, Pd, Ag, W, Pt, and Au. Initial positron energies were varied from 1 to 10 keV and the trajectories of 2,000 to 10,000 particles were studied. The mean implantation depth (as a function of incident energy) was calculated from the Monte Carlo data. The mean depths scale, albeit imperfectly, with the density of the host material, and it has been customary to assume that the density dependence could be factored out so that the mean depths would lie on a universal curve;

$$\langle z \rangle = \frac{A}{\rho} E^n \tag{1}$$

where E is the energy of the positrons (in keV), <z> the mean depth, ρ the density of the host material and A and n are constants. The values of these constants were found to be n=1.6 and A=400 μg/cm² (keV)^1.6, when the mean depth is expressed in angstroms.[20]

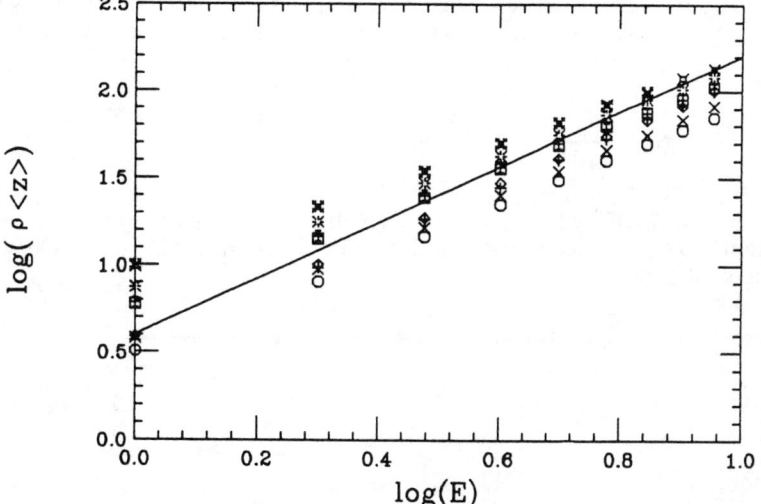

Fig. 1. Density scaled mean depths as a function of the implantation energy for Al◊, Pd□, Cu x, Si +, Pt ✗, W ✻, Au ✼, Ag ✚, Ni ○. The straight line represents the universal curve.

The density-scaled results of the mean depth from our simulations are plotted in Fig. 1. Our Monte Carlo results show that the fitted values of A and n are material dependent and can be correlated to the material density as is shown in Figs. 2(a) and 2(b), i.e. a simple inverse density dependence still leaves some residual density effects.

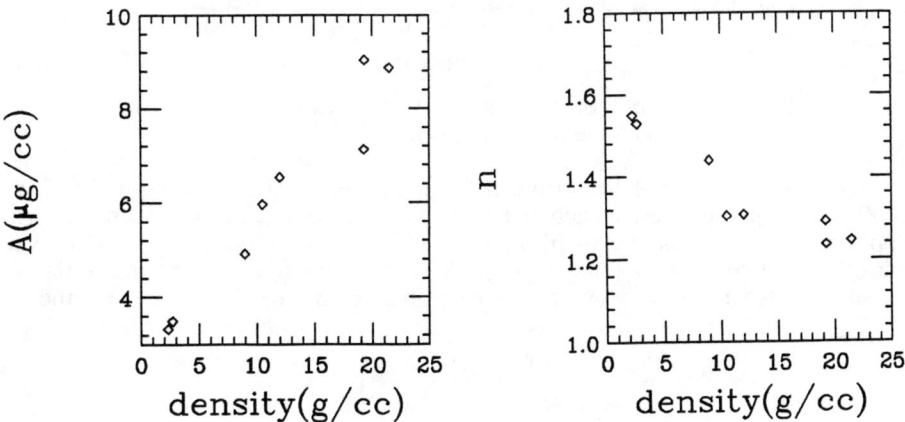

Fig. 2. (a) Variation of A with density z.
(b) Variation of n with density z.

B. Implantation Profiles

The Monte Carlo studies of Valkealahti and Nieminen[3,4] showed that the positron implantation profiles in a semi-infinite material were best represented by the derivative of a Gaussian. The derivative of a Gaussian profile is a special case of a more general class of profiles, the so-called Makhovian distribution 2b, where as a function of z, the probability P(z) that a given positron is implanted between z and z+dz is:

$$P(z) = N_0 m \frac{z^{m-1}}{z_0^m} \exp\left[-\left(\frac{z}{z_0}\right)^m\right] \tag{2}$$

with $m = 2$ and $z_0 = \langle z \rangle / \Gamma(1+1/m)$. N_0 is a normalization constant. For $m = 2$ the gamma function $= 0.8862$. Generally z_0 and m are used as adjustable parameters to fit the experimental data.

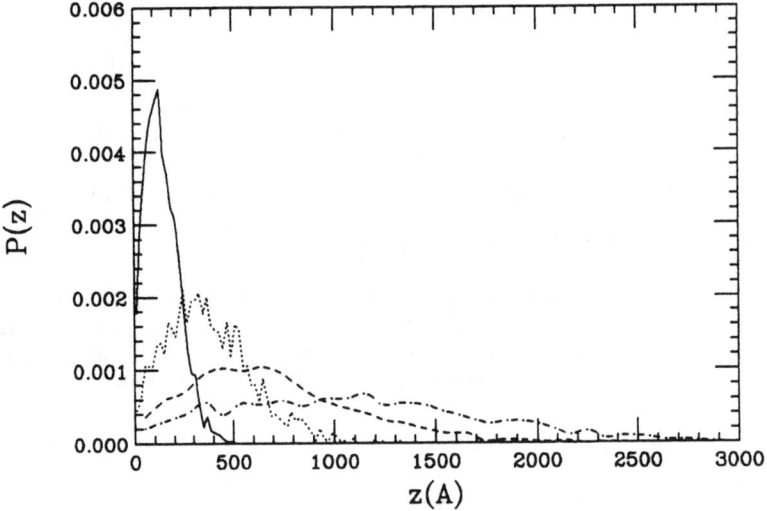

Fig. 3. Implantation profiles of positrons in Al for implantation energies of 1 keV (—), 2 keV (····), 3 keV (---) and 4 keV (·-·-).

The Monte Carlo profiles of positrons implanted in Al are plotted in Fig. 3 with 2000 positrons implanted at each energy. Since the mean depth $\langle z \rangle$ increases with implantation energy E_0, the profiles rapidly broaden with increasing E_0. However, this effect can be accounted for by a simple scaling procedure. The prescription for scaling profiles of different energies can be obtained by rewriting Eq. (2) in the form

$$\langle z \rangle P(z) = \frac{N_0}{c_m} \left(\frac{z}{c_m \langle z \rangle}\right)^{m-1} \exp\left[-\left(\frac{z}{c_m \langle z \rangle}\right)^m\right] \tag{3}$$

where, $z_0 = c_m \langle z \rangle$, with c_m a numerical constant which depends on m.

Figure 4 shows a plot of $\langle z \rangle P\langle z \rangle$ as a function of $z/\langle z \rangle$ for 10,000 positrons in Ni. This scaling has the advantage that the implantation profiles at different energies can be plotted on the same curve and one set of fitting parameters can describe all the profiles. This form of scaling also applied to more general forms of the profile than the Makhovian, as discussed below.

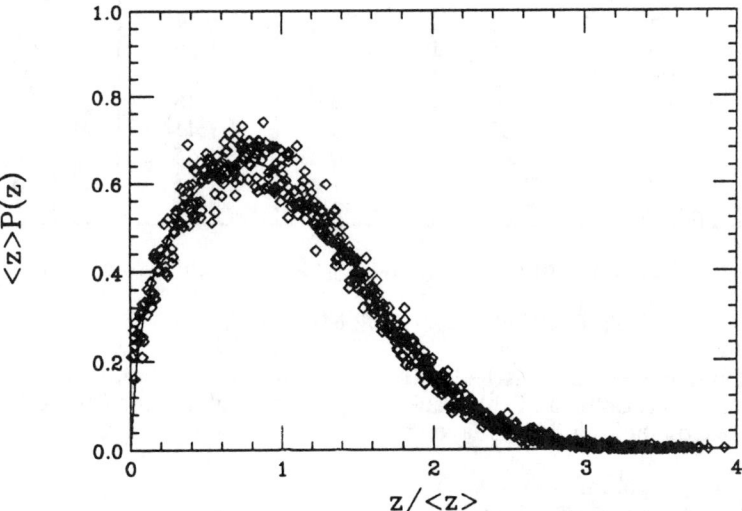

Fig. 4. Scaled implantation profiles for 1-10 keV positrons in nickel with 10,000 positrons implanted at each energy.

This universal scaling relationship, (i.e. $\langle z \rangle P(z)$ is the same function of $z/\langle z \rangle$, independent of the implantation energy,) was found to apply to all the materials that we studied (Al, Si, Co, Ni, Cu, Pd, Ag, W, Pt, and Au) for profiles with implantation energies of 1-10 keV.

Although our Monte Carlo profiles can be fit fairly well with a Makhovian using m and z_0 as adjustable parameters, we found, however, that even the best fit Makhovian underestimates the positron density at the surface. Monte Carlo results of Jensen and Walker[21] also showed similar results and led them to propose a multiplying factor which would adequately model the profile near the surface. We find that a good fit to our Monte Carlo data can be obtained with a function of the form

$$\langle z \rangle P(z) = N_{\ell m} \left(\frac{u}{c_{\ell m}} \right)^{\ell} \exp\left[-\left(\frac{u}{c_{\ell m}} \right)^{m} \right] \qquad (4)$$

where $u = z/\langle z \rangle$, $z_0 = c_{\ell m}\langle z \rangle$ and $P(u) = \langle z \rangle P(z)$. N_{ml} is the normalization constant and ℓ, and m are fitting parameters. $c_{\ell m}$ is a numerical constant obtained from the condition that $\langle z \rangle$ obtained from the distribution function equals that obtained from the Monte Carlo "data." (When $\ell = m-1$ this reduces to the Makhovian form.]

From the form of the fitting function Eq. (4) it is obvious that near the surface (i.e. for $u < 1$) the shape of the profile is determined by the value of ℓ, while the shape of the tail ($u > 1$) is dependent on the value of m. By not restricting $\ell = m-1$ we can model the shape of the profile with greater accuracy. The results of the fitting using Eq. (4) are listed in Table I. The fitting parameter m showed a systematic decrease from a value of 2.7 for Al to 1.78 for Pt. Values of ℓ ranged from 0.4 to 0.5 but there was no systematic z dependence and the values of c_0 varied from 1.30 to 1.55.

Element	Density (g/cc)	m	ℓ	$c_{m\ell}$	$N_{m\ell}$
Al	2.698	2.70	0.47	1.53	1.64
Si	2.329	2.52	0.54	1.46	1.71
Cu	8.96	2.55	0.40	1.56	1.57
Ni	8.96	2.53	0.43	1.53	1.60
Co	8.96	2.34	0.49	1.46	1.65
Ag	1.50	2.03	0.43	1.41	1.57
Pd	11.99	2.11	0.40	1.45	1.54
W	19.25	2.31	0.31	1.55	1.46
Pt	21.45	1.78	0.43	1.30	1.53

Table I. The values of m, ℓ, $c_{m\ell}$, and $N_{m\ell}$ for different elements.

MONTE CARLO STUDY OF MULTILAYERS

The two multilayer systems studied in detail were palladium-aluminum bilayers or overlayers of Al and Co on a Si substrate. For palladium-aluminum Monte Carlo studies were done on the following bilayers:

i) semi-infinite aluminum
ii) 100 Å layer of Al on Pd
iii) 300 Å layers of Al on Pd
iv) 500 Å layers of Al on Pd
v) semi-infinite Pd
vi) 100 Å Pd on Al
vii) 300 Å Pd on Al
viii) 500 Å Pd on Al

The variation of mean depth with energy for these bilayers is plotted in Fig. 5. The dashed and solid lines have been drawn to guide the eye and are not fits to any data. The top dashed line represents the mean depth for semi-infinite Al calculated from the Monte Carlo data; the bottom solid line represent semi-infinite Pd. The Monte Carlo results show that for low energies the multilayer system has the mean depth of the overlayer and that at higher energies the mean depth is close to that of the bulk material. The backscattered fraction of the Pd-Al system also showed a similar transition from overlayer to bulk properties with an increase in energy.

A 3-metal system consisting of 1000 Ang of Al on 968 Ang of Co on a silicon substrate was also modeled using the Monte Carlo code. X-ray reflectivity and variable-energy positron studies on this and similar systems have been done by Heald and Nielsen.[22] The values of the mean depth and backscattered fraction of the multilayer system were similar to those of Al at 1-3 keV, those of Co at 4-6 keV, and those of Si at 7-10 keV. The implantation profiles are plotted in Fig. 6 for incident energies of 1, 3, 5, 7 and 9 keV. At 1 keV the positrons are implanted primarily in the aluminum overlayer. At 3 and 5 keV some positrons are planted in the Al and some in Co, the fraction in Co going up at 5 keV. At 7 and 9 keV there is positron implantation in the Si, too. In fact, the profiles show the same trend as the backscattered fraction and the mean depth, showing that the results are quite consistent.

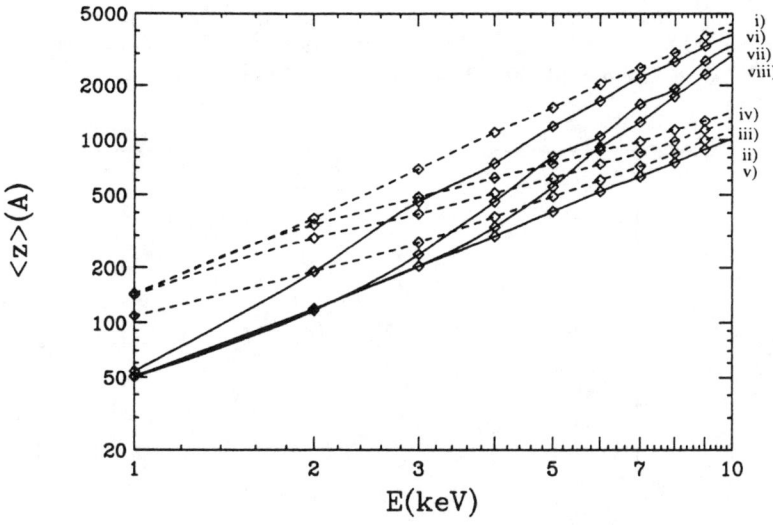

Fig. 5. Variation of mean depth with energy for the different palladium-aluminum bilayer systems.

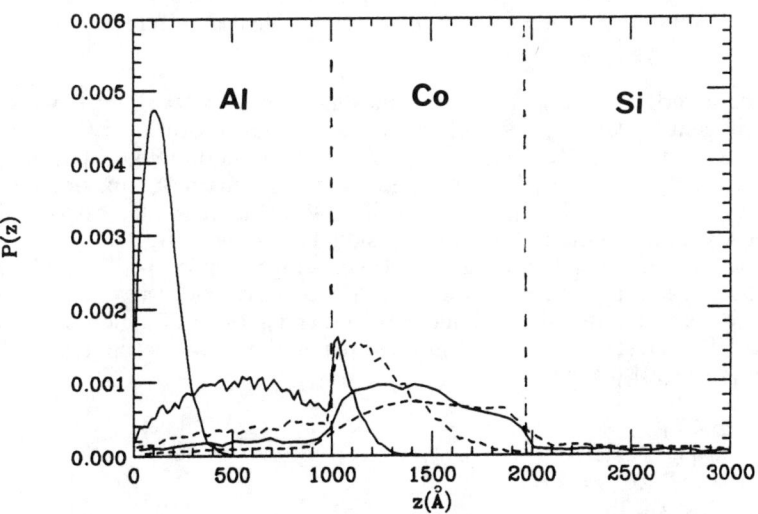

Fig. 6. Implantation profiles of positrons in Al/Co/Si. Solid lines represent 1, 3, 7 keV and dashed lines represent 5 and 9 keV profiles.

The differences in the density of the substrate and the overlayers have previously been modeled by a density-normalization procedure.[23] Density-normalized profiles of the Al/Co/Si system are plotted in Fig. 7. When density normalization was tried

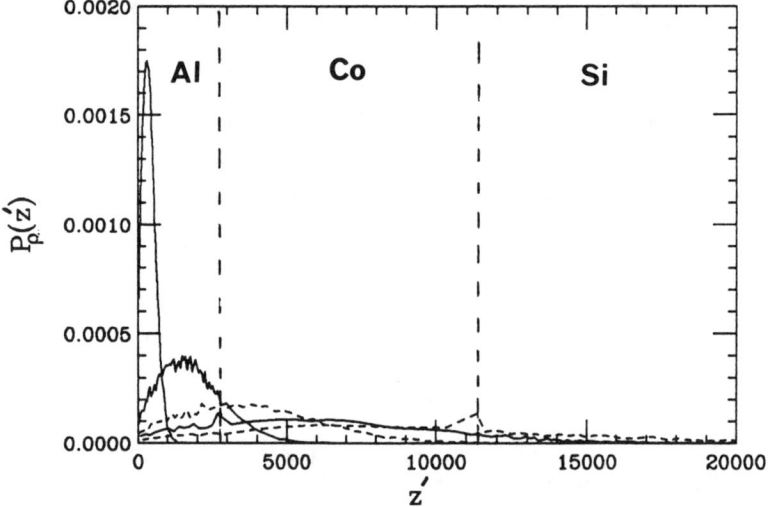

Fig. 7. Density normalized implantation profiles in Al/Co/Si. Solid lines represent 1, 3, 7 keV and dashed lines represent 5 and 9 keV profiles.

for copper overlayers on tungsten the procedure seemed to work fairly well and it should, in general, work fairly well when the densities of overlayer and substrate are not too different. However, in the case of palladium-aluminum, (Pd density = 11.95 g/cc., Al density = - 2.698 g/cc) we found that the density normalization did not adequately describe the implantation profiles. We note that in the case of Al/Co/Si also density normalization leaves some residual multilayer effects.

Since the mean-depth scaling had worked so well for semi-infinite elements we tried a similar scaling for multilayers as well. Let us consider a system of two overlayers, one of material 1 and thickness T_1, and another of material 2 and thickness T_2 on a semi-infinite substrate of material 3. Now define two new variables u_{eff} and $[<z>P(z)]_{eff}$ such that

for $z \leq T_1$
$$u_{eff} = z/<z_1>$$
$$[<z>P(z)]_{eff} = <z_1> P(z)$$
for $T_1 < z \leq T_2$
$$u_{eff} = (z - T_1)/<z_2> + T_1/<z_1>$$ (5)
$$[<z>P(z)]_{eff} = <z_2> P(z)$$
and for $z > T_2$
$$u_{eff} = (z - T_2)/<z_3> + T_2 - T_1)/<z_2> + T_1/<z_1>$$
$$[<z>P(z)]_{eff} = <z_3> P(z)$$

Here $\langle z_1 \rangle$, $\langle z_2 \rangle$ and $\langle z_3 \rangle$ are the mean depths for positron implantation in semi-infinite materials 1, 2 and 3. For a particular energy E these values are calculated using Eq. (1) with values of A and n obtained by fitting the Monte Carlo mean depths of semi-infinite elements. The results of this normalization procedure for Al/Co/Si are plotted in Fig. 8. Not only does this normalization scheme adequately take

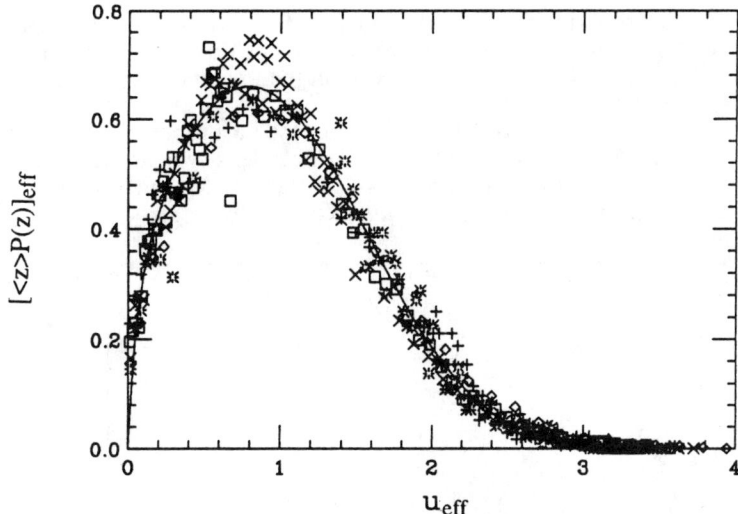

Fig. 8. Profiles for Al/Co/Si renormalized according to Eq. (5). The Monte Carlo data at 1 keV is represented by ◊, 3 keV by ×, 5 keV by □, 7 keV by +, and 9 keV by ※. The solid line represents the fitted curve.

into account the effect of the different densities of different materials but the profiles for different energies which look so different from each other all fall on the same universal curve. The normalized data from 1 to 10 keV can all be fitted with an expression analogous to Eq. (4) where the parameters ℓ, m, $c_{\ell m}$, $N_{\ell m}$ are all obtained for the normalized data:

$$[\langle z \rangle P(z)]_{eff} = N_{\ell m} \left(\frac{u_{eff}}{c_{\ell m}} \right)^{\ell} \exp\left[-\left(\frac{u_{eff}}{c_{\ell m}} \right)^{m} \right] \quad (6)$$

where u_{eff} is given by Eq. (5). The fitted profile for this multilayer system is compared in Fig. 9 to the fitted curves for elemental Al, Co and Si [i.e. Eq. (5) with T=c]. The parameters ℓ, m, etc. for the various profiles are compared in Table II.

Element or Multilayer	m	ℓ	$c_{m\ell}$	$N_{m\ell}$
Al/Co/Si	2.59	0.43	1.51	1.66
Al	2.70	0.47	1.53	1.64
Co	2.34	0.49	1.46	1.65
Si	2.52	0.54	1.46	1.71

Table II. Comparison of the fitting parameters of the Al/Co/Si multilayer with those of semi-infinite Al, Co and Si.

This scaling procedure was also tested for the various palladium-aluminum multilayer systems for incident energies of 1 to 10 keV and was found to work just as well.

CONCLUSION

From the Monte Carlo study of simple metallics we find that the energy variation of the mean depth can be expressed by the functional form of Eq. (1). However, A and n are not universal constants but density-dependent parameters whose density dependence needs to be explored further.

We also find that the Makhovian distribution is not the best representation of the implantation profile and that it underestimates the positron density near the surface as found by Jensen et al.[21] An alternate form, Eq. (4), similar to a Makhovian but somewhat more general, provides a better fit to our Monte Carlo results.

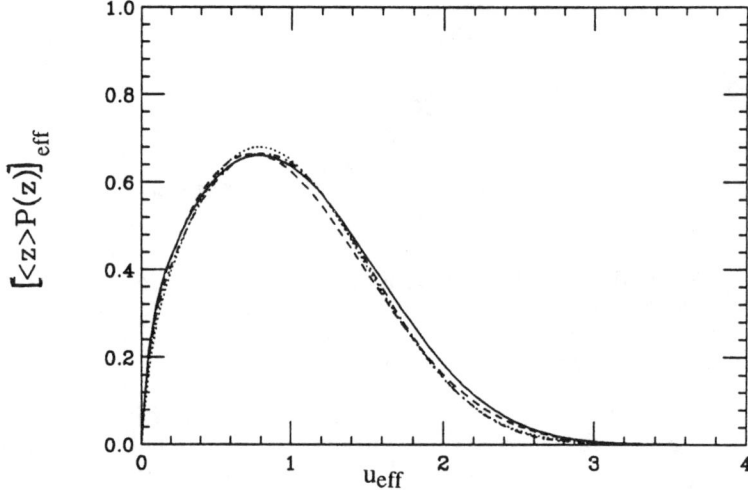

Fig. 9. Comparison of the renormalized Al/Co/Si profile (solid line) [Eq. (5)] with those of Al (---), Si (••••), and Co (•-•-•) [Eq. (5) with $T_1=\infty$].

The implantation profiles for different energies can be scaled in such a way that they all fall on a universal curve. This scaling procedure worked for positrons implanted at energies ranging from 1 to 10 keV in semi-infinite Al, Si, Co, Ni, Cu, Pd, Ag, W, and Au. It remains to be seen if these scaling relationships will still be valid for positrons implanted at higher energies and whether similar scaling can be done for electron implantation profiles.

The Monte Carlo studies of multilayer systems showed that the backscattered fraction and mean depth have values characteristic of the overlayer material at low energies and those of the substrate at higher energies. The implantation profiles also show the same trend; however, their detailed shapes are both complicated and energy-dependent. By suitably scaling these profiles, Eq. (5), we can get a profile independent of the implantation energy. This profile can be fitted with a Makhovian-like function [Eq. (6)] and it is very similar to the scaled profiles of the constituent metals. Whether this scaling process is applicable at higher energies and for different metal combinations remains to be seen, but for Pd/Al and Al/Co/Si it works well.

These scaled energy-independent profiles can be used as the initial condition in defect profiling programs like VEPFIT[23] for the analysis of s-parameter experimental data.

ACKNOWLEDGMENTS

This work was supported by the U.S. Department of Energy, Division of Materials Sciences, Office of Basic Energy Sciences under Contract DE-AC02-76CH00016. VJG received partial support from NEDO.

REFERENCES

1. K. A. Ritley, M. McKeown, and K. G. Lynn, Positron Beams for Solids and Surfaces, London, Ontario, Canada, 1990, P. J. Schultz, G. R. Massoumi, and P. J. Simpson, Editors: p. 3, American Institute for Physics, New York.
2. M. McKeown, K. G. Lynn, V. J. Ghosh, and D. O. Welch, (unpublished).
3. S. Valkealahti and R. M. Nieminen, Appl. Phys. A. 32, 95 (1983).
4. S. Valkealahti and R. M. Nieminen, Appl. Phys. A. 35, 51 (1984).
5. A. J. Green and R. C. G. Lecky, J. Phys. D 9, 2123 (1976).
6. L. Schiff, Quantum Mechanics, p. 117, McGraw-Hill, London, 1968.
7. M. Abramowitz and I. A. Stegun, Handbook of Mathematical Functions, p. 897, Dover, New York, 1965.
8. H. Kolbensvedt, J. Appl. Phys. 38, 4785 (1967).
9. J. M. Jauch and F. Rohrlich, The Theory of Photons and Electrons, p. 256, Addison-Wesley, Cambridge, 1954.
10. H. J. Bhabha, Proc. Roy. Soc. A 152, 559 (1935).
11. M. Gryzinski, Phys. Rev. A 138, 305, 322 and 336 (1965)
12. F. Rohrlich and B. C. Carlson, Phys. Rev. 93, 38 (1954).
13. D. R. Penn, J. Electron. Spect. 9, 29 (1976).
14. D. R. Penn, Phys. Rev. B 13, 5248 (1976).
15. J. Dhra, Phys. Rev. B 21, 4904 (1980).
16. A. Desalvo, A. Parisini, and R. Rosa, J. Phys. D 17, 1545 (1984).
17. J. J. Quinn, Phys. Rev. 12, 1453 (1962).
18. D. Isaacson, Compilation of R_s Values, Radiation and Solid State Laboratory, New York University, Internal Report.
19. K. A. Ritley, K. G. Lynn, V. J. Ghosh, and D. O. Welch (unpublished).
20. P. J. Schultz and K. G. Lynn, Rev. Mod. Phys. 60, 701 (1988).
21. K. O. Jensen, A. B. Walker, and N. Bouarissa, Positron beams for solids and surfaces, London, Ontario, Canada, 1990, P. J. Schultz, G. R. Massoumi, and P. J. Simpson, Editors, p. 17, AIP (New York).
22. S. M. Heald and B. Nielsen (to be published in same proceedings).
23. A. Van Veen, H. Schut, J. de Vries, R. A. Hakvoort, and M. R. IJpma. Positron beams for solids and surfaces, London, Ontario, Canada, 1990, P. J. Schultz, G. R. Massoumi, and P. J. Simpson, Editors, p. 171, AIP (New York).

CHARACTERIZATION OF AMORPHOUS SILICON

R.A. Hakvoort, A. van Veen and H. Schut,
Interfaculty Reactor Institute, Delft University of Technology, Mekelweg 15, NL–2629 JB Delft, The Netherlands,

M.J. van den Boogaard and A.J.M. Berntsen,
Utrecht University, Debye Institute, Dept. of Atomic and Interface Physics, P.O. Box 80000, NL–3508 TA Utrecht, The Netherlands,

S. Roorda[1] and P.A. Stolk,
FOM Institute for Atomic and Molecular Physics, Kruislaan 407, NL–1098 SJ Amsterdam, The Netherlands,

A.H. Reader,
Philips Research Laboratories, P.O. Box 80000, NL–5600 JA Eindhoven, The Netherlands.

ABSTRACT

S-parameter positron beam measurements have been done on several kinds of a-Si: Kr-sputtered a-Si, PECVD a-Si, MeV ion beam amorphized Si and a-Si grown in an MBE-system at a low deposition temperature. Kr sputtered a-Si becomes denser for higher Kr concentration. PECVD a-Si:H contains micro-cavities with a size depending on growth temperature. MeV ion beam amorphized Si contains 1.2 at.% small vacancies, which decreases upon annealing (relaxation) to 0.4 at.%. This effect can be mimicked by H-implantation and subsequent annealing, showing that at least some of the dangling bonds in a-Si are located at these vacancy-type defects. Finally positron measurements show that MBE-system grown a-Si contains large open-volume defects. The positron annihilation data are supplemented by data from some other techniques.

INTRODUCTION

The study of the annealing behaviour and the role of incorporated gases like H and noble gases in a-Si is interesting from both theoretical ('structure' of a-Si) and practical standpoint (solar cells, etc.). Dangling bonds in a-Si deteriorate the electrical properties of the material. The carriers are trapped at these bonds. On the other hand hydrogen may passivate these dangling bonds. Positrons are sensitive to electric charge and are therefore thought to be capable of probing these defects.

EXPERIMENTAL

Four types of a-Si have been examined:
  1) a-Si sputter deposited at 310 °C from a Kr-plasma at a pressure of 0.1 Pa. The Kr-Si atom ratio was varied, resulting in a-Si:Kr with differing Kr concentrations. The Kr concentration was determined by Rutherford Backscattering (RBS).
  2) a-Si grown by Plasma Enhanced Chemical Vapour Deposition (PECVD) at temperatures ranging from 50 °C to 200 °C and from a 2:1 mixture of $SiH_4:H_2$ (rf-power during deposition was 20 W at a frequency of 13.56 MHz).

---

[1]*Present address:* Département de Physique, Université de Montréal, Département de Physique, C.P. 6128, succ. "A", Montréal, Québec, H3C 3J7 Canada,

3) Ion beam amorphized silicon, prepared by implantation of 0.5 and 1 MeV Si^+-ions at -100 °C, resulting in a uniform 500 nm a-Si layer. The sample has been annealed at 500 °C, i.e. just below the recrystallization temperature. In a similar *as-implanted* sample an additional 50 keV H^+-implantation to a dose of 5×10^{16} cm^{-2} has been performed at room temperature. This sample was annealed in vacuum at 150 °C for 2, 6 and 25 hours respectively (see further [1]).

4) a-Si grown in a Molecular Beam Epitaxy (MBE) system. The first 10 nm was grown at room temperature, followed by 300 nm at 250 °C and a postgrowth anneal of 30 min. at 350 °C. Raman measurements show that at least 99 % of the overgrowth is amorphous.

Table 1. The fitted values of the S-parameters and the positron diffusion lengths of the Si-layers of the samples examined.

	scaled S	diff. length
monocrystalline Si		
c-Si	1.0000	245 nm
Kr-sputtered Si		
a-Si (0.55 % Kr)	1.0240	13.8 nm
a-Si (1.0 % Kr)	1.0244	12.6 nm
a-Si (2.7 % Kr)	1.0213	8.7 nm
a-Si (3.0 % Kr)	1.0209	7.0 nm
a-Si (3.9 % Kr)	1.0170	7.1 nm
a-Si (4.8 % Kr)	1.0183	8.0 nm
PECVD Si		
a-Si:H (50 °C)	1.0651	2.4 nm
a-Si:H (100 °C)	1.0534	2.4 nm
a-Si:H (150 °C)	1.0364	3.3 nm
a-Si:H (200 °C)	1.0273	6.2 nm
ion beam Si		
as-impl. a-Si (-100 °C)	1.0379	8.6 nm
annealed a-Si (500 °C)	1.0257	15.5 nm
ion beam a-Si + 50 keV H^+		
—	1.0391	6.5 nm
25 h anneal (150 °C)	1.0249	5.3 nm
MBE-system grown Si		
a-Si	1.1450	5.4 nm

All samples were examined with slow positrons, whereas the Kr-sputtered and 'MBE'-samples were also characterized with Raman-spectroscopy. Besides, the PECVD layers have been studied with Small Angle X-ray Scattering (SAXS) and the Ion beam samples by free carrier lifetime measurements.

The analysis of the data was done with the VEPFIT-program [2]. All S-values presented in this article, have been scaled to the fitted c-Si S-value and are therefore properly S/S_{c-Si}-values.

RESULTS AND DISCUSSION

The results of the measurements are shown in figs. 1–4 and the fit results are summarized in table 1.

The S-parameter of Kr-sputtered a-Si ranges linearly between 1.024 for low Kr-concentrations (0.5 and 1 %) to 1.018 for high concentrations (3–5 %). The positron diffusion length in the latter a-Si:Kr-layers is significantly lower than the one in a-Si with a low Kr-concentration.

Apparently, the 50 eV Kr creates Si self-interstitials which fill the grown-in vacancy-type defects, thereby reducing their size. On the other hand the electrically inactive Kr-atoms themselves fill the positron-traps. Both effects result in dense a-Si for a high Kr-concentration, corresponding to a lowering of the S-parameter. Nevertheless, the decrease of the diffusion length evidences more positron traps in the case of high Kr-concentrations. Though the Kr atoms and the Si self-interstitials have completely occupied the vacancies in the material, the oversized Kr atoms induce smaller positron traps due to the straining of the lattice.

This explanation is consistent with Raman-measurements, which show that the the *hwhm* of the Transverse Optical (TO) peak increases from 38.5 cm^{-1} for a-Si with

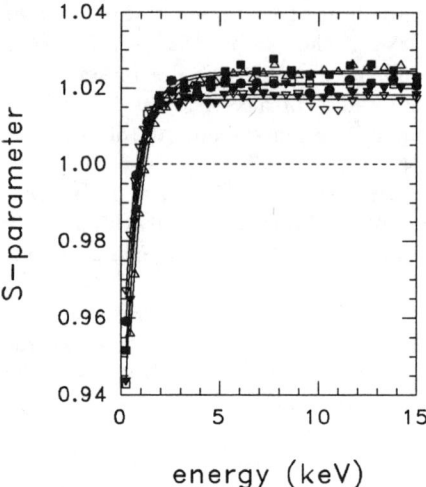

Figure 1. *S*-parameter *vs.* positron incident energy for Kr-sputtered a-Si containing 0.55 % (top curve) to 4.8 % Kr (lowest curve).

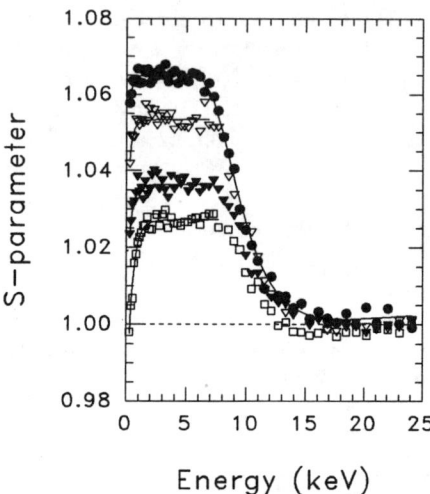

Figure 2. *S*-parameter *vs.* positron incident energy for PECVD a-Si grown at 50 °C (top curve), 100 °C, 150 °C and 200 °C (bottom curve).

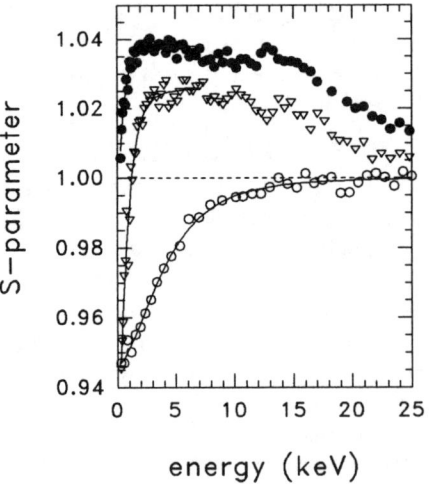

Figure 3. *S*-parameter *vs.* positron incident energy for c-Si (bottom) and *as-implanted* and *annealed* ion beam amorphized Si (top and middle curve).

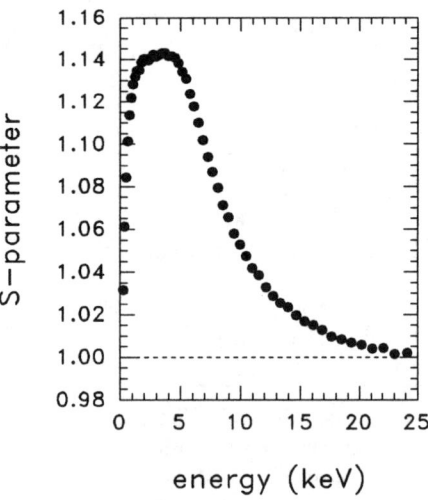

Figure 4. *S*-parameter *vs.* positron incident energy for MBE-system grown a-Si.

0.55 % Kr to 43.5 cm^{-1} for a-Si with 4.8 % Kr. Therefore, the mean bond angle distortion of a-Si is larger for a-Si with much Kr, corresponding to less ordering on the short range. As a reference: MBE a-Si has a TO$_{hwhm}$ of 37.2 cm^{-1}, while it contains large voids. We conclude: More Kr in the lattice leads to a larger bond angle distortion, which corresponds to less short-range order and results in a shorter positron diffusion length [3].

The S-parameter of PECVD a-Si:H ranges between 1.027 for material grown at 200 °C and 1.065 for 50 °C a-Si. The diffusion length decreases for these samples from 2.4 nm to 6.2 nm.

The high S-parameter indicates Positronium (Ps) formation, at least in the 50 °C sample, showing the a-Si:H contains large voids. Positrons trapped in small open-volume defects in a-Si have S-parameters of about 1.035 [4]. Extremely high S-values can only be the result of zero-momentum annihilations of para-Ps in (large) voids, which has indeed been observed in PECVD a-Si:H via both 2D-ACAR and positron lifetime measurements [5–6]. The rather low S-parameter for the 150 °C and 200 °C a-Si, even lower than the S-parameter for vacancy-type defects mentioned above, can be explained by the presence of Si–H bonds in the small voids. These bonds change the electron momentum density in the neighbourhood of the trapped annihilating positrons, apparently leading to a lower S. This explanation is consistent with the much higher S-parameter found for MBE-system grown a-Si, where no hydrogen is present (see below).

Small Angle X-ray Diffraction (SAXS) measurements have shown that the free volume fraction of this a-Si increases from a few percent for the 200 °C growth up to 12 % for the poor-quality 50 °C growth [7]. This is in accordance with our measurements, which show that both the size of the voids ($\sim S$) and the number of the voids ($\sim$diffusion length) reduce for higher deposition temperatures.

The S-parameter of ion beam amorphized Si is 1.038 after the implantation, but decreases upon annealing (500 °C) to 1.026. The diffusion length increases from 8.6 to 15.5 nm. After H-implantation in the *as-implanted* a-Si the S-parameter remains at 1.039 and decreases only after prolonged heating at 150 °C. The diffusion length slightly decreases from 6.5 to 5.3 nm, although these values are not completely beyond doubt, because the diffusion length is obtained from positron measurements in the surface region of the a-Si where there should be no implanted H.

By the Si-implantation defects are introduced in the Si, which eventually leads to a complete amorphization. The resulting network of structural defects of the a-Si contains many dangling and/or strained bonds. Positrons are trapped in the vacancy-type defects present in the lattice. Upon thermal annealing the strained bonds relax, while the dangling-bonds can be passivated by hydrogen. In the former case the number of defects decreases in accordance with the increase of the positron diffusion length. From the decrease of S it can be concluded that the mean radius of the vacancy-type defects lowers by the anneal.

Assuming the divacancy trapping rate of c-Si (3×10^{14}, [8]) to be a typical positron trapping rate in damaged Si, we can estimate the concentration of positron traps. The trap-density in the *as-implanted* a-Si would correspond to 1.2 at.%, while the defect density in 500 °C annealed a-Si is found to be 0.4 at.% [1].

H-implantation and subsequent annealing at 150 °C also leads to a lowering of the S-parameter. This temperature is too low for any significant structural changes to occur.

It is well-known that H passivates dangling bonds, which according to photocarrier lifetime measurements has also taken place in the H-implanted and annealed sample [1]. This passivation can be related to the measured decrease in S, probably due to reduced trapping in the larger 'voids'. It appears, therefore, that some of the vacancy-type defects which were able to trap positrons no longer serve as those traps after the dangling bonds have been passivated with H. This means that at least some of the original dangling bonds were located at vacancy-type defects.

Finally, the S-parameter of MBE-system grown a-Si amounts to 1.145, which is really very large, but in accordance with the value of 1.15 for MBE grown Si with large voids [4]. The only adequate explanation of this value is para-Ps formation in large cavities. Simpson *et al.* found voids of 3 to 6 nm diameter [4]. The diffusion length of 5.4 nm is rather small, indicating poor-quality material, although it is not as small as in some of the PECVD a-Si:H-layers. Raman-measurements show a rather small bond-angle distortion ($TO_{hwhm} = 37.2$ cm^{-1}).

In conclusion it has been shown that positrons are sensitive probes for a characterization of several types of a-Si. Changes in the S-parameter and diffusion length can be related to the presence of incorporated gases and voids or vacancy-type defects.

REFERENCES

[1] S. Roorda, R.A. Hakvoort, A. van Veen, P.A. Stolk and F.W. Saris, *submitted to J. Appl. Phys.*
[2] A. van Veen, H. Schut, J. de Vries, R.A. Hakvoort and M.R. IJpma, in: P.J. Schultz, et al.: *Slow Positron Beams for Solids and Surfaces*, Proc. 4th Int. Positron Workshop (AIP, New York, 1990), p.171.
[3] M.J.W. Greuter, L. Niesen, R.A. Hakvoort, J. de Roode, A. van Veen, A.J.M. Berntsen, W.G. Sloof, *to be published in:* Proc. 9th Int. Conf. on Hyperfine Interactions, August 17-21, 1992, Osaka, Japan.
[4] P.J. Simpson et al., in: P.J. Schultz, et al.: *Slow Positron Beams for Solids and Surfaces*, Proc. 4th Int. Positron Workshop (AIP, New York, 1990), p.125.
[5] V.G. Bhide, R.O. Dusane, S.V. Rjarshi, A.D. Shaligram and S.K. David, J. Appl. Phys., **62**, 108 (1987)
[6] R. Suzuki, Y. Kobayashi, T. Mikado, A. Matsuda, P. McElheny, S. Mashima, H. Ohgaki, M. Chiwaki, T. Yamazaki and T. Tomimasu, Jpn. J. of Appl. Phys, **30**, 2438 (1991).
[7] M.J. van den Boogaard, S.J. Jones, Y. Chen, D.L. Williamson, R.A. Hakvoort, A. van Veen, A.C. van der Steege, W.M. Arnold Bik, W.G.J.H.M. van Sark and W.F. van der Weg, *to be published* (MRS Spring meeting 1992).
[8] M. Shimotomai, Y. Ohgino, H. Fukushima, Y. Nagayasu, T. Mihara, K. Inoue and M. Doyama, in: *Defects and Radiation Effects in Semiconductors*, Inst. Phys. Conf. Ser., **59** (Institute of Physics, London, 1981), p.241.

Positron Stopping in Germanium

A. Halec, P. Maguire, P.J. Simpson and Peter J. Schultz
The Positron Beam Laboratory, Dept. of Physics, The University of Western Ontario
London, Ontario, CANADA N6A 3K7

G.C. Aers, T.E. Jackman and P. Marshall
Microstructural Sciences, The National Research Council of Canada
Ottawa, Ontario, CANADA K1A 0R6

ABSTRACT

The dependence of the mean penetration depth $<z>$ on the incident positron energy E is described by the empirical power law formula used originally for electrons and adopted by Mills & Wilson [1]. Doppler broadening as a function of incident positron energy is used to study amorphous Ge overlayers grown by thermal evaporation on GaAs substrates. It is shown that experimental data can be fitted using the power law dependence of $<z>$ on E, with parameters that are in acceptable agreement with MC simulations for Ge.

PACS codes: 68.35.-p, 61.80.Fe,

INTRODUCTION

The variable-energy positron beam (VEP) technique is becoming established as one of the most sensitive non-destructive probes available for structural point defects and their depth-resolved concentration.[1-4] The technique is based on the fact that energetic positrons implanted in a solid target will rapidly thermalize ($\leq 10^{-11}$ s), and then diffuse through the solid ($> 10^{-10}$ s) until they annihilate from either "free" or "trapped" environments. Quantitative analysis of the data requires a knowledge of the depth-distribution of the monoenergetic incident positrons *after* they have thermalized in the solid.

This distribution, known as the stopping or implantation profile $P(z,E)dz$, is defined as the probability that positrons injected into the solid with initial energy E reach a layer lying between z and $z+dz$. For materials with a constant mass density ρ (g/cm³) as a function of depth, $P(z,E)$ is believed to be well represented by a Makhovian distribution,[5] such as that shown in Fig.1 and given by:

$$P(z,E) = \frac{mz^{m-1}}{z_o^m} \exp[-(z/z_o)^m] \quad (1)$$

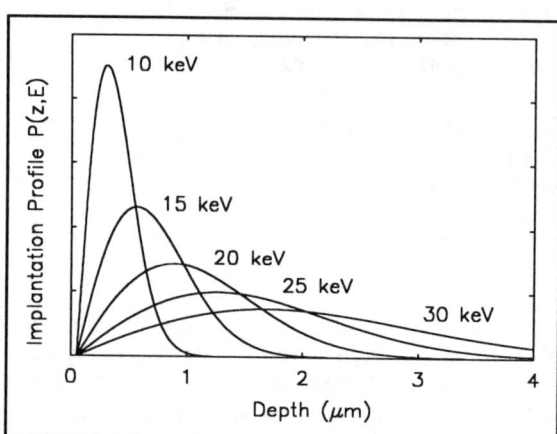

Figure 1 The Makhovian stopping profile, P(E,z), plotted against penetration depth for monoenergetic positrons incident on Ge (Eq. 1).

where
$$z_0 = \frac{\langle z \rangle}{\Gamma[(1/m)+1]}$$

and the mean penetration depth is:

$$\langle z \rangle = \frac{A}{\rho} E^n \qquad (2)$$

Both n and A are constants determined either by experiment[2,4,6-8] or by Monte Carlo simulation.[5,8,9] Typical values assumed for these constants are $n \sim 1.6$ and $A \sim 400$ (Å/keV^n).

In this communication we present data obtained for the depth dependence of the positron annihilation lineshape in layered Ge/GaAs samples. The similar mass densities ($\rho_{Ge} = 5.32$ and $\rho_{GaAs} = 5.36$ g/cm^3) allow us to assume that the simple Makhovian profile given in Eq. (1) is reasonable, and we use the program POSTRAP5 to model the positron data. POSTRAP5 is a modified version of POSTRAP4[10] that allows the analysis of multilayered structures with different diffusion coefficients and defect trapping rates. In an extension of this research we are also investigating stopping in layered materials with dissimilar mass densities. A preliminary report on the modelling of these data appears in these proceedings (see G.C. Aers et al.).

EXPERIMENT

Amorphous Ge overlayers of different thicknesses were grown by thermal evaporation on GaAs(100) substrates and studied with the variable-energy positron (VEP) beam. The S-parameter was measured as a function of incident positron energy. The S-parameter is defined as the relative number of counts in a fixed region ($\sim 50\%$) centred in the annihilation γ-ray energy spectrum. A larger S corresponds to a narrower lineshape.[1]

The thicknesses of the Ge overlayers were measured using a Philips EM430 transmission electron microscope (TEM), operating at 250 kV (with an accuracy of $\pm 1\%$). The cross sectional TEM samples were prepared by the small angle cleavage technique.[11] The thicknesses obtained were in very good agreement ($\pm 5\%$) with independent measurements made using a direct thickness profilometer (DECTAK).

RESULTS & DISCUSSION

Figure 2 shows the S-parameter data for all four samples studied, together with the results of fitting the data (solid curves) using POSTRAP5. By obtaining a good fit to the steep slope of the data in region 1, we find that the diffusion length in the Ge overlayers is $L_+ \sim 50\text{-}100$ Å. This is consistent with expectations, since there should be a high concentration of defects in the overlayers. A uniform product $Cv = 1.2 \times 10^{12}$ s^{-1} was satisfactory to fit all layers, where C is the defect concentration and v the specific positron trapping rate. This is ~ 3 orders of magnitude faster than the annihilation rate λ_B for positrons freely diffusing in undefected Ge,[12] and suggests that nearly all positrons were trapped by a concentration of more than 0.1% vacancies (assuming $v = 1 \times 10^{15}$ s^{-1}). The diffusion coefficient for the GaAs substrate was $D_+ \simeq 1.9$ cm^2/s, in good agreement with the previous results of Saarinen et al.[13]

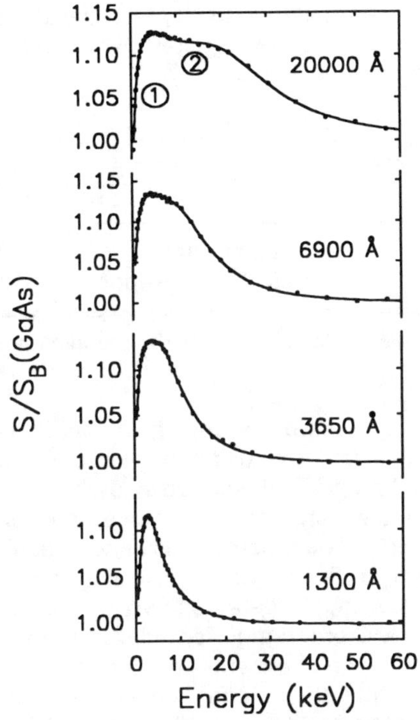

It can be seen in Fig. 2 that the S-parameter data have saturated at a maximum level for all of the thicker Ge layers (region 2). This is consistent with the high defect concentration in these layers. Nevertheless, the value is clearly not constant through the entire Ge layer, and it was necessary to allow a gradient in the S-parameter in order to obtain a good fit. This indicates that these layers (grown by thermal evaporation on "nominally" room temperature substrates) were not entirely uniform. This is not unreasonable for this growth process, where substrate temperature and, possibly, Ge flux may vary during growth of the layer. High resolution TEM (sufficient for lattice-images of the GaAs substrates) shows evidence of some sort of macroscopic defect feature in the Ge overlayers, which are otherwise amorphous. We do not have sufficient information to speculate on the nature or extent of these structures.

With the above considerations, the model for the entire set of samples profiled was simplified by various known or consistent features. The layer thicknesses were all well known, as were the constituent material densities. A consistent gradient was adopted for the S-parameter data through the Ge layers,

Figure 2: S-parameter and fit for defected Ge layers on GaAs(100). Ge layer thickness is shown beside each data set.

and diffusion and defect parameters were established as described above. The only remaining flexibility is in the stopping profile (Eqs. 1 & 2), and the parameters A, n and m were adjusted to obtain the best fit to *all* the data in Fig. 2. The results are listed in Table I.

	A (Å/keV^n)	n	m
Experiment	400	1.65	1.77
MC simulations	429	1.55	1.77

Table I: Parameters for the stopping profile of positrons in Ge/GaAs layered materials.

Positron stopping in Ge was also calculated using a Monte Carlo (MC) program (described elsewhere[14]), and the parameters are also listed in Table I. The parameters A and n result from a linear fit of the log($<z>$) versus log(E) data, shown in Fig. 3. The MC program is formulated to take advantage of experimental optical data through the Penn description of inelastic scattering, although it allows the option of employing a combination of random phase approximation (RPA) and Gryzinski cross section calculations for cases where accurate optical data do not exist. This was the case for Ge, so there may be some error associated with the deduced parameters. Generally, the Penn model

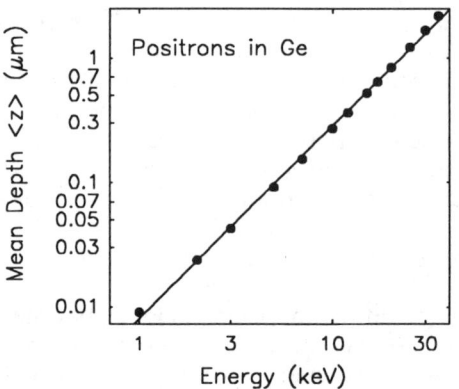

Figure 3 Mean stopping depth $<z>$ versus incident positron energy E, calculated using the Monte Carlo program.[14]

is superior to the RPA+Gryzinski combination.[14] The comparison was tested for the case of Si, where optical data are available. In this case, we found that both sets of MC calculations produced a similar dependence of $<z>$ on E, but the Penn model resulted in a $<z>$ that was uniformly ~10% lower than for the RPA+Gryzinski combination. This degree of precision is outside the bounds of the present theoretical data. By selectively fitting different regions of the MC data it is possible to get parameters ranging from $A \sim 463$ and $n \sim 1.47$ (for $E \leq 10$ keV) through to $A \sim 298$ and $n \sim 1.68$ (for $E > 10$ keV). Nevertheless, the agreement between calculated and measured stopping profiles is generally very good.

CONCLUSIONS

The results obtained for positron implantation in highly defected Ge overlayers on GaAs(100) substrates are consistent with those determined using MC calculations. The Makhov equation[5] routinely used to describe the stopping profile of positrons in solids is found to be satisfactory, with a shape parameter of $m = 1.77$ and a dependence of mean depth on energy given by $<z> = (400/\rho)E^{1.65}$, where ρ is the mass density of the material. This is identical to the relationship deduced empirically for the stopping of positrons in Si.[2,3]

We gratefully acknowledge the assistance of J. McCaffrey with TEM, useful discussions with K. Jensen, and technical assistance from T.C. Leung, P. Perquin, and I. Schmidt. We also gratefully acknowledge use of the Monte Carlo program written by K.O. Jensen and A.B. Walker. Research in the Positron Beam Laboratory is supported by grants from the Natural Sciences and Engineering Research Council of Canada.

REFERENCES

[1] P.J. Schultz and K.G. Lynn, Rev. Mod. Phys. **60**, 701 (1988).
[2] P.J. Schultz, E. Tandberg, K.G. Lynn, B. Nielsen, T.E. Jackman, M.W. Denhoff, and G.C. Aers, Phys. Rev. Lett. **61**, 187 (1988).
[3] P.J. Simpson, M. Vos, I.V. Mitchell, C. Wu and P.J. Schultz, Phys. Rev. B **44**, 12180 (1991).
[4] P.J. Simpson, P.J. Schultz, S.-Tong Lee, Samuel Chen, and G. Braunstein, J.Appl.Phys.**72**, 1799(1992).
[5] S. Valkealahti and R.M. Nieminen, Appl. Phys. A **35**, 51 (1984).
[6] A.P. Mills and R.J. Wilson, Phys. Rev. A **26**, 490 (1982).
[7] A. Vehanen, P. Huttunen, J. Mäkinen, and P. Hautojärvi, J.Vac.Sci&Technol. A **5**, 1142(1987).
[8] J.A. Baker, N.B. Chilton, K.O. Jensen, A.B. Walker, and P.G. Coleman, Appl.Phys.Lett. **59**, 23 (1991).
[9] J.A. Baker, N.B. Chilton, K.O. Jensen, A.B. Walker, and P.G. Coleman, J.Phys.: Condens.Matter **3**, 4109 (1991).
[10] G.C. Aers, p.162 in Positron Beams for Solids and Surfaces, ed. P.J. Schultz, G.R. Massoumi and P.J. Simpson (AIP Press, New York 1990).
[11] J.P. McCaffrey, Ultramicroscopy, **38**, 149 (1991).
[12] R. Würschum, W. Bauer, K. Maier, J. Major, A. Seeger, H. Stoll, H.-D. Carstanjen, W. Decker, J. Diehl, and H.-E. Schaefer, p. 671 in Positron Annihilation, ed. L.Dorikens-Vanpraet, M. Dorikens, and D. Segers (World Scientific, Singapore1989).
[13] K. Saarinen, P. Hautojärvi, J. Keinonen, E. Rauhala, J. Räisänen, and C. Corbel, Phys. Rev. B **43**, 4249 (1991).
[14] K.O. Jensen, A.B. Walker, and N. Bouarissa, p. 19 in Positron Beams for Solids and Surfaces, ed. P.J. Schultz, G.R. Massoumi and P.J. Simpson (AIP Press, New York 1990).

Slow Positron Beam Measurements on GaAs Grown by Molecular Beam Epitaxy at Low Temperatures

N. Hozhabri [1,2], S. C. Sharma [1], R. N. Pathak [2], and K. Alavi [2]

[1] Department of Physics

[2] NSF Center for Advanced Electron Devices and Systems, Department of Electrical Engineering

The University of Texas at Arlington, Arlington, Texas 76019

Variable energy positron beam spectroscopy is employed to study defects and arsenic precipitates in GaAs epilayers grown at low temperatures by Molecular Beam Epitaxy (MBE). We have measured the S-parameter as a function of positron implantation energy. We have obtained results for the depth profiles of the Ga monovacancy defects in unannealed LT-GaAs and Ga monovacancies and arsenic clusters related defects in annealed LT-GaAs.

Recently, a new type of GaAs epilayer has attracted the attention of researchers interested in semiconductor devices.[1-6] It is found that GaAs epitaxial layers grown at substrate temperatures that are much below the regular grown temperature of 600 °C in molecular beam epitaxy (LT-GaAs) have technologically important properties.[4] When GaAs is grown at a substrate temperature of 200 °C, it exhibits: 1) no measurable Photoluminescence (PL) signal, 2) greater than 1 at.% excess As, 3) a large EPR signal corresponding to 5% arsenic ionization. On the other hand, LT-GaAs's that have been grown above 250 °C, or have been subjected to post growth/anneal at normal MBE growth temperature of 600 °C exhibit : 1) a small PL signal , 2) larger than 1 at.% excess As, 3) uniformly semi-insulating character, and 4) no measurable EPR signal (resolution ~ 1 x 10^{18} cm^{-3}). Among the applications of these materials is the reduction of sidegating and backgating which are important parasitic problems associated with GaAs device technology. Due to a significantly short carrier lifetime

(≈0.3ps) in LT-GaAs, these structures have been used to fabricate subpicosecond photoconductors for ultrahigh speed optoelectronic switches and receivers.

In this paper, we present depth profiles of defects in MBE grown semi-insulating GaAs and LT-GaAs wafers using a variable energy positron beam. The Doppler broadening spectra of the gamma rays resulting from the annihilation of the implanted positrons in the LT-GaAs samples have been measured. These spectra have been analyzed in terms of the line shape parameter (S). The application of the positron beam technique in charactrization and defects profiling of the semiconductor materials has been discussed elsewhere.

The samples used in the present experiments are MBE grown normal GaAs and LT-GaAs. The normal GaAs wafer with a thickness of 1.5 µm was grown at a growth temperature of 680 °C as determined by thermocouple (T/C) . The LT-GaAs epilayers (thickness ≈ 1.5µm) were grown at 250 °C (T/C) on MBE grown buffer layers of 1.0 µm thick normal GaAs. One of the LT-GaAs wafers was annealed *in situ* at 652 °C (T/C) for one hour. The structures of these samples are given in Figure (1)

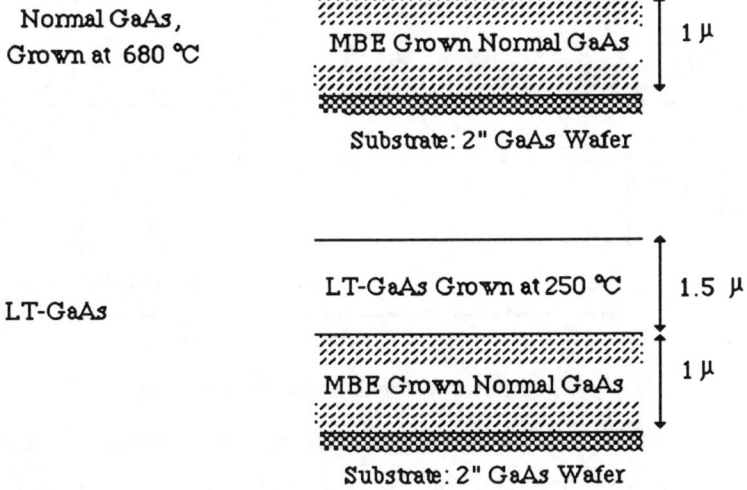

Figure 1. The strucures of the normal and LT-GaAs samples grown by molecular beam epitaxy used in the experiments.

The positron beam spectrometer, used in this experiment, consists of a magnetically guided positron beam, a high-purity Ge solid state detector with a resolution of 1.1 keV at 511 keV annihilation gamma-ray energy, and a multichannel analyzer. Positrons emitted from a ^{22}Na source are moderated by well annealed tungsten foils and then transported through a curved beam-line by a focusing magnetic field. The collimated positrons are accelerated to desired energies up to 25 keV prior to being implanted into the samples. Doppler broadening spectra were measured as a function of positron implantation energy. These spectra were analyzed in terms of the S-parameter. The S parameters of the Doppler broadening spectra measured for the normal GaAs, unannealed LT-GaAs and annealed LT-GaAs are shown as a function of positron implantation energy in Figure (2).

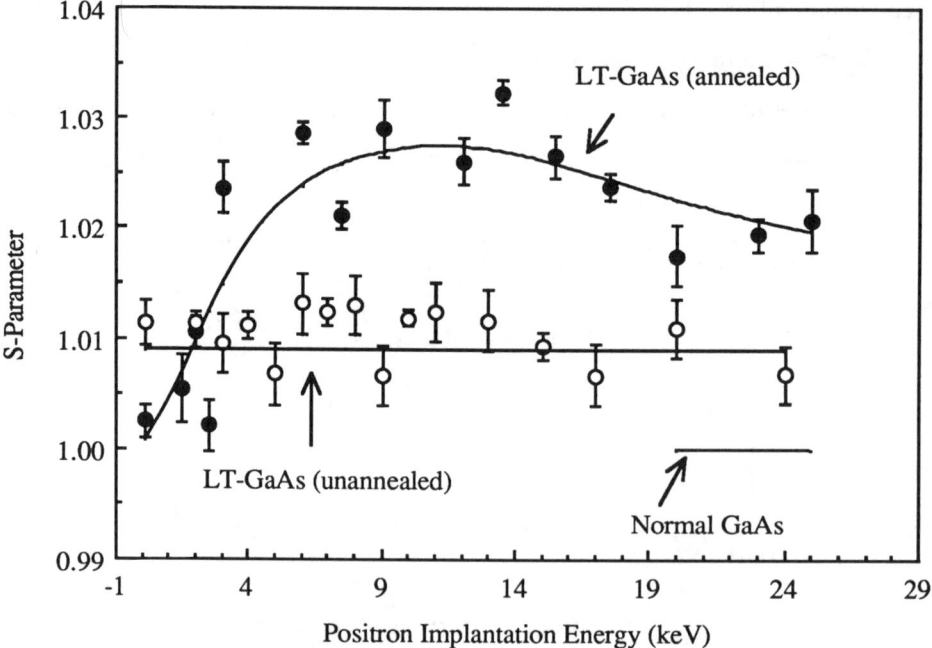

Figure 2. The S-parameter vs positron implantation energy, measured in annealed LT-GaAs, unannealed LT-GaAs, and normal GaAs. The error bars represent statistical uncertaities in the S parameter. The solid curve for the annealed LT-GaAs was obtained by fitting the positron diffusion equation.

The S-parameter values for the normal GaAs do not change with positron implantation energy (corresponding to an implantation range from surface down to ≈ 1 μm) and remains at an average value of 0.380 ± 0.001 (normalized to 1.0 in Fig.2). For the unannealed sample, again the S-parameter does not change with depth. However, the measured values are approximately 1 % higher than those measured for the normal GaAs. The higher value of the S-parameter, in the case of the unannealed LT-GaAs, indicates a much higher concentration of defects in this wafer. The constancy of the S-parameter, between the surface and the maximum implantation depth of about 1 μm, signals a uniform depth distribution of these defects in this wafer. The higher S values in the unannealed LT-GaAs are due to the presence of significant number of Ga monovacancies in the sample created during the growth. In contrast, the S-parameter of the annealed LT-GaAs increases from a normalized surface value of 1.0 to a value of 1.03 at the implantation energy of 5 keV (depth $\approx$ 80 nm). For positron energies from 5 keV to 17 keV (mean implantation depths of $\approx$ 80 nm to $\approx$ 570 nm), the S-parameter remains at 1.03. For mean implantations deeper than 570 nm, the S-parameter decreases slowly towards 1.01, which is characteristic of the unannealed wafer. These data clearly show that 1) as expected, the normal GaAs contains the least number of defects, 2) the unannealed LT-GaAs has a uniform defect distribution with a higher number of defects compared to that in the normal GaAs, and 3) the annealed LT-GaAs has a non-uniform depth distribution of defects. The observed changes in the S-parameter of the annealed LT-GaAs indicate a non-uniform spatial distribution of defects related to the presence of the arsenic clusters in LT-GaAs. The formation of the aresenic clusters has been suggested due to the following processes

$$As_{Ga} \rightarrow As \text{ (free)} + V_{Ga} \qquad (1)$$
$$n\ As \rightarrow As_n\text{-Cluster} \qquad (2)$$

which increase the number of the Ga vacancies after annealing the wafer.[7] The As_{Ga}'s defects are the dominant form of the defects in the unannealed LT-GaAs samples and they have a uniform spacial distribution. The density of these defects is about 10^{19}-10^{20} cm^{-3} (in agreement with 1-2% excess arsenics found in unanealed LT-GaAs)[8] which is relatively much higher than the Ga monovacancies[7]. The breakup of As_{Ga}'s defects following reaction (1) should result in an uniform but higher S-parameter in the annealed LT-GaAs sample. The non-uniformity of the S-parameter suggests that positrons 1) annihilate in the reshuffeld V_{Ga}'s and 2) annihilate from the trapped states in the distorted regions around arsenic clusters. The similarity between the S parameter in the annealed LT-GaAs and arsenic depth profile using XPS[9] suggests that the latter is more dominant in trapping the positrons. The trapping of the positrons in these strained regions, containing mostly low density conduction electrons, results in a narrower energy spectrum. The latter produces a higher S parameter. In order to obtain quantitative information, we have examined the S-parameter data by using VEPFIT[10] that solves the well-known one-dimensional positron diffusion equation,

$$D_+ \frac{d^2c(x)}{dx^2} - \frac{d[v_d\, c(x)]}{dx} - [k_t n_t + \lambda_b]\, c(x) + P(x,E) = 0 \qquad (3)$$

where $c(x)$ is the positron volume density at depth x, v_d is the field dependent positron drift velocity, k_t is the specific positron trapping rate, $n_t(x)$ is the atomic fraction of the defect density, λ_b is the positron annihilation rate in the bulk, and D_+ is the positron diffusion coefficient related to the positron diffusion length by $L_+ = (D_+/\lambda_{eff})^{1/2}$. The positron implantation profile is given by

$$P(x,E) = \frac{2x}{x_0^2} \exp[-(\frac{x}{x_0})^2] \qquad (4)$$

where $x_0 = (\sqrt{\pi}/2)\, \bar{x}\,(E)$. $\bar{x}\,(E) = [33.2/\rho](E)^{1.6}$ is the mean implantation depth (in nm) at energy E in KeV, and ρ is the density of the sample in g/cm^3. In the VEPFIT model, the measured S-parameters are represented by

$$S = F_s S_s + \sum_{k}^{n} F_k S_k \qquad (5)$$

where F_k is the fraction of the positrons in k^{th} layer, F_s is the fraction of the positrons returning to the surface, S_k is the value assigned to the kth layer, and S_s is characteristic of the surface.

The results of the VEPFIT analysis of the annealed LT-GaAs data show that 1) whereas the average positron diffusion length in the arsenic-rich region is (47±1) nm, it increases to (56±4) nm in the deeper regions (below 750 nm) of the wafer. A shorter diffusion length for the positrons in the top region of the wafer is due to a higher concentration of the positron trapping sites, i.e., arsenic clusters related defects and V_{Ga}.

In summary, we have obtained the depth profile of Ga monovacancies in unannealed LT-GaAs annealed wafer. We have also obtained the depth profile of the defects in the annealed LT-GaAs. From a comparison of the defect profiles, we have shown that annealed LT-GaAs contains a non-uniform and higher concentration of defects.

Research supported, in part, by a grant from the U. S. Army Research Office.

REFERENCES

1. F. W. Smith, A. R. Calwa, C. L. Chen, M. J. Manfra, and L. J. Mahoney, IEEE Electron Device Lett. 9, 77 (1988).
2. M. R. Melloch, N. Otsuka, J. M. Woodall, A. C. Warren, and J. L. Freeouf, Appl. Phys. Lett. 57, 1531 (1990).
3. Z.-Q. Fang and D. C. Look, Appl. Phys. Lett 61, 1438 (1992)
4. S. Gupta, J. F. Whitaker, A. Mouron, IEEE J. of Quantom Elec., V. 28, 2464 (1992)
5. S. Gupta, P. K. Bhattacharya, J. Pamulapati, and G. Mourou, Appl. Phys. Lett. 57, 1543(1990).
6. A. C. Warren, J. M. Woodall, J. L. Freeouf, D. Grischkowski, D. T. McInturff, M. R. Melloch, and N. Otsuka, Appl. Phys. Lett. 57, 1331 (1990).
7. F. W. Smith, Ph. D. Dissertation, MIT, Cambridge 1990
8. I. Ohbu, M. Takahama, and K. Hiruma, Appl. Phys. Lett. 61, 1679 (1992)
9. N. Hozhabri, A. R. Koymen, J. B. Bailey, R.N. Pathak, and K. Alavi, March Meeting of american Physical Society, 1993, to be published.
10. A. van Veen, H. Schut, J. de Vries, R. A. Hakvoort, and M. R. Ypma, in Positron Beams for Solids and Surfaces, edited by P. J. Schultz, G. R. Massoumi, and P. J. Simpson, American Institute of Physics Conference Proceedings (1990), Vol. 218

THE EFFECTS OF LOW-ENERGY SCATTERING ON POSITRON IMPLANTATION

K. A. Ritley
Department of Physics and Materials Research Laboratory
University of Illinois at Urbana-Champaign
1110 W. Green Street, Urbana, IL 61801

K. G. Lynn, V. Ghosh, and D. O. Welch
Department of Physics and Department of Applied Science
Brookhaven National Laboratory, Upton, New York 11973

ABSTRACT

Existing Monte-Carlo models are capable of simulating the behavior of positrons incident at keV energies, then following the energy loss process to arbitrary final kinetic energies of from 20 eV to 100 eV. In the present work we describe a Monte-Carlo simulation of the final stages of positron thermalization in Al, from 25 eV to thermal energies, via the mechanisms of conduction-electron and longitudinal acoustic phonon scattering. We show that the latter stages of thermalization can have important effects on the stopping profiles and mean depth. We describe a novel way to obtain information about positron energy loss by considering the time-evolution of a point-concentration (delta-function distribution) of positrons. We examine, for the first time in the context of a positron Monte-Carlo calculation, the effects of a positive positron work function. Finally, we discuss some issues relating to the agreement of Monte-Carlo calculations with experimental data.

I. INTRODUCTION

The need for reliable Monte-Carlo models of positron implantation and stopping is clear. A slow positron incident on a solid rapidly thermalizes ($\sim 10^{-12}$ s) and then diffuses at thermal energies, trapping in defects and interacting with the surface. Annihilation with electrons does not occur until much longer times ($\sim 10^{-10}$ s). The simplicity of this two-step "stopping-and-diffusion" process, at least to first order, is fortuitous, as it facilitates the analysis of experimental data using a straightforward application of the diffusion equation. Although the diffusion aspect of this problem is far from trivial, it is readily attacked and a variety of solutions are available. Unfortunately the problem of determining the stopping profiles (i.e., the initial conditions for diffusion) is less amenable to analytical treatments. These conditions are usually characterized by the implantation (or stopping) profile $P(z)$, where $P(z)dz$ is the probability of locating a positron between z and $z + dz$. Accurate experimental measurements of this quantity are difficult, and thus the experimentalists have had to rely quite heavily on the results of computer-based Monte-Carlo calculations.

Monte-Carlo calculations are able to track the progress of a positron incident at high kinetic energies, usually in the keV range, and follow its trajectory through the material as it scatters both elastically and inelastically, until it has reached an arbitrary endpoint energy, at which point the positron is considered to be stopped. In the Monte-Carlo programs to date, that endpoint energy has been typically 20 eV to 100 eV. At these energies, the inelastic process of positron-conduction-electron scattering will quickly deliver the positron to near-thermal energies, at which point phonon scattering becomes dominant and the diffusion process begins. Thus it has been argued that the final stages of thermalization, from ~ 20 eV to $\frac{3}{2}k_BT$, will have a negligible effect on the stopping profiles and other quantities generated by these programs.[1-4]

In the present work we have chosen to re-examine that assumption. Our motivation for this is essentially three-fold. First, although the time required for positrons to give up their final kinetic energy (~ 20 eV to $\frac{3}{2}k_BT$) is extremely short, the physical effects of scattering depend on the mean scattering rates and on the mean free paths, which as we discuss below are not negligible. Second, there is a growing need for information about positrons incident at low energies on multilayers and buried interfaces. The length scales associated with these systems are relatively short and even small changes to the stopping profiles made by low-energy scattering could be important. Additionally, positron work function differences between adjoining materials may have some effect on positrons incident on such systems. Even with a relatively large kinetic energy, the kinetic energy of the component of velocity normal to the surface may be of the order of the difference between work functions. Continuing the energy loss process to thermal energies allows us to examine possible work function effects. Third, these Monte-Carlo models have some utility as a straightforward and physically intuitive means for studying the crossover from positron stopping to positron diffusion. Although numerical solutions of the Boltzmann transport equation have been used to study such processes, it has been shown that stochastic approaches such as single-particle Monte-Carlo calculations are a direct and efficient way of numerically integrating the Boltzmann equation.[5,6]

This paper is organized in the following way: We briefly discuss in Sec. II the Monte-Carlo model and the scattering mechanisms it includes. We discuss in Sec. III a novel way to examine the energy loss process, by considering the time-evolution of a point concentration of positrons. In Sec. IV we discuss the effects of low-energy scattering on the stopping profiles and the mean depth, and in Sec. V we describe how a positive work function affects the stopping profiles. Finally, in Sec. VI, we discuss some important issues regarding the agreement of Monte-Carlo simulations with experimental data. Detailed results concerning backscattering and the energy distribution of the re-emitted positrons will appear in a forthcoming publication.[7]

II. MODEL AND MECHANISMS

In order to study the effects of low-energy inelastic scattering (in the ~ 20 eV to $\frac{3}{2}k_BT$ range), a single-particle Monte-Carlo model of the usual type was constructed.[1] The behavior of positrons contained within a semi-infinite sample is modelled by averaging a large number of single particle trajectories through the material. Using the notation of Ref. 14, a trajectory can be denoted by a series of states $(E_n, \mathbf{r}_n, \Omega_n)$, where E_n, $\mathbf{r}_n$ and Ω_n represent the energy, position vector and direction cosines of a particle after the nth scattering event. As a positron travels through the material, it scatters inelastically $(E_{n+1} < E_n)$, and its progress is followed until it is either backscattered out of the material $(r_z < 0)$ or until its energy has fallen below the threshold energy for the simulation. Full details of the calculation of the scattering states are given in Ref. 7.

The initial conditions of the present work are taken to be the results of a Monte-Carlo calculation using BNLPOS.[1,2] In that model an incident positron can be followed from a relatively high incident energy (keV) to an endpoint energy of 25 eV. An extensive complement of elastic and inelastic scattering processes is included; this is described in detail in Refs. 1 and 2.

The 25 eV endpoint implantation profiles generated by BNLPOS are then fit to Mahkovian distributions (see Sec. 5), and these are used as input parameters for the present work.[8] The scattering mechanisms considered are conduction- electron scattering, using the free-electron-gas formalism developed by Oliva; and longitudinal acoustic phonon scattering, using the Debye-model formalism developed by Nieminen and Oliva.[9,10] The combined mean free path per collision in the ~ 20 eV to $\frac{3}{2}k_BT$ energy range is typically 30Å or less in Al. We do not consider plasmon scattering or elastic scattering. The former mechanism does not apply (for reasons of energy-momentum conservation) for positrons with energy below approximately 20 eV. The latter mechanism does not apply as the positron nears thermal energy and quantum effects become important; the usual cut-off for elastic scattering is typically twice the width of the lowest positron energy band.

We account for the surface in the following way. In the case of materials with positive positron work functions ($\phi_+ > 0$), we consider the surface to be a simple energy step barrier of height ϕ_+. Positrons incident on this barrier (from within the material) for which the kinetic energy $E_\perp$ of the component of velocity normal to the surface is less than the positron work function, $E_\perp < \phi_+$, are specularly reflected. For positrons with $E_\perp > \phi_+$, a small fraction is considered undergo quantum mechanical barrier reflection, according to the usual reflection coefficient for scattering off a square energy step barrier. For negative work function materials ($\phi_+ < 0$), all positrons incident on the surface with the proper momentum are permitted to escape freely. More complicated surface effects (such as trapping) are presently not included in this model.

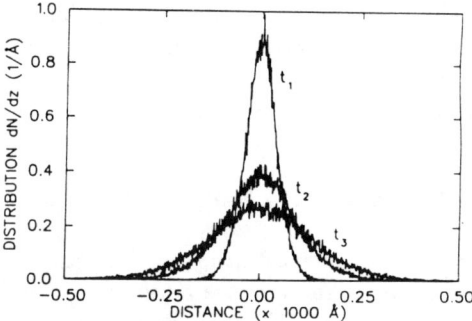 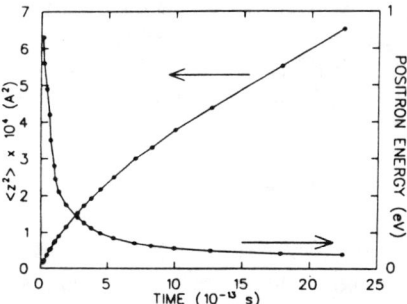

Figure 1. The spreading of an initial point concentration of positrons, at $t = 0.07$, 0.7, and 2.2 ps.

Figure 2. The positron energy and the mean-square-displacement, both as functions of time, of the profiles in Fig. 1.

III. STOPPING AND DIFFUSION

Our first step after constructing the Monte-Carlo model is to examine the process of positron stopping and the crossover from thermalization to diffusion. This is most easily accomplished by adopting for initial conditions a delta-function distribution of 25 eV positrons, $P(t_0) = \delta(\mathbf{r} - \mathbf{r_0})$, and then by tracking this distribution as the positrons scatter inelastically. The time-development of this distribution is depicted in Fig. 1, in which it is seen that the initial distribution leads to Gaussian-like profiles which broaden over time.

The energy loss rate of the initial positron "cloud" appears in Fig. 2. This energy loss curve exhibits three distinct features: an abrupt initial descent in which conduction-electron scattering quickly delivers the positron to near-thermal energies (stopping domain); a somewhat longer interval in which the positrons continue to lose energy but in which the principal scattering mechanism shifts from conduction-electron scattering to phonon scattering (thermalization domain); and the final, phonon-scattering-dominated regime in which the energy loss is negligible (beginning of diffusion domain).

Also presented in Fig. 2 is the mean-square displacement of the positrons as a function of time, $<z^2>$. Two interesting features are immediately apparent. First, $<z^2>$ becomes linear in the diffusion domain (i.e., $t > 1.0$ ps), as expected. Inasmuch as the final broadening of the positron cloud can be described by a simple diffusion equation with a constant, thermal-energy diffusion coefficient, D_+, we expect the mean-square displacement to vary linearly over time, $<z^2> \sim D_+ t$. This permits us to calculate the positron diffusion coefficient; in the case of Al, we find $D_+ = 1.1(1) \text{cm}^2/\text{s}$, in good agreement with the experimental value of $D_+ = 1.7(2) \text{cm}^2/\text{s}$, obtained by H. Huomo et al.[11]

Second, the values of $<z^2>$ at the point of crossover into the diffusion domain are not negligible. For Monte-Carlo simulations which terminate at higher energies (e.g., 25 eV), this distance represents the approximate length scale over which the neglect of the low-energy processes may become important. This has a direct bearing on the case of (narrow) stopping profiles for positrons incident at low kinetic energies; this is discussed in more detail below.

IV. STOPPING PROFILES AND MEAN DEPTH

We next explore the effects of low-energy inelastic scattering on the positron stopping profiles and mean depths. To accomplish this, we adopt for initial conditions the fitted Makhovian distributions for 25 eV-endpoint profiles generated by BNLPOS and reported in Ref. 13. We use these fitted profiles for computational simplicity and due to the near-universal use of the Makhovian distribution to date; however the program is quite general and actual particle information (location and angular distribution) from any Monte-Carlo simulation can be used. We then continue the energy loss process from 25 eV to 0.2 eV, the approximate onset energy for positron diffusion, according to the results of the previous section.

The initial 25 eV stopping profiles, together with the low-energy-broadened profiles, are reported in Fig. 3. Two qualitative features are immediately apparent. First, there is clearly a non-vanishing spatial probability density for positrons at shallow depths. This is in stark contrast to the Makhovian distribution, for which $P(z=0) = 0$. In cases in which other stopping profiles are taken for initial conditions, such as the results of Jensen, Walker, and Bouarissa (which predict a non-zero surface density of positrons at 100 eV endpoint energies), we expect the effect of the low-energy processes will be to increase the density at the surface.[15]

Second, we observe that the greatest relative effect obtained by extending positron transport from 25 eV to thermal energies occurs for positrons initially incident at low kinetic energies. Above ~3 keV incident energy, we see the low-energy scattering has little effect on the initial profiles. This is in agreement with the results of the previous section, in which we calculated the mean-square displacement of an initial delta-function distribution. The resulting low-energy stopping profiles are, in a broad sense, a convolution of the initial profiles with the 100Å-wide broadened delta-function distribution. For incident energies above ~3 keV, the 25 eV profiles in Al are sufficiently wide that the additional low-energy broadening has little relative effect.

To report these data quantitatively, we have attempted to fit the profiles with an analytical function. Neither the Makhovian distribution nor a recent modification of this provided an adequate fit. We did find, however, that the data were exceptionally well-described by the functional form proposed in the recent experimental study by Baker et al.[12],

$$P(z) = -\frac{d}{dz} \exp\left\{ -\left[\frac{z}{z_0}(1 + \frac{z}{z_0})^2\right]^m \right\} \qquad (1)$$

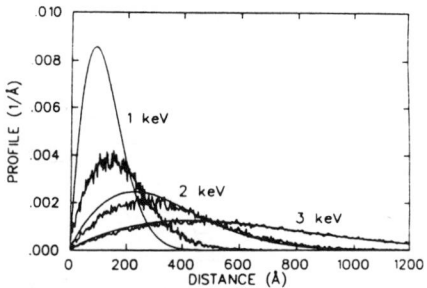
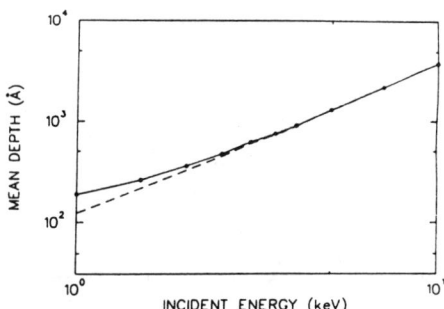

Figure 3. Comparison of 25-eV-endpoint profiles in Al, and profiles for positrons at thermal energy.

Figure 4. Comparison of mean depth for positrons at 25 eV (dashed line) and positrons at thermal energy.

where we find m (shape parameter) and z_0 (depth parameter) to be both material and energy dependent. In the case of Al, we find $<m>=1.15(1)$ and $<z_0/z_{1/2}>=2.62(0)$. These values are in excellent agreement with the experimental values, but the importance of this agreement will be further discussed in Sec. VI. We also note this form adequately describes stopping profiles for systems with positive work functions, as discussed below.

The mean implantation depths are reported in Fig. 4. As can be seen, the effect of the low-energy scattering is to cause a deviation from the power-law form predicted by BNLPOS. In particular, there is a relative increase in the mean depth at low incident energies, resulting in the introduction of a slight positive curvature. This is in accord with the expected behavior: low-energy scattering results in a substatial egress of positrons from the solid in this case (zero work function), thereby depleting the overall concentration near the surface, and resulting in increased penetration and, therefore, increased mean depth. The positive curvature in the energy-dependence of the mean depth has been reported in a recent experimental study by Baker et al.[13]; we comment further on this in Sec. VI.

V. WORK FUNCTION EFFECTS

The continuation of the energy loss process to thermal energies permits us to examine, for the first time, the effect of a positron work function on the implantation profile. We consider the case of a positive work function ϕ_+, and assume that this can be described by a square energy step barrier at the surface of height ϕ_+. Recent theoretical work regarding positron-surface interactions suggests that this simple model may not be optimal, but it serves as a first step in examining work function effects.[14] As discussed earlier, positrons incident on this barrier (from within the solid) for which the kinetic energy $E_\perp$ of the component of velocity normal to the surface is less than the positron work function, $E_\perp < \phi_+$, are specularly reflected; for positrons with

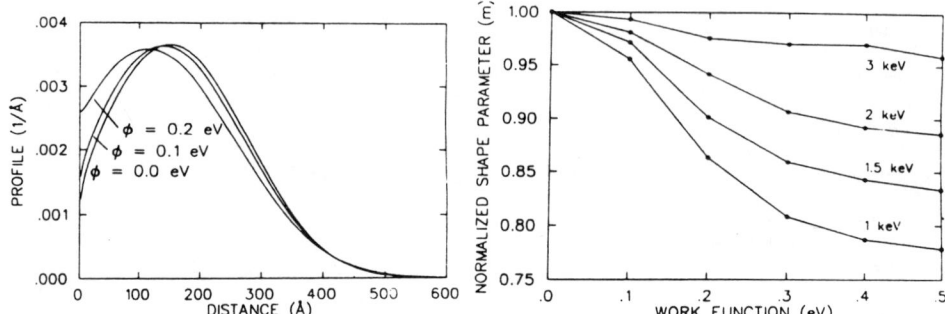

Figure 5. Stopping profiles for positrons incident at 1 keV in Al, for different values of the positron work function.

Figure 6. Values of the shape parameter (m) for fits of Eq. (1) to the stopping profiles in Al.

$E_\perp > \phi_+$, a small fraction is considered undergo quantum mechanical barrier reflection.

Stopping profiles for 1 keV incidence energy in Al are presented in Fig. 5, for assumed work functions of 0 eV though 0.2 eV. The number density of positrons near the surface increases as the magnitude of the work function, as expected: as the height of the energy step barrier increases, fewer positrons are able to overcome this barrier and escape via reemission.

We have found these profiles to be well-described by Eq. (1). In Fig. 6, we present the dependance of the shape parameter m on the assumed work function, obtained by fitting the stopping profiles to Eq. (1), for positrons incident at several kinetic energies on Al. Two qualitative features are immediately apparent. First, the greatest effect of the work function occurs for positrons incident at the lowest kinetic energies. In this case, the mean depth is small and the majority of positrons are within scattering distance of the surface. Second, as the work function increases, the resulting effect on the profiles dimishes and eventually vanishes, as indicated by the levelling of the curves in the direction of increasing work function. As the size of the energy step barrier increases, few positrons are able to overcome this barrier and escape. Increasing the barrier further thus has little effect, as positrons unable to overcome a small barrier are certainly unable to overcome a larger one.

The ability to examine work function effects may have application to the study of multilayers and buried interfaces. It has already been shown by independent Monte-Carlo calculations that the high-endpoint-energy stopping profiles may change dramatically across an interface, particularly an interface between materials of different density.[15,16] The effect of low-energy scattering will be to either increase or decrease the profile differences across the interface, depending on the gradient of the work function at the interface.

VI. SUMMARY AND DISCUSSION

In the present work we have continued the positron energy loss process from 25 eV (the usual cut-off for Monte-Carlo models of positron stopping) to near-thermal energies in Al, via conduction-electron and acoustic phonon scattering. We have detailed the energy loss process, and we have determined a diffusion coefficient for Al which is in good agreement with experiment. We have shown that the final stages of thermalization can have important effects on the stopping profiles and mean depth, particularly for positrons incident at low kinetic energies. We have also shown that a positive positron work function can have important effects on the stopping profiles. We have obtained agreement between our stopping profiles and those reported in the experimental study by Baker and co-workers. We have also shown that low-energy inelastic scattering can lead to a positive curvature in the incident-energy-dependence of the mean depth, also in good agreement with the experimental results of these authors.

It is the chief reward of theoretical physics to obtain agreement with experimental data. In the present work, we are not displeased with the agreement between our calculations and the recent experimental findings. In the case of Monte-Carlo simulations, however, particularly those in which there are many material-dependent variables to consider, agreement between theory and experiment is only a first step. The ultimate aim of this work is the construction of a model to provide accurate information about a wide variety of systems and which can be used in support of experimental data analysis or further theoretical studies.

For all who await this goal, there is an important point to consider. In every Monte-Carlo model of positron stopping constructed thus far, *there are many adjustable parameters.* In some cases, such as the present work, these adjustable parameters are obvious and take the form of physical quantities (such as the Debye temperature or the deformation potential), experimental measurements of which are not in full agreement with each other, or else are in disagreement with the results of theoretical calculations. In other cases, the adjustable parameters are far more subtle: for example, exactly how the electronic structure of atoms should be partitioned for elastic scattering calculations, or exactly when the optical mean free path model can be applied. Fortunately, the situation for positron Monte-Carlo models is in better shape than for many other models, and we need not contend with free parameters such as the inner-potentials or imaginary-potentials as are found in many coherent-wave calculations.

Our understanding of individual positron scattering mechanisms is probably sufficient to permit us to expect that accurate models of implantation can be constructed. Whether such models now exist, or whether more complicated models are necessary, such as those which can account for particle channeling at higher incident energies, is still a topic for research. Therefore, exhaustive benchmarking and comparisons between the results of Monte-Carlo calculations and experimental data must continue, so that a *consistant set of material-dependent parameters* can be attained; and so that, if necessary, the theoretical descriptions of positron-solid interactions or the Monte-Carlo models themselves can be refined.

ACKNOWLEDGEMENTS

It is a pleasure to thank J. Oliva, R. Nieminen, T. McMullen, and D. Britton for useful discussions. We gratefully thank the New Energy and Industrial Technology Development Organization (NEDO) for financial support, and DOE contracts DE-AC02-76CH00016, DE-FG02-91ER45439.

REFERENCES

1. Ritley, K. A., M. McKeown, and K .G. Lynn, *Positron Beams for Solids and Surfaces*, ed. by P. Schultz, G. R. Massoumi, P. J. Simpson, (American Institute of Physics, New York, 1990), pp. 3-18, and references therein.
2. Ghosh, V. J., Welch, D. O., and K. G. Lynn, pre-print.
3. Jensen, K., A. Walker, and N. Bouarissa, *Positron Beams for Solids and Surfaces*, ed. by P. Schultz, G. R. Massoumi, P. J. Simpson, (American Institute of Physics, New York, 1990), pp. 19-28.
4. Valkealahti, S. and R. M. Nieminen, *App. Phys. A* **32**, 95-106 (1983); and Valkealahti, S. and R. M. Nieminen, *App. Phys. A* **35**, 51-59 (1984).
5. Jensen, K. O., A. B. Walker, *J. Phys. Condensed Matter* **2**, 9757-9775 (1990).
6. Jacoboni, C., Lugli, P., *The Monte-Carlo method for semiconduction device simulation*, (Springer-Verlag, Berlin, 1989).
7. Ritley, K. A., K. G. Lynn, V. J. Ghosh, D. O. Welch, preprint.
8. Asoka-Kumar, P., and K. G. Lynn, *App. Phys. Lett.* **57**, 1634-1636 (1990).
9. Oliva, J., *Phys. Rev. B* **21**, 4909 (1980).
10. Nieminen, R. M. , and J. Oliva, *Phys. Rev. B* **22**, 2226-2247 (1980).
11. Huomo, H., E. Soininen, and A. Vehanen, *App. Phys. A* **49**, 647-658 (1989).
12. Baker, J. A., N. B. Chilton, and P. G. Coleman, *App. Phys. Lett.* **59**, 164-166 (1991).
13. Baker, J. A., N. B. Chilton, K. O. Jensen, A. B. Walker, and P. G. Coleman, *J. Phys. Condensed Matt.* **3**, 4109-4114 (1991).
14. Kong, Y., K. G. Lynn *Phys. Rev.* **44**, 13109-13111 (1991).
15. Ghosh, V. J., D. O. Welch, and K. G. Lynn, in these proceedings.
16. Aers, G. C., K. O. Jensen, A. B. Walker, "Calculation of positron diffusion in layered systems," SLOPOS5, August, 1992.

THE EFFECT OF CHANNELING ON THE DEFECT DEPTH DISTRIBUTION IN 110 keV Rb IMPLANTED POLY-W

H. Schut, A. van Veen, R.A. Hakvoort
Interfaculty Reactor Institute, Delft University of Technology
Mekelweg 15, NL-2629 JB Delft, The Netherlands

M.J.W. Greuter, L. Niesen
Materials Science Centre, University of Groningen
Nijenborgh 18, NL-9747 AG Groningen, The Netherlands

ABSTRACT

The positron annihilation Doppler broadening technique has been used to study the effect of annealing of poly-crystalline tungsten foils irradiated with 110 keV $^{85}Rb^+$ ions with doses up to 10^{12}, 10^{14}, 10^{15} and 10^{16} $^{85}Rb^+/cm^2$, and 10^{17} $^{83}Rb^+/cm^2$. The defect depth distribution after irradiation and subsequent annealing for 15 minutes at temperatures from 300 to 1600 K in steps of 100 K has been monitored by measuring the S-parameter as a function of the positron incident energy. The shape of the S-curves of the as-irradiated samples is explained by trapping of positrons in the near surface region (10 nm) containing Rb-defect clusters and positron trapping in simple defects created at larger depths due to channeling of Rb. Annealing of the "deep" defects in the tail of the distribution is observed as an increase of S at 15 keV in the 400-600 K temperature interval followed by a decrease of S between 1000 and 1300 K. Above these temperatures S rises again. These observations are assigned to trapping of vacancies at the small Rb-vacancy complexes causing their growth followed by the release of trapped vacancies. When at still higher temperatures the damage in the tail is recovered positrons diffuse back to the near surface region where large Rb-vacancy clusters or bubbles with correspondingly high S-parameter value have been formed.

INTRODUCTION

The preference of positrons for open-volume defects makes them a valuable tool to study dislocations, vacancies, vacancy clusters and voids in e.g. metals. Moreover variable energy positron analysis can be applied successfully for defect depth profiling of these defects. The interest of this study is based on the use of ^{83}Rb as source for Mössbauer studies in Rb implanted metals. Preliminary Mössbauer studies on 110 keV ^{83}Rb implanted tungsten show small but significant effects after annealing at 1200 K for half an hour, probably due to the growth of bubbles[1]. Furthermore alkali-metal interactions play an important role in the crystallisation behaviour of filaments in incandescent lamps[2].

EXPERIMENTAL

From 25 μm thick poly-crystalline tungsten (99.95%) foil 7 samples of 15x15 mm² were cut. Five samples were implanted with ^{85}Rb with doses up to 10^{12}, 10^{14}, 10^{13}, 10^{15} and 10^{16} Rb/cm². One sample was implanted with ^{83}Rb to a dose of 10^{17} Rb/cm² and one was kept as a reference sample. All samples were examined with slow positrons at the Delft Variable Energy Positron beam facility[3,4]. *In situ* heating of the samples was achieved by electron bombardment. The temperature of the samples was varied from 300 K to 1600 K in steps of approximately 100 K and was monitored by a W/Re thermocouple and an optical pyrometer. All heat treatments were carried out for 15 minutes. For each sample the so-called S(hape)-parameter was measured as a function of the positron incident energy (100 eV to 25 keV). The sample with dose 10^{13} Rb/cm² was not measured at elevated temperatures. S quantifies the Doppler broadening of the 511 keV photo peak and reflects the momentum distribution of electrons at the positron-electron annihilation site.

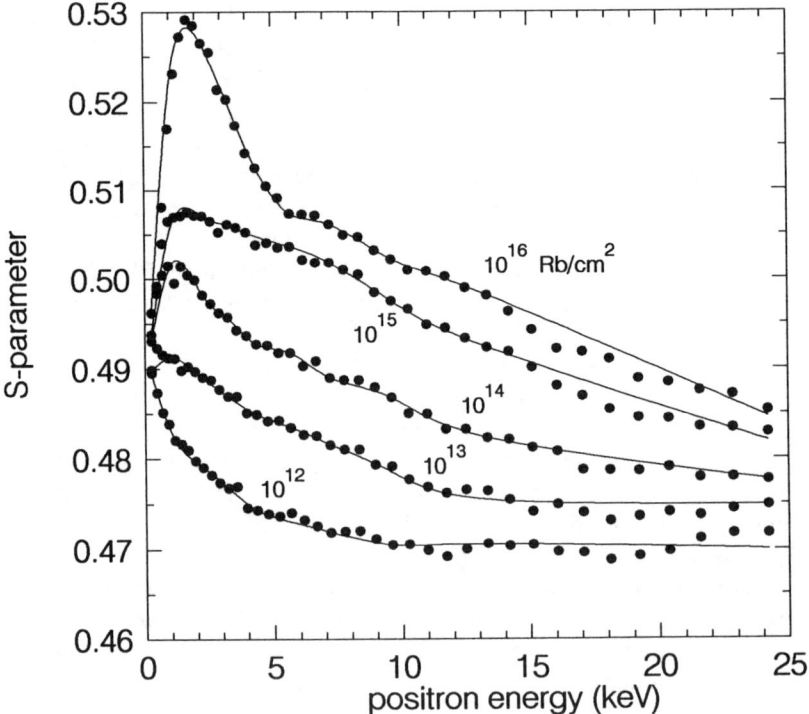

Figure 1. The S-parameter as a function of the positron incident energy for the poly-crystalline W foils implanted with 10^{12} to 10^{16} ^{85}Rb/cm².

RESULTS AND DISCUSSION

The results of the Doppler broadening measurements on the as-implanted samples are shown in figure 1. With increasing Rb-dose the S-parameter develops its maximum at 2 keV positron implantation energy. This energy corresponds to a mean positron implantation depth in tungsten of 6 nm. TRIM[5] calculations for 110 keV Rb in randomly orientated W yield a Rb range of 7 nm. Above 5 keV positron energy the S-parameter slowly decreases with energy. Even at 25 keV implantation energy the S-values are well above the S-parameter value of the polycrystalline reference sample (S_{ref} =.458). Apparently different types of defects are created by the implantation of 110 keV Rb. Both the strong increase of S in the near surface region with increasing Rb-dose and the results of the TRIM calculations indicate the formation of Rb-defect clusters. The relatively high S-parameter values at higher positron energies is explained by the presence of deeper lying simple defects, such as vacancies and small vacancy clusters, created by channelled Rb atoms.

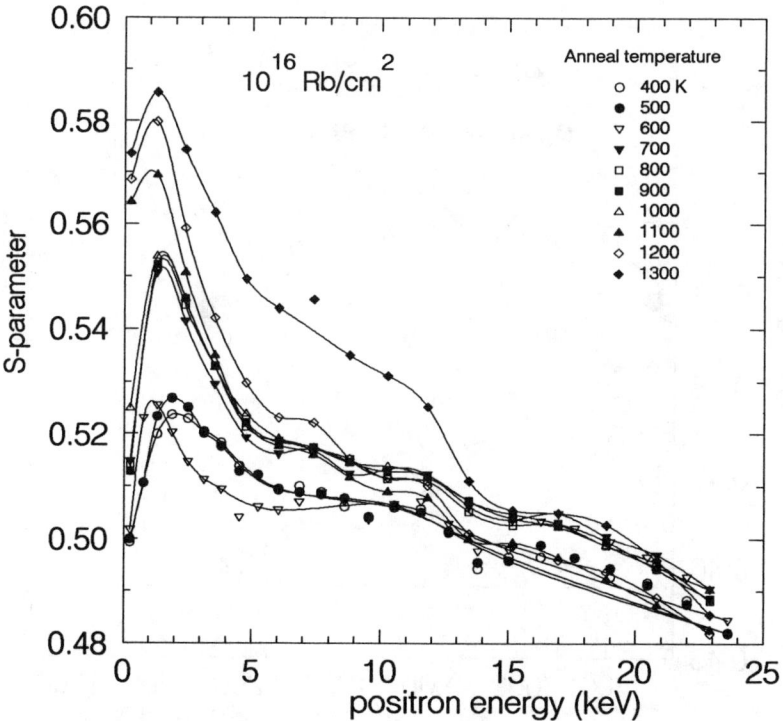

Figure 2. S-parameter curves measured at the 10^{16} Rb/cm^2 sample after annealing to the indicated temperatures.

In figure 2 the results of S-parameter measurements on the 10^{16} Rb/cm^2 sample after several anneal steps are shown. The effect of annealing of the near surface defects is seen as an increase of the S-parameter at low positron implantation energies. Although less pronounced, this behaviour is also observed for the samples with lower Rb implanted dose. The extremely high S-value found in the case of the 10^{16} Rb/cm^2 implantation indicates the presence of large cavities with an S/S_{bulk} ratio of 1.15 like found for cavities in silicon[6.]
The effect of annealing on the deep defects is derived from the S-parameter values measured at 15 keV positron energy (mean implantation depth of 160 nm). In figure 3 these values are shown as a function of the anneal temperature for all implanted samples. Annealing of the deep defects is observed as an initial increase of S in the temperature range from 400 to 600 K followed by a decrease between 1000 and 1300 K. Above these temperatures S increases again. This behaviour might be ascribed to the trapping of additional vacancies at the small Rb-vacancy clusters. However, it is not clear what causes the mobility at such low temperatures. Usually vacancy mobility is found around 700 K.

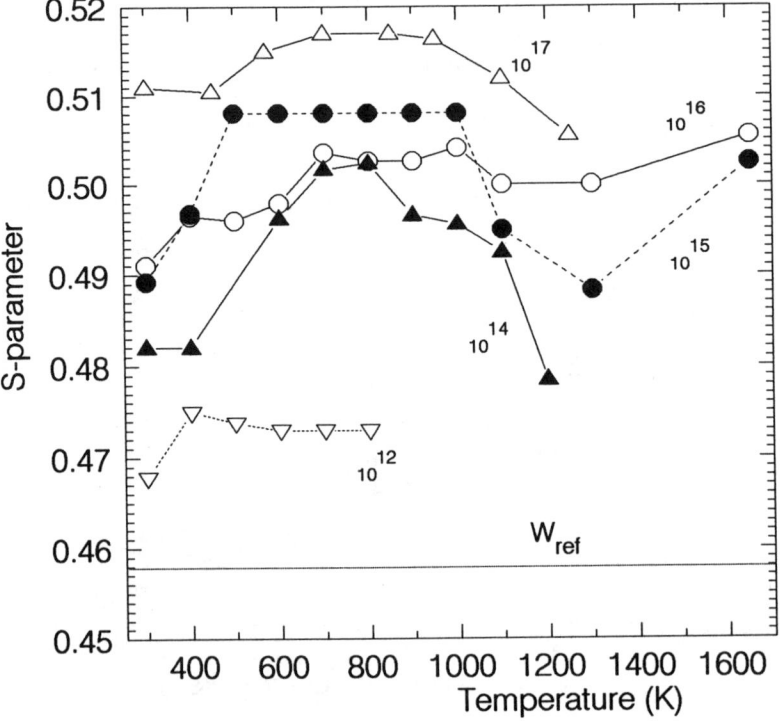

Figure 3. S-parameter as a function of anneal temperature measured at 15 keV positron energy. Mean positron implantation depth is 160 nm.

Trapping of vacancies enlarges the size of the positron trapping site which generally results in a higher S-parameter value. The subsequent decrease of S is ascribed to the release of trapped vacancies from the Rb-vancancy cluster.

A similar sequence is observed in PAC and THDS studies on silver and indium implanted tungsten[7]. Finally at temperatures above 1300 K the damage in the tail is recovered, thereby enabling positrons to diffuse back to the near-surface region where at these temperatures large Rb-vacancy clusters with correspondingly high S-parameters have been formed.

REFERENCES

1. M. Greuter and L. Niesen, to be published.
2. K.T. Tim and G. Welsch, Materials Letters, 9, (1990), 295-301.
3. H. Schut, "A variable energy positron beam facility with applications in materials science", Thesis, Delft University of Technology, Delft (1990).
4. A. van Veen, J. Trace and Microprobe Techniques, 8, (1990), 1-29.
5. TRIM-87, The transport of Ions in Matter, J.F. Ziegler, J.P. Biersack and U. Littmark, Pergamon Press, New York, 1985.
6. R.A. Hakvoort, A. van Veen, H. Schut, M.J. van den Boogaard, A.J.M. Berntsen, S. Roorda, P.A. Stolk, and A.H. Reader, these proceedings
7. G.J. van der Kolk, K. Post, A. van Veen, F. Pleiter and J.Th.M. de Hosson, Rad. Eff. 84, (1985), 131-158.

POSITRON EXPERIMENTS WITH CADMIUM MERCURY TELLURIDE

C Smith, P C Rice-Evans
Physics Department, Royal Holloway, University of London,
Egham, Surrey, TW20 0EX, England

N Shaw
D.R.A. Electronics Division, St Andrews Road,
Malvern, Worcs, WR14 3PS, England

ABSTRACT

The mixed crystal $Cd_xHg_{1-x}Te$ has been studied with positrons. Beam measurements reveal that the Doppler broadening is a good indicator of the hole concentration. Bulk studies in the temperature range 10-300K have been analysed on the basis of a temperature dependance of the negative charge state of Hg vacancies resulting in a variation in the trapping rate. Ionisation energies for the singly and doubly ionised vacancies were estimated to be 14 and 42meV respectively.

INTRODUCTION

The mixed crystal cadmium mercury telluride, $Cd_xHg_{1-x}Te$ (CMT) is an important infrared detector material. From the viewpoint of the device physics it is in many respects an ideal substance. By altering x, the mole fraction of CdTe, the forbidden energy gap can be varied continuously from that of the semi metal HgTe(-0.3 eV at 0 K) to that of CdTe (1.6 eV at 0 K) and within this bandgap range detectors can therefore be tailored for a specified spectral response with optimum performance. Most interest is in detectors operating in the atmospheric windows lying between the spectral wavelengths 8-14μm and 3-5μm requiring CMT with x-values around 0.2 and 0.3 respectively.

We have conducted experiments on the near surface region of CMT with our variable energy positron beam XENOPHON, and also on the bulk properties. In both cases we have employed Doppler-broadening measurements on the annihilation radiation.

BEAM MEASUREMENTS

The use of a low-energy positron beam provides a non-destructive means of probing surface and near-surface defects[1]. In such a facility, positrons from a ^{22}Na source are first slowed down to a few electron Volts in a tungsten moderator, after which they can be accelerated to various energies before penetrating a sample. By varying this energy, the depth to which positrons probe can be controlled according to a Makhovian stopping profile given by[2]

$$P(z, E) = -\frac{d}{dz}\exp\left(-\frac{z}{z_o}\right)$$

where $z_o = A/\rho E^n$, with $A = 4.5\mu gcm^{-3}$ and ρ the material density in gcm^{-3}. As grown and annealed samples of CMT have been investigated using the low- energy positron beam at Royal Holloway. Samples 1, 2 and 3 are bulk CMT grown by the cast recrystallize anneal (CRA) process with $x = 0.2$ and sample 4 is epitaxial CMT on a GaAs substrate (grown by MOVPE) with a thickness of 12μm and x=0.21. Samples 1

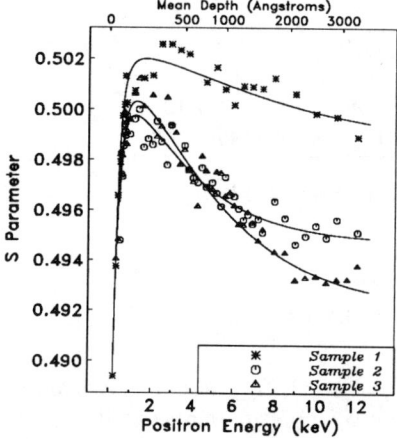
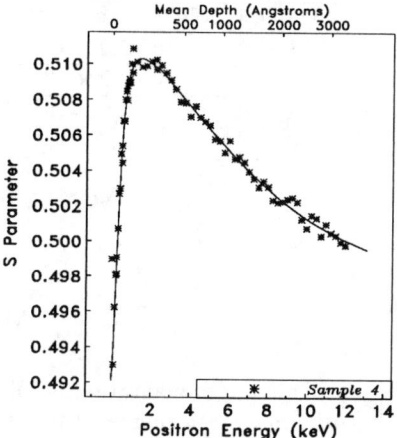

Figure 1. The S Parameter as a function of energy for samples 1,2 and 3.

Figure 2. The S Parameter as a function of energy for sample 4.

and 4 are unannealed while sample 2 has been annealed in a vacuum for 48 h at 503 K and sample 3 has been annealed in a mercury atmosphere for 24 h at 623 K. The hole concentrations of the samples were measured by the Hall effect at 77 K.

The photon spectra were recorded with a germanium detector situated directly behind the sample and were taken over a range of positron implantation energies 0-12 keV. The lineshape parameter S was defined as the ratio of the counts in the central eleven channels of the 511 keV photopeak to the total counts in the peak (40 channels) after background subtraction[3].

The results in figs. 1 and 2 have been analysed using a time-independent diffusion model. After coming to a virtual stop, diffusion takes place and the final distribution $n(z)$ of positrons in the sample is given by

$$D_+ \frac{dn(z)}{dz} + (\lambda_b + \kappa C)n(z) + P(z, E) = 0$$

where D_+ is the diffusion coefficient, λ the annihilation rate and κ the positron specific trapping rate. Three rates have to be accounted for: annihilation in the bulk, annihilation at the surface, and epithermal scattering[4]. Thus the parameter S may be written as,

$$S(E) = S_{ep}J_{ep} + (1 - J_{ep})(J_s S_s + (1 - J_s)S_b)$$

where S_{ep}, S_s and S_b are the S parameters which correspond to 100% of the positrons annihilating in the epithermal case, the surface state and in the bulk state respectively. The fractions J_{ep} and J_s are the probabilities of a positron annihilating in the states S_{ep} and S_b respectively and are given by,

$$J_{ep} = \int_0^\infty P(z, E)\exp(-z/l_+)dz$$

Sample	Hole Concentration (cm^{-3})	S_b	S_s	L(Å)
1	2.0×10^{17}	0.4986	0.5024	1800 ± 80
2	2.0×10^{16}	0.4946	0.5007	580 ± 80
3	1.0×10^{15}	0.4918	0.5015	1010 ± 80
4	1.0×10^{17}	0.4968	0.5117	1430 ± 80

Table 1: *Summary of Diffusion Model Analysis.*

and

$$J_s = \int_0^\infty P(z, E)\exp(-z/L_+)dz$$

where P(z,E) is the implantation profile given by eqn (1), L_+ the diffusion length and l_+ the epithermal scattering length. A summary of the diffusion model analysis is given in table 1.

S_b varies with hole concentration in a similar manner to that obtained by Krause et al[5] who related positron lifetimes to hole concentration. Krause claimed a sensitivity range of $10^{15} - 10^{18}$cm^{-3} and our results are in agreement with this, but both conflict with those of Gély et al[6] who concluded that positron trapping saturates at a hole concentration of 1.7×10^{16}cm^{-3}. The cause of this disagreement cannot be due to the growth process or the composition since both Krause and Gély used bulk CMT grown by the travelling-heater method with similar x values(0.2). It probably arises because Gély rapidly quenched the samples after annealing to fix the high-temperature state.

BULK MEASUREMENTS AT LOW TEMPERATURES

It is accepted that mercury vacancies are doubly ionised at 300 K, but there is some doubt as to whether they are singly or doubly ionised at 77 K[7,8].

An important related question is whether the trapping rate for positrons in semiconductors is independent of temperature; Puska et al[9] estimate different rates for V^{2-}, V^{1-} and V^0, and for the negative vacancies predict a $T^{-1/2}$ dependence.

We have conducted a study of the bulk properties at low temperature with a traditional sandwich geometry of 2 pieces of CMT embracing a 20μCi ^{22}NaCl source enclosed in a 3μm aluminium envelope. The Doppler broadening of three samples was measured over the temperature range 10 - 300 K. Sample (A) had been annealed at 873 K for one day and had a hole concentration of 10^{14}cm^{-3}; sample (B) had been annealed in a mercury atmosphere at 493 K for ten days and had a hole concentration of 1.1×10^{18}cm^{-3}; and sample (C) had the same processing as sample (B) but had been rapidly quenched in water, and its hole concentration was 5×10^{18}cm^{-3}. Figure (3) shows the results. Sample (A) with its low vacancy concentration shows a low value of S which implies only a low amount of positron trapping. Sample (C) has a high value of S, indicating much trapping, and the lack of a temperature dependence suggests a saturation of trapping.

Sample (B) is most interesting. It has an intermediate value of S, and displays a striking temperature dependence. The first thing to be noted is that although the vacancies are expected to be negative there is no sign of a $T^{-1/2}$ dependence with its negative gradient. The important question is whether the slope of the curve for (B) is indicating the variation of the charge state of the vacancies as the temperature changes. We might expect the actual trapping rates for V^0, V^{1-} and V^{2-} to be different, and we have conducted a 2-state analysis on the curve for (B) with this assumption. In a semiconductor where defects or impurities can donate or accept more than one electron a number of energy levels corresponding to the ionisation energies E_n of the flaw will be introduced into the band gap. For a given temperature and Fermi level E_F, the ratio of the number of flaws N_n with charge n to N_{n+1} is known to be[10]

$$\frac{N_n}{N_{n+1}} = g_n^{-1} \exp\left(\frac{E_n - E_F}{kT}\right)$$

where g_n^{-1} is the spin degeneracy. Then the probability of a flaw being in the charge state n is $P_n = N_n / \sum^n N_n$, where for a divalent acceptor like CMT n can have values 0, 1 or 2. Hence the probability of the vacancy being in a doubly negative-charge state is

$$P_2 = \frac{g_2 \exp\left(\frac{E_f - E_2}{kT}\right)}{1 + g_2 \exp\left(\frac{E_f - E_2}{kT}\right) + g_1^{-1} \exp\left(\frac{E_1 - E_f}{kT}\right)}$$

and for the singly negative charge state it is,

$$P_1 = \frac{1}{1 + g_2 \exp\left(\frac{E_f - E_2}{kT}\right) + g_1^{-1} \exp\left(\frac{E_1 - E_f}{kT}\right)}$$

the probability of the vacancy being neutral is then,

$$P_0 = 1 - (P_1 + P_2).$$

Now

$$E_f = kT \ln\left(\frac{N_v}{p}\right)$$

where p is the hole concentration and N_v is the density of states in the valence band given by,

$$N_v = 2 \left(\frac{m_v kT}{2\pi \hbar^2}\right)^{3/2} = 4.83 \times 10^{21} \, (Tm_v/m_0)^{3/2} \, m^{-3}$$

where m_v is the effective mass of a hole in the valence band and m_0 the rest mass of the electron.

With a total vacancy concentration C_v given by $C_0 + C_1 + C_2$ and $p = C_1 + 2C_2$, the equations lead[11,12] to

$$p = \frac{C_v \left[2 + g_2^{-1} f(E_1)\right]}{1 + g_2^{-1} f(E_2) + g_1^{-1} g_2^{-1} f(E_2) f(E_1)}$$

with $f(E_1) = \exp\left(\frac{E_1 - E_f}{kT}\right)$ and $f(E_2) = \exp\left(\frac{E_2 - E_f}{kT}\right)$. Rewriting with,

$$p = N_v \exp\left(-\frac{E_f}{kT}\right)$$

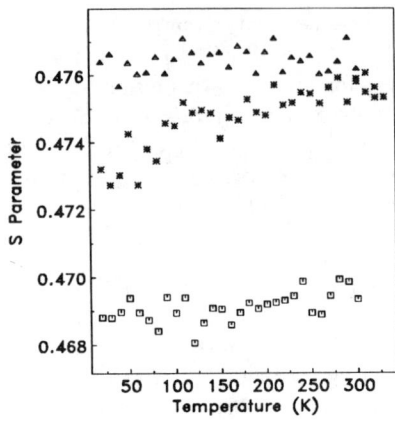 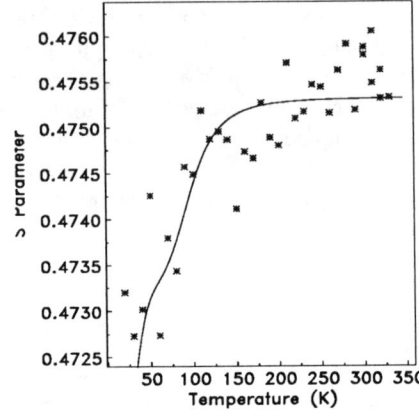

Figure 3. The S parameter as a function of temperature. (A) squares; (B) stars; (C) triangles.

Figure 4. The S parameter as a function of temperature for sample B with a theoretical fit.

the equation assumes the following form,

$$x^3 + 2\alpha x^2 - \alpha(1-\beta)x - \alpha\beta = 0$$

where

$$x = \frac{p}{2C_v}$$

$$\alpha = \frac{N_v}{4g_1^{-1}C_v}\exp\left(-\frac{E_1}{kT}\right)$$

and

$$\beta = \frac{N_v}{g_2^{-1}C_v}\exp\left(-\frac{E_2}{kT}\right).$$

We have analysed the positron S-parameter measurements for sample B with the conventional two state trapping model[13], with

$$C_v = \frac{\lambda_b}{\mu}\left(\frac{S - S_b}{S_v - S}\right)$$

where S_b and S_v correspond with 100% annihilations in the free and trapped states. We have assumed that S_v does not depend on the charge state of the vacancies, but that the trapping rate μ will. We therefore write $\mu = (\mu_0 P_0 + \mu_1 P_1 + \mu_2 P_2)$. The mercury vacancy concentration, C_v, has been obtained with Hall effect measurements assuming the vacancies are singly ionised at 77 K (Elizarov et al, 1990), and the bulk annihilation rate λ_b has been taken as 275ps (Krause et al, 1989). For CMT we take values of g_1 and g_2 as 3/2 and 4 respectively. Solving the cubic equation for p allows the probabilities P_0, P_1 and P_2 at any temperature to be calculated and these can then be employed in the fitting for S.

A least-squares fit to the data is shown in Figure 4 where it is seen that the data lie plausibly on the best calculated line. The best values for the ionisation energies of the singly and doubly ionised vacancies E_1 and E_2 are 14 meV and 42 meV respectively and the estimated trapping rates μ_0, μ_1, μ_2 are 0.1, 1.1 and $3.0 \times 10^{-8} cm^3 s^{-1}$.

It is clear from the figure that the data do not justify firm conclusions, and further work is required to improve the statistics and test the assumptions.

ACKNOWLEDGEMENTS

This work was supported by the Science and Engineering Research Council and the DRA Malvern.

REFERENCES

[1] Schultz P J and Lynn K G, 1988, Reviews of Modern Physics 60,701
[2] Makhov A F, 1960, Sov. Phys. Solid State 2,1934
[3] Chaglar I, Rice-Evans P C, El-Khangi F A R and Berry A A, 1981, Nucl. Instr. Meth 187,581
[4] Britton D T and Rice-Evans P C, 1988, Nucl. Instr. Meth. 57,3
[5] Krause R, Klimakow A, Kiessling F M, Polity A, Gille P and Schenk M, 1990, J. Cryst. Growth 101,512
[6] Gély C, Corbel C and Triboulet R J, 1990, J.Phys. Condens. Matter 2,4763
[7] Vydyanath H R, 1981, J. Electrochem. Soc. 128,2609
[8] Elizarov A I, Bogoboyashchii V V and Berchenko N N, 1990 Sov. Phys. Semicond. 24,278
[9] Puska M J, Corbel C and Nieminen R M, 1990, Phys.Rev. B41,9980
[10] Shockley W and Last J T, 1957, Phys.Rev. 107,392
[11] Champness, 1956, Proc.Phys.Soc London B69,1335
[12] Cohen M M, 1972, 'Introduction to the Quantum Theory of Semiconductors' Gordon and Breach, London.
[13] West R N, 1973, 'Positron Studies of Condensed Matter' Taylor and Francis, London.

CHARACTERIZATION OF THIN FILMS BY A PULSED POSITRON BEAM

R. Suzuki, T. Mikado, H. Ohgaki, M. Chiwaki,
T. Yamazaki, K. Awazu and A. Matsuda
Electrotechnical Laboratory, 1-1-4 Umezono, Tsukuba, Ibaraki 305, Japan

Y. Kobayashi
National Chemical Laboratory for Industry, 1-1 Higashi, Tsukuba,
Ibaraki 305, Japan

A. Uedono
Research Center for Advanced Science and Technology, University of Tokyo, 4-6-1
Komaba, Meguro-ku, Tokyo 153, Japan

S. Tanigawa
Institute of Materials Science, The University of Tsukuba, 1-1 Tennoudai, Tsukuba,
Ibaraki 305, Japan

ABSTRACT

Positron lifetime spectroscopy with an intense pulsed positron beam has been used to characterize defects in hydrogenated amorphous silicon films, porous silicon, diamond films, ion implanted SiO_2, and metal-oxide-semiconductor samples. Both long-lived component and short-lived component of the positron lifetime spectra strongly depend on the deposition condition, annealing temperature, and other conditions. The relationship between positron lifetime spectra and microscopic structures is discussed.

§1. INTRODUCTION

Positron lifetime spectroscopy is known to be a powerful method for the study of microstructural defects in solids. Since the positron lifetime (τ) depends strongly on the size of the defect, it is possible to distinguish different defect species with this technique to some extent. However, it is difficult to investigate thin films of less than several micrometers in thickness by means of conventional positron methods because positrons emitted from the β^+ sources penetrate too deeply into the sample.

In recent years, slow or variable-energy positron beams have been developed to measure positron lifetime spectra at a desired depth from the surface of the solid.[1-4] Recent progress of the variable-energy pulsed positron beam at the ETL linac facility enables us to measure positron lifetime spectra with high resolution, high count rate, high peak to background ratio, and wide measurable time range.[4] We have carried out positron lifetime experiments on several materials with the pulsed positron beam and found that this measurement is a very sensitive method to study microstructural defects and electronic structure in surface or near surface regions.[6-9] In this paper, we present positron lifetime study of several thin film samples, which are important for both technological application and basic research.

§2 EXPERIMENTS

The positron lifetime measurements were carried out with the variable-energy pulsed positron beam at the ETL. A description of the apparatus can be found elsewhere.[4,5] All the experiments presented in this paper were performed at room temperature under high-vacuum ($\sim 1 \times 10^{-7}$ Torr) conditions. The annihilation γ-ray was detected by a scintillation detector which consists of a BaF_2 crystal (50.8 mm in diameter and 25.4 mm in thickness) and a photomultiplier tube (Hamamatsu H3378). The lifetime spectra were obtained by measuring the time interval between the timing signal from the pulsing system and the timing signal of the annihilation γ-ray. For each spectrum, 1×10^5-3×10^5 counts were accumulated.

§3. RESULTS AND DISCUSSION

3.1 HYDROGENATED AMORPHOUS Si FILMS

Figure 1(a) shows lifetime spectra of hydrogenated amorphous Si (a-Si:H) films (930 nm - 960 nm in thickness) at an incident energy (E_i) of 6 keV corresponding to the mean penetration depth $\bar{x}$ of about 300 nm. These films were deposited on single crystal Si by a Xe-dilution plasma-enhanced chemical-vapor deposition (PECVD) method under different rf-power densities (0.03 W/cm^2, 0.13 W/cm^2, 0.51 W/cm^2, and 0.76 W/cm^2).[10] The lifetime spectra show large differences among the films while no significant difference was observed in the TO Raman spectra among these films.[6] This suggests that the formation of microstructural defects is strongly influenced by the rf-power density during the PECVD process.

The lifetime spectrum of the film prepared at 0.03 W/cm^2 has nearly a single component of 319 ps. This lifetime is close to the lifetime of divacancies in crystalline silicon.[11] This indicates that positrons are rapidly trapped at divacancy-like defect. The lifetime spectra of the other films have significantly longer-lived components, which can be attributed to annihilation at vacancy clusters or voids. This result is consistent with the experimental results of the refractive index, thermal effusion, NMR, and electron spin resonance—all of these shows the concentration of large vacancy type defects in the film of 0.03 W/cm^2 is lowest.[10]

In the films prepared at 0.13 W/cm^2 and 0.51 W/cm^2, both the long-lived o-Ps component ($\tau_3 \simeq$ 9 ns) and the intermediate component (420 ps $\leq \tau_2 \leq$ 440 ps) are observed.[6] The long lived component of $\sim$9 ns indicates the existence of voids of 1-2 nm in diameter.[6] A small angle X-ray scattering experiment[12] revealed that a structural inhomogeneity of nanometer size exists in the film of 0.51 W/cm^2. The results of the other experiments, for example, the hydrogen thermal effusion experiment and the NMR measurement, suggest indirectly that the nanometer-size structures are voids, but it could not be directly confirmed by these experiments. On

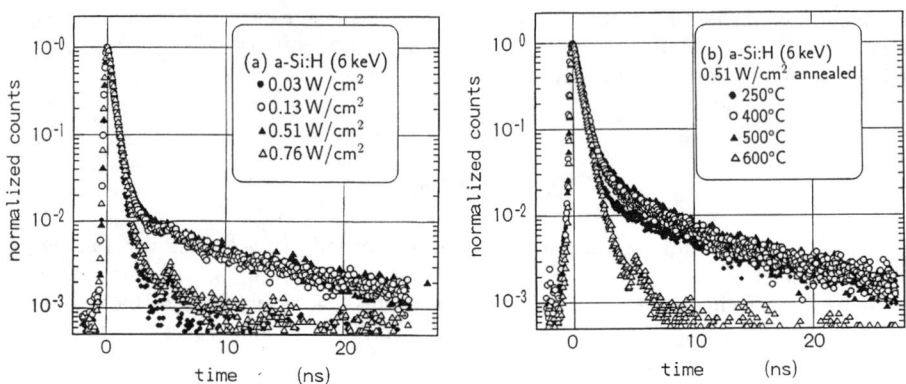

Fig. 1. Positron lifetime spectra of a-Si:H films. (a) Four films prepared under rf-power densities of 0.03, 0.13, 0.51 and 0.76 W/cm², (b) Annealing behavior of the a-Si:H film prepared at 0.51 W/cm².

the contrary, the present study directly confirmed that the structures are voids since positron lifetime experiments only gives information on vacancy type defects.

It should be noted that the film prepared at 0.03 W/cm² shows typical photo-degradation property which hinders the realization of inexpensive large-scale solar cell applications, while the film at 0.51 W/cm², which contains a high concentration of voids, does not show the photo-degradation during the experiment of photo-degratation.[10] Therefore, the present study directly confirmed that the nanometer-size voids do not contribute photo-degradation.

Figure 1(b) shows positron lifetime spectra of the film prepared at 0.51 W/cm² under different annealing temperatures. Both the intermediate component and long-lived component are drastically changed with the annealing temperature. In particular, the long-lived component was suddenly decreased at between 500°C-600°C. This suggests that some structural relaxation and/or hydrogen effusion were occurred in the film in the annealing process, and these would reflect to the lifetime spectra.

3.2 POROUS SILICON

Porous silicon, obtained by anodization of Si single crystals in HF acid solutions, has attracted much attention because of its unique properties, e.g., photoluminescence, high chemical reactivity, etc. The structure and properties of porous silicon have been found to be very dependent on the Si crystal characteristics and on the anodization conditions. Cruz and Pareja[13] first applied positron lifetime technique to porous silicon samples, and they found that the intermediate lifetime, τ_2, strongly depends on the anodization conditions, annealing temperatures, and silicon crystal properties. Recently, Itoh, Murakmi and Kinoshita[14] found that there exists a very long-lived (> 10ns) o-Ps component and that the component is depend on the sample conditions. However, both of these experiments use a conventional lifetime technique and a high intensity of the lifetime component of bulk silicon was included in their spectra.

Thus, we measured positron lifetime spectra in porous silicon using the variable-energy pulsed positron beam. Figure 2 shows the lifetime spectra of porous silicon obtained by 2 min anodization in a mixture of 50% HF solution and C_2H_5OH with a current density of 25 mA/cm^2. As shown in Fig. 2, a o-Ps long-lived component of $\sim$ 30 ns was observed. In porous Si, it is established that there are nanometer-size pores around nanometer-size Si structures. The o-Ps lifetime is considerably longer than that of the a-Si:H films of Fig. 1 which indicates existence of 1-2 nanometer voids. The long lifetime would be due to the o-Ps atoms which effuse from small pores to larger pores since these pores are connected to each other and the larger pores have larger biding energy. Two short-lived components of 240 ps·44% and of 530 ps·49% were resolved in the spectrum. The 530 ps lifetime component can be attributed to the positrons annihilated at vacancy clusters or the surface of the pores, and the 240 ps lifetime can be attributed to the mean lifetime of p-Ps, non-trapped positrons, and/or positrons trapped at small vacancy-type defects.

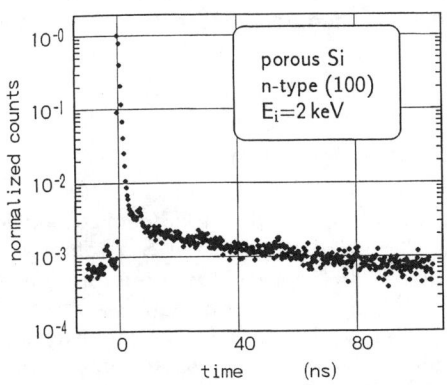
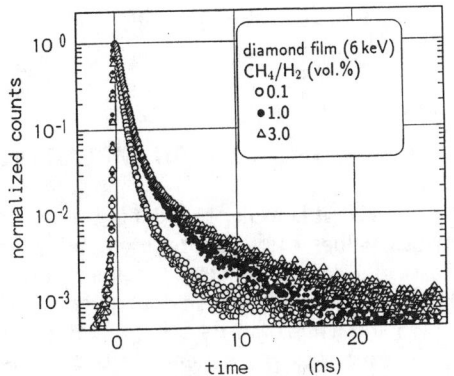

Fig. 2. Positron lifetime spectrum of porous Si at the incident positron energy of 2 keV.

Fig. 3. Positron lifetime spectra of diamond films at the incident positron energy of 6 keV.

3.3 DIAMOND FILMS

Figure 3 shows the positron lifetime spectra of polycrystalline diamond films (0.7 μm – 1.4 μm in thickness) synthesized on single-crystal silicon substrates (0.01 Ω·cm, n-type) by the microwave PECVD method using three different H_2/CH_4 mixtures.[9] The incident positron energy is 6 keV corresponding to the depth of $\sim$200 nm. Figure 3 shows clearly that the mean lifetime, τ_M, increases in order of CH_4/H_2 ratio.

The results of the Raman experiment indicate that a peak of 1333 cm^{-1}, which is due to the well-defined diamond structure, appears more clearly as the ratio of CH$_4$ is decreased.[15] Therefore, we may conclude that the mean lifetime decreases with the improvement of crystalline perfection. The detailed results and discussions are reported in ref.9. The positron lifetime spectrum of the diamond film prepared at the lowest CH$_4$/H$_2$ ratio of 0.1% has significantly longer-lived components than that of the bulk diamond (~100 ps). This indicates that vacancy-type defects still exist with high concentration in the film prepared at 0.1%. This result is consistent with the result of the Doppler broadening measurement on the same films.[15]

For all the diamond films, a component of around 400 ps lifetime and one or two long-lived ($\tau > 1$ ns) components were observed.[9] The component of around 400 ps lifetime can be attributed to the positrons annihilated at large vacancy clusters.[17] The long-lived components can be attributed to the annihilation of o-Ps atoms which formed in large vacancy clusters or voids. These voids presumably exist at the crystalline boundaries. It should be noted that the intensity of o-Ps components is considerably higher than that of previously reported experiments on various carbon materials which contain voids with high density.[16] The Ps formation is presumably due to the hydrogen termination of the void surface.

Comparing the spectra of the a-Si:H films with those of the diamond films, the separation of the short lived component and the long lived o-Ps component is better in the case of the a-Si:H films than in the case of the diamond films. This suggests that the size distribution of the voids in the diamond films is broader than that of the a-Si:H films.

3.4 ION IMPLANTED AMORPHOUS SiO$_2$

Several Doppler broadening measurements using slow positron beams observed various values of S-parameter in SiO$_2$ films.[19-22] In particular, the S-parameter is drastically decreased after the ion implantation.[21-22] However, the interpretation of this phenomenon has been controversial. Using the pulsed positron beam, we observed a high intensity of o-Ps component in SiO$_2$ films, and that both the intensity and lifetime are strongly depend on the deposition condition.[8] Thus, we speculated that the difference of S-parameter is due to the difference of positronium formation. To confirm this speculation, we measured lifetime spectra of ion implanted SiO$_2$ with the pulsed positron beam.

Figure 4 shows the 400 keV Xe-ion implanted and unimplanted SiO$_2$ at the positron incident energy of 2.6 keV which corresponds the mean depth of implanted ions. The amorphous SiO$_2$ specimen was obtained by a vaporphase axial deposition method.[18] The spectrum of the unimplanted specimen is similar to that of thermally grown SiO$_2$ film.[7,8] As shown in Fig. 4, o-Ps component of ~1.5 ns lifetime is drastically decreased and 450-500 ps component is increased with increasing the dose of ions. This result and results of Doppler broadening experiments on ion implanted SiO$_2$ suggests that the positronium formation is inhibited by defects produced in the ion implantation process. Similar results were observed in SIMOX (separation by implantation of oxygen) samples, and detailed discussion is presented in another contribution.[22]

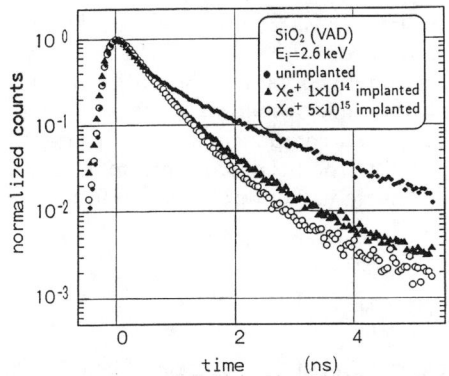

Fig. 4. Positron lifetime spectra of Xe-ion implanted SiO_2.

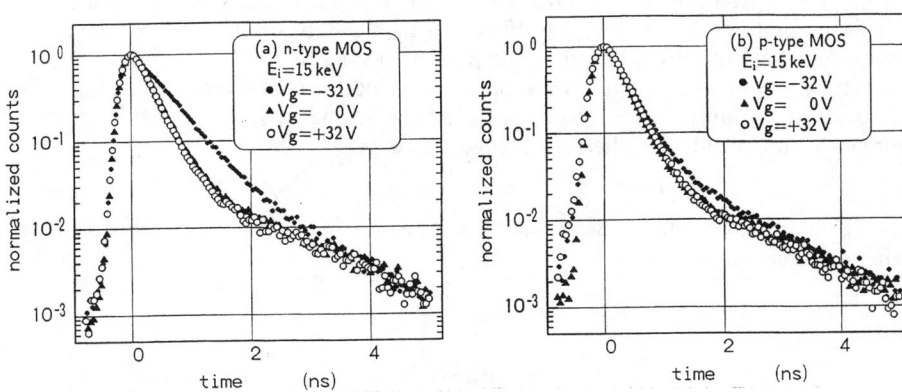

Fig. 5. Positron lifetime spectra of (a) n-type MOS and (b) p-type MOS at the incident positron energy of 15 keV.

3.5 MOS

In recent years, Uedono, et al.[23,24] and Leung, et al.[25,26] have studied MOS (metal-oxide-simiconductor) samples using Doppler broadening measurements with variable-energy positron beams, and they found that positrons implanted to Si substrate are drifted to Si-SiO$_2$ interface by the electric field of the depletion layer and trapped at Si-SiO$_2$ interface effectively when the sample is n-type MOS with negative gate bias voltages. Thus, in order to deduce the density and/or size of defects at the Si-SiO$_2$ interface, we applied the positron lifetime technique to MOS samples.

Figure 5 shows the lifetime spectra of n-type MOS and p-type MOS, both of which are same as the samples of the Doppler broadening experiment, performed by Uedono, et al.[23]; the gate is 100 nm poly Si and the oxide thickness is 400 nm. The incident positron energy is 15 keV corresponding to the mean depth of about

1.5 μm. Drastic change was observed in the lifetime spectrum of n-type MOS at the negative gate voltage. This is due to the positrons drifted to the Si-SiO$_2$ interface. The lifetime of this component is 460 ps obtained with a conventional fitting program. This lifetime is different from both the bulk Si lifetime (220 ps) and the SiO$_2$ lifetime which has a long-lived ($\tau=1.5$ ns) o-Ps component of a high intensity (~60%). This result and low S-value[23] suggest that the positrons are trapped at vacancy clusters of more than 5 vacancies and annihilate without positronium formation.

§4. CONCLUSION

Positron lifetime spectra were measured on a-Si:H films, diamond films, porous Si, ion implanted SiO$_2$, and MOS samples. We observed the long-lived o-Ps component in most of the lifetime spectra and found that both long-lived component and short-lived component are strongly influenced by the deposition condition, annealing temperature, and other conditions. The present results demonstrate the potential of the positron annihilation technique as a non-destructive characterization tool for the technologically important materials. Some of the studies presented in this paper are preliminary and detailed studies are being carried out.

The authors would like to thank Drs. T. Noguchi, S. Sugiyama and K. Yamada for helpful assistance.

REFERENCES

1. K.G. Lynn, W.E. Frieze, and P.J. Schultz, Phys. Rev. Lett. 52, 1137 (1984).

2. W. S. Crane and A. P. Mills Jr, Rev. Sci. Instrum. 56, 1723 (1985).

3. Schödlbauer, D., Sperr, P., Kögel, G. and Triftshäuser, W., Nucl. Instr. and Meth. B, 34, 258 (1988).

4. R. Suzuki, Y. Kobayashi, T. Mikado, H. Ohgaki, M. Chiwaki, T. Yamazaki and T. Tomimasu, Proc. Int. Conf. Evolution in Beam Applications, (Radiation Application Development Association, Tokyo) p.357, 1992.

5. R. Suzuki, Y. Kobayashi, T. Mikado, H. Ohgaki, M. Chiwaki, T. Yamazaki and T. Tomimasu, "Positrons at Metallic Surfaces", ed. A. Ishii, (Trans Tech Publ. Ltd), to be published.

6. R. Suzuki, Y. Kobayashi, T. Mikado, A. Matsuda, P. J. McElheny, S. Mashima, H. Ohgaki, M. Chiwaki, T. Yamazaki and T. Tomimasu, Jpn. J. Appl. Phys., 30, 2438 (1991).

7. R. Suzuki, Y. Kobayashi, T. Mikado, H. Ohgaki, M. Chiwaki, T. Yamazaki and T. Tomimasu, Materials Science Forum, 105-110, 1993 (1992).

8. R. Suzuki, Y. Kobayashi, T. Mikado, H. Ohgaki, M. Chiwaki, T. Yamazaki and T. Tomimasu, Materials Science Forum, 105-110, 1459 (1992).

9. R. Suzuki, Y. Kobayashi, T. Mikado, H. Ohgaki, M. Chiwaki, T. Yamazaki, A. Uedono, S. Tanigawa and H. Funamoto, Jpn. J. Appl. Phys., 31, 2237 (1992).

10. A. Matsuda, S. Mashima, K. Hasezaki, A. Suzuki, S. Yamasaki and P. J. McElheny, Appl. Phys. Lett. 58, 2494 (1991).

11. W. Fuhs, U. Holzhauer and F.W. Richter, Appl. Phys. 22, 415 (1980).

12. S. Mashima, K. Hasezaki, A. Suzuki, P. J. McElheny and A. Matsuda, Mat. Res. Soc. Symp. Proc. vd.123 (MRS Anaheim, CA, 1991).

13. R.M. de la Cruz and R. Pareja, Positron Annihilation, eds L. Dorikens-Vanpraet, M. Dorikens and D. Segers, (World Scientific, Singapore, 1989) p. 702.

14. Y. Itoh, H. Murakami, A. Kinoshita, Hyperfine Interactions, to be published.

15. A. Uedono, S. Tanigawa, H. Funamoto, A. Nishikawa and K. Takahashi, Jpn. J. Appl. Phys. 29, 555 (1990).

16. T. Iwata, H. Fukushima, M. Shimotomai and M. Doyama, Jpn. J. Appl. Phys. 20, 1799 (1981).

17. G. Kögel, D. Schödlbauer, W. Triftshäuser and J.Winter, J. Nucl. Mater. 162-164, 876 (1989).

18. P.C. Schultz, Fiber Optics, eds B. Bendow and S.S. Mitra (Plenum Press, New York and London, 1979) Vol.3.

19. K.G. Lynn and P. Asoka-Kumar, Materials Science Forum, 105-110, 359 (1992).

20. D.L. Smith, C. Smith, P. Rice-Evans, J.H. Evans and H.E. Evans, Materials Science Forum, 105-110, 1451 (1992).

21. A. Uedono, L. Wei, Y. Tabuki, H. Kondo, S. Tanigawa, J. Sugiura and M. Ogasawara, Materials Science Forum, 105-110, 1479 (1992).

22. A. Uedono, S. Watauchi, Y. Ujihira, L. Wei, S. Tanigawa, R. Suzuki, H. Ohgaki, T. Mikado, H. Kametani, H. Akiyama, Y. Yamaguchi and M. Koumaru, in this volume.

23. A. Uedono, S. Tanigawa and Y. Ohji, Phys. Lett. A 133, 82 (1988).

24. A. Uedono, L. Wei, Y. Tabuki, H. Kondo, S. Tanigawa and Y. Ohji, Materials Science Forum, 105-110, 1475 (1992).

25. T.C. Leung, Z.A. Weinberg, P. Asoka-Kumar, B. Nielsen, G.W. Rubloff and K.G. Lynn, J. Appl. Physics, 70 2874 (1992).

26. T.C. Leung, P. Asoka-Kumar, B. Nielsen and K.G. Lynn, J. Appl. Physics, to be published.

POINT DEFECTS IN AS-GROWN AND ION IMPLANTED GaAs
PROBED BY A MONOENERGETIC POSITRON BEAM

A. Uedono
Research Center for Advanced Science and Technology,
University of Tokyo, 4-6-1 Komaba, Meguro-ku,
Tokyo 153, Japan

S. Fujii
Opto-electronics R&D Lab., Sumitomo Electric Industries
Ltd., 1 Taya-cho, Sakae-ku, Yokohama 244, Japan

L. Wei and S. Tanigawa
Institute of Materials Science, University of Tsukuba,
Tsukuba, Ibaraki 305, Japan

ABSTRACT

Defects in ion implanted GaAs were studied by monoenergetic positron beams. From measurements of Doppler broadening profiles of the positron annihilation as a function of incident positron energy, it was found that a species of defects strongly depends on a species of implanted ions. Native defects in GaAs wafers were also studied. The present investigation shows that positrons provide a sensitive and nondestructive probe for the detection of both vacancies and interstitials.

1. INTRODUCTION

Ion implantation into GaAs has long been studied because of its technological importance. Si is frequently used impurities in order to fabricate n-type GaAs layer. Recently, Se has emerged as a potential candidate for n-type dopants in MBE-grown GaAs, because of its role in the reduction of DX centers.[1] In ion implantation technology, therefore, Se seems to be desirable dopant for the formation of a thin channel layer because of its heavy mass and single site substitution. However, Se is known as the dopant of low activation efficiency.[2] Oxygen is widely used in order to form an isolation layer between GaAs transistors. Implanted oxygen atoms were known to occupy substitutional sites and form deep electron traps after annealing treatments.[3] In the present paper, we report on the application of the monoenergetic positron beam technique to the study of defects in GaAs introduced by Si^+-, Se^+- and O^+-ion implantation. Native defects in GaAs wafers grown by the horizontal Bridgman (HB) method and the liquid encapsulated Czochralski (LEC) method were also studied.

2. EXPERIMENTAL

Undoped semi-insulating (s.i.) LEC-GaAs wafers were used as substrates for ion implantation. 70-keV Si^+-, 200-keV Se^+- and 50-keV O^+-ions were implanted into the wafers up to a dose of 3×10^{13} ion/cm^2. After ion implantation, the specimens were annealed with encapsulates of SiN or SiO_2 at 800 °C (25 min). The SiN film was deposited on the specimens by using the plasma enhanced CVD technique under SiH_4/N_2 gas at 320 °C. The SiO_2 film was deposited by the atmospheric-CVD technique under SiH_4/N_2O gas at 460 °C. The capless annealing under AsH_3 ambient was also performed at 850 °C (15 min). After these annealing treatments, the encapsulates were removed by using a buffered hydrofluoric acid. The transmission line method (TLM) was used in order to measure the sheet resistivity. Doppler broadening profiles were measured as a function of incident positron energy by using a monoenergetic positron beam line constructed at University of Tsukuba.[4] The spectrum with a total count of 10^6 was measured for each incident positron energy. The lifetime spectra for the Se^+-implanted specimen were measured by using a pulsed monoenergetic positron beam line constructed at Electrotechnical Laboratory (Tsukuba).[5]

Native defects in GaAs wafers cut from various single crystals were also studied by using the monoenergetic positron beam and by the conventional positron annihilation techniques. Characteristics of these specimens are listed in Table I. The positron lifetime spectra were measured by a fast-fast system with the time resolution of ~210 ps (FWHM). Doppler broadening profiles of the annihilation radiation were measured by a Ge detector. All measurements were performed at room temperature.

Table I. Characteristics of the GaAs specimens.

specimen	growth method	dopant	carrier concentration (/cm^3)
HB-N-A	HB	Si	8.4×10^{16}
HB-N-B	HB	Si	3.0×10^{17}
HB-N-C	HB	Si	7.2×10^{17}
HB-N-D	HB	Si	2.5×10^{18}
HB-P-A	HB	Zn	1.0×10^{18}
HB-P-B	HB	Zn	5.9×10^{18}
HB-P-C	HB	Zn	1.7×10^{19}
HB-P-D	HB	Zn	5.0×10^{19}
HB	HB	undoped	$10^{12} \sim 10^{14}$ (n-type)
LEC	LEC	undoped	s.i.

3. RESULTS AND DISCUSSION
3.1 DEFECTS IN Si⁺-IMPLANTED GaAs

Figure 1 shows the S parameter as a function of incident positron energy E for the 70-keV Si⁺-implanted GaAs specimens before and after annealing treatments. The distribution of Si atoms on the basis of LSS theory[6] is also shown in Fig. 1. The observed increase of S in the S-E plots for the as-implanted specimens can be attributed to the annihilation of positrons trapped by vacancy-type defects. The S-E relations for the annealed specimens with the SiN film and without the encapsulate were found to be similar to the S-E relation for the unimplanted specimen. This fact shows the recovery of vacancy-type defects by the annealing treatments. For the annealed specimen with the SiO₂ film, however, the value of S was found to rapidly decrease and to almost saturate above $E=5$ keV. This means that the diffusion length of positrons L for the SiO₂-cap annealed specimen is lower than that for the another annealed specimen. The value of L was obtained as 43 nm by using the one-dimensional diffusion model of positrons. By using the value of L for the unimplanted specimen (290 nm), the annihilation rate from the free state (4.3×10^9 s⁻¹) and the specific trapping rate for monovacancies in Si ($\mu_d = 2 \times 10^{-9}$ cm³s⁻¹, ref. 7), the concentration of defects C_d was derived as 1×10^{20} cm⁻³ for the SiO₂-cap annealed specimen. The obtained value of C_d suggests that almost all positrons are trapped by defects in the SiO₂-cap annealed specimen. For this specimen, however, no increase in S was observed. Thus the species of defects can not be attributed to vacancy-type defects. Since interstitial-type defects cause the smaller value of S than S_{bulk} (see section 3.3), the dominant species of defects is considered to be antisite defects. It has been reported that for a phosphosilicate glass and SiO$_x$N$_y$ cap annealed GaAs specimen, an out-diffusion of Ga atoms from the substrate into the film

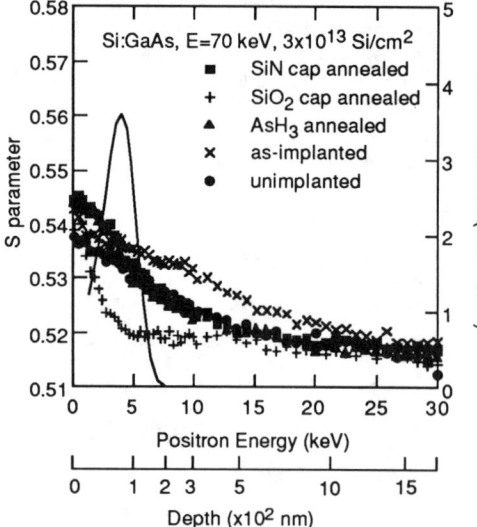

Fig. 1. The S-E relationships for the 70-keV Si⁺-implanted GaAs specimens with a dose of 3×10^{13} Si/cm².

enhances the electrical activation of implanted Si, whereas the opposite phenomenon was observed for the SiO_2 film in another experiment.[8,9] The sheet resistivities for the present specimens are listed in Table II. The lowest activation of implanted Si was observed for the SiO_2-cap annealed specimen. The out-diffusion of Ga atoms introduces Ga-vacancies (V_{Ga}) in the substrate. Thus, following two kinds of reactions can explain the lack of the increase in S and the low activation of implanted Si,

$$V_{Ga} + As_{As} \rightarrow As_{Ga} + V_{As} \tag{1}$$

$$V_{As} + Si_i \rightarrow Si_{As}, \tag{2}$$

where a positron is scattered by As-antisites (As_{Ga}) and Si_{As} decreases the activation efficiency of Si. The effect of V_{Ga} introduced by the SiO_2-cap annealing was also found for the O^+-implanted GaAs specimen.

Table II. Sheet resistivities of the Si^+-implanted GaAs specimens after various annealing treatments.

Annealing method	Sheet resistivity (Ω)
SiN-cap	222±5
SiO_2-cap	443±4
capless	355±5

3.2 DEFECTS IN Se^+-IMPLANTED GaAs

Figure 2 shows the S-E relations for the 200-keV Se^+-implanted GaAs specimens. It can be seen that the values of S in the defected region drastically increased after the annealing treatments. The sheet resistivities for these specimens are listed in Table III. From Fig. 2 and Table III, the largest value of S and the highest activation efficiency were observed for the SiN-cap annealed specimen. In turn, the smallest value of S and the lowest activation efficiency were observed for the capless-annealed specimen. Thus it can be concluded that the observed increase in S is attributed to the annihilation of positrons trapped in vacancy-type defects introduced by the activation of implanted Se-ions. In order to know the species of defects, lifetime spectra for the SiN-cap annealed specimen were measured. The spectra were resolved into an one-component or two-components with a time resolution of 250 ps (FWHM). The results are listed in Table IV. The lifetime at $E=20$ keV is longer than the bulk-lifetime in GaAs (235 ps). This is due to the effect of the positron annihilation in the defected region (0 keV $< E <$ 10 keV). The values of τ_2 at $E=2$ and 3 keV (770~780

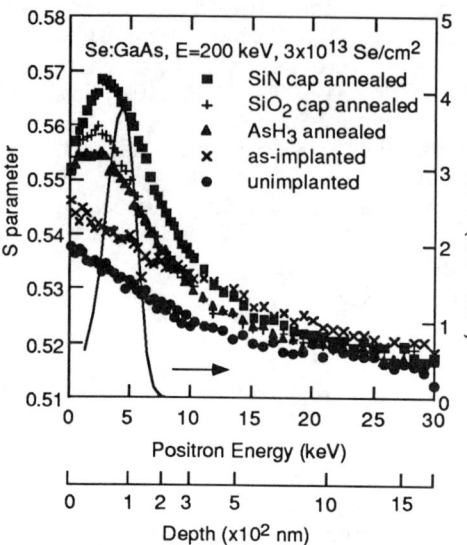

Fig. 2. The S-E relationships for the 200-keV Se^+-implanted GaAs specimens with a dose of 3×10^{13} Se/cm^2.

ps) show the trapping of positrons by vacancy-type defects. Such long lifetime, however, was not normally observed for vacancy-type defects in GaAs specimens. For example, the lifetime of positrons trapped by vacancy clusters introduced by a plastic deformation was reported as ~500 ps.[10,11] Thus the observed value of τ_2 shows the size of defects are larger than such clusters. The defects in the Se^+-implanted specimen are considered to be introduced by the change of the Fermi level position due to the activation of Se atoms and/or by the mismatch stress introduced by Se atoms located at the As site.

Table III. Sheet resistivities of the Se^+-implanted GaAs specimens after various annealing treatments.

Annealing method	Sheet resistivity (Ω)
SiN-cap	1004±13
SiO$_2$-cap	1227±37
capless	insulating

Table IV. The results of the lifetime experiments for the Se^+-implanted specimen after the SiN-cap annealing.

E (keV)	τ_1 (ps)	τ_2 (ps)	I_2 (%)
2	317±8	770±60	12±2
3	335±7	760±40	15±2
20	267±5	-	-

3.3 DEFECTS IN O^+-IMPLANTED GaAs

Figure 3 shows the S-E relations for the O^+-implanted GaAs specimens. It can be seen that the value of S was decreased after the SiO$_2$-cap annealing. Dannefaer et al.[12] reported measurements of positron lifetime spectra

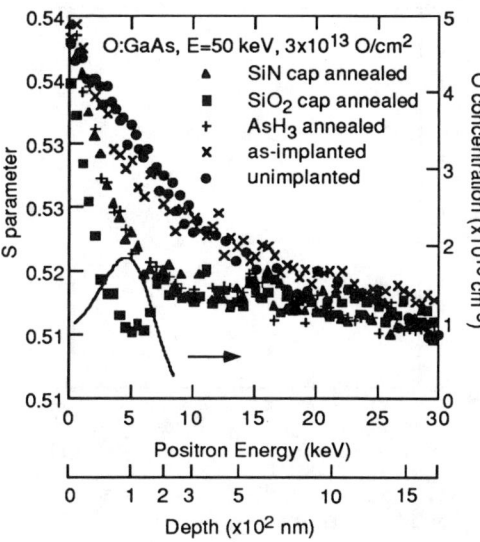

Fig. 3, The S-E relationships for in the 50-keV O+-implanted GaAs specimens with a dose of 3x10¹³ O/cm².

and Doppler broadening profiles for Czochralski-grown Si. They found that positrons can be trapped by interstitial oxygen clusters yielding the positron lifetime of about 100 ps and a small value of S. Thus the observed decrease in S also can be attributed to the annihilation of positrons trapped by oxygen clusters. When oxides are formed by following reactions,

$$2GaAs+3O \rightarrow Ga_2O_3+2As \quad (3)$$
$$2GaAs+3O \rightarrow As_2O_3+2Ga, \quad (4)$$

interstitial atoms are emitted into the lattice. If excess vacancies exist, the oxygen precipitation is considered to be enhanced by following reactions,

$$2GaAs+3O+vacancy \rightarrow Ga_2O_3 \quad (5)$$
$$2GaAs+3O+vacancy \rightarrow As_2O_3. \quad (6)$$

Therefore, it can be concluded that the introduction of V_{Ga} due to the out-diffusion of Ga atoms into the SiO_2 film enhances the oxygen precipitation.

3.4 NATIVE DEFECTS IN GaAs

Figure 4 shows the S parameter for HB-GaAs with different carrier concentrations and undoped LEC-GaAs obtained by the conventional positron annihilation technique. For n-type GaAs, the value of S was found to be larger than that for undoped GaAs. This fact can be attributed to the annihilation of positrons trapped by vacancy-type defects. The lifetime spectra for n-type GaAs and undoped HB-GaAs were decomposed into two-components. However, only a single lifetime was resolved for p-type GaAs and undoped LEC-GaAs. Figure 5 shows the lifetimes for HB-GaAs and LEC-GaAs. For n-type GaAs, the obtained value of τ_2 for HB-N-A (297 ps) is in good agreement with the lifetime of divacancies.[13] For HB-N-B~HB-N-D and HB, the values of τ_2 were obtained as ~276 ps.

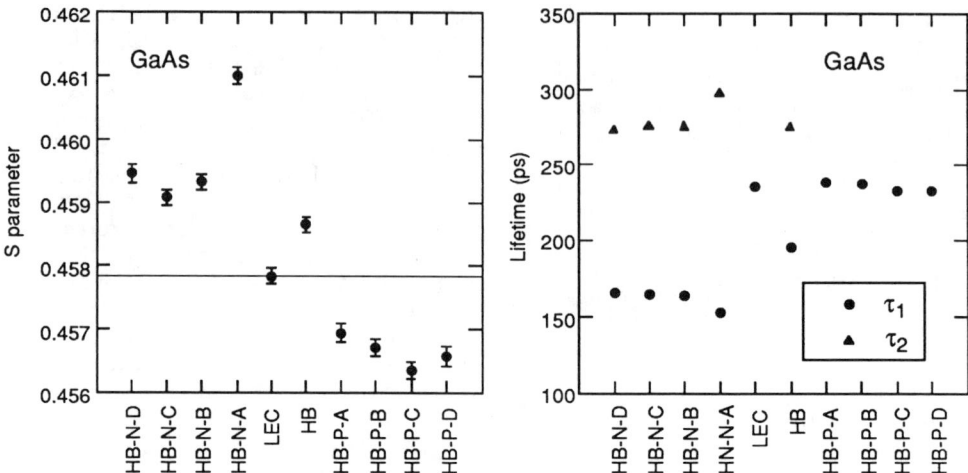

Fig. 4, The S parameter for HB-GaAs with different carrier concentrations and undoped LEC-GaAs.

Fig. 5. The lifetimes of positrons for HB-GaAs with different carrier concentrations and undoped LEC-GaAs.

These values are close to the lifetime of monovacancies.[13] Thus the major type of defects can be identified as monovacancies. Since it is generally accepted that V_{As} is positively charged and V_{Ga} is negatively charged,[14] only V_{Ga} is detectable by the positron annihilation technique. Thus Dannefaer et al.[13] concluded that V_{Ga} is main defects in n-type GaAs. The present results are in good agreement with their conclusion.

For the undoped HB-GaAs, the obtained value of L (57±13 nm) was found to be smaller than that for the undoped LEC-GaAs (290±5 nm). The obtained short diffusion length of positrons means that the concentration of defects for HB-GaAs is higher than that for LEC-GaAs. The typical diffusion length of positrons for Si is ~250 nm. Since the observed diffusion length for LEC-GaAs (290 nm) is close to that for Si, almost all positrons in LEC-GaAs are considered to annihilate from the free state. Thus, the value of S and the lifetime of positrons for the undoped LEC-GaAs can be identified as the characteristic values of the positron annihilation in the bulk GaAs.

In Figs. 4 and 5, the values of S and the lifetimes for the p-type GaAs specimens are lower than those for LEC-GaAs. It is generally accepted that positively charged point-defects do not act as the trapping centers of positrons. Candidate of negatively charged point-defects is Ga_{As}, because As_{Ga} as well as both As interstitials (I_{As})

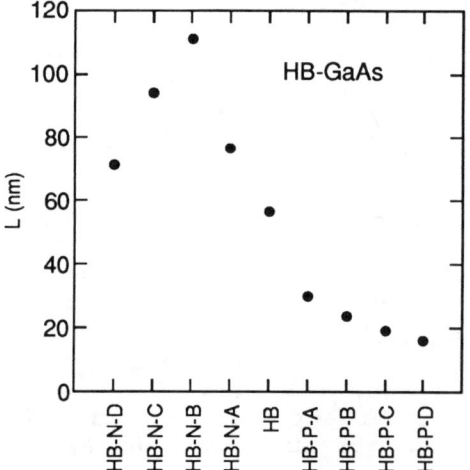

Fig. 6. The diffusion length of positrons for the HB-GaAs specimens with different carrier concentrations.

and Ga interstitials (I_{Ga}) are usually believed to exist in positive or neutral charge states.[14] Saarinen et al.[15] reported the trapping of positrons by negative point defects (probably Ga_{As}) below 125 K. In the present experiments, however, such localization is not expected because all measurements were performed at room temperature. Therefore the observed decrease in S for p-type HB-GaAs can be attributed to the annihilation of positrons trapped by interstitial clusters.

Figure 6 shows the diffusion length of positrons for the HB-GaAs specimens. It can be seen that the value of L for the n-type specimens is larger than that for the p-type ones. This can be attributed to a bending of the band in the subsurface region and a resultant enhanced diffusion of positrons. For the n-type specimens with a carrier concentration above 3.0×10^{17} /cm^3, the value of L decreased with increasing the carrier concentration. This fact is due to the trapping of positrons by V_{Ga}. For the p-type specimen, the value of L decreased with increasing the carrier concentration. This can be attributed to not only the band bending but also the trapping of positrons by interstitial clusters.

4. CONCLUSIONS

We have presented a study of defects introduced by ion implantation in LEC-GaAs. For the Si$^+$-implanted GaAs after the SiO$_2$-cap annealing, the dominant species of defects was identified as antisite defects. For the Se$^+$-implanted specimens, however, vacancy clusters were introduced by the activation of Se atoms. For the O$^+$-implanted specimens, oxygen clusters were introduced by the SiO$_2$-cap annealing. Native defects in GaAs were also studied. For n-type HB-GaAs, dominant species of defects was identified as V_{Ga}, however, divacancies were found in the specimen with low carrier concentration. For p-type HB-GaAs, main species of defects was identified as interstitial clusters.

ACKNOWLEDGMENTS

The authors would like to thank Dr. R. Suzuki (Electrotechnical laboratory) for the measurements of lifetime spectra of the Se^+-implanted specimen.

REFERENCES

1. T. Ishikawa, T. Maeda and K. Kondo, J. Appl. Phys., 68, 3343 (1990).
2. A. Lindow, J.F. Gibbons, V.R. Deline and C.A. Evance, Jr., J. Appl. Phys., 51, 4130 (1980).
3. P.N. Favennec, J. Appl. Phys., 47, 2532 (1976).
4. A. Uedono, L. Wei, C. Dosho, H. Kondo, S. Tanigawa and M. Tamura, Jpn. J. Appl. Phys., 30, 1597 (1991).
5. R. Suzuki, Y. Kobayashi, T. Mikado, H. Ohgaki, M. Chiwaki, T. Yamazaki and T. Tomimasu, Jpn. J. Appl. Phys., 30, L532 (1991).
6. J. Lindhard, M. Scharff and N.E. Schiott, K. Dan. Vidensk. Selsk. Mat.-Fys. Medd. 33, No 14 (1963).
7. S. Dannefaer, G.W. Dean, D.P. Kerr and B.G. Hogg, Phys. Rev. B, 14, 2709 (1976).
8. S. Singh, F. Baiocchi, A.D. Butherus, W.H. Grodiewcz, B. Schwartz, L.G. Van Vitert, L. Yesis and G.J. Zydzik, J. Appl. Phys., 64, 4194 (1988).
9. M. Kuzuhara, T. Nozaki and T. Kamejima, J. Appl. Phys., 66, 5833 (1989).
10. K. Saarinen, C. Corbel, P. Hautojärvi, P. Lanki, F. Pierre and D. Vignaud, J. Phys. Condensed Matter, 2, 2453 (1990).
11. P. Mascher, S. Dannefaer and D. Kerr, Can. J. Phys., 69, 298 (1991).
12. S. Dannefaer and D. Kerr, J. Appl. Phys., 60, 1313 (1986).
13. S. Dannefaer, P. Mascher and D. Kerr, J. Phys. Condensed Matter, 1, 3213 (1989).
14. G. A. Baraff and M. Schlüter, Phys. Rev. Lett., 55 1327 (1985).
15. K. Saarinen, P. Hautojärvi, A. Vehanen, R. Krause and G. Dlubek, Phys. Rev. B, 39, 5287 (1989).

CHARACTERIZATION OF SiO_2 BY MONOENERGETIC POSITRON BEAMS

A. Uedono, S. Watauchi and Y. Ujihira
Research Center for Advanced Science and Technology,
University of Tokyo, 4-6-1 Komaba, Meguro-ku,
Tokyo 153, Japan

L. Wei and S. Tanigawa
Institute of Materials Science, University of Tsukuba,
Tsukuba, Ibaraki 305, Japan

R. Suzuki, H. Ohgaki and T. Mikado
Electrotechnical Laboratory, 1-1-4, Umezono, Tsukuba,
Ibaraki 305, Japan

H. Kametani
Advanced Process Research Laboratory, Mitubishi Electric
Corporation, 1-1, Tsukaguchi-Honmachi, Amagasaki, Hyogo
661, Japan

H. Akiyama, Y. Yamaguchi and M. Koumaru
LSI Research and Development Laboratory, Mitsubishi Electric
Corporation, 4-1, Mizuhara, Itami, Hyogo 664, Japan

ABSTRACT

Vacancy-type defects in silicon-on-insulator (SOI) structures fabricated by the separation by implanted oxygen (SIMOX) process were studied by monoenergetic positron beams. From measurements of Doppler broadening profiles of the positron annihilation as a function of incident positron energy, it was found that a dominant defect species in as-implanted specimens was identified as oxygen-vacancy complexes. The formation of oxide films in the SOI structure were also studied by a pulsed monoenergetic positron beam.

INTRODUCTION

The silicon-on-insulator (SOI) technology has attracted an interest because of its high potential for achieving the device structure of ULSIs.[1] Separation by implanted oxygen (SIMOX) is the most promising technique in order to fabricate desired SOI structures. The conventional SIMOX process needs an implantation of high-energy (~200 keV) oxygen-atoms with a high-dosage (~10^{18} O/cm^2) and a subsequent high temperature annealing of over 1300 °C. A top Si layer formed by the SIMOX process is a homogeneous single crystal. Threading dislocations in the top Si layer, however, are very stable and difficult to remove even by the high-temperature annealing.[2] In the present paper, we applied monoenergetic positron beams in order to probe defects in Si introduced by 200-keV O^+-ion implantation up to doses of $1\sim2\times10^{18}$ O/cm^2.

EXPERIMENTAL

The specimens used in the present experiments were Czochralski-grown (Cz) Si wafers (p-type, 10 Ωcm) with a (100) orientation. The implantation of 200-keV O^+-ions was performed at 550°C up to a dose of 1×10^{18} O/cm^2. After ion implantation, the specimen was subsequently annealed at 1350 °C. The thickness of an oxide layer grown on the Si substrate by this thermal treatment was 90 nm. Then the implantation was performed again after the removal of the oxide layer. Thus a total implantation dosage was 2×10^{18} O/cm^2. Finally the specimen was annealed again. The thickness of the oxide layer grown on the Si substrate by the final annealing treatment was 110 nm. The density of dislocations in the top Si layer was below 1×10^4 $/cm^2$. The obtained thicknesses of the top Si layer and that of the oxide film were 150 nm and 870 nm, respectively.

The characterization of oxide films was performed by using a pulsed monoenergetic positron beam line constructed at Electrotechnical Laboratory (Tsukuba). The detail of the system was described elsewhere.[3] An intense monoenergetic positron beam (10^7 e^+/s) was produced by an electron beam with a energy of 75-MeV and with a pulse width of 1 μs (50 pulse/s). The obtained positrons were stored in a linear-storage section and a stretched positron beam was used for the pulsing system. The width of bunched positrons was about 150 ps and the counting rate of about 200 counts/s was achieved at the average electron current of ~2 mA. The lifetime spectra were obtained by measuring the time interval between the timing signal derived electrically from the pulsing system and the 511-keV annihilation γ-ray detected by a BaF_2 scintillation detector. The incident positron energy was varied from 0.5 keV to 24 keV. For each spectrum, about 2×10^5 counts were accumulated. The lifetime spectra were analyzed by RESOLUTION.[4] The analysis included a "pseudo-source" component with a fixed intensity (13 %) which resulted from both reflected positrons from the surface of the specimen and an imperfect operation of the pulsing system. The lifetime spectra were resolved into two-components with a time resolution of 280 ps. Frieze et al.[5] developed the positron trapping model combined with the spatial motion of implanted positrons prior to the annihilation. Because of a low statistics of lifetime spectra obtained in the present experiments, however, this model was not included in the analysis.

Doppler broadening profiles of the positron annihilation were also measured by using a monoenergetic positron beam line installed at the University of Tsukuba.[6] The spectrum with a total count of 5×10^6 was measured for each incident positron energy. The annihilation spectrum was characterized by the S parameter where the central region of the spectrum was defined from 510.5 keV to 511.5 keV.

RESULTS AND DISCUSSION

Figure 1 shows the S parameter as a function of incident positron energy for the 200-keV O^+-ion implanted Si(100) specimens with doses of 1×10^{18} and 2×10^{18} O/cm^2. Before the measurements, the specimens were dipped into 10 % HF solution for 5 min in order to remove native oxides formed on the surface of the specimens. From Fig. 1, it can be seen that the values of S at 5 keV < E < 10 keV drastically decrease and these values are lower than the value of S for the positron annihilation from the free state in Si, S_{bulk} (0.535). After the high-temperature annealing treatment, it was found that the values of S in this region increased again. An additional ion implantation decreased the values of S, however, these were still higher than those for the as-implanted specimen with a dose of 1×10^{18} O/cm^2. The high value of S was obtained again after the final annealing treatment. Similar annealing behavior of the S parameter for the oxygen implanted Si specimens was already reported by Nielsen et al..[7] We have reported studies of defects in B^+-, P^+-, As^+-ion implanted Si specimens.[8-11] For these specimens, the values of S corresponding to the positron annihilation in the defected region were larger than S_{bulk}. It is generally accepted that the characteristic value of the S parameter for "pure" vacancy-type defects is larger than S_{bulk}. From a study of electron irradiated Cz-Si specimens, however, the characteristic value of S for a complex of a monovacancy

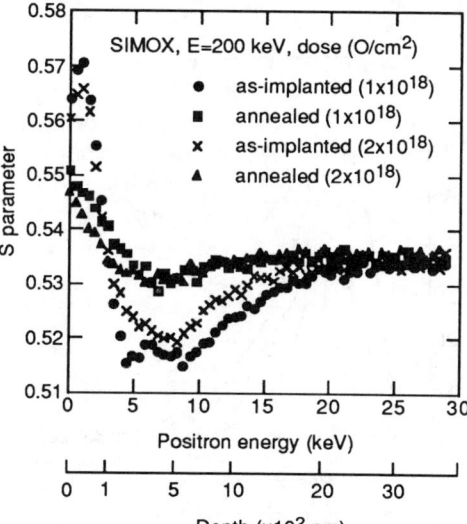

Fig. 1. The S parameter as a function of incident positron energy E for the 200-keV O^+-ion implanted specimens as implanted and after subsequent annealing treatment.

and two oxygen atoms (VO_2) was found to be smaller than S_{bulk}, while the characteristic value of S for VO is larger than S_{bulk}.[12] The positrons trapped by the such complexes have a probability of the annihilation with electrons of oxygen atoms. Because of a high momentum distribution of electrons of oxygen atoms, the characteristic value of S for oxygen-vacancy complexes is considered to be decreased with increasing a number of oxygen atoms coupled with vacancy-type defects. Therefore, in the present experiments, implanted oxygen atoms are considered to be trapped by vacancy-type defects and form stable oxygen-vacancy complexes.

Figures 2 and 3 show the value of the longest lifetime τ_2 and its relative intensity I_2 as a function of incident positron energy for the O^+-ion implanted Si specimens before and after annealing treatment, respectively. The very long lifetime of τ_2 (~1.5 ns) directly shows the pick-off annihilation of ortho-positronium (o-Ps). From the conventional lifetime experiments for amorphous SiO_2 specimens, o-Ps was found to annihilate with the intensity of 50~70 % (τ=0.9~1.8 ns).[13] Therefore, the observed increase in the value of I_2 (Fig. 3) can be attributed to the formation of o-Ps in the SiO_2 film formed by the high-temperature annealing treatment. Because of the removal

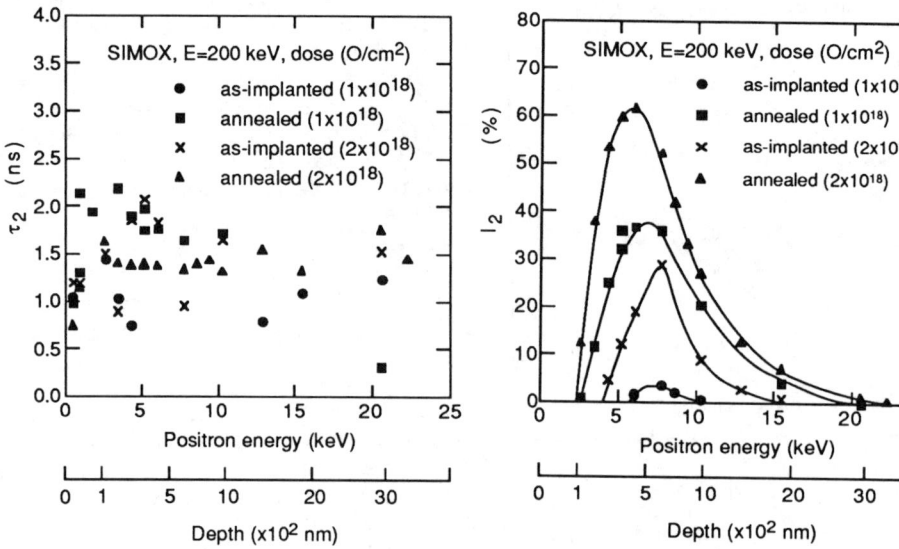

Fig. 2. The τ_2-E relations for the 200-keV O^+-ion implanted specimens as implanted and after annealing treatment.

Fig. 3. The I_2-E relations for the 200-keV O^+-ion implanted specimens as implanted and after annealing treatment.

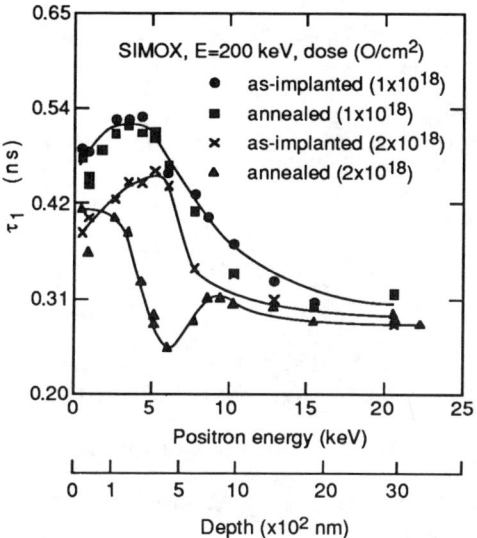

Fig. 4. The τ_1-E relations for the 200-keV O^+-ion implanted specimens.

of the oxide film grown on the Si substrate by chemical etching, the depth of the Si/SiO_2 interface seems to shift towards the surface. Thus the shift of the maximum value of I_2 towards the surface observed in the I_2-E plots corresponds to the removal of the oxide films. Annihilation from the para-Ps (p-Ps) state produces two γ-rays with very sharp energy width. Thus, the increase in the value of S observed after annealing treatment (Fig. 1) can be attributed to the formation of Ps in the oxide films.

Figure 4 shows the shortest lifetime τ_1 as a function of incident positron energy. The values of τ_1 are attributed to not only the annihilation of p-Ps but also the annihilation of positrons from trapped states or delocalized states. For the annealed specimen with a dose of 2×10^{18} O/cm^2, the sudden decrease in τ_1 was found at $E \cong 6$ keV. This correlates with the increase in the formation probability of o-Ps shown in Fig. 3. For the as-implanted specimens, the value of τ_1 was approximately 500 ps at 1 keV $< E <$ 6 keV. This lifetime is close to the characteristic lifetimes for $V_5 \sim V_6$ in Si.[14] Since the lifetime of positrons trapped by VO_2 was found to be similar to that of V (270 ps),[12] the dominant defects in the SOI region can be identified as vacancy-clusters such as V_5 or V_6. The drastic increase in the value of S at $E \cong 2$ keV (Fig. 1) can be attributed to the positron annihilation in such defects.

CONCLUSION

We have presented the study of defects and the characterization of the oxide films in the SOI structure fabricated by the SIMOX technique. The monoenergetic positron beam technique proved to be sufficiently sensitive for the detection of defects in SIMOX wafers. The relative intensity of o-Ps was found to be sensitive for the formation of the SiO_2 film by the high temperature annealing. This suggests that the microstructure of the

SiO_2 film can be studied by measurements of lifetime spectra. For the as-implanted specimens, the dominant defects in the SOI region was identified as the vacancy-clusters such as V_5 or V_6. The present investigation shows the potential of monoenergetic positron beams as a nondestructive probe for the study of the microstructure of SiO_2 and point defects in SIMOX wafers.

REFERENCES

1. A.H. Van Ommen, Nucl. Instrum. & Methods B, <u>39</u>, 194 (1989).
2. J.P. Colinge, Electron. Lett., <u>22</u>, 187 (1986).
3. R. Suzuki, Y. Kobayashi, T. Mikado, H. Ohgaki, M. Chiwaki, T. Yamazaki and T. Tomimasu, Jpn. J. Appl. Phys., <u>30</u>, L532 (1991).
4. P. Kirkegaard and M. Eldrup, Comput. Phys. Commun., <u>7</u>, 410 (1974).
5. W.E. Frieze, K.G. Lynn and D.O. Welch, Phys. Rev. B, <u>31</u>, 15 (1985).
6. A. Uedono, L. Wei, C. Dosho, H. Kondo, S. Tanigawa and M. Tamura, Jpn. J. Appl. Phys., <u>30</u>, 1597 (1991).
7. B. Nielsen, K.G. Lynn, T.C. Leung, B.F. Cordts and S. Seraphin, Phys. Rev B., <u>44</u>, 1812 (1991).
8. A. Uedono, S. Tanigawa, J. Sugiura and M. Ogasawara, Jpn. J. Appl. Phys., <u>28</u>, 1293 (1989).
9. A. Uedono, S. Tanigawa, J. Sugiura and M. Ogasawara, Jpn. J. Appl. Phys., <u>29</u>, 1867 (1990).
10. A. Uedono, L. Wei, C. Dosho, H. Kondo, S. Tanigawa, J. Sugiura and M. Ogasawara, Jpn. J. Appl. Phys., <u>30</u>, 201 (1991).
11. A. Uedono, L. Wei, S. Tanigawa, J. Sugiura, M. Ogasawara and M. Tamura, Radiation Effects and Defects in Solids, <u>124</u>, 31 (1992).
12. A. Uedono, Y. Ujihira, A. Ikari and H. Haga, accepted for the publication in Hyperfine Interactions.
13. A. Uedono, S. Watauchi and Y. Ujihira, submitted to Hyperfine Interactions.
14. S. Dannefaer, Defect Control in Semiconductors, ed. K. Sumino (Elsevier Science B. V., North-Holland, 1990) p.1561.

DEFECT IN AMORPHOUS SILICON PREPARED BY ION IMPLANTATION

L. Wei and S. Tanigawa,
Institute of Materials Science, University of Tsukuba, Tsukuba,
Ibaraki 305, Japan

Y. Hiroyama, T. Motooka and T. Tokuyama
Institute of Applied Physics, University of Tsukuba, Tsukuba, Ibaraki 305, Japan

ABSTRACT

Amorphization and structural relaxation in Si-implanted Si has been studied using slow positrons and Raman spectroscopy. The S parameter in the defect region decreases slightly with the annealing temperature. However, no significant change of S parameter was observed with 200°C and 450°C isothermal annealing. One interesting point is that S parameter holds lower value in the amorphized region, but such a behavior cannot observed by the W parameter. The change of TO peak half width in the Raman spectra (480 cm^{-1}) was found to correlate with the change of S parameter. The defect configuration is expected to alter during isothermal annealing.

Ion implantation is a increasingly important technology for fabrication of doped layers in semiconductor micro-electronic devices as devices sizes are decreased. The temperature dependence of ion-beam-induced amorphization has been generally considered to be due to competition between defect accumulation in an energetic collision cascade and out-diffusion of the defects from the cascade,[1] however, the amorphization mechanism is still controversial and not yet fully understood.[2] Furthermore, the complete understanding of recovery of defects and the recrystallizations of amorphized Si by subsequent heat treatments is urgently required.

In this presentation, we have investigated the defect in amorphized Si prepared by Si self-implantation, using a slow positron beam and Raman spectroscopy.

The substrates used in this study were optically flat 6~8 Ωcm p-type Si (100) wafers. Ion implantations were performed at room temperature with the substrates 5°C off-normal to the incident ion beam for suppression of channeling effects. ^{30}Si ions were implanted at energies of 100 keV and 150 keV in doses from 10^{14} to 5×10^{15} ions/cm^2. Isothermal annealing were performed in the furnace at 200°C and 450°C under nitrogen gas flow for 150-keV Si implanted specimens with a dose of 5×10^{14} cm^{-2}. The annealing time was set from 30 min to 180 min by a increasing step of 30 min for 200°C isothermal annealing, and from 10 min to 60 min by a step of 10 min for 450°C isothermal annealing. The thickness of amorphized layers were 180 nm for 100-keV Si-implanted specimen and 200 nm for 150-keV Si-implanted specimens, as determined by Rutherford backscattering spectrometry (RBS) and by transmission electron microscopy (TEM), respectively.

The Doppler broadened spectra of annihilation radiations were measured using a slow positron beam line constructed at University of Tsukuba. A high-purity Ge detector with an energy resolution of 1.1 keV at 512 keV line of ^{106}Ru was used. For each Doppler broadened spectrum, about 6×10^5 counts were accumulated. The fixed central region of the spectra was chosen from 510.5 keV to 511.5 keV and the total region from 508 keV to 514 keV. The incident positron energy was adjusted from 100 eV to 30 keV. All the measurements were performed at room temperature.

Structural relaxation of a-Si layer was observed by using Raman scattering spectroscopy. 488 nm Ar ion laser was used as a light source with the diameter of illumination ≈ 0.1 mm on the specimen surface and a total power of 120 mW. Scattering light was focused and passed through a SPEX single monochromator (controlled with SPEX DM3000 photon counting system) using a 1800 lines/mm grating and detected by a photomultiplier.[3] Measurements were repeated three times for each specimens with small displacements of the laser irradiation point. The TO peak in the Raman spectra (480 cm^{-1}) was observed in the following experiments with a accuracy of ≈ 2 cm^{-1}.

Figure 1 shows the S parameter as a function of incident positron energy, i.e., the S(E) response, for Si-implanted specimens with energy of 100 keV and doses

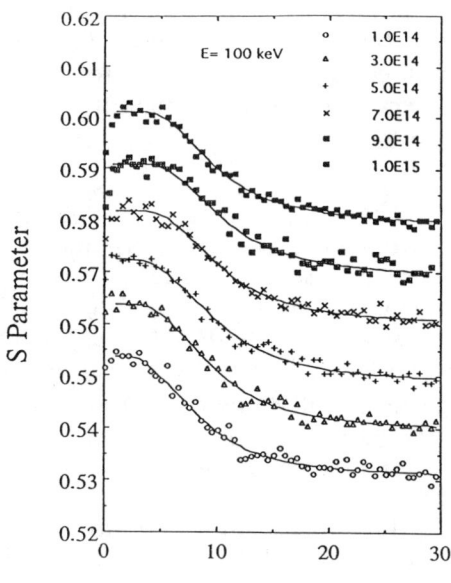

Fig. 1. S(E) response for Si-implanted specimens with energy of 100 keV and doses of $1.0 \times 10^{14} \sim 1.0 \times 10^{15}$ cm^{-2}. The solid lines indicate the fitting of positron diffusion model to the experimental data. The data are displaced upwards by 0.01 for clarity.

of $1.0 \times 10^{14} \sim 1.0 \times 10^{15}$ cm^{-2}. The data are displaced upwards by 0.01 for clarity. The solid lines indicated the fitting of positron diffusion model to the experimental data. However, no significant change was observed in S(E) response within the implantation doses though the RBS results indicate the increasing of damage with the implantation dose.[4]

The data have been modeled using a variation of an analysis successfully used in the previous investigation of Si.[5-7] The implantation profile P(x, E) of positrons may be described by a Makhovian profile[8]

$$P(x, E) = -\frac{d}{dx}[\exp(-\frac{x}{x_0})^m], \quad (1)$$

$$x_0 = \bar{x}/\Gamma(1+1/m), \quad (2)$$

$$\bar{x} = (40/\rho) E^{1.6}, \quad (3)$$

where m is selected as 1.9 and $\bar{x}$ is the mean penetration depth of positrons. The Doppler response S(E) is found by summing the integral of P(x, E) and weighting by the characteristic S value for different depth of specimen. Thus, the energy-dependent Doppler response S(E) can be defined as

$$S(E) = S_s F_s(E) + S_d F_d(E) + S_b F_b(E), \quad (4)$$

where S_s, S_d and S_b indicate the characteristic values of S parameter for the annihilation at surface, in the damaged region and in the defect-free substrate, respectively.

The values of D (damaged depth) were estimated to be around 500 nm, larger than the thickness of amorphized region obtained by RBS measurement. Since the values of S_d/S_b, which are about 1.043, are larger than those for divacancies, the main defect species can be identified as vacancy clusters.

The 150 keV Si-implanted specimen with dose of 5×10^{15} cm^{-2} were isothermally annealed at 200°C and 450°C, respectively. The S(E) response for isothermally annealed specimens are shown in Fig. 2. The S(E) for as-implanted specimen is also included. The S parameter in the damaged region decreased with the annealing time slightly, implying the recovery of defects in the damaged region.

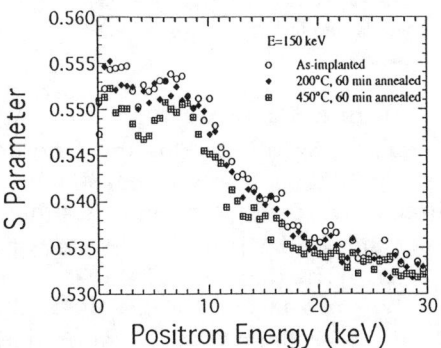

Fig. 2. S(E) response for isothermally annealed specimens and for as-implanted one. Only two data for annealed specimens are included for clarity of the graph.

However, no further significant variation of S parameter was observed when the annealing time increased. The interest point is that the S parameter holds the lower value around 200 nm from the surface. The remeasured S(E) responses corresponding to Fig. 2 are shown in Fig. 3 in detail. Data are also upwards shifted for clarity in Fig. 3. The decrease of S parameter around 200 nm from surface within statistical deviation are also found in the other post-annealed specimen. However, such a manner of S parameter could not recognized by the W parameter as shown in Fig. 4. The reason why S and W parameter show different manners remains elusive.

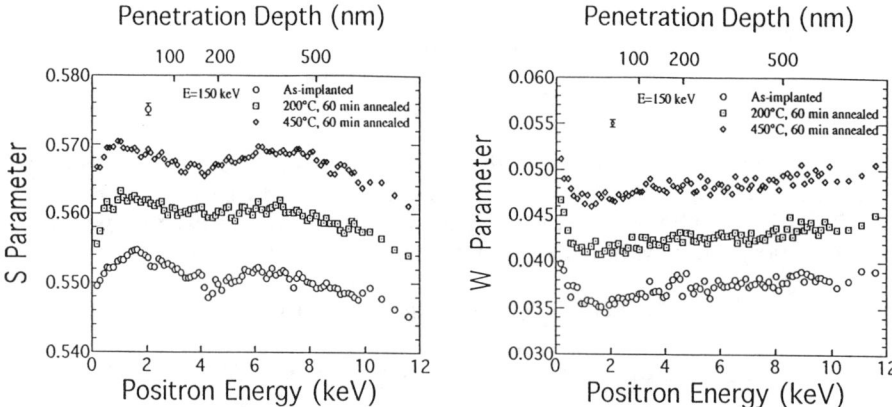

Fig. 3. S(E) response for isothermally annealed specimens and for as-implanted one measured in detail. Data are displaced upwards for clarity.

Fig. 4. W parameter for isothermally annealed specimens and for as-implanted one measured in detail. Data are displaced upwards for clarity.

Figure 5 shows a typical example of Raman spectra of a-Si (formed by 150 keV, 5×10^{15} cm^{-2} Si$^+$ implantation) after furnace annealing for 360 min at 450°C. The TO peak half width decrease and the peak position shifted towards higher wavenumber after thermal treatment. At the same time the LA peak located at wavenumber ≈300 cm^{-1} became slightly stronger. The decrease of the TO peak half width can be attributed to the reduction of the bond angle deviation which describes the structural relaxation. Beeman et al.[9] proposed the following linear relationship between the TO peak full width of the half maximum Γ (cm^{-1}) and bond angle deviation $\Delta\theta$ (degree) in a-Si,

$$\Gamma/2 = 7.5 + 3\Delta\theta \ . \tag{5}$$

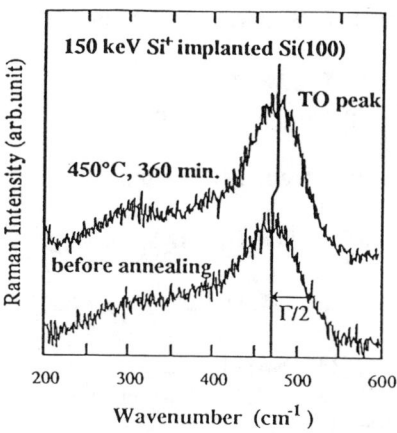
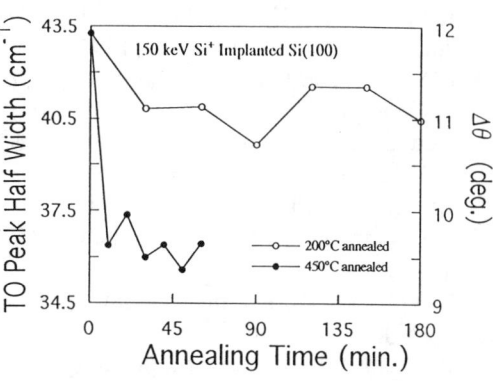

Fig. 5. Typical Raman spectra from Si (100) specimens implanted with 5×10^{15} cm^{-2} Si$^+$ before and after annealing.

Fig. 6. Changes in $\Gamma/2$ and $\Delta\theta$ during isothermal annealing at 200°C and 450° with different annealing times. The TO peak half width is calculated by $\Gamma/2 = 7.5 + 3\Delta\theta$.

Fig. 6 show the a-Si TO peak half width $\Gamma/2$ of Si-implanted Si as a function of time of isothermal annealing at 200°C and 450°C. The bond angle deviation, $\Delta\theta$ generally decreased with annealing but detailed observation of $\Delta\theta$ shows a small increase after the initial decrease. The behavior is similar to the observation in the relaxation process of a-Si:H.[10] From a comparison between the results of Raman spectra and the results of positron annihilation, one can find that the decrease of $\Delta\theta$ corresponds to the decrease of S parameter. The defect configuration is expected to alter during isothermal annealing. It is worthwhile to investigate the recovery and recrystallization of amorphized Si in aspect of structural relaxation. The detailed studies, including cross-sectional TEM measurement are in progress now.

This work was supported in part by a Grant-in-Aid for Scientific Research from the Ministry of Education, Science and Culture, and by a NEDO grant for international joint research.

REFERENCES

1. F. F. Morehead, Jr. and B. L. Crowder, Proc. 1st Int. Conf. on Ion Implantation, ed. L. T. Chadderton and F. H. Eisen (Gordon and Breach, New York, 1971) p. 25.
2. R. G. Elliman, J. Linnros and W. L. Brown, Mat. Res. Soc. Proc. 100, 363 (1988).

3. T. Motooka, F. Kobayashi, P. Fons, T. Tokuyama, T. Suzuki and N. Natsuki, Jpn. J. Appl. Phys. 30, 3617 (1991).
4. T. Motooka and O. W. Holland, Appl. Phys. Lett. 58, 27 (1991).
5. A. Uedono, L. Wei, C. Dosho, H. Kondo, S. Tanigawa, J. Sugiura and M. Ogasawara, Jpn. J. Appl. Phys. 30, 201 (1991).
6. A. Uedono, L. Wei, C. Dosho, H. Kondo, S. Tanigawa and M. Tamura, Jpn. J. Appl. Phys. 30, 1597 (1991).
7. P. J. Simpson, M. Vos, I. V. Mitchell, C. Wu and P. J. Schultz, Phys. Rev. B44, 12180 (1991).
8. A. Vehanen, K. Saarinen, P. Hautojärvi and H. Huomo, Phys. Rev. B35, 4606 (1987).
9. D. Beeman, R. Tsu and M. F. Thorpe, Phys. Rev. B32, 874 (1985).
10. R. Tsu, Disordered Semiconductors, ed. M. A. Kastner, G. A. Thomas and S. R. Obvsbinsky (Plenum, New York, 1987) p. 479.

INVESTIGATION OF VACANCY-RELATED DEFECTS IN HEAVILY PHOSPHORUS-DOPED Si:P GROWN BY PLASMA CHEMICAL VAPOR DEPOSITION

L. Wei and S. Tanigawa
Institute of Materials Science, University of Tsukuba, Tsukuba, Ibaraki 305, Japan

Y. Jia, A. Yamada and M. Konagai
Faculty of Engineering, Tokyo Institute of Technology, Tokyo 152, Japan

ABSTRACT

Investigation of vacancy-related defects in heavily phosphorus-doped Si:P grown by plasma chemical vapor deposition (CVD) has been carried out using variable-energy positrons. Modeling and fitting of S(E) responses indicate that a high concentration of vacancy-related defect ($>10^{-3}$ /atom) exists in both as-deposited and annealed specimens. The 600°C annealed one has a lowest carrier concentration, highest sheet resistivity and highest concentration of vacancies. The vacancy-related defects were assigned for the electrical deactivation. At higher annealing temperature (900°C), the precipitation of phosphorous may also play an important role in the electrical deactivation. The thermal equilibrium calculation obtained the defect concentration which coincides with results of positron annihilation in the same order.

INTRODUCTION

Development of low-temperature Si epitaxial technology and the fabrication of heavily doped Si films with ultra-low resistivity are of great importance in microelectronics in order to further reduce device geometries as required by the very-large-scale integration (VLSI). Conventional Si epitaxy with chemical vapor deposition (CVD) [1] technique uses processing temperatures exceeding 1000°C and causes impurity redistribution and thermal stress. To minimize autodoping and solid state diffusion effects for an abrupt impurity profile, several low temperature approaches including molecular-beam epitaxy (MBE) [2,3] and CVD techniques [4-6] have been proposed. Among these techniques, plasma-CVD appear to be one of the most suitable process due to its high throughput and low cost. In this paper, we demonstrate the investigation of vacancy-related defects in heavily doped Si films with doping level of 10^{21} cm^{-3} deposited by plasma-CVD at low temperature of 250°C.

EXPERIMENTAL

A plasma-CVD apparatus was used in this study. The reactant gases SiH_4, SiF_4, H_2 and PH_3 are introduced into the reactor with a base pressure of 10^{-7} Torr, and the RF glow discharge (13.56 MHz) is employed to decompose the mixed gases. The substrate used was a (100)-oriented Si wafer which has a resistance over 1000 Ωcm. The substrate temperature was 250°C during the growth and the annealing was performed at 600°C and 900°C in N_2 ambient for 1 hr. Two series of specimens with different epilayer thickness (200 nm and 340 nm) were obtained. The dopant concentration ($\sim 10^{21}$ cm^{-3}) was verified by SIMS measurement. Hall effect measurements were made on van der Pauw specimens at room temperature for a magnetic flux density of 0.84 T following a standard procedure. Table I shows the measured electrical properties.

The Doppler broadened spectra of annihilation radiations were measured using a slow positron beam line constructed at University of Tsukuba. A high-purity Ge detector with an energy resolution of 1.1 keV at 512 keV line of ^{106}Ru was used. For each Doppler broadened spectrum, about 6×10^5 counts were accumulated. The fixed central region of the spectra was chosen from 510.5 keV to 511.5 keV and the total region from 508 keV to 514 keV. The incident positron energy was adjusted from 100 eV to 20 keV. All the measurements were performed at room temperature.

Table I. Electrical properties for Si:P.

Samples	Thickness (nm)	Carrier concentration (cm^{-3})	Mobility (cm^2/V·sec)	Resistivity (Ω·cm)
As-deposited Si/Si	200	1.11×10^{21}	14.76	3.82×10^{-4}
	340	1.10×10^{21}	13.02	5.52×10^{-4}
600°C annealed	200	5.76×10^{19}	60.16	1.08×10^{-3}
	340	2.75×10^{19}	57.43	3.94×10^{-3}
900°C annealed	200	3.50×10^{20}	89.30	2.01×10^{-4}
	340	3.21×10^{20}	57.09	3.40×10^{-4}

RESULTS AND DISCUSSION

Positron annihilation

Figure 1 shows the S parameters as a function of incident positron energy, i.e., the S(E) response, for as-deposited, 600°C annealed and 900°C annealed Si (340 nm)/Si specimens. The solid lines indicate the best fitting of positron trapping model. The remeasured S(E) responses after the removal of native oxide on the surface of Si epitaxial layer are show in Fig. 2. The correspondent plot for Si (200 nm)/Si is not included here because of its similarity with Si (340 nm)/Si. The S(E) response in Fig. 1 for as-deposited Si (340 nm)/Si differs significantly from that of defect-free Si wafer. The high S parameter in the epitaxial layer indicates the vacancy-type defects were introduced during the growth of epitaxial Si layer. From Figs. 1 and 2, one can find that the S parameter for the 600°C annealed specimen increased in comparison with that of as-deposited one, indicating the increase of vacancy-type defects after annealing at 600°C. Furthermore, the S parameter for 600°C annealed specimen holds highest value in the interfacial region between epitaxial layer and the Si substrate. However, by annealing at 900°C, the drastic decreasing of S parameter was observed, indicating the

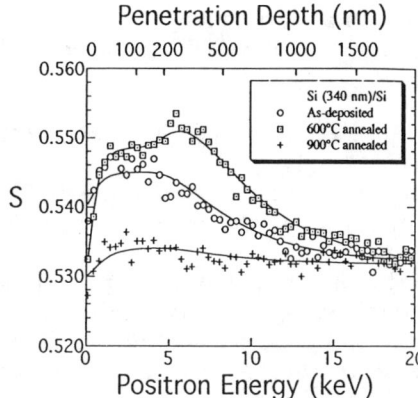
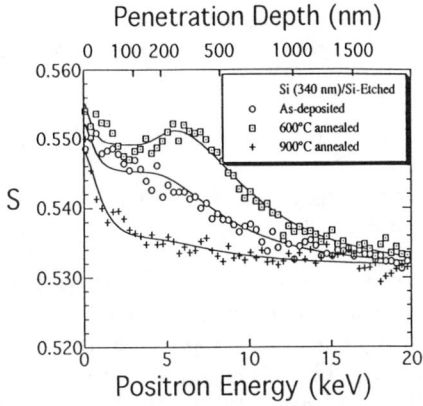

Fig. 1. S parameter versus positron energy for as-deposited, 600°C annealed and 900°C annealed Si (340 nm)/Si. Solid lines denote the fits to the experimental data.

Fig. 2. S parameter versus positron energy for as-deposited, 600°C annealed and 900°C annealed Si (340 nm)/Si after removal of native oxide on the surface of specimens. Solid lines denote the fits to the experimental data.

recovery of vacancy-type defects.

The data have been modeled using a variation of an analysis successfully used in the previous investigation of Si and GaAs.[7-9] The implantation profile P(x, E) of positrons may be described by a Makhovian profile[10]

$$P(x, E) = -\frac{d}{dx}[\exp(-\frac{x}{x_0})^m], \qquad (1)$$

$$x_0 = \bar{x}/\Gamma(1+1/m), \qquad (2)$$

$$\bar{x} = (40/\rho)E^{1.6}, \qquad (3)$$

where m is selected as 1.9 and $\bar{x}$ is the mean penetration depth of positrons. The Doppler response S(E) can be expressed by summing the integral of P(x, E) and weighting by the characteristic S value for different depth of specimen.

The area analyzed in the case of the Si:P was divided into three regions: (i) the surface; (ii) the epitaxial layer with a considerable amount of defects ($0 < x \le D$); and (iii) the defect-free region ($D \le x < \infty$). For the 600°C annealed specimen the interfacial state was also taken into account. In the second region, positrons can diffuse forth towards the Si substrate or diffuse back towards surface of specimen depending upon the implanted position and the mean diffusion length within the defective region. But the positrons implanted into the Si substrate could not diffuse back to the surface, because after penetrating into the defective region, they would be trapped by defects and eventually annihilate with the environmental electrons.

Therefore, the energy-dependent Doppler response S(E) can be defined as

$$S(E) = S_s F_s(E) + S_{epi} F_{epi}(E) + S_i F_i(E) + S_b F_b(E), \quad (4)$$

where S_s, S_{epi}, S_i and S_b indicate the characteristic values of S parameter for the annihilation at surface, in the epitaxial layer, at the interface and in the defect-free substrate, respectively.

For Si, it is convenient to determine the diffusion length of positrons in the subsurface region using $\Delta S(E)$ [11] because S_s is easily modified by etching the surface native oxide regardless of other parameters such as $F_s(E)$, $F_{epi}(E)$, $F_b(E)$, S_{epi} and S_b. The difference in the value of S parameter before and after removal of native oxide is expressed as

$$\Delta S(E) = [S_s|_{\text{after etching}} - S_s|_{\text{before etching}}] F_s(E)$$
$$= \Delta S_s F_s(E), \quad (5)$$

where $S_s|_{\text{before etching}}$ and $S_s|_{\text{after etching}}$ are characteristic values of S_s for the specimens before and after etching of native oxide, respectively. By fitting the experimental data using Eq. 5, one can simply obtain the diffusion

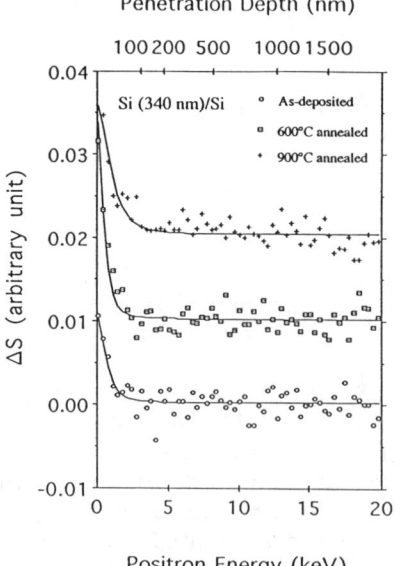

Fig. 3. ΔS(E) responses for as-deposited, 600°C annealed and 900°C annealed Si (340 nm)/Si with fitting curves. The arbitrary unit is used in the vertical axis.

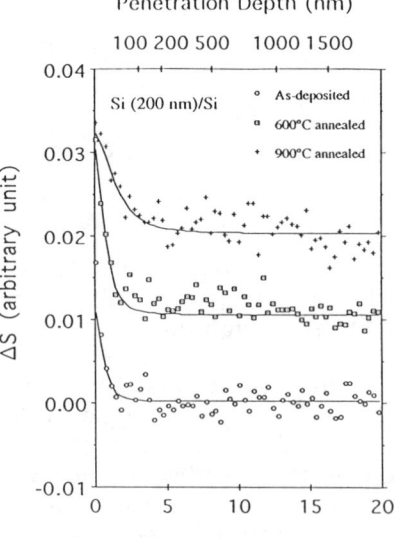

Fig. 4. ΔS(E) responses for as-deposited, 600°C annealed and 900°C annealed Si (200 nm)/Si with fitting curves. The arbitrary unit is used in the vertical axis.

length of positrons in the subsurface region. In Figs. 3 and 4, shown are ΔS(E) responses for Si (340 nm)/Si and Si (200 nm)/Si, respectively. The solid lines are the best fitting of Eq. 5 to the experiment data. The obtained parameters by fitting of Eqs. 4 and 5 are summarized in Table II. From Table II, one can find that the ratio of S parameter for epitaxial layer, S_{epi}, to that for a annihilation in Si substrate are 1.025, 1.034 and nearly 1 for as-deposited, 600°C annealed and 900°C annealed specimens, respectively. The value of 1.034 is close to the characteristic value of positron annihilation in divacancy, implying the production of divacancy-like defects in the epitaxial layer during annealing at 600°C. Furthermore, the S_i is more higher than the saturated value of annihilation in divacancy. Therefore, the agglomeration of vacancies at the interface should be considered. By annealing at 900°C, the S parameter drastically decreased, however, the diffusion length of positrons remains significantly short as listed in Table II. The precipitation, which may be responsible for the shortening of diffusion length of positrons, is expected to occur after annealing at 900°C, since the dopant concentration extended the solid solubility of phosphorous in Si prior to the annealing treatment.[12] We roughly estimated the concentration of vacancies by taking $\lambda_f=4.55\times10^9$ s^{-1} [Ref. 13], $L_{free}=180$ nm and $\mu=3\times10^{-14}$ s^{-1} [Ref. 14], as listed in Table II.

Table II. Summary of obtained parameters by fitting the positron diffusion equation to the experimental data.

	Samples	S_s/S_b	S_{epi}/S_b	S_i/S_b	L (nm)	C_v (/atom)
As-deposited	Si (200 nm)/Si	1.008	1.026	–	8	7.6×10^{-3}
	(Etched)	1.038	1.025	–		
	Si (340 nm)/Si	1.017	1.026	–	10	4.8×10^{-3}
	(Etched)	1.040	1.026	–		
600°C annealed	Si (200 nm)/Si	1.004	1.028	1.071	12	1.7×10^{-3} *
	(Etched)	1.043	1.035	1.079		
	Si (340 nm)/Si	1.001	1.034	1.074	7	5.0×10^{-3} *
	(Etched)	1.047	1.034	1.073		
900°C annealed	Si (200 nm)/Si	0.992	1.006	–	39	3.0×10^{-3}
	(Etched)	1.016	1.008	–		
	Si (340 nm)/Si	0.997	1.005	–	21	1.3×10^{-3}
	(Etched)	1.032	1.007	–		

*) For divacancy.

Vacancy complex model

Pandey et al.[15] showed by *ab initio* total-energy calculations that a new defect complex (V-As$_4$), a vacancy surrounded by four arsenic atoms, is responsible for explaining electrical deactivation when heavily As-doped Si was annealed. The chemical property of phosphorus is similar to arsenic, so we supposed that a defect complex (V-P$_4$), a vacancy surrounded by four phosphorus atoms, is also formed by annealing the heavily P-doped Si films. Furthermore, calculations for the total number of complex have been carried out from the thermodynamical point of view. The calculation can be illustrated as follows:

The multiplicity W of distinguishable distribution of V-P$_4$ complex is given by the formation of

$$W = {}_{N_t+N_c}C_{N_c} \times {}_{N_t-4N_c}C_{N_p-N_c}, \tag{6}$$

where N_t, N_p and N_c are the total number of lattice site, the phosphorus atoms and the total number of V-P_4 complex, respectively. The free-energy F of the system could be get from

$$F = -EN_c - kT\ln W, \quad (7)$$

where E, k and T are the formation energy of the V-P_4, Boltzmann constant and the temperature, respectively. The free energy must be minimized subject to the equilibrium constraint

$$dF/dN_c = 0. \quad (8)$$

The electron concentration at various temperature is determined by

$$N_e = N_p - 4N_c. \quad (9)$$

During the calculation, the independent parameter is only E, which was 1.4 eV for V-As_4 as determined by Pandey et al. Other analogous vacancy-complex, such as V_2-P_6 and V-P_2, may calculated by the similar way.

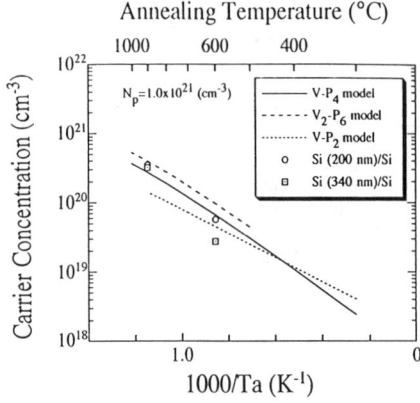
Fig. 5. Carrier concentration as a function of annealing temperature. Open circles (o) and dotted squares (□) are experimental data obtained from the Hall measurement.

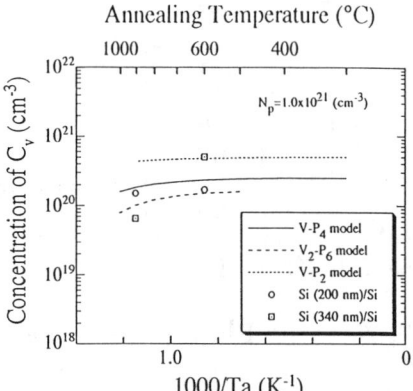
Fig. 6. Concentration of vacancies as a function of annealing temperature. Open circles (o) and dotted squares (□) are data obtained from the positron annihilation.

Figure 5 shows the carrier concentration as a function of annealing temperature. The good agreement between the calculation using V-P_4 complex model and the electrical measurement was obtained. The defect concentration obtained from the calculation coincides with the results of positron annihilation in the same order as shown in Fig. 6. However, the specification of defects by positron annihilation is discrepant with the calculation of the complex model. The configuration of vacancy complex may be more complicated. Nevertheless, it is acceptable, from the present experiment, that the vacancy complex results in the electric deactivation in Si:P after heat treatment. The precipitation may also play an important role in the electric deactivation when annealing at high temperature.

ACKNOWLEDGMENTS

This work was supported in part by a Grant-in-Aid for Scientific Research from the Ministry of Education, Science and Culture, and by a NEDO grant for international joint research.

REFERENCES

1. T. Yamazaki, H. Hiroshi and T. Ito, App. Phys. Lett. 55, 879 (1989).
2. Y. Ohta, J. Appl. Phys. 51, 1102 (1980).
3. J. C. Bean, G. E. Becher, D. J. Petroff and T. E. Scidel, J. Appl. Phys. 48, 907 (1977).
4. T. Uematsu, S. Matsubara, M. Kondo, M. Tamura and T. Saito, Jpn. J. Appl. Phys. 27, 193 (1988).
5. S. Nishida, T. Shiimoto, A. Yamada, M. Konagai and K. Takahashi, Appl. Phys. Lett. 49, 79 (1986).
6. A. Yamada, Y. Jia, M. Konagai and K. Takahashi, Jpn. J. Appl. Phys. 28, L2284 (1989).
7. A. Uedono, L. Wei, C. Dosho, H. Kondo, S. Tanigawa, J. Sugiura and M. Ogasawara, Jpn. J. Appl. Phys. 30, 201 (1991).
8. L. Wei, Y.-K. Cho, C. Dosho, T. Kurihara and S. Tanigawa, Jpn. J. Appl. Phys. 30, 2863 (1991).
9. L. Wei, S. Tanigawa, M. Uematsu and K. Maezawa, Jpn. J. Appl. Phys. 31, 2056 (1992).
10. A. Vehanen, K. Saarinen, P. Hautojärvi and H. Huomo, Phy. Rev. B35, 4606 (1987).
11. S. Shikata, S. Fujii, L. Wei and S. Tanigawa, Jpn. J. Appl. Phys. 31, 732 (1992).
12. G. F. Cerofolini, M. L. Polignano, F. Nava and G. Ottaviani, Thin Solids Films 97, 363 (1982).
13. P. J. Simpson, M. Vos, I. V. Mitchell, C. Wu and P. J. Schultz, Phys. Rev. B44, 12180 (1991).
14. S. Dannafaer, Phys. Status Solidi A 102, 481 (1987).
15. K. C. Pandey, A. Erbil, G. S. Cargill, R. F. Boehme and D. Vanderbilt, Phys. Rev. Lett. 61, 11 (1988).

SECTION 2
DEFECTS—BULK

POSITRON LIFETIME MEASUREMENTS IN γ-IRRADIATED POLYETHYLENE UNDER DIFFERENT CONDITIONS

G. Brauer
Forschungszentrum Rossendorf, O-8051 Dresden, Germany

Th. Daniel
Phys.-Chem. Institut, Universität Giessen,
W-6300 Giessen, Germany

W. Faust, H. Schneider
Strahlenzentrum, Universität Giessen,
W-6300 Giessen, Germany

Z. Michno
Technical University Opole, Opole, Poland

ABSTRACT

Positron lifetime measurements were performed on polyethylene probes of different densities under various atmospheric conditions within a dose range from 0 kGy up to 1460 kGy. Four components could be resolved in each lifetime spectrum. The components consist each of a lifetime τ and an intensity I. The behaviour of the lifetime parameters of the longest-lived and the second long-lived component, respectively, may be attributed to the γ-induced recrystallization and the creation of hydroperoxides. The average free volume as a function of the applied γ-dose was estimated.

INTRODUCTION

The study of molecular solids by positron annihilation techniques is a very broad subject, and polyethylene has been the subject of detailed investigations [1-6]. Of the four resolved positron lifetimes the longest one ($\tau_4 \approx$ 2,5 ns) is associated with ortho-positronium (o-Ps), the shortest one ($\tau_1 \approx$ 0.1-0.2 ns) is - at least partially - associated with para-positronium (p-Ps), and an intermediate one ($\tau_2 \approx$ 0.3 ns) with free positrons not forming positronium (Ps). The origin of the third component ($\tau_3 \approx$ 1 ns) has been under debate, i.e. it should be due either to a positron or to o-Ps bound to a molecule.

Radiation resistance is one of the most important properties for the practical use of polymers, especially polyethylene, as construction elements. Therefore this property was chosen for our investigations. The work presented here can be divided into two parts. Part I describes results obtained due to a search for a real-time determination of material properties under irradiation. Part II describes present experiments non destroying the sample on polyethylene of high density by using the positron lifetime technique to study the irradiation induced effects within a dose range from 0 kGy up to 1460 kGy.

RESULTS AND DISCUSSION - Part I

Several investigations were performed on polyethylene of about 170 mg cm^{-2} thickness irradiated with dose rates of 0.1 and 12.5 kGy h^{-1}, respectively, up to doses of 1000 kGy. Positron lifetimes measurements [7] enabled us to conclude on the existence of four lifetime components, which all where of the order cited above from the literature [1-6]. Several constraints were tested in order to improve the calculations, including the suggestion that the third lifetime component may also be due to an o-Ps state. In our opinion the best fits were obtained from the following constraints: (i) τ_1 = 125 ps, (ii) τ_3 = 1085 ps, (iii) $3I_1 = I_3$. Constraints (i) and (iii) mean that only these two states are due to the annihilation of Ps. Fixing τ_1 was necessary because of our time resolution (350 ps FWHM) and the relatively low intensity of I_1. The value τ_3 = 1085 ps was calculated from averaging the τ_3 results from the constraint-free analysis of the lifetime spectra. Due to a bound state - of unknown origin - we expect at least very small or negligible changes of this lifetime. The final results of our lifetime calculations with the cited constraints are given in table I.

We have shown [7] that logarithmic functions represent the most natural approximations of an effect measurable over some decades of radiation dose. These functions tend towards a saturation value which might be reached if a further increase of radiation dose would not change anymore the annihilation scheme. The test of other correlation functions gave no better results. The statistical error of all parameters never exceeds 2% of the actual value.

Results of oxygen permeation measurements [8], and determination of physical and chemical characteristics [9] of the same samples have shown:
- a more or less definite, and generally non-linear change of the physical and chemical properties with increasing dose; exponential (permeability, elongation at break), or power-law functions (density, carbonyl index, gel content),
- a strong effect of dose rate on property changes.

Positron annihilation measurements are non-destructive and possible on real construction elements, even in place on operating equipment, or can be carried out on very small test samples (some 20 mg) of arbitrary shape. All the other methods cited either require a certain geometry of the specimen or are destructive test methods. Therefore it was suggested to estimate material properties in real-time non-destructively by positron annihilation measurements and correlation analysis [10]. In principle, thereby material properties due to an expected load can be estimated too.

RESULTS AND DISCUSSION - PART II

The measurements were performed with a ^{22}Na-β^+-source, and a positron lifetime spectrometer based on BaF$_2$-scintillators (Harshaw Chemie, Wermelskirchen, Germany) and XP2020Q (Valvo, Hamburg, Germany) photomultipliers. The time resolution of the spectrometer was about 400 ps FWHM with ^{60}Co and energy windows set for ^{22}Na.

Table I. Experimental values for different measuring parameters at different irradiation conditions.

Parameter	unirradiated state	irradiated state			
		dose (kGy)			
		1	10	100	1000
I_4 (%)	18.5	17.4	16.8	16.0	15.2 *
		17.4	16.8	17.0	16.7 +
τ_4 (ps)	2546	2541	2460	2425	2395 *
		2565	2616	2547	2476 +
I_3 (%)	11.8	11.8	10.0	10.3	9.4 *
		12.8	12.7	11.7	10.8 +
I_2 (%)	63.6	65.1	67.6	68.4	70.4 *
		64.1	65.0	65.6	66.9 +
τ_2 (ps)	268	271	278	278	283 *
		267	278	275	275 +
		Dose rate: * 0.1 kGy h^{-1}, + 12.5 kGy h^{-1}.			

The sample (50 mm · 50 mm · 5 mm) of polyethylene (Lupolen 4261 A, BASF AG, Ludwigshafen, Germany) with a density (unirradiated) of 0.942 g/cm^3 and a cristallinity (unirradiated) of 59 % have been cut of plates (0.5 m^2) and placed as a sandwich at the positron source (3.7 MBq ^{22}Na between two 7.5 μm thick Ni-foils). The ^{60}Co source of the Strahlenzentrum of the Justus-Liebig-Universität Giessen was used to irradiate the polyethylene samples in air at room temperature with a dose rate of 0.54 kGy/h. Each lifetime spectrum has been measured to achieve more than 1 · 10^6 counts at least. All measurements were carried out at room temperature. We used a separate pair of samples for each dose.

The spectra have been analysed using the programme POSFIT [11], which is running interactively on a PC and is able to fit the prompt curve (two Gaussians) simultaneously with the lifetime parameters. For the final analysis we used here the physically justified constraints $I_1 = (I_3+I_4)/3$ and $\tau_1 = 120$ ps. A comparison with the programme POSGAUSS [12] was performed.

Due to the γ-irradiation in the polymer chains there may be produced radicals by kicking off H-atoms [13]. Subsequently different chemical reactions (recristallisation, creating of hydroperoxides, crosslinking, degradation) are induced. Also immediately after the beginning of irradiation the oxygen, dissolved in polyethylene, may be bound forming hydroperoxides at the polymer chains in the amorphous regions. More oxygen for this reaction is supplied by diffusing from the air into the sample [14]. The lifetimes and intensities of the components vary with the degree of irradiation. Especially the lifetime of the fourth component (τ_4) depends on the free volume [15].

The behaviour of the reciprocal density ρ^{-1} (measured by BASF AG), is shown in fig. 1a, and of the intensity I_4 in fig. 1b as a function of the ^{60}Co-γ-dose.

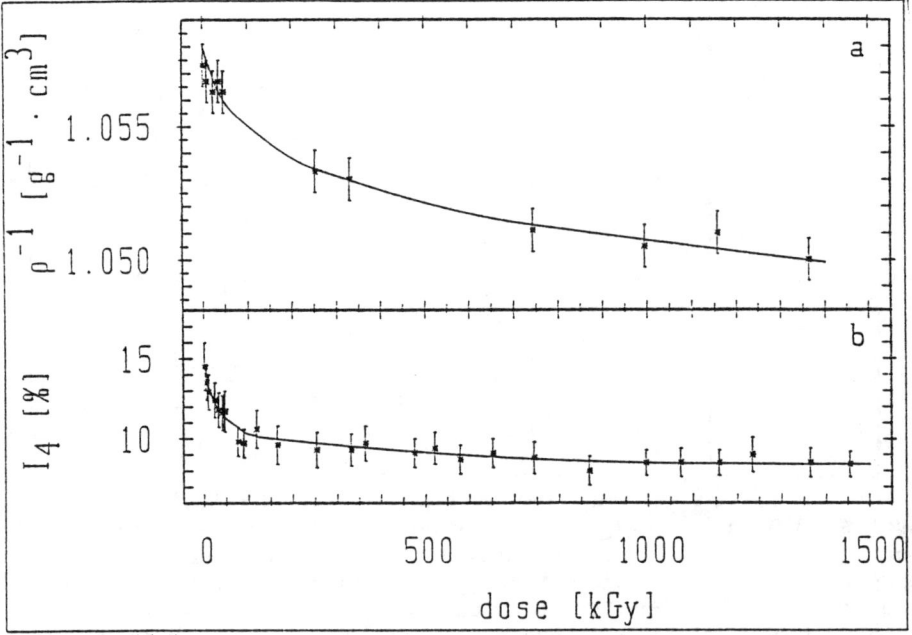

Fig.1. Behaviour of the reciprocal density ρ^{-1} (a) and of the intensity I_4 (b) (relative units) as a function of the ^{60}Co-γ-dose.

ACKNOWLEDGEMENTS

One of us (Z.M.) participated in this work as a scholar of the WE-Heraeus-Stiftung, Hanau, Germany. This work was supported by the Deutsche Forschungsgemeinschaft, Bonn-Bad Godesberg, Germany. We are thankful for providing the polyethylene to the BASF, the analysing programmes to Dr. B. Lèvay, Karlruhe, Germany and to P.D. Dr. H. Sormann, Graz, Austria.

REFERENCES

1. M. Eldrup, in: Positron Solid-State Physics, Proc. Internat. School of Physics "Enrico Fermi", Course LXXXIII, Varenna on Lake Como 1981, Ed. V. Brandt, North Holland, Amsterdam 1983, p. 644.
2. P. Kindl, Habil. thesis, university of Graz (1983).
3. P. Kindl, G. Reiter, phys. stat. sol. (a) 104, 707 (1987).
4. G. Reiter, Ph D thesis, university of Graz (1987).
5. J. Ch. Abbé, G. Duplâtre, J. Serna, Positron Annihilation, Ed. by L. Dorikens-Vanpraet, M. Dorikens, D. Segers, 796 (1989).
6. A. Badia, J. Ch. Abbé, G. Duplâtre, ICPA-9, 9th Int. Conf. on Positron Annihilation, Szombathely, August, 26th-31st, 1991, D10.
7. G. Brauer, W. Hertle, A. G. Balogh, phys. stat. sol. (a) 105, K7 (1988).
8. G. Brauer, W. Hertle, E. Müller, W. Tzscheutschler, Radiat. Phys. Chem. 33, 307 (1989).
9. G. Brauer, W. Hertle, 'Kunststoffe' 78, 424 (1988).
10. G. Brauer, W. Hertle, E. Müller, W. Tzscheutschler, Report ZfK-678 (1989), Zentralinstitut für Kernforschung Rossendorf.
11. B. Lèvay, private communication.
12. H. Sormann, private communication.
13. E. Köhnlein, in Kunststoffe, 65th year, no. 9 (1975).
14. H. U. Voigt, F. Wizcnerowicz, Energiewirtschaftl. Tagesfragen (ET) 11, 524 (1974).
15. Y. C. Jean, Microchem. J. 42, 72 (1990); Nucl. Instr. and Meth. B56/57, 615 (1991).

Free Volume Model for Dielectric Constant of Polymer Films

Abe Eftekhari[1]
Anne St. Clair, Diane M. Stockly, Danny R. Sprinkle,
and Jag J. Singh[2]

ABSTRACT[3]

A slow positron flux generator reported in another paper at this conference was used to measure positron lifetime in a series of especially developed fluorine containing thin polyimide films. The positron lifetime spectra was analyzed into 2-components using a standard least square routine. No evidence for positronium formation was observed in any of test films studied.

The trapped positron lifetimes were used to calculate the radii of the shallow trap sites. Equating the total volume occupied by the traps with the saturation moisture content of Kapton (reference) films, free volume fractions (f) were calculated in all the samples. These free volume fractions affect the dielectric constants (ϵ) of the test films as follows:

$$\frac{1}{\epsilon} = \frac{(1-f)}{\epsilon_R} + \frac{f(1-d)}{\epsilon_{Air}} + \frac{fd}{\epsilon_{Water}}$$

Where, ϵ_R is the dielectric constant of the trap-free medium, ϵ_{Air} is the dielectric constant of air, ϵ_{Water} is the dielectric constant of water, and d is the moisture uptake inhibition factor. Several examples illustrating the applicability of this model to various types of polymers will be presented.

1- Department of Physics, Hampton University, Hampton, Virginia.
2- NASA Langley Research Center, Hampton, Virginia.
3- For more information refer to the paper "A Low Energy Flux Generator for Microstructural Characterization of Thin Polymer films" by: J.J. Singh, A. Eftekhari, & T. L. St. Clair in this book.

PRELIMINARY RESULTS OF A SLOW POSITRON STUDY ON AN EPOXY POLYMER

Y.C. Jean, G.H. Dai, H. Shi
University of Missouri-Kansas City
Kansas City, MO 64110, USA

R. Suzuki
Electrotechnical Laboratory
Tsukuba, Japan

Y. Kobayashi
National Chemical Laboratory for Industry
Tsukuba, Japan

ABSTRACT

The positron annihilation lifetime spectra and positronium energy spectra of an epoxy polymer with a 5:2:3 equivalent ratio of DGEBA/DDA/DAB epoxy (T_g = 52°C) are measured as a function of temperature and of positron incident energies (0-5 keV). Preliminary results from these experiments show: (1) a 1-9% of long-lived o-Ps are emitted from the epoxy surface, (2) the lifetime of o-Ps near the surface is about three times longer than that in the bulk, (3) the positron lifetime spectra obtained from the incident positron energy exceeding 2 keV are essentially the same as those in the bulk, and (4) the energies of emitted Ps are distributed from thermal to about 100 eV. These results are discussed in terms of the potential uses of slow positrons in probing the free-volume hole properties of thin films and near surfaces of polymeric materials.

INTRODUCTION

In the recent years, positron annihilation spectroscopy (PAS) has been developed as a useful tool in probing the microscopic properties of polymeric materials[1]. One of the great successes in this line of research is the direct determination of free-volume hole properties at an atomic scale (2-20 Å) in polymers. In a series of experiments[2-6], we have demonstrated that PAS is able to determine the free-volume hole size[2,3], distribution[4], fraction[5], and anisotropic structure[6] of polymers. The high sensitivity of PAS in probing free-volume properties arises from the fact that the positronium atom (Ps, an atom consists of a positron and an electron) is preferentially trapped (localized) in atomic-scale free-volume holes. The free-volume theory has been postulated more than 4 decades[7] and been widely used to explain many important physical and mechanical properties of polymeric materials[8]. As far as we know, PAS is the only physical probe that gives direct experimental free-volume information in polymers.

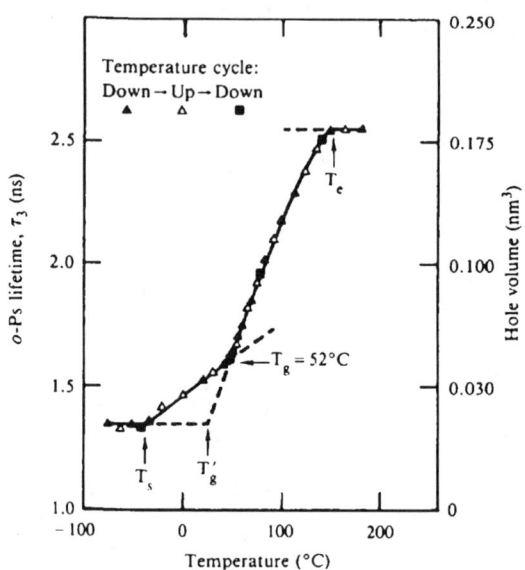

Fig. 1. o-Ps lifetime vs. temperature in a DGEBA/DDA/DAB epoxy by using fast positrons (ref. 2). The intensity of o-Ps varies from 22-25 %.

For example[2], in Fig. 1, the measured o-Ps (the triplet state of Ps) lifetimes are plotted as a function of temperature in an epoxy polymer. As shown in Fig.1, the temperature variation is very dramatic and the largest change of o-Ps lifetime coincides with the well-known glass transition temperature T_g (= 52°C). Below T_g, there exists a temperature dependence with a smaller slope than that above T_g. At a very low temperature T_s (about $T_g - 100°C$), we observe a constant lifetime which indicates the frozen character of molecular structures including a rotational motion. On the other hand, at very high temperatures, above T_e (about $T_g + 100°C$), we also observe a constant lifetime which indicates the formation of a Ps-bubble as those found in liquids. The observed lifetime-temperature coefficients in polymers above T_g are about 1-2 orders of magnitude larger than the volume expansion coefficients measured by other methods[8]. Similar large variations are also observed as a function of pressure[3] and of physical aging[9]. All existing results give direct evidence that Ps is a unique probe for free-volume holes. The time scale of the probe covers any molecular motions longer than 10^{-10} s.

Along the path of developing PAS as a free-volume microprobe, there are still fundamental problems which are not well understood, such as, Ps and positron diffusion constants, their capture cross sections to free-volume holes, and formation mechanisms of Ps in polymeric materials. Furthermore, applications of PAS by using the conventional positron sources from radioisotopes are limited to bulk properties due to a relatively large penetration depth of the energetic positrons, i.e. in the order of 100-200 μm. Both the fundamental problems and practical applications can be eliminated by employing a slow positron beam with a well-defined energy. In this paper, we report the preliminary results of a slow positron study in an epoxy polymer which has been thoroughly investigated by using fast positrons in the past by us.

EXPERIMENTS

The epoxy samples used in this study are the same as those reported in our previous studies by using fast positrons[2-6]. They are amine-cured epoxies with a chemical composition of DGEBA (diglycidyl-ether of bisphenol A), DDH (N,N'-dimethyl-1,6-diamino-hexane), and DAB(1,4-diaminobutane) (supplied from Imperial Chemical Inc. Americas) at an equivalent ratio (5:3:2). A T_g of 52°C was determined by DSC

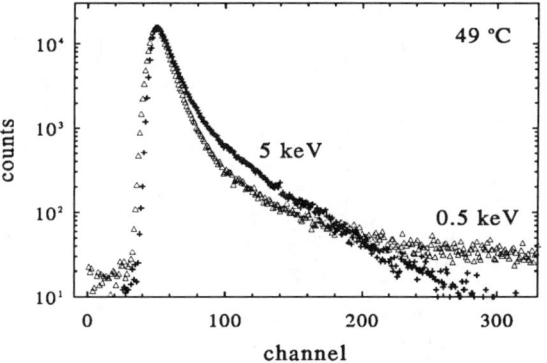

Fig. 2. PAL spectra of a DGEBA/DDA/DAB epoxy by using slow positrons. The time calibration was 0.0466 ns/ch. Total counts = 2.5 × 10^5.

(Differential Scanning Colorimetry) and by PAS (as shown in Fig. 1) measurements. The samples were cured under a dry N_2 environment and span at a speed of 1000 rpm on an aluminum plate during curing. The cured epoxy sample (thickness 25-35 μm) coated on an aluminum plate was cut (2.5 cm × 2.5 cm) and then mounted in the beam line for the positron measurements. For each experiment, the sample was first heated to 130°C for 30 min and then slowly cooled down to the measurement temperature. The vacuum of the sample chamber was about 10^{-6} and 10^{-8} torr at high and low temperatures, respectively.

The positron annihilation lifetime (PAL) spectra were recorded at the intense slow positron facility in the Electrotechnical Laboratory (ETL)[10]. The slow positrons were generated with the electron linac (75 MeV, 1-4 μs, 50-100 pps) by the use of a Ta converter and a W moderator. The generated slow positrons were temporarily stored in a linear storage section and then extracted to the pulsing system as a quasi-continuous beam. The positron pulsing system consists of three stages under a magnetic field of

0.008 T; a reflection type chopper, a sub-harmonic pre-buncher, and a double harmonic buncher. The finally pulsed positron beam has a cross section of 1 cm diameter with a variable energy from 50 eV to 30 keV. The positron lifetime was monitored by a start timing from the finally pulsed signal (150 ps width) and by a stop timing from the annihilation photons detected by a BaF_2 scintillator. The lifetime resolution is 250 ps at a counting rate of 100 cps (1 μA of electron linac current). Each PAL was collected at a period about 50 min with a total statistics approximately 3×10^5 counts.

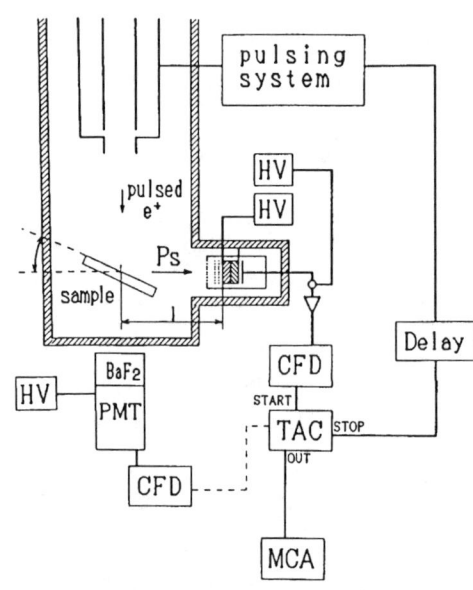

The obtained PAL spectra were χ^2 fitted into three or four lifetimes by using PATFIT programs[11]. Two series of PAL spectra were acquired; short-gated (80 ns) and long-gated spectra (800 ns) for the search of the surface Ps and of the free Ps in vacuum, respectively. The PAL spectra from the short- and long-gated experiments were fitted into three- and four-lifetimes, respectively. Typical PAL spectra with a short gate obtained by using slow positrons are shown in Fig. 2.

A Ps time-of-flight (TOF) was setup to measure the energy distributions of Ps emitted from the surface. A sketch of the Ps-TOF system is shown in Fig. 3. The Ps-TOF spectra were timed by the pulsing signals of positrons (start) and by the signals detected at the multichannel plate (MCP). The sample was situated 45° from the beam direction and the MCP was 90° from the beam line. The distance between the sample and the MCP was 4.8 cm. Any charged particles emitted from the surface were rejected by both positive and negative bias in front of the MCP detector. Only Ps signals were detected in a Ps-TOF spectrum. This design is similar to the Ps-TOF system used by Sferlazzo et al.[12] The background counts were obtained by measuring a Ps-TOF spectrum in the same sample rotated 90° from the detector. The Ps energy spectra were calculated from the obtained Ps-TOF spectra after the

Fig. 3. A diagram of a Ps-TOF setup used in measuring Ps energy spectra.

background subtraction and after taking the o-Ps decay into account.

RESULTS AND DISCUSSIONS

The obtained PAL spectra were fitted either in three- or four-lifetimes for the short-gated and for the long-gated spectra, respectively. In the three-component fits, the short lifetime $\tau_1 \approx$ 0.12 ns, and the intermediate lifetimes $\tau_2 \approx$ 0.38 ns are attributed to the p-Ps (singlet Ps) and the positron annihilation, respectively. The long lifetime, τ_3=1.5-4.5 ns, is attributed to the o-Ps annihilation. The short lifetimes do not vary while the intermediate lifetime vary slightly with the temperature and positron incident energy. The results of o-Ps lifetimes τ_3 and intensities I_3 are main useful information in the interpretation of free-volume hole properties. In Figs. 4 and 5, we plot these results vs. the sample temperatures at different positron incident energies.

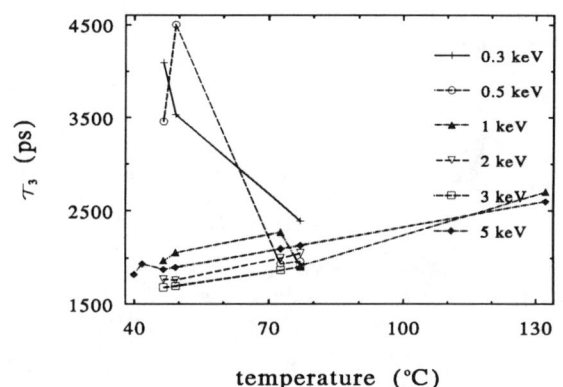

Fig. 4. o-Ps lifetime in an epoxy vs. sample temperature at various positron incident energies.

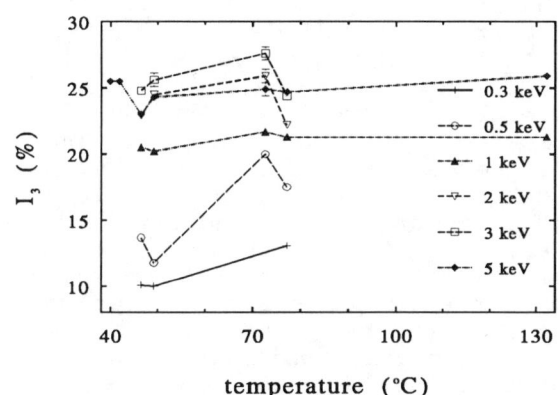

Fig. 5. o-Ps intensity in an epoxy vs. temperature at various positron incident energies.

From the o-Ps lifetime results (Fig. 4), we have the following observations: (1) at low temperatures, the observed o-Ps lifetime decreases from 4.5 ns to 1.7 ns as the positron incident energy increases from .3 to 5 keV, (2) at low energies, the lifetime decreases as a function of positron energy while at higher energies, it increases, (3) the o-Ps lifetimes and their temperature variations for

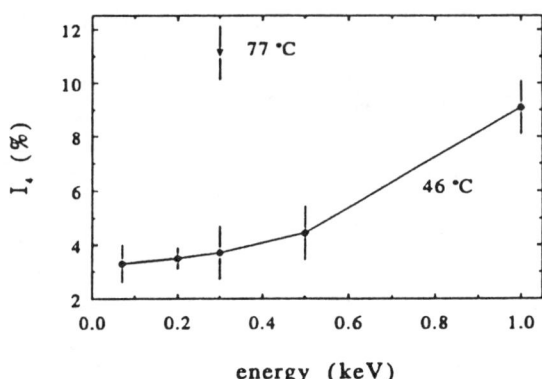

Fig. 6. Free o-Ps intensity (τ = 140 ns) emitted from an epoxy surface vs. positron incident energy.

energies exceeding 2 keV are essentially the same as those obtained by using fast positrons (as shown in Fig. 1).

The larger τ_3 value at low positron energies is due to the contribution of o-Ps annihilation near and on the surface. The penetrated slow positrons interact with the bulk and then slow down to certain energies that are capable to capture electrons in forming Ps in a very short time, i.e. in the order of 10^{-12} s. The formed Ps is either trapped and annihilated in the free-volume holes or diffused back to the surface. A surface can be considered as a hole with an infinite radius. The mean hole size near the surface is found to be larger than that in the bulk. The o-Ps lifetime has a direct correlation with the free-volume hole size[13]: a larger hole results a longer lifetime. The observed o-Ps lifetime is longer from the lower positron energies because the formed Ps has a higher probability to diffuse to the surface. Since the current slow positron beam can not be pulsed to the energies below 0.2 keV for a good time resolution, we estimated the o-Ps lifetime near and on the surface from the 0.3 keV result to be about 4.8 ns. The extrapolated surface o-Ps lifetime (4.8 ns) is about three times longer than that in the bulk lifetime (1.7 ns) of the epoxy. This result is consistent with the longer surface lifetimes observed on the surface than in the bulk of molecular systems, such as zeolites[14].

At higher positron energies, the temperature variation of o-Ps lifetime follows the similar trend as those found in the bulk. A large increase of lifetime at higher temperature is resulted from the expansion of free-volume holes. It is interesting to observe when the positron energy exceeds 2 keV, the PAL spectra are essentially the same as those in the bulk. This implies that the diffusion length of Ps in polymeric materials is relatively short. We estimated it to be in the order of hundred Å instead of thousand Å as in crystalline systems[15]. A shorter Ps diffusion length in polymers is due to the existence of free-volume holes which are the effective trapping sites for Ps. From this result, one will be able to obtain bulk information of polymers with an incident energy about 2 keV. This will be an advantage in using slow positrons to probe thin film polymers with hundreds to thousands Å thickness.

The variations of o-Ps intensity I_3 as shown in Fig. 5 are also very interesting. In applications of PAL to polymers, we use I_3 as a measure for the free-volume fraction[1]. The increase of I_3 as a function of incident energy can be explained due to an increase of o-Ps annihilation in the bulk. This explanation is consistent with the converging of τ_3 to the bulk lifetime as discussed above. As seen in Fig. 5, the values of I_3 also converge to the I_3 (25%) as observed in using fast positrons[2]. At low energy, o-Ps intensities as a function of temperature show an increase of the free-volume fractions both in the bulk and near the surface.

The low values of I_3 at low energies can be contributed to the back diffusion of positrons and Ps to the surface. The emission of Ps from the surface to the vacuum can be determined by measuring the long-lived o-Ps probability from PAL experiments. We have performed a series of PAL experiments by setting the timing gate of the spectra to 800 ns in order to search for free o-Ps. The obtained PAL spectra were fitted into four lifetimes. Besides three lifetimes as described above, we obtained the fourth component with a lifetime 140±5 ns which corresponds to free o-Ps annihilation in the vacuum. The absolute fractions of I_4 are calibrated to the I_3 obtained between the short- and long-gated experiments at high incident energies. The results of the fractions of free o-Ps emitted from the epoxy surface are plotted in Fig. 6.

As shown in Fig. 6, we observe an increase of free o-Ps fraction as the positron incident energy increases. The lower fraction of o-Ps emission at a lower incident energy is due to the higher probability of positrons diffusing back to the surface. This is similar to the result reported in the ice at low positron incident energy[15]. In principle, one should observe a decrease of I_4 at certain positron incident energy due to the diffusion of Ps from the bulk to the surface. Unfortunately, the current slow positron facilities can not be performed at an energy exceeding 1 keV for the long-gated PAL experiments. This part of experiment will be performed in the near future. From the current results of Figs. 5 and 6, we observe a large increase of free positrons emitted from the surface, i.e., I_2, as the incident positron energy decreases.

Next we present the Ps-TOF spectra for o-Ps emitted from the epoxy surface with the incident positron energies of 70, and 200 eV in Figs. 7 and 8, respectively. The Ps-TOF spectra were performed at three temperatures. As shown in Figs. 7 and 8, the time of flight spreads widely from the zero time (the left-hand peaks) to about 120 ns. From these TOF results, we calculate the kinetic energies of

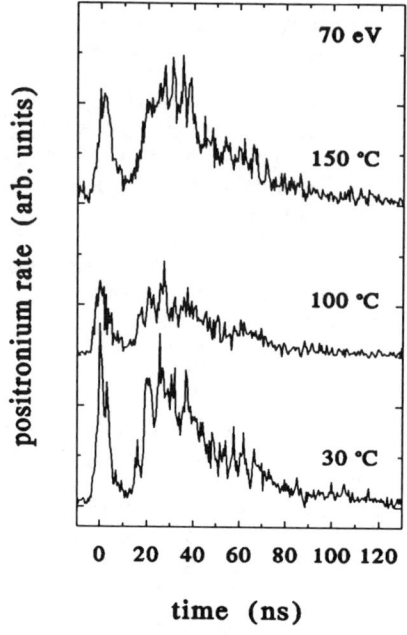

Fig. 7. Ps-TOF spectra of an epoxy surface at three temperatures. The total counts for each measurement are 2×10^4.

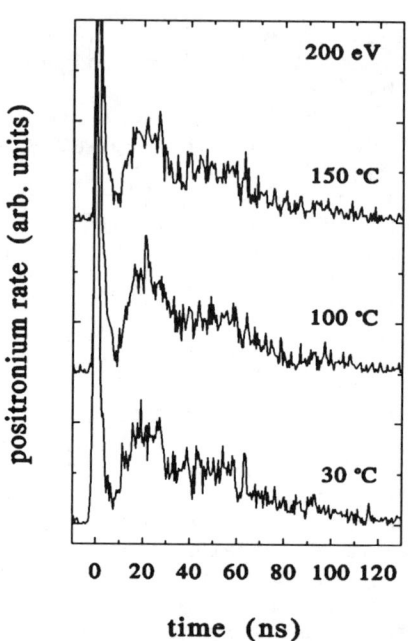

Fig. 8. Ps-TOF spectra of an epoxy surface at three temperatures. The total counts for each measurement are 2×10^4.

emitted Ps after subtracting the background and taking the o-Ps decay into account. The resulting Ps energy distributions are plotted in Figs. 9 and 10 for the incident energies 70, and 200 eV, respectively. As shown in Figs. 9 and 10, the Ps energies are distributed from thermal states to nearly 100 eV. This is very different from the existing Ps energy spectra emitted from metal[16] or oxide surfaces[12] where narrow energy distributions are observed at low Ps energies (a few eV). Besides the difference between a polymeric surface and those relatively clean surfaces, the Ps formation mechanism might be quite different in polymeric materials. In polymers Ps can be formed at the energies between the thermal state to nearly hundreds eV as judged from the broad distributions of Ps energy shown in Figs. 9 and 10. Another possible mechanism is a direct abstraction of electrons on the surface by the incident positrons without penetrating and thermalized into the bulk. This is seen by comparing the energy spectra between 70 and 200 eV incident positrons. We found that Ps is distributed more to a higher energy for the 200 eV than for the 70 eV of incident energy. On the other hand, it is

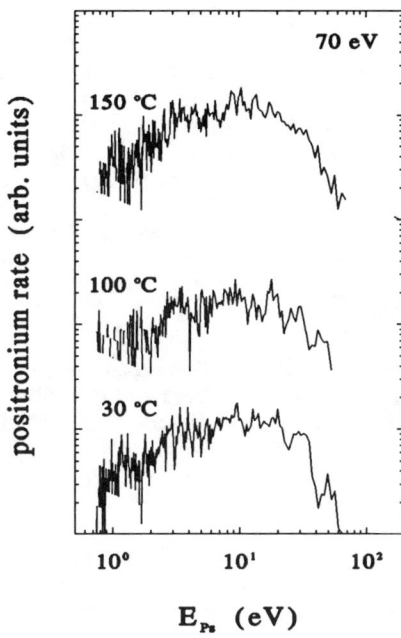

Fig. 9. Ps energy spectra emitted from an epoxy surface. Energies of Ps were calculated from the data shown in Fig. 7.

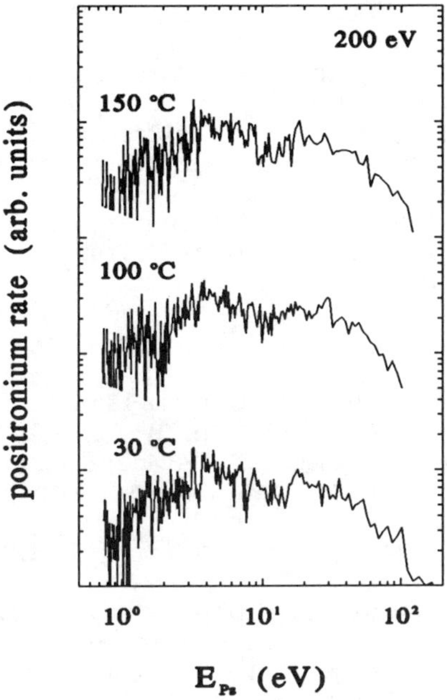

Fig. 10. Ps energy spectra emitted from an epoxy surface. Energies of Ps were calculated from the data shown in Fig. 8.

hard to see any significant difference in the Ps energy distributions at different temperatures. This might need a much better statistics to detect any difference in the energy distributions at the temperatures above and below T_g.

CONCLUSION

We have reported a series of experimental PAL data for an epoxy surface as a function of temperature and of incident positron energy. We also reported the Ps-TOF spectra from an epoxy polymer. We found that slow positrons can be a powerful probe for the investigations of free-volume hole and defect properties for a thin film and for near the surface of polymeric materials. The bulk properties of polymers can be determined by using the positron energy exceeding 2 keV. These PAL results also show a significant fraction of free Ps emission from the epoxy surface. This indicates that Ps has a negative work function in epoxy polymers. The free-volume holes near the surface are found to be larger than in the bulk. The emitted Ps energy distributions are ranging from thermal energies to nearly 100 eV. This indicates that Ps can be formed in various mechanisms by: (1) direct abstracting electrons from the surfaces, (2) attracting the ionized electrons while the positron is still at a hot stage, and (3) combing the surrounding electrons after the positron is thermalized. Further experiments in various polymeric surfaces will be very useful in the understanding the fundamental Ps behaviors in polymers as well as for the practical applications of PAS to determine the polymer properties.

ACKNOWLEDGEMENT

We are specially thankful to Dr. T.C. Sandreczki in preparing the thin film samples of epoxy polymers for this experiment. This research has been supported by the NSF grants (DMR-9004083, and INT-89-15060-ENGR-177).

REFERENCES

1. For examples, see, Y.C. Jean, Microchem. J. **42**, 72 (1990); Y. Kobayashi, W. Zheng, E.F. Meyer, J.D. McGervey, A.M. Jamison, and R. Simha, Macromolecules **24**, 2302 (1989).
2. Y.C. Jean, T.C. Sandreczki, and D.P. Ames, J. Polym. Sci. B **24**, 1247 (1986).
3. Q. Deng, C.S. Sundar, and Y.C. Jean, J. Phys. Chem. **96**, 492 (1992).
4. Q. Deng, F. Zandiehnadem, and Y.C. Jean, Macromolecules **25**, 1090 (1992).
5. H. Nakanishi, Y.C. Jean, E.G. Smith, and T.C. Sandreczki, J. Polym. Sci. B **27**, 1419 (1989).
6. Y.C. Jean, H. Nakanishi, L.Y. Hao, and T.C. Sandreczki, Phys. Rev. B **42**, 9705 (1990).
7. A.K. Doolittle, J. Appl. Phys. **22**, 1471 (1951).
8. For example, see, J.D. Ferry, "Visco-elastic Properties of Polymers," J. Wiley and Sons, N.Y. (1980).
9. T.C. Sandreczki, H. Nakanishi, and Y.C. Jean, in "Positron Annihilation Studies of Fluids," Ed. S.C. Sharma, World Scientific, Singapore (1988), p. 200.
10. R. Suzuki, Y. Kobayashi, T. Mikado, H. Ohgaki, M. Chiwaki, T. Yamazaki, and T. Tomimasu, Jap. J. Appl. Phys. **30**, L532 (1991).
11. PATFIT-88 package (1989), purchased from Risø Nat. Lab., Denmark.
12. P. Sferlazzo, S. Berko, and K.F. Canter, Phys. Rev. B **35**, 5315 (1987).
13. H. Nakanishi, S.J. Wang, Y.C. Jean, in "Positron Annihilation Studies of Fluids," Ed. S.C. Sharma, World Scientific, Singapore (1988), p. 292.
14. K. Venkateswaran, K.L. Cheng, and Y.C. Jean, J. Phys. Chem. **89**, 3001 (1985).

15. M. Eldrup, A. Vehanen, P.J. Schultz, and K.G. Lynn, Phys. Rev. B $\underline{32}$, 7048 (1985).
16. R.H. Howell, I.J. Rosenberg, and M.J. Fluss, Phys. Rev. B $\underline{35}$, 5303 (1987).

On the interpretation of ACPAR spectra with respect to electron momentum density in real metals

G. Kontrym-Sznajd and A. Rubaszek

W. Trzebiatowski Institute of Low Temperature and Structure Research,
Polish Academy of Sciences, Wroclaw, P.O.Box 937 (Poland)

The influence of the positron distribution and electron-positron interactions on the momentum density of annihilation quanta in real metals is discussed. The role of momentum dependence of two-particle electron-positron correlations is set forth.

The average densities determined by electron-positron correlations are presented for valence electrons in simple metals. Momentum-dependent enhancement factors $\epsilon(p)$ describe the deviation of momentum density of annihilation quanta from the electron momentum density in the material investigated. The necessity of reconstruction of experimental ACPAR spectra while determining $\epsilon(p)$ is substantiated.

I. Basic formulas.

The angular correlation of positron 2γ annihilation radiation (ACPAR) technique has been established for studies of the electronic structure of materials, particularly for electron momentum spectroscopy[1]. In real metals, however, (due to strong electron-positron correlations as well as to deviation of positron distribution from the uniform one) the investigated electron momentum density (EMD),

$$\rho^e(\mathbf{p}) = \frac{1}{\Omega}(-i)^2 \int_\Omega \int_\Omega e^{-i\mathbf{p}\cdot(\mathbf{r}-\mathbf{r}_1)} \cdot G_e(\mathbf{r}t; \mathbf{r}_1 t^+) d\mathbf{r} d\mathbf{r}_1 =$$

$$\frac{1}{\Omega} \sum_i n(i) \left| \int_\Omega e^{-i\mathbf{p}\cdot\mathbf{r}} \cdot \psi_i^e(\mathbf{r}) d\mathbf{r} \right|^2 \quad , \quad (1a)$$

differs from the momentum density of the 2γ annihilation quanta,

$$\rho^{2\gamma}(\mathbf{p}) = \frac{\pi r_0^2 c}{\Omega}(-i)^2 \int_\Omega \int_\Omega e^{-i\mathbf{p}\cdot(\mathbf{r}-\mathbf{r}_1)} \cdot G_{ep}(\mathbf{r}t, \mathbf{r}t; \mathbf{r}_1 t^+, \mathbf{r}_1 t^+) d\mathbf{r} d\mathbf{r}_1 =$$

$$\frac{\pi r_0^2 c}{\Omega} \sum_i n(i) \left| \int_\Omega e^{-i\mathbf{p}\cdot\mathbf{r}} \cdot \psi_i^e(\mathbf{r}) \cdot \left[\psi_+(\mathbf{r}) \cdot \frac{\psi_i^{ep}(\mathbf{r},\mathbf{r})}{\psi_i^e(\mathbf{r}) \cdot \psi_+(\mathbf{r})} \right] d\mathbf{r} \right|^2, \quad (1b)$$

by a momentum-dependent factor

$$\epsilon^{2\gamma}(\mathbf{p}) = \rho^{2\gamma}(\mathbf{p})/\rho^e(\mathbf{p}). \quad (1c)$$

Here G_e and G_{ep} are the zero-temperature electron and electron-positron Green's functions, respectively, $\psi_i^{ep}(\mathbf{r}_e, \mathbf{r}_p)$ denotes the pair wave function of the electron in the *initial* state i located at $\mathbf{r}_e$ and thermalized positron at $\mathbf{r}_p$, and ψ_+ and ψ_i^e are the positron and *unperturbed* electron wave functions, respectively.

Quantities r_0, c, Ω and $n(i)$ in Eqs.(1) denote the classical electron radius, velocity of light, volume of unit cell and occupation numbers of electronic states i, respectively.

Within the independent particles model (IPM) $\psi_i^{ep}(r_e, r_p) = \psi_i^e(r_e) \cdot \psi_+(r_p)(G_{ep} = G_e \cdot G_p)$ and Eq.(1b) leads to well-known IPM ACPAR formula

$$\rho^{IPM}(p) = \frac{\pi r_0^2 c}{\Omega} \sum_i n(i) \left| \int_\Omega e^{-i p \cdot r} \cdot \psi_i^e(r) \cdot \psi_+(r) dr \right|^2. \quad (2a)$$

The corresponding factor

$$\epsilon^{IPM}(p) = \rho^{IPM}(p)/\rho^e(p) \quad (2b)$$

provides information about the influence of the shape of the positron wave function on the resulting ACPAR spectrum[2]. The function $\epsilon^{2\gamma}(p)$ may be written as the product of $\epsilon^{corr}(p)$ and $\epsilon^{IPM}(p)$. The enhancement factor (most often investigated in theoretical and experimental works),

$$\epsilon^{corr}(p) = \rho^{2\gamma}(p)/\rho^{IPM}(p), \quad (2c)$$

describes mainly electron-positron correlation effects. Positron and unperturbed electron wave functions (Green's functions), used in Eqs. (1a) and (1c) may be obtained within standard band-structure calculations. It is not so with functions ψ_i^{ep} (or G_{ep}) of Eq. (1b). For determining G_{ep} or the corresponding correlation factors,

$$f_i(r) = |\psi_i^{ep}(r, r)|^2 / |\psi_i^e(r) \cdot \psi_+(r)|^2, \quad (3)$$

in real metals, various approximations have been used, leading to alternative results for $\epsilon^{corr}(p)$[3-6]. Of particular interest there is the local density approach (LDA) developed in a series of papers[5,6], which approximates $f_i(r)$ by correlation functions in jellium of local electron density characterized by the parameter $r_s(r)$, $\epsilon^{corr}_{jell}[i, r_s(r)]$.

It should be noted here that if within LDA $f_i(r)$ is assumed to be independent of r (i.e. $f_i^{LDA}(r) \equiv \epsilon^{corr}_{jell}(i, r_s)$), the approach reduces to the average density approximation (AED)[4]. In a series of works, developing approaches of Refs. 6, the state-selectivity of $f_i(r)$ is neglected. The next paragraph is devoted to the influence of state-dependence of $f_i(r)$ on the resulting annihilation characteristics.

2.Electron-positron enhancement factors inside the Fermi surface and applicability of AED.

In this paragraph the influence of positron distribution as well as of local correlation effects $f_i(r)$ on the resulting momentum density $\rho^{2\gamma}(p)$ is discussed. Let us take into account the contribution of electronic Bloch states

$$\psi^e_{kj}(\mathbf{r}) = \frac{1}{\sqrt{\Omega}} e^{i\mathbf{k}\cdot\mathbf{r}} \cdot \left[\sum_{\mathbf{G}} u_{kj}(\mathbf{G}) e^{i\mathbf{G}\cdot\mathbf{r}}\right] \quad (4)$$

to EMD and $\rho^{2\gamma}(\mathbf{p})$. For this purpose it is convenient to check how the EMD,

$$\rho^e(\mathbf{p} = \mathbf{k} + \mathbf{G}^*) = \sum_j n(kj) \mid u_{kj}(\mathbf{G}^*) \mid^2 \quad (5a)$$

changes after perturbing functions ψ^e_{kj} by the periodic functions (cf. expression in brackets on the right-hand-side of Eq.(1b))

$$h_{kj}(\mathbf{r}) = \psi_+(\mathbf{r}) \cdot [f_{kj}(\mathbf{r})]^{1/2} = \sum_{\mathbf{G}} h_{kj}(\mathbf{G}) e^{i\mathbf{G}\cdot\mathbf{r}}, \quad (5b)$$

where $f_{kj}(\mathbf{r})$ are the two-particle correlation functions given by Eq.(3) (for more details see Ref.7). The momentum $\mathbf{k} = \mathbf{p} - \mathbf{G}^*$ in Eq.(5a) is in the first Brillouin zone (1BZ), $\mathbf{p}$ is in the extended zone scheme and the vectors $\mathbf{G}$ are the reciprocal lattice vectors.

This approach includes all the cases under study; if $f_{kj}(\mathbf{r})$ is approximated by unity, we get IPM; f_{kj} independent of positron position $\mathbf{r}$ leads to AED[4]; local correlation functions $f[r_s(\mathbf{r})]$ independent of initial electronic state kj represent formalism of Refs.6; $f_{kj}[r_s(\mathbf{r})]$ dependent on local electron density parameter $r_s(\mathbf{r})$ and state kj provide LDA of Refs.5; if any approximations to $f_{kj}(\mathbf{r})$ are made, the exact form of $\rho^{2\gamma}(\mathbf{p})$ is investigated. The corresponding "generalized" momentum density $\rho(\mathbf{p})$ is given by the expression (see Ref.7)

$$\rho(\mathbf{p} = \mathbf{k} + \mathbf{G}^*) = \sum_j n(kj) \mid u_{kj}[\mathbf{G}(j)] \mid^2 \mid h_{kj}[\mathbf{G}^* - \mathbf{G}(j)] \mid^2 \mid 1 + \alpha(kj, \mathbf{G}^*) \mid^2 =$$

$$\mid u_{kj^*}(\mathbf{G}^*) \mid^2 h_{kj^*}(0) \mid^2 \mid 1 + \alpha(kj^*, \mathbf{G}^*) \mid^2 \left[n(kj^*) + \sum_{j \neq j^*} n(kj) \mid \beta(kj, \mathbf{G}^*) \mid^2 \right], (6a)$$

where

$$\alpha(kj, \mathbf{G}^*) = \{h_{kj}[\mathbf{G}^* - \mathbf{G}(j)]\}^{-1} \sum_{\mathbf{G} \neq \mathbf{G}(j)} h_{kj}(\mathbf{G}^* - \mathbf{G}) \frac{u_{kj}(\mathbf{G})}{u_{kj}[\mathbf{G}(j)]}, \quad (6b)$$

and

$$\beta(kj, \mathbf{G}^*) = \frac{u_{kj}[\mathbf{G}(j)]}{u_{kj^*}(\mathbf{G}^*)} \frac{h_{kj}[\mathbf{G}^* - \mathbf{G}(j)]}{h_{kj^*}(0)} \frac{1 + \alpha(kj, \mathbf{G}^*)}{1 + \alpha(kj^*, \mathbf{G}^*)}. \quad (6c)$$

Vectors $\mathbf{G}(j)$ in Eqs.(6) provide the highest of amplitudes $\mid u_{\mathbf{kj}}(\mathbf{G}) \mid$ of the expansion (4), i.e.

$$\mid u_{\mathbf{k}j}[\mathbf{G}(j)] \mid \geq \mid u_{\mathbf{k}j}(\mathbf{G}) \mid \text{ for } \mathbf{G} \neq \mathbf{G}(j), \qquad (7a)$$

while

$$\mid u_{\mathbf{k}j^*}(\mathbf{G}^*) \mid \geq \mid u_{\mathbf{k}j}(\mathbf{G}^*) \mid \text{ for } j \neq j^* \qquad (7b)$$

and

$$\mathbf{G}(j^*) = \mathbf{G}^*. \qquad (7c)$$

For momenta $\mathbf{p} = \mathbf{k} + \mathbf{G}^*$ inside the central Fermi surface (FS), the state $\mathbf{k}j^*$ is occupied, i.e. $n(\mathbf{k}j^*) = 1$ in Eq.(6a).

The valence electrons in simple metals are described by the nearly-free electron (NFE) model reasonably well. Except for the states $\mathbf{k}j$ close to the BZ boundary, within the NFE there is only one "leading" Fourier coefficient in the representation (4) and the other (Umklapp) components of $\psi^e_{\mathbf{k}j}(\mathbf{r})$ are very small, i.e. $\mid u_{\mathbf{k}j}[\mathbf{G}(j)] \mid \gg \mid u_{\mathbf{k}j}(\mathbf{G}) \mid$ for $\mathbf{G} \neq \mathbf{G}(j)$. This relation gives (see Eq.(6b) and Appendix A in Ref.7) $\mid \alpha(\mathbf{k}j, \mathbf{G}^*) \mid \ll 1$. Moreover[7], within IPM and LDA $\mid h_{\mathbf{k}j}(\mathbf{G}) \mid \ll \mid h_{\mathbf{k}j}(0) \mid$ for $\mathbf{G} \neq 0$, that provides $\mid \beta(\mathbf{k}j, \mathbf{G}^*) \mid \ll 1$. As the result, the NFE momentum density and enhancement factors inside the FS may be approximated by the expressions[7]

$$\rho_{NFE}(\mathbf{p} = \mathbf{k} + \mathbf{G}^*) \cong \mid u_{\mathbf{k}j^*}(\mathbf{G}^*) \mid^2 \mid h_{\mathbf{k}j^*}(0) \mid^2 \qquad (8a)$$

and

$$\epsilon_{NFE}(\mathbf{p} = \mathbf{k} + \mathbf{G}^*) \cong \mid h_{\mathbf{k}j^*}(0) \mid^2. \qquad (8b)$$

From formula (8b) it clearly follows that the momentum-dependence of the electron-positron enhancement factors $\epsilon(\mathbf{p})$ inside the FS is due to the state-selectivity of two-particle correlation functions $f_{\mathbf{k}j}(\mathbf{r})$ only. Neither the form of the positron wave function[2], nor state-independent correlations[6] $f(\mathbf{r})$ can change visibly the shape of $\rho^{2\gamma}(\mathbf{p})$ inside the FS with respect to EMD for delocalized electronic populations (e.g. valence electrons in simple metals). This fact is illustrated in Fig.1, where the relative enhancement factors $\epsilon(\mathbf{p})/\epsilon(0)$ in Na and Mg are presented for $\mathbf{p} \in FS$. Dashed line displays both IPM results and those following from use of LDA state-independent correlation functions $\epsilon^{corr}_{jell}[0, r_s(\mathbf{r})]$. Application of state-selective correlations $\epsilon^{corr}_{jell}[E_{\mathbf{k}j}/E_F, r_s(\mathbf{r})]$ leads to results shown by the solid line. Jellium enhancement factors $\epsilon^{corr}_{jell}(p, r_s)$ were taken from Ref.8.

Within LDA Eq.(8b) leads to the AED[4b)] result

$$\epsilon_{NFE}(\mathbf{p}=\mathbf{k}+\mathbf{G}^*) \cong \left| \int \psi_+(\mathbf{r})\{\epsilon_{jell}^{corr}[E_{\mathbf{k}j}*/E_F, r_s(\mathbf{r})]\}^{1/2}d\mathbf{r} \right|^2 \cong$$

$$\int |\psi_+(\mathbf{r})|^2 \, \epsilon_{jell}^{corr}[E_{\mathbf{k}j}*, r_s(\mathbf{r})]d\mathbf{r} =$$

$$\epsilon_{jell}^{corr}[E_{\mathbf{k}j}*/E_F, r_s(\mathbf{r}^*)]. \qquad (8c)$$

In Fig.2 the effective electron densities $r_s(r_0), r_s(r_1)$, and $r_s(r_\lambda)$ are presented, where r_0 and r_1 are obtained according to Eq.(8c) for momenta $|\mathbf{p}| = 0$ and $|\mathbf{p}/\mathbf{p}_F| = 1$, respectively, and ϵ_{jell}^{corr} is the jellium result of Ref.8. Density parameter $r_s(r_\lambda)$ was obtained in the same way, by comparing the correlation factor for total valence annihilation rates in real metals, $\gamma^{corr} = \lambda_{val}/\lambda_{val}^{IPM}$, extracted from Ref.9 with its jellium analog $\gamma_{jell}^{corr}(r_s)$, parametrized in Ref.10. It is visible, that the values of $r_s(r_0), r_s(r_1)$, and $r_s(r_\lambda)$ are very close each other, determining the effective electron density parameters r_s^{eff} describing electron-positron correlations for valence electrons in simple metals. These effective densities differ from the average valence electron densities in the Wigner-Seitz cell, described by r_s^{free}(marked by stars in Fig.2). This agreement of $r_s(r_0), r_s(r_1)$ and $r_s(r_\lambda)$ is the great advantage in the studies of EMD by positron annihilation method; knowing the total annihilation rates in real metals, λ_{val} and λ_{val}^{IPM} we are able to determine $r_s(r_\lambda) \cong r_s^{eff}$ and read EMD for $p \in FS$ directly from $\rho^{2\gamma}(\mathbf{p})$ as $\rho^e(\mathbf{p}) \cong \rho^{2\gamma}(\mathbf{p})/\epsilon_{jell}^{corr}(p, r_s^{eff})$, avoiding laborious calculations of individual electronic wave functions.

For localized electronic populations (d-electrons, rare-gas-cores or metal surfaces), neither $\alpha(\mathbf{k}j^*)$ nor $\beta(\mathbf{k}j^*)$ can be neglected in the formula (6a). These functions are strongly dependent on the electronic state $\mathbf{k}j$, even if $f_{\mathbf{k}j}$ are not state-selective, as $u_{\mathbf{k}j}(\mathbf{G})$ change as functions of the state $\mathbf{k}j$. Thus, the assumptions about the form of the positron distribution[2] may have a crucial influence on the resulting momentum densities $\rho^{IPM}(\mathbf{p})$ and $\rho^{2\gamma}(\mathbf{p})$. Similarly, application of state-independent two-particle correlations[6] in the ACPAR formula leads to momentum-dependent enhancement factors. Nevertheless, neglect of state-selectivity of $f_{\mathbf{k}j}(\mathbf{r})$ for more delocalized d-electrons is not recommended[7].

3. Methods of extracting enhancement factors from experimental ACPAR data.

In quite a few experimental and theoretical works the real form of $\epsilon(\mathbf{p})$ is approximated by the biparabolic formula[11]

$$\epsilon(p) = a + b \cdot p^2 + c \cdot p^4. \qquad (9)$$

This approximation, however, is not valid for momenta close to the Fermi momentum, even in such a simple model as an electron gas, as discussed in details in Ref.12. Experimental enhancement factors are usually determined based on formula (9) by fitting parameters b/a and c/a to measured ACPAR data within the X^2 test. This procedure is performed either with e.g. the one-dimensional $(1D)$ spectra

$$N(p_z) = \int_{p_z}^{\infty} p\epsilon(p)dp,$$

or with the $(3D)$ density reconstructed from the experiment, $\rho^{rec}(p)$. For momenta close to the Fermi momentum the differences between resulting enhancement factors are dramatic! This fact is illustrated in Fig.3(a), where $\epsilon(p)$ obtained based on the reconstructed density ρ^{rec} (stars) and resulting from $N(p_z)$ (dots) are compared with $\epsilon_{jell}^{corr}(p, r_s = 2)$ (solid line), which was used in the calculations of both ρ^{rec} and $N(p_z)$. The features of enhancement factors presented in Fig.3(a) follow from the fact that the biparabolic approximation (9) falls down near the Fermi momentum. The reconstruction method leads to appreciably smaller difference between the starting "true" enhancement factor and the fitted one (than fitting them to $N(p_z)$) for momenta up to $0.75 p_F$. However, when p approaches the FS, $N(p_z)$ provides much better fit.

In Fig.3(b) theoretical and experimental enhancement factors in simple metals are presented. The considerable differences between experimental results are observed. These deviations of experimental data can be attributed to the following reasons:
1) different methods of data analysis
2) various statistics and resolution functions of the apparatus.

As a conclusion, we suggest to perform analysis of ACPAR data and reconstructed densities simultaneously. In order to diminish the influence of experimental errors on the results, we recommend the expansion of ACPAR data into a series of orthonormal Chebyshev polynomials. This series has the least-squares approximation properties and hence it properly takes experimental errors into account.

Acknowledgments

AR is grateful to the Stefan Batory Foundation (Warsaw, Poland), Idaho State University, and INEL (Idaho, USA) for financial support of participation in SLOPOS5. Work of GKS is done under grant of State Committee for Scientific Research, $N^o 204049101$.

References

[1] For reviews, see, e.g., S.Berko, in *Positron Solid State Physics*, edited by W.Brandt and A.Dupasquier (North-Holland, Amsterdam, 1983), p.102
[2] H.Sormann and M.Sob, Phys.Rev.B **41**, 10 529 (1990); H.Sormann, Phys.Rev.B **43**, 8841 (1991)
[3] For rewievs of jellium results, see, e.g., Ref.7
[4] (a)P.E.Mijnarends and R.M.Singru, Phys.Rev.B **19**, 6038 (1979); (b)B.Chakraborty, ibid. **24**, 7423 (1981); (c)M.Sob, J.Phys.F **12**, 571 (1982)

[5] S.Daniuk, G.Kontrym-Sznajd, J.Mayers, A.Rubaszek, H.Stachowiak, P.A.Walters, and R.N.West, J.Phys.F **17**, 1365 (1987)
[6] J.Arponen, P.Hautojarvi, R.Nieminen, and E.Pajanne, J.Phys.F **3**, 2092 (1973); T.Jarlborg and A.K.Singh, Phys.Rev.B **36**, 4460 (1987)
[7] G.Kontrym-Sznajd and A.Rubaszek, *to be published*
[8] A.Rubaszek and H.Stachowiak, phys.stat.sol.(b) **124**, 159 (1984); Phys.Rev.B **38**,3846 (1988)
[9] S.Daniuk, M.Sob, and A.Rubaszek, Phys.Rev.B **43**, 2580 (1991)
[10] E.Boronski and R.M.Nieminen, Phys.Rev.B **34**, 3820 (1986)
[11] S.Kahana, Phys.Rev. **129**, 1622 (1963)
[12] A.Rubaszek and H.Stachowiak, J.Phys.F **15**, L231 (1985)
[13] G.Kontrym-Sznajd and J.Majsnerowski, J.Phys.:Cond.Matter **2**, 9927 (1990)
[14] (a)S.M.Kim and A.T.Stewart, Phys.Rev.B **11**, 2490 (1975); P.Kubica and A.T.Stewart, Phys.Rev.Lett. **34**, 852 (1975); (b)L.Oberli et al., in *Positron Annihilation*, edited by P.C.Jain et al., (World Scie., Singapore 1985), p.251; (c)J.Arponen and E.Pajanne, J.Phys.F **9**, 2359 (1979); G.Kontrym-Sznajd and J.Majsnerowski, see Ref.13; P.Kubica and A.T.Stewart, Can.J.Phys. **61**, 971 (1983); H.Nakashima, T.Kubota, Y.Murahami, H.Kondo, and S.Tanigawa, in *Positron Annihilation* (1992, in press)

Figure Captions

Fig.1. Relative enhancement factors $\epsilon(p)/\epsilon(0)$ in Na and Mg. ϵ^{IPM} and ϵ^{corr} are displayed by dashed and solid lines, respectively.

Fig.2. Effective electron density parameters in Mg, Li, Na, and K following from comparison of

(a) $\epsilon^{corr}(0)$ with $\epsilon^{corr}_{jell}(0, r_s)$(solid lines)

(b) $\epsilon^{corr}(p_F)$ with $\epsilon^{corr}_{jell}(p_F, r_s)$(dashed lines)

(c) $\lambda_{val}/\lambda_{val}^{IPM}$ with $g(r_s) = \lambda_{jell}(r_s)/\lambda_{jell}^{IPM}(r_s)$(dotted lines).

Values of ϵ^{corr} and λ_{val} in real metals were obtained within LDA (Refs. 13 and 9, respectively) while ϵ^{corr}_{jell} and $g(r_s)$ are results for an electron gas (Refs. 8 and 10, respectively). The average densities of valence electrons, r_s^{free}, are marked by stars.

Fig.3. (a) enhancement factor $\epsilon(p, r_s)$ in an electron gas for $r_s = 2$(solid line) compared with its biparabolic approximation, where the parameters b/a and c/a are fitted to reconstructed density $\rho^{rec}(p)(stars)$ and to $N(p_z)(points)$.

(b) relative enhancement factors in jellium as a function of r_s for $p = p_F$(full line), $p = 0.99 p_F$(broken line), and $p = 0.95 p_F$(chain line). Parameters $(b+c)/a$ fitted to $\rho^{rec}(p)(stars)$ and $N(p_z)(points)$ are compared with experimental values of $(b+c)/a$ extracted from Refs 14(a), (b), and (c) (circles, triangles, and points, respectively).

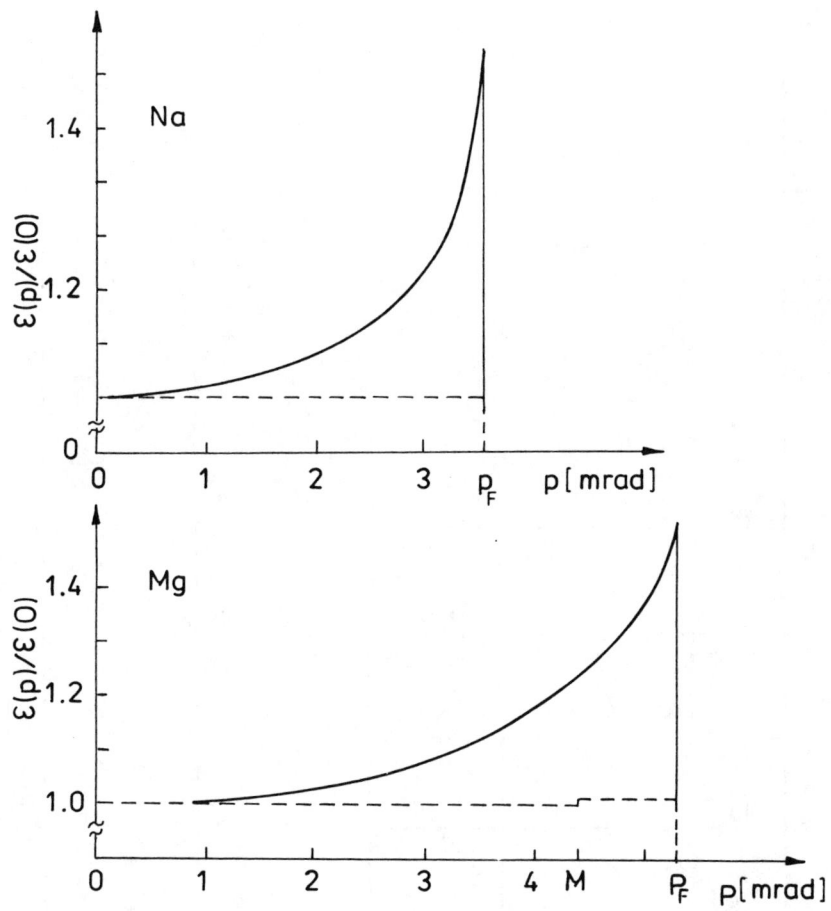

Figure Captions
*Fig.1.*Relative enhancement factors $\epsilon(p)/\epsilon(0)$ in Na and Mg. ϵ^{IPM} and ϵ^{corr} are displayed by dashed and solid lines, respectively.

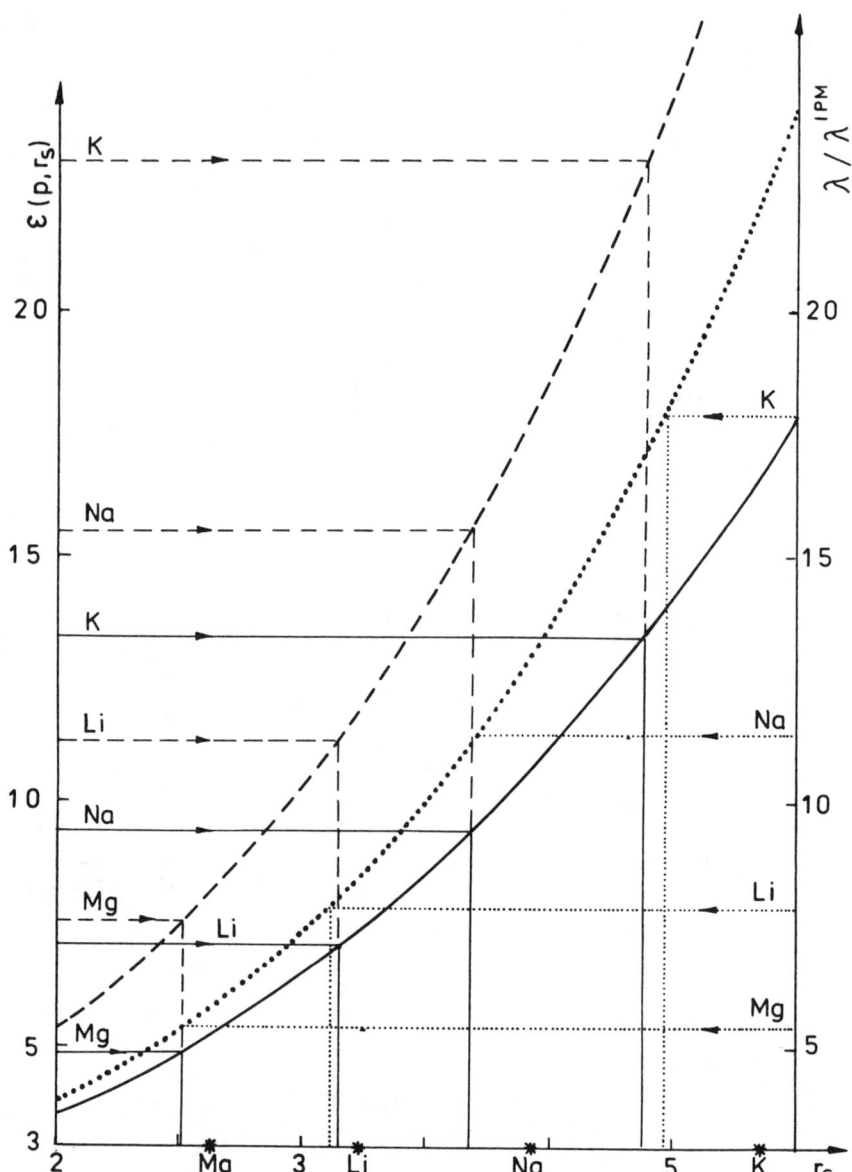

*Fig.2.*Effective electron density parameters in Mg, Li, Na, and K following from comparison of
(a) $\epsilon^{corr}(0)$ with $\epsilon^{corr}_{jell}(0, r_s)$ (solid lines)
(b) $\epsilon^{corr}(p_F)$ with $\epsilon^{corr}_{jell}(p_F, r_s)$ (dashed lines)
(c) $\lambda_{val}/\lambda^{IPM}_{val}$ with $g(r_s) = \lambda_{jell}(r_s)/\lambda^{IPM}_{jell}(r_s)$ (dotted lines).

Values of ϵ^{corr} and λ_{val} in real metals were obtained within LDA (Refs. 13 and 9, respectively) while ϵ^{corr}_{jell} and $g(r_s)$ are results for an electron gas (Refs. 8 and 10, respectively). The average densities of valence electrons, r_s^{free}, are marked by stars.

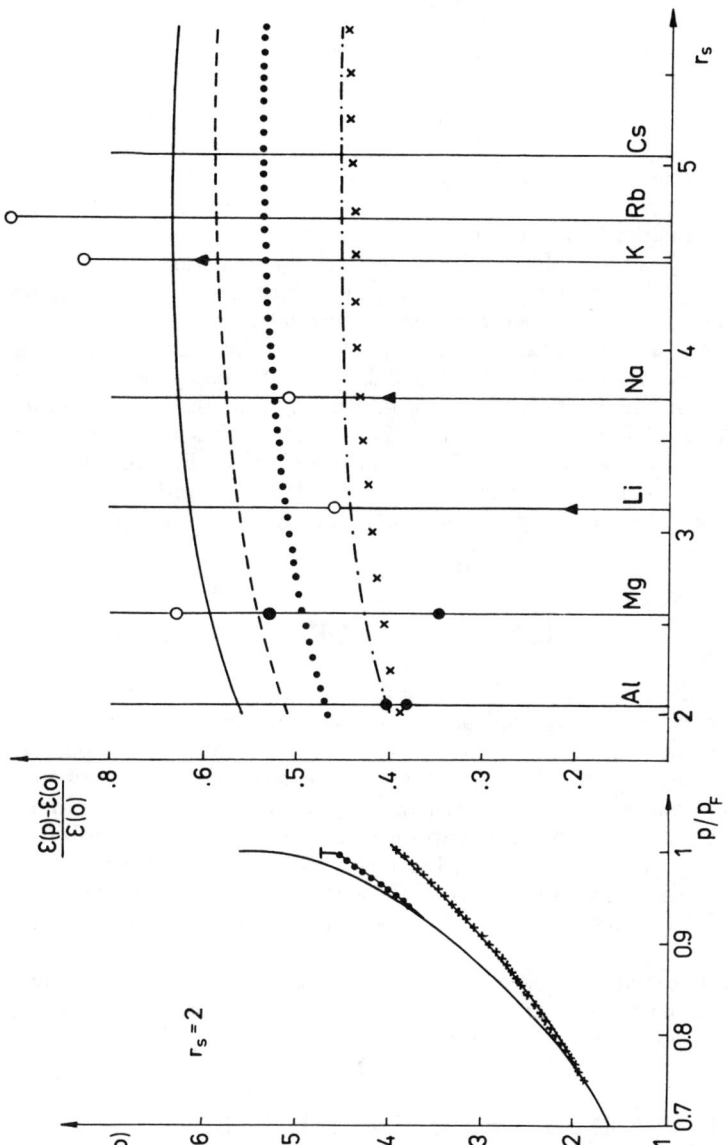

Fig. 3.(a) enhancement factor $\varepsilon(p, r_s)$ in an electron gas for $r_s = 2$(solid line) compared with its biparabolic approximation, where the parameters b/a and c/a are fitted to reconstructed density $\rho^{rec}(p)(stars)$ and to $N(p_z)(points)$.
(b) relative enhancement factors in jellium as a function of r_s for $p = p_F$(full line), $p = 0.99p_F$(broken line), and $p = 0.95p_F$(chain line). Parameters $(b + c)/a$ fitted to $\rho^{rec}(p)(stars)$ and $N(p_z)(points)$ are compared with experimental values of $(b+c)/a$ extracted from Refs 14(a), (b), and (c) (circles, triangles, and points, respectively).

THE CORRELATION BETWEEN LIFETIME AND MOMENTUM OF $e^+ - e^-$ PAIR

K. Krištiaková, J. Krištiak, O. Šauša, M. Morháč, P. Bandžuch
Institute of Physics, Slovak Academy of Sciences, 84228 Bratislava, Czecho-Slovakia

ABSTRACT

The triple coincident measurements of the Doppler broadening of annihilation line and the positron lifetime have been performed on the polymer (polypropylene PP/polyethylene PE) and the high-temperature superconductor ($YBa_2Cu_3O_7$) samples. Two time spectrometers have simultaneously been used with the time resolution of 320 ps and 620 ps, respectively. The energy resolution of common HP Ge detector was 1.5 keV at 570 keV. Preliminary results show that correlation between the centroid of annihilation line and the time channel has been observed in both types of materials. The energy shifts up to +30 eV (for polymer) and up to ~ -200 eV for $YBa_2Cu_3O_7$ have been found. The existence of correlation between the intensity of long-lived component of ~ 1.9 ns and a momentum interval of annihilating $e^+ - e^-$ pair can be concluded in the case of polymer sample.

MOTIVATION OF THE WORK

There are still not completely understood the various modes of positron annihilation in polymers as well as the relation of these modes to the details of polymer morphology [1], e. g. the longest positron lifetime component is attributed to the decay of ortho-Ps by pick-off an electron with a momentum distribution. There is a suggestion that positrons may annihilate with long lifetime without forming Ps. Anyway, we may suppose that the different annihilation modes are reflected in a momentum dependence of annihilation lifetime.

As we know the interesting question, how the momentum spectrum of electrons in a large defect (free volumes, etc.) looks like, is also unanswered by an experiment.

The work reported here is trying to find the answers to the above-mentioned questions using two-parameter, momentum-lifetime measurements of the positron annihilation.

EXPERIMENTAL TECHNIQUE AND RESULTS

The experimental 2-dimensional set-up is shown in Fig. 1. The momentum selection has been done with a HP Ge spectrometer using the Doppler broadening of the 2 γ - annihilation photon line. The time parameter was obtained using conventional fast - fast technique. The coincidence three parameters (τ_1, τ_2, E_γ) event has been formed by the CAMAC system and the home -

developed software. The positron source was ^{22}NaCl sealed between KAPTON foils of 8 μm thickness. This source was sandwiched between a pair of specimens. The Pilot U scintillators of different thickness limited the time resolution to 320 ps and 680 ps, respectively. HP Ge detector was ∼ 50 cc detector with a resolution of 1.5 keV at 570 keV.

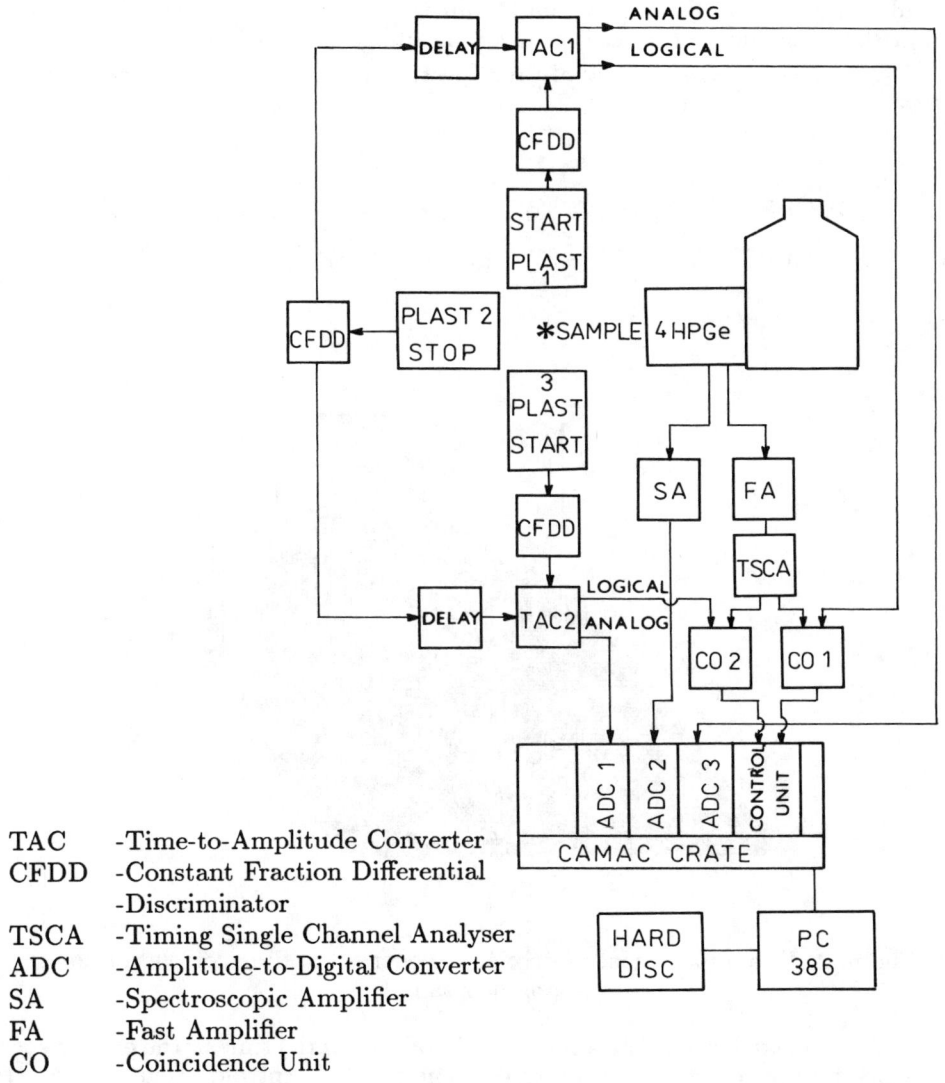

TAC — Time-to-Amplitude Converter
CFDD — Constant Fraction Differential Discriminator
TSCA — Timing Single Channel Analyser
ADC — Amplitude-to-Digital Converter
SA — Spectroscopic Amplifier
FA — Fast Amplifier
CO — Coincidence Unit

Figure 1: Experimental setup

The measurements have been performed with two differents kinds of materials. The polymer polypropylene + polyethylene (PP/PE   1:1) has been used for the study of positronium. The high-temperature superconductor

$YBa_2Cu_3O_{7-x}$ has been used for an investigation of bulk and defects annihilation.

Data were accumulated for 6 days on an 8k x 1k x 1k matrix at several runs. Plots of the reduced matrix are shown in Figs. 2 and 3 for the polymer and YBaCuO sample, respectively. The relevant windows on E_γ or T axis allow a reconstruction of the time spectra or the γ-ray spectra. The time spectra were analyzed by PATFIT - 88 program [2] but the statistics in a window was low so that the assignment of intensities has a meaning in the polymer sample only.

The analysis of integral time spectra provide the following characteristics:

PP/PE   125 ps (21 %)   380 ps (54 %)   2400 ps (25 %)

YBaCuO   181 ps (88 %)   299 ps (12 %)

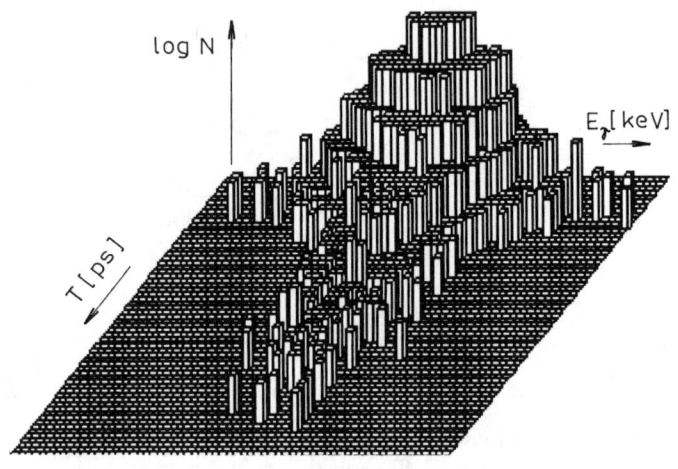

Figure 2: Two-dimensional matrix of the energy vs. time correlation for the polymer sample

Due to low statistics the centroid of each of the time or the energy spectra has been calculated and discussed. The shift of centroid of energy spectra means that the momentum distribution changes its shape and / or its maximum. The centroid of complex time spectra has no such simple interpretation but its change can indicate a dependence of relative intensities I on the momentum of annihilating pair.

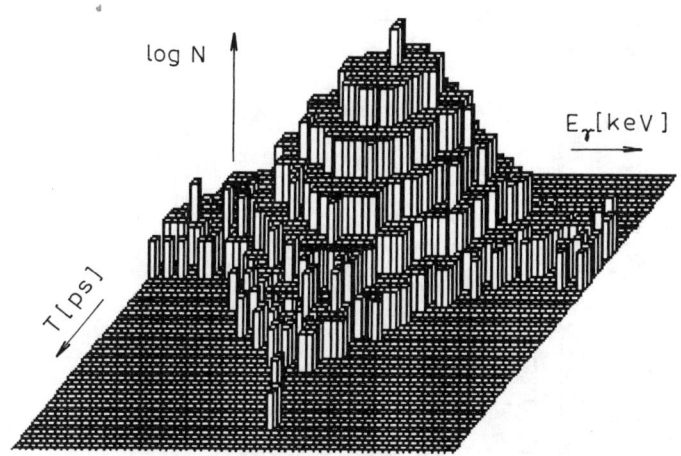

Figure 3: Two-dimensional matrix of the energy vs. time correlation for the high temperature superconductor $YBa_2Cu_3O_7$ sample

POLYMER RESULTS

In Fig 4. the centroid of the annihilation line is plotted for several time intervals. It indicates the change of the electron momentum distribution in place where o - Ps is sitting relative to the short - and mediate - component.

This conclusion is supported by Fig. 5 where a shift of the centroid of long - lived time window is shown. This result is similar to the result published by Mac Kenzie and Mc Kee [1].

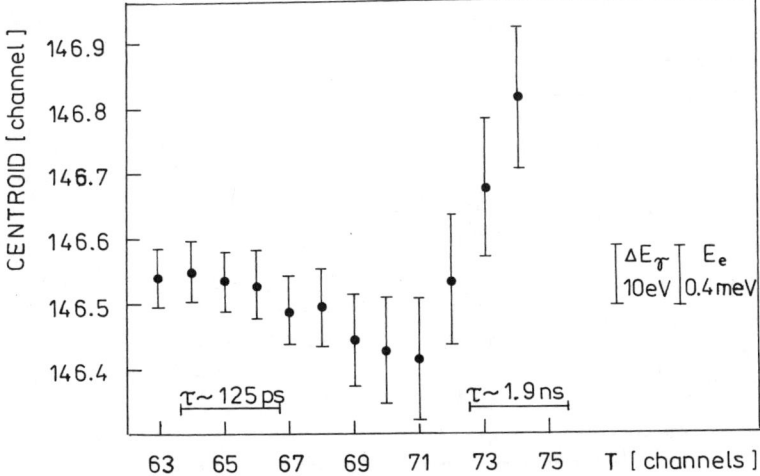

Figure 4: The position of centroid of the annihilation line for the various time channels ($\Delta T = 105$ ps), polymer sample

154 Lifetime and Momentum of e^+-e^- Pair

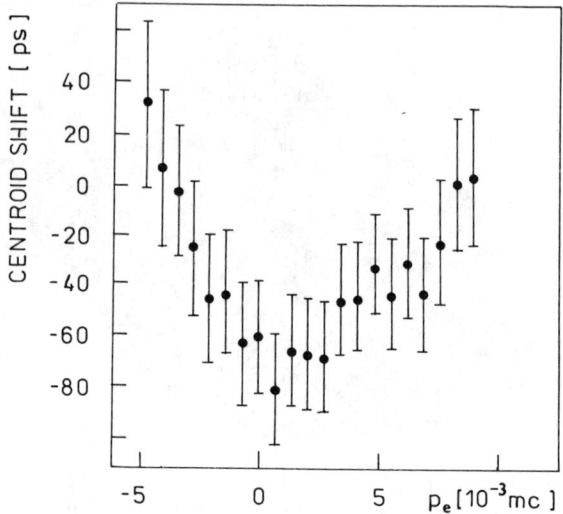

Figure 5: The position of centroid of the time component ~1.9 ns vs. momentum of e^+ - e^- pair normalized to zero for wings of annihilation line

In Fig. 6 we plotted the intensity of the long component I_3 as a function of the momentum interval calculated from the Doppler shift. It shows that the intensity of I_3 depends on momentum. It means that the electron surroundings of annihilation places is different.

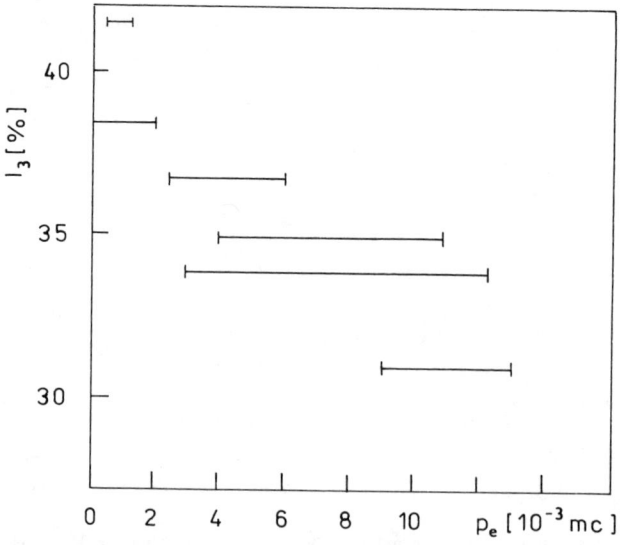

Figure 6: The dependence of the intensity of the third component (τ_3 ~1.9 ns) on a momentum of e^+ - e^- annihilating pair

YBaCuO RESULTS

In Fig. 7 the shift of annihilation line centroid as a function of the time is shown. It once more indicates that the positron annihilating from the metal vacancies and from interfacial defects is feeling a quite different electron momentum distribution.

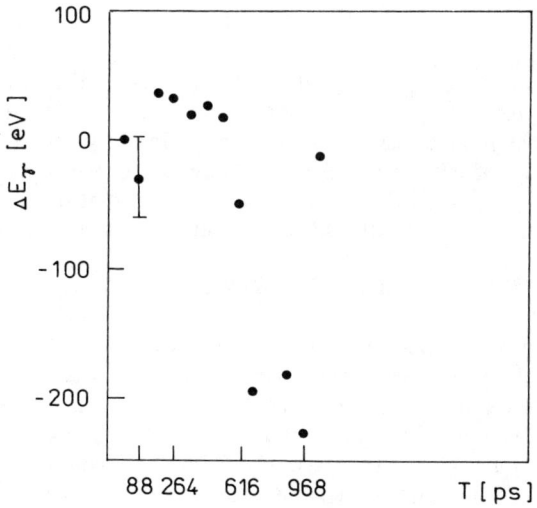

Figure 7: A shift of annihilation line centroid vs. various time channels, YBaCuO sample

CONCLUSION

Our preliminary results indicate that 2-parameter measurements are able to provide information on the changes of the electron momentum distribution in a bulk as well as in defects. In such case it would be very interesting to look for structure of defects near the surface and in multilayer systems. Such experiments can be done with a positron beam from a radioactive source.

REFERENCES

[1] I. K. Mac Kenzie, B. T. A. Mc Kee, Appl. Phys. **10**, 245 (1976).

J. R. Stevens, in "Methods of Experimental Physics", vol. 164, (Academic Press, N.Y. 1980).

[2] P. Kirkegaard, N. Y. Pedersen, M. Eldrup, RISØ-M-2740.

CREATION AND EVOLUTION OF VACANCY CLUSTERS IN SOLIDS UNDER IRRADIATION. THEORY AND COMPUTER SIMULATION.

Alexander I. Melker
Department of Metal Physics, Physics & Mechanics Faculty
St.Petersburg State Technical University
St. Petersburg, 195 251, Russia

ABSTRACT

Radiation damage induced by neutral and charged particles is decomposed into four stages: dynamic, relaxation, diffusion, and slow fracture. For the first stage formulae describing a space structure of collision cascades, the energy of splitting a cascade into subcascades, formation probabilities for vacancy clusters are considered. Other stages are analyzed by molecular dynamics. The primary defects created at the dynamic stage transform into voids, stacking fault tetrahedra, dislocation loops, and cracks.

INTRODUCTION

In an explicit form, radiation fracture is observed when plasma is interacting with the first wall of a fusion reactor. Fracture is also induced by pulsed electron and laser irradiation. The different types of radiation fracture can be classified as follows: sputtering, blistering, swelling as well as neutron-, electron-, and laser-induced fracture. General features and peculiarities of these types of fracture are considered in Reference 1. It is difficult to investigate these phenomena because they cover a wide time interval from 10^{-15} to 10^6 sec. They also encompass regions with dimensions ranging from an interatomic distance to a macroscopic one which is commensurate with target thickness. Nevertheless, one can distinguish four time stages in radiation damage[1] as follows:

1) dynamic stage. At this stage, collision cascades appear in solids, caused by primary knock-on atoms (neutron induced fracture), implanted ions, or relativistic electrons. A collision cascade, in its turn, creates interstitials, vacancies, and athermal vacancy clusters. The time scale of the dynamic stage is approximately r_o/v where r_o is the interatomic distance in solid and v is the velocity of a primary knock-on atom.

2) relaxation stage. Here the damaged crystalline or amorphous structure relaxes. As a result, the defect configurations are transformed into embryos of dislocations, voids, cracks, and new phases. The time scale of the relaxation stage is equal to a period of atomic vibrations in a solid.

3) diffusion stage. This stage can be divided into two substages of fast and slow time. The interstitials begin to migrate first. The mobile defects, having a larger migration energy, e.g. implanted ions, vacancies, divacancies, are annealed at the slow time substage. The time scale of each substage is defined by the jump time of the most mobile defect occurring therein.

4) slow fracture. Here one observes accretion of cracks and voids. The time scale is roughly equal to the interval between two successive jumps of a crack or void.

It should be noted that sometimes a smaller number of stages materializes, e.g. in neutron-, electron-, and laser-induced fracture. In this paper, the dynamic and relaxation stages of radiation fracture are considered in detail. A glimpse of the diffusion and fracture stages is also given.

DYNAMIC STAGE

Cascade Structure

During 14 MeV neutron irradiation the maximum PKA (primary knock on atoms) energies for solid targets are in the 0.5 to 1.0 MeV range. Radiation damage produced by such PKA is usually simulated by bombarding a material with its own ions. Observations by TEM (transmission electron microscopy) have shown that solids irradiated by their own ions or 14 MeV neutrons contain defect clusters whose appearance is attributed to splitting of collision cascades into subcascades.

The high-energy PKA collisions in solids can be divided into two classes. For an impact parameter $\rho < r_o$ where r_o is the interatomic distance in a solid, the moving atom scatters at large angles. In contrast, if $\rho \leq r_o$, scattering occurs at small angles. The scattering type determines the subsequent trajectories of colliding atoms and, as a consequence, the radiation damage structure.

Let us represent a collision cascade as a graph whose vertices are the cascade branching points in which large-angle scattering takes place, and whose arcs are the trajectories of bombarding and knock-on atoms. Each arc is connected with a high-energy branch of a cascade. Each high-energy branch has the following structure: the momentum of any knock-on atom produced by small-angle scattering of the high-energy atom is normal to the high-energy trajectory. The vacancy created by the high-energy particle constitutes the take-off point of a knock-on atom and the initiation of a low-energy branch of the cascade. This scheme agrees well with numerous results of computer simulation made by molecular dynamics (Fig. 1).

Now we have the mathematical model of the radiation damage and can try to describe it quantitatively. First we seek the distribution function for graph arcs, the probability $P(\ell)$ that the atom moving through a solid with initial energy E traverses a path of length ℓ without large-angle scattering. This probability can be found from the equation

$$- P'(\ell) / P(\ell) = \eta(\ell) \tag{1}$$

where $\eta(\ell)$ is the function which is known in reliability theory as the failure danger. In our case it is the danger of large-angle scattering. Usually, one assumes that

$$\eta(\ell) = \eta_o = \text{constant} \tag{2}$$

resulting in the exponential scattering law

$$P(\ell) = \exp(-\eta_o \ell) \tag{3}$$

with the constant mean free path

$$1/\eta_o = \lambda(E_o) = 1/\eta_o \, \sigma(E_o) \tag{4}$$

Here η_o is the number of atoms per unit volume and $\sigma(E_o)$ is the large-angle total cross section.

However, the exponential law is rather rough if we deal with the high-energy branches of the collision cascade where inelastic losses are large. Let us take into account the inelastic collision energy losses in the form

$$- dE/d\ell = \text{const. } E^{\frac{1}{2}} \tag{5}$$

and approximate the interatomic potential for the sake of simplicity by the function

$$\phi(r) = \text{const}/r^2 \tag{6}$$

This is a reasonable approximation for the PKA energy interval considered. In this case the energy depends on the distance ℓ as

$$E(\ell) = E_o (1 - \alpha \ell)^2 \tag{7}$$

and we can get the concise danger of large-angle scattering by

$$\eta(\ell) = \eta_o /(1 - \alpha \ell)^2 \tag{8}$$

where $\alpha = \text{const}/E_o^{\frac{1}{2}}$.

Contrary to the old simplified danger of large-angle scattering, the new danger is based upon the more strict consideration of the charged particle transmission and gives birth to the more precise scattering law

$$P(\ell) = \exp(-\eta_o \ell /(1 - \alpha \ell)) \tag{9}$$

with the mean free path $<\lambda>$ which is rather a complicated function.[2] This new scattering law resembles partly the Weibull distribution

$$P(\ell) = \exp(-\eta_o \ell^{\gamma}) \tag{10}$$

where η_o and γ are parameters given in Reference 3. The Weibull distribution is widely used in reliability theory because it has two parameters and therefore is more flexible than the exponential law.[3] However, the Weibull failure danger is a monotonic function

$$\eta(\ell) = \gamma \eta_o \ell^{\gamma-1} \tag{11}$$

The new distribution is even more flexible than the Weibull one because a nonmonotonic character of the new failure danger (Equation 8) agrees better with experience.[3] Besides, it gives the fixed boundary $\ell_m = 1/\alpha$ for particle penetration (lifetime in the reliability theory).

Using the new scattering law based on the concept of large inelastic losses, we deduced the equation for the mean number of the high-energy branches:[2]

$$\beta'(\ell) / \beta(\ell) = -2q(\ell) \tag{12}$$

where $q(\ell) = Q(\ell)$ is the collision probability density and $Q(\ell) = 1 - P(\ell)$ is the collision probability. The equation has the solution

$$\beta(\ell) = \exp 2[Q(\ell_s) - Q(\ell)] \tag{13}$$

Here the distance ℓ_s is the splitting distance which corresponds to the PKA minimum energy below which the collision cascade does not split. On averaging $\beta(\ell)$ in the interval $(0,\ell_s)$ one can obtain the mean number of the subcascades generated by the PKA with the initial energy E_o

$$\overline{\beta} = \tfrac{1}{2} [\exp 2Q(\ell_s) - 1] \tag{14}$$

For $Q(\ell_s) = 1$, we get $\overline{\beta} = 3.2$. This agrees well with the experimental value, $\beta_{exp} = 3.3$, which is observed for vanadium after irradiation by 14 MeV neutrons.[4]

In general, the subcascade size depends on the energy of the PKA, its free path during large-angle scattering collisions, and the number of subcascades in a collision cascade. Strictly speaking the subcascade size distribution must be obtained as a folding of the distributions of these quantities. However, such an evident approach leads to big mathematical difficulties even when the cascade does not split and coincides with a subcascade.[5] Therefore, we consider a simpler way to estimate the size distribution.

All the possible graphs, the arcs of which are the high-energy branches of subcascades, define the ways of cascade splitting into subcascades. For example,[2] there are only two different graphs if $\beta = 5$, whereas for $\beta = 7$ the number of different graphs is five. Let us write down the formal expression

$$\overline{q} = \Sigma \, q(\beta) \cdot \beta / \Sigma \, \beta \tag{15}$$

where the summation is extended over all the elements of the set of β. This expression, when written as a function of ℓ defines the subcascade density distribution. The detailed analysis shows[2] that $q(\ell)$ does not depend on the PKA energy. That agrees well with the experimental results published for gold.[6] $q(\ell)$ can be approximated by the formula

$$q(\ell) = \eta_s / (1-\alpha_s \ell)^2 \, \exp(-\eta_s \, \ell / (1-\alpha_s \ell)) \tag{16}$$

where $\eta_s = n \, \sigma(E_s)$, $\alpha_s = \text{const}/E_s$, and E_s is the PKA minimum energy below which the collision cascade does not split.

Except for a few metals, nothing is known about the experimental value of this energy. However, the splitting threshold can be calculated from the conservation of kinetic energy for a head-on collision of a PKA with a solid atom.[7,8] The result is

$$E_s \approx 1.8 \, Z^2 \, a_o/a \, (a/r_o)^2 \; [\text{keV}] \tag{17}$$

where Z is the charge number, $a_o = 0.0529$ nm is the radius of the first electron orbit in a hydrogen atom, a is the screening constant for the Lindhard-Nielsen-Scharff potential, and r_o is the interatomic distance in a solid. Strictly speaking, the resulting formula contains the adjustable parameter $\delta = (6/r_o)^2 \ll 1$, where 6 is the collision diameter. Taking the experimental value of the splitting energy for gold, $E_s \approx 50$ keV, as a benchmark, one finds good agreement with the theoretical estimate calculated for $\delta = 5\%$. The splitting energies calculated with the 5% tolerance for Si, Fe, Ga, As, Mo, and U are equal to 4.7, 12, 13, 11, 22, and 79 keV, respectively.

In order to describe the cascade structure completely, one must also take into account temperature. It is found by computer simulation [9] that the temperature of a material does not influence the high-energy trajectories of knock-on atoms. At the same time, increasing the temperature decreases dimensions of the low-energy branches which are generated by the secondary knock-on atoms with momenta normal to the high-energy trajectories. References 7 and 8 suggest the following simple and effective method to calculate the temperature influence. First we present the mean square root displacement of an atom in a solid at zero temperature and at temperature $T>\Theta_D$:

$$\langle u_o^2 \rangle = \hbar^2/m\, K\, \Theta_D \qquad \langle u_T^2 \rangle = 9\hbar^2\, T/m\, K\, \Theta_D^2 \qquad (18)$$

Here $\hbar = h/2\pi$, h and k are the Planck and Boltzmann constants, Θ_D is the Debye temperature, m is the atom mass. From these formulae, it follows that we can realize the displacement $\langle u_T^2 \rangle$ even at zero temperature if the atom mass changes as

$$m(T) = (\Theta_D/9T)\, m(0) \qquad (19)$$

where m(T) is the effective mass of a target atom. Using the temperature-dependent mass, we can reformulate the expressions which describe the collision process, and get the temperature-dependent magnitudes, such as the collision diameter, mean free path, etc. For example,

$$\lambda(T) = 2\, \lambda/(1 + 9T/\Theta_D) \qquad (20)$$

which agrees well with the computer simulation results obtained for the crowdion in α-iron.[8,9]

Vacancy Clusters at the Dynamic Stage

It is well known that the properties of irradiated solids depend not only on the quantity of radiation defects but also on the structure of damaged regions created by primary knock-on atoms. Computer simulation shows[9] that the development of any collision cascade is accompanied by the appearance of not only single vacancies but also of vacancy clusters. The share of vacancies which enter into clusters reaches two thirds on an average.[9]

The simplest vacancy cluster is a divacancy. In the binary-collision approximation an atom moving through a target can produce a divacancy in two ways: (i) by successively knocking two adjacent atoms out of lattice sites, (ii) by knocking out of a lattice site an atom which in turn displaces another atom out of the neighboring lattice site.

As before, we can use a graph to represent a displacement cascade which produces isolated vacancies and vacancy clusters in a solid (Fig.2). In this case, each arc corresponds to a part of a trajectory of a moving atom and each arrow points out a direction of movement. Each vertex coincides with a site in a crystal lattice and each symbol over an arc denotes a moving atom. The number of a site is marked by a figure under a vertex. The length of the first arc (0,1) is arbitrary. The length of the second arc (1,2) equals the interatomic distance in a crystal lattice. If vertex 1 is encircled, it means that the knock-on atom 1 continues the process of defect formation. By joining the divacancy diagrams to the monovacancy ones, we obtain all the processes leading to formation of trivacancies which have a chain structure. However, there is the fifth diagram representing a new mechanism of the parallel defect formation. Here atom A produces a vacancy in site 1 and then together with atom 1 makes two vacancies in the neighboring sites 2 and 3. As a result, a trivacancy of a dendritic structure is formed. By combining all the formation diagrams of monovacancies and trivacancies one can construct the formation diagrams of tetravancies and so forth.

The structure of the diagrams is such that they compile the set of all the formation processes of a vacancy cluster which can be separated into the uncoupled subsets whose elements are the formation processes of clusters of lesser dimensions. Considering each subset as a group of correlated processes independent of the processes of other subsets, one can get the group representation of the athermal formation processes for multivacancies. For example, the group representation for multivacancies up to a hexavacancy can be written as follows[10]

$$\begin{aligned} V_2 &= V_1 \cdot 2V_1 \\ V_3 &= V_1 \cdot 2V_2 + L_3 \\ V_4 &= V_1 \cdot 2V_3 + L_3 \cdot 4V_1 \\ V_5 &= V_1 \cdot 2V_4 + L_3 \cdot (4V_2 + 2V_1 \cdot 2V_1) + 2L_5 \\ V_6 &= V_1 \cdot 2V_5 + L_3 \cdot (4V_3 + 2V_1 \cdot 2V_2 + 2V_2 \cdot 2V_1) + 2L_5 \cdot 6V_1 \end{aligned} \quad (21)$$

Here V_n and L_n denote the subsets with the chain and branching structure, respectively.

The mean number of Frenkel pairs created by an atom moving with the energy E can be gotten from the Kinchin-Pease equation

$$v = E / 2E_d \quad (22)$$

where E_d is the displacement energy. If $v = 1$, the moving atom produces only one vacancy, but if $v = 2$, it can make either two isolated vacancies or one divacancy. The distribution of the vacancies can be found from the normalizing condition

$$p_1 + p_2 = 1 \quad (23)$$

If $v > 2$, then the analogous reasoning leads to the conclusion that

$$\Sigma \, p_v = 1 \quad (24)$$

Suppose that the probability of placing a vacancy cluster into the vth subset is determined by the group representation of the formation processes for the corresponding cluster. Denoting $p_1 = p$, one can get the following equations for the probability

$$\begin{aligned} v &= 1, \; p = 1 \\ v &= 2, \; p + 2p^2 = 1 \\ &\quad\cdots\cdots\cdots\cdots \\ v &= 5, \; p + 2p^2 + 5p^3 + 14p^4 + 45p^5 = 1 \end{aligned} \quad (25)$$

By averaging p(v) over the energy spectrum of knock-on atoms in the interval $v = 1$ to n, one can get the vacancy cluster distribution which is in a good agreement with the computer simulation[9] as well as with the experimental results obtained by field-ion microscopy [11] and electron-positron annihilation.[12,13]

The distribution obtained is of importance not only for bulk studies but also for surface investigation. For example, in order to estimate damage to a first wall of a fusion reactor, one must know what kinds of radiation defects are produced (their type and number) in the surface layer of a material. It is also necessary to know the distribution of the imbedded ions with distance, since these ions also strongly influence how the damaged region develops.

Usually the depth distribution of the defects produced by the ion which was incident at the target surface with the initial energy E_o is written in the form

$$C(x;E_o) = \int f(E,x;E_o) \, \nu(G(E,x),x) \, dE \tag{26}$$

where $f(E,x;E)$ is the energy-space distribution of the ions, $G(E,x)$ is the energy share of elastic collisions, $\nu(G(E,x),x)$ is the number of displaced atoms or the cascade function. Taking into consideration the different types of defects one must rewrite the distribution in the form

$$C(n,x;E_o) = \int f(E,x;E_o) \, \nu(n,G(E,x),x) \, dE \tag{27}$$

Here the additional argument n characterizes the formation of vacancy clusters. Unfortunately, no reliable analytical methods are currently available for solving this equation. However, we can use Monte Carlo methods on an individual collision scheme in which the integrand functions are calculated from first principles on the basis of the elementary events statistics. This directly gives the depth distribution of point defects, their clusters, and imbedded particles.[14] It is interesting to note that the imbedded light ions distribution can be approximated with good accuracy by the formula which defines the concise danger of large-angle scattering. Moreover, using these results as the input data for kinetic equations it is possible to describe quantitatively, and therefore to explain, the nucleation of voids due to athermal vacancy clusters[15] and the diffusion stage of radiation damage leading to blistering.[16]

Thus far we have not taken account of the crystal structure in an explicit form. However, it is important sometimes, e.g. for ion implantation in semiconductor crystals, to know what kinds of vacancy clusters are formed. In this situation, one may superimpose the formation diagrams of vacancy clusters on a crystal structure. Consider the simplest vacancy cluster, namely divacancy. In this case there is no need for the full crystal structure: it suffices to include only the crystallographic direction. Assume that the two atoms collided can be approximated by hard spheres with diameter $\mathfrak{b}$. Using classical mechanics one gets the following cross section for divacancy formation:[17]

$$\sigma(V_2,\theta) = \sum_{j=1}^{2} \left\{ 1 - \cos\left[\frac{\mathfrak{b}}{D_{hk\ell}}\left(1 - \frac{1}{\epsilon \cos^2\theta - \delta_{j,2}}\right)^{1/2}\right] \right\} \sigma(V_1) \tag{28}$$

Here Θ is the angle between the [hkℓ] direction and the direction of the incident atom momentum, $D_{hk\ell}$ is the interatomic distance in the [hkℓ] direction, $\epsilon = E/E_d$, where E_d is the threshold for the displacement of an atom from a lattice site, $\delta_{j,2}$ is the Kronecker symbol, j is equal to 1 or 2 for the first and second mechanism, respectively, and $\sigma(V) = \mathfrak{b}(1-1/\epsilon)$ is the total cross section for vacancy formation.

This expression is valid for $\Theta \leq \Theta_d$ where $\Theta_d = \arccos[(2-\delta_{j,2})/\epsilon]$ is the maximum displacement angle for the divacancy formation. When integrated over the displacement solid angle Ω_d, divided by the number of the different orientations of the divacancy in a crystal, $Z_{hk\ell}$, the expression gives the effective total cross section for the divacancy formation

$$\sigma(V_2, hk\ell) = \frac{Z_{hk\ell}}{4\pi} \int_{\Omega_d} \sigma(V_2, \theta) \, d\Omega \qquad (29)$$

To take account of the dependence of the collision diameter δ on the energy of the incident atom, it is necessary to describe the atomic interaction by an appropriate potential, e.g. by the Born-Mayer potential. In this case

$$\delta = a \, \ell n \, (2A/E) \qquad (30)$$

where a and A are the constants.

The calculated values for the divacancy formation are in a good agreement with those found by molecular dynamics both for metals[17] and for semiconductors[18] in the energy range to $E \leq 1$ keV. However, for high-energy cascades the hard-sphere approximation is incorrect because it assumes that all the scattering angles are equally probable, whereas with increasing energy the incident atom begins to scatter predominantly through small angles. Nevertheless, using the momentum approximation instead of the hard-sphere one, it is possible to develop the expression which characterizes the divacancy formation on small-angle scattering[19]

$$\sigma(V_2, hk\ell) = Z_{hk\ell} \, D^2_{hk\ell} \, G(\rho_d/D_{hk\ell}) \, / \, 2 \qquad (31)$$

Here ρ_d is the displacement impact parameter which corresponds to the displacement momentum $p_d = (2mE_d)^{1/2}$ and substitutes for the diameter δ in the expression for $\sigma(V_1)$. The function $G(x)$ is tabulated in Reference 19. The two expressions (29) and (31) cover all the energy range for the divacancy production in metals[17,19,20] and semiconductors[18,21] which can be realized experimentally.

We have considered the simplest vacancy cluster, a divacancy, rather carefully. For larger clusters, there is no need to investigate the collision mechanics what is, in general, a tedious procedure; instead, one can proceed as follows. Let us write down the effective cross section of the divacancy formation

$$\sigma(V_2, hk\ell) = w_{12}(hk\ell) \cdot \sigma(V_1) \qquad (32)$$

Here $w(hk\ell)$ is the probability of the transition from a monovacancy to the divacancy along the $\langle hk\ell \rangle$ direction. We put each divacancy in a correspondence with the displacement cone which has the following features: the cone top coincides with one of the vacancies which constitute the divacancy, the symmetry axis coincides with the $\langle hk\ell \rangle$ direction, and the top angle is equal to $2\Theta_d$ where

$$\Theta_d = \text{arc sin} \, (2\rho_d/D_{hk\ell}) \qquad (33)$$

Draw a sphere around the cone top. The displacement cone cuts the displacement solid angle Ω_d out of the sphere so we can write the transition probability in the form

$$w_{12}(hk\ell) = Z(hk\ell) \cdot (\Omega_d/4\pi) \cdot \langle w \rangle \qquad (34)$$

Here $\langle w \rangle$ is the probability that the beam particle hit the displacement cone.

Now we can substitute any multivacancy for a diverging bundle of the divacancies which are equivalent to the displacement cones having a common top. For example, in the general case any trivacancy configuration can be changed for a combination of two displacement cones, any tetravacancy configuration is equivalent to three cones and so on. If these cones overlap, it means that a projectile creates an athermal multivacancy. Therefore we need only to calculate a solid angle of overlapping, which is a more simple geometric problem than the initial rather sophisticated problem of collision mechanics.[22]

And what is left of physics? Physics is present in the total cross section of vacancy formation

$$\sigma(V_1) = \pi \rho_d^2 \tag{35}$$

where the displacement impact parameter ρ_d is a complicated function of the projectile energy E, the displacement threshold and the interaction potential, the latter factor being the most important.[22] For gold, as an example, no monovacancies or divacancies are generated at all. The majority of trivacancies are 120 and 60 ones, and the chain tetravacancies prevail over other configurations[22] whereas in GaAs divacancies dominate over trivacancies.[21] Here we classify trivacancies by the largest angle of an isosceles triangle.

RELAXATION AND GROWTH

One central problem of the radiation physics of solids is the influence of point defects on the properties of materials. Single defects are usually those mainly considered in this respect. However, as is clear from the previous section, the majority of radiation vacancies occur as clusters and the proportion of these becomes predominant as the radiation damage evolves further.[23] It is shown that athermal vacancy clusters can turn into the germs of the dislocations, voids, or cracks in the course of evolution.[1] Currently the evolution of point defects and their clusters is evidenced mainly by computer simulation because analytical methods are inefficient for such an investigation. As for experimental methods they can be used to identify a defect if its size exceeds approximately 1 nm. For example, in f.c.c. metals such as nickel, one can distinguish dislocation loops, stacking fault tetrahedra as well as voids.[24-31] On the other hand, molecular dynamics makes it possible to explore nucleation and growth of defects if their dimensions are less than approximately 0.5 nm. Usually, these defects are small vacancy clusters. Consequently, computer simulation allows to close the size gap between analytical and experimental methods.

There is no need to discuss the calculation procedure of molecular dynamics and its application to solids which is thoroughly considered elsewhere.[32] Instead, we review the most interesting results which illustrate the possibilities of computer simulation with respect to evolution of a radiation damage structure.[33,34] We shall restrict our consideration to two metals: nickel and aluminium because of great importance to industry being the basis for commercial alloys.

Stacking Fault Tetrahedra and Dislocation Loops

According to the atomistic calculations[35] stacking fault tetrahedra arise in f.c.c. metals due to condensation in a (111) plane if the atoms from one side of this plane displace to a greater extent than the opposite side atoms (asymmetric crystal relaxation). Consider the growth of a stacking fault tetrahedron. The germ of any stacking fault defect is a tetrahedral trivacancy (Fig. 3a). Such a configuration can be formed if a f.c.c. crystal has three vacant sites being the nearest neighbors to each other, so that they form a 60 vacancy triangle. During the asymmetric relaxation one of the atoms, equidistant to the three vacant sites, displaces strongly to the center of the vacancy triangle. The final configuration represents a four-vacancy tetrahedron, the center of which is occupied by the strongly displaced atom.

One can take this defect as an elementary bricklet and build more complex defects by joining such bricklets. For example, the least stacking fault tetrahedron consists of the four tetrahedral trivacancies or four bricklets (Fig. 3b). It can be formed during the asymmetric relaxation of a crystal containing six adjacent vacancies which lie in a (111) plane and compose an equilateral triangle. The details of this process are given in Reference 36. The next size stacking fault tetrahedron (Fig. 3c) can be made from a ten-vacancy equilateral triangle or ten bricklets. If this scheme is true, one can find the energy of a stacking fault tetrahedron analyzing the geometric structure of the tetrahedron.

Let the equilateral tetrahedra be numbered 0,1,2,...,i. Denote the quantity of the strongly displaced atoms by m, the binding energy of the vacancy cluster which contains n vacancies by E^B_{nv} and the specific binding energy by ε^B_{nv}. Then it is not difficult to get the following relations

$$E^B_{nv} = m\ E^B_{3v} = 3\ m\ \varepsilon V^B_{3v}$$

$$\varepsilon^B_{nv} = E^B_{nv}/n = 3\ (m/n)\varepsilon^B_{3v}$$

$$m_{i+1} = m_i + n_i, \quad m_o = 1$$

$$n_{i+1} = n_i + i + 2, \quad n_o = 3 \tag{36}$$

Here ε^B_{3v} is the specific binding energy of the tetrehedral trivacancy which equals to 0.24 eV for nickel.[36] Using these formulae one gets the specific binding energies 0.47 and 0.72 eV for the six and ten-vacancy tetrahedron, respectively, what is in a good agreement with the values calculated by molecular dynamics. Knowing only the binding energy of a small stacking fault tetrahedron and the vacancy formation energy E it is possible to calculate the stacking fault energy from the expression

$$\gamma_{sf}\ \ell^2\sqrt{3} = nE^F_v - E^F_{nv} \tag{37}$$

where ℓ is the length of the tetrahedron edge. Taking $E^F_v = 1.76$ eV,[37] we get $\gamma_{sf} = 350$ and 200 mJ/m for n = 6 and 10, respectively. These values agree well with the experimental results, viz. $\gamma_{sf} = 125$-400 mJ/m.[38,39]

In contrast to stacking fault tetrahedra, dislocation loops are formed as a consequence of symmetric crystal relaxation which embraces both sides of an initial vacancy cluster lying in one of the (111) planes (Fig. 4). The specific binding energy of the hexagonal loop becomes larger if it grows, but this increase is less pronounced than for the stacking fault tetrahedron. We have investigated the growth of these defects by removing the nearest atoms what is equivalent to vacancy absorption. In the course of cluster development there appear mixed configurations which contain the elements both of stacking faults and of dislocation loops. Their specific binding energy oscillates between the values for the stacking fault tetrahedra and dislocation loops. For example, the strong lattice relaxation around the vacancy rhombs lying in a (111) plane creates the configuration consisting of two adjacent stacking fault tetrahedra (Fig. 5).

It should be noted that the perfect configurations of stacking fault tetrahedra and dislocation loops can be created only for certain number of vacancies which form a regular structure: an equilateral triangle, rhomb, or hexagon.

Voids

It is known[13] that there are two essentially different kinds of vacancy clusters: two and three dimensional. This leads to dramatic consequences. Increasing the volume of a material, three dimensional volume clusters of vacancies induce swelling which changes mechanical properties and construction dimensions. In contrast to two-dimensional defects, only a moderate body of work is devoted to investigating atomic mechanisms of void nucleation.

An elementary vacancy cluster which has a compact volume configuration is a tetravacancy in the form of a tetrahedron (Fig. 6). This cluster can be gotten athermally during irradiation by knocking the inner atom out of the tetrahedral trivacancy shown in Fig. 4. The crystal relaxation around this void is small and its binding energy is negative. We have investigated the growth of this void adding vacancies, i.e. removing atoms 1, 2, and 3 successively (Fig. 6). The first two configurations got (Figs. 6b, 6c) are also unstable, the atomic displacements around these configurations are small as before. Nevertheless, adding the third vacancy changes the configuration radically. Rotating around its center, the triangle of the nearest atoms (4, 5, and 6) enters the empty zone and forms a stacking fault (Fig. 6d). This configuration is very stable. If to remove the stacking fault atoms in series, another group of the nearest atoms (7, 8, 9, and 10 in Fig. 6d) displaces into the empty zone and forms the elementary stacking fault tetrahedron (Fig. 6e) the edge length of which is near to an interatomic distance of a perfect lattice. The binding energy of the configuration got is high, the displacements of surrounding atoms are small.

One can consider the empty-full cube as a void with a nucleus inside which stabilizes its space form. The vacancy cube can be inscribed into a truncated octahedron shown in Fig. 7 which determines a void form. Such vacancy voids are found in irradiated f.c.c. metals and alloys (e.g. Ni, Cu, Al, and stainless steels).[24] It seems that to have a stabilizing nucleus is a necessary condition for survival of a vacancy cluster in a volume form.

To verify this hypothesis, we have investigated the growth of different stable volume clusters using the procedure described and received analogous results (Figs. 8 and 9). These results not only confirm the hypothesis of a stabilizing nucleus but lead to the following conclusion: in order to get a stable void, it is not necessary to start off an elementary vacancy embryo as it is accepted presently in almost all the theories of radiation swelling.[40] Under irradiation, a void embryo can be easily got athermally from small vacancy clusters which transformed into dislocation-type defects during crystal lattice relaxation.[41]

The concept of a stabilizing nucleus consisting of its own atoms permits us to look at vacancy-type-dislocation-loop embryos from another point of view. We can consider the strongly displaced atoms on the both sides of a loop plane as a stabilizing nucleus for a loose zone which is not so dense as a surrounding crystal. Suppose that we have removed some stabilizing atoms without collapse. Then we get an ellipsoidal void with a lesser number of stabilizing atoms. We have checked this possibility and come to the conclusion that in irradiated metals the embryos of voids and vacancy-type dislocation loops can convert into each other.[41] This computer simulation result can explain experimental data. Really, in irradiated metals and alloys, there appear sometimes voids which are stretched in one direction.[23] Usually these voids are situated near dislocations. One can think that embryos of such voids were dislocations.

Now we can propose the swelling mechanism for pure irradiated metals which is based on the peculiarities of void growth discovered by computer simulation. This mechanism incorporates four postulates:

1. Any stable void embryo has a stabilizing nucleus of its own atoms.
2. During incubation period there arise void embryos which have no explicit character of voids.
3. When adjoining void embryos form a common void, their stabilizing nuclei interact with each other inducing quick relaxation growth of the common void.
4. At the final stage, a large void can grow absorbing vacancies in an isotropic way.

The first assertion explains why pure metals swell at all. The second one makes clear the existence of incubation time for swelling. Consider the other postulates. It is known[40] that swelling is proportional to irradiation time to some power m, which is at first larger, and later less, than unity. The third postulate means that small void embryos absorb vacancies not only in a diffusional manner. In principle, if one knows the form of all the void embryos, their quantity, and the number of atoms which enter into stabilizing nuclei, one can estimate the coefficient m. For a void embryo in the form of a truncated cube, which consists of 13 vacancies, and has a stabilizing nucleus of 6 atoms, we found[41] that m = 1.35. The fourth postulate explains saturation: here, it is not difficult to see that m = 1/3.

Cracks

As mentioned above, the vacancy clusters can convert not only into dislocations or voids but also into cracks. Consider this possibility. Usually, the boundary atoms which surround a plane vacancy cluster displace during crystal relaxation into an empty zone, and the vacancy cluster collapses transforming into a dislocation loop. However, when applying a tensile force normal to a vacancy cluster plane, the boundary atoms displace less and at some force compensating the crystal relaxation, no displacements occur. Further increasing the force leads to growth of an empty zone volume, thus stimulating crack nucleation.

Let us consider, for example, a twelve-vacancy cluster lying in a (001) plane of aluminium (Fig. 10). The displacements of atoms surrounding the cluster are shown in Fig. 11. One can see two critical stains, the lesser E_{cr}^{min} refers to the displacement of atom C which is in the vicinity of a future crack tip, and the larger ones E_{cr}^{max} correspond to the other atoms displacements. The lesser critical strain resembles a crack tip opening which is very popular in fracture mechanics.[42] The larger critical strains are similar to that introduced by Griffith[43] and has just the same dependence on a crack length. It separates two regions: a region in which a plane cluster can be considered as a dislocation, and a region where the cluster becomes a crack.

Considering a plane vacancy cluster as a possible fracture nidus, it must be born in mind that the vacancy cluster can occupy various crystallographic planes and have different binding energies. Molecular dynamics shows that vacancy clusters collapsed in (011) or (111) planes are being opened in just the same manner as a (001) plane vacancy cluster, i.e. the cluster transformation begins in a future crack tip and ends in the middle of the cluster. However, at larger strains, a (111) crack does not propagate. Instead, there form alternating groups of the boundary atoms which displace in opposite directions. The final configuration resembles a void surrounded by stacking fault tetrahedra. These results are in a good agreement with the experimental observations[44] according to which cracks propagate in aluminium in (001) and (011) planes only.

In fracture mechanics, there are three idealized modes of cracks corresponding to the different orientation of an external force with respect to a fracture plane.[45] In Mode 1, the opening mode, the force is a tensile force normal to a cleavage plane. In Modes 2 and 3, the shear modes, the force is in a crack plane normal to or along a crack line, respectively. The first mode is the only one leading to physical fracture. Other modes are assumed to produce a physical crack if enough Mode 1 is present to separate a cleavage plane.

Indeed, with Modes 2 and 3 only, large shear strains transform a (001) plane vacancy cluster into a group of hexagon dislocation loops lying in closed-packed (111) planes.[34] Combination of Modes 1 and 2 (or 3) can be interpreted as an incline of the principal axes of a stress tensor with respect to a vacancy cluster plane. In this case, we also observed the crack cleavage: however, the cleavage surface does not lie in one and the same plane.[46]

Now consider one of the most interesting phenomena discovered by computer simulation.[47] It is known[38] that in aluminium multi-layer dislocations loops are observed. Their appearance is ascribed to vacancy condensation. We have investigated the behavior of the double layer of vacancies lying in two adjacent (001) planes. In Fracture Mode 1, the cluster transforms at first into a void, then cleavage begins. Another volume vacancy cluster composed of two plane clusters which lie in (001) and (011) planes normal to each other, collapses, if unloaded, forming a (011) dislocation loop. In Fracture Mode 1, the dislocation loop is opened out forming a void (Fig. 12). However, if a tensile force is applied parallel to both vacancy planes, the cluster transforms into a sharp crack (Fig. 12). The crack nucleation is facilitated if the second tensile force is applied normal to the first one. In this case, we have a penny crack. Consequently, one and the same vacancy cluster can be converted into a dislocation loop, void, or crack. This phenomenon of cluster shaping explains the experiments performed on ductile metals for which fracture, including radiation fracture, is accompanied by appearance of large number of microcracks and microvoids.[48,49]

CONCLUSION

In summary, it could be said that at present the first stage of radiation damage is understood quite satisfactorily. The most unclear now is the relaxation stage. New facts discovered at this stage are gradually changing the postulates which the analytical description of the diffusion stage is based upon. The latter is a subordinate stage because the kinetic equations describing it use as input data microscopic values and processes detected during the relaxation. As for the fracture stage, its processes are analyzed mostly by continuum mechanics the approach of which to fracture does not embrace many important features identified by computer simulation.

REFERENCES

1. A. I. Melker, Dr. Sc. Thesis, Leningrad Polytechnical Institute, 1987.

2.   A. I. Melker, S. N. Romanov, and N. L. Tarasenko, *Phys. Stat. Sol.(b) 132*, 425 (1985)

3.   B. V. Gnedenko, Yu. K. Belyev, and A. D. Solovyev, *Mathematical Methods in Reliability Theory* [in Russian], Nauka, Moscow 1965.

4.   T. Aruga, Y. Katano, K. Suzuki, Y. Ikeda, T. Nakamura, K. Shiraishi, *J. Nucl. Mater. 134-135*, 667 (1985)

5.   D. K. Holmes and G. Leibfried, *J. Appl. Phys. 31*, 1046 (1965)

6.   P. P. Pronko and K. L. Merkle, in: *Applications of Ion Beams to Metals*, Eds. S. T. Picraux, E. P. EerNisse, and F. L. Vook, Plenum Press, New York 1974

7.   A. I. Melker, in: *Radiation Defects in Metals* [in Russian], Ed. A. T. Lukyanov, Nauka, Alma-Ata 1981

8.   A. I. Melker, S. N. Romanov, and N. L. Tarasenko, *Phys. Stat. Sol. (b) 133*, 111 (1986)

9.   V. V. Kirsanov, *Computer Experiment in Atomic Materials Science* [in Russian], Energoatomizdat, Moscow 1990

10. A. I. Melker and S. N. Romanov, *Phys. Stat. Sol. (b) 126*, 133 (1984)

11. A. L. Suvorov, in: *Radiation Defects in Metals* [in Russian], Ed. A. T. Lukyanov, Nauka, Alma-Ata 1981

12. K. Hinoda, S. Tanigawa, H. Kumakura, M. Doyama, K. Shiraishi, *J. Phys. Soc. Japan 45*, 1858 (1978)

13. P. Hautojarvi, L. Pollanen, A. Vehanen, J. Yli-Kaupilla, *J. Nucl. Mater. 114*, 250 (1983)

14. A. I. Melker and S. N. Romanov, *Zh. Tekh. Fiz. 52*, 1362 (1982)

15. A. I. Melker, S. N. Romanov, and V. S. Tokmakova, in: *Computer Simulation of Defects Kinetics in Crystals* [in Russian], FTI Acad. Sci. USSR, Leningrad 1985

16. A. I. Melker, S. N. Romanov, and V. S. Tokmakova (unpublished)

17. S. N. Romanov and A. I. Melker, *Zh. Tekh. Fiz. 51*, 1252 (1981)

18. B. V. Klimovich and V. V. Nelayev in: *Computer Simulation of Structure Defects in Crystals* [in Russian], FTI Acad. Sci. USSR, Leningrad 1988

19. A. I. Melker and S. N. Romanov, *Phys. Stat. Sol. (b) 122*, K17 (1984)

20. S. N. Romanov, A. I. Melker, and D. B. Mizandrontsev, *Phys. Stat. Sol. (b) 153*,

K5 (1989)

21. Z. V. Basheleishvili, S. N. Romanov, Ye. P. Romanova, and A. I. Melker in: *Computer Simulation of Struture-dependent Properties of Crystal Materials* [in Russian], FTI Acad. Sci., USSR, Leningrad 1986

22. A. I. Melker and S. N. Romanov, *Phys. Stat. Sol.(b) 126*, 517 (1984)

23. Yu. V. Konobeev, in: *Computer Simulation of Defects in Crystals* [in Russian], FTI Acad. Sci. USSR, Leningrad 1979

24. D. J. Mazey and J. A. Hudson, *J. Nucl. Mater. 37*, 13 (1970)

25. V. N. Bykov and A. G. Vakhtin, *Fiz. Tverd. Tela 15*, 910 (1973)

26. V. N. Bykov, G. G. Zdorovtseva, V. A. Troyan, and V. S. Khaimovich, *Kristallografiya 19*, 896 (1974)

27. W. J. Yang, R. A. Dodd, and G. L. Kulcinski, *J. Nucl. Mater. 64*, 157 (1977)

28. H. Akimichi, K. Yoshio, H. Schozo, and S. Kenzaki, *Kakuugu Kenku 43*, 69 (1980)

29. M. I. Robinson and M. I. Jenkins, *Phil. Mag. (a) 43*, 999, (1981)

30. S. I. Rudnev, *Fiz. Metal. Metalloved. 58*, 411 (1984)

31. K. Niwase, T. Ezava, F. E. Fujita, H. Kusanagi, and H. Takaku, *Rad. Eff. 106*, 65 (1988)

32. A. I. Melker, *Modelling Experiment* [in Russian], Znanie, Moscow 1991

33. V. V. Sirotinkin, Ph.D. Thesis, Leningrad Polytechnical Institute, 1988

34. O. A. Shenderova, Ph.D. Thesis, Leningrad State Technical University, 1991

35. R. D. Dokhner, *Fiz. Tverd. Tela 12*, 3556 (1970)

36. A. A. Vasilyev, V. V. Sirotinkin, and A. I. Melker *Phys. Stat. Sol. (b) 131*, 537 (1985)

37. A. I. Melker and V. V. Sirotinkin (unpublished)

38. J. P. Hirth and J. Lothe, *Theory of Dislocations*, McGraw-Hill, New York 1969

39. F. Kroupa and A. Machova, *Fyzika kovu (Teorie Dislokaci)*, Edicni Stredisko CVUT, Praha 1988

40. Sh. Sh. Ibragimov, V. V. Kirsanov, and Yu. S. Pyatiletov, *Radiation Damage in*

Metals and Alloys [in Russian], Energoatomizdat, Moscow 1985

41. A. I. Melker, V. V. Sirotinkin, and O. A. Shenderova (unpublished)

42. S. N. Atluri and A. S. Koboyashi, in: *Computational Methods in the Mechanics of Fracture,* Ed. S. N. Atluri, North-Holland, Amsterdam 1986

43. A. A. Griffith, *Phil. Trans. Roy. Soc. (a) 221,* 163 (1920)

44. D. N. Fager, M. V. Hyatt, and H. T. Diep, *Scr. Met. 20,* 1159 (1986)

45. R. Thomson, *Solid State Physics 31,* 1 (1986)

46. A. I. Melker and O. A. Shenderova (unpublished)

47. A. I. Melker and O. A. Shenderova, in: *Euromech-291, Macro and Micromechanical Aspects of Fracture, 23-27 June 1992, St. Petersburg, Russia*

48. H. G. F. Wilsdorf, *Mater. Sci. Eng. 59,* 1 (1983)

49. A. I. Melker, I. L. Tokmakov, D. A. Chumichev, and T. I. Makarenko, *Poverkhnost, 1,* 123 (1985)

172 Creation and Evolution of Vacancy Clusters

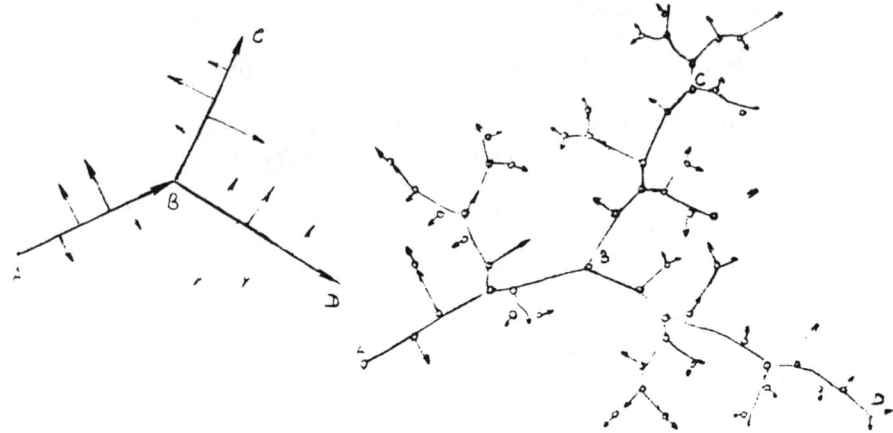

Fig.1 Structure of the collision cascade composed of three high-energy branches: scheme (a), real cascade in - iron initiated by PKA with the energy 0.1 MeV [9] (b)

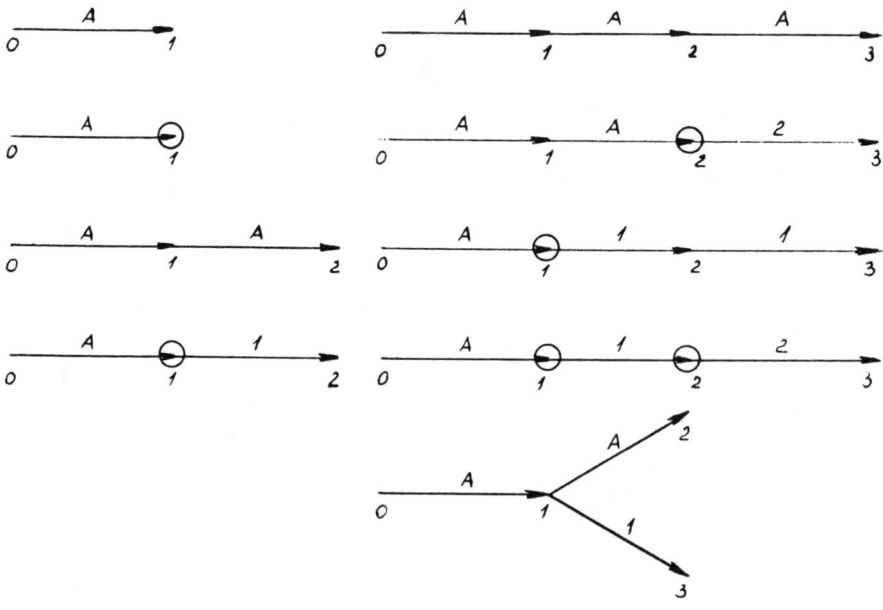

Fig.2 Fomation diagrams for mono, di, and trivacancies.

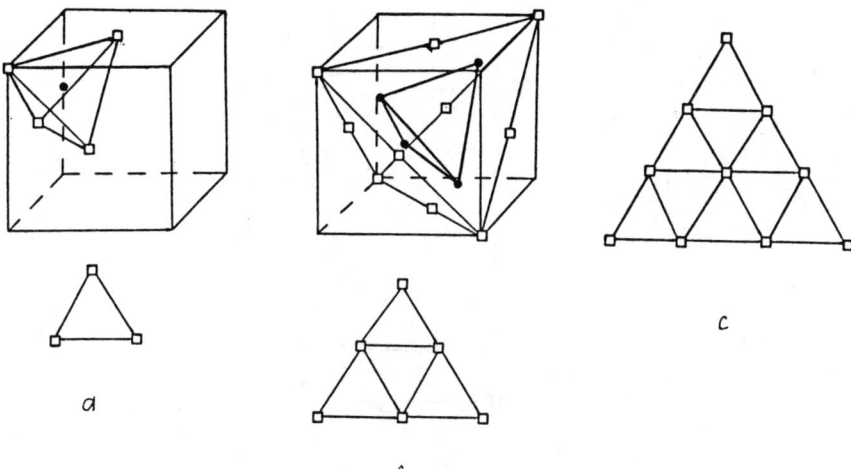

Fig.3 Stacking fault tetrahedra in nickel. For larger defects, the initial positions of vacancies are only shown. vacancy, strongly displaced atom
Binding energy (eV): 0.71(a), 2.82(b), 7.17(c)

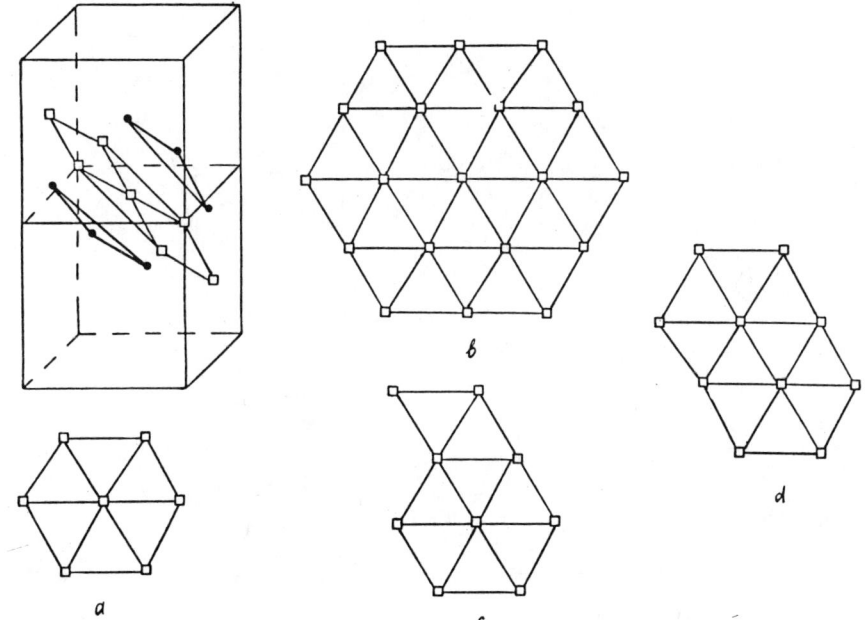

Fig.4 Vacancy clusters in a (111) plane transforming into dislocation loops. Binding energy (eV): 2.55(a), 10.27(b), 3.82(c), 4.86(d)

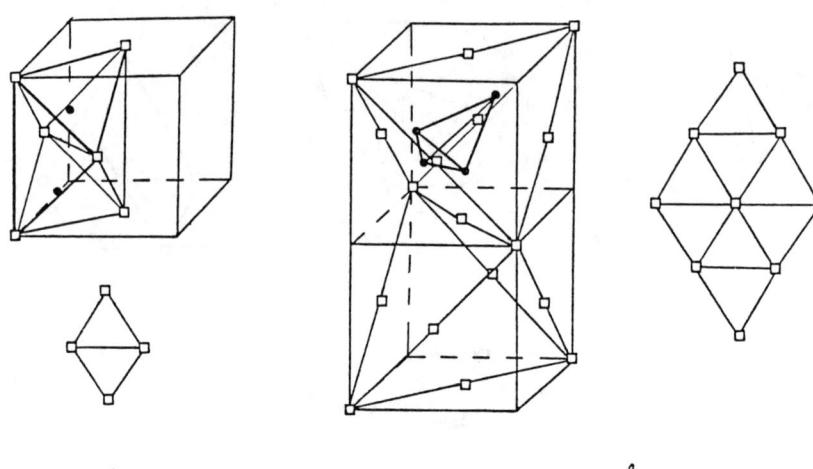

Fig.5 Adjacent stacking fault terahedra. Binding energy (eV): 1.23(a), 4.68(b)

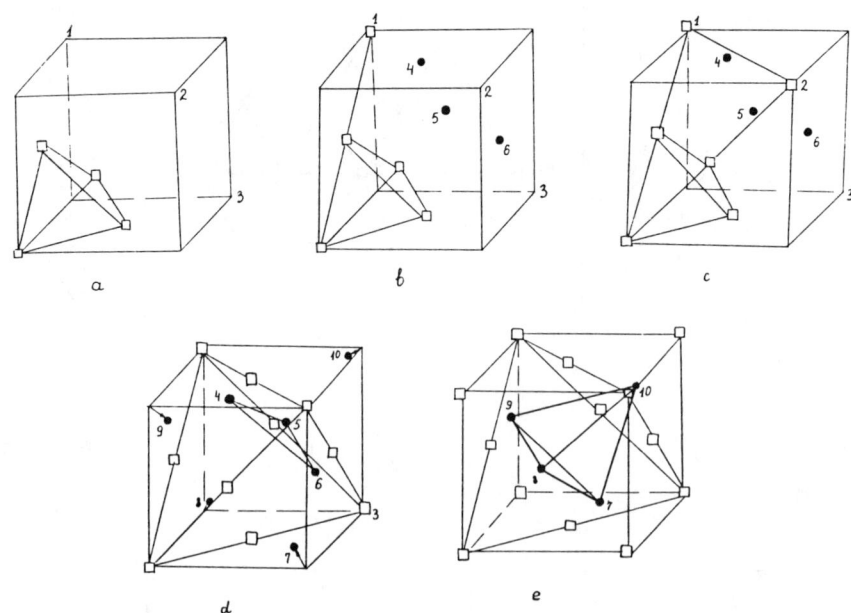

Fig.6 Growth of a tetrahedral void. Binding energy (eV): -0.24(a), -0.11(b), 0.08(c), 2.23(d), 2.27(e)

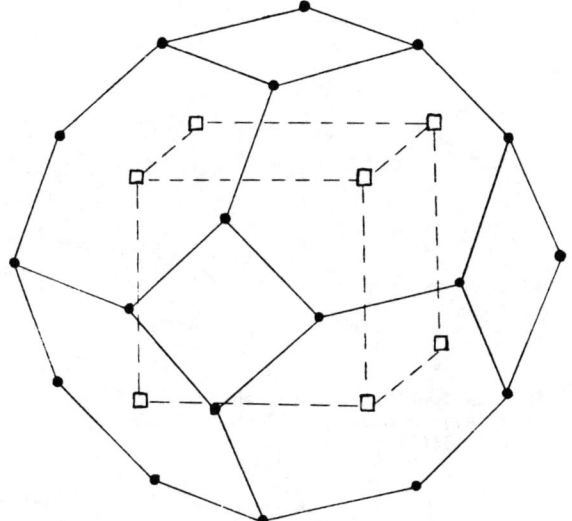

Fig.7 Vacacy cube inside a truncated octahedron.

176 Creation and Evolution of Vacancy Clusters

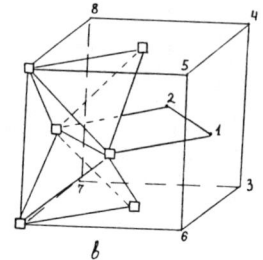
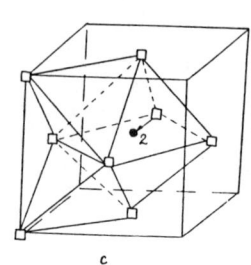
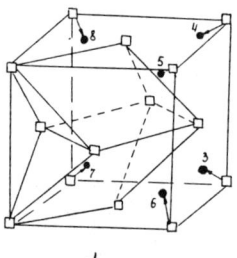
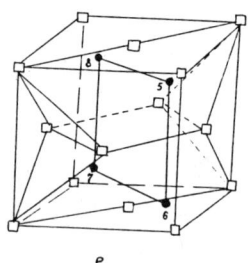

Fig.8 Growth of two adjacent vacancy tetrahedra.
Binding energy (eV): 1.01(a), -0.24(b),
2.42(c), 2.00(d), 1.67(e)

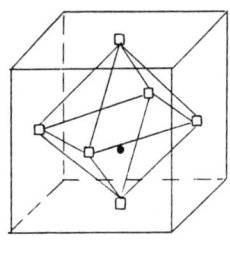
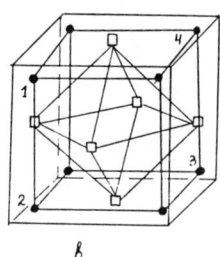
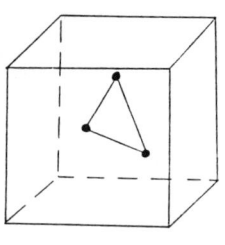
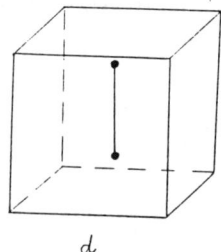
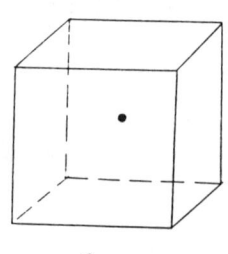

Fig.9 Growth of a vacancy octahedron. Binding energy

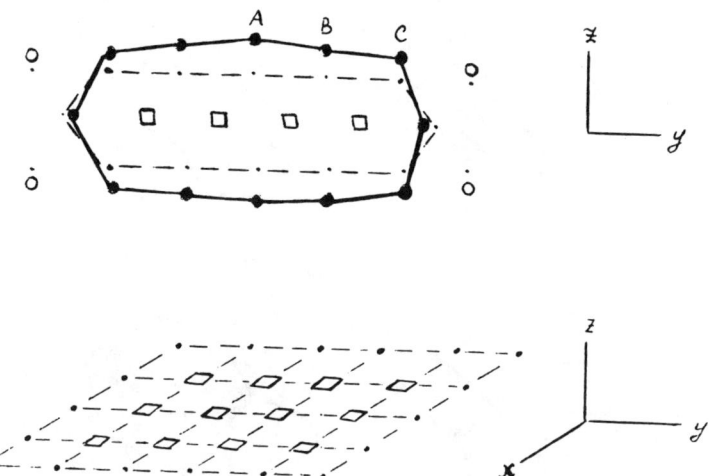

Fig.10 Plane vacancy cluster in stretched aluminium: atom positions before (dotted line) and after stretching (solid line).

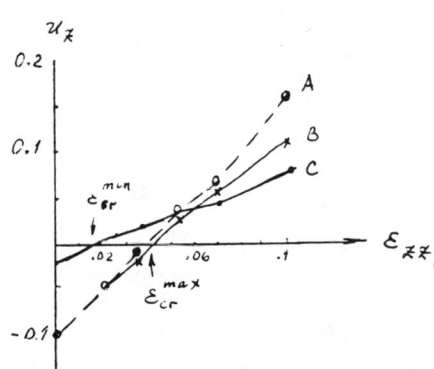

Fig.11 Displacement of boundary atoms surrounding the vacancy cluster shown in Fig.10

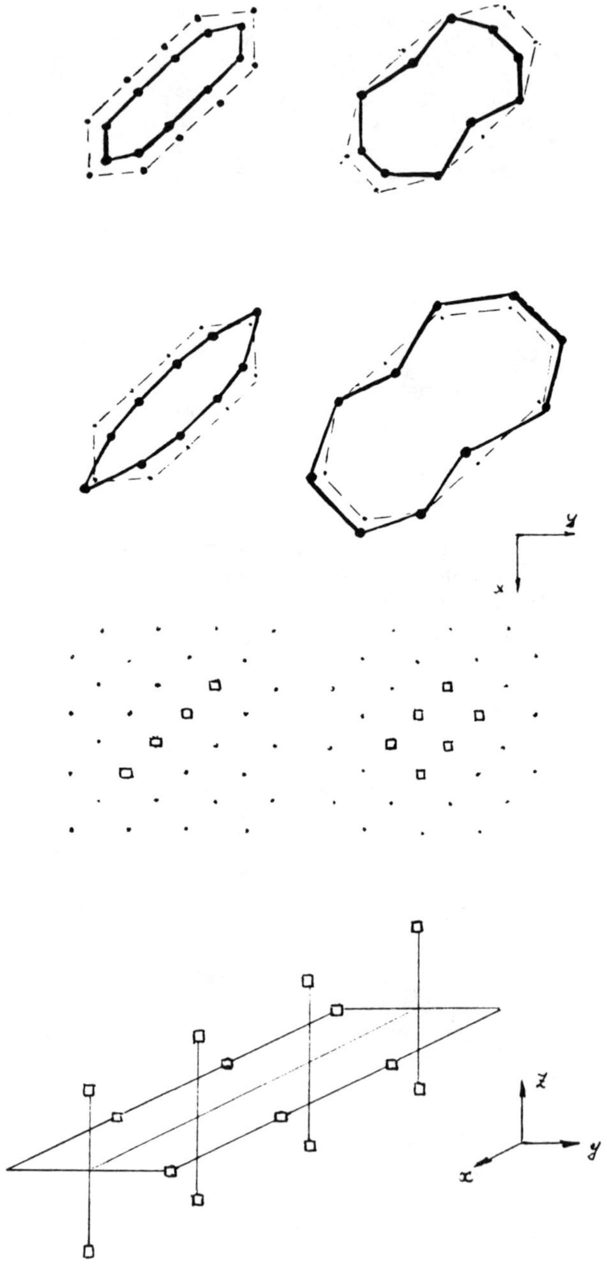

178 Creation and Evolution of Vacancy Clusters

Fig.12 Transformation of a vacancy cluster into a void (a) and a sharp crack (b)

BEAM-BASED AGE–MOMENTUM CORRELATION STUDIES OF POSITRONIUM SPIN CONVERSION IN PARAMAGNETIC SOLUTIONS AND OF POSITRON TRAPPING AT DEFECTS IN DIAMONDS

H. Stoll, M. Koch, U. Lauff, K. Maier*, J. Major, A. Seeger, P. Wesołowski
Max-Planck-Institut für Metallforschung, Institut für Physik, Postfach 800665,
D 7000 Stuttgart 80 , Germany

I. Billard, J. Ch. Abbé, G. Duplâtre
Centre de Recherches Nucléaires, Laboratoire de Chimie Nucléaire, Strasbourg, France

S. H. Connell, J. P. F. Sellschop, E. Sideras-Haddad
Schonland Research Centre, University of the Witwatersrand, Johannesburg, South Africa

K. Bharuth-Ram, H. Haricharun
Physics Department, University of Durban-Westville, Durban, South Africa

ABSTRACT

Correlated measurements of the lifetime and of the Doppler broadening of the 511 keV annihilation radiation of positrons (Age–Momentum Correlation, AMOC) using an MeV positron beam have become a powerful tool for investigating reactions of positrons or positronium as a function of time. The room-temperature reaction rate of the spin conversion of positronium in methanol induced by the presence of a paramagnetic solute (HTEMPO) has been found to be proportional to the HTEMPO concentration with a reaction-rate constant of $(22.5 \pm 0.5) \cdot 10^9 \, \mathrm{l\,mol^{-1}\,s^{-1}}$ up to the highest concentration investigated ($C \leq 0.1 \mathrm{mol/l}$).

By means of the AMOC technique it was shown that a large Doppler broadening in natural and synthetic diamonds is correlated with a very short lifetime component. At higher positron ages a second positron state with a narrower momentum distribution of the positron–electron pair indicates positron trapping at defects with concentrations ranging from 10^{-7} to 10^{-6} in different types of diamonds.

1 AGE–MOMENTUM CORRELATION MEASUREMENTS WITH AN MEV POSITRON BEAM

By measuring individual positron lifetimes (= *positron age*) together with the Doppler shift ΔE ($\rightarrow$ *momentum* of the annihilating positron-electron pair) of the energy of one of the annihilation quanta in a triple-coincidence set-up (Age-Momentum Correlation, AMOC)[1-5] time-resolved information on the evolution of the positron states (e.g., trapping of positrons at defects or chemical reactions of positronium) may be obtained. The application of this technique was retarded, however, because of the rather slow data accumulation in conventional (source-based) $\gamma\gamma\Delta E$ measurements resulting from the requirement of triple coincidence in three *gamma* detectors.

The beam-based $\beta^+\gamma\Delta E$ AMOC technique implemented at the Stuttgart MeV positron facility[6-9] offers substantial advantages over the conventional $\gamma\gamma\Delta E$ coincidence technique. The unity detection efficiency of the β^+ start detector permits faster data accumulation and thus better statistics and/or reduced measuring times. Pile-up pulses in the β^+ start scintillator can be used efficiently to discriminate against random coincidences due to positrons arriving almost simultaneously. This allows AMOC spectra with a very large peak-to-background ratio

---

*now: Universität Bonn, Institut für Kern- und Strahlenphysik, Nußallee 14 - 16, D 5300 Bonn, Germany

to be obtained even at high coincidence rates. In addition, the beam-based technique often considerably simplifies specimen preparation and experimental set-up. E.g., MeV positrons may be released into air and/or be deeply implanted into solids or liquids including high-temperature melts[10,11].

In this paper beam-based AMOC studies of spin conversion of positronium in methanol induced by a paramagnetic solute (HTEMPO) (Sect. 2) and of positron trapping in different types of diamond (Sect. 3) are presented. Both experiments were performed with a positron beam energy of 4 MeV.

2 AGE–MOMENTUM CORRELATION MEASUREMENTS OF POSITRONIUM SPIN CONVERSION IN PARAMAGNETIC SOLUTIONS

The systems investigated were methanol and methanol with different additions of the paramagnetic solute HTEMPO (4-hydroxy-2,2,6,6-tetramethylpiperidine-1-oxyl). The unpaired electron of the nitrosyl group (Fig. 1) is responsible for the paramagnetism of this substance. All samples were carefully degassed and then transferred into cylindrical glass vessels under argon pressure in order to avoid oxidation of positronium. The 4 MeV positrons were implanted through a 0.2 mm thin glass window. Owing to the size of the vessels (3 cm in diameter, 3 cm long) almost all positrons annihilated inside the liquid.

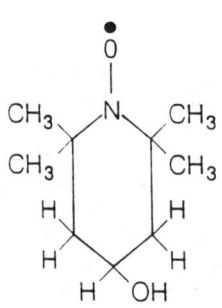

The two-dimensional AMOC spectrum of pure methanol measured at room temperature is shown in Fig. 2. The coincidence counts are plotted on a logarithmic scale versus the positron age and the energy of one of the annihilation quanta. The planes of constant energy represent lifetime spectra at different momenta of the electron–positron pairs. The planes of constant positron age represent energy spectra at different times after the e+ implantation. From these energy spectra a time-dependent centroid lineshape parameter $S_t(t)$ can be calculated as a function of positron age (Fig. 3). Low values of the S_t parameter correspond to large Doppler broadening of the annihilation quanta, hence to wide momentum distributions of the positron–electron pairs.

Fig. 1: Structure formula of HTEMPO. The dot indicates the unpaired electron.

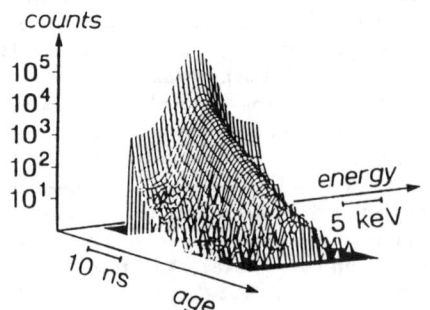

Fig. 2: $\beta^+\gamma\Delta E$ AMOC spectrum of methanol at room temperature.

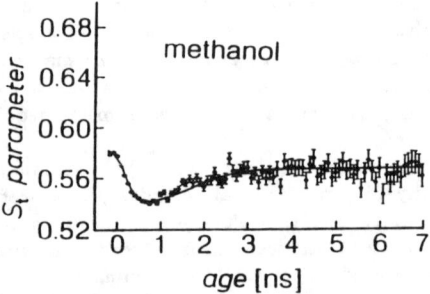

Fig. 3: Lineshape parameter S_t vs. positron age for methanol at room temperature.

A simple model for the variation of the S_t parameter as a function of positron age is given in the Appendix. Positrons are assumed to annihilate from three different positron states: the free positron state, the para-positronium (p-Ps), and ortho-positronium (o-Ps) state. The solid line in Fig. 3 represents a fit using this model. The fit parameters, given in Table I, are the positron lifetimes $\tau_i = \lambda_i^{-1}$, the Doppler-lineshape parameter S_i, and the normalized initial population

$$\tilde{n}_i(0) = n_i(t=0)/\sum_i n_i(t=0) \tag{1}$$

of each of the three individual states (for definitions of λ_i, S_i, and n_i see Appendix, Eqns. A.1 and A.2). The global variation of the S_t parameter as a function of positron age is determined by the sum of the lineshape parameters of the individual states S_i (see Table I) weighted by their contributions $\lambda_i n_i(t)$ to the total annihilation rate (see Appendix, Eqn. A.2).

Table I: Values of the fit parameters $\tau_i = \lambda_i^{-1}, S_i$, and $\tilde{n}_i(0) = n_i(t=0)/\sum_i n_i(t=0)$ of the three different positron states (p-Ps, free e$^+$, and o-Ps) according to the solid lines in Figs. 3 and 4 (for definitions see Appendix). These parameters are independent of the HTEMPO concentrations investigated. The dependence of the only additional parameter K_{conv} (see Appendix Eqn. A.12) on the HTEMPO concentration is given in Fig. 5.

i	τ_i	S_i	$\tilde{n}_i(0)$
p-Ps	125ps	0.75	6.7%
free e$^+$	450ps	0.54	73.2%
o-Ps	3.5ns	0.57	20.1%

The S_t parameter observed in pure methanol at positron ages exceeding 3 ns (Fig. 3) is dominated by the longest-lived positron state, i.e., o-Ps in which the positron annihilates predominantly in a "pick-off" process with a "foreign" electron of opposite spin. The lower value of the S_t parameter in the time interval of 0.5 ns to 3 ns is due to a larger contribution of free-positron annihilation, which shows a slightly broader momentum distribution than the pick-off annihilation of o-Ps. At ages lower than 0.5 ns the contribution of p-Ps annihilation causes the S_t parameter to increase again, since the annihilation of p-Ps gives not only the shortest lifetime of all three positron states but also a very narrow momentum distribution.

The time-dependent lineshape parameter $S_t(t)$ changes dramatically, in particular at higher positron ages, when 0.1 mol/l of HTEMPO is added (Fig. 4a, upper data points). Other HTEMPO concentrations were also measured, such as 0.05 mol/l (Fig. 4b), 0.01 mol/l (Fig. 4c), and 0.005 mol/l (Fig. 4d). The change in the lineshape parameter $S_t(t)$ is attributed to the spin-conversion reaction

$$\begin{aligned} \text{o-Ps} + M &\longrightarrow \frac{1}{4}\text{p-Ps} + \frac{3}{4}\text{o-Ps} + M \\ \text{p-Ps} + M &\longrightarrow \frac{1}{4}\text{p-Ps} + \frac{3}{4}\text{o-Ps} + M \end{aligned} \tag{2}$$

induced by the paramagnetic solute M. Such a reaction tends to counteract the depletion of the p-Ps population and to retain the initial 1:3 ratio between p-Ps and o-Ps population. The spin-conversion reaction causes the short-lived p-Ps with its narrow annihilation line to make a significant contribution to the AMOC spectrum also at high positron ages. The values of the lineshape parameter S_t (Fig. 4a-d) at positron ages exceeding 3 ns are determined by the ratio of the annihilation of o-Ps and spin-converted p-Ps and thus by the HTEMPO concentration.

The preceding interpretation is strongly supported by the fact, that the data on all four HTEMPO concentrations investigated (Fig. 4a-d) can be described by the simplest conceivable rate-equation model (cf Appendix) allowing for the above reaction (2). Only one additional (concentration-dependent) parameter, the *reaction rate of spin conversion* K_{conv} (for definition see Appendix, Eqn. A.12), is needed to obtain reasonable fits (solid curves in Fig. 4 a-d).

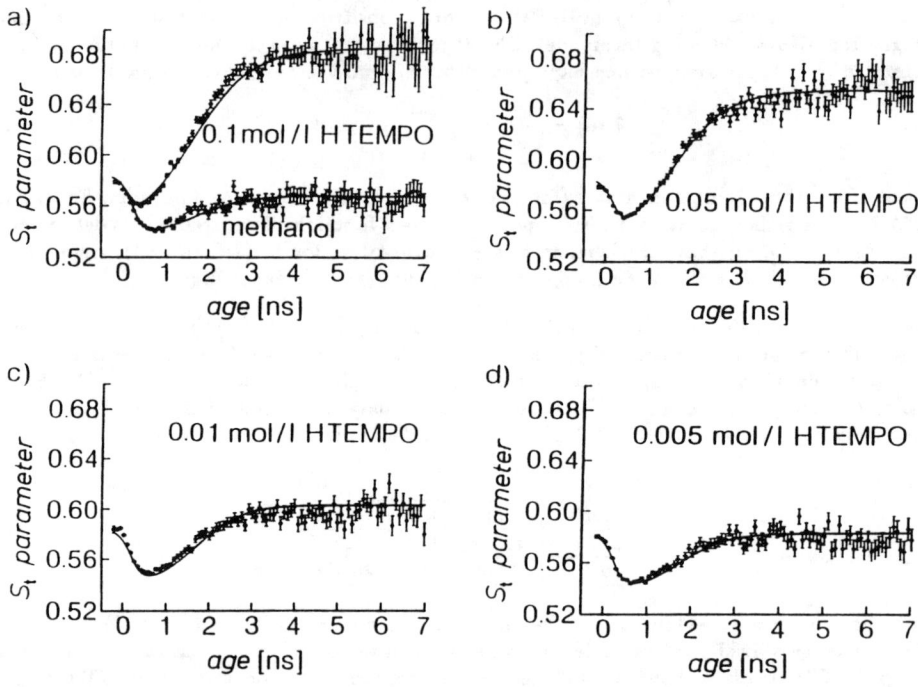

Fig. 4a-d: Lineshape parameter S_t vs. positron age for methanol and different concentrations of HTEMPO in methanol at room temperature.

All other parameters have been kept at the values determined from the fit of the pure solvent (see Table I). In the concentration range investigated it was not necessary to postulate a significant effect of the solute on the ratio of the initial populations of free positrons and positronium.

The spin-conversion reaction rates obtained in this way as a function of HTEMPO concentration C are shown in Fig. 5. In the concentration range ($C \leq 0.1\,\mathrm{mol/l}$) examined in the present work we obtain a linear relationship,

$$K_{\mathrm{conv}} = k_{\mathrm{conv}} \cdot C, \tag{3}$$

between the reaction rate K_{conv} of the spin conversion and the HTEMPO concentration C with a room-temperature reaction-rate constant (specific reaction rate)

$$k_{\mathrm{conv}} = (22.5 \pm 0.5) \cdot 10^9 \, \mathrm{l\,mol^{-1}\,s^{-1}}. \tag{4}$$

This is in fairly good agreement with previous results ($k_{\mathrm{conv}} = 19.2 \cdot 10^9\,\mathrm{l\,mol^{-1}\,s^{-1}}$) obtained from uncorrelated lifetime spectroscopy and Doppler-broadening measurements[12].

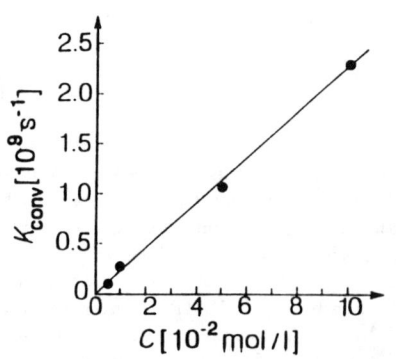

Fig. 5: Room-temperature reaction rate of spin conversion vs. HTEMPO concentration.

3 POSITRON TRAPPING AT DEFECTS IN NATURAL AND SYNTHETIC DIAMONDS

Diamond is a very interesting candidate for positron spectroscopy studies owing to its many unique and extreme properties as the simplest single-element molecular solid. Major recent technological advances lend additional interest to the study of diamond. The controlled doping of diamond is now a reality[13], and there have also been tremendous improvements in growing synthetic diamond both in the stable (high-pressure, high-temperature) regime and in the metastable regime (in particular, growth by chemical vapor deposition, CVD [14,15]). Defect studies on diamond now have a similar motivation as they have, e.g., on silicon.

Measurements of positron lifetimes and of positron–electron momentum distributions on diamond[16,17,18] have revealed an anomalously large Doppler broadening together with a spread of lifetimes. One of the lifetimes, usually the most intense one, is among the shortest measured on any material. The data tend to scatter rather strongly, reflecting the complex (and little known) defect structure of the different diamond samples.

It should be borne in mind that in the majority of diamond samples many impurities are present in concentrations above the limit of saturation trapping for positrons. In natural diamond hydrogen, nitrogen, and oxygen are the dominant impurities. Apart from them, natural diamonds display a rich and distinctive trace element chemistry which reflects the composition of the Earth's mantle material in which the diamond had its genesis[19]. In synthetic diamonds, it is normal to find appreciable concentrations of the light volatiles mentioned above, as well as inclusions of the metal solvent used in the synthetic crystal-growth process. Structural defects such as monovacancies, multiple vacancies, dislocations, and complexes of these are also present. More recently, synthetics grown slowly in high-pressure–high-temperature presses but seeded from CVD material have been shown to be extremely pure and defect free.

The present section reports on *age–momentum correlation measurements* on different types of diamond. Typical spectra obtained at room temperature are shown in Fig. 6a-c for three different types of diamond samples, viz. synthetic, natural type Ia, and natural type IIa. These samples differ mainly as follows: In type–Ia stones, nitrogen is present in a variety of aggregated forms at levels of many hundred ppm or more, while in type IIa the nitrogen level is less than 20 ppm. The dislocation density in type IIa is higher, and a mosaic spread of 0.1° is typical. Type Ia showed clearly visible fine dark streaks, probably due to amorphous carbon. The synthetic diamond is classed as type Ib; it is also rich in nitrogen, though dominantly in the form of single substitutional atoms. The mosaic spread is lowest in the case of the synthetic samples.

The AMOC spectra (Fig. 6a-c) reveal at least two different positron states. The giant Doppler broadening mentioned earlier is associated with the short positron lifetime. Long-lived positrons undergo a transition to a second state with a narrower e^+e^- momentum distribution. Apart from these general features, the spectra differ quantitatively in, amongst other things, the transition rate to and the Doppler broadening of the second state, evidencing a sensitivity of the measurements to differences in the defect structures of the samples. A model for the first state must satisfy the following three requirements imposed by the measurements: It must be occupied by almost the entire young positron ensemble, its lifetime must be remarkably short and its Doppler broadening very large. We have previously speculated[16] that the first state could be p-Ps that is quenched by pick-off annihilation with the 1s electrons of the carbon atoms. However, this would not be consistent with a large initial population. An alternative could be annihilation of positrons from intra-bond sites. Support for this conjecture is found in the calculated bulk lifetime[20] $\tau_{bulk} = 114$ps, which is similar to the observed value for the first state. This theoretical value is dominated by the contribution from the sp^3 valence electrons. The Doppler broadening at the intra-bond site should be large as can be inferred from measurements and calculations for positron Doppler spectroscopy of the hydrocarbons[21].

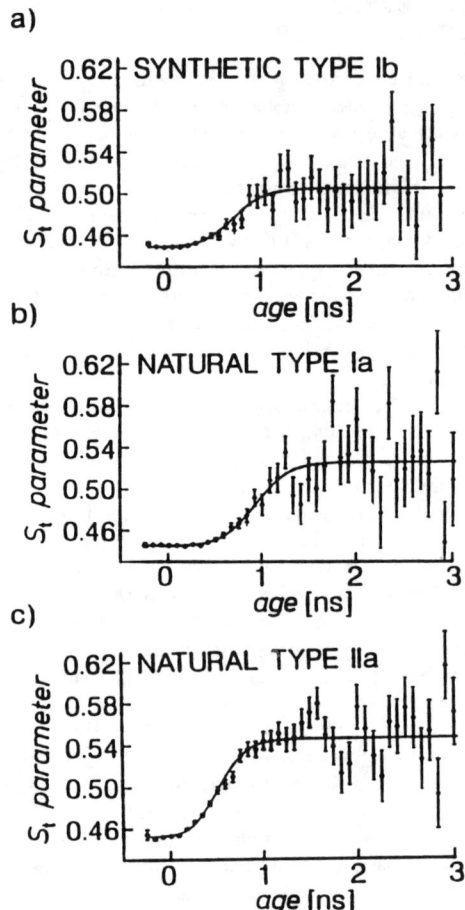

Fig. 6 a-c: Lineshape parameter S_t vs. positron age for different types of diamonds at room temperature.

Relating intra-bond parameters derived from small hydrocarbon clusters to the diamond macromolecule matrix is usually found to be surprisingly reliable. Annihilation from the intra-bond site may still be consistent with the expected long thermalization time of positrons in diamond by invoking a fast quasi-adiabatic diffusion mechanism, with the positrons hopping between intra-bond sites. The intra-bond site was found for the positive muon in diamond[22]. Of course, other speculations are also possible at this stage.

The narrowing of the Doppler spectrum with increasing positron age may be interpreted as positron trapping at defects in the diamonds. The solid lines in Fig. 6 a - c represent fits according to the two-state trapping model outlined in the Appendix. De-trapping has been neglected ($K_{\text{detrap}} = 0$). The fit parameters are collected in Table II. The positron lifetimes given in Table II may have systematic errors as it is very difficult to evaluate the time resolution of the spectrometer precisely enough to deconvolute such very short lifetime components.

The positron lifetimes and occupation probabilities deduced from $S_t(t)$ as given in Table II are consistent with the analysis of the integral lifetime spectra. This supports the interpretation of the observation as positron trapping at defects. Due to both the plethora of possibilities for the defects and the positron initial state configuration the precise nature of the defects is not yet known. There is some indication based on results on electron-irradiated diamond that vacancies and vacancy-complexes may trap positrons and give rise to longer lifetimes[17].

Table II: Values of the positron lifetimes $\tau_{\text{free}} = \lambda_{\text{free}}^{-1}$, $\tau_{\text{trap}} = \lambda_{\text{trap}}^{-1}$, of the occupation probabilities N_1, N_2, and of the lineshape parameters S_1 and S_2 for different types of diamonds according to the solid lines in Fig. 6. For definitions see Appendix. To calculate the trap concentration C_{trap}, a specific trapping rate σ_{trap} of 10^{15} s^{-1} has been assumed.

type	τ_{free} [ps]	τ_{trap} [ps]	N_1 [%]	N_2 [%]	S_1	S_2	C_{trap}
synth.	108	277	93	7	0.45	0.50	$3 \cdot 10^{-7}$
Ia	113	319	97	3	0.45	0.53	$1 \cdot 10^{-7}$
IIa	130	394	69	31	0.45	0.55	$1 \cdot 10^{-6}$

In order to further illuminate the possible physical models for the positron behaviour in diamond, we intend to perform a new series of measurements on different (well-characterised) natural and synthetic diamonds at different temperatures. Angular-correlation (ACAR) and Doppler-broadening measurements to detect crystal symmetry in the momentum distribution of the e^+e^- system are also planned or underway.

4 CONCLUSIONS AND OUTLOOK

Age–Momentum Correlation with an MeV positron beam ($\beta^+\gamma\Delta E$ AMOC) has become a routine technique, allowing us to follow kinetic processes directly by time-domain observations of the populations of different positron states tagged by the Doppler-broadening characteristics of the 511 keV annihilation radiation. The present paper illustrates this by a room-temperature study of the spin conversion of positronium in the system HTEMPO/methanol as a function of HTEMPO concentration. The same technique can be applied to other systems and other chemical reactions in positronium chemistry, e.g., oxidation, complex formation, and inhibition of positronium as well as to the dependence of these reactions on external parameters, e.g., temperature.

$\beta^+\gamma\Delta E$ AMOC measurements on diamonds show a narrowing of the Doppler broadening at longer positron ages that is consistent with a two-state trapping model with different concentrations of positron traps in different types of synthetic and natural diamonds. Hopefully, future measurements on well-characterised diamonds will help to unravel the physical or chemical nature of positron traps in diamond.

With the rather weak $3 \cdot 10^8$ Bq (8mCi) ^{22}Na source in the high-voltage terminal of the Stuttgart pelletron used in the present measurements the time required to obtain a $\beta^+\gamma\Delta E$ AMOC spectrum is about 20 hours. The measuring time will be reduced to about 2 hours per $\beta^+\gamma\Delta E$ AMOC spectrum when the present positron source is replaced by a $3.7 \cdot 10^9$ Bq (100mCi) ^{22}Na source.

The rate of data accumulation and the time resolution can be further improved by the *positron-clock technique*. Here, the positron beam is circularly deflected in the electric field of a pair of radio-frequency (1.25 GHz) resonators, moving over a *position sensitive* ring detector like the hand of a clock, and subsequently refocussed onto the sample. First lifetime measurements using the positron-clock principle have already been published[23]. When combined with secondary-electron detectors the positron-clock principle may also be applied to keV positron beams. Since the positron clock can handle much higher positron fluxes than a β^+ start detector without deterioration of its excellent time resolution, it will have a particularly strong impact when *positron factories* with positron fluxes exceeding those hitherto available will become operational.

We aim at increasing the positron beam flux at the Stuttgart pelletron by replacing the present tungsten moderator with a more efficient solid-rare-gas moderator. Different solid-rare-gas moderators were tested at various temperatures in an electrostatically guided slow positron (keV) beam set-up[24]. Under deliberately chosen poor vacuum conditions ($\approx 10^{-4}$ Pa) that simulated those in accelerator terminals a solid-krypton moderator operating at 37.5 K showed a moderator efficiency about four times higher than that of a monocrystalline tungsten moderator. The combination of solid-rare-gas moderation with the positron-clock technique should allow us to accumulate AMOC spectra with good statistics within about 30 minutes.

ACKNOWLEDGEMENTS: We should like to thank A. Bonnenfant (Strasbourg) and J. Sprater (MPI für Festkörperforschung, Stuttgart) for preparing the methanol and HTEMPO samples and Messrs. De Beers Industrial Diamonds (Pty) Ltd for their ongoing support and encouragement. We are grateful to H.-D. Carstanjen, W. Decker, J. Diehl, and H.-E. Schaefer for co-operation and helpful discussions. The help of I. Schemminger, H. Schneider, and A. Siegle in preparing of the manuscript is gratefully acknowledged. S. H. Connell is indebted to the Alexander-von-Humboldt-Stiftung for the award of a fellowship.

APPENDIX: ANALYSIS OF AMOC SPECTRA

In this appendix we give a quantitative model of the AMOC spectra for *spin conversion* of positronium and for *trapping* and *detrapping* of positrons at defects.

Positrons implanted into the sample are assumed to annihilate in different individual positron states (e.g., free positrons, trapped positrons, p-Ps, o-Ps) characterized by their annihilation rates λ_i and their momentum distributions $P_i(p)$. The two-dimensional AMOC spectrum $F(t,p)$ can then be written as

$$F(t,p) = \sum_i P_i(p) \lambda_i n_i(t), \qquad (A.1)$$

where $n_i(t)$ denotes the time-dependent population of the i-th positron state.
The time-dependent S_t-parameter reads

$$S_t(t) = \frac{\sum_i S_i \lambda_i n_i(t)}{\sum_i \lambda_i n_i(t)}, \qquad (A.2)$$

where S_i is the lineshape parameter of the i-th positron state. For comparison of (A.1) and (A.2) with measured AMOC spectra, $P_i(p)$ and S_i have to be convoluted with the energy resolution of the spectrometer, whereas $\lambda_i n_i(t)$ has to be convoluted with the time resolution of the set-up.

If in (A.2) the populations $n_i(t)$ are replaced by the occupation probabilities N_i, which are defined as

$$N_i = \frac{\int_0^\infty n_i(t)\, dt}{\sum_i \int_0^\infty n_i(t)\, dt}, \qquad (A.3)$$

we obtain the time-averaged lineshape parameter

$$\bar{S} = \frac{\sum_i S_i \lambda_i N_i}{\sum_i \lambda_i N_i} \qquad (A.4)$$

familiar from conventional Doppler-broadening measurements.

The parameters λ_i, $P_i(p)$, and S_i of the individual positron states as well as all reaction rates (see below) are assumed to be time-independent. Then the time dependence of the S_t parameter defined in Eqn. (A.2) is determined entirely by the time dependence of the populations $n_i(t)$. In the case of a two-state system with interchanging populations, e.g., spin conversion of o-Ps and p-Ps, or trapping and detrapping of positrons at defects, the time dependence of the two populations n_1 and n_2 is determined by two coupled rate equations of the form

$$\begin{aligned}\frac{dn_1}{dt} &= -\Lambda_1 n_1 + K_2 n_2, \\ \frac{dn_2}{dt} &= -\Lambda_2 n_2 + K_1 n_1.\end{aligned} \qquad (A.5)$$

In the following, the subscript 1 stands for p-Ps in the case of spin conversion of positronium and for free positrons in the case of positron trapping, whereas the subscript 2 stands for o-Ps or trapped positrons.

POSITRON TRAPPING

With the initial condition

$$n_2(0) = 0 \tag{A.6}$$

we obtain as solutions of the rate equations (A.5)

$$n_1(t) = n_1(0) \left[1 + \frac{\Lambda_2 - \Lambda_1}{\gamma} \tanh \frac{\gamma}{2} t\right] \cosh \frac{\gamma t}{2} \exp \frac{-\alpha t}{2},$$

$$n_2(t) = n_1(0) \frac{2 K_1}{\gamma} \sinh \frac{\gamma t}{2} \exp \frac{-\alpha t}{2}, \tag{A.7}$$

where

$$\alpha = \Lambda_1 + \Lambda_2 \quad \text{and} \quad \gamma = \sqrt{(\Lambda_1 - \Lambda_2)^2 + 4 K_1 K_2}. \tag{A.8}$$

With the identification

$$\begin{aligned}\Lambda_1 &= \lambda_{\text{free}} + \sigma_{\text{trap}} C_{\text{trap}}, & K_2 &= K_{\text{detrap}}, \\ \Lambda_2 &= \lambda_{\text{trap}} + K_{\text{detrap}}, & K_1 &= \sigma_{\text{trap}} C_{\text{trap}},\end{aligned} \tag{A.9}$$

Eqns. (A.7) describe the change in the populations of free positrons (n_1) and positrons trapped at defects (n_2), where σ_{trap} denotes the specific trapping rate, C_{trap} the trap concentration, and K_{detrap} the detrapping rate. λ_{free}, λ_{trap} denote the annihilation rates from the free and trapped states, respectively.

POSITRONIUM CHEMISTRY

With the initial condition

$$n_2(0) = 3 n_1(0) \tag{A.10}$$

the solution of the rate equations (A.5) yields

$$n_1(t) = n_1(0) \left[1 + \frac{\Lambda_2 - \Lambda_1 + 6 K_2}{\gamma} \tanh \frac{\gamma t}{2}\right] \cosh \frac{\gamma t}{2} \exp \frac{-\alpha t}{2},$$

$$n_2(t) = 3 n_1(0) \left[1 + \frac{\Lambda_1 - \Lambda_2 + \frac{2}{3} K_1}{\gamma} \tanh \frac{\gamma t}{2}\right] \cosh \frac{\gamma t}{2} \exp \frac{-\alpha t}{2}, \tag{A.11}$$

with α and γ given in (A.8). With the identification

$$\begin{aligned}\Lambda_1 &= \lambda_{\text{p-Ps}} + \frac{3}{4} K_{\text{conv}}, & K_2 &= \frac{1}{4} K_{\text{conv}}, \\ \Lambda_2 &= \lambda_{\text{o-Ps}} + \frac{1}{4} K_{\text{conv}}, & K_1 &= \frac{3}{4} K_{\text{conv}},\end{aligned} \tag{A.12}$$

Eqns. (A.11) describe the change of the p-Ps population (n_1) and the o-Ps population (n_2) for a spin-conversion process with a reaction rate K_{conv},

$$\begin{aligned}\text{o-Ps} + \text{M} &\longrightarrow \frac{1}{4} \text{p-Ps} + \frac{3}{4} \text{o-Ps} + \text{M}, \\ \text{p-Ps} + \text{M} &\longrightarrow \frac{1}{4} \text{p-Ps} + \frac{3}{4} \text{o-Ps} + \text{M},\end{aligned} \tag{A.13}$$

where M denotes a paramagnetic solute and $\lambda_{\text{p-Ps}}$, $\lambda_{\text{o-Ps}}$ are the annihilation rates of the p-Ps and o-Ps, respectively. A third rate equation

$$\frac{dn_{e+}}{dt} = -\lambda_{e+} n_{e+} , \qquad (A.14)$$

with the solution

$$n_{e+}(t) = n_{e+}(0) e^{-\lambda_{e+} t} , \qquad (A.15)$$

has to be considered if a finite fraction of the implanted positrons does not form positronium but annihilates from the free positron state with an annihilation rate λ_{e+} (see Sect. 2).

REFERENCES

1. F. H. H. Hsu, C. S. Wu, Phys. Rev. Lett. 18, 889 (1967).

2. J. D. McGervey, V. F. Walters, Phys. Rev. B2, 2421 (1970).

3. I. K. McKenzie, B. T. A. McKee, Appl. Phys. 10, 245 (1976).

4. I. K. McKenzie, P. Sen, Phys. Rev. Lett. 37, 1296 (1976).

5. Y. Kishimoto, S. Tanigawa, in: Positron Annihilation eds. P. G. Coleman, S. C. Sharma, L. M. Diana (North Holland, Amsterdam, 1982) pp. 404, 790, 815.

6. W. Bauer, K. Maier, J. Major, H.-E. Schaefer, A. Seeger, H.-D. Carstanjen, W. Decker, J. Diehl, H. Stoll, Appl. Phys. A43, 261 (1987).

7. W. Bauer, J. Briggmann, H.-D. Carstanjen, S. Connell, W. Decker, J. Diehl, K. Maier, J. Major, H.-E. Schaefer, A. Seeger, H. Stoll, E. Widmann, Nucl. Instr. and Meth. B50, 300 (1990).

8. H. Stoll, M. Koch, K. Maier, J. Major, Nucl. Instr. and Meth. B56/57, 582 (1991).

9. H. Stoll, P. Wesolowski, M. Koch, K. Maier, J. Major, A. Seeger, Mat. Sci. Forum 105 - 110, 1989 (1992).

10. R. Würschum, W. Bauer, K. Maier, A. Seeger, H.-E. Schaefer, J. Phys.: Condens. Matter 1, SA 33 (1989).

11. H.-E. Schaefer, W. Eckert, J. Briggmann, W. Bauer, J. Phys.: Condens. Matter 1, SA 97 (1989).

12. I. Billard, J. Ch. Abbé, G. Duplâtre, J. Phys. Chem. 88, 2071 (1991).

13. J. F. Prins, Mat. Sci. Reports, 7, 271 (1992).

14. R. C. De Vries, Ann. Rev. Mater. Sci., 17, 161 (1987).

15. J. C. Angus and C. C. Hayman, Science, 241, 913 (1988).

16. M. Koch, K. Maier, J. Major, A. Seeger, J.P. F. Sellschop, E. Sideras-Haddad, H. Stoll, S. H. Connell, Mat. Sci. Forum 105 -110, 671 (1992).

17. S. Dannefaer, P. Mascher, D. Kerr, Diamonds and Related Materials 1, 407 (1992).

18. Xue-Song Li, S. Berko, A. P. Mills, Jr., Mat. Sci. Forum 105 - 110, 739 (1992).

19. J. P. F. Sellschop, in: The Properties of Diamonds, ed. J. E. Field (Academic Press, New York, 1991).

20. M. J. Puska, S. Mäkinen, M. Manninen, R. M. Nieminen, Phys. Rev. B 39, 7666 (1989).
21. S. Y. Chuang, W. H. Holt, B. G. Hogg, Can. Journ. Phys. 46, 2309 (1968).
22. T. L. Estle, S. Estreicher, D. S. Morguick, Phys. Rev. Lett. 58, 1547 (1987).
23. P. Wesolowski, K. Maier, J. Major, H. Stoll, T. Grund, M. Koch, Nucl. Instr. and Meth. B68, 468 (1992).
24. T. Grund, K. Maier, A. Seeger, Mat. Sci. Forum 105 - 110, 1879 (1992).

SECTION 3
SURFACE STUDIES

POSITRON RE-EMISSION STUDIES OF THE GROWTH AND ANNEALING PROPERTIES OF EPITAXIAL PALLADIUM OVERLAYERS ON Cu(100)

G.W. Anderson, K.O. Jensen*, T.D. Pope, K. Griffiths,
P.R. Norton and P.J. Schultz
Departments of Chemistry and Physics, The University of
Western Ontario, London, Ontario, Canada N6A 3B7

ABSTRACT

The growth and annealing properties of Pd overlayers on Cu(100) have been investigated using re-emitted positron spectroscopy (RPS), electron work function, low energy electron diffraction (LEED) and Auger electron spectroscopy (AES). Work function measurements indicate that by 1.5 ML coverage the surface of the film resembles a pure Pd layer. Two changes in the overlayer growth mode have been observed at 0.5 ML and 1 ML Pd coverages, corresponding to the completion of the first alloy layer and the beginning of the growth of bulk Pd. The bulk Pd film does not grow epitaxially and contains ~1% of vacancy-type defects. For 0.5 ML Pd overlayers the annealing of defects associated with the alloy surface has been observed at 353K to have a characteristic time constant of $\tau = 82\pm5$ min.

INTRODUCTION

Currently there is great interest in the potential applications of metal epitaxy on metal substrates. Unique structural phases, unstable in the bulk material, can be grown as thin films by epitaxial growth on a suitable substrate. These structural phases when combined with the two-dimensional nature of these systems often result in unique chemical, electronic and magnetic properties. In a manner analogous to strained layer superlattices in the semiconductor industry, it is hoped that by selecting appropriate material and growth conditions it will be possible to engineer metallic hetero-systems with specific properties.

The potential of slow positrons as a probe for metal overlayer systems has recently been realized [1-5]. The work function of a positron in a solid material can be defined in a manner analogous to the electron work function. However, the effect of the surface dipole contribution is reversed (due to the positron charge) and the positron work function of a material can be negative. A positron diffusing in such a negative work function material may follow one of three branches if it encounters the surface: spontaneous re-emission with energy equal to the work function, scattering into a surface state or positronium formation. In RPS the re-emitted positrons are examined to provide information about the emitting surface.

In the present study positron re-emission results are correlated with a variety of other surface science techniques to allow the investigation of the growth, morphology and annealing of ultrathin Pd overlayers on Cu(100). Previous studies of this system

[6-9] have shown that the first 0.5 ML of Pd forms a well ordered 50/50 Pd/Cu surface alloy with a c(2×2) LEED pattern.

EXPERIMENTAL

Two separate UHV systems (base pressures $<1\times10^{-10}$ torr) equipped with Pd evaporation sources were utilized in this study. One of the systems (described elsewhere [10]) is equipped with a four grid retarding field analyser (RFA) for LEED and AES experiments and is interfaced to a variable energy, magnetically guided positron beam for positron re-emission measurements. The other chamber is equipped with a similar RFA and a Kelvin probe for electron work function measurements.

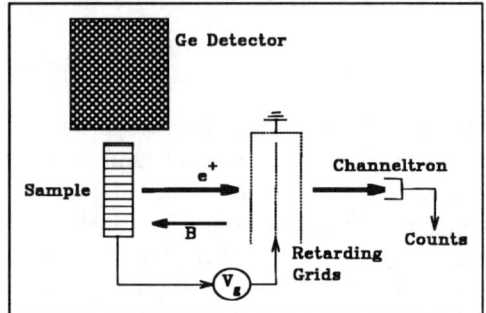

Fig. 1. A schematic representation of the apparatus for the positron experiments.

The Pd overlayers were deposited from a sublimation source onto the Cu(100) substrate at 295K. The sample preparation and cleaning procedures have been described previously [9]. All positron experiments were performed using a 3 keV positron implantation energy.

In the positron re-emission experiments, schematically illustrated in Figure 1, only the component of the re-emitted positron energy along the beam direction can be measured, due to the magnetic field used to steer the positron beam. This results in the measured energy of the re-emitted positrons being $E\cos^2\alpha$, where E is the total energy and α is the angle between the sample normal and the incident positron beam [11]. By measuring the RPS spectra at angles of $\alpha=0°$ and $\alpha=60°$ the positron work function can be determined according to

$$-\phi^+ = \frac{4}{3}(E_0 - E_{60}) \quad (1)$$

where E_α is the energy of the elastic re-emission peak in the RPS spectra.

The absolute positron work function of clean Cu(100) was measured to be $\phi^+=-0.24 \pm 0.04$ eV. While this result is in disagreement with a recent theoretical prediction of $\phi^+=+0.22$ eV [12], it is in reasonable agreement with related measurements of ϕ^+ for Cu(110) and (111) [13] and the known differences in surface dipole as determined from ϕ^- measurements [14].

Figure 2 displays the change in positron and electron work functions (the latter measured using the Kelvin probe) with Pd coverage. The solid curve is an empirical fit to the data utilized in subsequent modeling.

The work function change saturates after approximately 1.5 ML of Pd has been deposited, indicating that at 1.5 ML the surface layer resembles a complete Pd layer.

In the electron work function data there appears to be a plateau around 0.5 ML coverage, which likely coincides with the completion of the alloy layer and a subsequent change in the growth mode. Although evidence for this plateau is limited to a few data points, four separate experiments displayed the same structure.

Figure 3 shows the dependence of the yield of re-emitted positrons (Y^+) on Pd coverage. Y^+ is determined by comparing the number of gamma photons (measured with a Ge detector) emitted by positron annihilations when positrons are being re-emitted (γ_r) versus when they are confined to the sample (γ_c) by an appropriate bias [4,5].

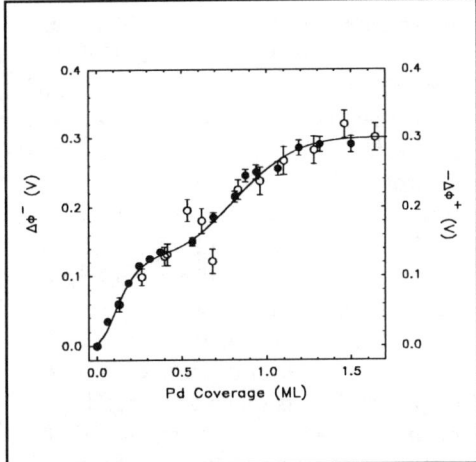

Fig. 2. The change in positron work function (open circles) and electron work function (closed circles) with Pd coverage.

$$Y^+ = \frac{\gamma_c - \gamma_r}{\gamma_c} \quad (2)$$

Below 1 ML, Y^+, like $-\phi^+$, increases with Pd coverage, showing evidence of a more pronounced plateau near 0.5 ML than in the work function results. This behavior is expected, as a work function decrease has been observed to correlate with an increase in Y^+ for other absorbate systems [15]. However, at approximately 1 ML Pd coverage Y^+ passes through a maximum and begins to decrease. This decrease is not due to changes in the work function, which is still increasing in this region.

The data are fit with a modified form of the density of final-states (DOFS) model of Gullikson et al [15]:

$$Y^+(\theta) = \frac{Y_o(\phi^+(\theta))^{1/2}}{(\phi^+(\theta))^{1/2} + b} \left(1 - \frac{N^t}{1 + e^{-(\theta-c1)/w1}}\right) \quad (3)$$

where θ is the Pd coverage and b accounts for the surface branching ratios. The prefactor Y_o was constrained at 100% and $\phi^+(\theta)$ is modeled with the curve shown in Figure 2. Equation (3) was found to provide a good fit to the data above 0.5 ML, with the sigmoidal loss term of amplitude N^t accounting for the decrease in Y^+ above 1 ML. However to fit the data for all θ we had to modify equation (3) to account for the fact that Y^+ does not track ϕ^+ exactly below 0.5 ML. We elected to do this by replacing the constant b with $b(\theta)=b_0+b_1\exp((\theta-c)/\sqrt{2}w)^2$ to model the variation of the surface branching ratios with θ. The solid curves in Figure 3 represent this fit, while the dashed curves are the original DOFS model.

The offset between the two experiments is found to be due to different surface branching ratios: $b_0=1.02\pm0.01$ for the upper curve and $b_0=1.20\pm0.04$ for the lower curve. These values are in good agreement with the values of 1.16 and 1.12 found for Ni(100) and Cu(111) respectively [16]. The higher value of b_0 for the lower data is indicative of a less perfect substrate, resulting in more positrons scattering into the surface state and/or positronium formation.

The gaussian modifiction to b_0 is centred at a coverage of $c = 0.35 \pm 0.04$ ML with a width of $w = 0.11 \pm 0.05$ ML. This is attributed to a surface effect, most likely due to the evolution of surface defects (such as steps) during the initial stages of the growth of the first alloy layer and their subsequent removal as the layer is completed.

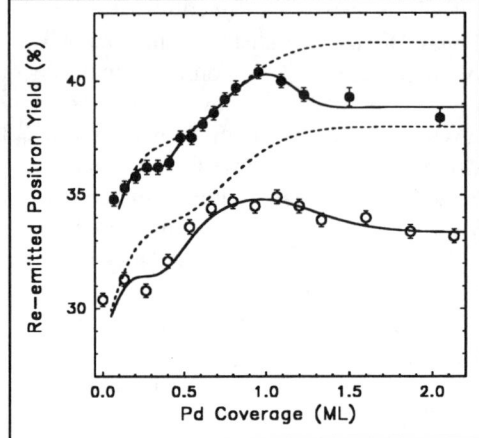

Fig. 3. The change in re-emitted positron yield with Pd coverage for two experiments.

While it has not yet been established conclusively that surface defects trap positrons [16,17] it has been demonstrated experimentally [18] that they do effect the positron surface branching ratios, and therefore in all likelyhood the positron re-emission probability.

The decrease in Y^+ modeled by the sigmoidal term is centred at a coverage of $c1 = 1.05 \pm 0.09$ ML with a width of $w1 = 0.22 \pm 0.09$ ML. This effect persists as the coverage increases further, resulting in the decrease in Y^+ observed above 1 ML coverage. This effect is attributed to positron trapping due to the overlayer defects and/or a change in the surface branching ratios due to differences in the morphology of the overlayers. While it is not possible to unambiguously distinguish between these two causes we associate the effect with a change in growth mode from the intial interfacial alloy layers to the growth of bulk Pd at a coverage of 1 ML.

If this effect is assumed to be solely due to vacancy defects, then the positron trapping fraction of $N^t = 0.09 \pm 0.03$ corresponds to a defect concentration of the order of $C^v = N^a N^v {v^+}/\mu d = 3 \times 10^{14}$ cm^{-2}. This result is obtained by assuming the positrons are moving through the overlayer of thickness d with a velocity of $v^+ = (2\Delta^+/m)^{1/2}$, where $\Delta^+ \approx 0.5$ eV [19,20] is the difference between the Pd and Cu positron affinities, m is the positron mass, $N^a = 1.53 \times 10^{15}$ cm^{-2} is the atomic density of a Cu(100) plane and μ is the specific trapping rate, which for vacancy

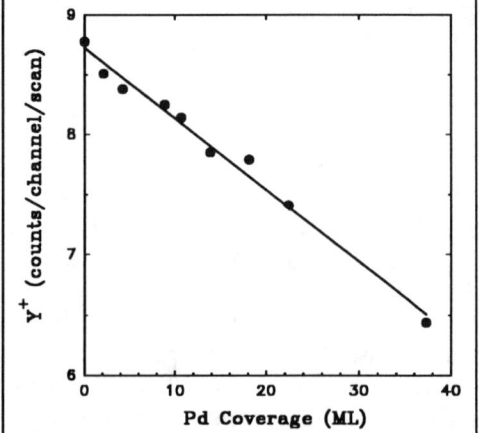

Fig. 4. The change in re-emitted positron yield as a function of Pd coverage. The decrease in yield is due to positron trapping at defects in the bulk Pd film.

type defects in metals is on the order of $10^{15} s^{-1}$ [21,22]. This number, on the order of 18% of a monolayer seems unrealistically high, implying that both depletion mechanisms discussed are likely playing a role.

The Y^+ versus coverage data is extended to a coverage of 37 ML in Figure 4. For these data, Y^+ was measured by counting the number of positrons re-emitted per scan during an RPS experiment. The yield is observed to decrease linearly with coverage, which is attributed to trapping at vacancy-type defects in the bulk Pd film. Fitting the data gives a trapping fraction of $N^t=0.007$, which corresponds to a defect concentration of $C^v=2\times10^{13}$ cm^{-2} ML^{-1}. This defect concentration, on the order of 1% of a monolayer, is consistent with the observation that for coverages above 1 ML no LEED pattern is observed, indicating that the bulk Pd film does not grow epitaxially, but is either amorphous or has extremely small crystallites.

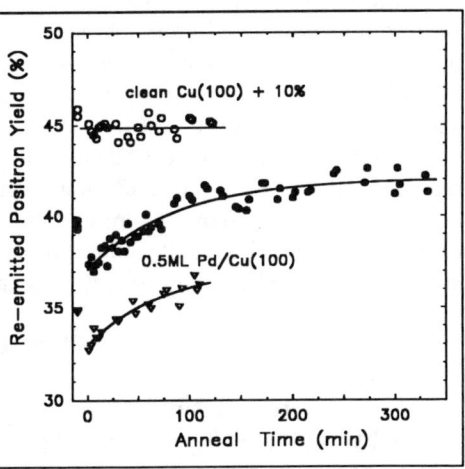

Fig. 5. Re-emitted positron yield as a function of annealing time at 353K for clean Cu(100) (offset by +10% for display purposes) and 0.53 ML Pd/Cu(100).

The effects of annealing upon Y^+ for the 0.53 ML Pd/Cu(100) alloy surface are illustrated in Figure 5. Several samples were prepared at 300K and annealed to 353K at a heating rate of approximately 1 K/s. Y^+ was measured before the anneal (shown at t =-10 min) and as a function of time at 353K.

For clean Cu(100) there is an initial drop in Y^+ after heating to 353K, then the yield remains constant over 120 min. For 0.53 ML Pd/Cu(100) overlayers a similar initial drop in yield occurs upon heating. As the samples anneal, Y^+ is observed to increase by ~5% over a 6 hr anneal. The increase in yield during the anneal has been fit with a function of the form $(1-e^{-t/\tau})$, giving a characteristic annealing time of $\tau = 82 \pm 5$ min. At the end of one of the experiments the sample was cooled to 295K

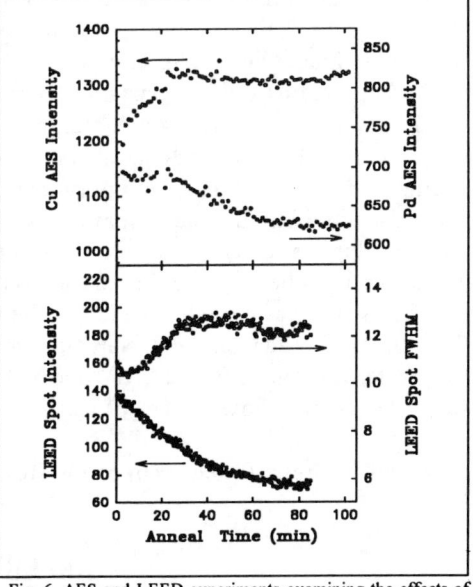

Fig. 6. AES and LEED experiments examining the effects of annealing at 353K.

and Y^+ was found to further increase by an amount equivalent to the initial decrease. This indicates that the source of the initial decrease is a reversible temperature effect, likely due to changes in the positron diffusion constant or the positron surface branching ratios.

Figure 6 illustrates the results of LEED and AES experiments performed under similar annealing conditions. In the LEED experiment the 1/2 order beams (due to the c(2×2) alloy layer) are observed to broaden for times <30 min and then sharpen slowly. They also decrease in intensity during the anneal. The AES result shows a decrease in the Pd (330+326 eV) signal and an increase in the Cu (60 eV) signal. These results are attributed to the dissolution of some of the Pd into the bulk and its replacement by Cu. However, during the anneal the positron work function is observed to be constant. Therefore the change in Y^+ is not associated with work function changes, but is attributed to annealing of defects associated with the alloy surface, facilitated by the dissolution of a small amount of Pd.

CONCLUSIONS

In summary, we have studied the structure and defects of Pd overlayers on Cu(100) grown at 295K. Independent positron and electron workfuction measurements are in good agreement, and show that by 1.5 ML Pd coverage the surface resembles a pure Pd layer. Two changes in growth mode have been observed in re-emitted positron yield measurements: the completion of the alloy layer at 0.5 ML and the change from the intial interfacial layers to the growth of bulk Pd at 1 ML. The bulk Pd film has been observed to contain 1% of vacancy-type defects. Annealing of the 0.53 ML alloy surface has been shown to result in an improvement in the positron re-emission yield. This is attributed to the annealing of defects associated with the alloy surface facilitated by the dissolution of a small amount of Pd into the bulk, and occurs with a characteristic time constant of $\tau = 82 \pm 5$ min at 353K.

These measurements have highlighted the usefulness of the positron re-emission yield as a probe for studying metal epitaxial systems. Due to its sensitivity to overlayer defects and morphology (unobservable by other techniques), great potential exists for providing unique insight into such systems. Also of particular interest is the potential applications to the positron re-emission microscope [23,24], since differences in surface morphology should lead to a clear re-emission contrast.

This work was partially funded by the Natural Sciences and Engineering Research Council of Canada and the Network of Centres of Excellence in Molecular and Interfacial Dynamics (CEMAID), one of fourteen Networks of Centres of Excellence supported by the Government of Canada.

* Present Address: Department of Physics, University of Essex, Colchester C04 3SQ, UK.

REFERENCES:

1. P.J. Schultz, K.G. Lynn, W.E. Frieze and A. Vehanen, Phys. Rev. B **27**, 6626

(1983).
2. D.W. Gidley and W.E. Frieze, Phys. Rev. Lett. **60**, 1193 (1988).
3. D.W. Gidley, Phys. Rev. Lett. **62**, 811 (1989)
4. J.G. Ociepa, P.J. Schultz, K. Griffiths and P.R. Norton, Surf. Sci. **225**, 281 (1990).
5. P.J Schultz, K.G. Lynn, Rev. Mod. Phys. **60**, 701 (1988).
6. G.W. Graham, Surf. Sci. **171**, L432 (1986).
7. S.C. Wu, S.H. Lu, Z.Q. Wang, C.K.C. Lok, J. Quinn, Y.S. Li, D. Tian, F. Jona and P.M. Marcus, Phys. Rev. B **38**, 5363 (1988).
8. G.W. Graham, P.J. Schmitz and P.A. Thiel, Phys. Rev. B **41**, 3353 (1990)
9. T.D. Pope, G.W. Anderson, K. Griffiths, P.R. Norton and G.W. Graham, Phys. Rev. B **44**, 11518 (1991).
10. P.J. Schultz, Nucl. Instrum. Methods B **30**, 94 (1988).
11. C.A. Murray and A.P. Mills, Jr., Solid State Commun. **34**, 789 (1980).
12. O.V. Boev, M.J. Puska and R.M. Nieminen, Phys. Rev. B **36**, 7786 (1987).
13. C.A. Murray, A.P. Mills, Jr. and J.E. Rowe, Surf. Sci. **100,** 647 (1980).
14. J.Hölzl and F.K. Schulte, in *Solid Surface Physics: Springer Tracts in Modern Physics No. 85*, edited by G. Höhler (Springer-Verlag, Berlin, 1979) p 88.
15. E.M. Gullikson, A.P. Mills, Jr. and C.A. Murray, Phys. Rev. B **38**, 1705 (1988).
16. R.M. Nieminen and M.J. Puska, Phys. Rev. Lett. **50**, 281 (1983).
17. A.P. Brown, K.O. Jensen and A.B. Walker, Surf. Sci. **211/212**, 173 (1989).
18. A.R. Köymen, D.W. Gidley and T.W. Capehart, Phys Rev. B **35**, 1034 (1987).
19. G.W. Anderson, T.D. Pope, K.O. Jensen, K.Griffiths, P.R. Norton and P.J. Schultz, to be published.
20. M.J. Puska, P. Lanki and R.M. Nieminen, J. Phys. Cond. Matt. **1**, 6081 (1989).
21. T. McMullen and M.J. Stott, Phys. Rev. B **34**, 8985 (1986).
22. M.J. Puska and M. Manninen, J. Phys. F **17**, 2235 (1987).
23. J. Van House and A. Rich, Phys. Rev. Lett **61**, 488 (1988).
24. G.R. Brandes, K.F. Canter and A.P. Mills, Jr., Phys. Rev. Lett. **61**, 492 (1988).

Verification of Focusing from a Hemispherically Shaped Surface.

Benjamin L. Brown, Tamara S. Andrew,
Margaret S. Clarkson, C. Sean Sutton
Mount Holyoke College
South Hadley, MA 01075

Pedro Encarnación
University of Michigan
Ann Arbor, MI

Art Denison and Henry Makowitz,
EG&G INEL
Idaho Falls, ID

Kermit Bundy
Idaho State University
Pocatello, ID

Abstract

The proposed intense slow positron source (ISPS)[1] at Idaho National Engineering Laboratory (INEL) requires focusing from a large area source to a small (1 cm) remoderator foil. A diameter reduction of >30 is desirable, and a gridless design appears to accommodate the design requirements. Experimental confirmation of the gridless focusing concept is presented here together with SIMION (simulated ion tracking) computer simulations, which indicate a diameter reduction of ~50x.

Prepared for U. S. Department of Energy through the EG&G Idaho, Inc. LDRD program. Under DOE-ID Contract DE-AC07-76ID01570

The proposed intense slow positron source at INEL will have a maximum of 10^{13} slow positrons emanating from a large area source dish. In order to focus the positrons efficiently to the foil remoderator for brightness enhancement, an appropriately designed electrostatic focusing geometry has been studied. The design features gridless focusing, thus avoiding the complexity and inefficiency of grid optics, and making the experimental goal of a >30x diameter reduction much easier.

The SIMION program uses relaxation calculations to obtain solutions to Laplace's equation with specified boundary conditions, and the potentials in our case are calculated to ~1% accuracy. In addition, an analytical solution to a nearly identical geometry indicates agreement with both the SIMION simulations and the PMT measurements. A separate measurement of a uniformly illuminated photocathode shows linearity well above a total current of 1 μa, indicating no unforeseen problem at the anticipated positron beam currents.

After some initial design considerations involving a grid focused system,[2,3] a gridless design appears to be superior for the ISPS beam. The gridless design has the following advantages: increased efficiency (x2) with the grid and grid mounting structure eliminated; increased focusing possible due to the elimination of grids that can produce significant defocusing; total insensitivity of alignment in the gridless geometry as opposed to the very fine tolerances and external adjustments needed in the grid focused beam; easy adaptability to a cryogenic gas moderator system; less sensitivity to surface roughness than the grid case; and ease of construction. A typical hemispherical configuration for focusing is shown in Fig. 1. The radius of the hemisphere is 25 cm. The positrons are attracted to a central disk that is biased at -5kV, relative to all other surfaces which are at or near ground potential.

A photomultiplier tube manufactured by Hammamatsu (Fig. 2) has a geometry similar to the geometry that we wish to use with the ISPS at INEL (Fig.3). In the work presented here, we analyze the focusing of low energy photoelectrons from the hemispherical surface of the photomultiplier. All measurements were taken with the PMT in a light tight box with a single green LED as a light source. The LED is positioned externally. The photon count rate was recorded, as the distance from the LED to the PMT was varied both laterally (A) and longitudinally (Z). The measurements show, with a small correction for a systematic variation with position Z, the distance of the LED from the PMT, there is no significant variation in the photon count as a function of arc distance (A in cm) from the central axis. Refer to Fig. 4. This demonstrates a diameter reduction of 8:1 for the first stage, limited only by the large (2.5 cm) acceptance diameter of the first dynode stage.

A SIMION simulation of the PMT geometry is shown in Fig. 3. The

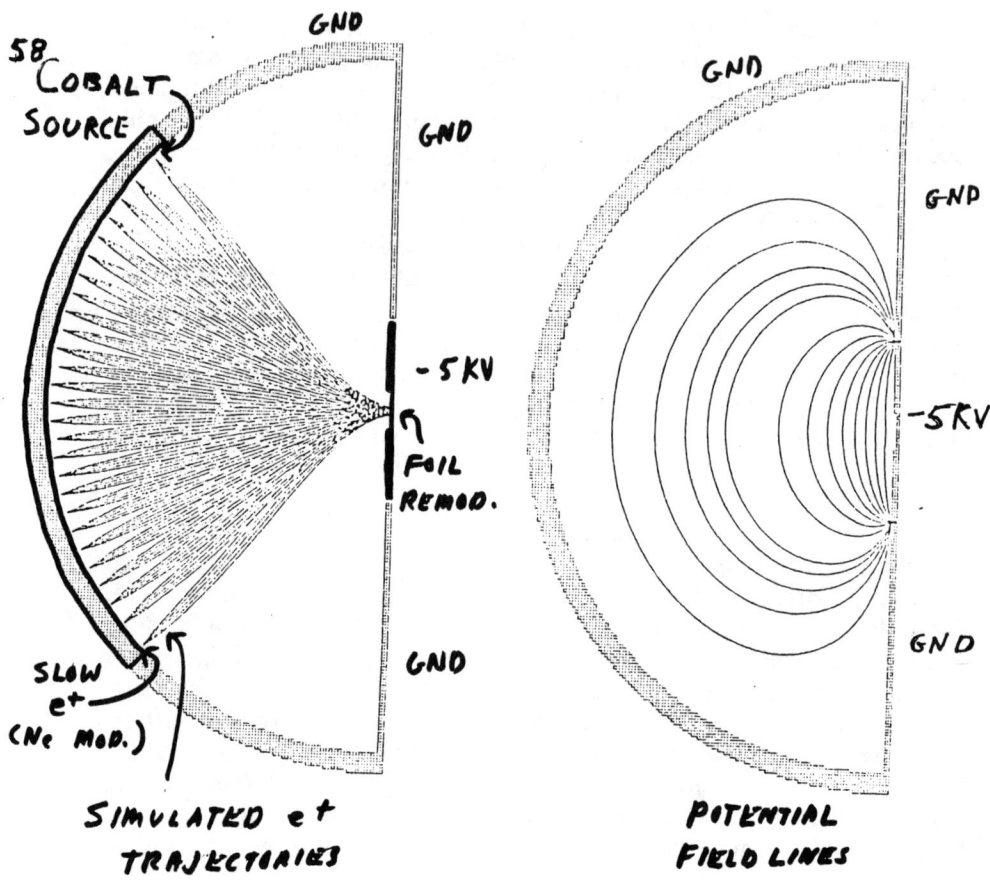

FIGURE 1. Simion simulation of the first stage of an intense positron beam source (ISPS) proposed to be built at INEL. The positron trajectories are shown on the left, with positrons launched at 1 eV with an angle varying to 45 degrees from the surface. Equipotential curves are also shown on the right. The surface shown is a hemisphere which mates to an annular plate. A small disk at the center of the plate is biased to -5 kV. The scale of the proposed beam corresponds to a 25 cm radius of the hemisphere.

internal structure of the tube is shown and the figure represents a cross section with cylindrical symmetry. A second dynode voltage of 800 volts was applied, followed by a multi-dynode chain for amplification (not shown) located in the grounded central cylinder. The focusing in the simulation appears to be dramatic, and it corresponds to the experimental PMT results. A trick was used in the SIMION simulation, and in most of our other similar simulations, to minimize the rough surface effects due to the finite mesh at the hemispherical boundary. The trick is the following: the positrons were launched along an equipotential surface a few mesh units away from the rough surface where the potentials are likewise rougher. This gives a much smoother hemispherical potential surface from which to launch the positrons. The smoothness is easily attainable in practice. With the gridless focusing, the trick improves the focusing slightly (<10% with the finest mesh size used). In simulations of some grid focusing geometries where a grid is placed very near the hemispherical surface,[3] the trick improves the simulation focusing dramatically (x2 or more). In grid designs where the grid is nearer the target and farther from the hemisphere,[2,4] the design is less sensitive to surface roughness and alignment considerations. The trick does not improve the simulation focusing as much for this design.

An analytical solution which is also close to the current geometry appears to independently confirm the above results. Using a Bessel function solution to Laplace's equation, it is not difficult to solve the problem with the boundary conditions of a biased disk set into an infinite grounded plane.[5] This problem is substantially similar to our proposed geometry. From the Bessel function solution involving the potential, the E field can be obtained from the gradient of the potential, and the characteristic focusing determined. Positrons are assumed to be launched on an equipotential surface which approximates our hemispherical surface of emission at a distance of several disk radii from the disk center. The hemispherical emission surface is thus at a slightly more negative potential (for positrons) than the ground potential. This equipotential surface can be correspondingly biased in our SIMION calculations (biasing the hemisphere slightly negative in Fig. 1 for example) for comparison to the Bessel function solution. The analytical solution further supports the simulations and the PMT measurements.

Calculations of space charge effects show no significant effects in the positron beam current range of 1 μA ($\sim 10^{13}$ e$^+$/s), which represents a static density of ~ 100 e$^+$/cm^3. We also examined the Vlassov equation and determined that the E field external, is $>10^5$x the induced E field, and thus collective effects in transport are negligible. Indeed, the plasma oscillation time is an order of magnitude larger than the transit time from the dish to the remoderator. Possible charging of the Ne moderator must be investigated in detailed tests using an intense source.

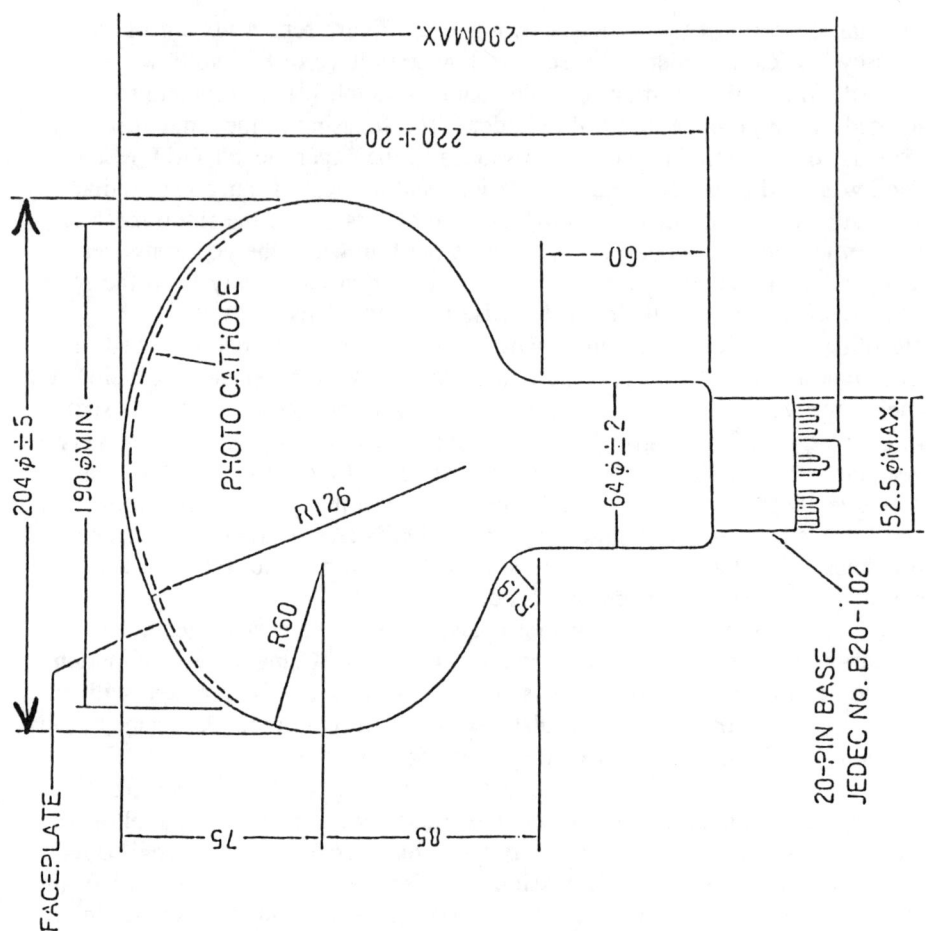

FIGURE 2. Hammamatsu phototube R-1408. This tube uses the same geometry for focusing as the proposed ISPS beam.

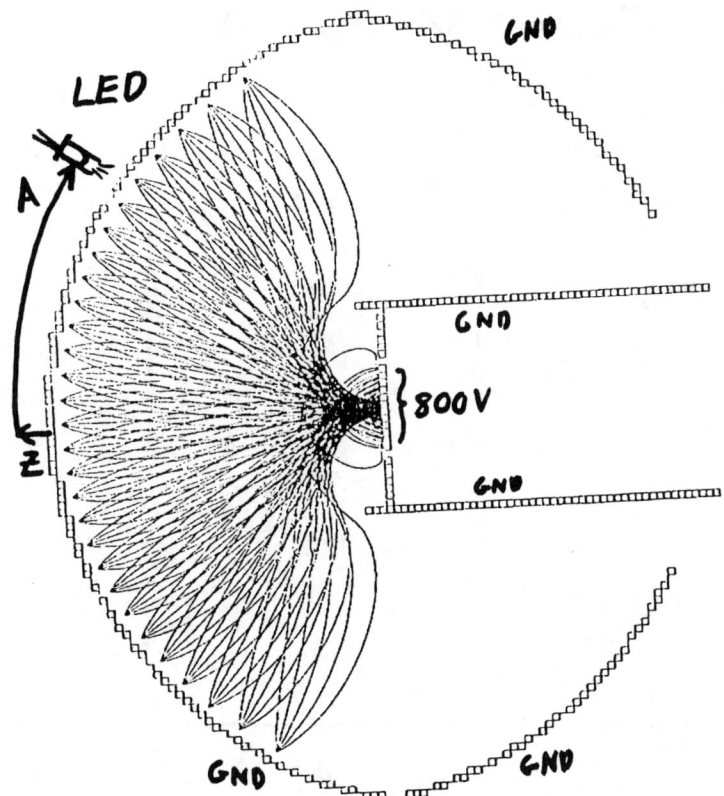

FIGURE 3. SIMION simulation of the PMT focusing from ~20 cm to ~1 cm. The electrons at extreme angles to the central axis (large values of A) tend to fall outside the central 0.6 cm spot.

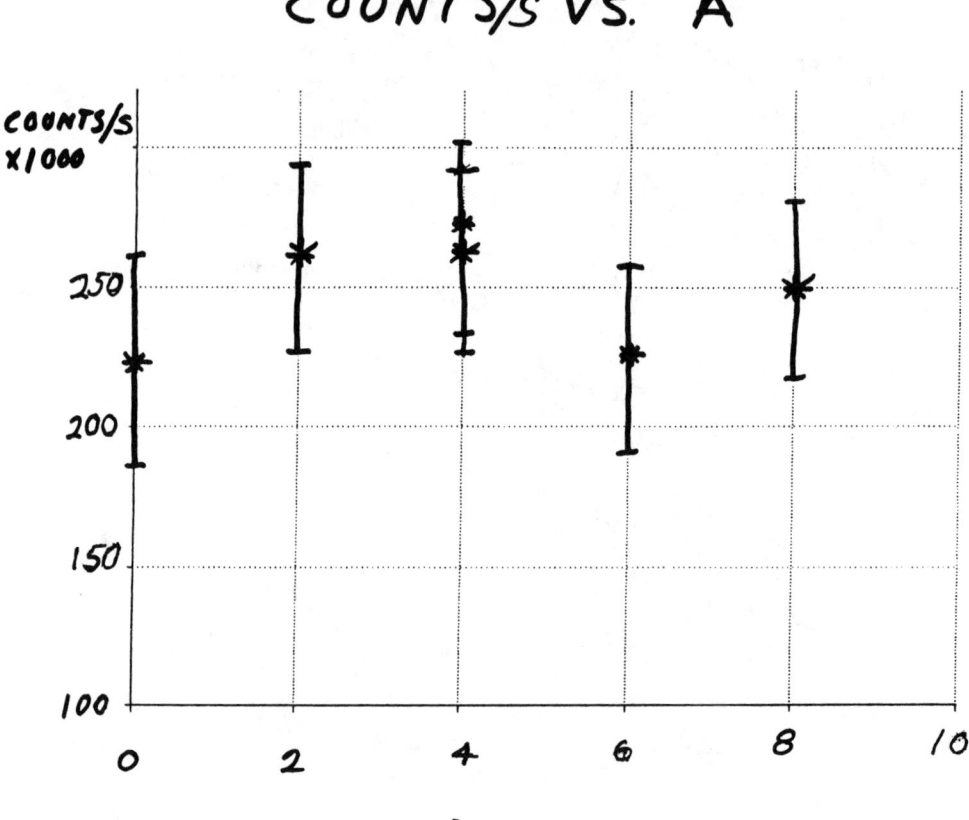

FIGURE 4. PMT response in Counts/S (photoelectron per second) vs. distance A(cm), the position of the LED measured from the PMT central axis. There is no significant change in the variation in counts as the position A is varied. The data has been corrected for a small systematic effect (15%) due to the distance variation of the LED from the PMT surface. The SIMION calculations and the experimental verification with a Hammamatsu phototube both indicate that a hemispherical emission geometry can be used to focus positrons in the proposed beam.

As an additional check of the focusing predictions of the SIMION simulations, a transverse magnetic field was applied with a Helmholtz coil in the laboratory. The PMT response was compared to the SIMION result, with a similar magnetic field strength and geometry. Although the trajectories were quite complicated, there was an unmistakable similarity in the SIMION and the PMT measurement results. For example, the strength of the magnetic field to cut the photocurrent in half was measured to be 80 G in the experiment and 95 G for the SIMION simulations. This leads us to believe that the SIMION calculations are indeed reliable for this class of hemispherical geometries, and a diameter reduction of 50:1 has been demonstrated convincingly.

In the study of magnetic field effects, we found a small sensitivity to magnetic fields at the ~ 3 G level. Large diameter, low current coils could easily be incorporated to cancel stray fields, if they were of that magnitude.

The test of higher beam current was performed with a uniform light source illuminating the photocathode, and a United Detector Technology 40x opto-meter monitoring the irradiance (W/cm^2). The estimated number of electrons leaving the surface was made using a photoefficiency of $\sim 15\%$ given by the manufacturer for the wavelength of light used. With an illuminance of 100 nW/cm^2 a current at the first dynode was measured directly to be 1.6μA. The response was linear up to 10μW/cm^2.

In conclusion, the focusing without a grid seems to work well, and it appears to have many advantages over the design involving a grid. Barring any unforeseen problem, this concept will be incorporated into the ISPS beam in the demonstration Phase I beam.[1] We are confident that the focusing, as predicted by SIMION for this class of gridless spherical geometries, will have a diameter reduction of 50:1.

1. See B. L. Brown elsewhere in this volume.

2. Grid focusing of positrons from a curved surface appears in Eric H. Ottowitte, "High-Intensity Positron Beam Via 200 MW, Large Volume, Test (Fission) Reactor", Proceedings of the BB Factory Workshop, January 27, 1988 (World Press).

3. See H. Makowitz elsewhere in this volume.

4. A. Zecca and R. S. Brusa, Nuclear Inst. and Meth. in Phys. Research **A313**, 337(1992).

5. J. D. Jackson, *Classical Electrodynamics*, Vol 2, problem 3.11.

A Model for the PAES Cu M23VV Signal versus Cs Coverage on the Cu(100) Surface

N.G. Fazleev [a], J.L. Fry, J.H. Kaiser, A.R. Koymen,
T.D. Niedzwiecki and Alex Weiss
Department of Physics, The University of Texas at Arlington,
Arlington, Texas 76019-0059

ABSTRACT

Positron-annihilation-induced Auger-electron spectroscopy (PAES) employs positrons trapped at the surface to create core-holes and so initiate the Auger process in atoms in the topmost layer of the surface. This technique has already proved itself by providing the surface layer chemical composition of surface alloys and chemisorbed systems. Recent experimental investigations of the attenuation of the Cu M23VV Auger peak with Cs coverage on Cu (100) at 163 K using PAES revealed that the normalized peak intensity for Cu remains nearly constant at the clean surface value until the Cs coverage reaches approximately 0.7 physical monolayer at which point the peak intensity drops precipitously. We present a semi-phenomenological analysis of this unusual behavior. Our model treats the positron as trapped in a double well potential in the direction perpendicular to the surface: one well is associated with the Cu substrate and the other with the Cs adsorbate. The sharp drop in the PAES intensity which occurs over a small change in the Cs coverage is attributed to a structural transition from a disordered to a hexagonal close-packed ordered structure, as suggested by LEED spectra.

INTRODUCTION

Like electrons, positrons have quantum-mechanical states at surfaces on a number of metals. These states are critically dependent on the short-range "correlation well" in the proximity of the surface atoms and are the consequence of the interplay between repulsion from the surface ionic cores and the attractive electron-positron correlations just outside the surface. The existence of positron surface states has been demonstrated by the observation that the positron could be thermally desorbed from clean metal surfaces at elevated temperatures as positronium (Ps).[1] While the trapping of electrons at metal surfaces may be interpreted on the basis of a simple long-range image potential, truncated at the surface, there is still much controversy on the modeling of the positron surface state. This is due to the fact that positrons reside so close to metal surfaces that the electron-positron correlations strongly affect the nature of the positron surface states.[2]

Recently, the positron-surface interactions have become the subject of extensive experimental studies by positron-annihilation-induced Auger electron spectroscopy (PAES).[3] This novel technique employs the positron trapped at the surface state to create core-hole excitations and so initiate the Auger processes in atoms in the topmost layer of the surface. Since in PAES the Auger signal intensities are quite sensitive to the spatial distribution of the positron density at the surface, PAES experiments can provide detailed information for clarifying the nature of the positron surface states. In addition, because these positron surface states are the same from which the Ps are thermally excited,[3] the PAES signal intensities should be consistent with the experimental Ps desorption data.

The measurements of the attenuation of the Cu $M_{2,3}VV$ Auger peak with Cs coverage on Cu(100) using PAES[4] were aimed at studies of the effect of the alkali-metal adsorption on the localization of the positron surface states.

The Nieminen - Jensen (NJ) theory[5] predicts that the charge rearrangement that leads to the lowering of the electron work function causes the positron to become localized in the region between the substrate and the alkali-metal overlayer up to the coverage of one physical monolayer, producing an increase in the positron binding energy. Thus, according to this theory the PAES Cu signal should remain close to the clean-surface value after deposition of Cs. The reduction in the PAES Cu signal up to about 55% of the clean surface value at the Cs coverage of approximately 0.9 physical monolayer would be expected due mostly to the attenuation caused by the inelastic scattering of outgoing Auger electrons as they traverse the thin Cs overlayer.[4] Calculations based on the NJ theory indicate that the decrease in the Cu core annihilation probability caused by a decrease in the overlap of the positron wavefunction with the Cu substrate should have a comparatively smaller effect (less than 16%).[4]

However, the Cu PAES signal intensity in these experiments at 163 K drops sharply almost to zero when the Cs coverage reaches approximately 0.7 physical monolayer, significantly deviating from the predictions of the theory.[5,6] The calculations of the Cu PAES signal intensity based on the NJ theory[4] are in reasonable agreement with the experimental data only below the critical coverage of approximately 0.7 physical monolayer, but cannot reproduce the sharp drop in the PAES intensity observed for higher coverages.

We attribute the failure of the NJ theory[4] to explain the observed PAES Cu intensity dependence on the Cs coverage to the fact that the calculations[4] of the positron surface state on alkali-metal-(Cs) covered Cu surfaces were not performed self consistently and did not take fully into account the correlations between the positron and the electrons in the system. Electronic structure calculations show that consideration of the above mentioned factors may significantly change the final theoretical results.[7] In addition, their calculations were performed for certain assumed regular periodic surface structures in the alkali-metal overlayer, which is not the case even for low Cs coverages.[8,9]

The observed behavior of the PAES Cu $M_{2,3}VV$ signal intensity normalized to the clean Cu value, I_{PAES}, correlates strongly with the Cs coverage dependence of the positronium fraction, f_{Ps} as it should do: for coverages at which the Ps fraction is changing rapidly the suppression of the desorption of Ps causes a corresponding increase in the Cu PAES signal and when the Ps fraction increases most rapidly I_{PAES} drops most sharply.

In contrast to the PAES results, the intensity of the electron- induced Cu Auger electron signal decreases linearly to about 60% of the clean surface value at one physical monolayer of Cs coverage consistent with attenuation due to inelastic scattering of the outgoing Auger electrons in the Cs overlayer.

The purpose of this work is to present a model for a semi-phenomenological analysis of the unusual behavior of the PAES Cu $M_{2,3}VV$ signal versus Cs coverage on the Cu(100) surface.

THE MODEL

Computer simulations of the distribution of Cs atoms on a substrate as they go down at random on the Cu surface were performed on the basis of a simple sticking model in which the possibility of one Cs atom landing on top of another was excluded. These simulations show that at approximately 60% alkali coverage of the overlayer area (relative to close packed coverage) no more Cs atoms can be deposited on the Cu surface without crowding some Cs atoms together. Simple calculations show that for the Cs coverages from 60% up to 100% (the latter is adequate to one physical monolayer) the alkali overlayer has an increasing percentage of its area that has hexagonal close-packed structure: at least 25% of the overlayer area should have hexagonal close-packed structure with the nearest Cs - Cs distance on Cu (100) of 5.26 Å for the 70% Cs coverage, and at least 50% of the overlayer area should be close packed for the 80% Cs coverage. As a consequence, the alkali adsorbate is expected at high coverages (from 60% up to 100%) to form islands increasing in area with the locally hexagonal close packed structure of the Cs atoms. As it follows from calculations, the areas with close-packed structure of the Cs atoms in the overlayer appear mostly between the 60%-80% Cs coverage leading to the disorder-order structural transition in the Cs adsorbate in the Cs/Cu(100) system.

The results of our computer simulations of the Cs distribution on the Cu(100) surface are supported by the low-energy electron diffraction (LEED) observations.[8,9] Studies of the deposition of Cs on a Cu(100) surface for coverages varying between 0 and one monolayer by LEED at low temperatures revealed that the Cs adatoms occupy hollow sites with fourfold symmetry for coverage up to 0.7 monolayer. Above 0.7 monolayer coverage, the Cs overlayer is formed by two kinds of domains with quasi-hexagonal meshes rotated by 90°.

As indicated by studies of the spectrum of collective and single-particle excitations of alkali adsorption systems using electron-energy-loss spectroscopy (EELS)[10] the electronic structure of the Cs adsorbate appears to change in a gradual manner: a small change in coverage produces a small change of excitation energy due to the rather uniform distribution of adatoms. As a consequence, there should be no areas with close-packed Cs atoms at coverages below the critical coverage of 60% which is consistent with the performed computer simulations of the distribution of Cs atoms. The EELS studies and investigations of coverage-dependent binding energies by ultraviolet photoelectron spectroscopy (UPS)[10] reveal that a characteristic feature of the Cs overlayer is the large difference in the effect of the adsorbate at low and high coverage. Studies of the charge transfer between adsorbate and substrate at different Cs coverages by measuring the threshold energy for alkali core-level excitations using EELS[10] revealed that the threshold for Cs 5p excitations decreases from 13.2 eV for the more ionic Cs adsorbate at low coverage to 11.6 eV for the nearly neutral adsorbate at high coverage. Such changes in the character of the Cs adsorbate with coverage correlate with the Cs coverage dependence of the electron work-function. A depletion of electronic charge on the vacuum side of the adsorbate and increase of electronic charge in the Cs/Cu interface region at low coverages produces the rapid initial drop of the electron work function. The change in the electron work function at low coverage is equated with the potential drop across a dipole layer at the surface created by the polarized adatoms. The variation of the work-function with the Cs coverage suggests that the alkali overlayer is essentially neutral at coverages close to a full monolayer. Similar results for the work function changes due to adsorption of alkali-metal overlayer on a transition-metal surface are obtained in the theoretical calculations by Lang[11] within a jellium model and by Wimmer et al.[7] and Ning [12] for the Cs/W system.

Thus, at low coverages the Cs adsorbate can be regarded as an array of polarized adatoms, whereas at high coverage its behavior is that of a metal. As the Cs coverage reaches the value of approximately 60% the areas with a hexagonal close-packed structure of Cs atoms appear in the Cs overlayer due to the structural phase disorder-order transition. In these close packed areas, Cs atoms lose their atomic character and form the two-dimensional metallic adsorbate.

In our model we assume that, as the Cs overlayer becomes metallic, positron surface states appear on the vacuum side of the Cs overlayer. As a consequence of this assumption, we treat the positron as trapped in a double well potential in the direction perpendicular to the surface: one well is associated with the Cu substrate and the other with the Cs adsorbate. Both the positron surface state on the vacuum side of the clean Cu surface and the positron surface state at the Cs/Cu interface are characterized by the binding energy E_1. (For the purposes of simplicity we neglect the possible energy difference between the binding energies of these positron surface states.) The positron surface state on the vacuum side of the areas of alkali adsorbate with the hexagonal close-packed structure of Cs atoms is characterized by the binding energy E_2.

The relative positions of the energy levels of the positron surface states between different materials in contact are determined by the positron affinity, A_+[13,14]: the difference between the lowest positron energies on the different sides of the interface is given by the difference in the positron affinities. The self consistent calculations of the positron affinity performed for bulk Cu and Cs give -4.81 eV and -6.94 eV, respectively.[14] However, it is necessary to take into consideration that the positron energy level in bulk is higher than in vacuum since the surface dipole is slightly larger in magnitude than the positron chemical potential.[13,14] In addition, the electronic structure of the Cs overlayer on the Cu substrate may differ considerably from that of the bulk. Thus, the calculated affinity values for the bulk can not be used to determine the relative positions of the positron binding energies in the Cs adsorbate and Cu substrate. The calculated[6] value for the positron binding energy E_b for the clean Cu(100) surface is equal to 2.77 eV, the calculated value for the positron surface binding energy for the Cs overlayer is not available.

In our model we assume also that the energy level E_2 associated with the positron surface state on the vacuum side of the Cs overlayer lies lower than the energy level E_1 associated with the positron surface states on the clean Cu surface and at the Cs/Cu interface. As a result, as soon as the positron surface state appears on the vacuum side of the alkali overlayer with close-packed structure of Cs atoms, it would become occupied by positrons. Since according to the results of the computer simulations the formation of areas with a hexagonal close-packed structure of Cs atoms occurs mostly at the 60%-80% Cs coverage, the relative number of positrons that can occupy the surface state on the vacuum side of the close packed islands of Cs can be regarded to be a sharp function of the Cs coverage:

$$f(c) = 1/[1 + e^{-\alpha(c-c^*)}] \tag{1}$$

where $c^* \sim 70\%$, and $\alpha \sim 81$. The parameter c^* defines the Cs coverage at which the positron surface state appears on the vacuum side of the Cs overlayer, its value is determined from computer simulations and the LEED spectra.[8,9] The parameter α defines the sharpness of the coverage dependence, its value is determined from the fitting of the obtained theoretical curve for the Cs coverage dependence of the normalized PAES Cu intensity to the experimental data. The number of positrons that can occupy the positron states at Cs/Cu interface is taken to be constant. Then the normalized PAES Cu intensity would be proportional to the probability of the positron to occupy the state at the Cs/Cu interface, which is given by the following expression:

$$f = 1/[1 + f(c) \, e^{-\Delta E/kT}] \tag{2}$$

where $\Delta E = E_2 - E_1$ is the difference between the binding energies of the positron states associated with the Cs adsorbate and the Cu substrate, E_2 and E_1, respectively. (For simplicity we neglect the possible Cs coverage dependence of ΔE.)

The results of the theoretical calculations for the Cs coverage dependence of the normalized PAES Cu intensity multiplied by an attenuation factor determined from the attenuation of the EAES intensity, to take into consideration inelastic scattering of the outgoing Auger electrons, are presented in Fig. 1.

As can be seen from Fig. 1, the model correctly describes the behavior of the experimental PAES Cu signal with the Cs deposition. The calculated normalized PAES Cu intensity remains nearly constant at the clean surface value until the Cs coverage reaches approximately 0.7 physical monolayer at which the theoretical curve drops sharply in agreement with the experimental results. According to the proposed model this drop occurs due to the fact that above the critical Cs coverage a structural phase disorder-order transition takes place in the Cs overlayer. As a consequence of this structural transition the Cs overlayer becomes metallic and the positron surface state appears on the vacuum side of the Cs overlayer.

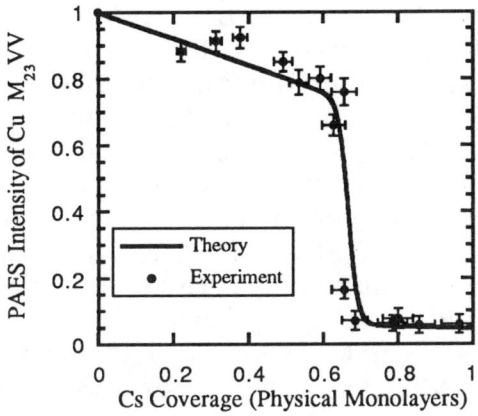

Fig. 1 Normalized Cu M$_{23}$VV PAES intensity plotted as function of Cs coverage

Due to the fact that the energy of this state is lower than the energy level of the positron state at the Cs/Cu interface, the positron surface state initially localized in the Cs/Cu interface becomes depopulated as a result of occupation of positrons of the surface state localized on the vacuum side of the Cs overlayer. This leads to a sharp drop in the PAES Cu intensity. The nonzero value of the PAES Cu intensity at Cs coverages exceeding the critical coverage is due to partial population of the positron state at the Cs/Cu interface.

The predicted core annihilation probabilities are very small for the Cs core levels. For example, calculation of bulk core annihilation rates[6] indicate that of the levels giving Auger transitions in the experimental range currently available, the two highest probabilities are 0.078% for 4p and 0.26% for 4d as compared to ~ 6% for the 3p level of Cu.[6] As a consequence of this it is difficult to observe the PAES Cs signal.

CONCLUSIONS

The PAES data obtained at 163 K indicate that above the critical Cs coverage there is a structural transition from disordered to the hexagonal close-packed ordered structure of the Cs overlayer. This is supported by a computer stimulation of the distribution of the Cs atoms on the Cu substrate performed within the static sticking model and by observation of the evolution of the LEED pattern. Thus, PAES can serve as a method for investigating structural transitions in the topmost layer.

A simple model which treats the positron as trapped in a double well potential in the direction perpendicular to the surface (one well located just outside the Cu substrate and the other located on the vacuum side of the Cs adsorbate), explains the observed behavior of the normalized Cu $M_{23}VV$ PAES intensity at 163 K with Cs coverage. A sharp drop in the normalized Cu $M_{23}VV$ PAES intensity at 163 K for the system Cs/Cu(100) which occurs over a small change in the Cs coverage at the critical Cs coverage of approximately 0.7 physical monolayer is attributed to a rapid growth of population of the positron surface state as it appears on the vacuum side of the Cs overlayer. Inclusion of the known attenuation of the EAES signal with Cs coverage provided very good fit of the theory to the experiment.

PAES has provided clear evidence for the migration of positrons trapped initially at the Cs/Cu interface to the vacuum side of the overlayer as a function of the Cs coverage, suggesting that PAES can become a powerful tool in investigating structural phase transitions and epitaxial growth.

ACKNOWLEDGMENTS

We would like to thank A.P. Mills Jr., Yuan Kong and K.G. Lynn for very useful and stimulating discussions. This work was supported by The Robert A. Welch Foundation, The National Science Foundation (DMR 910 6238), Texas Advances Research Program and by a Fulbright grant.

(a) Permanent address: Department of Physics, Kazan State University, Kazan 420008, Russia.

REFERENCES

1. A.P. Mills, Jr., Solid State Commun. **31**, 623 (1979); K.G. Lynn, Phys. Rev. Lett. **43**, 391 (1979); C.H. Hodges and M.J. Stott, Solid State Commun. **12**, 1153 (1973). K.G. Lynn and D.O.

Welch, Phys. Rev. B **22**, 99 (1980); I.J. Rosenberg, A.H. Weiss, and K.F. Canter, J. Vac. Sci. Technol. **17**, 253 (1980).
2. P.M. Platzman and N. Tzoar, Phys. Rev. B **33**, 5900 (1986); D.W. Gidley, A.R. Koymen and T.W. Capehart, Phys. Rev. B **37**, 2565 (1988); P.J. Schultz and K.G. Lynn, Rev. Mod. Phys. **60**, 701 (1988).
3. A.Weiss et al., Phys. Rev. Lett. **61**, 2245 (1988); Alex Weiss et al., in Proceedings of the Eighth International Conference of Positron Annihilation, 1988 (World Scientific, Singapore, 1989); D. Mehl et al., Phys. Rev. B **41**, 799 (1990); R. Mayer, A. Schwab and A. Weiss, Phys. Rev. B **42**, 1881 (1990).
4. A.R. Koymen et al., Phys. Rev. Lett. **68**, 2378 (1992).
5. R.M. Nieminen and Kjield O. Jensen, Phys. Rev. B **38**, 5764 (1988).
6. Kjield O. Jensen and A. Weiss, Phys. Rev. B **41**, 3928 (1990).
7. E. Wimmer et al., Phys. Rev. B **28**, 3074 (1983).
8. C.A. Papageorgopoulos, Phys. Rev. **B25**, 3740 (1982).
9. Cousty, R. Riwan and P. Soukiassian, Surface Science **152/153**, 297 (1985).
10. S.A. Linddgren and L. Wallden, Phys. Rev. B **22**, 5967 (1980).
11. N.D. Lang, Phys. Rev. B **4**, 4234 (1971).
12. Wang Ning et al., Phys. Rev. Lett. **56**, 2759 (1986).
13. O.V. Boev, M.J. Puska and R.M. Nieminen, Phys. Rev. B **36**, 7786 (1987).
14. M.J. Puska, P. Lanki and R.M. Nieminen, J. Phys.: Condens. Matter **1**, 6081 (1989).

WORK-FUNCTION AND EPITHERMAL POSITRON EMISSION FROM SURFACES

P.G. Coleman, A. Goodyear and A.P. Knights
School of Physics, University of East Anglia, Norwich NR4 7TJ, U.K.

ABSTRACT

Energy spectra of positrons re-emitted from metallic surfaces have been measured using an electrostatically-focussed positron beam system incorporating a hemispherical energy analyser. A summary of the performance characteristics of the apparatus is presented, together with preliminary measurements of re-remitted positron energy spectra from polycrystalline tungsten for positrons incident with energies between 100eV and 3keV. The spectra exhibit features attributable to work-function and epithermal positrons, and an energy loss peak whose relative intensity increases as the incident positron energy decreases.

INTRODUCTION

An important contribution to the understanding of the interactions experienced by positrons in the surface and subsurface regions of solids is the measurement of the energy and angular distributions of positrons emitted from a surface bombarded with a well-characterised positron beam. A particular point of interest has been whether the energy spectra of positrons emitted from some metals are broadened by energy loss processes at the surface or by measurement of the axial components of momenta of positrons emitted with a broad angular distribution.

A number of studies have been carried out in this area; in 1974 Pendyala et al observed interesting double-peaked energy spectra for positrons emitted from polycrystalline tungsten and other metallic surfaces under high vacuum conditions.[1] Nine years later Wilson and Mills demonstrated the existence of a significant tail on the low-energy side of the re-emitted positron energy spectrum for clean W(111).[2] A similar tail, attributed to energy loss, was observed by Chen et al for W(100)[3], in the high-resolution energy spectroscopy measurements of Fischer et al for Ni(100),[4] and in the measurements by the Aarhus group and by Willutski et al[5] on positron re-emission from W foils.

As well as low-energy tails, the emission of epithermal positrons is of significance, both in fundamental and applied studies. An example of the latter is the distortion of S-parameter vs incident energy plots, typical of positron implantation studies of subsurface defect profiles, at low energies (where epithermal positron emission, and fast positronium formation, becomes significant). In 1986 Nielsen et al [6] reported their observation of epithermal positrons emitted from Al(111) bombarded by 50 and 500eV positrons, exhibiting significant high-energy tails whose relative intensity was higher at the lower incident energy. Spectra of similar form were reported by Baker et al[7] for Cu(110) bombarded by 10eV positrons. In 1991 Jensen and Walker published their first Monte Carlo simulations of epithermal positron distributions.[8]

In this short paper our first measurements of work-function and epithermal positron energy distribution from polycrystalline tungsten are reported, as well as observations of an energy-loss spectrum, as a function of incident positron energy in the range 0.1-3keV.

EXPERIMENTAL METHOD

The development of the UHV electrostatic positron beam system at the University of East Anglia (UEA) has been described by Goodyear et al[9] and Goodyear and Coleman.[10] Positrons are focussed to a small (~mm) spot on the sample, which is positively biased, and re-emitted positrons are transported to a hemispherical energy analyser. A simplified schematic of the centre of the system is shown in figure 1.

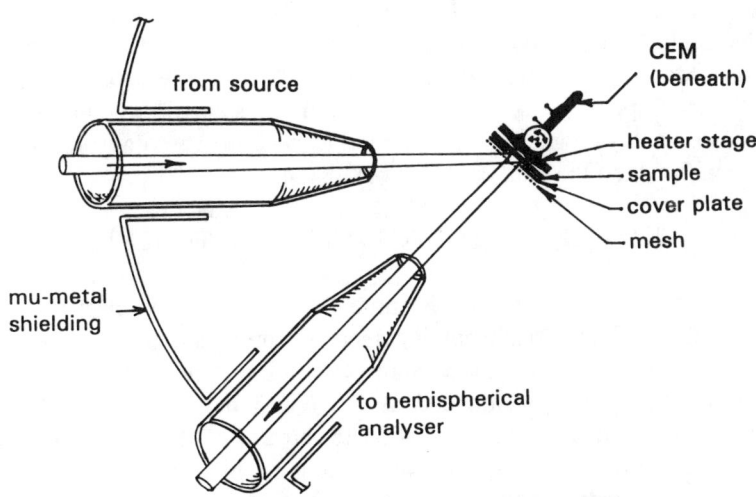

Fig. 1 Schematic diagram of the sample region

Recent modifications include the replacement of the trapped diffusion pump by a turbomolecular pump, and the installation of guiding coils around the incident positron beam arm. The latter addition allows the focusing of the incident beam on to the same position on the sample at each energy used. The sample holder, originally a simple plate on to which an annealed tungsten foil was clamped, has been replaced by a compact stage incorporating a filament for electron-beam heating. The sample itself is behind a hole in a coverplate which is held at a different potential, so that positrons emerging from the coverplate cannot appear on the energy spectrum of the sample. Incident beam optics were first simulated using a transfer-matrix program, and later the Simion code, and fine-tuned experimentally by profiling the beam cross-section using a moveable CEM detector.

The resolution of the system is limited to about 800meV FWHM by the need to obtain re-emitted spectra of reasonable intensity - in the experiments described below this meant using a fixed analyser transmission energy of 22eV, for a sample bias of +150V. The system resolution can be improved in a number of ways, including narrowing the entrance slits, increasing the radius of the hemispheres, and reducing the ripple on power supplies to the analyser elements. However, for the measurements described herein the current resolution is adequate.

The angular acceptance of the entrance optics to the hemispherical analyser is 2°, so that the viewed area of the sample is essentially defined by the analyser slits. It is clear that the number of re-emitted epithermal positrons per unit solid angle entering the analyser slit will depend on their intrinsic energy of emission, because of the +150V bias on the sample (which, with the 2.7eV work function, accelerates the positrons normally to the surface). For example, 100% of positrons approaching the surface isotropically from within with energies of up to 0.2eV will eventually have trajectories at angles less than 2° to the normal, whereas this is true for only 13% of positrons approaching with 1eV and 5.5% with 2eV. As a consequence the recorded energy spectrum is distorted, with the systematic attenuation increasing with increasing energy, and this would have to be folded into any future theoretical simulation with which the data could be compared.

RESULTS FOR POLYCRYSTALLINE TUNGSTEN

Figure 2 shows the measured re-emitted positron energy spectra for 0.1, 0.2, 0.5, 1, 1.5, 2 and 3 keV positrons incident on polycrystalline tungsten foil which had been annealed in an external vacuum system and transported through air to the spectrometer. The angle of incidence was 45°. The relative shapes of the spectra are demonstrated in figure 2 by normalisation to the top of the "work-function" peak. Reliable normalisation of the spectra to incident beam flux could not be done and was thus not attempted; although careful profiling at each incident energy provided information on the spatial position of the beam, the percentage overlap of the beam with the area of the sample viewed by the analyser could not be unambiguously determined. In the future this problem will be overcome by adjusting the incident lenses to produce beams with similar profiles at each energy, and then steering them to the same spatial position, so that the same percentage of the incident positrons will interact with the viewed area of the sample.

There are three principal features of the spectra in figure 2; the work-function peak at 2.7eV, the higher-energy epithermal tail, and a lower-energy distribution

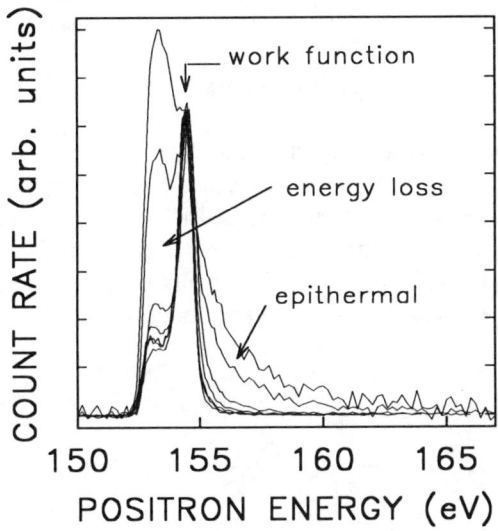

Fig. 2 Re-emitted positron energy spectra from W

which can be attributed to energy loss. At 3keV the work-function peak dominates the spectrum, with essentially no epithermal tail, and a broad energy-loss shoulder comprising about 23% of the total spectrum. As the incident energy decreases the fraction of the re-emitted spectrum attributable unambiguously to epithermal positrons increases monotonically to at least 14% at 100eV; the relative size of the work-function peak appears to decrease steadily with decreasing energy, but the overlap of the three elements of the spectra in this central region means that deconvolution into constituent parts is impossible and firm conclusions should not be drawn.

The surprising feature of the spectra is the energy-loss peak. The fraction of the spectrum at energies below the work function remains approximately constant as the incident energy decreases, until at energies below 500eV it jumps to about 30%, and indeed dominates the 100eV spectrum. Although similar distributions have been reported in the past for "clean" tungsten surfaces, it is difficult to explain the increasing magnitude of this part of the spectra at very low incident energies without invoking interactions of the positrons with an overlayer on the tungsten surface. It may be that at the higher incident energies, when most positrons returning to the surface are thermalised, a constant fraction of the latter undergo energy loss collisions as they pass through the surface/overlayer region. At 100-200eV, however, the majority of positrons may interact principally with the adsorbate, and the re-emitted positron energy spectrum would then be largely characteristic of the overlayer rather than the tungsten. This model is substantiated by measurements performed after the tungsten sample was heated *in situ* at 800°C; the percentage of the spectrum in the "energy-loss" tail decreased significantly.

CONCLUSIONS

The energy spectra for 0.1-3keV positrons incident on polycrystalline tungsten foil proved to be more interesting than first expected; further studies on atomically clean, single-crystal tungsten will answer many of the questions raised by this work, particularly with respect to the origins and dependence on incident energy of the low-energy or energy-loss distribution. Measurements on clean single-crystal silver are underway; silver has a positive positron work function and therefore the spectra show only the epithermal distributions.

Acknowledgement This work is supported by SERC, Swindon, UK.

REFERENCES

1. S. Pendyala, D. Bartell, F.E. Girouard and J. Wm. McGowan, Phys. Rev. Lett. **33**, 1031 (1974)
2. R.J. Wilson and A.P. Mills, Jr., Phys. Rev. B. **27**, 3949 (1983)
3. D.M. Chen, K.G. Lynn, R. Pareja and B. Nielsen, Phys. Rev. B **31**, 4123 (1985)
4. D.A. Fischer, K.G. Lynn and D.W. Gidley, Phys. Rev. B**33**, 4479 (1986)
5. P. Willutski, J. Störmer, D.T. Britton, G. Kögel, P. Sperr, R. Steindl and W. Triftshäuser, paper elsewhere in this volume.
6. B. Nielsen, K.G. Lynn and Yen-C Chen, Phys. Rev. Lett. **57**, 1789 (1986)
7. J.A. Baker, M. Touat and P.G. Coleman, J.Phys.C **21**, 4713 (1988)
8. K.O. Jensen and A.B. Walker, Mat. Sci. Forum **105-110**, 317 (1992)
9. A. Goodyear, I.R. Farthing and P.G. Coleman, in *Positron beams for Solids and Surfaces*, eds. P.J. Schultz, G.R. Massoumi and P.J. Simpson (AIP, New York) 239 (1990)
10. A. Goodyear and P.G. Coleman, Mat. Sci. Forum **105-110**, 1867 (1992)

POSITRONIUM AT A NITRIC OXIDE MONOLAYER ON GRAPHITE

C E Haynes and P C Rice-Evans
Physics Department, Royal Holloway, University of London
Egham, Surrey TW20 0EX

ABSTRACT

Positronium production as a function of monolayer coverage of nitric oxide on a graphite substrate is described. In spite of its unpaired electron, the formation characteristics appear to be different from those for physisorbed oxygen.

INTRODUCTION

Physisorption concerns the laying down of a thin layer of gas atoms or molecules onto a solid surface. The layer will be maintained in equilibrium with the gas pressure above the surface. As the temperature of the substrate is decreased, more gas molecules will condense down onto the surface. The coverage (n) and the pressure (P) are related by the equation[1].

$$P = \left(\frac{2\pi m k^3}{h^2}\right)^{1/2} n T^{3/2} \exp\left(\frac{-\varepsilon_0}{kT}\right)$$

where ε_0 is the binding energy of the gas onto the substrate, and m the adatom mass. The substrate used in the present experiment is exfoliated graphite as it offers a huge surface area (20 m^2 g^{-1}) for the adsorption of gas molecules. To a first approximation we assume the substrate presents a uniform continuum and that there are no preferential adsorption sites. Thus on reducing temperature it is possible to produce a clean, two-dimensional monolayer of varying density.

In the past we have conducted various experiments looking at the positronium formation at the carbon surface in the presence of various gases including oxygen, nitrogen, argon[2] and methane[3]. Oxygen has proved unique among the gases tried. Other gases typically show a sharp peak in positronium production at 50% monolayer coverage, whereas in oxygen, positronium production rises to a maximum and remains high until the bilayer is built up.

Oxygen has been shown strongly to quench positronium from the ortho (↑↑) to the para (↑↓) state. The ground state of oxygen has two unpaired electrons ($3\Sigma_g^-$) so the rôle of such unpaired electrons needs to be examined.

Here we report on an experiment employing the gas nitric oxide which has one unpaired electron. It is known that nitric oxide quenches positronium even more strongly than oxygen[4].

EXPERIMENT

The grafoil was first annealed at 420K for 12 hours in vacuum to drive off water and other absorbed gases. Nitric oxide was then introduced into the sample chamber which is held in a liquid nitrogen cryostat. The vapour pressure at 77.4K was found to be 2.0 Torr. Continuous one hour measurements were taken at temperatures between

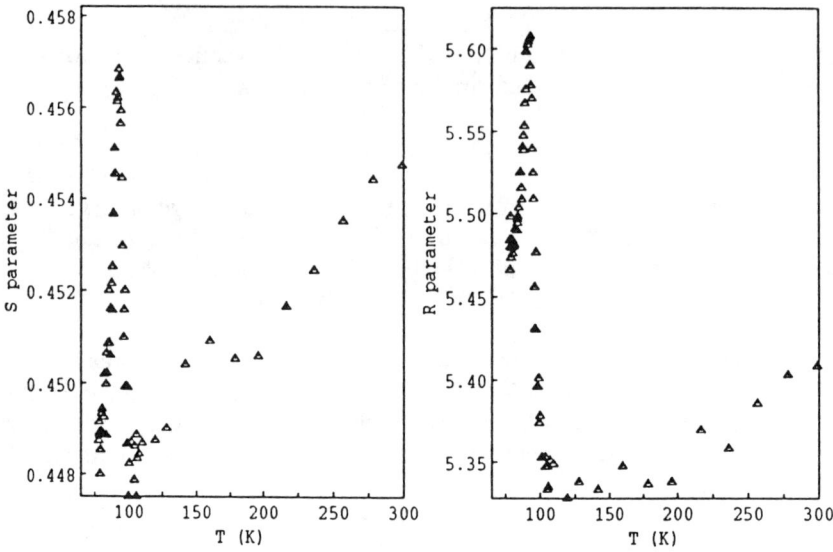

Figure 1: Results for Condensation of Nitric Oxide on grafoil; the parameters S and R indicate para-Ps and ortho-Ps annihilation respectively.

300K and 77K using a germanium photon detector, with the whole spectrum being recorded.

Figure 1 shows the analysis for nitric oxide; S is the conventional line-height parameter used to describe the Doppler-broadening of the 511 keV annihilation line. As the temperature is lowered a gentle decline in the parameter is observed, but below 100K a sharp peak is observed. In common with previous studies we attribute this line-narrowing to the growth in the parapositronium component, at the expense of the normal 2-γ annihilation with carbon electrons.

Figure 1 also shows the R parameter for nitric oxide which is essentially the ratio of the counts in the spectrum below 400 keV to the counts in the 511 keV line. Changes in R thus represent variations in the $3\gamma/2\gamma$ ratio, which indicates changes in the annihilation intensity of the orthopositronium. Again we see the sharp rise below 100K with a peak occurring at 90K in agreement with S. It is noticeable that at the lowest temperature, R does not decline to its original value, but we believe this is due to increased Compton scattering as multilayers of NO are laid down.

For comparison we also show previously obtained curves for oxygen (Fig. 2). The rise in positronium production with monolayer development is easily seen, but a plateau appears rather than a sharp peak. We have argued elsewhere[5] that maximum Ps production is maintained up to complete monolayer coverage but is eliminated by the growth of the bilayer.

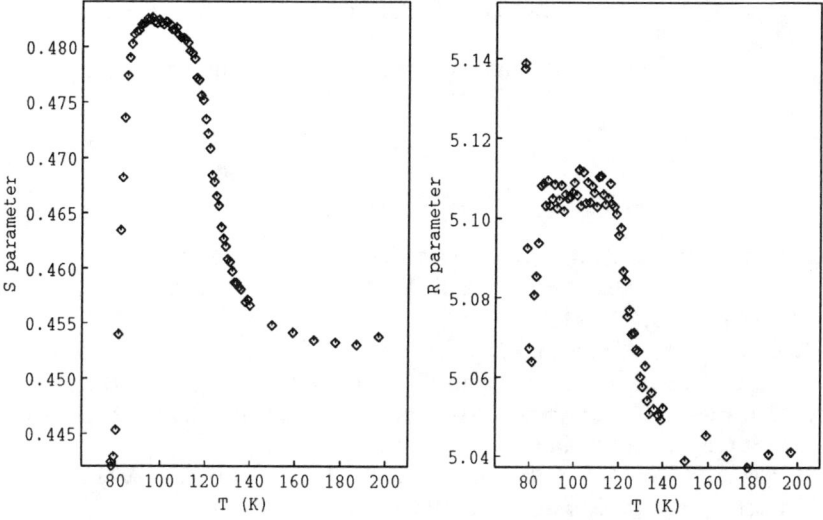

Figure 2: Results for Condensation of Oxygen on grafoil ; the parameters S and R indicate para-Ps and ortho-Ps annihilation respectively.

DISCUSSION

Positronium is not formed at a graphite surface at room temperature. However, Sferlazzo et al[6] found Ps at higher temperatures and suggested that phonons satisfied conservation requirements by absorbing the surplus electron parallel momentum. In our studies with condensed monolayers we have suggested that atoms physisorbed on graphite may recoil and thereby absorb any excess momentum.

With methane we found a sharp peak in Ps with a maximum at 50% coverage, and approximately zero at 0% and 100%. At the lower coverages, Ps production at the surface is proportional to the possibility of there being a free adatom in the vicinity; at higher coverages the freedom of recoil is inhibited by close neighbours.

Nitrogen, argon, krypton also give sharp peaks, and now we find that this is true of nitric oxide. We therefore conclude that the anomalous behaviour of oxygen is not just the result of there being unpaired electrons.

The plateau we see for positronium production must be as a result of some unique feature of oxygen which allows the absorption of excess momentum. The vibrational levels of oxygen and nitrogen are similar[7] and hence are unlikely to be the explanation. Oxygen has a low energy first excited state at 1eV compared with 5eV for NO[7] and we are inclined to suggest this state is involved in the positronium production for O_2 with spin exchange via the excitation of the 1eV state. This implies excited molecules of oxygen are able to recoil into their neighbours in two dimensions.

ACKNOWLEDGEMENT

We thank the Science and Engineering Research Council for supporting this work.

REFERENCES

1. J G Dash, Films on solid surfaces, (Academic, New York, 1975).
2. P Rice-Evans, M Moussavi-Madani, K U Rao, D T Britton and B P Cowan, *Phys Rev B* **34** 6117, (1986).
3. P Rice-Evans and K U Rao, *Phys Rev Lett* **61** 581, (1988).
4. S J Tao, S Y Chuang and J Wilkenfield, *Phys Rev A* **6** 1967, (1972).
5. P C Rice-Evans, C E Haynes, I Al-Qaradawi, F A R El Khangi, H E Evans and D L smith, *Phys Rev B* (in press).
6. P Sferlazzo, S Berko, K G Lynn, A P Mills, Jr., L O Roellig, A H Viescas and R N west, *Phys Rev Lett* **60** 538, (1988).
7. G Herzberg, Spectra of Diatomic Molecules (Van Nostrand, New York, 1950).

Theory of Surface Adsorbate Analysis by Positronium Formation

Akira Ishii

Faculty of General Education, Tottori University, Koyama, Tottori 680, Japan

INTRODUCTION

Atomic structure determination of an adsorbate on a surface is one of the most significant problems in surface science. for ordered adsorbates, we can use the LEED technique or X-ray diffraction technique for the determination. However, for disordered adsorbates, we could not apply the ordinary diffraction technique. The STM image would be very helpful for us to understand atomic structure of the adatom, but quantitative resolution is still limited by the uncertainty of the shape of the tip. Though the EXAFS technique is very useful to determine the local atomic structure, it cannot be applied to low Z atoms.

Positronium formation is one of the most surface-sensitive phenomena, because Ps formation occurs only at a topmost atomic layer or outside of the solid, for metals and semiconductor [1]. Thus, we can obtain some information for surface adsorbates with Ps formation phenomena. since Ps formation is caused by a positron which is diffracted by a crystal, Ps information cross section contains information of positron diffraction. In other words, we could consider that a surface adatom would be a detector of diffracted positron beam just at the surface. [2,3,4]

Therefore, in principle, we can get information of surface adatom by measuring Ps formation phenomena due to electron of the adatom. There are two ways to observe Ps formation; Ps TOF spectrum and desorption of the adatom. For both ways, measurement of Ps formation from the adatom without distortion by Ps formation due to substrate electrons is possible.

The purpose of this paper is to discuss the sensitivity of Ps formation cross sections to surface atomic structure, especially to the position of surface adatoms.

2. MODEL

The Ps formation matrix element is considered to be as follows.

$$M = \iint \psi(r_+, r_-) V(r_+, r_-) \phi_+(r_+) \phi_-(r_-) dr_+ dr_-$$

where $\Psi(r_+, r_-)$ is the wave function of Ps, ϕ_+ is the wave function of the incident and diffracted positron, and ϕ_- is the wave function of the electron. $V(r_+, r_-)$ is an interaction between the positron and the electron. In the following calculation, V is fixed to be simple isotropic screened Coulomb potential.

This matrix element has been confirmed by some calculations which agree with experiments [5,6]. In fig. 1, we show an example of the calculation.

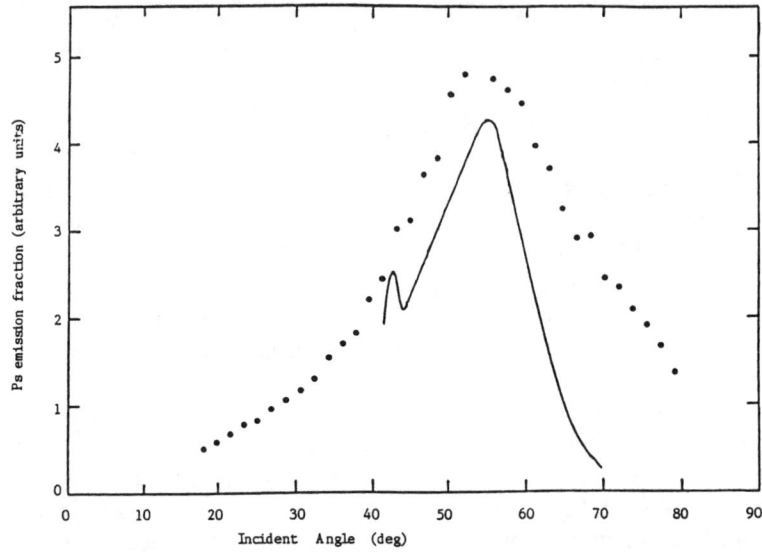

Fig.1 Comparison of calculated Ps fraction with experimental data measured by S.Tang[7]. The jellium model is assumed in this calculation. Ps scattering angle is 100° measured from the incident positron angle. Incident positron energy is 30eV.

If we concentrate our attention to Ps formation with electrons of the surface adsorbates, the electron wave function ϕ_- should be molecular orbitals of the surface adatom, which is localized on the surface. Thus, by matrix element (1), we can observe diffraction effects in the positron wavefield ϕ_+, projected onto the positions of the adatoms. According to the Ps diffraction experiment for Cu[7], diffraction effects for the Ps wavefield; $\Psi(r_+,r_-)$, can be neglected.

Therefore, because of the diffraction effect in the positron wavefield, Ps formation intensity is changed by changing the incident positron energy. The variation of the Ps intensity is analogous to the I-V curve analysis in Low Energy Electron Diffraction. Since the Ps formation with the adatom occurrs at the adatom position, the variation of the Ps intensity contains enough information to determine the position of the adatom on the surface. (This would be a new spectroscopy to determine surface atomic structure.)

There are at least two methods to observe such Ps formation only with the surface adsorbates: one is to use angle-resolved Ps spectrum and the other is to observe the ion desorbed due to Ps formation [2,3,4]. The angle-resolved Ps formation spectrum (ARPsFS) is very similar to the angle-resolved ultraviolet photoemission spectroscopy (ARUPS)[8,9,10]. Thus, like ARUPS, we can measure the Ps formation intensity by measuring the intensity of an extra peak due to the adatom state in ARPsFS spectra, as we show in fig. 2. Such angle-resolved PsFS experiments are now possible, for example, by using a short-pulse intense positron beam line of ElectroTechnical Laboratory in Tsukuba, Japan.[11,12].

Desorption of adatoms due to Ps formation would be possible in some cases. After Ps formation occurs with an adatom on the surface, the adatom is ionized by losing an electron. The desorption occurs if the ionized adatom feels a repulsive force from the substrate surface. Measuring the intensity of the desorbed adatoms is the best way to determine the amount of Ps formation with the electron of the adatom.

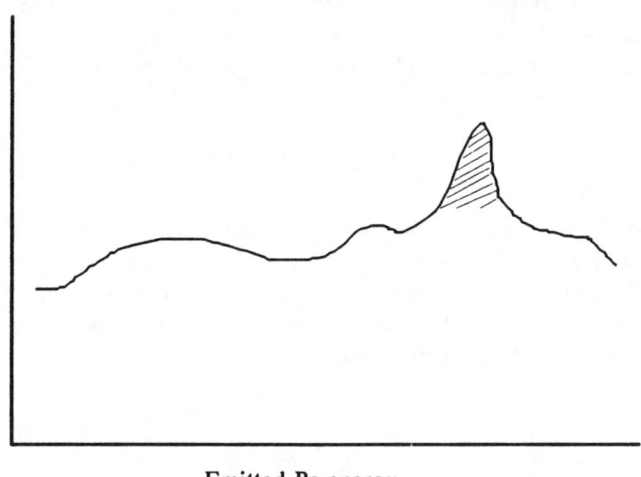

Fig.2 Schematic picture of expected angle-resolved Ps energy spectrum. The hatched area means contribution from surface adatom.

3. CALCULATION

In this paper, we present calculations of the Ps formation cross section. We consider hydrogen atoms on a Si(100)2x1 surface. Since the electronic state of the hydrogen atom on the s-p band metal, Al(111), is made up only from s-states [13], the electronic state of the hydrogen on Si(100) surface is mainly constituted from 1s and 2s orbitals. Thus, we assume here that the hydrogen has only the 1s state. Even if we assume the 2s state, the qualitative features of the following calculation is the same. For the substrate, we take the atomic positions of Si(100)2x1 assymmetric dimer structure obtained by K.Inoue et al. using Car-Parrinello-like molecular dynamics [14].

The muffin-tin potential of Si atom and empty sphere potential at every interstitial sites calculated be Ikeda and Terakura [15] are used in the calculation.

The matrix element of the Ps formation with hydrogen 1s orbital has been presented in ref. 4 and 16, where the positron wavefield has been calculated using a LEED program with potentials for positron. The significant point of the matrix element is that it depends clearly on the position of the adatom. Thus, similar to LEED I-V curve analysis [17] or diffuse LEED [18], we can use the position dependence to determine the adatom position.

In fig. 3, we show the Ps formation cross section with an electron of a hydrogen adatom on Si(100)2x1 surface as a function of the incident positron energy. The positron beam is considered to be normal incident. Each of the panels of figure 3 corresponds to a different site for the hydrogen adatin, as defined in fig. 4. As we can see in the figures, the Ps cross section is very sensitive to the site of the hydrogen atom. Thus, by measuring such curves, we can determine the adsorption site of hydrogen atom on a surface.

4. CONCLUSION

Theoretical investigation of the Ps formation cross section with surface adsorbates is presented. Since the Ps formation cross section is very sensitive to the position of the adatom, it can be used to determine atomic structure of adsorbed atom on crystal surface. Especially for low coverage cases, this method would be very powerful in the determination adsorption sites.

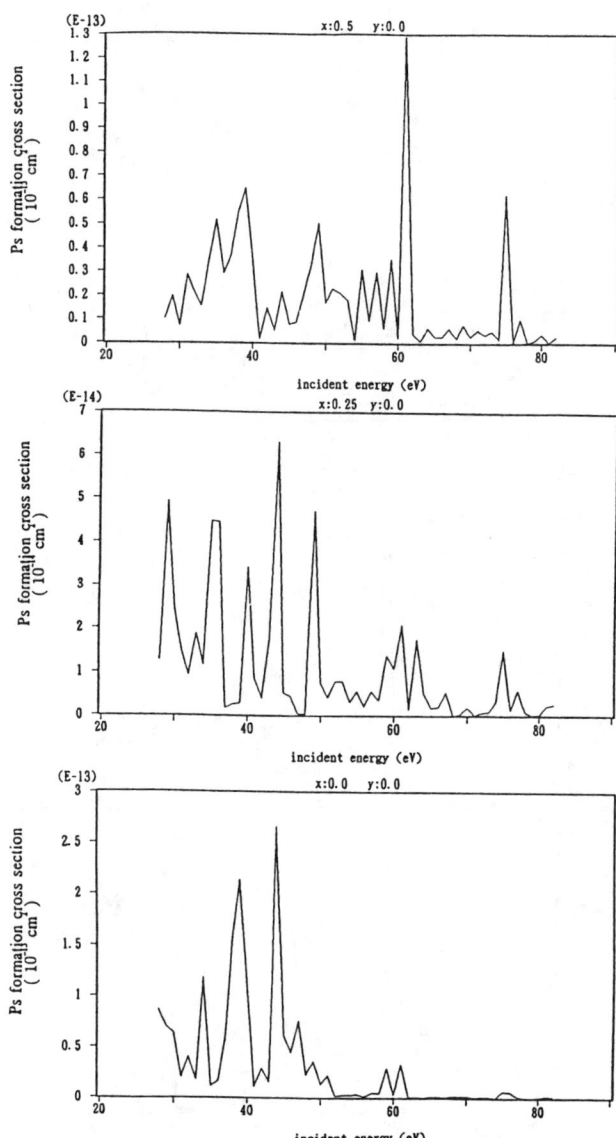

Fig.3 Theoretical prediction of incident positron energy dependence of Ps total cross section due to a hydrogen atom on a Si(100)2x1 surface. Contribution from the substrate is not included. The position of the hydrogen atom on the surface unit cell is shown at the top of the each figures: x and y correspond to the axis indicated in fig.4.

232　Surface Adsorbate Analysis

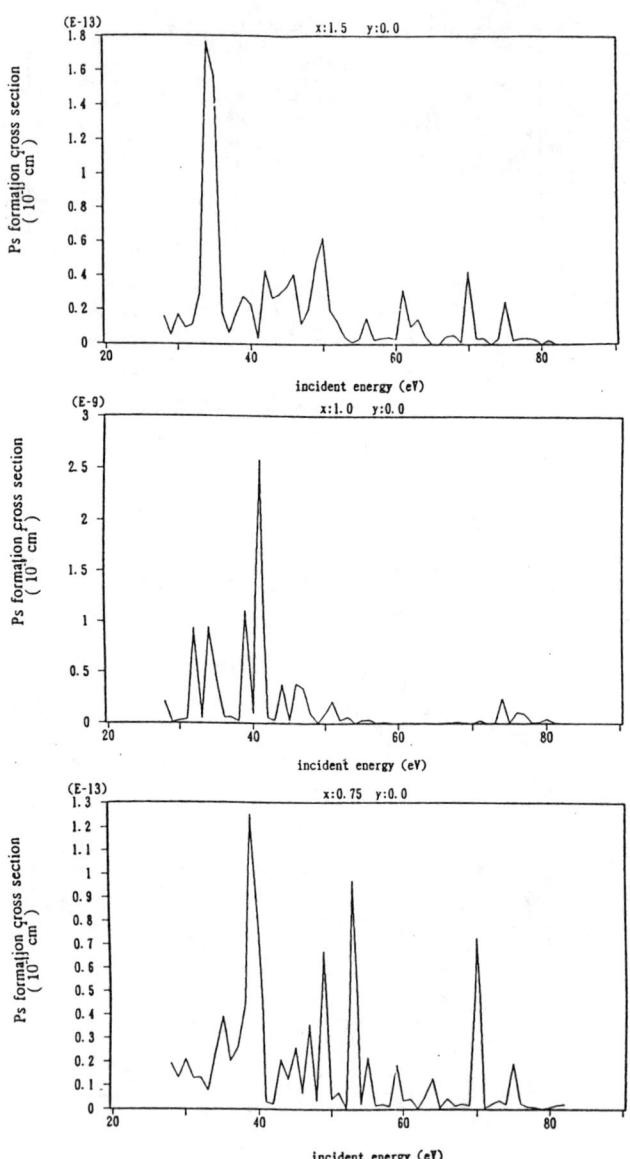

Fig. 3 Continued.

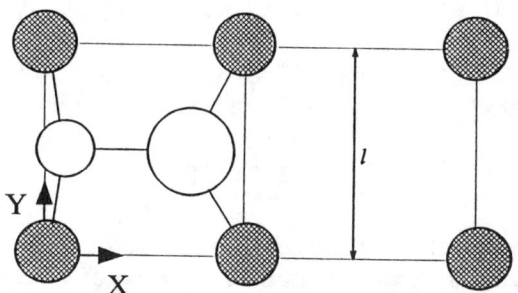

Fig.4 The surface unit cell of Si(100)2x1. The hydrogen adsorption sites indicated in fig.3 are defined by the axis, x and y. The unity of the axis is l shown in the figure.

ACKNOWLEDGMENT

The author is grateful to the Computer Center of the Institute for Molecular Science at Okazaki, Japan for their support to use the main frame computer, Hitachi M680H. The author thanks also to Profs. Y. Murata, K. Komori and K. Fukutani of the Institute for Solid State Physics of the University of Tokyo for helpful discussions.

References.
[1] A. Ishii, in *Positron at Metallic Surfaces* edited by A. Ishii (Trans Tech Publishers, Aedermannsdorf, 1992)
[2] S. Ishii, Nucl. Instrum. Methods B67, 509, (1992)
[3] A. Ishii and Y. Murata, Material Science Forum 105-110, 297, (1992)
[4] A. Ishii and Y. Murata, Surface Sci. 273, 442, (1992)
[5] A. Ishii and S. Shindo, Surface Sci. 242, 256, (1991)
[6] A. Ishii, Radiation Effects and Defects in Solids, 117, 245, (1991)
[7] S. Tang, doctoral thesis (The City University of New York, 1990)
[8] A. P. Mills, Jr., L. Pfeiffer and P. M. Platzman, Phys. Rev. Lett. 51, 1085, (1985)
[9] A. Ishii, Surface Sci. 209, 1, (1989)
[10] A. Ishii and J. B. Pendry, Surface Sci. 209, 23, (1989)
[11] R. Suzuki, Y. Kobayashi, T. Mikado, H. Ohgaki, M. Chiwaki, T. Yamazaki and T. Tomimatsu, in *Positrons at Metallic Surfaces* edited by A. Ishii (Trans Tech Publishers, Aedermannsdorf, 1992)
[12] R. Suzuki, contribution in this proceedings.
[13] A. Ishii, Surface Sci. in press.
[14] K. Inoue, M. Nakayama, Y. Morikawa, K. Terakura and K. Kobayashi, unpublished
[15] M. Ikeda and K. Terakura, private communication
[16] A. Ishii and Y. Murata, Surface Sci. in press
[17] J. B. Pendry, Low Energy Electron Diffraction (Academic, London, 1974)
[18] J. B. Pendry and D. K. Saldin, Surface Sci. 145, 33, (1984)

The Temperature Dependence of the Atomic Composition of the Surface of Cu(100) after Deposition of Submonolayer Films of Au

J.H.Kim, G.Yang, S.Yang, K.H.Lee, A.R.Koymen, and A.H.Weiss
Department of Physics, University of Texas at Arlington, TX. 76019

ABSTRACT

Positron Annihilation Induced Auger Electron Spectroscopy (PAES) was used to observe the surface composition of vapor-deposited Au films on a Cu(100) substrate as a function of temperature. PAES intensities were measured at three different coverages of Au : 0.05ML, 0.1ML, and 0.15ML and at seven different temperatures, ranging from 198K to 348K. At all three Au coverages, the PAES signal decreased from 198K until 273K and remained constant from 273K to 348K. This implies that Au atoms are alloying with the Cu atoms in the top layer of the Cu substrate to form a Au-Cu surface alloy structure whose stochiometry is stable from 273K to 348K. EAES spectra showed only a small decrease in the Au Auger intensities as the sample was warmed. This demonstrates that PAES has significantly more surface selectivity than EAES. The irreversibility of intermixing between the Au and Cu(100) substrate was checked in another experiment in which Au was deposited at 198K, then heated up to 348K, and then cooled back down to 198K.

INTRODUCTION

Positron Annihilation Induced Auger Electron Spectroscopy (PAES),[1] Electron Induced Auger Electron Spectroscopy (EAES), and Low Energy Electron Diffraction (LEED) have been used to study surface composition, surface alloying and surface morphology of submonolayers of Au vapor-deposited on a Cu(100) substrate.[2] Wang et al.[3] showed that at a Au coverage of 1/2ML, the C(2x2) structure of the LEED pattern was formed due to an ordered surface alloy layer with every other Cu atom occupied with Au atom in the room temperature. Hansen and Tobin[4] studied the temperature dependence of surface alloying using angle resolved photoemission spectroscopy (ARPES) in the Au/Cu(100) system. Lee et al.[2] performed the first PAES measurements on Au/Cu(100); measuring PAES intensities at three temperatures and fifteen coverages. In this paper, extend the work of Lee et al., and consider in more detail the temperature dependence of the structure and atomic composition of the top layer at low-coverages (0.05ML, 0.1ML, and 0.15ML) of Au on Cu(100) by measuring PAES intensities for a larger number of more closely spaced temperatures.

EXPERMENT

The experimental apparatus, which consists of a magnetically guided slow positron beam and trochoidal energy analyzer attached to an ultra-high vacuum chamber, has been described previously.[5] The positron beam, sample, and particle detecting system all reside in an UHV chamber, equipped with Electron Induced Auger Electron Spectroscopy (EAES), LEED, and an ion sputter gun. A Cu single crystal was oriented to within 1º of (100) face by X-ray diffraction and then polished. Sample cleaning in the vacuum system was accomplished by repeated neon ion sputtering followed by annealing at 973K. After cleaning the sample, Au was deposited on the Cu (100) substrate at 198K using a thermal evaporation source. The Au deposition rate was measured using a quartz micro-balance thickness monitor. EAES was taken both before and after the experiment to determine the contamination of the sample. LEED observations were used as a means to monitor the surface structure. A clean Cu EAES spectrum and sharp P(1x1) LEED pattern were observed after annealing the sample. Positron Annihilation Induced Auger Electrons were detected by a Micro Channel Plate (MCP) in coincidence with annihilation gamma rays detected by any of three NaI scintillators. The energy of the low energy positron beam was 25eV and the sample was biased at -3V with respect to the ground reference of the energy spectrometer to attract the slow remitted positrons back to the sample. A series of PAES data were obtained while the films deposited at 198K were warmed up from 198K to 348K. The kinetics of interdiffusion of Au and Cu were then studied by monitoring the Au and Cu Auger signals from the top-surface versus temperature.

RESULTS AND DISCUSSION

In fig. 1. we present PAES spectra for the Au depositions of 0.05, 0.1, and 0.15ML on Cu(100) at 4 of the 7 temperatures measured. The x-axis represents the energy of the Auger electron and the y-axis corresponds to the number of electron-gamma-ray coincidence counts. Spectrum(a) was obtained for 0.05ML of Au deposition, spectrum (b) and (c) were obtained for depositions of 0.1ML and 0.15ML respectively. The primary PAES peaks (Au O_{23}VV and Cu M_{23}VV) are shown in these spectra. The relative PAES intensities of Au and Cu were obtained from the data by using a two parameter least square fit.[6] The Au and Cu PAES intensities were taken to be the values of the parameters determined from a least square fit. The sharp rise in counts below about 25eV is due to collisionally excited secondary electrons. The ratio of the Au (42eV) to Cu (60eV) PAES intensity as determined from the fits changes from 45:50 to 17:85 upon warming the film deposited at 198K to 348K at a coverage of 0.05ML (fig.1(a)). The Au/Cu ratios show a similar trend at the coverages of 0.1ML and 0.15 ML (fig.1 (b) and (c)). The PAES intensities of the Au peak (a) and Cu peak (b) as a function of the sample temperature are shown in Fig.2. The Cu and Au PAES intensities were determined from fits to PAES spectra obtained at each temperature (12 of which are shown in Fig. 1). The total data acquisition time at each temperature for a given coverage was 150 minutes. The PAES intensity versus temperature curves for both Au and Cu have a sharp change in slope at a temperature of about 273K (see Fig.2). The intensity of the Au peak decreased from 198K to 273K and remained constant from 273K to 348K. The intensity of Cu peak increased between 198K and 273K and remained at a constant value above the temperature of 273K.

As seen in Fig.2, the intensities from low coverages of Au are even higher than what would be expected if the positrons were sampling the entire surface uniformly. For example, the PAES intensity at 0.05ML of Au is about 45% of that from 1ML of Au. This is probably due to positron trapping at Au islands formed even 198K. We interpret the gradual decrease in the Au PAES intensity from 198K to 273K as due to the intermixing of Cu atoms with the Au to form a surface alloy. We interpret the plateau at about 273K in the PAES intensities versus temperature curve as an indication of the predominance of islands of Au/Cu alloy whose stochiometry is stable up to 348K. This is consistent with previous room temperature measurements using ARPES[7], LEED[8], AES[9], XPS[10], ISS[4] and STM[11] which have shown the Au-Cu surface alloy monolayer forms was created at Au deposition of 0.5ML at 300K.

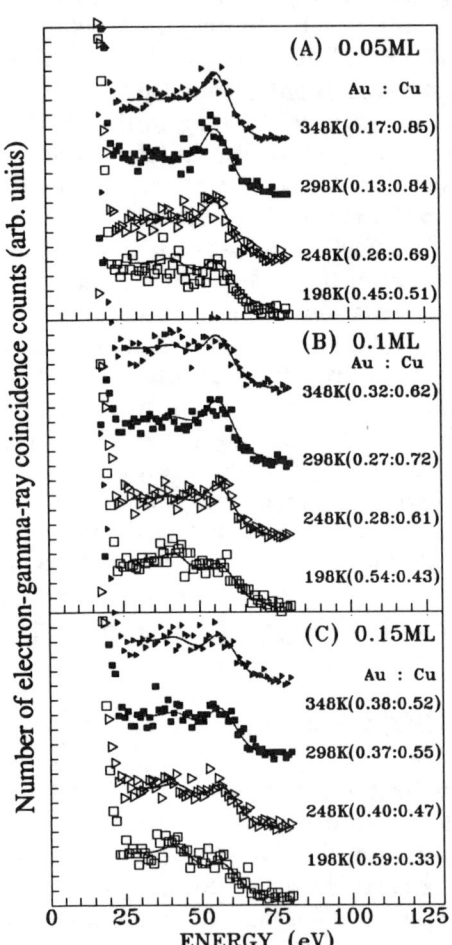

Fig. 1. PAES spectrum obtained for three coverages of Au on Cu(100) : 0.05ML, 0.1ML, and 0.15 ML at four different temperatures. The solid lines were obtained from fits to a linear combination of pure Au and pure Cu spectra. The intensities for Au and Cu (indicated in parentheses) were obtained from the fitting parameters. The spectra have been shifted along the y-axis to aid visibility.

J. H. Kim et al. 237

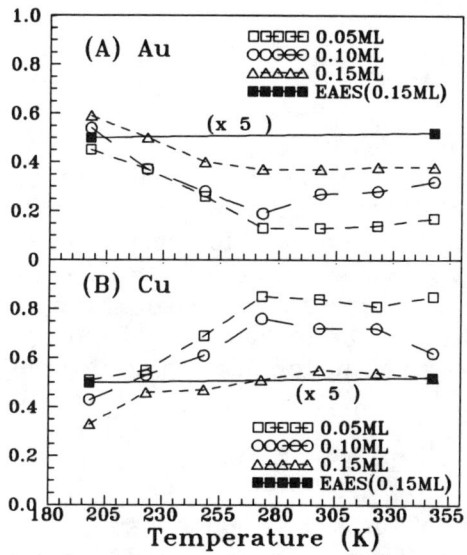

Fig. 2. PAES intensity of Au(A) and Cu(B) as a function of temperature (dashed lines) and the ratio (x 5) of the Au(69eV) to Cu(920eV) EAES intensities at 198K and 348K (solid line in both 2A and 2B)).

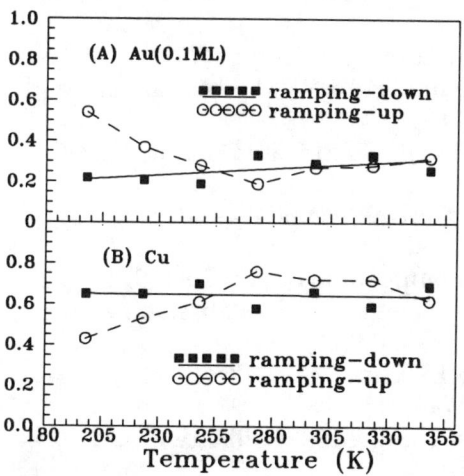

Fig. 3. The PAES intensity Au(3A) and Cu(3B) as a function of temperature during ramp-up (circle) and ramp-down (solid square).

Unlike PAES, EAES showed only small changes in the Au $O_{23}VV$ to Cu $M_{23}VV$ Auger signals upon warming (see Fig. 2). Specifically, the ratio,

R = [Au $O_{23}VV$] / [Cu $M_{23}VV$],

changed from R = 0.106 at 198K to R = 0.108 at 348K (the square brackets indicate peak to peak values taken from derivative spectra). The Au PAES signal is always higher than the actual coverage even at high temperatures (see Fig.2). This may be due to the localization of positrons at alloy islands which were formed on the terraces. A comparison between the extent of changes EAES and PAES intensities demonstrates that PAES is more surface sensitive than EAES and can be used to characterize the changes in the content and structure of the topmost atomic layer.[5,12]

Another experiment was performed to confirm that the formation of the Au-Cu surface alloy is an irreversible process. The data points in fig.3 were collected at seven different temperatures while the sample was being heated from 198K to 348K and then cooled back to 198K. The rate of temperature change was about 1K/min. The PAES intensities of Cu and Au changed significantly as the sample was heated from 198K to 348K (fig.3 (a) Au, (b) Cu, open circle) but during cooling from 348K to 198K, the intensities remained approximately constant (fig. 3, solid square). The above results demonstrate their reversibility of Au/Cu(100) surface alloy formation process.

This is in agreement with the work of Hansen et al. using photoelectron diffraction and photoelectron spectroscopy.[4]

CONCLUSION

We have discussed measurements in which PAES was used to examine the temperature dependence of the structure of submonolayer films of Au deposited on Cu(100). The PAES and LEED results taken together indicate that the Au-Cu surface alloy starts to be formed even at low temperatures (198-273K). The plateau in the PAES intensities above 273K indicates that the Au-Cu surface alloy is stable and that the Au atoms are not dispersed deeply into the bulk up to 340K. Our results clearly indicate the ability of PAES to monitor the elemental content and structure of the topmost atomic layer in mixed metal systems.

ACKNOWLEDGEMENTS

This work performed at University of Texas at Arlington is supported by the Welch Foundation, Texas Advanced Research Program, and National Science Foundation (Grant No. DMR-9106238).

REFERENCE

1. A.Weiss, R.Mayer M. jibaly, C.Lei, D.Mehl, and K.G. Lynn, Phys. Rev. Lett. 61, 2245, 1989
2. K.H.Lee, D.Sc. Thesis, University of Texas at Arlington(1992)
3. Z.Q. Wang, Y.S. Li, C.K.C. Lok, J.Quinn, F.Jona and P.M.Marcus, Solid State Communications 62, 181(1987).
4. J.C. Hansen, M.K.Wagner and J.G.Tobin, Solid State Commn. 72,319 (1989)
5. Chun.Lei, D.Mehl, A.R.Koymen, F.Gotwald, M.Jibaly and A.Weiss Rev.Sci.Instrum.60(1989)3656.
6. Temperature dependent top layer composition of Pd/Cu(100) A.R.Koymen, K.H. Lee, G. Yang, H. Zeou and A.H. Weiss, to be published.
7. J.C. Hansen and J.C. Tobin, J.Vac.Sci.Technol.. A7.2475(1988)
8. G.W. Graham, Surf. Sci. 184, 137(1986)
9. J.C.Hansen, J.A. Benson, W.D.Clendening, M.T.McEllistrem, and J.G.Tobin,Phys.Rev. B36,6186(1987)
10. B.J.Knapp, J.C. Hansen, J.A. Benson, and J.G. Tobin, Surf. Sci. 188, L675(1987)
11. D.D.Chambliss, R.J.Wilson, S. Chiang, Surf. Sci. 264, L187 (1992)
12. D.Mehl, A.R.Koymen,K.O.Jenson,F.Gotwald and A.Weiss Phys. Rev., B41(1990),799

STUDY OF SUBMONOLAYER FILMS OF Au ON Cu(100) USING POSITRON ANNIHILATION INDUCED AUGER ELECTRON SPECTROSCOPY

K. H. Lee, Gimo Yang, A.R. Koymen, and A. H. Weiss
Department of Physics, University of Texas at Arlington, P. O. Box 19059, Arlington, TX 76019

ABSTRACT

Positron Annihilation induced Auger Electron Spectroscopy (PAES)[1], Electron induced Auger Electron Spectroscopy (EAES) and Low Energy Electron Diffraction (LEED) have been used to study ultrathin film of Au on Cu(100). Temperature induced changes in the top surface compositions in Au/Cu(100) are directly observed in PAES spectra while EAES spectra indicate only minor changes as the sample is warmed from 173K to 423K. A sharp kink at ≈ 0.06 ML in a plot of Au PAES intensity versus coverage at 173K indicates that the positrons are preferentially sampling Au atoms. Saturation of the Au PAES intensity is observed near 1monolayer of Au demonstrating the high top layer selectivity of PAES. The surface defect dependence of the PAES intensity is interpreted in terms of the positron localization at surface defects.

INTRODUCTION

In early studies of the Au/Cu(100) system, Palmberg and Rhodin[2] were the first to observe a C(2x2) LEED pattern at submonolayer coverages attributing it to an ordered surface alloy layer with every other Cu atom replaced with a Au atom. Graham demonstrated that the ordered surface structure of the Au/Cu(100) was formed with an Au coverage estimated to be 1/2 ML, and closely resembling the (100) surface of the ordered Cu_3Au. LEED intensity analysis[3] also confirmed that the C(2x2) LEED pattern was formed at a coverage of 1/2 ML of Au with the layer buckled, and with the Au atoms located 0.1Å outwards from the Cu substrate. Hansen and Tobin[4] studied the temperature and concentration dependencies of electronic structures in the Au/Cu(100) system and consolidated previous results of surface alloying studies using angle resolved photoemission spectroscopy (ARPS).

Previous studies have indicated that PAES (Positron Annihilation induced Auger Electron Spectroscopy) is extremely sensitivity to the composition of the top most layer[1]. In this paper we report the results of measurements in which the surface selectivity of PAES was exploited in the study of the dynamics of Au overlayer growth and interdiffusion on Cu(100). A more complete paper which will include a discussion of theoretical aspects of this work is currently in preparation. The PAES measurements were performed in conjunction with complimentary Electron induced Auger (EAES) and LEED measurements. The relative PAES intensities of Au and Cu were measured at different substrate temperatures with varying amounts of Au deposition in order to determine the extent of surface alloying and the Au surface coverage.

We also present evidence suggesting the trapping of positrons at surface defects. Such trapping has been suggested by previous studies.[6] This could imply that the surface information derived from PAES is representative of the atomic species at the vicinity of the defects, and thus may be different from averages over the whole surface.

EXPERIMENTAL

The experiments were performed using the UT Arlington PAES system which has been described previously[1]. The PAES system consists of a magnetically guided positron beam and a trochoidal energy analyzer. A micro channel plate (MCP) is used to detect Auger electrons emitted from the sample and three NaI(Tl) detectors were used to detect annihilation gamma rays emitted in coincidence with the annihilation induced Auger electrons. The PAES system is mounted on a UHV (base pressure 2.0×10^{-10} Torr) sample chamber equipped with EAES, LEED and a sputter ion gun. The sample holder is equipped with a resistance heater and a LN_2 cooling system.

The sample temperature was regulated to ($\pm$ 1°C) using a custom built temperature control unit. The sample, a Cu(100) single crystal, was mechanically polished using standard techniques and oriented parallel to the (100) face to within $\pm$1degree by Laue X-ray diffraction. The sample was cleaned by multiple Ne+ sputtering and annealing (973K) cycles. A Au thermal evaporator, which consists of an aluminum oxide tube which was filled with Au and wrapped with a W heating wire, was used to deposit Au on Cu(100). Prior to deposition, the Cu substrate was cooled to 173K to reduce the interdiffusion between Au and Cu atoms during deposition. Data were then taken at 173K using PAES, EAES and LEED. Subsequently the sample was warmed to first 303K and later 423K without changing the Au coverage and data were obtained at these two temperatures.

RESULTS AND DISCUSSION

PAES spectra obtained from a Cu(100) sample with a coverage of 0.68ML (as determined by PAES) are presented in Fig.1. This coverage corresponds to the coverage which produced the sharpest C(2x2) LEED pattern (at 303K). The x-axis represents the electron energy and the y-axis represents the number of e⁻- γ coincidence counts. The data is the sum of 5 loops in which the energy was increased in 1eV steps with an accumulation time of 30 seconds per point for each loop (adding up to a total of 150 seconds per point). The spectrum shown in Fig. 1(a) was obtained at 173K directly after deposition. The spectra shown in Figs. 1(b) and 1(c) were obtained after the sample was heated to 303K and 423K respectively. The primary PAES peaks (Au $O_{23}VV$ (42eV), Cu $M_{23}VV$ in Au/Cu(100)) are indicated in these spectra. The Auger peaks from other transition that originate from more tightly bound core levels are predicted to be much smaller than the primary peaks and are not observable in our measurement.[5] A two parameter fit to the data was used to extract the contributions from Au and Cu to the spectra.[5]

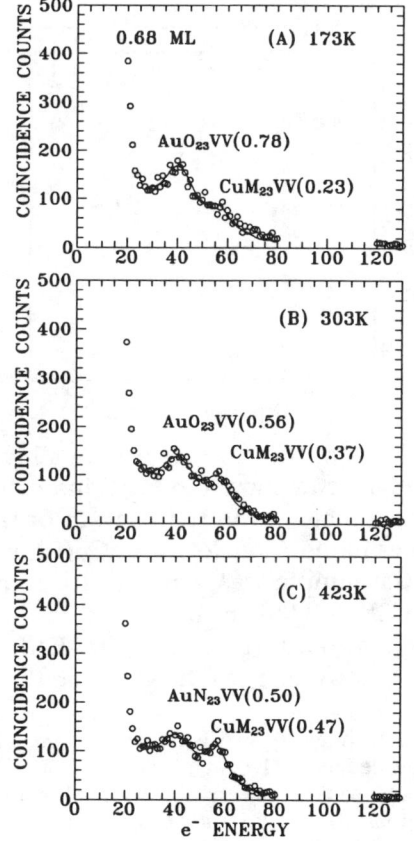

Fig. 1 PAES spectra obtained from a coverage of 0.68ML Au on Cu(100), at (A) 173K, (B) 303K, and (C) 423K. Solid lines are obtained from a two parameter least square fit to a linear combination to Au and Cu reference spectra.

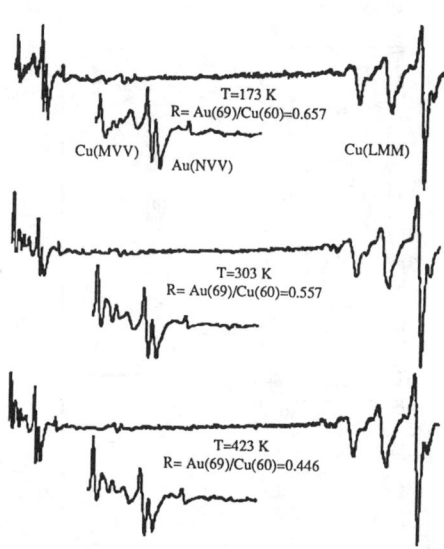

Fig.2. EAES spectra of Au/Cu(100) at the coverage of 0.58ML taken at 173K (A), 303K (B), and 423K (C).

Temperature induced changes in the PAES spectrum can be clearly seen in Fig.1(a)-1(c). As the temperature is increased, the Cu peak can be seen to grow while the Au peak decreases. The ratio of Au to Cu PAES intensity changed from 3.4±0.8 to 1.5±0.2 upon warming from 173K to 303K and then to 1.06±0.15 upon further warming to 423K. These changes correspond to temperature induced changes in surface compositions associated with

Fig.3. EAES intensity versus deposition time for Au deposition on Cu(100) at (a) 173K, and (b) 303K. Solid lines represent fits to function of the form $A+Be^{Ct}$.

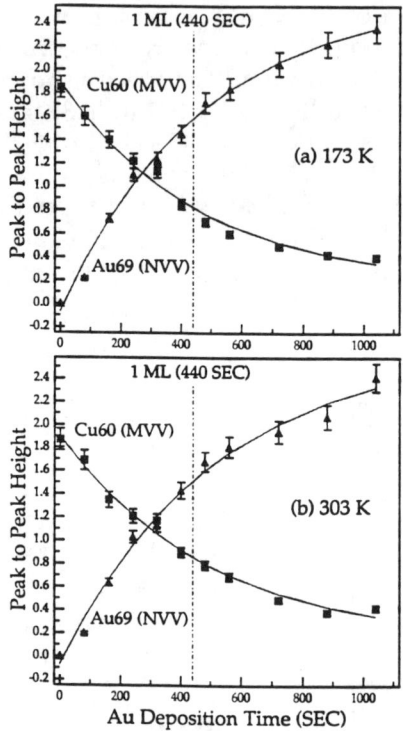

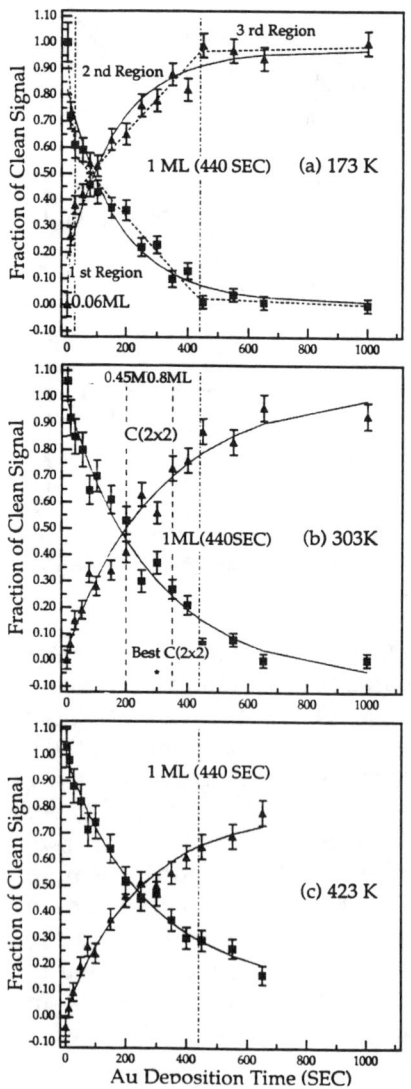

the formation of the ordered Au-Cu alloy. The EAES spectra shown in Fig.2 indicate only a small changes in the peak to peak amplitudes in the Au (69eV) and Cu(60eV) peak as the sample was warmed up from 173K to 303K and then to 423K.

Fig.3(a) and (b) show the EAES intensity (peak to peak height) of the

Fig.4. PAES intensity versus deposition time of Au deposited on Cu(100) at (a)173K,(b) 303K, (c) 423K. Solid squares correspond to measured Cu PAES intensities, solid triangles to Au PAES intensities.,1st region:0-0.06ML, 2nd Region: 0.06-1ML, 3rd Region: above 1ML. Solid lines are result of a fit to a function of the for $A+Be^{Ct}$ and dashed line are piecewise fits to straight lines. In (b), the star symbol indicates the coverages at which the best C(2x2) LEED pattern was observed. A C(2x2) LEED pattern was observed over the region between the dashed dot lines corresponding to coverages between 0.45 and 0.8ML.

Au(69eV) and the Cu(60eV) derivative peaks versus the Au deposition time at 173K and 303K respectively. The EAES intensity versus deposition time plot shows smooth and continuous curves without a clear change in slope. These curves are ambiguous in their meaning and the determination of 1ML Au coverage is not clear.

In contrast, PAES intensity versus deposition time plots show clear changes in slope. Fig.4 shows the PAES intensity versus deposition time for Au deposited on Cu(100) at 173K(a), 303K(b) and 423K(c). Referring to Fig. 4(a) it can be seen that the Au PAES intensity increased very rapidly at low coverages reaching 45% of its thick layer value at 0.06ML. The intensity then increases approximately linearly attaining its thick layer value at about 1ML. The calibration of 1ML was taken to be the point at which the PAES intensity saturated as determined by fitting the PAES data with three straight lines. The fact that the PAES intensity of Au grows to approximately 45% of the pure Au signal at a coverage of about 0.06ML of Au indicates that the positrons sample the Au adatom preferentially.

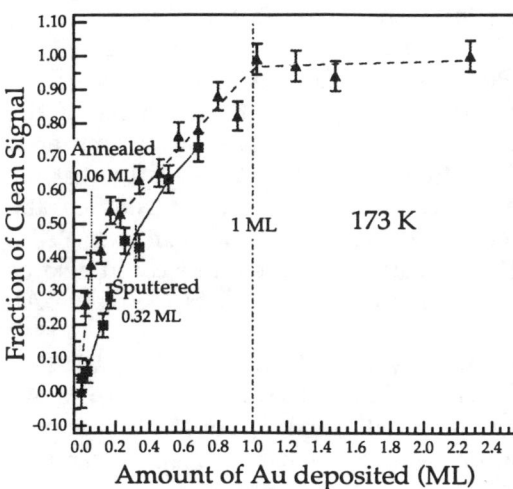

Fig. 5 PAES intensity versus coverage for a sputtered Cu(100) sample at 173K. Solid straight lines represent a fit to the Au PAES data (solid squares) from the sputtered sample and dashed straight lines represent a fit to the Au PAES data (solid straight) from the annealed sample.

Two factors which may be important in determining the way in which overlayer atoms are sampled are the distribution of the adatoms on the surface and the localization of positrons at defect sites. Positrons may trap at the same surface defects that are favorable sites for trapping an overlayer atom. Thus the PAES intensity of the overlayer atom will be enhanced to the extent that overlayer atoms are trapped at the defect sites.

In addition, positrons may trap at the edge or on top of Au islands on the Cu substrate which would again lead to an enhancement of the Au signal at low coverages. In Fig4(b) and (c), the PAES intensity versus coverage plots do not exhibit a clear break the slope at submonolayer coverages. This is probably because Au atoms form an ordered Au-Cu alloy at 303K and some of the Au adatoms diffuse into the bulk at 423K preventing the positron from preferentially sampling the Au atoms. A plot of PAES intensity versus deposition time indicates that at 303K there is an approximately linear relationship between the Au deposition time and the PAES intensity up to 1ML. At 423K, PAES signals are smaller than at 303K presumably due to the diffusion of some of the Au adatoms into the bulk.

The PAES intensity was found to be strongly dependent on surface defects introduced by ion sputtering. We explored the sensitivity of PAES to surface defects by comparing PAES intensities for a sputtered and annealed sample with one which was just sputtered. The Cu(100) sample was sputtered for 5 minutes with 3 KeV Ne ions at 173K until the LEED spot profiles were visibly broadened. Fig. 5 shows the PAES intensity of the sputtered Cu(100) substrate as compared with the PAES intensity of the annealed Cu(100). The PAES intensity versus coverage plot of the sputtered Cu substrate has a much smaller slope than that of the annealed substrate presumably because of the presence of more positron trapping sites in the sputtered Cu substrate as compared to the annealed Cu. The first PAES break is shifted from 0.06ML to 0.32ML coverage. These effect could also be due to the affect of sputter induced defects on the growth mode of the Au overlayer.

Surface structure was also examined using LEED. At 173K, a P(1x1) LEED pattern was observed for the clean Cu(100) substrate. As the Au coverage was increased to 1 ML, several small weak spots start to appear between the spots of the P(1x1) pattern and a weak center spot was also observed in the center of the P(1x1). Above 1ML, the center spots disappear and a few weak spots were observed between the spots of the P(1x1) pattern at 173K. At 303K, a C(2x2) LEED pattern was observed at the coverage from 0.45ML to 0.8ML and the best C(2x2) LEED pattern was found at the Au coverage of 0.68ML. Previous LEED studies indicated that the Au/Cu(100) is arranged in a structure with a C(2x2) LEED pattern at a Au coverage near 0.5 ML.[4] The discrepancy with our results may indicate an error in our coverage calibration due to our assumption that the point at which the PAES signal saturates was 1ML.

CONCLUSIONS

In this paper we have presented the results of a PAES study of ultrathin films of Au deposition on a Cu(100). Two important findings are that we can observe temperature induced changes in PAES intensity corresponding to surface alloying and that the PAES intensity saturates above 1ML demonstrating the extremely high top layer selectivity of PAES. One of the most interesting features of the data is the fact that the PAES intensity of Au grows sharply to approximately 45% of the saturation value at a coverage of 0.06ML. This sharp increase in PAES intensity at low coverages indicates that positrons are annihilating preferentially at sites near Au adatoms. In addition, the large dependence of PAES intensity on sputter damage provides evidence for positron localization and trapping at surface defects.

REFERENCE

1. A. Weiss, R. Mayer, M. Jibaly, C. Lei, D. Mehl, and K. G. Lynn, Phys. Rev. Lett. 61, 2245 (1988)
2. P. W. Palmberg and T. N. Rhodin, The J. Chem. Phys. 49, 134 (1968)
3. Z. Q. Wang, Y. S. Li, C. K. Lok, J. Quinn, F. Jona and P. M. Marcus, Solid State Commun. 62, 181 (1987)
4. J. C. Hansen and J. G. Tobin, J. Vac. Sci. Technol. A 7, 2475 (1988)
5. K. H. Lee, A. R. Koymen, D. Mehl, K. O. Jensen, and A. Weiss, Surf. Sci. 264, 127 (1992)
6. A. R. Koymen, D. W. Gidley and T. W. Capehart, Phys. Rev. B, 35, 1034 (1987)

Low Energy Electron and Positron Diffraction from Surfaces. What you Learn. How they Differ.

D. L. Lessor, Pacific Northwest Laboratory[a], Richland, WA 99352
K. F. Canter, Brandeis University, Waltham, MA 02254
C. B. Duke, Xerox Webster Research Center, Webster, NY 14580

Electron and positron beams of energy 30< E < 250 eV have a number of similarities as probes of material surfaces: 1) Wavelengths for both on the order of 0.5 to 2.5 Angstroms give coherent diffraction from an ordered solid; 2) attenuation lengths on the order of five Angstroms make them sensitive to the outer (≤5) surface layers; 3) the probing particles respond to a strongly screened, strong Coulomb potential; 4) because of low momentum of the probing particle, few atomic dislocations are produced; 5) patterns and symmetries in the discrete diffracted beams give qualitative information about clean crystal surfaces and ones with ordered overlayers; and 6) the energy dependence of the intensities in the discrete diffracted beams gives quantitative atom position information, which can be related through models to electronic structure and to the nature of the atomic bonds. Nevertheless, low energy electron diffraction (LEED) and low energy positron diffraction (LEPD) exhibit a number of differences: 1) There are no excluded final states for positron inelastic processes, i.e., no antisymmetry requirement with electrons of the material, thus giving somewhat shorter attenuation lengths for positrons than for electrons; 2) electrons entering a solid fall into a potential well of order 10 eV, but positrons see a potential between ≈2 ev attractive and 3 ev repulsive; and 3) positrons experience no exchange interaction with the electrons of the solid. The most important difference[1], however, is that positrons see an atomic scattering center as a repulsive nucleus whose potential is screened away by attractive atomic electrons, whereas electrons see an attractive nucleus whose potential is screened away by repulsive atomic electrons. Consequences of these differences[2] are: 1) greater calculability of positron diffraction, 2) even greater surface sensitivity for

© 1994 American Institute of Physics

LEPD than for LEED, 4) less difference between the scattering factors of highly dissimilar atoms in LEPD than in LEED, and 3) a less highly multiple scattering process for LEPD.

Preliminary investigations show that lower multiplicity of scattering for positrons results in less sharp minima for the figures of merit for theory-experiment agreement at the correct surface structure in LEPD than in LEED, at least in the case of diatomic crystals whose atomic species have scattering factors of similar magnitude. Less sharpness of minima can contribute to lower spatial resolution for LEPD than for LEED, but the greater accuracy with which LEPD multiple scattering can be calculated may fully compensate for this seeming lower spatial resolution of atom positions. Also, LEPD may offer an advantage when the atomic species present differ greatly in their Z values.

The repulsive potential seen by positrons at the atomic scattering centers gives little sensitivity of positron scattering to either atomic potential or electron density at points inside the classical turning points. By contrast, scattering of low order spherical waves of electrons is quite sensitive to details of core potential and electron density. Consequently, scattering factors for positrons show less variation with energy and less variation between different atom species. The first of these reductions of variation contributes to the more accurate calculability of positron multiple scattering, while the latter offers a potential advantage in resolving surface structures of polyatomic crystals whose atoms differ greatly in Z values. We speculate that having atomic scattering factors of similar magnitude enhances interference effects in LEPD over those in LEED, and that this in turn offers better spatial resolution of surface structures from LEPD. An investigation is under way to determine if LEPD indeed offers greater position sensitivity than does LEED for polyatomic materials having atoms quite different in size. The results are planned for publication elsewhere. LEPD has been used successfully for surface structure determinations on CdSe[3,4], GaAs[2], and InP[5] crystals.

(a) Operated for the U. S. Department of Energy by Battelle Memorial Institute under contract DE-AC06-76RLO 1830.

[1] C. B. Duke and D. L. Lessor, Surf. Sci. 225, 81 (1990).

[2] D. L. Lessor, C. B. Duke, X. M. Chen, G. R. Brandes, K. F. Canter, and W. K. Ford, J. Vac. Sci. Technol. A **10**, 2585 (1992).

[3] C. B. Duke, D. L. Lessor, T. N. Horsky, G. Brandes, K. F. Canter, P. H. Lippel, A. P. Mills, Jr., A. Paton, and Y. R. Wang, J. Vac. Sci. Technol. A **7**, 2030 (1989).

[4] T. N. Horsky, G. R. Brandes, K. F. Canter, C. B. Duke, S. F. Horng, A Kahn, D. L. Lessor, A. P. Mills, Jr., A. Paton, K. Stevens, and K. Stiles, Phys. Rev. Lett. **62**, 1876 (1989).

[5] X. M. Chen, K. F. Canter, C. B. Duke, A. Paton, D. L. Lessor, and W. K. Ford, to be submitted to Phys. Rev. B.

Annihilation characteristics for positrons trapped at the surfaces of simple metals

A. Rubaszek[†], A. Kiejna[‡] and S.Daniuk[†]

[†] W.Trzebiatowski Institute of Low Temperature and Structure Research, Polish Academy of Sciences, 50-950 Wroclaw 2, P.O.Box 937 (Poland)

[‡] Institute of Experimental Physics, University of Wroclaw, 50-205 Wroclaw, ul.Cybulskiego 36 (Poland)

Angular correlation of annihilation radiation (ACAR) spectra for positrons trapped in the surface state are calculated within an approach developed by the authors for simple metals. Positron annihilation parameters (momentum-dependent enhancement factors and lifetimes) are presented. Characteristics of host material (electron and positron work functions) are discussed. The role of electron-positron correlations in theoretical studies of surface properties by positron annihilation is emphasized.

Introduction.

A beam of mono-energetic slow positrons is a useful tool to probe electronic properties of metal surfaces[1]. Positron lifetime experiment evidences the presence of the positron surface state (SS) at the Al (110) surface[2]; the measured SS component amounts to 580 psec., i.e. about 15% more than the spin-averaged free-positronium value. The extended angular correlation of annihilation radiation (ACAR) technique[1] has been successfully applied to investigations of the electron and positron SS in Al[3], Cu, Si, Ni, Pb and graphite[4]. In contrast to more complicated metals[4], the SS components of the two-dimensional (2D) ACAR spectra for any of low index surfaces of Al [(100), (110), and (111)] are nearly isotropic and face independent[3]. It seems to be interesting to check to what extent the properties of the SS annihilation characteristics in Al can be attributed to the nearly-free character of valence electrons in bulk aluminium. In the present work we try to answer this question; the lifetimes, the ACAR spectra, and momentum-dependent enhancement factors for positrons trapped at the surface of simple metals, characterized by nearly-parabolic valence bands in the bulk, are calculated.

Another interesting problem is what kinds of effects must be taken into account in theoretical studies of the SS annihilation parameters. As pointed out in Refs. 5 and 6, both the distributions of individual electronic states and electron-positron correlations must be treated very carefully in the near-surface region.

Neglecting electron-positron enhancement effects[7] or including them within the local density approximation (LDA)[8,9] leads to underestimation of the value of the positron SS lifetime τ. For this reason conclusions about the corresponding ACAR spectra need some deal of caution. Electron-positron correlations obtained within the weighted density approximation (WDA)[10] provide correct theoretical value of lifetime for positrons trapped at the Al surface[11], and therefore allow for more confidence in the corresponding ACAR spectrum[6].

Invalid assumptions on the form of electron wave functions in the near-surface region applied in the ACAR formula[8] can lead to wrong conclusions about the shape of the spectrum and the role of the electron-positron correlation effects, as shown in Refs. 5 and 6.

In this work the approach developed in Refs. 5 and 6 is applied to calculation of annihilation characteristics in simple metals. Electron wave functions in the host material are determined in the way proposed in Ref.12, while electron-positron correlations are obtained within WDA. In the positron model the energy levels in the bulk following from the band structure calculations[13] are taken into account.

Calculations and results.

At the metal surface the ACAR formula may be approximated by the expression[5,6]

$$\rho(\mathbf{p}) = \frac{\pi r_0^2 c}{8\pi^3} \sum_{i \text{occ}} \left| \int e^{-i\mathbf{p}\cdot\mathbf{r}} \psi_+(\mathbf{r}) \cdot \psi_i^e(\mathbf{r}) \left[1 + \frac{\Delta\rho(\mathbf{r};\mathbf{r})}{n_{el}(\mathbf{r})}\right]^{1/2} d\mathbf{r} \right|^2, \quad (1)$$

where $\psi_i^e(\mathbf{r})$ and $n_{el}(\mathbf{r})$ are the *unperturbed* electron wave functions and density, respectively, in the material investigated, $\psi_+(\mathbf{r})$ is the positron wave function at the surface, and $\Delta\rho(\mathbf{r}_e;\mathbf{r}_p)$ denotes the conditional density of the electronic screening charge at $\mathbf{r}_e$, assuming that the positron is located at $\mathbf{r}_p$ (for more details see Ref.6). Summation in Eq.(1) is over all occupied initial electronic states i, and r_0 and c are the classical electron radius and velocity of light, respectively. The corresponding lifetime τ is related to $\rho(\mathbf{p})$ according to the formula

$$1/\tau = \lambda = \int \rho(\mathbf{p})d\mathbf{p} = \pi r_0^2 c \int |\psi_+(\mathbf{r})|^2 \cdot n_{el}(\mathbf{r}) \left[1 + \frac{\Delta\rho(\mathbf{r};\mathbf{r})}{n_{el}(\mathbf{r}))}\right] d\mathbf{r}. \quad (2)$$

In this work the electron wave functions $\psi_i^e(\mathbf{r})$ are obtained following the self-consistent scheme of Monnier and Perdew[12]. The electron density $n_{el}(\mathbf{r})$ is constructed from the one-electron wave functions ψ_i^e which satisfy the Schrödinger equation with the effective electron potential $V_-(\mathbf{r})$ being the sum of electrostatic and electron-electron exchange-correlation contributions. The electrostatic potential V_C in turn is the sum of the potential coming from the positive background of ions, $n_+(\mathbf{r})$, in the semi-infinite metal filling the half-space $z \leq 0$, and the Coulomb potential from the other electrons. Electron-electron correlations are taken into account in the way developed in Ref.12; the classical image-potential limit is matched self-consistently to the local exchange-correlation potential of the Ceperley-Alder form.

As follows from band structure calculation results, the nearly-free electron model is quite adequate for valence electrons in the bulk of simple metals (cf.

also Ref.15). In the present work we apply the jellium model of the surface, where the ions are thought of as forming a constant positive background, i.e. $n_+(r) = n_0 \Theta(-z)$, where $n_0 = 3/(4\pi r_s^3)$ is the average electron density in the bulk and $\Theta(x)$ is the unit-step function. This model may be corrected in order to reproduce properties of real metals, by adding to the electrostatic electron potential $V_C(r)$ the difference potential $\delta v_-(r) = <\delta v_-> \Theta(-z)$ (for more details see Ref.12). This procedure, however, does not affect considerably the resulting electron wave functions ψ_i^e at the surface.

Electron-positron correlation function $g(r) = 1 + \Delta\rho(r;r)/n_{el}(r)$ and corresponding positron correlation potential $V_{corr}(r)$ were obtained within WDA[11,6], based on the electron density in the host material, $n_{el}(r)$ (for more details see Ref.6). Positron wave function is the eigenfunction of the Schrödinger equation with the effective potential $V_+(r)$ consisting of the electron Coulomb (with the opposite sign), $-V_C(r)$, and the electron-positron correlation, $V_{corr}(r)$, parts. In order to switch from the jellium model to the real surface, the potential $\delta v_+(r)$ should be added to the positron potential $V_+(r)$ inside the metal, where $\delta v_+(r)$ is the difference between the potential of the discrete lattice of ions and the uniform background $n_+(r) = n_0\Theta(-z)$. In this work $\delta v_+(r)$ was approximated by a constant value $<\delta v_+>$ for $z \le -d/2$ and by a linear function $-2<\delta v_+> z/d$ for $-d/2 \le z \le 0$, where d is the interplanar distance averaged over main crystallographic directions. The value of $<\delta v_+>$ should reproduce the positron work function ϕ_+, i.e.

$$-V_C(-\infty) + V_{corr}(-\infty) + <\delta v_+> = -\phi_+. \qquad (3)$$

As the first step we determined the positron work functions in Al, Cd, Mg, Li, Na, and K (the experimental values of ϕ_+ in simple metals are available, according to our best knowledge, only for Al, while for Cd it is known to be positive) according to the relation[14]

$$\phi_+ = -(\mu_+ + \mu_-) - \phi_-, \qquad (4)$$

where μ_+ and μ_- are the positron and electron chemical potentials in the bulk and ϕ_- is the electron work function. Electron and positron energy levels in the bulk, providing the values of μ_+ and μ_- (cf. Refs.14), were obtained based on the band structure calculations[13] performed relativistically using the linear-muffin-tin-orbital method within the atomic spheres approximation, and applying the solid-state configurations. The details concerning the electron and positron models can be found in Ref.13. In Fig.1 the resulting positron work functions are presented as a function of an electron density parameter r_s. Solid line connects the values of ϕ_+ obtained according to Eq.(4) with ϕ_- following from the jellium model of the surface. The open squares correspond to ϕ_+ obtained basing on the experimental electron work functions, ϕ_-^{exp}. Theoretical values of ϕ_+ in Al, Na, and K extracted from Refs.14(b) and (c) are quoted in Fig. 1 for comparison.

Positron work functions obtained within both models appear to be increasing functions of r_s. Except for Al, they take positive values, that suggests positron tunneling into the vacuum. Since the positron energy eigenvalues E_+ at the surface are lower than the positron work functions ϕ_+, the positron tunneling into the metal occurs either. This fact is illustrated in Fig. 2, where the positron distributions at the surfaces of Al and Cd(positron work functions close to zero) are compared with those in Li and Na(values of ϕ_+ considerably positive). Due to the appreciably negative values of the positron potential V_+ inside the metal (V_+ is scaled to zero in vacuum, i.e. $V_+(\infty) = 0$), the portion of positron wave function in Li and Na, found in the metal side of the surface plane is not negligible, in contrast to Al and Cd.

The positron distribution at the surface is reflected in the SS lifetimes, calculated according to Eq.(2). In Fig.3 the values obtained in the present work for electron work functions, ϕ_-^{jell}, following from the jellium model as well as from experiment, ϕ_-^{exp}, are compared with bulk experimental data.

In Al and Cd the values of τ exceed 500 psec., i.e. the spin-averaged free-positronium lifetime. Theoretical result for Al is in excellent agreement with experiment[2]. As we could not find in the literature any experimental lifetime data for positrons trapped at the surface of simple metals, except Al, we would like to encourage experimentalists to perform measurements, at least for Cd. With decreasing bulk electron density, the differences between the SS and bulk lifetimes decrease. This fact, in our opinion, may be attributed to the increase of positron work function, as discussed above.

One-dimensional projections of ACAR spectra obtained according to Eq.(1) within quasi-IPM (which assumes $\Delta\rho(\mathbf{r};\mathbf{r}) = 0$; this approach is discussed in Ref.6) and using WDA electron-positron correlation function $g(\mathbf{r})$, are presented in Fig.4 (parts (a) and (b), respectively). In any of metals under consideration, including electron-positron correlations causes narrowing of ACAR spectra with respect to quasi-IPM and reverses anisotropy direction, providing normal component of $\rho(\mathbf{p})$ narrower than the parallel one. Each of spectrum calculated within the enhanced model is almost isotropic, suggesting that the nearly-free character of valence electrons in the bulk may be responsible for the nearly-isotropic shape of experimental SS ACAR spectra in Al[3]. With increasing values of r_s(increasing the positron work function ϕ_+), the shape of the SS spectrum becomes more and more similar to the classical inverted parabola (dotted lines in Fig.4).

Electron-positron enhancement factors $\epsilon(\mathbf{p}) = \rho(\mathbf{p})/\rho^{IPM}(\mathbf{p})$ (for more details see Refs. 6, 13, and 15) are displayed in parts (c) of Fig.4. These parameters show similar momentum-dependence in simple metals, with parallel component considerably higher than the normal one.

Conclusions

The theoretical annihilation characteristics for positron trapped at the surface of simple metals are studied. The positron work function is found to be a decreasing function of electron density in the bulk. Application of WDA to description of electron-positron correlation effects at the surface leads to the SS lifetimes in Al and Cd higher than the spin-averaged free-positronium value. Theoretical result for Al is in very good agreement with experiment[2], while experimental lifetime data in other simple metals are not available.

Electron-positron enhancement effects cause narrowing of ACAR spectra and in any of metals under study reverse the anisotropy direction (cf. remarks in Ref.3, concerning SS components of experimental data for Al). These spectra are almost isotropic with normal component slightly narrower than the parallel one. With increasing electron density parameter r_s both the positron lifetime and ACAR spectra become to approach to the bulk ones, due to the increase of the positron work functions, which causes the positron to be trapped in the metal side of the surface plane.

Present calculations were performed within the structureless jellium model of the metal surface. The method, however, may be applied to studies of face-dependent SS annihilation characteristics in simple metals by using in the electron model the face-dependent step difference electron potential $< \delta v_- >$ in the way presented in Ref.12. The face-dependence of SS annihilation parameters in jellium-like metals is an interesting problem for further studies.

Acknowledgments

Anna Rubaszek is grateful to the Stefan Batory Foundation (Warsaw, Poland), for covering the cost of the travel to SLOPOS5 and to the Idaho State University and INEL (Idaho, USA) for the Conference grant.

References

[1] For reviews, see, e.g., A.P.Mills,Jr., in *Positron Solid State Physics*, edited by W.Brandt and A.Dupasquier (North-Holland, Amsterdam, 1983), p.432; K.G.Lynn, *ibid.*, p.609; A.Dupasquier and A.Zecca, Riv. Nuovo Cimento 8, 1 (1985); P.J.Schultz and K.G.Lynn, Rev.Mod.Phys. 60, 701 (1988), and references cited therein

[2] K.G.Lynn, W.E.Frieze, and P.J.Schultz, Phys.Rev.Lett. 52, 1137 (1984)

[3] K.G.Lynn, A.P.Mills,Jr., R.N.West, S.Berko, K.F.Canter, and L.O.Roeling, Phys.Rev.Lett. 54, 1702 (1985); D.M.Chen, S.Berko, K.F.Canter, K.G.Lynn, A.P.Mills,Jr., L.O.Roelling, P.Sferlazzo, M.Weinert, and R.N.West, *ibid.* 58, 921 (1987); Phys.Rev.B 39, 3966 (1989)

[4] see citations 5-9 in Ref.6

[5] A.Rubaszek, J.Phys.Condens.Matter, 1, 2141 (1989)

[6] A.Rubaszek, Phys.Rev.B 44, 10 857 (1991)

[7] B.Rozenfeld, W.Swiatkowski, and K.Jerie, Acta Phys.Pol.A 64, 93 (1983); Lou Yongming, Phys.Rev.B 38, 9490 (1988)

[8] R.M.Nieminen, M.J.Puska, and M.Manninen, Phys.Rev.Lett. 53, 1298 (1984); R.M.Nieminen and M.J.Puska, *ibid.* 50, 281 (1983); A.P.Brown, A.B.Walker, and R.N.West, J.Phys.F 17, 2491 (1987); A.P.Brown, K.O.Jensen, and A.B.Walker, *ibid.* 18, L141(1988)

[9] A.Rubaszek and J.Lach, J.Phys.Condens.Matter 1, 9243 (1989); Surf.Sci. **211/212**, 227 (1989)
[10] O.Gunnarson, M.Johnson, and B.I.Lundqvist, Phys.Rev.B **20**, 3136 (1979)
[11] K.O.Jensen and A.B.Walker, J.Phys.F **18**, L277, (1988)
[12] A.Kiejna, Phys.Rev.B **43**, 14 695 (1991), and references cited therein
[13] S.Daniuk, M.Sob, and A.Rubaszek, Phys.Rev.B **43**, 2580 (1991)
[14] (a) G.Fletcher, J.L.Fry, and P.C.Pattnaik, Phys.Rev.B **27**, 3987 (1983); (b) M.Farjam and H.Schroe, *ibid.* **36**, 5089 (1987); (c) O.V.Boev, M.J.Puska, and R.M.Nieminen, *ibid.*, 7786 (1987)
[15] G.Kontrym-Sznajd and A.Rubaszek, *this Conference; to be published*

Figure captions

Fig.1. Positron work functions obtained according to Eq.(4) using ϕ_- calculated in this work (circles connected by a solid line) and experimental electron work functions (open squares) compared with those extracted from Refs.14(b) (Li, MG) and 14(c) (black squares).

Fig.2. Positron wave functions at the surfaces of Al, Cd, Li and Na.

Fig.3. Positron SS lifetimes as a function of r_s obtained within model using jellium electron work function (black circles connected by a solid line) and experimental one (open circles connected by dashed line). Lower solid line represents experimental lifetime values in the bulk.

Fig.4. 1D ACAR spectra (normalized to the same peak height) obtained according to Eq.(1) within quasi-IPM [parts(a)] and within the enhanced model [parts (b)]. Solid and dashed lines correspond to parallel and normal components, respectively. Dotted line represents the classical inverted parabola. 1D enhancement factors $\epsilon(p_z) = \rho(p_z)/\rho^{IPM}(p_z)$ and $\epsilon(p_x) = \rho(p_x)/\rho^{IPM}(p_x)$ (normalized to the same peak height) are displayed in parts (c) by dashed and solid lines, respectively.

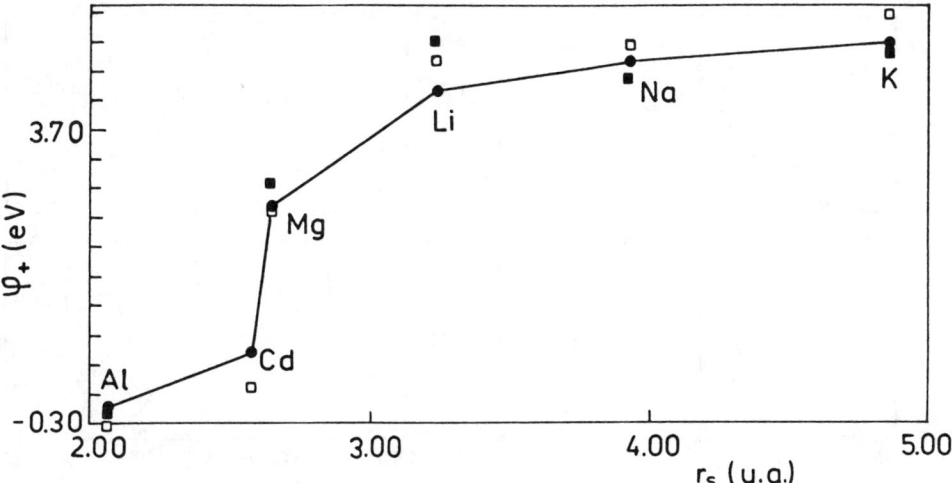

Fig. 1

256 Annihilation Characteristics for Positrons

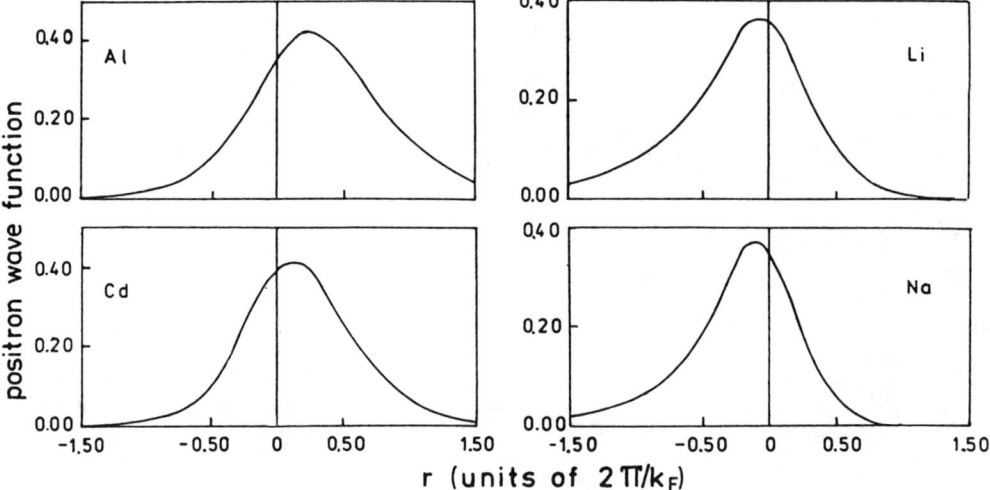

Fig. 2

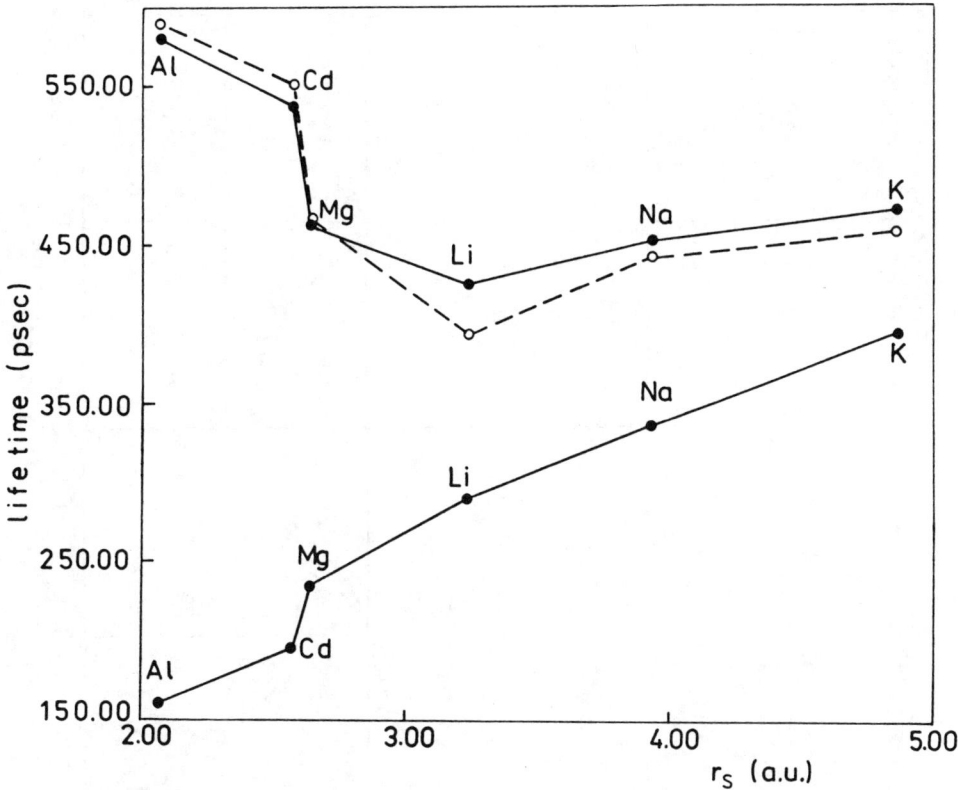

Fig. 3

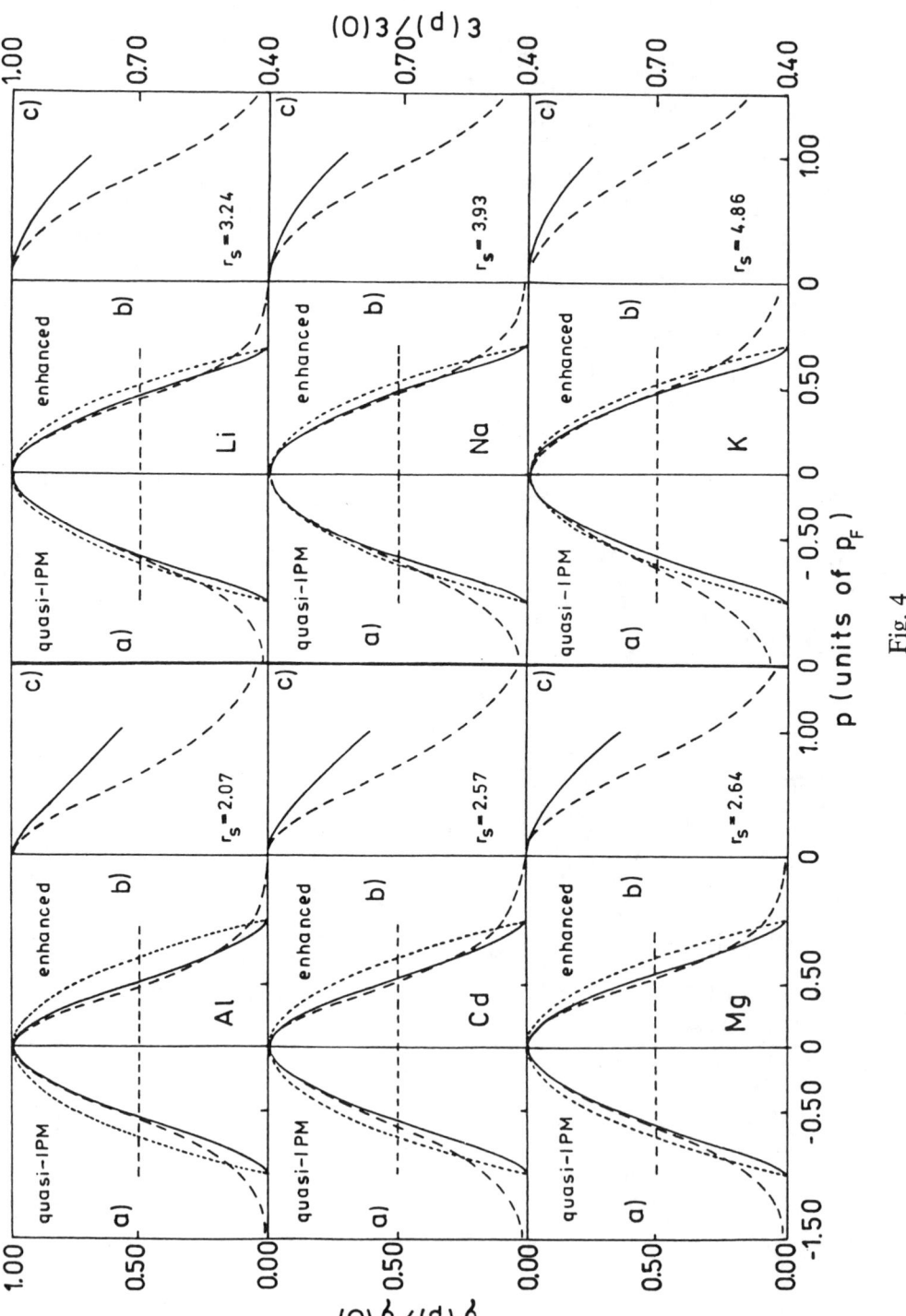

Fig. 4

PHYSISORBED SURFACES STUDIED BY POSITRON ANNIHILATION SPECTROSCOPY

S.J.Wang

Department of physics, Wuhan University, Wuhan 430072, P.R.China

Physisorbed nitrogen and methane surfaces on grafoil with large specific surface area were studied by positron annihilation. The positron lifetime and Doppler broadening S-parameter were measured as a function of coverage and as a function of temperature at a fixed fraction of coverage. The activation energy for surface defect created by adsorption was estimated and detailed surface phase transitions were observed.

1. Introduction

Physical adsorption of atomic or molecular layers on solid substrates by the relatively weak Van Der Waals force is one of the most rapidly growing research fields in surface science today. Physisorbed surface is an ideal system for detailed quantitative investigations about various fundamental problems in condensed matter physics, such as melting, welting and microstructural phase transitions. Various experimental techniques have been utilized to study physisorption phenomena. The positron annihilation spectroscopy (PAS) has been known as a powerful tool for material science for defect and electronic structure. In the recent years, the development of slow positron beam has indicated that the positron is a potential new probe for surface science. However this technique requires ultra-high vacuum and clean surface. On the other hand, one can use fast positrons from radioactive source for surface studies. In this case, the system studied must have high specific surface area, such as grafoil and porous materials etc. The physisorption results of positron lifetime measurements for argon and nitrogen on grafoil was first reported by Jean et al[1]. The surface inhomogeneity created by adsorption can be viewed as a defect site by the positron. The unique trapping and diffusion properties of the positron and positronium (Ps) on the surface are responsible for the enhanced annihilation signals from the surface as compared with bulk. In this paper, we reported the continued research on physisorption on grafoil.

2. Experiment

It is very important for positron scientists to know how to make submonolayer and multilayer surfaces. The graphite is chosen as a substrate because: 1) the surface has well been studied by other techniques; 2) the surface can prepared relatively clean; 3) a class of different surface area can be obtained by graphitization process. The substrate was grafoil (from Union Carbide Co.) which is an exfoliated recompressed sheet with a specific surface area of about 20 m^2/g. A wide class of adsorbates and moleculars can be chosen as system of study. Preparation of submonolayer and multilayer can be achieved by the so-called isotherm method which maintains at a constant temperature of 2D phase and uses the equilibrium pressure between the 2D and 3D molecules to form a certain fraction of surface coverage. Fig. 1 shows a schematic diagram of the isotherm experiment for nitrogen physisorption[2]. The grafoil sample was baked at 700°C under 10^{-5} Torr vacuum for 6-8 hours. The baked grafoil was transferred to an aluminum sample cell under dry helium atmosphere. A positron source was embedded in the grafoil. The sample cell was evacuated under 10^{-5} Torr for 6h prior to isotherm measurements. The gas manifold system consisted of standard glass chamber V_0, two absolute pressure gauges P_1 and P_2, a liquid nitrogen bath and diffusion pump. The sample cell was immersed in liquid nitrogen bath during the measurement at 77K. Fig. 2 shows an isotherm plot for methane on grafoil at 77K[3]. As shown each sudden jump corresponds to monolayer, bilayer, trilayer, etc. Other experimental details concerning about PAS can be found in our previous papers[3,4].

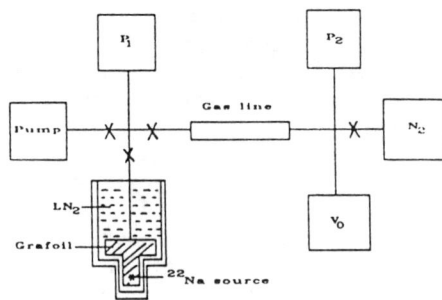

Fig.1 Schematic diagram of the gas manifold system. P1 (0-10Torr) and P2 (0-800Torr) are absolute pressure gauge.

3. Results and Discussions

1). Monolayer

The structural phase transformation of nitrogen and methane monolayers have been studied by PAS. In the methane monolayer on grafoil system, positron lifetime and Doppler broadening measurements were performed as a function of temperature between 13K and 95K at fixed fraction of coverage submonolayers. Three positron lifetimes were resolved: a long-lived component $\tau_3=2.5\pm0.5$ns, $I_3=0.4\pm0.1\%$ is attributed to o-Ps annihilation in the inter-grain regions and on the surface which is nearly independent of temperature; $\tau_1=0.18\pm0.02$ns and $I_1=69\pm2\%$ is attributed to the positron annihilation in the bulk graphite; the intermediate component $\tau_2=0.32$ns corresponds to the positron annihilation on the

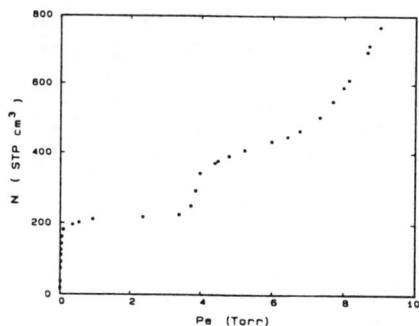

Fig.2 Isotherm plot for methane adsorption on grafoil at 77K. N is the adsorption volume at STP and Pe is the equilibrium pressure.

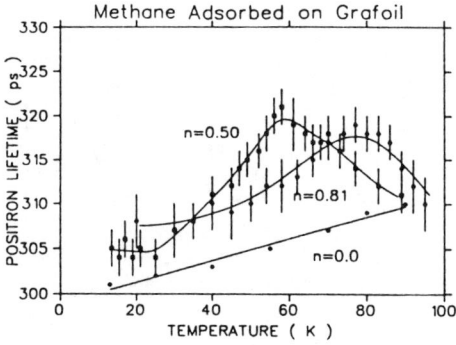

Fig.3 Surface positron livetime vs temperature for submonolayer methane adsorbed on grafoils. n is the fraction of complete monolayer adsorption.

physisorbed surface of graphite. The assignment of τ_2 to the surface state has been verified by the chemical quenching (O_2) and magnetic quenching experiments[1]. The surface positron lifetime varies as a function of temperature and fraction of coverage. In Fig. 3, we show the variation of τ_2 with temperature for n (fraction of coverage) =0, 0.50 and 0.81 respectively[4]. As shown, surface positron lifetime is sensitive to the solid-solid, solid-liquid and liquid-hypercritical fluid phase transitions. The maximum values of positron lifetime correspond to the known 2D melting temperatures, i.e. 69K and 72K for n=0.50 and 0.81 respectively. The sigmoidal behavior of S vs T has been employed to deduce the activation energy for the defect formation on 2D surface. The obtained energies of 0.017±0.002 ev (n=0.50) and 0.007±0.002 ev (n=0.81) are new informations in surface science.

The Doppler broadening S-parameters have been measured as a function of coverage for nitrogen on grafoil at 77K[2]. The S-parameter shown in Fig. 4 starting from zero coverage increases with the fractional coverage till a maximum value at about half a monolayer and then decreases. This can be interpreted by "island formation" mechanism. The structure of the 2D nitrogen physisorption is a commensurate phase. But islands and defects structure may be created due to the presence of molecular adsorbates. The maximum structural inhomogeneity occurs at about half a monolayer. Beyond one monolayer, further adsorption will fill into these defect sites, the physisorbed surface has more homogeneous structure and S-parameter reaches a value lower than that of clean grafoil surface.

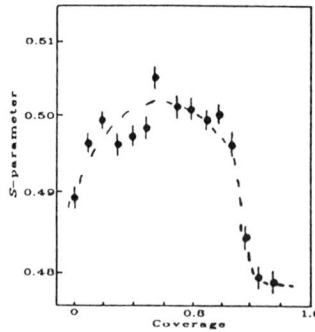 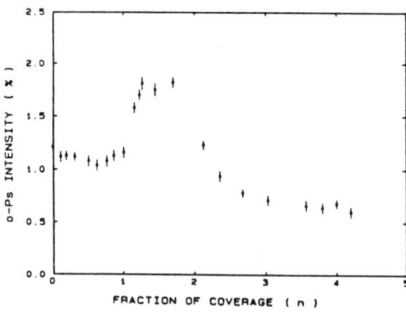

Fig.4 The S-parameter vs the fraction of nitrogen coverage at 77K.

Fig.5 The o-Ps intensity vs the fraction of methane coverage at 77K

2). Multilayer

The positron lifetime and Doppler broadening experiments were performed as a function of coverage from n=0 to 4.2 monolayer of methane on graphite at 77K[4]. The most interesting results are the variations of o-Ps lifetime and its intensity. The o-Ps intensity shown in Fig.5 is nearly constant below monolayer, but significantly increases above n=1. The higher o-Ps formation probabilities between n=1 and n=1.8 reveals the structural transition from one to two layers. This fact suggests that new free volume or open spaces such as voids are created due to methane adsorption when n>1. Considering that the weight fraction of methane at 1.5 monolayer is only 0.5% and an increase of 0.6% o-Ps intensity was observed, adsorption of methane must be a very non-uniform process, one of the best candidates for this process is island formation, where multilayers form at the same positions on the surface while other position may still has a low coverage. Further adsorption of methane beyond n=2 results in a slight decrease of o-Ps intensity. The level-off value at high coverage is 0.6% less than that of the clean grafoil surface. This may be explained by that at high coverage the amount of free space available for o-Ps formation is reduced. It might be associated with formation of more thicker film when n>2.

The value of o-Ps lifetime gives a quantitative value of the size of the free volume or the defects where o-Ps localized and annihilated. If we fit the relationship between o-Ps lifetime and free volume, we obtain the sizes of free volumes to be 230$Å^3$, 330$Å^3$ and 180$Å^3$ for n=0, n=1.5 and n>2 respectively[3]. If we further assume that the free volume space is spherical, we estimate the size of these spaces on the surface to be 7.8Å, 8.6Å and 7.0Å for n=0, n=1.5 and n>2 respectively.

4. Conclusion

The utilization of PAS in studying physisorbed surface of monolayer and multilayer is promising. New informations about the surface defect formation due to the adsorption, phase transition and defect activation energy, are uniquely obtained from PAS data. Further exploits in deferent adsorbates and substrates including using a slow positron beam are suggested.

References:
1. Y.C.Jean et al. Phys.Rev.B 32,4313(1985)
2. H.Liu and S.J.Wang, Chinese Phys.Lett. 9,217(1992)
3. S.J.Wang, D.M.Zhou and Y.C.Jean, Chem.Phys. 129,503(1989)
4. S.J.Wang and Y.C.Jean, Phys.Rev.B 37,4869(1988)

RE-EMITTED POSITRON SPECTROSCOPY OF COBALT AND NICKEL SILICIDE FILMS

B.D. Wissman, W.E. Frieze, and D.W. Gidley
Department of Physics, University of Michigan, Ann Arbor, Michigan 48109

ABSTRACT

The techniques of re-emitted positron spectroscopy (RPS) have been employed in the first systematic investigation[†] of the positronic properties of the various stoichiometric phases (M_2Si, MSi, and MSi_2) of Co and Ni silicide films grown *in situ* on Si substrates. The positron work function is found to be negative for all of the different phases; thus implanted positrons may be re-emitted. The energy of the re-emitted positrons is found to have a surprisingly large variation for the different phases. This feature should provide the image contrast necessary to observe each phase on a microscopic scale using the positron re-emission microscope (PRM). The positron deformation potential, $E_d^+ \equiv V(\partial\Sigma/\partial V)$, was determined for $CoSi_2$ films; it can be used to estimate the size of the positron diffusion constant, which is found to be comparable to that of other metals. Thus the short positron diffusion length (of order 150 Å) determined from depth-profiling measurements of $CoSi_2$ films must be a result of positron trapping in either the film or at the interface with the Si substrate. RPS results considered as a function of film thickness support the conclusion that defects in the film (misfit dislocations and/or vacancies) represent the major source of positron trapping.

INTRODUCTION

There is a great deal of interest in the formation of metal silicide films due to their many device applications, such as Schottky barriers, ohmic contacts, low resistivity interconnects, and metal base and permeable base transistors.[1,2,3] A number of the metal silicides are known to grow epitaxially, and of these, $CoSi_2$ and $NiSi_2$ are of particular interest due to their small lattice mismatches with Si (1.2 % and 0.4 %, respectively[4]). These small lattice mismatches permit the growth of silicide films with nearly perfect epitaxial structure on Si substrates, which, under certain conditions, may be pseudomorphic.[1,5,6] In addition to their technological applications, the possibility of growing silicide films which have nearly perfect interfaces with Si substrates makes them ideal systems in which to study the basic physics of metal-semiconductor junctions.[4] Such films are attractive systems for studying the diffusion, drift[7], and trapping of positrons in the vicinity of the Schottky *well* and the associated depletion region. However, very little is known about the properties that determine positron transport in these materials. The only study[8] is of $CoSi_2$, in which it was discovered that a thin film grown on a Si(111) substrate has a negative positron work

---

[†] B.D. Wissman, W.E. Frieze, and D.W. Gidley, to be published in Phys. Rev. B.

function. $CoSi_2$ therefore re-emits positrons; thus the techniques of re-emitted positron spectroscopy (RPS) can be applied to this and other silicide systems.

In RPS one measures the energy distribution of positrons which are re-emitted after being implanted with incident energies of order several keV, thermalizing, and subsequently diffusing back to the surface. Due to the presence of contact potentials, the energy of elastically emitted positrons is determined[9] by a parameter Σ, where

$$\Sigma \equiv \mu^- + \mu^+ = -(\phi^- + \phi^+). \tag{1}$$

Here μ^- and μ^+ are the bulk electron and positron chemical potentials, and ϕ^- and ϕ^+ are the electron and positron work functions, respectively. Note that Σ (sometimes referred to as the positron affinity) is negative, and that a more negative value indicates a stronger affinity of a particular material for positrons.[10] Positrons which have been implanted into a multilayer structure may be re-emitted with well-defined energies which are characteristic of each material. By identifying the peak energy and any shifts thereof, RPS can be used to probe heterogeneous growth systems, e.g. interdiffusion alloying[9] and pseudomorphism.[11] In particular, in the present work we can observe the thermal reaction of Co and Ni layers deposited on Si to form the various silicide phases. RPS spectra are investigated as a function of: the initial thickness of the metal overlayer, the annealing/reaction temperature, and the incident positron energy.

EXPERIMENTAL TECHNIQUE

Co and Ni silicides were grown *in situ* by thermal reaction of thin (5 – 200 Å) Co and Ni films which had been deposited by evaporation from a heated W filament onto Si substrates which were at room temperature. All work was performed in a UHV surface analysis chamber, with base pressure 1×10^{-10} Torr. The background pressure remained below 2×10^{-7} Torr at all times during the deposition and annealing. Silicide growth was monitored by both Auger electron spectroscopy (AES) and low-energy electron diffraction (LEED), with the various silicide phases obtained by varying the annealing time and temperature.[12] The substrates were cleaned by sputtering with 1 – 2 keV Ar ions, followed by a quick (< 30 sec) heat to 1100 – 1200 °C. Samples cleaned in this manner exhibited good LEED patterns, and had negligible surface contaminations of C. In the case of the (111) substrates, the 7x7 surface reconstruction was typically observed.

RPS studies were done using a variable energy (0.5 – 4.0 keV) monoenergetic beam. At these beam energies the mean implantation depth[13] $\bar{z}$ varies from approximately 25 – 750 Å. The energies of the re-emitted positrons were measured with a double-pass cylindrical mirror analyzer operating in a constant pass-energy mode.[9] The value of Σ is determined by the energy of the elastic peak relative to that of a Ni(100) reference crystal, with Σ_{Ni} taken to be -3.8 eV, as discussed below. The absolute energy scale is calibrated by defining

ϕ^- (and hence the zero positron energy cutoff) of the clean Ni(100) crystal to be 5.2 eV.[14] $|\phi^+|$ is determined by the energy difference between the elastic peak and the corresponding zero energy cutoff, minus a correction due to the resolution of the analyzer. On an absolute scale our measurements of ϕ^+ and Σ are uncertain at the ±0.1 eV level. However, by measuring Σ relative to a reference crystal, the relative values of Σ can be determined to an accuracy of approximately ±20 meV. Relative values of ϕ^- may be determined to an accuracy of approximately ±50 meV. The clean Ni(100) reference crystal is found (after a resolution-shift correction of 0.25 eV) to have $\phi^+ = -1.4$ eV, therefore from Eq.(1) $\Sigma_{Ni} = -3.8$ eV.

RPS OF SILICIDES

A. Ni Silicides on Si(100)

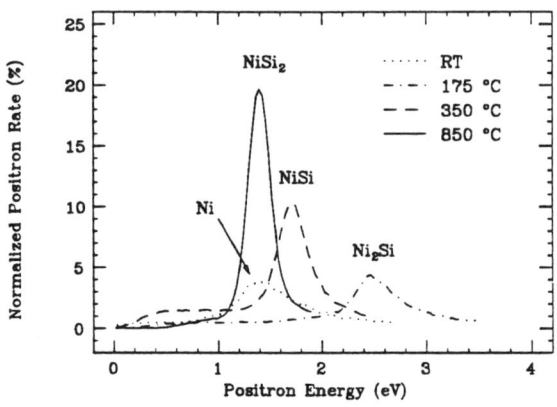

Figure 1: RPS Spectra of 50 Å Ni/Si(100). The spectra, acquired at RT using 1 keV positrons, show the progression from un-reacted Ni (as deposited at RT), to Ni$_2$Si, to NiSi, to NiSi$_2$ with increased annealing temperature. The spectra are normalized such that the vertical scale represents the peak rate as a fraction of the rate (at 1 keV) of the Ni(100) reference crystal.

A Ni film of thickness approximately 50 Å was deposited on a n-Si(100) substrate ($\rho \approx 10$ $\Omega \cdot$ cm) and annealed at progressively higher temperatures (as noted in Fig. 1) to produce the different silicide phases: Ni$_2$Si, NiSi, and NiSi$_2$, in order of increasing reaction temperature. Assuming uniform layer growth, this amount of Ni would produce silicide film thicknesses of approximately 75, 100, and 175 Å, respectively.[15] The corresponding mean implantation depths for 1 keV positrons range from approximately 50–80 Å. At annealing temperatures below 175 °C the RPS spectrum indicated the presence of only un-reacted Ni, consistent with AES. A distinct Ni$_2$Si peak appeared after annealing at 175 °C. After annealing at 250 °C a distinct NiSi peak appeared. The transition to NiSi$_2$ began after reaction at 650 °C, with LEED indicating a highly-ordered surface. After annealing at 850 °C a well isolated peak corresponding to NiSi$_2$ was observed. All three Ni silicide phases are clearly distinguishable in Fig. 1, and it is interesting to note that their peaks are equal or higher in energy than that of pure Ni (i.e. despite the fact that Σ_{Si} has a low value, the Ni silicides all have $\Sigma \geq \Sigma_{Ni}$).

By increasing the incident positron beam energy, and hence the mean implantation depth, we can sample progressively deeper below the surface. This simple form of depth-profiling was used to demonstrate the transition from a NiSi film of thickness approximately 100 Å to a NiSi$_2$ film of thickness approximately 175 Å as the annealing temperature of a 50 Å Ni film was increased from 650 °C (Fig. 2) to 750 °C. As can be seen in Fig. 2, the slightest hint of NiSi$_2$ is observed for the lowest implantation energy ($\bar{z} \approx 25$ Å), indicating the presence of NiSi$_2$ near the surface, consistent with AES and LEED observations. After annealing at 750 °C, a 4 keV spectrum ($\bar{z} \approx 750$ Å) shows a NiSi peak, but the spectrum acquired using less penetrating 1 keV incident positrons ($\bar{z} \approx 75$ Å) shows only a slight mounding at that energy, indicating the presence of deep-lying NiSi, thus indicating that the the final transformation occurs deep in the film.

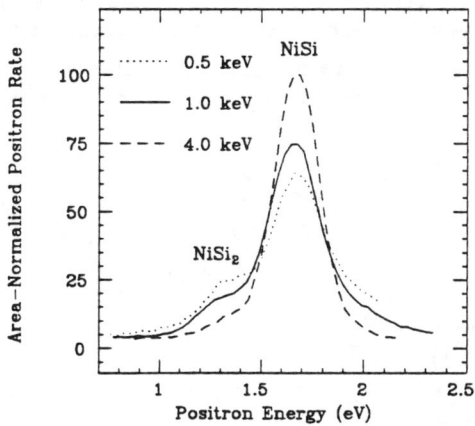

Figure 2: RPS Spectra as a Function of Incident Energy. RPS spectra of a 50 Å Ni film annealed at 650 °C. The spectra, acquired at RT, are normalized to equal areas.

B. Co Silicides on Si(111)

A Co film of thickness 100 Å was deposited on a n-Si(111) substrate ($\rho \approx 0.13$ $\Omega \cdot$ cm) and annealed at progressively higher temperatures to produce the different silicide phases: Co$_2$Si, CoSi, and CoSi$_2$, in order of increasing reaction temperature. Assuming uniform layer growth, this amount of Co would produce silicide film thicknesses of approximately 150, 200, and 350 Å, respectively.[15]

The corresponding mean implantation depths for 1 keV positrons range from approximately 50 – 80 Å. At annealing temperatures below 300 °C the RPS spectrum indicated the presence of only un-reacted Co, consistent with AES. After annealing for 5 minutes at 300 °C the RPS spectrum had distinct Co and Co_2Si peaks, plus a mound at the energy corresponding to CoSi. An isolated Co_2Si peak was never seen - it was always accompanied by Co and CoSi peaks. This was also the case for silicide films grown from Co films of initial thicknesses of 150 and 200 Å, consistent with the observation that the growth of Co_2Si and CoSi occurs simultaneously.[16,17] Further annealing at 300 °C resulted in a transition from Co_2Si to CoSi, with the size of the CoSi peak increasing relative to that of Co_2Si with increasing annealing time. After annealing for 10 minutes at 350 °C the Co_2Si peak disappeared, leaving a lone CoSi peak. The transition from CoSi to $CoSi_2$ occurred after annealing near 450 °C. At reaction temperatures above 550 °C, a lone peak corresponding to $CoSi_2$ was observed, with LEED indicating a highly-ordered surface.

C. Discussion of Results

The results of the above-mentioned spectra are presented graphically in Fig. 3. All of the different silicides are found to have negative ϕ^+. In all cases the Σ values are large enough that the silicide films present an energy barrier of several electron-volts to positrons which have thermalized in the Si substrate (and thus would not be appropriate as electrical contacts for a field-assisted positron moderator). Of course, if the predicted[10] value of Σ_{Si} shown in Fig. 3 is correct, then it is very unlikely that any metal contact would be suitable.

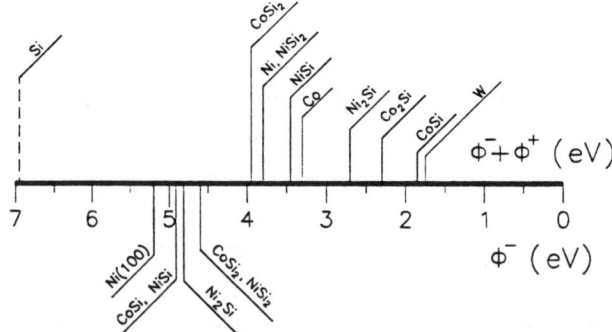

Figure 3: Measured RPS Peak Energies and Cutoffs. The value for Si is a theoretical prediction.[10] The energy scale is reversed so that positron energy increases to the right. The absolute energy scale is set by defining ϕ^- for Ni(100) to be 5.2 eV.[14]

Several other interesting features can be seen in Fig. 3. There is a relatively wide variation in Σ for the different silicides, particularly for the Co silicides. It is this property that allows each phase to be very easily distinguished in the

RPS spectra. There is a correspondingly large variation in $|\phi^+|$ (0.6 - 3.0 eV) with relatively little variation in ϕ^- (4.5 - 4.9 eV). In addition, the value of Σ is surprisingly large compared with that of the pure metal. The Ni silicide peaks all have Σ equal to, or greater than, that of pure Ni even though they rank in order of increasing Σ from Si-rich to Ni-rich. Naively, one might have expected (on the basis of alloying results[9]) that the silicides would lie between Si and Ni in Fig. 3, with the Ni-rich phase approaching Ni from the left. This is clearly not the case, and thus it would be interesting to have detailed calculations of μ^- and μ^+ for the silicides(as per Ref. 10 for the elements). Another interesting feature is the difference in the ordering of the Σ values for the Co silicides as opposed to the Ni silicides. This may be explained by the fact that while both metal-rich phases have the $PbCl_2$ structure, and both Si-rich phases have the CaF_2 structure, the structures of the intermediate phases are different; NiSi has the orthorhombic MnP structure, whereas CoSi has the cubic FeSi structure.[15] As a result, the atomic density of CoSi is slightly larger than that of NiSi, which may account for the fact that its Σ value is larger (i.e. less negative) than that of the other silicides. Nonetheless, we were surprised to find such a large value of Σ (-1.85 eV), and such a large and negative ϕ^+ (-3.02 eV) for CoSi, comparable to such extreme values as those for W.

Another feature, clearly evident in Fig. 1, is that the re-emitted positron yield in the elastic peaks of all of the silicides are small, ranging from approximately 4 – 20 % for the Ni silicides at 1 keV (and correspondingly 1 – 10 % for the Co silicides), where the yields are given as a fraction of the elastic peak yield (at 1 keV) of a clean, well-annealed, single crystal Ni reference. The total (energy-integrated) yield at 1 keV of re-emitted positrons shows much less variation, ranging from approximately 15 % for the metal-rich and intermediate phases to 25 – 30 % for the Si-rich phase, where the yields are given as a fraction of the total yield (at 1 keV) of the Ni reference crystal. These results are not surprising for the metal-rich and intermediate phases since the non-epitaxial nature of their growth (indicated by the lack of a LEED pattern) presumably leads to highly defective films. The low peak rates relative to that of the Ni reference indicate that the positrons are emitted with an angular distribution that is much broader than the angular acceptance of our energy analyzer, thus suggesting a rough or faceted surface. The higher re-emission rates from the Si-rich phase are certainly due to their better epitaxy. Nevertheless, their relatively low total yields may well be due to the presence of misfit dislocations or other positron-trapping defects, as discussed below.

To further investigate trapping defects, the total yield of re-emitted positrons from $CoSi_2$ films (thickness $\approx$ 700 Å) grown by the deposition of a 200 Å Co film, with subsequent annealing at 600 °C and 850 °C are shown in Fig. 4 as a function of incident positron energy. Note that the total yield drops rather sharply with increasing incident energy, and hence implantation depth. The data were fitted to a function of the form[18] $f = f_0 [1 + (\frac{E}{E_0})^{1.6}]^{-1}$. The fits yield E_0 values of 1.30 keV and 1.36 keV for the 600 °C and 850 °C data, respectively. These low values indicate that the film has a very short positron diffusion length, of order 150 Å. This is in reasonable agreement with the result obtained by Gullikson et al. for a film grown by MBE.[8] There are either positron-trapping defects (misfit dislocations and/or vacancies) in the film, or the Si interface/Schottky

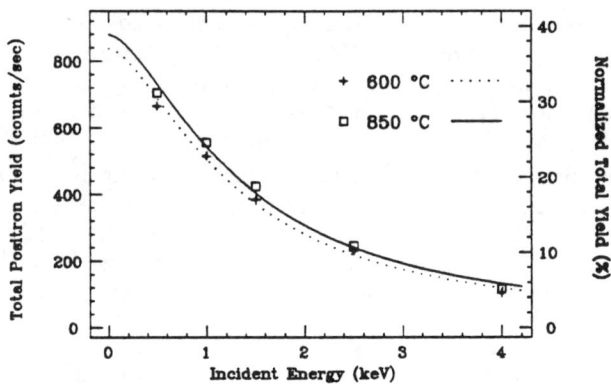

Figure 4: Total Positron Yield vs. Incident Energy. Total yield at RT from a Co film annealed at 600 °C and 850 °C to form $CoSi_2$. The right-hand scale represents the total yield as a fraction of the yield (extrapolated to zero incident energy) of the Ni(100) reference crystal. The dotted and solid curves represent fits to the 600 °C and 850 °C data, respectively.

well is trapping most of the positrons in the film. There is evidence for positron-trapping defects, as will be discussed in the following sections.

POSITRON DEFORMATION POTENTIAL OF $CoSi_2$

The positron deformation potential relates changes in Σ to variations in the bulk atomic density[10], and is defined as $E_d^+ \equiv V\,(\partial\Sigma/\partial V)$. The temperature dependence of Σ was measured for several $CoSi_2$ films grown by depositing Co films of thickness 100 - 200 Å on Si(111) substrates and annealing at 850 °C. This temperature dependence appears in the RPS spectrum in the form of a shift in the elastic peak energy. The peak energy was measured as a function of temperature as the samples were radiatively heated. Incident positrons of energy 1 keV were used. The peak shift was found to be linear in the range from 25 – 275 °C, with the slope $d\Sigma/dT = -0.22 \pm 0.03$ meV/ K. This value can be related to E_d^+ using the linear coefficient of thermal expansion, α.[11] Using $\alpha = 10.1 \times 10^{-6}$ K^{-1},[19] the measured temperature dependence then yields $E_d^+ = -7.3 \pm 1.0$ eV, where the error quoted is due to the statistical error in determining $d\Sigma/dT$.

The deformation potential is a measure of the strength of the positron-phonon coupling, and therefore may be used to estimate the size of the positron diffusion constant, D^+, due to acoustic phonon scattering.[20] The diffusion constant is directly related to the positron diffusion length[13], which determines many of the positron transport properties of a material. Using the measured values of the elastic constants[21], and taking the effective mass of the positron to be

$m^* = 1.5\, m_e$, which represents a reasonable compromise between theoretical and experimental estimates for most materials[20], the above value of E_d^+ yields the relatively large value of 2.9 cm^2/sec for D^+ at 300K (typical metals have diffusion constants of order $0.1-1.0$ cm^2/sec[13]). In the absence of positron-trapping defects, the resulting positron diffusion length is of order several thousand Angstroms, comparable to that of Ni. Thus the relatively low yield of re-emitted positrons from the silicides indicates that a significant number of positrons are trapping. We will consider this point further in the following section.

ULTRATHIN CoSi$_2$ FILMS

As mentioned earlier, all of the silicides exhibit relatively small re-emitted positron yields. We concluded earlier that in all likelihood the metal-rich and intermediate phases have a high density of open-volume defects that trap positrons, but that the situation for the Si-rich phase was less clear. In order to distinguish trapping in bulk defects from trapping at the Si interface/Schottky well, we measured the positron peak rate and total yield of ultrathin CoSi$_2$ films grown by a multiple-step deposition and 850 °C reaction technique. Co films were deposited with rates of order 0.2 Å/sec, with the background pressure in the chamber remaining below 2×10^{-9} Torr at all times during the deposition and annealing. These data (acquired using 1 keV incident positrons with $\bar{z} \approx 80$ Å) are plotted in Fig. 5, as a function of the initial Co film thickness. We find that both the peak rate and total yield approach the thick film (700 Å) values of $8-10$ % and $25-30$ %, respectively, for Co film thickness of order 40 Å (corresponding to 140 Å CoSi$_2$ thickness). This is consistent with a *bulk* diffusion length of order 150 Å (in agreement with our depth-profiling results). It is not consistent with a long *bulk* diffusion length (e.g. no bulk defects) and trapping only at the interface. If this were the case, a sizable increase in re-emission (of order 50 %) would be expected when the film thickness is increased from 140 Å to 700 Å. Thus the Si interface/Schottky well is not the major source of positron trapping, and we therefore conclude that the positrons are mainly trapping in misfit dislocations/and or vacancies.

CONCLUSION

All of the different phases of Co and Ni silicides re-emit positrons, with $|\phi^+|$ ranging from 0.6 to 3.0 eV. As there is little variation in ϕ^-, the parameter Σ (which represents the positron energy level in a particular material) therefore also has a comparably large variation. In general, Σ increases with increasing atomic density, which in turn tends to decrease with the silicide reaction temperature as the film is transformed from the metal-rich to the Si-rich phase. The rate of positron re-emission in the elastic peak increases in going from the metal-rich to the Si-rich phase. This feature, together with the widely separated and thus easily distinguishable peaks in the RPS spectra for each silicide phase, should provide the necessary image contrast for observing each phase on a microscopic scale using the positron re-emission microscope (PRM).[22,23] Depth-profiled PRM images may provide a unique perspective on the dynamics of the

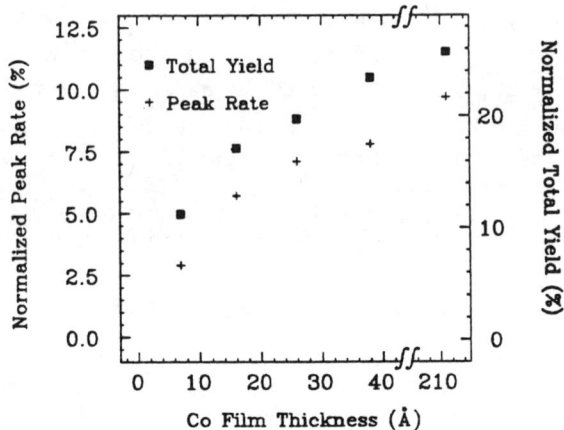

Figure 5: Peak Rate and Total Yield vs. Co Film Thickness. $CoSi_2$ films were grown by sequential steps of Co deposition followed by annealing at 850 °C. The peak rate and total yield (at 1 keV) are expressed as fractions of those (at 1 keV) of the Ni(100) reference crystal.

silicide growth as it proceeds through the various phases by diffusion and/or nucleation. In the particular case of $CoSi_2$ films, a PRM with the predicted lateral resolution of order 10 Å [22,23] could readily be used to observe the formation of non-emitting Si pin-holes, which have lateral dimensions larger than 100 Å.[4] Such pinholes play a strong role in determining the electron transport properties of $Si/CoSi_2/Si$ metal and permeable base transistors.[2,3]

Our measurement of the positron deformation potential, along with a reasonable estimate of the positron effective mass, can be used to deduce that the positron diffusion constant in $CoSi_2$ is comparable to that of typical metals. Thus the short positron diffusion length (of order 150 Å) determined in depth-profiling measurements cannot be attributed to a small diffusion constant. Positrons must be trapping in the Schottky well, or in defects in the film or at the interface with the Si substrate. Our RPS measurements, considered as a function of film thickness, distinguish defects in the film (presumably misfit dislocations and/or vacancies) as the dominant source of positron trapping. We cannot distinguish any significant trapping at the Si interface/ Schottky well. It would be interesting to employ depth-profiled Doppler broadening spectroscopy[13] on a thick $CoSi_2$ film to provide further confirmation of this conclusion.

ACKNOWLEDGMENTS

We thank S.M. Yalisove, and members of the Michigan positron group for helpful discussions. This work is supported by the National Science Foundation, grant DMR-9003987, with shared equipment assistance from grant PHY-9119899.

REFERENCES

1. R.T. Tung and J.M. Gibson *Mat. Res. Soc. Symp. Proc.* **67**, 211 (1986).
2. A.F.J. Levi, R.T. Tung, J.L. Batstone and M. Anzlowar *Mat. Res. Soc. Symp. Proc.* **102**, 361 (1988).
3. R.T. Tung, A.F.J. Levi and J.M. Gibson *Appl. Phys. Lett.* **48**, 635 (1986).
4. R.T. Tung in *Silicon-Molecular Beam Epitaxy, Vol. II*, edited by E. Kasper and J.C. Bean (CRC Press, Boca Raton, 1988).
5. J.L. Batstone, J.M. Phillips, and J.M. Gibson *Mat. Res. Soc. Symp. Proc.* **91**, 445 (1987).
6. J.L. Batstone, R.T. Tung, J.M. Phillips and J.M. Gibson *Mat. Res. Soc. Symp. Proc.* **102**, 253 (1988).
7. T.C. Leung et al. *Appl. Phys. Lett.* **58**, 86 (1991).
8. E.M. Gullikson, A.P. Mills, Jr. and J.M. Phillips *Surf. Sci.* **195**, L150 (1988).
9. D.W. Gidley and W.E. Frieze *Phys. Rev. Lett.* **60**, 1193 (1988).
10. M.J. Puska, P. Lanki and R.M. Nieminen *J. Phys.: Condens. Matter* **1**, 6081 (1989).
11. D.W. Gidley *Phys. Rev. Lett.* **62**, 811 (1989).
12. K.N. Tu and J.W. Mayer in *Thin Films - Interdiffusion and Reactions*, edited by J.M. Poate, K.N. Tu and J.W. Mayer (Wiley, New York, 1978).
13. P.J. Schultz and K.G. Lynn *Rev. Mod. Phys.* **60**, 701 (1988).
14. B.G. Baker, B.B. Johnson and G.L.C. Maire *Surf. Sci.* **24**, 572 (1971).
15. M.A. Nicolet and S.S. Lau in *VLSI Electronics: Microstructure Science, Vol. 6*, edited by N.G. Einspruch and G.B. Larrabee (Academic Press, New York, 1983).
16. C.D. Lien, M.A. Nicolet, C.S. Pai and S.S. Lau *Appl. Phys. A* **36**, 153 (1985).
17. C. d'Anterroches *Surf. Sci.* **168**, 751 (1986).
18. B. Nielsen, K.G. Lynn, A. Vehanen and P.J. Schultz *Phys. Rev. B* **32**, 2296 (1985).
19. C.W.T. Bulle-Lieuwma, A.H. Van Ommen and J. Hornstra *Mat. Res. Soc. Symp. Proc.* **102**, 377 (1988).
20. O.V. Boev, M.J. Puska and R.M. Nieminen *Phys. Rev. B* **36**, 7786 (1987).
21. G. Guénin, M. Ignat and O. Thomas *J. Appl. Phys.* **68**, 6515 (1990).
22. J. Van House and A. Rich *Phys. Rev. Lett.* **61**, 488 (1988).
23. G.R. Brandes, K.F. Canter and A.P. Mills, Jr. *Phys. Rev. Lett.* **61**, 492 (1988).

STUDY OF THE STRUCTURE OF THE Rh/Ag SURFACE USING POSITRON ANNIHILATION INDUCED AUGER ELECTRON SPECTROSCOPY (PAES)

G. Yang, S. Yang, J. H. Kim, K.H. Lee, A. R. Koymen, and A. H. Weiss

Department of Physics, University of Texas at Arlington, TX. 76019

G. A. Mulhollan

Department of Physics;University of Texas at Austin;TX. 78712-1081

ABSTRACT

Positron Annihilation induced Auger Electron Spectroscopy(PAES), Electron induced Auger Electron Spectroscopy(EAES), and LEED have been used to study the sandwich-like structure resulting from vapour-deposition of Rh on Ag(100). Schmitz et al found using AES, ISS, and TDS that the Ag layer diffuses to the Rh surface upon annealing to form a structure which Rh is sandwiched between a Ag cap layer and the Ag substrate. In this work, the top layer selectivity of PAES was utilized to study the mechanism of formation of the sandwich structure. The shapes and peak positions of PAES spectra for as-deposited(173K) and annealed (573K) Rh films confirm that Ag diffuses to Rh surface upon annealing. The temperature at which Ag starts to diffuse to the surface was directly monitored by PAES as the films deposited at 173K were heated to 573K.

INTRODUCTION

In recent years, there has been widespread interest in the physics and chemistry of metal films on metals. Fundamental interests in this area lie in the kinetics of film growth, the electronic properties, and surface chemistry of metal films. Furthermore the technological importance of bimetallic materials as catalysts[1] has also driven much attention to this field. To fully understand overlayer-substrate systems, it is necessary to investigate the structure of the overlayer and it's influence on the substrate. Structural changes associated with overlayer growth range from surface segregation and reconstruction, to substrate faceting[2]. In turn, an understanding of the growth process may make it possible to produce equilibrium or non-equilibrium structures which exhibit novel physical properties.[3] Among metal on metal systems, the Rh/Ag(100) configuration needs special attention because the equilibrium structure of this system cannot be predicted by any of three traditional growth modes (Frank-Van der Merwe, Stranski-Krastanov, or Volmer-Weber). P.J.. Schmitz et al [4] demonstrated that the equilibrium film structure of Rh/Ag(100) is that of a Ag-Rh-Ag sandwich in which a Ag layer diffuse to the Rh surface upon annealing, and their result was rationalized thermodynamically by the difference in surface free energies between Ag and Rh and kinetically by the high mobility of Ag atoms. In addition to the unique characteristic of growth mode, the magnetic properties of this system have also spurred interests and controversy. Various theories[5,6] predicts the existence of 4d ferromagnetism in ultra thin Rh on Ag(100) films. However, the SMOKE study [7] of Mulhollan et al. failed to identify ferromagnetism even at low temperatures. One possible explanation for this discrepancy is the presence of a diffuse interface between the Rh film and Ag substrate or the Ag-Rh-Ag sandwich structure in the case of room-temperature growth.[7] In this paper we present the results of measurements in which the top layer sensitivity[9] of Positron Annihilation induced Auger Electron Spectroscopy (PAES) was exploited to examine the mechanism of the formation of the Ag-Rh-Ag sandwich structure. Our results confirm the work of Schmitz, while suggesting the possibility of the occurrence of the migration of Ag to the Rh film at temperature as low as 173K.

EXPERIMENTAL

The measurements were performed using UTA PAES system which has been described previously[10]. The apparatus can be divided into three subsystems: (1) a magnetically guided low-energy positron beam; (2) a sample chamber for surface preparation and analysis; and (3) a trochoidal energy analyzer. Positrons are emitted from a 10mCi ^{22}Na radioactive source and moderated by a W foil to form a monochromatic low-energy beam. The sample chamber is equipped for electron induced Auger electron spectroscopy(EAES) and low energy electron diffraction(LEED). The sample is mounted on a stage capable of maintaining regulated temperatures ranging from 170K to <1400K which is maintained at a pressure of $<2 \times 10^{-10}$ Torr. The Ag(100) surface was cleaned by repeated sputter and anneal cycles, resulting in a clear P(1 x 1) LEED pattern and no detectable surface contamination as determined by EAES. Rh is deposited on the Ag(100) substrate using a thermal evaporation source following a design of DeCooman and Vook[11]. The evaporator consists of a resistively heated Tungsten wire wrapped with 0.25-mm-diam Rh wire. and enclosed in a liquid-nitrogen-cooled coil covered by a Ta shield with an 1 cm orifice directed toward the substrate. Before the Rh source was installed in the sample chamber, it was outgassed thoroughly in a separate chamber.

RESULTS AND DISCUSSION

The surface of the Rh/Ag(100) was investigated as a function of temperature and Rh exposure. The amount of Rh deposition was measured using a quartz microbalance (1 ML = 1.2×10^{15} atoms/cm^2, equal to the atomic density of the Ag(100) surface). A clean Ag(100) surface and a Ag(100) surface with a relatively thick layer (5.5 ML) of vapor deposited Rh were first examined at 173K and 573K PAES spectra at 173K for these two surfaces provided a basis for compositional calibration of PAES. Auger spectra from clean Ag(100) has two peaks (corresponding to Ag $N_3N_4N_5$ transition at ~44 eV and Ag NVV transitions at ~80 eV) in the 1 to 115 eV energy range. On the other hand, the Auger spectra from the thick layer of Rh on Ag has one peak at ~42 eV (Rh $N_2O_1O_1$ transition) and a second peak at ~67 eV (Rh $N_1N_{45}O_1$ transition). It is not possible to resolve the principle peaks for the two elements (the Ag $N_3N_4N_5$ and the Rh $N_2O_1O_1$) with the ~14 eV resolution of our spectrometer. However, the line shapes for clean Ag and for Ag with a thick layer of Rh grown at 173K can easily be distinguished, : the high energy side of Rh $N_2O_1O_1$ peak falls off more rapidly than the counterpart Ag

N3N4N5 peak, that is to say, the Ag peak is much broader. The spectra from the clean Ag warmed to 573K is uniformly attenuated while keeping line shape at 173K. This attenuation is due to an increase in the number of positrons that leave the surface as positronium before annihilation with atoms at the surface. On the other hand, the spectra from the sample with a thick layer of Rh annealed to the same temperature undergoes a number of changes. In addition to broadening of the peak at 43 eV, the appearance of the Ag peak at 80 eV and the disappearance of the Rh peak at 67 eV imply this diffusion of Ag atoms to the surface upon annealing. EAES also provides an indication of the diffusion. The peak to peak height ratio of Rh_{222} eV(MNN) to Ag_{356} eV(MNN) transitions changes from 2.23 at 173K to 0.169 at 573K indicating the existence of Ag as the surface species. The restoration of a very diffuse p(1 x 1) LEED pattern after annealing is consistent with previous work[5] which ascribe the diffuseness to a large population of defects.

Successive annealings of the films deposited at 173K provided more insight into the mechanism of Ag migration to the surface. Data was obtained for each of three different Rh coverages, 0, .75, and 1.5 ML, at four different temperatures: 273K, 373K, 473K, and 573K. After deposition at 173K, the sample was warmed successively from lower to higher temperature. It was maintained at each temperature for two and a half hours while data was taken. For each coverage the PAES spectra from Rh/Ag sample was compared to that of the clean Ag at the same temperature. Fig. 1 shows a

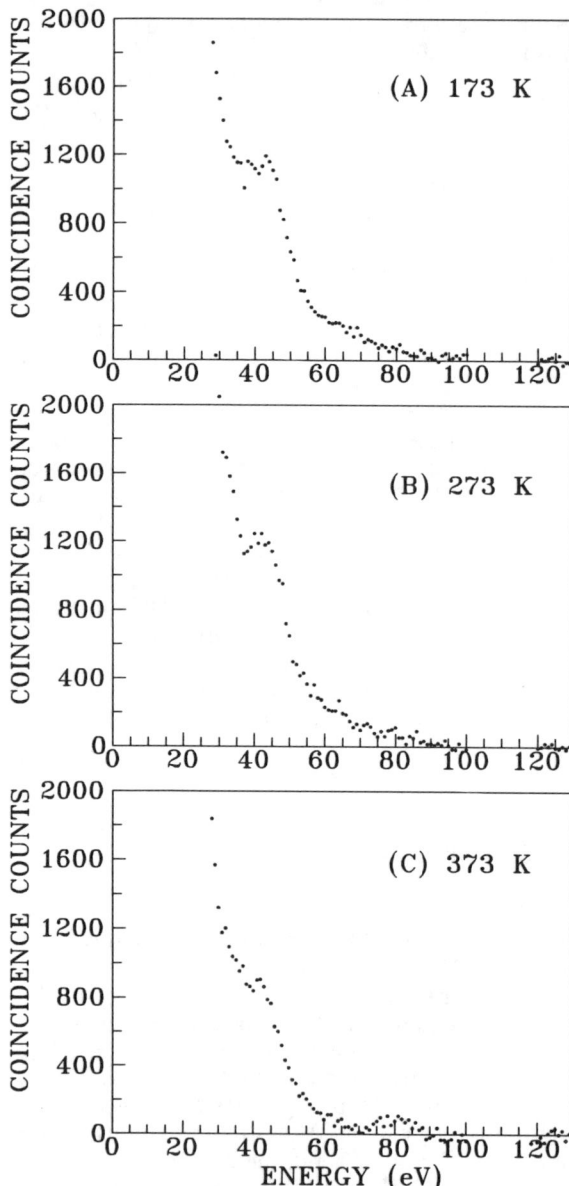

Fig.1 PAES spectra obtained at three different temperatures from Ag(100) with 1.5 ML Rh deposited at 173K.

series of these spectra for a Rh deposition of 1.5 ML. As expected, the spectrum at 173K (Fig.1 (a)) has two main features, one at 44 eV and one at 67 eV, which correspond to Rh Auger transitions. Besides these two, there exists a weak peak at 80 eV which correspond to the NVV transition of Ag. This indicates that some Ag already exist on the surface of the film even before annealing and is consistent with the results of the depth profiling measurement of Schmitz at al.[5](It should be noted however that their measurements were performed at 300K). The spectra obtained after raising the temperature to 273K (Fig. 1 (b)) clearly indicates a decrease in the peak at 67 eV and the growth of the peak at 80 eV. Raising the temperature to 373K produces a disappearance of the feature at 67 eV and more growth of the

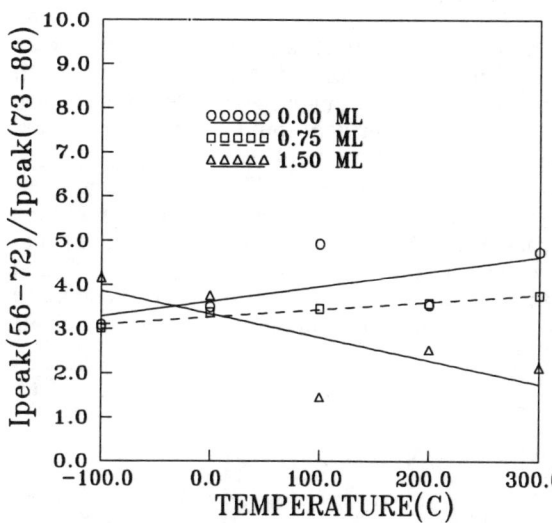

Fig. 2 The variation of the ratio of the area under the spectra from 56-72 eV to the area from 73 -86 eV for different Rh coverages as a function of temperature. Solid and dashed lines were determined from least square fits.

peak at 80 eV (fig.1(c)). These changes imply the completion of at least one monolayer of Ag by the migration of Ag to the surface. Schmitz et al[5] proposed two different migration mechanisms for low and high Rh coverages : i) simple lateral migration of Ag atoms from bare patches of the substrate at low coverages, ii) diffusion through defects in the overlayer at higher coverages. Our PAES results are consistent with Schmitz's mechanisms. Fig. 2 shows the variation of the ratio of the area from the Auger spectra 56 - 72 eV corresponding to Rh to the area from 73 - 86 eV corresponding to Ag for three different Rh coverages as a function of temperature. A similar trend of this ratio at the two coverages 0 and .75 ML suggests that , at 0.75 ML coverage, Rh is already capped by Ag atoms even at 173K(This result represents an average over the 2.5 hours needed to obtain one PAES spectrum). This is consistent with mechanism i. On the other hand, a gradual decrease of this ratio for 1.5 ML coverage correspond to the growth of the peak at 80 eV and the decay of the peak at 67 eV as a function of temperature suggesting thermally activated diffusion of Ag atoms through defects to the surface. This is consistent with mechanism ii. Unfortunately, the amount of Ag atoms which arrive at the surface as a function of temperature could not be quantized in this work. The closeness of the two main peaks along with changes in lineshape due to surface contamination at high temperatures led to physically unreasonable value when we applied a fitting routine based on a linear superposition of spectra obtained from elemental standards. Surface contamination problems (even at pressure $<2 \times 10^{-10}$ torr) are suggested by the fact that spectra with high counts and good statistics were obtained at 573K when the surface is heated to that temperature directly. However, when the sample is heated to 573K after a total of 7.5 hours at three intermediate temperatures, the spectra have fewer counts. Therefore, a stronger positron source which can reduce data-taking-time is needed.

CONCLUSIONS

PAES measurements performed on a Ag(100) substrate as functions of temperature and the amount of Rh deposited shows that Ag atoms covered by a Rh layer diffuse to the surface. Our results confirm previous work. In addition, we found that the migration of Ag atoms to the surface could proceed at lower temperatures than has been suggested previously[5]. What is worth noting is that, even at 173K, the PAES spectra from a 1.5 ML Rh covered surface has a small feature at 80 eV which corresponds to the Ag NVV transition. This fact could be a clue in resolving the controversies concerning the existence of 4d ferromagnetism in ultra thin Rh films on Ag (100). It will be our next step to study the diffusion behavior of Ag atoms to the surface as function of time at different deposition temperatures. The recent purchase of a strong positron source should make this possible by reducing the data-taking time.

REFERENCES

1. A D. Logan, M.T. Paffett, J. Catal. 133, 179(1992)
2. T. E. Madey, K-J. Song, C-Z. Dong, Surf. Sci. 247, 175(1991)
3. A. Y. Cho, J. Cryst. Growth 95, 1(1989)
4. P.J.Schmitz,W-Y.Leung,G.W.Graham,,P.A.Theil, Phys. Rev. B40, 11477(1989)
5. O.Eriksson,R.C.Albers,A.M.Boring,Phys. Rev. Lett. 66, 1350(1991)
6. M.J.Zhu,D.M.Bylander,L.Kleinman,Phys. Rev. B 43,4007(1991)
7. G.A. Mulhollan;R.L. Fink;J.L. Erskine;Phys. Rev. B44;2393(1991)
8. A.H.Weiss,R.Mayer,M.Jibaly,C.Lei,D.Mehl,K.G.Lynn, Phys. Rev. Lett. 61, 2245(1988)
9. D.Mehl,A.R.Koymen,K.O.Jensen,F.Gotwald,M.Jibaly,A.H.Weiss Phys. Rev. B41, 799(1990)
10. C.Lei,D.Mehl,A.R.Koymen,F.Gotwald,M.Jibaly,A.H.Weiss, Rev. Sci. Instrum. 60, 3656(1989)
11. B.C.De.Cooman,R.W.Vook, J. Vac. Sci. Technol. 21, 899(1982)

DESIGN OF AN ELECTROSTATIC POSITRON BEAM FOR BACKGROUND-FREE HIGH-RESOLUTION AUGER LINESHAPE STUDIES

H.Q. Zhou, S. Yang, A.R. Koymen, and A.H. Weiss
*Physics Department, The University of Texas at Arlington,
Arlington, TX 76019*

ABSTRACT

Previous measurements of positron annihilation induced Auger electron spectroscopy (PAES) using a magnetically guided positron beam have demonstrated the ability of PAES to produce essentially background free Auger spectra. In this paper we will describe the design of an electrostatic positron beam which will be used in conjunction with a large cylindrical mirror analyzer (CMA) to perform high resolution PAES measurements on clean and adsorbate covered surfaces. The positrons were traced through the beam optics and into the field free region using the SIMION charged-particle trajectory program with space charge effects neglected. The results of ray tracing will be described along with details of design parameters and projected system performance.

INTRODUCTION

Auger lineshapes contain information pertaining to the chemical environment at the surface, the surface density of states, and details of the Auger transition mechanism. In addition, an accurate determination of the Auger lineshapes is of great practical importance in the correct interpretation of conventional Auger electron spectra. However the necessity of subtracting a background which is in general significantly larger than the Auger signal itself has posed a major problem in determining Auger lineshapes. Although background subtraction techniques have been developed, the Auger lineshapes often cannot be definitively ascertained using conventional Auger electron spectroscopy (AES). This is because the spectral intensity due to intrinsic energy loss overlaps with that of extrinsic loss, as well as with other portions of secondary electron background, and cannot be unambiguously separated out.[1]

In positron annihilation induced Auger electron spectroscopy (PAES), core electrons are removed by matter-antimatter (positron-electron) annihilation rather than collisional ionization as in conventional electron induced Auger electron spectroscopy (EAES). Previous PAES measurements[2,3] have demonstrated that the large secondary electron background present in conventional AES can be eliminated using positron beam energies less than the Auger electron energy. The low background permits the determination of Auger lineshapes without the ambiguity introduced by having to subtract a large secondary electron background whose function form is unknown.

To date, PAES experiments have been performed using magnetically guided positron beams and trochoidal energy spectrometers. These systems are limited by practical considerations to a relative poor energy resolution ($\Delta E/E \sim 0.1$) which is too broad for detailed studies of Auger lineshapes. We are therefore in the process of building a new electrostatic positron beam which will be coupled with a large CMA. Details of the design of this new system which will be capable of achieving significantly high energy resolution ($\Delta E/E < 0.01$) will be presented in this paper.

PAES SYSTEM

The new PAES apparatus (see Fig. 1) can be divided into three sub-systems: 1) an electrostatic slow positron beam; 2) a large CMA; and 3) an UHV chamber for sample preparation. The considerations and results for the design of the positron beam will be presented in the next section.

A CMA was chosen as the Auger electron energy analyzer because of its high energy resolution and high luminosity.[4] To avoid the complications and loss of flux associated with beam brightness enhancement techniques, we use a CMA[5] which has physical dimensions much larger than those of commercially available systems. The CMA consists of two coaxial cylinders enclosed in a vacuum tank. The diameters for the inner and outer cylinders are 0.18 and 0.46 meters, respectively, and the distance from the sample to the detector is 0.6 meters. The UHV sample preparation chamber will be equipped with an ion sputter gun, a low energy electron diffraction (LEED) system and a long-travel sample manipulator with heating, cooling and multi-axis rotational abilities.

POSITRON BEAM DESIGN

A cross sectional view of the positron beam optics is shown in Fig. 1. Positrons from a planar moderator are extracted and accelerated by a Soa-type positron gun and a four-element field lens, bent by a parallel-plate analyzer, transported by an einzel lens, and finally focused onto the target by a four-element zoom lens. There are two sets of deflection plate located before and after the parallel-plate analyzer for varying the beam angle. The positron beam is mounted perpendicular to the CMA axis, while the sample is positioned with its normal at 45° to both the positron beam and the CMA axis.

Initially a ^{22}Na source will be used in conjunction with a single crystal tungsten foil moderator with a diameter of 8 mm. A similar source/moderator arrangement in a magnetically guided system produced an effective low energy positron source diameter of ~ 3 mm with an energy spread of < 1 eV and a flux of $\sim 4 \times 10^4$ e+/sec from a ^{22}Na source of 15 mCi. To

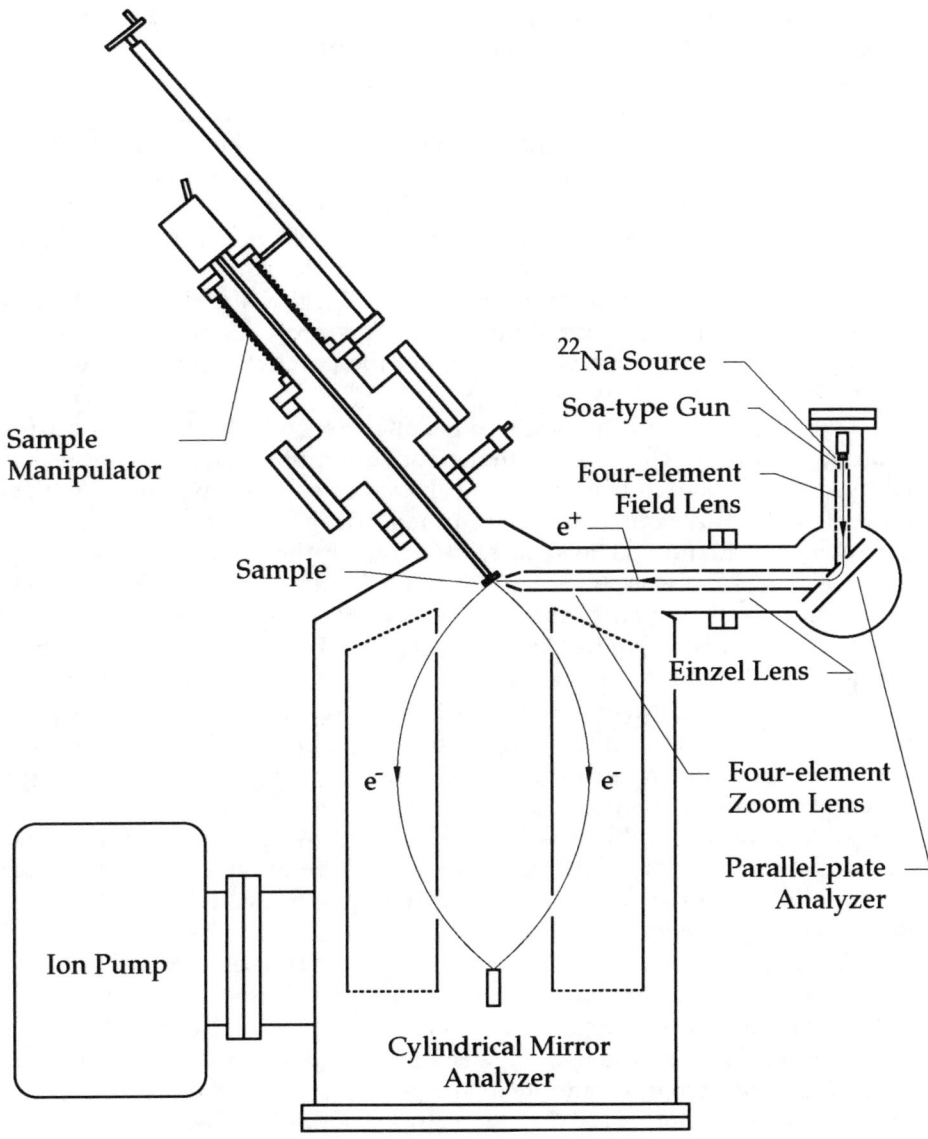

Figure 1 Cross-sectional view of the new PAES system with an electrostatic positron beam and a large cylindrical mirror analyzer.

mimic the positron beam in a SIMION simulation, we chose seven trajectories (ranging from -15° to 15°) for each of eleven points of origin which are evenly spaced out to a diameter of 6 mm.

The first stage of the positron beam, as shown in Fig. 2(a), consists of a Soa-type positron gun and a four-element field lens. The positron gun is an adaptation of the design of Canter et al.[6] with some parameter changes to suit our particular beam requirements. The overall dimension of the Soa gun has been reduced by a factor of two to make the beam more compact. The choice of the potentials for the Soa-type positron gun elements (Cathode, Welnelt and Soa tube and Anode) were guided by the SIMION simulations. Attached to the Soa gun is a four-element field lens. The purposes of the field lens are to collimate the positrons before they enter the parallel plate analyzer and to accelerate them from 2.5 eV to 1000 eV.

The second stage is a parallel-plate analyzer, which bends the positron beam by 90° in order to prevent the unconverted positrons and γ–rays emitted by the ^{22}Na source from reaching the sample. Basic parameters for the parallel-plate analyzer were determined using the design principle of Harrower and Eland.[7] The separation between the lower plate and upper plate (see Fig. 2(b)) was set equal to $0.3x_0$, where x_0 is the distance between the entrance and exit slits; and the deflecting voltage v_d was set equal to $0.6v_0$, where v_0 is the initial kinetic energy of the incoming positron beam. The potential of the lower plate was set at the same potential as the last element of the first stage. Two plates (of 0.02" thickness) will be placed between the lower and upper plates as field guard-plates. The field aberration due to the finite thickness of the plates should be negligible. The openings of the entrance and exit slits of the lower plate are a more serious source of aberration, and the problem will be minimized by spot-welding a Ta-mesh over each slit. Two sets of deflection plates will be provided before and after the parallel plates to correct for residual magnetic fields and possible misalignment in the optical components.

The third stage, as shown in Fig. 3(a), consists of essentially two einzel lenses, whose purpose is to transport the beam into the CMA chamber without significantly changing the beam energy and angles. Finally, as illustrated in Fig. 3(b), the last stage is a four-element zoom lens[8] which will be used to both decelerate and focus the positron beam. Ray tracing with the SIMION program indicates that we can decelerate the beam to 10 eV and focus the beam to a spot of 6 mm on the sample which is 10 mm away from the end of the zoom lens. The deceleration is done by the second and third elements of the zoom lens, while the first element is attached to the last element of the third stage. A cone shape, as shown in Fig. 3(b), was chosen for the last element of the zoom lens for two reasons: 1) to avoid the obstruction of the path of Auger electrons from the sample to the CMA; and 2) to reduce the penetration of the electric field due to the zoom lens into the region between the cone and the sample.

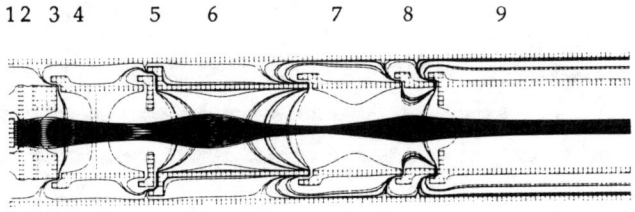

(Electrode voltages: 1: 10 V, 2: 9 V, 3: 9.75 V, 4: -15 V, 5: -70 V, 6: 10 V, 7: -390 V, 8: 450 V, 9: -990 V)

(a) First Stage

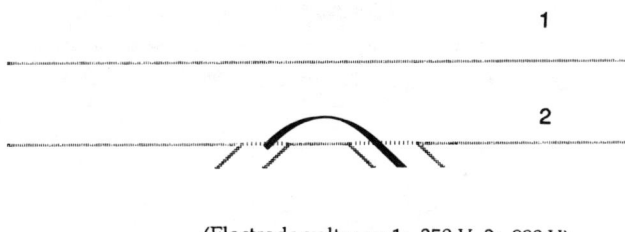

(Electrode voltages: 1: -350 V, 2: -990 V)

(b) Second Stage

Figure 2   Ray tracing simulations (with the SIMION charged-particle trajectory program) of the first and second stages of a electrostatic positron beam.

284 Design of an Electrostatic Positron Beam

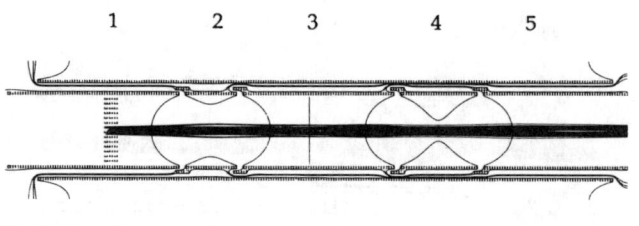

(Electrode voltages: 1: -990 V, 2: -300 V, 3: -990 V, 4: -3000 V, 5: -990 V)

(a) Third Stage

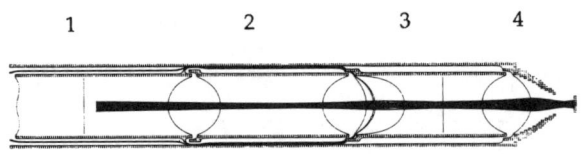

(Electrode voltages: 1: -990V, 2: -4000 V, 3: -500 V, 4: 0 V)

(b) Last Stage

Figure 3 Ray tracing simulations (with the SIMION charged-particle trajectory program) of the third and fourth stages of a electrostatic positron beam.

SUMMARY

In this paper, we have presented the design considerations and the results of computer simulations for an electrostatic positron beam which will be used in conjunction with a large cylindrical mirror analyzer to perform high resolution PAES studies. SIMION ray tracing analysis indicated that, this design, which makes use of a Soa-type positron gun, a parallel plate analyzer, two einzel lenses, and a four-element zoom lens, is capable of efficiently transmitting ~ 80% positrons emitted from the moderator to a spot of 6 mm diameter at the sample with an energy as low as 10 eV. It is estimated that the large CMA used in conjunction with the positron beam will permit PAES spectra to be obtained with an energy resolution ($\Delta E/E$) of less than 1%.

ACKNOWLEDGEMENTS

This research is supported by the Welch Foundation, Texas Advanced Research Program, and NSF(DMR9106238). We wish to express our thanks to Dr. P. Citrin, AT&T Bell Labs for lending us the large CMA. And also we would like to acknowledge useful discussions with Dr. P. Citrin, Dr. P. Lippel and Professor K. Canter.

REFERENCES

1. C.J. Powell and M.P. Seah, J. Vac. Sci. Technol. A **8**, 735 (1990).
2. A.H. Weiss, R. Mayer, M. Jibaly, C. Lei, D. Mehl and K.G. Linn, Phys. Rev. Lett. **61**, 2245 (1989).
3. Alex Weiss, David Mehl, Ali R. Koymen, K.H. Lee and Chun Lei, J. Vac. Sci. Technol. A **8**, 2517 (1990).
4. P.H. Citrin, Jr. R.W. Shaw and T.D. Thomas, *Proceedings of the International Conference on Electron Spectroscopy*, Pacific Grove, Calif., p. 105 (North-Holland, 1972).
5. On loan from AT&T Bell Labs., Murry Hill, New Jersey.
6. K.F. Canter, P.H. Lippel, W.S. Crane and Jr. A.P. Mills, *Positron Studies of Solids, Surfaces, and Atoms-A Symposium to celebrate Stephan Berko's 60th Birthday*, Brandeis University, p. 199 (World Scientific, 1984).
7. G.A. Harrower, Rev. Sci. Instrum. **26**, 850 (1955).
8. I.J. Rosenberg, Ph.D. Thesis, *Positron Emission and Low-energy Positron Diffraction from Metal Surfaces* (Brandeis University, 1981).

SECTION 4

INTENSE POSITRON BEAM FACILITIES

A PROPOSED INTENSE SLOW POSITRON SOURCE BASED ON ^{58}Co

Benjamin L. Brown
Mount Holyoke College

Art Denison and Henry Makowitz
INEL

Dave Gidley, Bill Frieze, Henry Griffin, and Pedro Encarnación
University of Michigan

Abstract

Positron beams have proven very useful for condensed matter and surface research. The highest intensity of the current operating positron beams is $\sim 10^9$ slow e$^+$/second. The goal of our proposal is to build an Intense Slow Positron Source (ISPS) demonstration beam (Phase I) of unprecedented brightness at the Idaho National Engineering Laboratory, INEL (up to 10^{10} slow e$^+$/s at 5 keV over a <0.03 cm. diameter). This Phase I beam will prove the principles necessary to build a larger facility scale ISPS Phase II beam which will have a potential of 10^{13} e$^+$/s, or $> 10^{12}$ e$^+$/s over 0.03 cm. The INEL is an ideal location for the ISPS because of the fast breeder reactor EBR-II, which is perfectly suited to creating the positron emitting isotope ^{58}Co, and the excellent radioactive materials handling capability and expertise. Sufficient scientific expertise is available at INEL for the construction and operation of a user facility (Phase II).

Prepared for U. S. Department of Energy through the EG&G Idaho, Inc. LDRD program. Under DOE-ID Contract DE-AC07-76ID01570

ISPS BEAM OVERVIEW

The principles associated with the INEL Intense Slow Positron Source have already been demonstrated on a "small" scale (i.e., ≤ 1 cm^2) by the positron physics community and by commercial isotope producers in the case of the ^{58}Co production. The INEL proposes to scale up this technology, by several orders of magnitude, using available hot cells, etc., and to build a special demonstration beam facility. Refer to Figure 1. Concurrent to this three-year ISPS device development effort, a "Super-Microscope" will be designed by the Michigan Group. The INEL Experimental Breeder Reactor (EBR II) has ample fast neutrons and is ideally suited to efficient ^{58}Co production. INEL also has previous experience in handling high levels of radioactive material.

We are currently proposing a demonstration scale (Phase I) intense slow positron source, ISPS, device which will achieve a monoenergetic positron beam of $\sim 10^{10}$ e$^+$/s at 5 keV on a ≤ 0.03-cm-diameter target using a ^{58}Co. This trial beam will represent a significant advancement over beams now available. One preeminent purpose of the positron beam will be the development of a super microscope and user facility which will obtain the ultimate possible resolution of a positron reemission microscope. We shall refer to this demonstration beam as Phase I ISPS, to distinguish it from Phase II beam (including a microscope/microprobe) that would be designed based on what is learned during the Phase I project.

The positron microscope[1-3] opens up the possibility of a high resolution electron type microscope that has several differing qualities relative to the electron counterpart. An advantage to positrons is the extreme sensitivity to lattice imperfections and voids which can be exploited using Doppler broadening and angular correlation techniques. Another novel technique involves the examination of surface conditions using reemission microscopy.[4-5] The potential resolution is 10 Å and the ultimate positron beam strength is ~ 1 µA. A combination electron and positron microscope would find applications in the basic research and technological development of solid materials, interfaces, surfaces, liquids, and a host of other applications including the study of biological molecules. A scanning intense positron beam for angular correlation of annihilation radiation would find applications in materials research, defects studies, surfaces and interfaces. The microelectronics industry has expressed strong interest in this facility and other industries will benefit from this technology. A recent meeting, held by the Department of Energy in Palm Springs CA, in September 1992, was devoted to exploring various applications of intense positron beams. Several additional possible

ISPS Phase I Beam

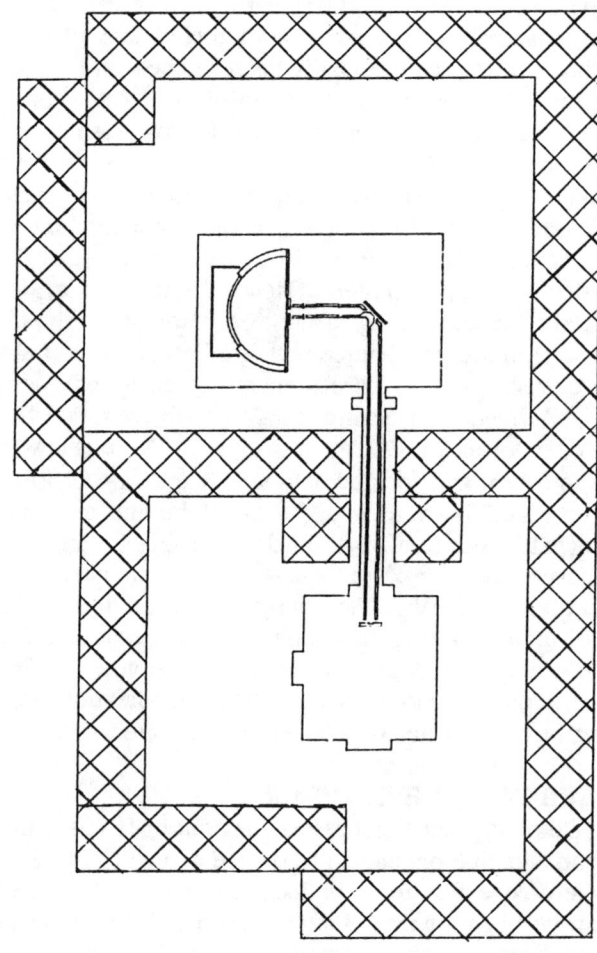

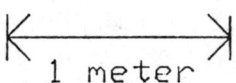

1 meter

FIGURE 1 INTENSE SLOW POSITRON SOURCE (ISPS DEVICE) BASED ON ^{58}Co. THE SOURCE IS PLATED ON A HEMISPHERICAL SURFACE, AND Ne IS DEPOSITED ON THE SURFACE TO PROVIDE EFFICIENT MODERATION OF THE POSITRONS.

applications were detailed at this meeting. There is no doubt that an intense positron beam would be of considerable interest not only to the positron community[6] but to many outside the present community. The Japanese positron physics community has managed to find several techniques that are adaptable to industrial and process control applications. We hope this type interface with industry can eventually be developed in this country.

Positrons are created by beta decay of ^{58}Co plated[7] on the inner side of a spherical surface (spherical segment) that forms a dish shape. We refer to this as the source dish. Refer to Figure 2. The beam, in the latest design configuration, consists of the following: a 30 cm diameter positron source; a neon solid moderator; a W thin crystal remoderator; and appropriate electrostatic optics.[8] ^{58}Ni bombarded by neutrons in a hard flux nuclear reactor creates ^{58}Co, a β^+ emitter. Chemical separation of the ^{58}Co, from the ^{58}Ni, followed by electroplating of the ^{58}Co onto part of the dish area (10's of cm^2 for Phase I) produces a source with a minimum of self absorption.[7] The positrons are produced with a broad energy spectrum with an endpoint energy of 474 keV. A fraction of positrons ($\leq 10^{-2}$) are reemitted from the neon rare gas solid moderator at low energy at ~ 1 eV. These "slow" positrons are accelerated to 5 keV and focused onto a W single crystal foil remoderator.[8] A large fraction (≤ 0.35) are reemitted through the thin moderator at low energy, due to the negative positron work function. The solid angle of emission is somewhat narrow and the positrons emerge with the characteristic W work function energy, ~ 2.5 eV. The energy loss in the foil increases the brightness of the beam (brightness enhancement[1]) enabling a subsequent reduction in beam size. A beam diameter reduction of ~ 35x is achieved for each remoderation step and the acceleration remoderation step is repeated until the resulting beam diameter is reduced to the desired size.

ISPS PHASE I BEAM DEVELOPMENT

The Idaho National Engineering Laboratory (INEL), the University of Michigan and Mount Holyoke College have agreed to jointly pursue a Super Positron Microscope Facility to be located at the INEL. The beam development will begin with a 3 year Phase I proof of principle program.

During Year One of the program, in collaboration with Michigan, a ≤ 15 Ci ^{58}Co room temperature source will be developed for use with the existing microscope in Ann Arbor. The University of Michigan research reactor will be used to irradiate Ni samples and the separation chemistry and plating technology will be explored. This will allow the INEL to gain experience in the chemical separation of ^{58}Co and demonstration ISPS device. Mount Holyoke College will develop the cryogenic Ne thin film technology. Design of the cryogenic system for eventual Ne moderation will take place at Mount Holyoke. The choice of the isotope ^{58}Co has been made partly because

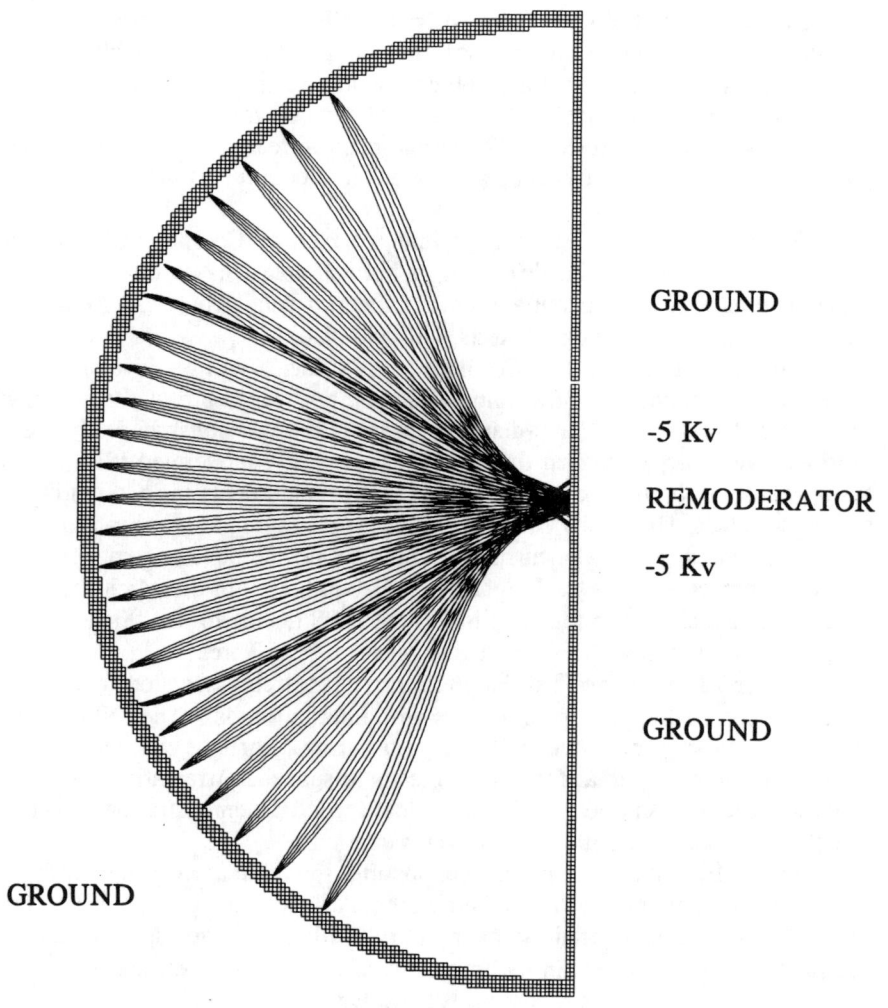

FIGURE 2 Positron emission from a hemispherical surface, as simulated by SIMION. The disk on the right is biased at -5 kV; all other surfaces are grounded. The positrons are assumed to be emitted at high energies and then moderated by a thin film of Ne condensed directly on the source.

of its excellent potential yield from Ni in the INEL EBR-II fast breeder reactor, and partly due to the high specific activity attainable through chemical separation. The EBR-II reactor combined with the excellent hot cell and engineering capabilities, make INEL an ideal location for the ISPS. The possibility of using the INEL ATR thermal neutron reactor with other isotopes such as Kr, Cu, or F will also be studied as an alternative to the ^{58}Co approach.

In Year Two the INEL will continue work on a ^{58}Co chemical separation and deposition capability inside one of its hot cell areas, based on in-house experience as well as research and development in Ann Arbor. At Mount Holyoke, preliminary cryogenic tests will take place.

During Year Three the ^{58}Co Phase I ISPS device will be developed, tested, and operated. A "slow" monoenergetic positron beam of $\sim 10^8$ to 10^{10} e$^+$/s at 5 keV on a ≤ 0.03-cm-diameter target will be demonstrated. In the following year, depending on funding, a cryogenically moderated ISPS device at 10^{11} e$^+$/s will be pursued at INEL. Recent developments in ^{58}Co plating techniques at the University of Michigan are encouraging and indicate that this goal is not unrealistic.[7] Concurrent to this ISPS device development effort, a "Super-Microscope" will be designed in collaboration with the Michigan Group. Issues critical to the coupling of the ISPS device to the "Super" Positron Microscope will also be investigated in Year Three.

The INEL has supported the preliminary research associated with the ISPS concept for the past several years with internal funds.[9] The INEL intends to continue its support of the ISPS program during Year One through Three with supplemental internal funds and intends to subsidize irradiation, hot material handling, and hot cell costs, so that the ISPS demonstration program goals can be achieved within the budget requested.

The INEL ISPS concept has one possible spinoff that may benefit the entire positron community: that of providing Curie sources of ^{58}Co at low cost. This is by no means a certainty, as the INEL will have to decide if this is practical and desirable within its limitations of mission, budget, and potential benefit to the INEL. It is a possibility, however, that should not be overlooked when evaluating this concept in light of other concepts for intense beams, such as the proposed ^{79}Kr reactor based source.[10] The used ^{58}Co from the main dish, which would normally be discarded, may be extremely useful to some research groups.

CONCLUSION

The INEL ISPS is a promising contender for a major positron facility which could serve the needs for the next generation of slow positron experiments. Using essentially a waste product produced in the reflectors of the EBR-II fast breeder reactor, the beam is economically attractive and is capable of producing extremely high slow positron rates, up to 10^{13} slow e$^+$/s, or 10^{12} slow e$^+$/s over a 0.03 cm diameter. This intensity will allow a facility with multiple user ports for a variety of possible experimental arrangements that can be tailored to user needs and specifications.

1. A. P. Mills, Jr., Appl. Phys. **23**, 189(1980).

2. J. Van House and A. Rich, Phys. Rev. Lett. **61**, 488 (1988).

3. G. R. Brandes, K. F. Canter, T. N. Horsky, and A. P. Mills, Jr., Appl. Phys. **46**, 335 (1988).

4. L. D. Hulett, J. M. Dale, and S. Pendyala, Mater. Sci. Forum **2**, 133 (1984).

5. G. R. Brandes, K. F. Canter, and A. P. Mills., Jr., Phys. Rev. B **43**, 10103(1991).

6. Refer to other papers on high intensity beams and applications in this volume.

7. H. Griffin et al. elsewhere in this volume.

8. B. L. Brown et al. elsewhere in this volume.

9. Eric H. Ottewitte, "High-Intensity Positron Beam Via 200 MW, Large-Volume, Test (Fission) Reactor", Proceedings of the BB Factory Workshop, January 27, 1988 (World Press). See also H. Mokowitz elsewhere in this volume.

10. A. P. Mills, Jr. Nuc. Sci. and Eng. **110**, 165(1992).

POSITRONS AT CEBAF

W. J. Kossler and A. J. Greer
College of William and Mary, Williamsburg, VA 23187-8795[*]

L. D. Hulett, Jr.
Oak Ridge National Laboratory, Oak Ridge, TN 37831-6056[†]

It is estimated that about 2.5×10^{10} slow positrons per second might be produced if 1.0 mA of 400 MeV electrons were incident on target. This estimate and some problems of target design are discussed. The electron beam structure is 1 to 2 ps pulses with a repetition rate of 7.5 MHz which is appropriate for a Free Electron Laser (FEL) which is being proposed for CEBAF. The electron beam would exit the FEL and would otherwise be transported to a beam dump.

INTRODUCTION

We consider the production of slow e^+ by an electron beam at the Continuous Electron Beam Accelerator Facility (CEBAF) in Newport News, Virginia. These positrons would be useful for material science, atomic physics, and also as a source for reinjection into CEBAF for further acceleration. It is also possible to produce muons for μSR, but we will only quote the estimated yield for these particles. The electron beam would have 400 MeV, 1.0 mA average current, in pulses of 1 to 2 ps duration and a repetition rate of 7.5 MHz. This beam has been proposed for a Free Electron Laser (FEL) at CEBAF. The electron beam after leaving the UV FEL has nearly the same quality (e.g. $\leq 1mm$ diameter) as before it enters the FEL, and if it is not used for anything else will proceed to a beam dump.

YIELD

An estimate of beam intensity can be easily made for the slow positron yield by noting that the yield is proportional to beam power once the electron energy is over 20 MeV or so. Thus one can scale from the Livermore results using the same conversion efficiency as quoted in the review article on positron physics by Schultz and Lynn.[1] This scaling has recently been checked by Monte Carlo calculations discussed below using the Egs4[2] computer program. The result is a slow positron rate of $2.5 \cdot 10^{10}$ from vane tungsten foils of identical configuration to those used at Livermore.[3] The spot size at this point would be about 1/2 cm in diameter (again from the Monte Carlo calculations, but also in agreement with the results at Livermore[3]).

To summarize, the expected results are:

---

[*]Supported by a Grant from the Center for Innovative Technology of the Commonwealth of Virginia.
[†]Supported by the US DOE.

© 1994 American Institute of Physics

Beam	Intensity	Pulse width	Rep. Rate
electrons	$1mA = 2/3 \cdot 10^{16} s^{-1}$	1ps	7.5MHz
slow e^+	$\approx 2.5 \cdot 10^{10} s^{-1}$	??	7.5MHz
surface μ^+	$\approx 2.5 \cdot 10^{6} s^{-1}$	$\tau_\pi \approx 26ns$	7.5MHz

MONTE CARLO CALCULATIONS

The computer program Egs4,[2] a Monte Carlo code which treats electron-gamma showers, has been used to study the positrons which could be produced by the 400 MeV electrons at CEBAF and also, for comparison, 100 MeV electrons, which energy corresponds to that of the Livermore Linac.

The first set of calculations were with a tungsten target geometry which was quite thick longitudinally and radially. From this we obtained the general characteristics of the stopping positron spatial distributions and also the energy deposition distributions. For the present calculations 1 MeV lower cutoffs were assumed for both $e^\pm$ and γs. Thus a positron was assumed to have stopped if its energy fell below 1 MeV. The residual range of a 1 MeV electron or positron is less than 1 mm.

The general stopping behavior can be seen from Fig. 1 in which the stopping positions are indicated for positrons produced by 100 electrons of 400 MeV. One can see that the lateral spread is not too great. Most of the positrons stay within a radius of 1/2 cm. This is not much different from that which we found for 100 MeV electrons by calculation and for that matter from the 100 MeV electron experimental results of Livermore.[3]

In order to determine the amount of energy deposited in the target a plot was made of total energy deposited within a given radius as a function of depth into the target. This is shown in Fig. 2 for which 1000 incident electrons were used. The energy deposited is given per electron. Each longitudinal slice was 1 mm and the radii differed by .5 mm, and the lowest was the .5 mm radius. One sees that most of the energy is deposited by about the 5th radial ring (2.5 mm). By integrating out to about 1.2 cm depth, which is roughly the target thickness which we show later to be optimal, one finds that about 80 MeV is deposited totally. The remainder exits into whatever is past the 1.2 cm point. This power level is within current water cooled rotating anode X-ray loads.

As a final point on the solid target calculations, it had been noted by Howell[3] that the angular spread of low energy positrons is very great. This may be seen from the Monte Carlo results also by plotting the directions of low energy positrons (less than 20 MeV) at the 10th 1 mm slice interface. This is shown in Fig. 3. The line lengths are proportional to positron energy and their directions and positions are the positrons' directions and positions at the interface. That there are some backward follows from this being in the midst of a solid target. Only the low energy positrons have such a wide angular spread. The higher energy positrons are much more forward.

The next series of Monte Carlo calculations were performed with a tungsten vane moderator system which approximates the geometry used by Livermore. This is shown in Fig. 4. The vanes are roughly 1/4 inch deep, 1/8 inch apart and 10 mils thick.

A series of calculations were made for differing tungsten block thickness for both 100 MeV and 400 MeV electrons. The stops in the foils as seen end on is shown in Fig. 5. It is seen that the diameter of intense stops is about 1/2 cm. The shift to one side is a reflection of the electron beam being assumed to enter

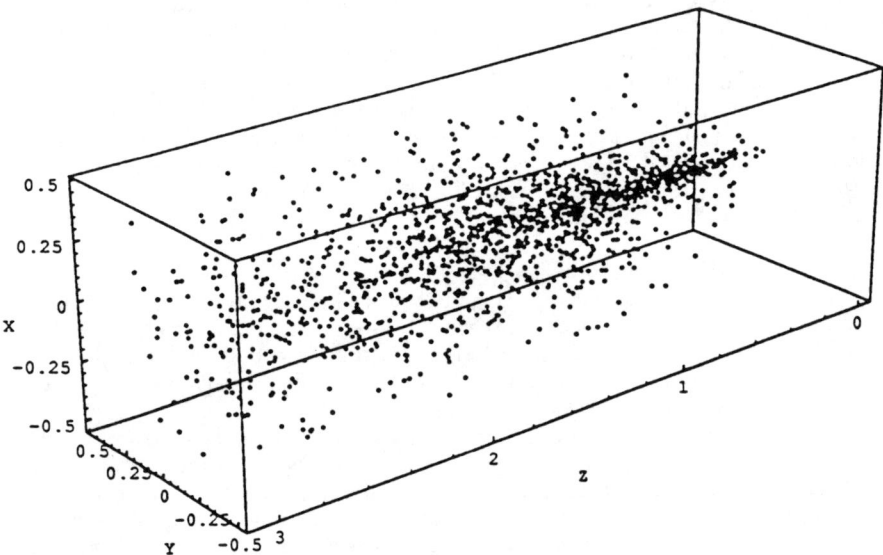

Figure 1: Positrons stopping in a block of tungsten. Stopping is defined as at or below 1 MeV.

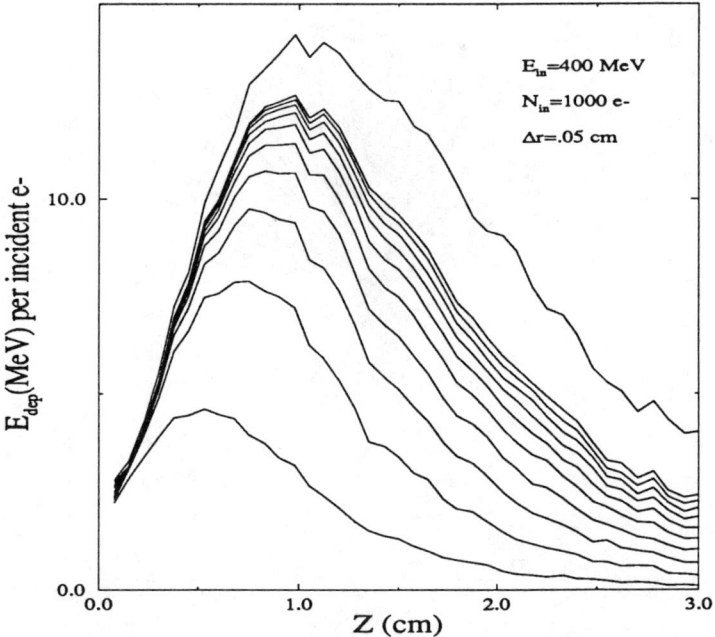

Figure 2: Energy deposited in the tungsten target within a given radius as a function of length z along the incident electron direction.

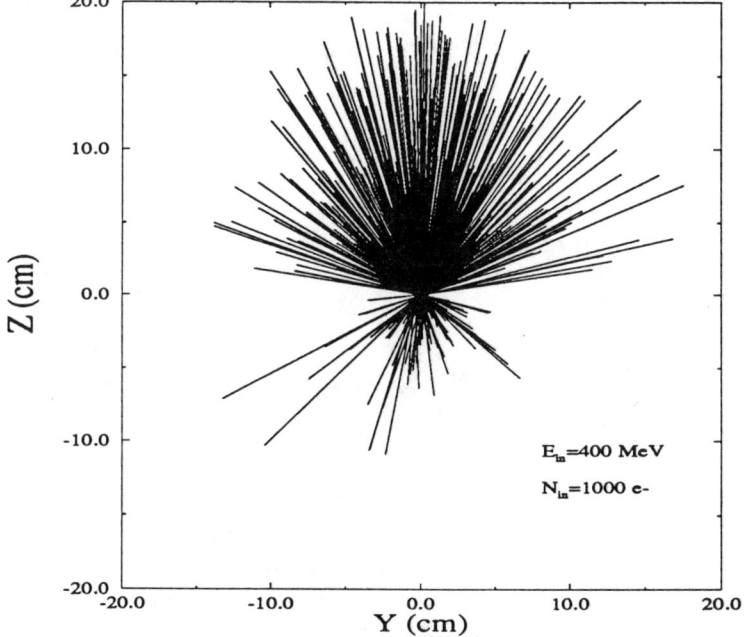

Figure 3: Positrons leaving the target at roughly 1 cm in z. Lengths are proportional to energy; maximum energy shown is 20 MeV.

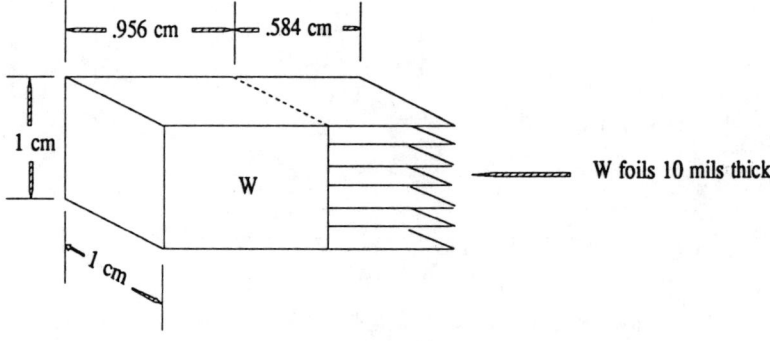

Figure 4: Geometry of the second target used in EGS4 calculations.

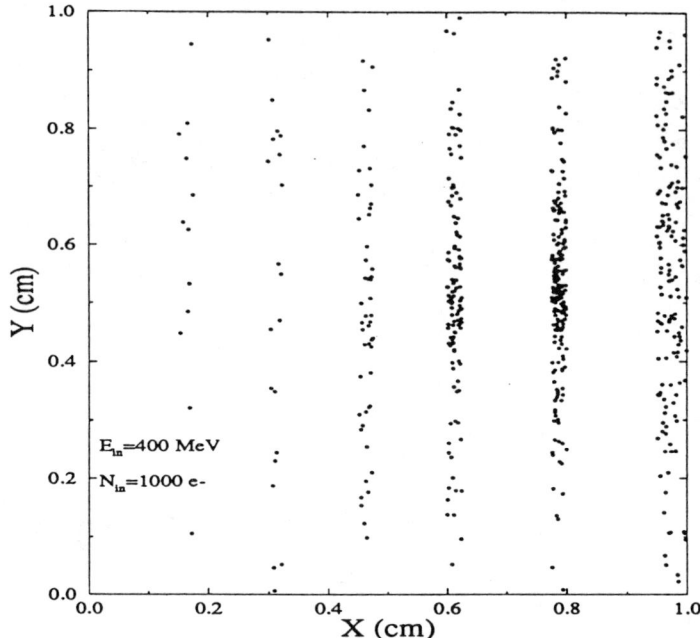

Figure 5: End-on view of positrons stopped in the tungsten foils. Counts are shown for stops along the whole length of the foils in z.

at the center of the block, but with a 10 degree offset from the target-moderator axis. The number of positrons stopping in the moderator foils as a function of thickness of the bulk tungsten block is shown in Fig. 6. The actual thickness used by Livermore is 1 cm, which Howell had determined to yield the most, while the calculations show a thinner target would yield more. This discrepancy may reflect either the beam offset or that the actual target is surrounded by solid tantalum and backscatter may favor a thicker target.

A point of considerable practical importance can be seen from the ratio of peak yields. This ratio is nearly exactly the factor of 4 expected from the energy ratios. One could have worried that the higher energy electron beam would have produced positrons that were more spread out and thus yielding fewer positrons in any given moderator system. That this is not the case is fortunate for CEBAF.

PROBLEMS OF THE TARGET

400 kW is the power in the electron beam and the spot diameter on entering the target might be less than 1 mm. This power and size are such as to burn holes in any material if it is left uncooled in the path of the beam. Fortunately, as seen from the Monte Carlo calculations, only 80 MeV is deposited in a tungsten block of appropriate dimensions to maximize positron stops in moderator vanes as used at Livermore; and that by the depth at which the linear energy deposition density peaks the area of the beam is on the order of 10 mm^2. Still, the heating is a problem.

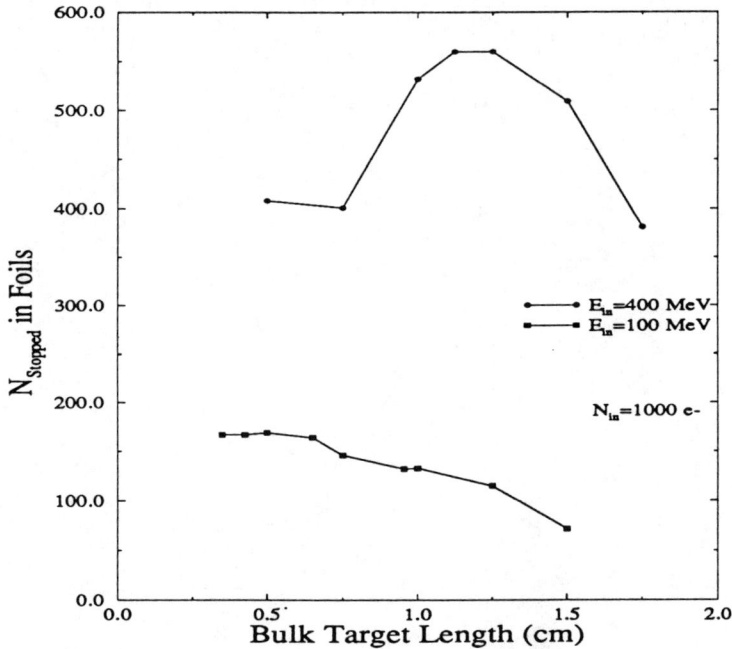

Figure 6: The number of positrons stopped in the foils as a function of thickness of the bulk tungsten block for both 100 and 400 MeV incident electron energies.

We have calculated that a 1 meter diameter water-cooled target rotating at about 1500 rpm can easily handle the heating, and shown that this wheel is mechanically feasible. As a measure of the practicality of an 80 kW target, one should note that 90 kW rotating anode X-ray sources are commercially available.

Further work is going into a first order design of a complete target system so that its cost can be estimated by engineers at CEBAF.

REFERENCES

1. Peter J. Schultz and K. G. Lynn. Interaction of positron beams with surfaces, thin films and interfaces. *Reviews of Modern Physics*, **60**:70, 1988.

2. Walter R. Nelson, Hideo Hirayama, and David W. O. Rogers. The egs4 code system. Technical Report SLAC-Report-265, Stanford Linear Accelerator Center, 1985.

3. R. H. Howell, 1992. private communication.

Submitted for publication in the Proceedings of the Fifth International Workshop on Slow-Positron Beam Techniques for Solids and Surfaces, Jackson Hole, Wyoming, USA, August 6–10, 1992.

THE INTENSE SLOW POSITRON SOURCE CONCEPT: A THEORETICAL PERSPECTIVE ON A PROPOSED INEL FACILITY[†*]

Henry Makowitz, James D. Abrashoff, William H. Landman, and Richard K. Albano
Idaho National Engineering Laboratory, EG&G Idaho, Inc.
Idaho Falls, Idaho 83415 USA

Toshiki Tajima, Physics Department
University of Texas at Austin
Austin, Texas 78712 USA

James Daniel Larson
Independence, Missouri 64052 USA

ABSTRACT

An analysis has been performed of the INEL Intense Slow Positron Source (ISPS) concept. The results of the theoretical study are encouraging. A full-scale device with a monoenergetic 5 keV positron beam of $\geq 10^{12}$ e$^+$/s on a $\leq$0.03-cm-diameter target appears feasible and can be obtained within the existing infrastructure of INEL reactor facilities. A 30.0-cm-diameter, large area source dish, moderated at first with thin crystalline W films and later by solid Ne, is proposed as the initial device in order to explore problems with a facility scale system. A demonstration scale beam at $\geq 10^{10}$ slow e$^+$/s is proposed using a ^{58}Co source plated on a 6-cm-diameter source dish insert, placed in a 30- cm adaptor.

[†] We would like to acknowledge Dr. H. Bakhru, Director, Accelerator Laboratory and Professor of Physics at the State University of New York at Albany, USA, and Dr. J. H. Broadhurst, Director, Accelerator Laboratory and Professor of Physics at the University of Minnesota, Minneapolis, Minnesota, USA, for their contributions and advice during the very early stages of this work.

[*] Work supported by the INEL LDRD Program under DOE Contract No. DE-AC07-76ID01570.

In this paper we discuss the theoretical basis for the ISPS in the context of both computer calculations for a point design and general scaling relationships developed from parametric simulation studies. We present the ISPS concept in the context of a 20-year program, which has the goal of achieving a positron plasma (hot or cold) at 10^{14} e$^+$/cm^3.

INTRODUCTION

This paper is intended as both a historical summary of certain recent research directions in the general area of intense positron beam production at the INEL, as well as a starting point for explaining present directions. In the first part of this paper we discuss the evolution of the NEAR (Nuclear E-Plus Accelerator Reactor) concept.[1,2,3] We then discuss the "Umbrella" concept and how it has evolved to the INEL ISPS (Intense Slow Positron Source).[4,5] Finally we discuss in detail "Concept A" for the INEL ISPS.[4]

SECTION A—THE NEAR CONCEPT

The NEAR Concept[1,2,3] was conceived as a method of producing fast positron beams by direct extraction. The basic ideas for the NEAR are illustrated in Figure 1. Early analysis[2,3] indicated that the NEAR could produce $\sim 10^{17}$ fast e$^+$/s for a five to ten assembly system. Neutron Monte Carlo studies for a High Temperature Gas Reactor (HTGR) NEAR were performed based on the model shown in Figures 2 and 3 and the system obtained criticality (i.e., $k_{eff} \sim 1.0$) for a central neutron thermal flux of $\leq 10^{15}$ n/cm^2/s. We believe that a beryllium/water/enriched uranium system of similar design could be configured for a central neutron thermal flux of $\leq 5 \times 10^{16}$ n/cm^2/s. However this is only a speculation at this time. Positron beam extraction technology was investigated and appears to be feasible.[1] The main technological issues associated with the NEAR are the high voltage breakdown characteristics of accelerator components in a high neutron flux environment. It was believed at the time these studies were concluded that the initial INEL program (yet to be formulated) should not have as its focus research on high voltage materials breakdown because of high flux neutron damage. It was also believed that since the ultimate NEAR reactor would cost $\sim$\$5 billion, pursuing such a program would be premature. Even a single NEAR assembly capable of producing $\leq 10^{16}$ fast e$^+$/s utilizing the INEL Advanced Test Reactor (ATR) seemed to be too expensive a first step (in the range of \$10 to \$20 million).

Idealized fast positron electromagnetic extraction computer code calculations were performed, based on INEL developed software, for various NEAR geometries. Both cylinders and disks were studied. We summarized these studies below.

Positive beta particle trajectories were calculated numerically using a fourth order Runge-Kutta method to integrate the relativistic equations of motion in static E and B fields.

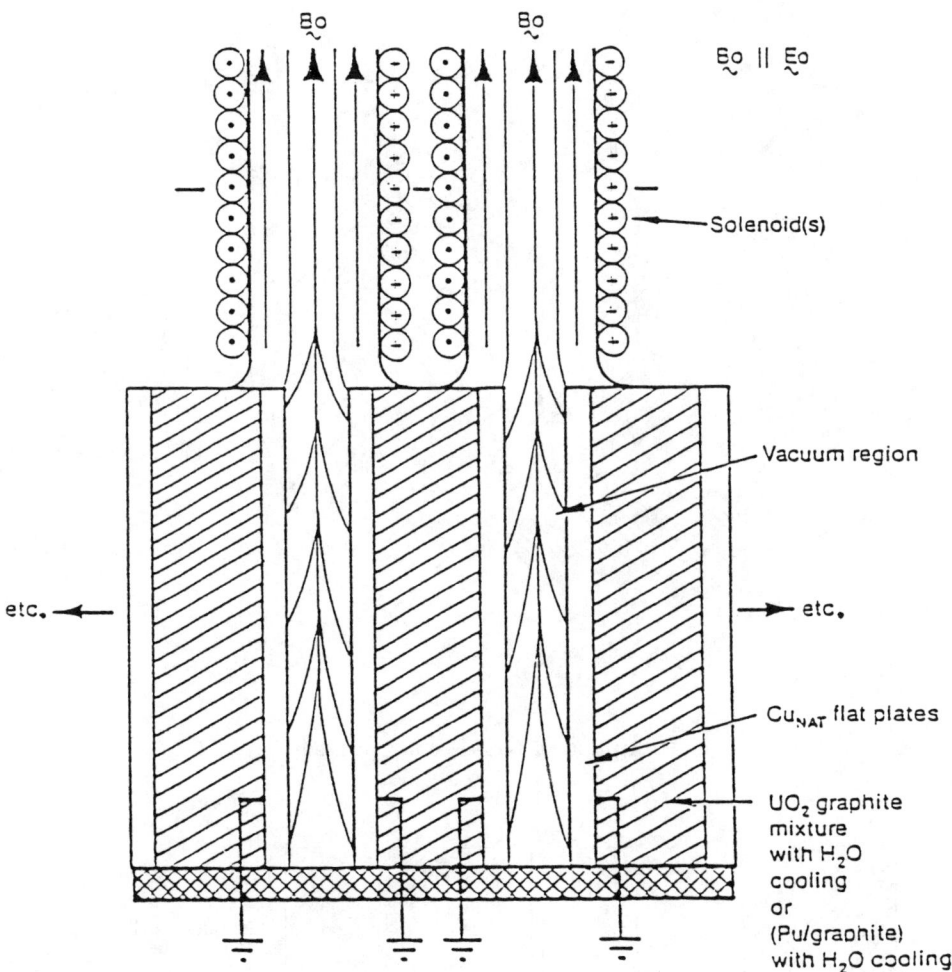

Fig. 1. NEAR assemblies—conceptual.

The geometry consists of a cylinder of radius, b, and length, d. The cylinder axis is oriented along the z-axis beginning at $z = 0$ and extending to $z = -d$. A current loop of radius **a** lies in the x-y plane at $z = 0$.

The magnetic field components produced by the current loop are evaluated via elliptic integrals except near the z-axis where they are supplemented by an expression for B_z holding along the z-axis. The B-field is scaled to its value at the center of the loop, B_0. A uniform electric field was applied along the z-direction.

The positrons were assumed to be emitted isotropically from the interior cylinder walls at $\rho = b$ and along the base disk at $z = -d$. The computed trajectories

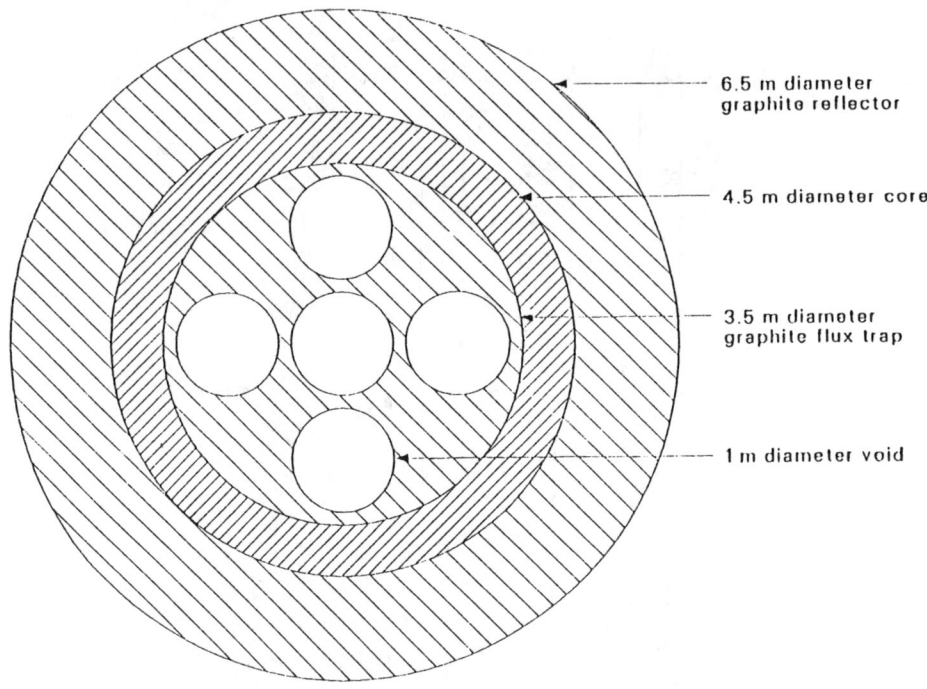

Fig. 2. Cross-sectional description of NEAR concept model.

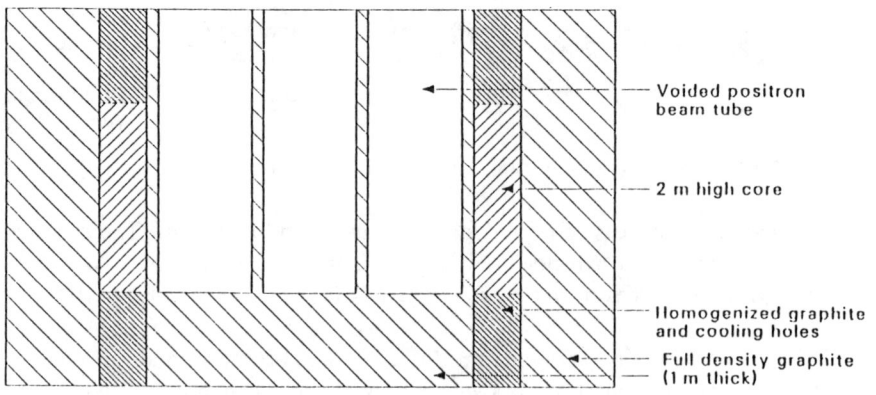

Fig. 3. Axial description of NEAR concept model.

were assigned weights based on the positron emission energy spectrum of Cu^{64}. Particles were followed in time until one of the following conditions were met:

1. The particle escaped the cylinder through the exit aperture at the top of the cylinder.

2. The particle was lost by making a collision with the cylinder wall.

3. The particle was lost by making a collision with the base-disk of the cylinder.

4. The particle survived the maximum time allowed for a trajectory without encountering the cylinder wall.

The particles were launched from discrete bins in velocity and configuration space. Cylindrical symmetry was assumed so that all particles could be launched from the x-axis.

Cylinder Walls—The cylinder walls were binned into 0.15 m segments with trajectories launched from the center of each segment. The initial launching angles were binned in the polar angle ϕ from 90.0 to 270 degrees in steps of 10 degrees referred to the x-axis and in the polar angle θ from zero to 180 degrees in steps of 10 degrees referred to the z-axis. The actual angle used for the launching was the center of each bin.

Base Disk—The base disk was binned radially into five segments and particles launched from the midpoint of each segment. The azimuthal angle was binned from $\phi = 0$ to 360 degrees in steps of 20 degrees. The polar angle was binned from $\theta = 0$ to 90 degrees in increments of five degrees.

Contour plots were drawn showing the phase space distribution of the particles reaching the exit aperture. Since a 20 by 20 contour grid was specified, some of the plots are rather coarse. Because of the coarse binning used in this preliminary study, a more detailed investigation involving either finer binning or Monte Carlo analysis is probably warranted.

The contour plot axes are specified in units of the perpendicular and parallel components of the dimensionless vector $\gamma\beta$ where $\gamma = (1 - \beta^2)^{-1/2}$ and $\beta = v/c$ with c the speed of light. The particle kinetic energy at any point in phase space is given by $E = m_0c^2(\gamma-1)$ where $m_0c^2 = 0.511$ MeV. The quantity $\gamma\beta = p/(m_0c)$ is a dimensionless relativistic momentum and is related to the total particle energy by $E_T = m_0c^2(\gamma^2 \beta^2 + 1)^{1/2}$.

Plots of relative number of escaping particles versus radius were made at the exit aperture. These are also binned in radius.

The energy range expected for the emitted particles was discretized into seven bins with energies 0.025, 0.1, 0.2, 0.3, 0.4, 0.5, and 0.6 MeV, respectively. Each computed trajectory was weighted according to the phase space volume corresponding to its discrete bin.

Each trajectory reaching the exit aperture is written to a file and subsequently weighted via Liouville's theorem to the phase space volume it initially occupied. Table 1 shows the fraction of particles escaping through the exit aperture per unit time. Each case corresponds to $B_0 = 10$ kG and a current loop with $a = 0.5$ m. The fraction of the total binned, weighted positron trajectories exiting the top of the cylinder was computed based on the total number of binned, weighted trajectories that were launched.

Phase space studies illustrated in Figures 4 and 5 indicated that ~10 to 15% total extraction efficiency for usable phase space exists for the NEAR geometries investigated, when fast positrons are the desired outcome. Obviously one could also utilize conventional slow positron moderation techniques for such a NEAR produced beam in an appropriate geometry.

We have also investigated cryogenic moderated discs in NEAR extraction geometry (Ar, Kr, Ne, etc.) and found 100% transport efficiency. These calculations are summarized below.

Low energy positrons were launched from a disk $a = 0.5$ m in radius and 1.5 m axially from the center of a wire loop producing a B-field of 10 kG at its center. The initial energies of the positrons ranged from zero to 5.5 eV in 11 discrete steps. Uniform E-fields of 1 and 5 MV/m were applied parallel to the disk axis.

Since the initial energies are very small compared to energy each particle gained in the E-field, all particles originating at a given radius on the disk exited the loop with practically the same energy. The gyro-radii of the launched particles were small enough to ensure that all trajectories reached the exit aperture without intersecting the cylinder wall at $r = 0.5$ m.

A table of trajectory results as a function of launching radius r_i follows for $E_0 = 1$ and 5 MV/m (Table 2). The radius at which the positron intercepts the exit aperture is r_f.

SECTION B—THE "UMBRELLA" CONCEPT

Our analysis of optimum NEAR positron extraction geometries lead us to the conclusion that a simple disc geometry would be very effective. Additionally, our

Table 1a. Disk emissions.

E_0 (MV/m)	Hole radius (m)	Depth (m)	Area (m^2)	Fraction exciting (%)	Area × % product
0	0.5	-1.5	0.785	1.3	0.010
1	0.5	-1.5	0.785	26	0.2
5	0.5	-1.5	0.785	97	0.76
1	0.5	-3.0	0.785	7.6	0.060
5	0.5	-3.0	0.785	34	0.27
0	0.08	-1.5	0.020	1.5	0.00030
1	0.08	-1.5	0.020	21	0.0042
5	0.08	-1.5	0.020	35	0.007
1	0.08	-3.0	0.020	1.6	0.00032
5	0.08	-3.0	0.020	34	0.0068

Table 1b. Cylinder emissions.

E_0 (MV/m)	Hole radius (m)	Depth (m)	Area (m^2)	Fraction exciting (%)	Area × % product
1	0.5	-1.5	4.71	38	1.8
5	0.5	-1.5	4.71	70	3.3
1	0.08	-1.5	0.754	13	0.098
5	0.08	-1.5	0.754	22	0.17
1	0.5	-3.0	9.42	27	2.5
5	0.5	-3.0	9.42	67	6.3
1	0.08	-3.0	1.51	10	0.15
5	0.08	-3.0	1.51	21	0.32
0	0.5	-3.0	9.42	5.3	0.50
0	0.08	-3.0	1.51	3.2	0.048

economics analysis indicated that an ≤$5 million cost was associated with an ex-situ disc [as compared to an in-situ NEAR assembly (~$10 to $20 million)]. Initial INEL interest was in using ^{64}Cu as the positron emitter. A one meter squared surface area disc was envisioned emitting ~10^{15} fast positrons/s. A proof-of-principle 30-cm-diameter device was studied. The source, rather than being a disc, was envisioned to be a spherical segment of 30-cm chord (25-cm radius). The source would fold up like an umbrella, and hence the name, and would be placed inside the INEL ATR for irradiation (inside a capsule). After irradiation the source would be

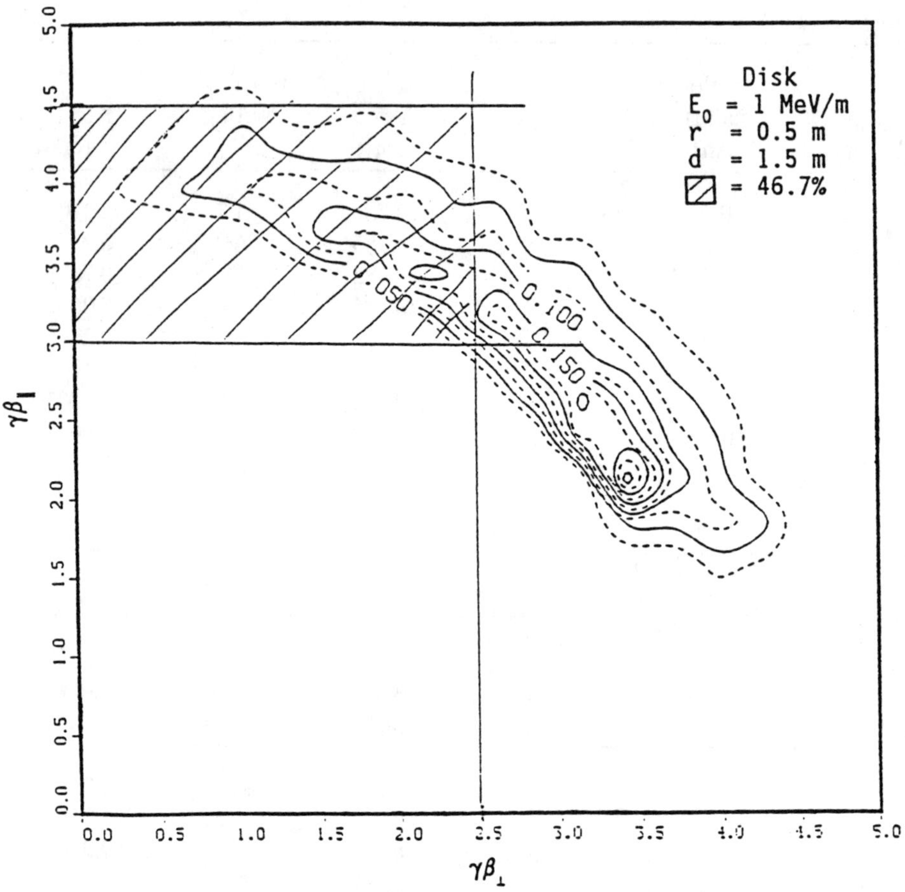

Fig. 4. Relative distribution for disk emission.

transferred via an interim cask (after removal from the irradiation capsule) and placed into a source insertion machine. The source insertion machine would transfer the hot source into position for beam generation in the ISPS prototype.[4]

Both a thin crystal moderator array and cryogenic moderation was analyzed (refer to Figures, 6, 7, and 8 for details of the "Umbrella" Concept).

The INEL also analyzed ^{58}Co, ^{22}Na, ^{18}F, and ^{79}Kr as isotope options. After much debate ^{58}Co was selected, utilizing the INEL EBR-II Reactor. Again, cost considerations were the deciding factors. Although chemical processing and isotope deposition are required for ^{58}Co, it was perceived by INEL personnel as a viable approach for a prototype ISPS device given INEL expertise in hot materials handling,

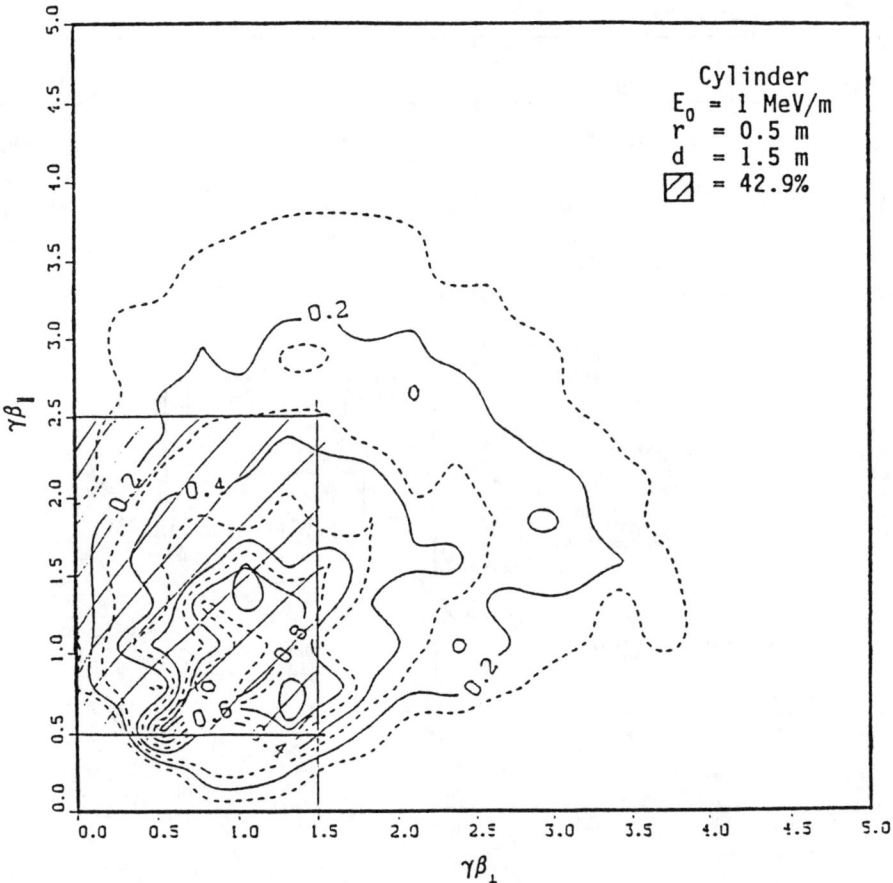

Fig. 5. Relative distribution for cylinder emission.

Table 2a. $E_0 = 1$ MV/m.

r_i(m)	$\gamma\beta_\perp$	$\gamma\beta_\parallel$	r_f/a
0.05	0.026258	3.8075	0.017840
0.15	0.080155	3.8069	0.053222
0.25	0.13857	3.8047	0.097713
0.35	0.20418	3.8020	0.12082
0.45	0.28004	3.7987	0.15214

Table 2b. $E_0 = 5$ MV/m

r_i(m)	$\gamma\beta_\perp$	$\gamma\beta_\parallel$	r_f/a
0.05	0.13264	15.962	0.017986
0.15	0.42635	15.654	0.05352
0.25	0.78354	15.635	0.088241
0.35	1.2565	15.598	0.12166
0.45	1.8728	15.575	0.15465

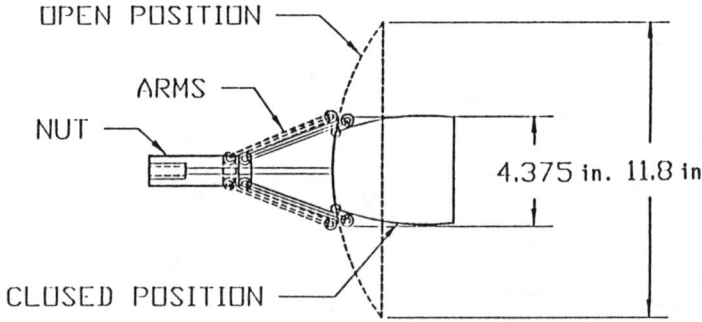

Fig. 6. ISPS source concept.

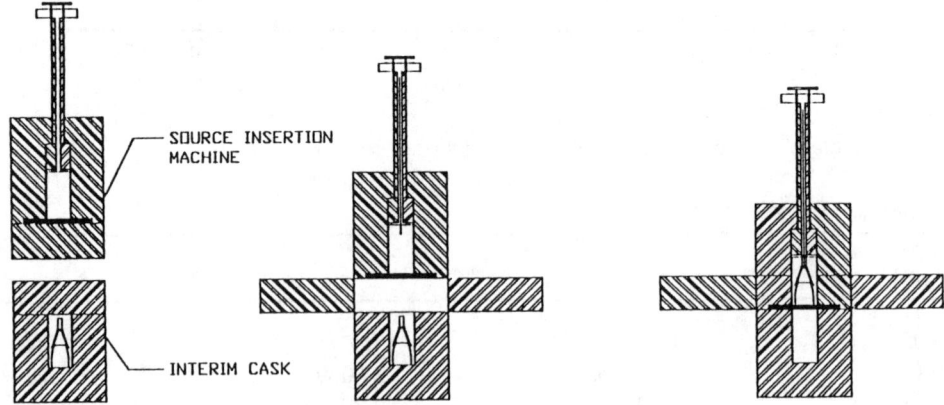

Fig. 7 Interim cask to source insertion machine transfer sequence.

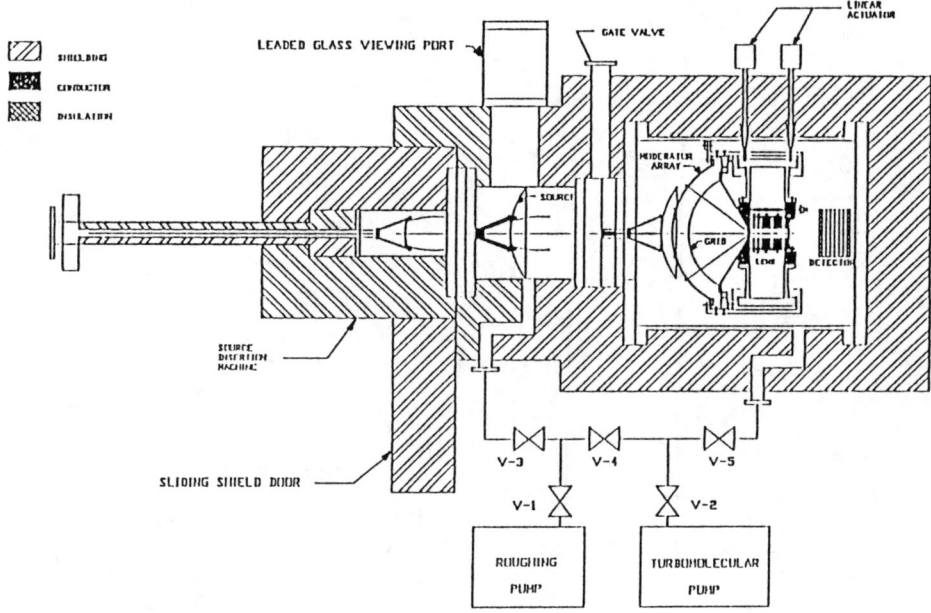

Fig. 8. ISPS cross section.

chemistry, and extensive hot cell facilities. An $\sim 10^{15}$ fast e^+/s, 30-cm chord (25-cm radius) ^{58}Co ISPS device was envisioned.

A twenty-year program was proposed with the ultimate goal of achieving a 10^{14} e^+/cm^3 plasma. Positron microscope applications were stressed for the early phases of the program (refer to Figure 9 for details).

Our attention then focused on the startup phase of our program. Two conceptual devices were investigated: "Concept A," a gridded design, and "Concept B," a gridless design. In this paper we will only discuss "Concept A" in detail.

SECTION C—BEAM OPTICS—"CONCEPT A" SYSTEM DESIGN

Our Concept A design has a spherical section onto which a neutron activated positron source is deposited (Co, Cu, F, Kr, etc.). Such a source produces a broad spectrum of positron energies ranging from a few to thousands of electron volts. An energy moderating material placed over the source slows down (thermalizes) the positrons and reemits a small fraction ($\sim 10^{-2}$ to 10^{-4}) with a nearly monoenergetic distribution near a few eV. These "slow" positrons are then reaccelerated and focused

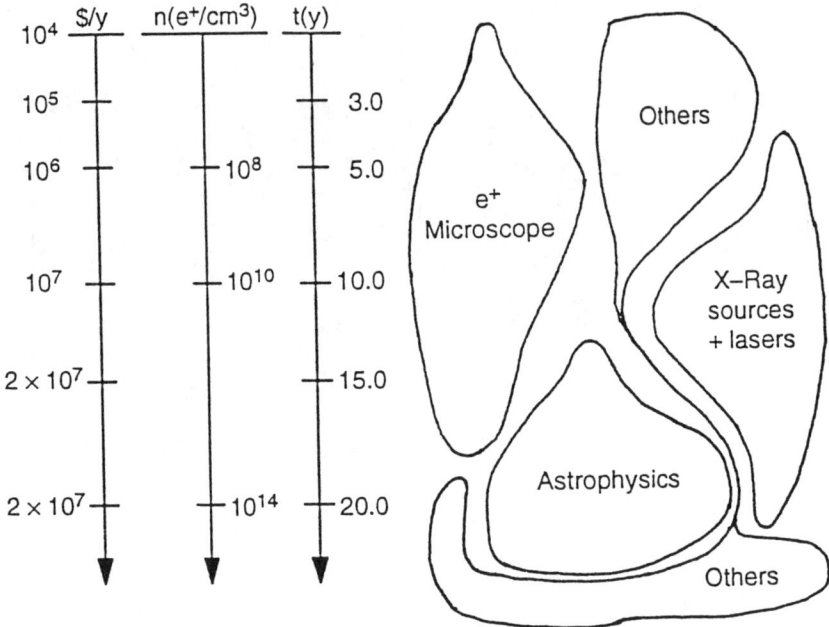

Fig. 9. INEL 20-year view of positron application.

to a smaller diameter using an electric field of several keV. This process of acceleration, focusing, and remoderation (known as brightness enhancement) is repeated until the beam diameter is reduced to the desired size. Brightness enhancement through remoderators circumvents Liouville's theorem as long as the theorem is obeyed within each stage segment [i.e., between the moderator (or remoderator) and the next remoderator]. Within stages

$$\Theta_1^2 d_1^2 E_1 = \Theta_2^2 d_2^2 E_2 \qquad (1)$$

where E_1 and E_2 are the particle energies exiting the first moderator (remoderator) and entering the second remoderator, respectively; d_1 and d_2 are the beam diameters at the first surface and second surface, and Θ is the angle of beam divergence at the corresponding surface. The prototype instrument design (Concept A) consists of the source and one stage of remoderation (see Figure 10). The design brings the 30-cm source diameter down to a 0.3-mm beam diameter in two stages.

The beam physics of the INEL intense slow positron source is relatively straightforward. To examine the optics design features in detail, positron trajectory codes have been run for a wide variety of configurations. The beam analysis followed the positron trajectory from a spherical segment through an initial moderator, acceleration and beam size reduction (Stage One), to a first remoderator, and then acceleration and refocusing to a second remoderator (Stage Two). Some preliminary calculations have also been done for the final beam extraction after the second remoderator. The bulk of the analysis has used two computer codes, SIMION and EGUN. SIMION, which was developed at the INEL and is well-known and used internationally, treats the positrons as independent particles and calculates their trajectories for a given set of boundary conditions over a specified geometry. EGUN, developed at the Stanford Linear Accelerator Center, is capable of more detailed calculations than SIMION because it can take into account collective effects and the effects of curved surfaces without introducing unspecified field errors. Space charge issues, which are important at high current densities, have been analyzed with a Particle In-Cell (PIC) technique. The details of the INEL Concept A design for a positron generator are described in the next few sections.

STAGE ONE ANALYSIS

From a detailed fine-mesh positron beam optics analysis with EGUN, we have obtained a set of empirical scaling relationships for the various physical effects.

Thermal Effect

The angle of positron emission from a surface, Θ_*, is determined by the properties and temperature of the moderator. To a good approximation, in tungsten (a well-studied moderator) the transverse beam energy is proportional to $k_B T$ (where k_B is the Boltzmann constant). Defining the angle of emission from a remoderator surface with

$$\tan \Theta_* = \left(\frac{E_\perp}{E_\parallel}\right)^{1/2} = \frac{|v_\perp|}{|v_\parallel|} \tag{2}$$

if $E_\perp = 4 k_B T$ (HWHM[a]), then

T (K)	$k_B T$ (mV)
300 (water cooling)	100
80 (liquid nitrogen)	30
5 (liquid helium)	2

This is a good approximation for tungsten between 300 and 77 K, although it breaks down below ~60 K remaining somewhere near 30 mV (HWHM) or 60 mV

---

a. HWHM = half width at half maximum.

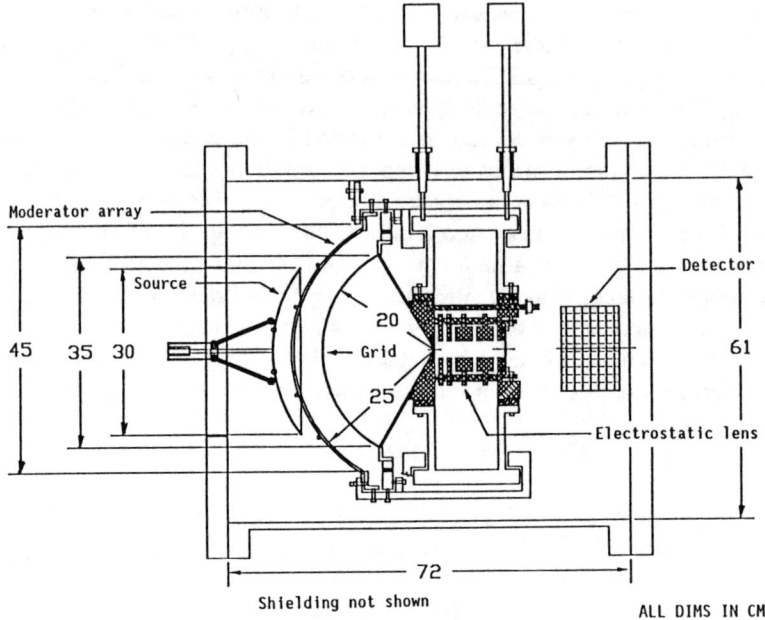

Fig. 10. Features of the Concept A design.

(FWHM[b]). The 5 K, 2 mV analysis serves as a theoretical lower bound for moderator and remoderator materials other than tungsten in this study.

Based on these observations and EGUN calculations, we obtain the relationship

$$\frac{\text{Thermal Error at Focus (mm)}}{\text{Radius (mm)}} = \pm 0.56 \times 10^{-2} \left[\frac{\text{Surface Ejection Energy (eV)}}{\text{Extraction Energy (eV)}} \times T_{\text{HWHM}} (K) \right]^{1/2} \quad (3)$$

This relationship has been used to predict the "smear" in focus for a number of cases. For tungsten (2.5 eV ejection energy), using an extraction energy of 5 keV and a temperature of 300 K, the thermal error at the focus is ±0.54 mm for a 250-mm-radius device.

---

b. FWHA = full width at half maximum.

Effect of "Flat" Moderator Crystal Plates and Placement

A design issue that needs to be addressed is the focus effect of flat moderator plates placed on or near a hemisphere (i.e., a spherical segment). Using a second scaling relationship developed from EGUN calculations, we have

$$\frac{\text{Error at Focus caused by Flat (mm)}}{\text{Radius (mm)}} = \pm 0.112 \left[\frac{\text{Length of Flat (mm)}}{\text{Radius (mm)}}\right] \left[\frac{\text{Length of Flat (mm)}}{\text{Extraction Gap Length (mm)}}\right]^{1/2} \quad (4)$$

For a typical case of 1 cm × 1 cm crystal of tungsten and an accelerator grid 5 cm away, the focus error is ±0.5 mm because of the flatness of the crystal plates for a 250-mm-radius device.

For the single crystal moderator flats, the misalignment error relative to the center of a crystal flat that acts as the pivot on a spherical surface results in the scaling relationship

$$\frac{\text{Alignment Error at Focus (mm)}}{\text{Radius (mm)}} = \pm 0.93 \frac{\text{Edge Displacement (mm)}}{\text{Length of Flat (mm)}} \left[\frac{\text{Length of Flat (mm)}}{\text{Extraction Gap Length (mm)}}\right]^{1/2} \times \text{Form Factor} \quad (5)$$

The form factor is ≤2.0 for parameters of interest to us. Considering the same tungsten flat as before (with the accelerating grid), a 0.1-mm edge displacement results in a focus misalignment of ±2.0 mm.

SIMION calculations, where applicable, were found to be in agreement within the accuracy of the code, for this set of empirical scaling relationships.

Effect of Grid and End Effects

As mentioned elsewhere,[5] detailed EGUN calculations were made to study the effect of the accelerating grid. Trajectories for a concentric grid have been calculated for several placements of such a grid, as well as for the case with no grid (i.e., biasing the first remoderator sufficiently to bring the majority of positrons to it). For such a concentric grid, the effects of the finite grid mesh size have also been examined. We calculated the defocusing effect, which was first analyzed by Davisson and Calbick,[6,7] and found smearing on the order of 1 mm^2 for a grid size of 1 mm. For our Concept A Stage One design, the error caused by the finite grid mesh size is given by

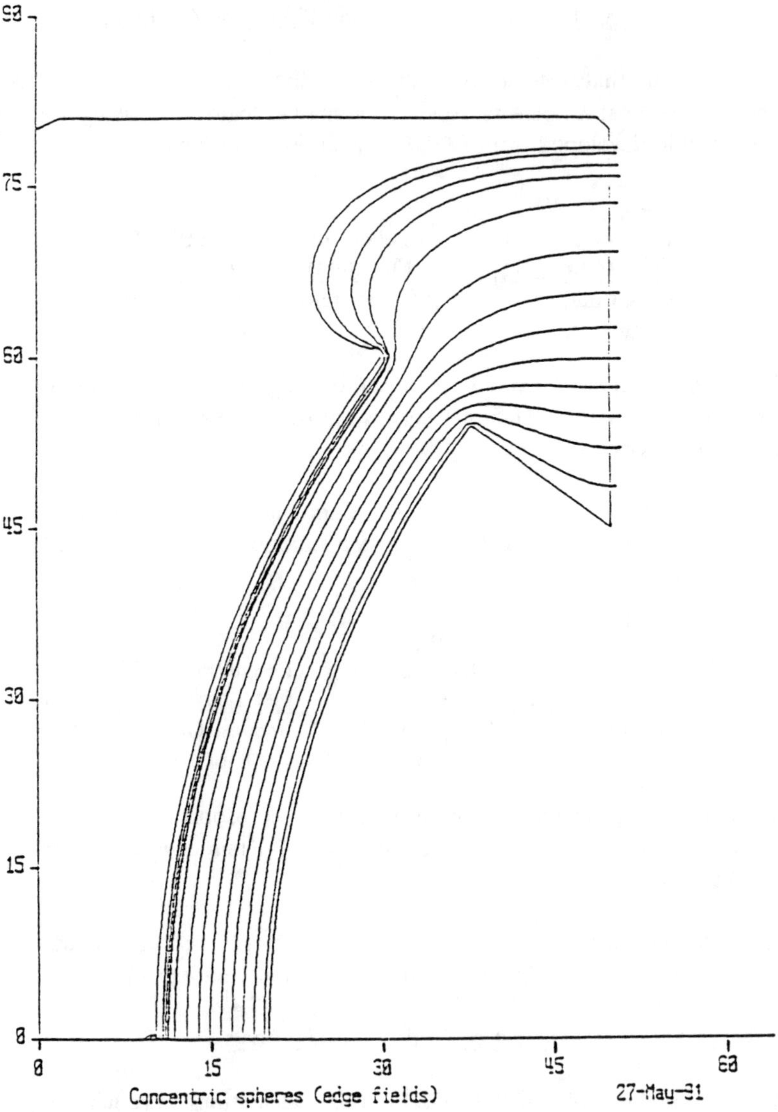

Fig. 11. End effect EGUN calculation for Stage One.

$$\frac{\text{Total Error caused by Grid Mesh at the Focus (mm)}}{\text{Mesh Spacing (mm)}} = \frac{\text{Radius (mm)}}{4 \times \text{Extraction Gap Length (mm)}} \qquad (6)$$

Other effects that can cause defocusing occur at the edges of the moderator and/or grid. These "end" effects arise because of the electric field gradients at these terminations. An example is shown in Figure 11. For a 5.0-cm gap between the grid and moderator array, the end-effect defocusing is ±0.50 mm when the concentric field region is extended by two gap lengths beyond the active positron transport region.

Summary

The final remaining issue for Concept A, other than space charge which will be discussed later, is global misalignment of the moderator array and the electrostatic grid. If these two elements are not exactly concentric (i.e., their radial vectors do not have the same focus), the entire beam spot will shift by an amount approximately equal to the displacement from the original focus of the radial vectors. The Intense Slow Positron Source (ISPS) Concept A device is designed so that if bench alignment of these two elements does not produce the same focal radial vectors and, hence, the beam spot is not on the optical axis of Stage Two, the entire Stage Two Einzel lens assembly can be displaced in the x-y plane (assuming $\hat{z}$ to be the positron beam axis) by a set of linear actuators. An electrostatic, x-y beam focusing field device can be incorporated into the exit aperture of Stage Two to compensate for this mechanical displacement.

Table 3 gives the maximum set of errors for a "well-constructed" Concept A device.

Examination of our Stage One, Concept A device physics suggests that a gridless design should have certain advantages. Equations (4) and (5) indicate that a factor of ~2x benefit exists for placing the grid at the origin, and hence reducing the effect of local field errors from the moderator surface. This could result in a factor of 4x brightness enhancement, relative to Concept A, if the end effects can be overcome. An additional factor of 2x in positron beam brightness can be obtained from elimination of beam loss to the grid and its support structure, and an overall reduction of complexity of the total design can be achieved. These ideas are presently under investigation both computationally and experimentally, in the context of the Concept B[8,9] optics design.

STAGE TWO ANALYSIS

The purpose of the second stage of positron acceleration is to reduce the positron beam delivered by Stage One from millimeter to submillimeter radius. Most of the calculations for the Stage Two design were performed with the SIMION code.

The size of the beam spot delivered by Stage One will be affected by various "parasitic" aberrations of mechanical (alignment) origin. Starting with a concave spherical emitter of 250-mm radius and 300-mm chord, there is good reason to expect (on the basis of preliminary beam trajectory calculations) that the converging positron

Table 3. Maximum set of errors.

Cause of Error	Error (mm)
Moderator tile positioning error of 0.1 mm	±2.0
Flat tiles 1 cm × 1 cm	±0.5
Thermal spreading (reemitted positrons)	±0.6
Grid wire mesh defocusing	±0.6
End effect (5-cm separation with 2x gap extension)	±0.5
Rms	2.3

beam will be smaller than 10 mm in diameter. If the structure is well-built, the focal spot will reduce to about 3 mm FWHM. This corresponds to a radial size reduction of 10^{-2} and an area reduction of 10^{-4}. Stage Two should be designed to provide comparable performance, i.e., to reduce an initial 3-mm-diameter spot to 0.3 mm or less. Preliminary calculations of a simple electrostatic focusing system suggest that this is indeed possible and, therefore, a realistic goal to pursue.

To avoid unnecessary geometric aberrations, the particle source for Stage Two (a remoderator) should have either a spherical (preferably concave for natural focusing) or flat emitting surface. Since the remoderator will most likely be a flat film or foil, a flat surface was assumed for model calculations. An applied electric field at the emitting surface accelerates a beam of particles away from the emitter.

An electrode having negative voltage must be placed near the emitting surface to generate the extraction field. [Although this field does not extract positrons from within the surface—positrons are energetically expelled in the absence of an external field—the term "extraction" will be used in the following discussion of the surface field and the electrode(s) which produce it to demarcate this region.] As a tentative (and simplistic) first design, this electrode was shaped as a moderately thick disk, placed parallel to the source plane with an extraction aperture (having rounded edges) surrounding the extraction region (see Figure 12). This combination of open electrodes (without grids) produces a longitudinal accelerating field that diverges away from the opening; consequently, it contributes radial forces that cause divergence or defocusing of the accelerated particles.

To focus this diverging beam back to the axis, an additional convergent lens is required. This second lens could be magnetic (solenoidal or "round" type used in electron microscopes) or electrostatic (Einzel lens). Einzel lenses can be accel/decel (meaning particles accelerate when entering and decelerate when exiting) or decel/accel. In very broad terms, decel/accel Einzel lenses require less operating voltage (and are much preferred for this reason) but produce greater focal aberrations. In either case, the focal length scales linearly with the lens diameter (i.e., a magnified image of lens and rays at fixed potentials does not change proportions) and the focal length is a nonlinear function of voltage. After examining a large set of cases, our analysis indicates that a decel/accel design gives the optimum result with an ~36x beam diameter reduction, including 300 K HWHM noise.

Calculations by EGUN for a Stage Two accel/decel type of lens indicate that space charge interactions for a beam focused down from 3-mm radius to 0.06-mm radius are not significant for currents less than 1 mA. At about 10 mA the spot size begins to enlarge and refocusing would be required. At about 100 mA it will not be possible to keep the spot below about 0.5-mm radius using lens adjustment.

Calculations for a similar beam envelope in a decel/accel lens were not performed. Although the decel/accel case can be anticipated to be less favorable, since the beam remains large except near the final "point" focus, the two results should not be too different for low currents (<1 mA). Based on the EGUN calculations, there should be no problem with space charge for any positron current that is likely to be produced in Stage Two. Space charge issues are discussed further in the next section.

Space Charge Analysis

Based on this analysis, experience, and recent simulations, it is generally possible to say, qualitatively, that if $\lambda_{De} \gg \ell$ (where λ_{De} is the Debye length and ℓ is the

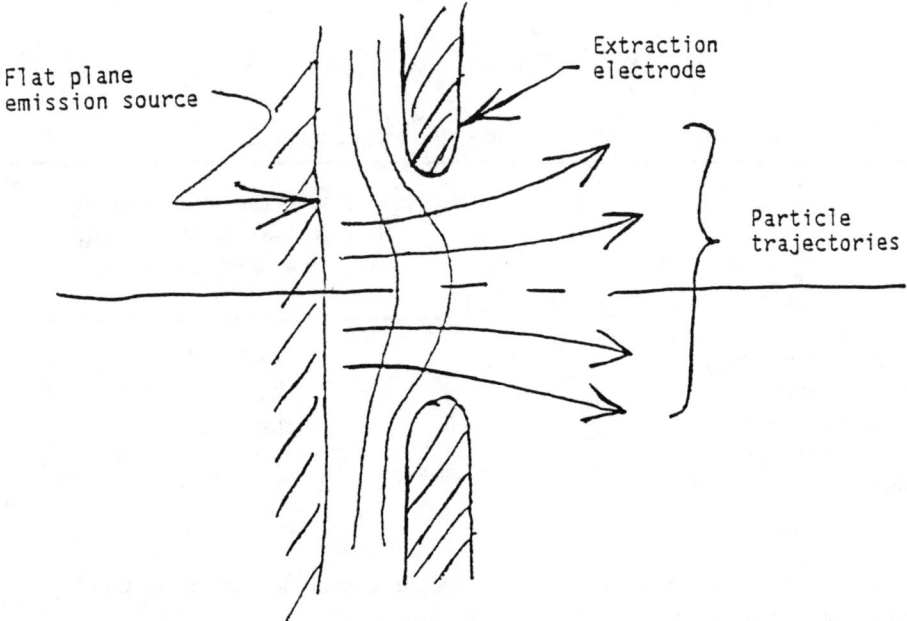

Fig. 12. Extraction region of Stage Two.

characteristic length), space charge effects can be neglected. On the other hand, if $\lambda_{De} \leq \ell$, space charge effects begin to occur. We will present, here, an order of magnitude analysis of the positron beam parameters expected for the ISPS device during various phases of the proposed experimental program (Table 4). Phase I is assumed to occur during the first three years of the experimental program.

The longitudinal space charge effect is severest at the surface of the final target. Therefore, we estimated the space charge effect at that point. As can be seen from Table 5, the most severe number for Phase II for the longitudinal Debye length is ~10 cm, while the transport length is of this order (or longer), and hence, the space charge effects are minimal. Thus, precise effects need to be looked at numerically. In Phase I it is safe to say that little to no longitudinal space charge effect exists. Note that numbers in Table 5 are based on the instantaneous electron density on the target, which is higher than the in-flight effective density. Table 5 results were obtained from the following calculation.

The longitudinal Debye length at the dynamical energy of 5 keV is given by

$$\lambda^l_{De} = \frac{2 \times 10^9 \text{ cm/s}}{6 \times 10^4 \sqrt{n_e}} \quad . \tag{7a}$$

The positron density on the target (where A is the area) for Phase I (very early) is given by

$$n_e = \frac{1.6 \times 10^7}{2 \times 10^9 \text{ A}} \cdot \frac{10^{-2}}{\text{A}} \text{ cm}^{-3} \quad . \tag{7b}$$

Table 4. ISPS physics parameters[a] used in space charge analysis.

	ϕ_e+ (5 keV) on exit remoderator of Stage Two Einzel lens with 0.03-cm-diameter unit area (e^+/s)
Phase I (very early)	1.6×10^7
Phase I (early)	8.2×10^9
Phase I (late)	1.6×10^{10}
Phase II (early)	2.0×10^{11}
Phase II (facility scale)	2.9×10^{12}

a. Stages One and Two as per previous analysis; one moderator (Stage One) surface and one remoderator (Stage Two) surface.

Table 5. Longitudinal space charge analysis.

	$\lambda^{\parallel}_{De}$ (cm)	
	$A = 10^{-1}$ cm^2	$A = 10^{-3}$ cm^2
Phase I (very early)	$\lambda^{\parallel}_{De} \approx 10^5$	$\lambda^{\parallel}_{De} \approx 10^4$
Phase I (early)	$\lambda^{\parallel}_{De} \approx 5 \times 10^3$	$\lambda^{\parallel}_{De} \approx 5 \times 10^2$
Phase I (late)	$\lambda^{\parallel}_{De} \sim 3 \times 10^3$	$\lambda^{\parallel}_{De} \sim 3 \times 10^2$
Phase II (early)	$\lambda^{\parallel}_{De} \sim 5 \times 10^2$	$\lambda^{\parallel}_{De} \sim 50$
Phase II (facility scale)	$\lambda^{\parallel}_{De} \sim 10^2$	$\lambda^{\parallel}_{De} \sim 10$

A = Area of target spot size.

The transverse space charge effect is most severe for the ISPS on the surface of the first remoderator, where the transverse temperature is expected to be at its lowest value. The precise temperature rise of the beam during flight from the surface of the remoderator to the target will be evaluated in later work. As can be seen in Table 6, the most severe numbers are again for Phase II, where the transverse Debye length is of the order of the focal size. Thus, a dynamical estimate becomes necessary for this case.

The initial transverse temperature of the positron beam emanating from the surface of the remoderator can, conservatively, be assumed to be as low as the temperature of the metal surface. If the remoderator is at room temperature, the beam transverse temperature is ≥300 K. If the remoderator is at a lower temperature, so is the positron beam transverse temperature. Therefore, the Debye length of the beam in the transverse direction is much smaller than that in the longitudinal direction. The transverse Debye length can be estimated (at 300 K) as

Table 6. Transverse space charge analysis.

	$\lambda^{\perp}_{De}$ (cm) at three values of T_e		
	300 K	70 K	4 K
Phase I (very early)	~10^2	~50	~10
Phase I (early)	~5	~3	~0.5
Phase I (late)	~3	~1.5	~0.3
Phase II (early)	~0.4	~0.2	~0.04
Phase II (facility scale)	~0.1	~0.05	~0.01

$$\lambda_{De}^{\perp} = \frac{4 \times 10^7 \, T_e^{1/2}}{6 \times 10^4 \, n_e^{1/2}} \approx \frac{10^2}{\sqrt{n_e}} \text{ cm} \quad , \tag{8a}$$

where T_e is in eV and n_e is in (cm)$^{-3}$. The positron density on the target for Phase I (very early) is estimated as

$$n_e \sim \frac{1.6 \times 10^7}{6 \times 10^7 \, A} \sim \frac{0.3}{A} \text{ cm}^{-3} \quad , \tag{8b}$$

If A is ~0.3 cm^2, $n_e = 1$ cm^{-3}.

The dynamical effect of the transverse space charge is estimated next. Since space charge effects become important only in Phase II at the first remoderator surface, that case is considered. The simplest way is to estimate the space charge acceleration time by evaluating the plasma time. For Phase II, the inverse plasma frequency is given by

$$\omega_{pe}^{-1} \sim \frac{1}{6 \times 10^4 \, \sqrt{n_e}} \text{ s} \approx \frac{\sqrt{2}}{6} \times 10^{-7} \sim \frac{1}{4} \times 10^{-7} \text{ s} \quad . \tag{9a}$$

On the other hand, the acceleration time for the beam to reach from ~2 eV to 5 keV over the length ℓ is

$$t = \frac{v}{a}, \text{ where } 2 \, a\ell = v^2 \quad . \tag{9b}$$

Therefore, $t \approx 6 \times 10^{-8}$ s if we set $\ell = 10^2$ cm (estimated maximum distance between the first remoderator and the target). Note that this time is comparable to or slightly longer than the plasma time. This means that space charge can eventually affect the dynamical orbits over the flight from the surface of the first remoderator to the target in the Phase II scenario. However, we believe at this time that this will be a small effect.

Examination of Table 6 indicates that for a certain range of parameters the transverse Debye length tends to be equal to, or smaller than, the beam transverse size. When the beam transverse size is greater than the transverse Debye length, we expect that the beam will begin to spread because of space charge effects. When it spreads, we expect that the beam will acquire an increased transverse temperature. On the other hand, the dynamical Debye length in the longitudinal direction is greater than, or at its worst about equal to, the longitudinal dimension, so that the space charge effect in the longitudinal direction is not serious.

When the transverse space charge effect increases the transverse temperature, the beam expansion (because of space charge) ceases when the heated beam acquires a transverse Debye length roughly equal to the beam transverse size. On the other hand, the time for growth of the space charge expansion is comparable with the beam drift time. Thus the e-folding of the instability is not large. This results in a dynamical situation for Phase II for $T_e = 0.03$ eV, where the transverse temperature begins to increase while the beam is longitudinally boosted. An accurate assessment of these cases thus requires dynamical (time-dependent) analysis and/or computer simulation (perhaps 2D). We estimate that the saturated temperature will be the original longitudinal temperature of ~2 eV. However, confirming this requires further theoretical and computational work.

A one-dimensional PIC simulation code developed at the Physics Department of the University of Texas was adapted for simulation of some aspects of the proposed experiments. Preliminary one-dimensional PIC simulations assumed a beam longitudinal energy of 5 keV (instantaneous boost from 2.5 eV to 5 keV) with a beam transverse temperature of 300 K and a 5-mm initial beam diameter. For Phase II (early) we have observed an ~40% increase in transverse velocity over the thermal velocity (because of space charge) over a time of 20 ω_{pe}^{-1} (this corresponds to a length of ~10 m). For Phase II (facility scale), we have observed an ~300% increase in transverse velocity over a time of 20 ω_{pe}^{-1}. Based on the results presented earlier and these simulations, we conclude that space charge effects will be negligible in Stages One and Two of the ISPS for Phase I of the program. Further work needs to be done to investigate the ISPS Phase II device, and the brightness enhanced beams beyond Stage Two, for Phases I and II. This work will be performed during the INEL ISPS demonstration program.

Beam Extraction

At this point in the ISPS device development, the INEL has not devoted as much effort to the design of a beam extraction system for the ISPS. It is our intent to design a beam extraction system during Phase I. We present some preliminary ideas below.

Further brightness enhancement steps with remoderators for Phases I and II require a careful analysis. Hence, the use of reflection or transmission remoderation steps over distances of several meters or more is not advisable unless additional brightness enhancement is not attempted. It is desirable to design a beam extraction system with at least two 90-degree bends so that the positron microscope's first remoderator is not in a direct line of sight from the ISPS source disk (because of the strong γ radiation field). The simplest approach is to (a) boost the beam rapidly to 5 keV off the second Stage Two remoderator; (b) use two 90-degree electrostatic (or magnetic) bending elements and allow the beam to thermally expand (because of the ~1/40 eV transverse beam energy) at 5 keV longitudinal beam drift energy between these elements; and then (c) repeat the Stage Two Einzel lens to reduce the beam back to submillimeter size. A more desirable approach is to compensate for the thermal beam expansion by placing focusing elements between the 90-degree bending sections. A third approach is to separate the Stage One and Stage Two ISPS elements. This approach has several operational advantages for a facility-scale ISPS device. We discuss several aspects of such a preliminary design below.

The main feature of the proof-of-principle ISPS Phase I design is two blockhouses, with a 90-degree bend for source shielding. A third blockhouse may also be desirable for safely loading the source. The optics includes a modified electrostatic lens just after the first remoderator. This is followed by an asymmetric Einzel lens, an electrostatic 90-degree turn, a second asymmetric Einzel lens, and then either a transmission remoderator or a reflection remoderator in Blockhouse 2. Placing the modified electrostatic lens just after the first remoderator has some advantages.

Engineering Design Considerations

Engineering studies have investigated both Cu and Co as positron sources for the ISPS device. Most of our detailed (early) studies were done for Cu.

Moderator Array

A cryogenic rare gas moderator structure and a W moderator array have been investigated. Our Concept A engineering analysis has emphasized the W moderator array. This array would be constructed of W foils 1,500 to 20,000 Å thick. Two design concepts have been identified for the array. The first, a tile design, requires a large number of foil "tiles" that would be attached to a hemispherical structure (Figure 13). The second concept, called the integral design, involves fabricating a hemispherical segment, applying a substrate, and coating it with W by sputtering or chemical vapor deposition, and then removing the substrate. It is anticipated that the structure will be fabricated from a readily machinable, nonmagnetic material. Ideally, the material should have a coefficient of thermal expansion similar to that of W. Matching the thermal expansion coefficients would eliminate concerns that the differential thermal growth would cause the foil to go slack (possibly wrinkling and losing focusing accuracy) or to be pulled too tightly (possibly damaging the foil).

In the tile design a number of W-foil devices, each consisting of a 1,500 to 20,000 Å W single crystal (or polycrystal) foil attached to a W frame (Figure 14), are assembled into a structure. The 0.5-mm-(0.020-in.) thick frames have an outside diameter of 1.0 cm (0.39 in.) and an inside diameter of 0.7 cm (0.28 in.). The hemispherical structure supports the frames and focuses them on the first remoderator target. This structure will have a number[c] of 0.7-cm-(0.28-in.-) diameter holes and

---

c. Over 600 in the Cu option and 10 to 20 for the Co option for Phase I.

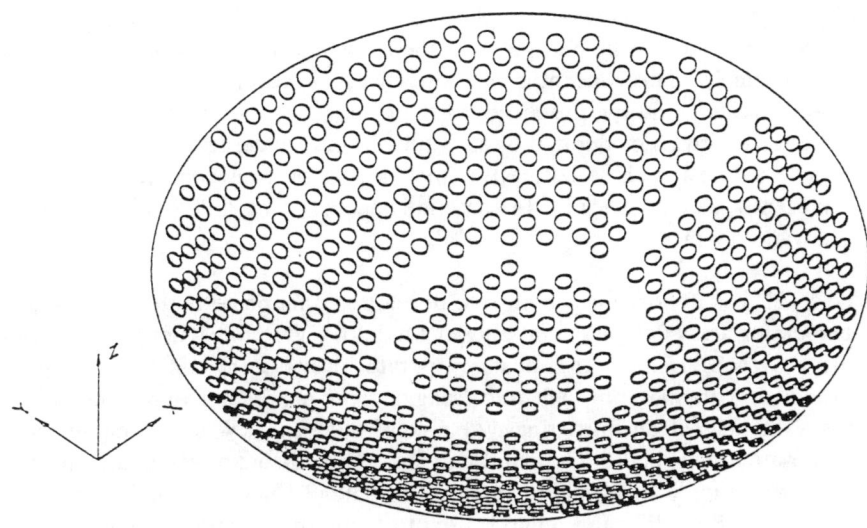

Fig. 13. Moderator array isometric view.

1.0-cm-(0.39-in.-) diameter by 0.5-mm-(0.020-in.-) deep spotfaces on the inner (concave) surface on a 1.07-cm-(0.42-in.-) triangular pitch. The spotfaces are used to align the foil devices. The desired alignment tolerance for each frame is ±50 μm (0.002 in.-) on the target; early prototype tolerances of ±1 mm are acceptable. Evaluation of foil preparation and mounting methods will be performed early in the design phase. The advantage of this tile concept is that it is proven—as stated in various articles in the positron literature. The major disadvantage is that a large number of foil devices are required and the assembly will be very labor intensive. However, the Co option significantly reduces the number of the W foils required (10 to 20) for the array for Phase I of the project.

The integral design starts with a spherical segment structure similar to that of the tile design but there are no spotfaces. The size of the holes depends on the structural properties of the W foil, so the optimum hole size will be determined based on the strength of the foil and the optical performance of the assembly. A substrate will be applied to the back of the structure, filling the holes. The front of the substrate/structure assembly will then be machined to obtain a smooth, spherical

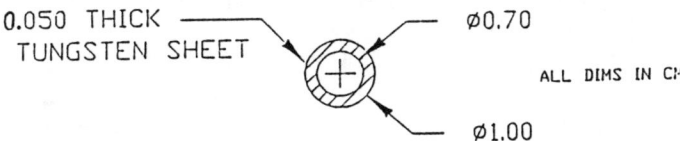

Fig. 14. Tungsten foil tile.

surface (Figure 15). A W film will be applied to this surface using a sputtering or chemical vapor deposition technique. Finally, the substrate will be dissolved, leaving a foil/structure assembly. This concept is very attractive. However, the technique has not been demonstrated, and the properties of the resulting foil are unknown, as are the properties of positrons moderated by such foils. Thus, some development is needed early in the design phase before the final moderator array concept can be selected. Presently, the tile design is the baseline concept. The "integral" design is a promising advanced concept.

The INEL has in-house capabilities to develop and fabricate both the tile and integral moderator array designs. Our Enerjet Model UCV-18/6 magnetron sputtering system with a CHA SR15-2 electron beam evaporator provides the capability to apply coatings of metals, ceramics, polymers, etc., ranging from nanometer to millimeter thicknesses. The magnetron sputtering system consists of a 10 kW dc power supply, 600 W RF power supply, and 6-, 3-, and 2-in. guns. The e-beam attachment consists of a 15 kW power supply and two e-beam guns, capable of simultaneous deposition. A LN_2 cryotrapped 8-in. diffusion pump is used to pump the vacuum chamber to the 10^{-7} torr range. Thin-film deposition rates and thicknesses are monitored by quartz crystal oscillators that are calibrated using standard atomic absorption techniques. Substrates (static or rotated) can be cooled to LN_2 temperatures or heated to 1200°C. Reactive gases, such as O_2, N_2, and C_2H_2, can be introduced into the chamber to deposit thin films of oxides, nitrides, and carbides, respectively. An experienced research and development staff of five people is available to aid in the moderator array development effort.

For the Phase II design, the maximum allowable moderator array temperature is on the order of 300 K, with a maximum allowable variation in temperature over the array surface of less than 30 K. Calculations indicate that this can be accomplished for the Cu concept with two water-cooling coils, one at a radius of one-half the outside radius and another at the outer radius. The temperature variation over the

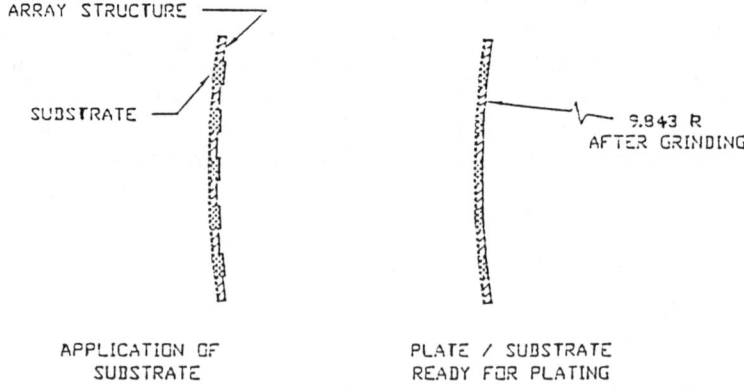

Fig. 15. Integral moderator array fabrication.

structure itself (i.e., the 3-mm thickness) will be less than 15 K. It is believed that a similar design will work for the Co concept, however, additional calculations need to be performed because of the higher specific activity of Co.

The temperatures of the foils attached to the moderator structure were also investigated for the Cu concept. In these calculations, the maximum calculated temperature difference between the center and edge of the foil was 25 K. Thus, the maximum temperature variation between the coolest part of the moderator array structure and the hottest point on the foil is 40 K (15 + 25). Although this is larger than the desired value of 30 K, given the conservative nature of the analysis and the early stage of the design the concept is certainly feasible.

Electrostatic Grid

The electrostatic grid accelerates the positrons emitted by the moderator array. The grid is a hemispherical segment with a radius of 20 cm (7.87 in.) and chord of 34 cm (13.5 in.) located concentric with the moderator array. In this design, it is maintained at −5.0 keV relative to the moderator array.

The precision of the focus of the positron beam is strongly affected by the dimensional alignment of the electrostatic grid. The current plan is to make the grid from 1-mm-square Cu mesh using a spherical form; a structural support will be attached to the mesh before it is removed from the form. Dimensional tolerances for the grid are a factor of two to three less severe than those of the moderator array.

Remoderators

The Stage Two remoderators focus and transmit the positron beam. Both remoderators are of single-crystal W foil and frame construction. The first remoderator is expected to be 1 cm in diameter. The second remoderator is expected to be 0.2 cm in diameter, although the beam diameter at the second remoderator could be as small as 0.03 cm.

Calculations were performed to determine the temperature distribution in the remoderator foils. In this analysis, the heat sources considered were thermalization of the positron beam and thermal radiation from the surroundings. Various beam impact diameters and remoderator operating temperatures were considered. The results indicate that the practical lower limit for the remoderator operating temperature is on the order of 20 to 30 K with He cooling.

Electrostatic Lens (Stage 2)

An electrostatic (Einzel) lens will be used to focus the positron beam emitted by the first remoderator onto the second remoderator. The lens, shown in Figure 16, consists of six coaxial electrodes. These electrodes will be separated by insulating sections; precise spacing between the electrodes will be achieved by machining the insulators. The electrodes and insulating spacers will be aligned prior to assembly of the lens in the moderator/grid/lens mounting structure. The penetrations through the vacuum vessel for the electrical connections will be made using standard high vacuum, high voltage, insulated feedthroughs.

Mounting Structure

The various components of the Concept A ISPS must be precisely aligned to each other to ensure optimum beam collection and focusing. The moderator array, electrostatic grid, remoderators, and electrostatic lens are all mounted on a common base. Where necessary, electrical insulation is provided between components with different electrical potentials. These devices will be assembled and aligned "on the bench" prior to installation in the vacuum vessel. After the source has been positioned next to the moderator array, the remoderators can be remotely aligned in the horizontal directions to optimize the beam brightness.

SECTION D—SUMMARY AND CONCLUSIONS

Various positron production concepts have been investigated at the INEL. We believe that NEAR technology can produce $\sim 10^{17}$ fast e^+/s for an ultimate device and $\leq 10^{16}$ fast e^+/s for a single assembly in the INEL ATR. ISPS technology can produce $\leq 10^{15}$ fast e^+/s. We have discussed the "Concept A" ISPS device in this paper. It has

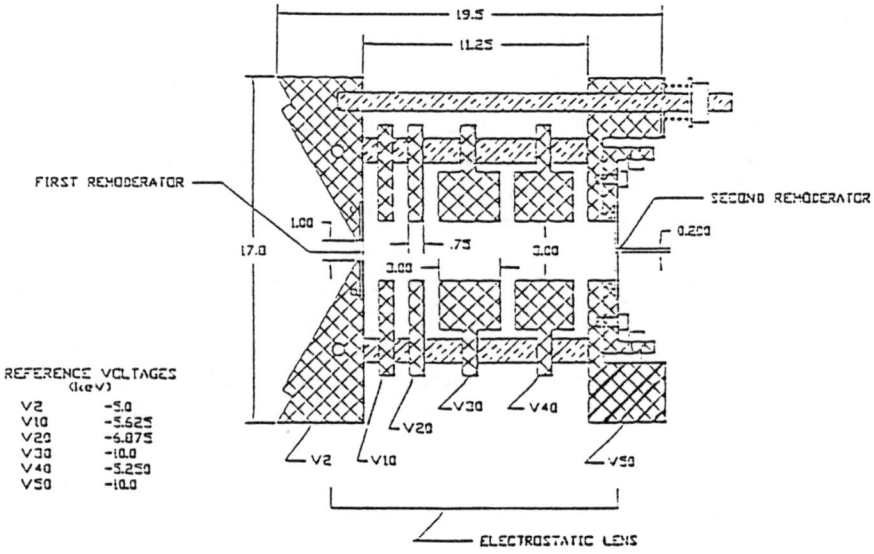

Fig. 16. Remoderator/lens assembly.

the highest theoretical confidence level at this time (utilizing thin crystal moderation or cryogenic Neon). Two other papers in these proceedings discuss our "Concept B" gridless cryogenic device.[8,9] The principle theoretical uncertainty with "Concept B" is the macroscopic physics of the device versus the microscopic physics of Concept A (i.e., gridless versus gridded). A proof-of-principle three-quarter-scale, gridless design experiment utilizing electrons has achieved ~100x beam diameter compression in one stage (large area source to first remoderator surface). Beam transport efficiency is currently under study for this Concept B device.[8]

The INEL believes that the ^{58}Co ISPS is the best first step it can take at the lowest cost to the positron research community. We also believe that the ISPS will have numerous application opportunities in both basic and applied physics (refer to Table 7 for a list of possible applications).

Table 7. Applications of the ISPS.

	National impact
Positron Microscope Facility • U. of Michigan and INTEL Corporation collaboration	Reduce material defects
Microelectronics Fabrication Technology • X-rays • Positrons • INTEL Corporation collaboration	Reduce integrated circuit size
Tunable channeling x-ray sources • Lithography • 10 keV for DOE-DP	Higher x-ray beam intensity Unique
Thin film NDE/NDT solid state R&D with positrons • U. of Michigan and INTEL Corporation collaboration	Density and reliability improvement
e$^+$ Accelerator and storage ring	Unique
e$^+$ Pure plasma and (e$^+$ e$^-$) equal mass plasma research	Basic research
Positronium QED measurements, Bose condensation of Ps, etc.	Basic research
Spin-off technologies • Material science for naval reactors • Aging work for NRC • In-service inspection for ATR	N/A

REFERENCES

1. H. Makowitz, "The NEAR Concept," Patent application submitted to DOE as EGG-PI-356, June 1, 1989, assigned Case Number S-70,066 by DOE, Chicago Office of Patent Council.

2. H. Makowitz, "The NEAR Concept," in *Proceedings of the 1987 Workshop on Intense Positron Beams, INEL, Idaho USA, June 18–19, 1987*, E. H. Ottewitte and W. Kells (eds.), World Scientific Publishing Co. Pte Ltd., pp. 140–143.

3. H. Makowitz, "The NEAR Concept," *Trans. Amer. Nuc. Soc.*, 60, pp. 209–210, 1989.

4. H. Makowitz, "Intense Slow Positron Source (ISPS) Concept," Patent application submitted to DOE as EGG-PI-444, January 9, 1991, assigned Case Number S-73,125 by DOE, Chicago Office of Patent Council.

5. H. Makowitz, "The Intense Slow Positron Source Concept," Presented at the 1992 March APS Meeting, March 16–20, 1992, *Bul. Amer. Phys. Soc.*, 37, No. 1, p. 453, 1992.

6. C. J. Davisson and C. J. Calbick, *The Physical Review*, 38, 2nd Series, July–December 1931, p. 585.

7. C. J. Davisson and C. J. Calbick, *The Physical Review*, 38, 2nd Series, October–December 1932, p. 580.

8. B. L. Brown, H. Makowitz, J. D. Larson, and P. A. Encarnacion, "An Improved Cryogenic Design for the INEL ISPS," Patent application submitted to DOE as EGG-PI-608, July 30, 1992, assigned Case Number S-77,404 by DOE, Chicago Office of Intellectual Property Council.

9. B. L. Brown et al. (to be found in these proceedings).

REPORT ON POSITRON SPECTROSCOPY FOR THE *BESAC* PANEL ON NEUTRON SOURCES

A. P. Mills, Jr.
AT&T Bell Laboratories, Murray Hill, NJ 07974-0636

[*Note to the reader: A meeting of the Walter Kohn BESAC panel on neutron sources requested input from positron users at a meeting held at Argonne, Ill on September 10, 1992. The following report was submitted to the panel and is included in this conference summary as a matter of general interest.*]

ABSTRACT

An unofficial consensus of a recent DOE-sponsored meeting on the prospects for materials studies with positrons is that, but for the present lack of intensity, positron spectroscopy of momentum densities and defects would realize its full potential as a unique and powerful probe of condensed matter. The applications range from the recent measurement of a part of the Fermi surface in $YBa_2Cu_3O_7$ that has not been seen by any other method to quantitative studies of defects in polymers, laminates, reactor materials, interfaces, and semiconductor multilayers. The relation to neutron sources is that the most intense source of positrons can be produced at a nuclear reactor. As intermediate steps it is recommended that there be improved instrumentation and other upgrades at existing facilities and tests of proposed high flux production methods that are relevant to the ANS.

INTRODUCTION

The probes used in the study of condensed matter range from feebly interacting microscopes and diffractometers to strongly perturbative impurities; from susceptibility measurements in weak applied fields to the layer-by-layer or sometimes total destruction of a sample. The complete state of most anything is obviously incomprehensible in its entirety, and various probes give pieces of the whole picture. Compared to neutrons and X-ray photons, positrons being charged tend to interact rather strongly with a sample. Because of the greater amount of theory sometimes needed for interpretation and because of the small source intensities, positron experiments have often been neglected in favor of other methods. Neutrons and photons are widely accepted because they scatter weakly and because the necessary facilities are available. In running a neutron or X-ray experiment, it is not necessary to be an expert in the production and distribution of the probe beam; the very opposite is the case for positron experiments at present.

Historically, a considerable amount of understanding was needed before it was realized that positrons can be the probe of choice for electron momentum density measurements. In addition, laboratory positron sources are not very strong, data rates are low and detectors have been slow to develop. In the 1950's it was demonstrated that the angular correlation of the annihilation radiation (ACAR) was different for different materials [1,2] and contained information about the magnitude and anisotropy of the Fermi momentum of the electrons in a metal. [3] By the 1980's the introduction of 2D detectors [4-6] made it possible to make a detailed comparison between theory and experiment in many interesting materials. And in the 1990's it was possible to exhibit a Fermi surface in a high temperature superconductor by a tour de force experiment: Fig 1 shows the Fermi surface discontinuity running parallel to Γ-X in YBCO. [7] All of these experiments were done using small radioactive positron sources and apparatus costing much less than 1M dollars.

Meanwhile it was found in 1967 that positrons are uniquely sensitive to monovacancies [8] and dislocations. [9] There is now a whole industry associated with such measurements on a variety of materials from polymers to industrial alloys, the result of which fills the voluminous pages of the triennial international positron meetings. [10-16] It is now obvious that an upgrading of the ACAR and defect spectroscopies far beyond the capabilities of a small laboratory could have benefits unusually great in proportion to the costs.

A parallel development has been the gradual emergence of techniques to control positrons in a manner analogous to the neutron moderation discovered in the 1930's. Again, progress has taken place over a time scale measured in decades. By the late 1970's the negative work function mechanism [17] was elucidated and the first efficient slow positron moderators were made. [18] By 1980 we not only knew how to make a high flux (10^{10} e$^+$ s^{-1}) positron beam, [19] but also a great number of interesting effects that would make it very useful had been discovered. [20]

Of course electrons are enormously useful and one may wonder why changing the sign of the charge would make positrons sufficiently different to justify their increased use in materials science. The answer is that the positive charge, freedom from the constraints of the exclusion principle and the existence of the annihilation channel result in positron interactions that are qualitatively different at low energies from those of electrons. Besides the ACAR and defect spectroscopies which could easily be expanded into having enviable spatial and momentum resolution and a high count rate, a number of surface spectroscopies based on positrons are ripe for incorporation into spatially-resolved techniques for imaging samples under a new and unique illumination.

WEAK SCATTERING

A particle that sees only a weak potential in a solid may be scattered to measure the structure factor $S(\mathbf{q},\omega)$ which, for a sufficiently weak potential is directly related to the diagonal part of the one-particle density matrix and the local single particle density $n(\mathbf{r})$: the electronic density if we are using X-rays or a combination of the magnetic moment and nuclear density if we are using neutrons. Thus neutrons and X-rays may be used to find the positions of the atoms in a crystal or the radial distribution function of the molecules of a liquid. For non-zero ω we may learn about the excitations of a material. When we combine the ease of interpretation with a penetration measured in tens of mm or tens of µm for neutrons and X-rays of typical energies it is easy to see why they are so useful.

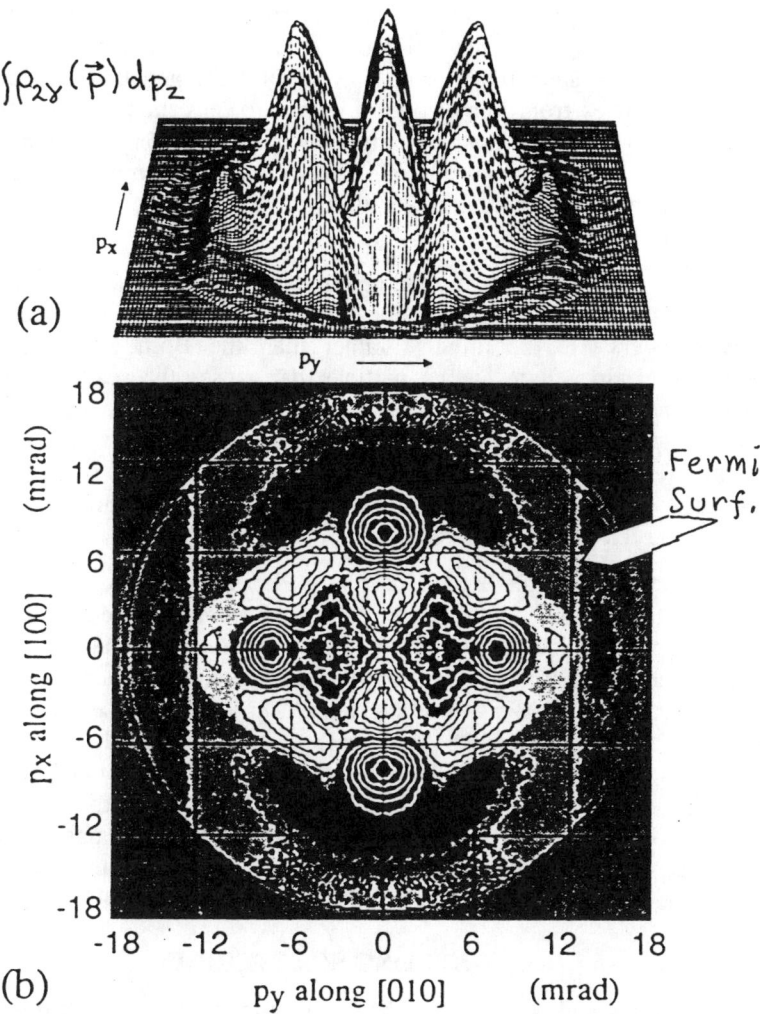

Figure 1. 2D projection on the ab plane of the 3D electron-positron momentum density of untwinned single crystal $YBa_2Cu_3O_7$. (a) Isometric and (b) gray-scale-plus-contour plots of the anisotropy of the final spectrum. [From Ref 7]

There is of course much more to a complete understanding of a material than is contained in the spatial distribution of atoms or in the spectrum of excitations. For a many particle system there is no way to find the actual wave functions of the individual particles from knowledge of $n(\mathbf{r})$ and thus much information is missing even at the single particle level. Completely complementary information is obtained if we measure also the single particle density in momentum space, i.e. the average number of particles with momentum $\mathbf{p}$, $n(\mathbf{p})$, which is the Fourier transform of the one-particle density matrix. [21]

Most neutron and X-ray studies involve small ω and thus give information about crystal structure, spatial densities and fluctuations. On the other hand for large energy transfers $\hbar\omega$ the impulse rather than the Born approximation becomes appropriate. If in a time ω^{-1} a particle that causes the scattering does not itself scatter from its neighbors, one may consider that the probe particle scatters from individual scattering centers rather than from large regions of the sample. For fixed momentum transfer $\mathbf{q}=\mathbf{q}_0$ the scattering intensity $S(\mathbf{q}_0,\omega)$ becomes the Hohenberg-Platzman-Compton profile [22] which is a one-dimensional projection of the single particle momentum density.

A famous example is the Hohenberg-Platzman [22] suggestion to measure the superfluid fraction in liquid He^4. [23] Unfortunately He-He scattering is not negligible so that much theory is needed to account for the final state effects [24] and to extract an estimate for the superfluid fraction. Another famous example is Compton scattering of X-rays which gives a 1D projection of the electron momentum distribution. In this case the final state effects are small, but there are severe limitations to the method since the X-rays scatter typically from all the electrons. [25] The third example is of course two-photon annihilation of positrons in a solid. The momentum of the pair of photons is precisely the center-of-mass momentum of the annihilating pair, and one can measure the 3D electron momentum distribution by studying the angular correlation of annihilation radiation (ACAR).

STRONG SCATTERING

Neutrons and X-rays are rather weakly interacting probes that are essentially free of initial state interaction effects, i.e. there is little perturbation of the sample. As we turn up the strength of the interaction, the probe particle becomes dressed by virtual excitations of the sample and the picture that we get will thus be modified. Turning the interaction up more we can get bound states like polarons and snowballs and the possibility exists for trapping of the probe particle. Contrary to one's first thoughts if one is used to neutrons, the existence of such effects can become an advantage under the appropriate circumstances. Trapping by defects is in fact highly desirable if we are searching for them, but one must be careful to establish that defects are not significant when measuring momentum distributions. In the case of ACAR, the repulsion of the positrons from ion cores causes the positrons to sample mostly the outer electrons, making it possible to study high-Z materials that are difficult to examine with Compton scattering. Fig 2a shows a comparison between the 1D ACAR [26] and Compton profiles [27] of Al; notice that the core contribution is almost negligible as seen by the positrons. Fig 2b shows how one can see the

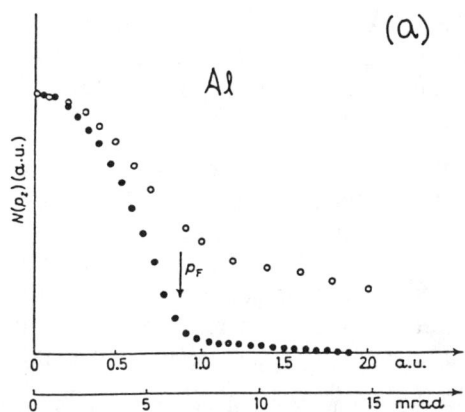

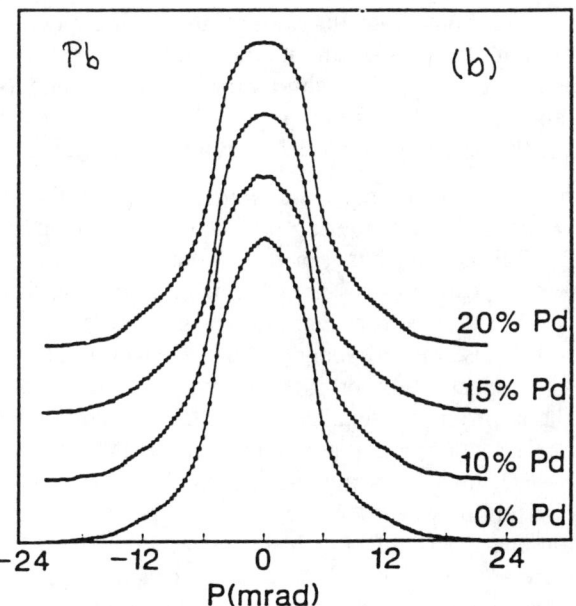

Figure 2. (a) Comparison of Compton-profile (open circles) and 1D ACAR (filled circles) for Al. [From Refs 26 and 27] (b) 2D ACAR of Pb-Pd alloys. [From Ref 28]

Fermi distribution by 2D ACAR in a non-dilute alloy of Pb and Pd. [28] Such a high atomic number material would be impossible to see by Compton scattering and the alloy concentration is totally beyond the capabilities of transport methods like the de Haas-van Alphen effect.

The Coulomb attraction of electrons to the positron causes the electron density in the neighborhood of the positron to be enhanced by as much as an order of magnitude in a metal. Such initial state effects do not change the shape of the momentum distribution very much basically because the attraction of two particles does not change the center-of-mass momentum. Since electron-positron annihilation is short-ranged, final state effects are negligible and the two-photon momentum density $\rho_{2\gamma}(\mathbf{p})$ is proportional to $n(\mathbf{p})$ as modified by the initial state effects.

The existence of the bound state of an electron and a positron, positronium, has been the subject of many studies that could be characterized as studies of the impurity rather than the material. However, the formation of positronium at a surface turns out to be a possible way to remove electrons from near the surface and thereby to measure the electron momentum distribution in a manner complementing angle-resolved photoemission. In this case the interaction that forms the positronium from a thermalized positron in a metal is short ranged and the impulse approximation appears to be valid. [29] Also, surprisingly, final state effects are not serious presumably because the positronium particle leaving the surface is neutral.

Another aspect of a strongly interacting probe is the interplay between multiple scattering and inelastic scattering. A notable example is low energy positron diffraction (LEPD) [30] which is superior to low energy electron diffraction (LEED) for determining the positions of the atoms on the surface of compound semiconductors. [31] The positron inelastic mean free path is shorter than the electron's because the absence of exchange for the positron allows scattering into a larger piece of phase space. Furthermore, the positron phase shifts are not wildly varying because the positrons are repelled from the insides of the atoms. As a result, there is less multiple scattering for positrons and the theory converges to a more accurate structural determination using simpler phase shifts. The upper panel of Fig 3 shows a comparison between the electron and positron measurements of the intensity vs. lepton energy for several diffracted beams from a CdSe surface. The lower panel shows that a measure of the goodness of fit (R_x) of the theory to the experimental data has a much lower minimum vs a structural parameter ω for LEPD than for LEED.

If the inelastic mean free path is much longer than the elastic mean free path then there can arise the phenomenon of dynamical scattering that can be seen with neutrons and X-rays and is at work in the bottling of ultra-cold neutrons. [32] In the rare gas solids positrons with energies below the band gap (see Fig 4a) exhibit the dynamical scattering Darwin top-hat [33,34] which has allowed a sensitive test of the theory of the positron rare gas atom interaction potential. [35] In the same materials, high energy positrons exhibit kinematic diffraction (see Fig 4b) which occurs when the inelastic mean free path is short compared to the elastic one.

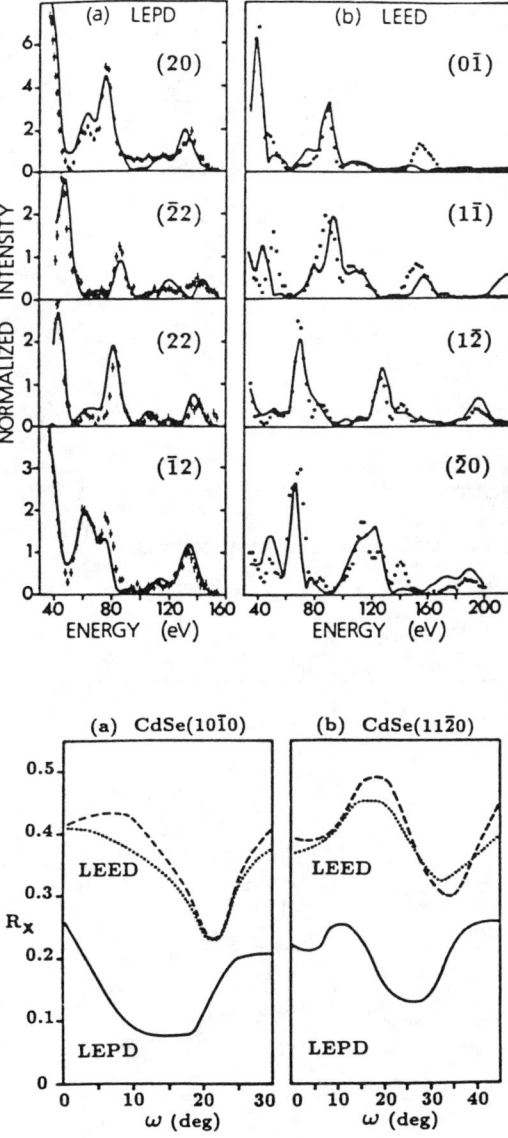

Figure 3. (Upper panel) Data (points) and theory (lines) for four beams diffracted from the CdSe(11-20) surface: (a) low energy positron diffraction (LEPD); (b) low energy electron diffraction (LEED). Intensities are given in absolute reflectivity multiplied by 10^4 in (a) and in arbitrary units in (b). (Lower panel) X-ray R factor R_x as a function of the bond-rotation angle ω for the (a) CdSe(10-10) and (b) CdSe(11-20) surfaces. [From Ref 31]

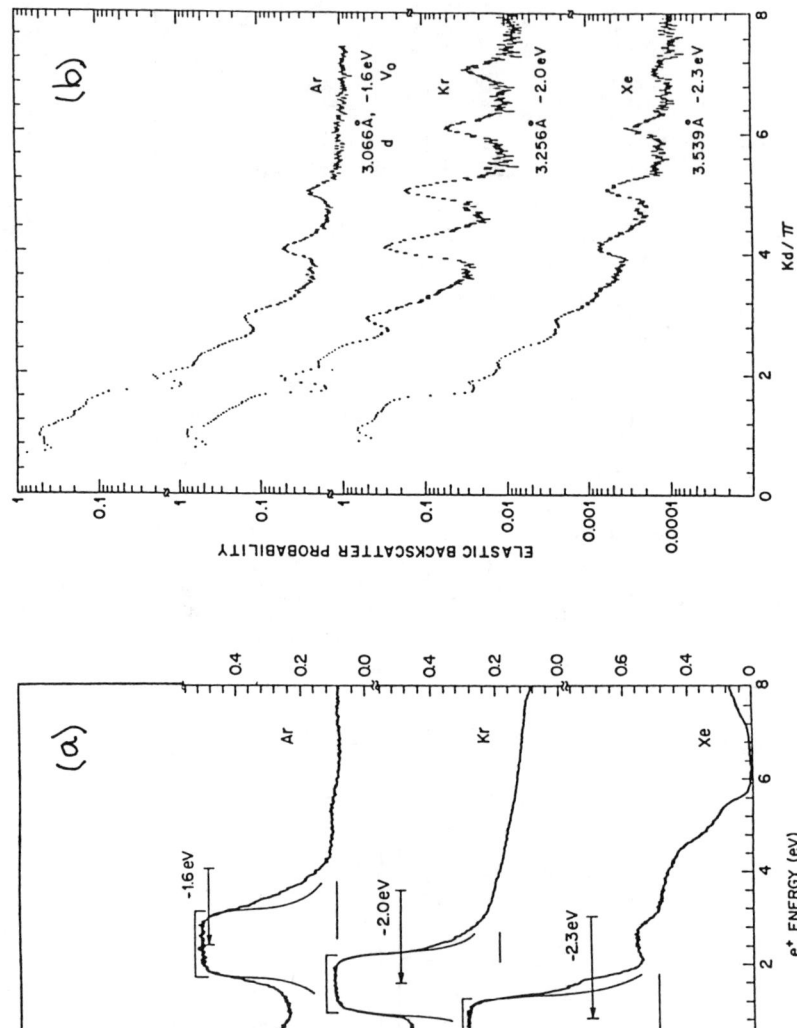

Figure 4. (a) Positron specular reflection probability versus positron energy for the (111) surfaces of Ar, Kr, and Xe. (b) Positron specular reflection intensity for the same surfaces as in (a), but plotted versus kd/π, where $\hbar k$ is the momentum of the positrons in the crystal, and d is the layer spacing of the planes of atoms parallel to the surface. [From Ref 34]

The point to be stressed is that a strongly interacting probe like the positron is not necessarily to be shunned. There follows a few examples of experiments suggested by various people that would be useful for materials science and would be made possible by a high flux of positrons. There are also several fundamental atomic physics and many-positron experiments that could be done as a spin-off from an intense positron facility. We should of course remember that, as in other fields, part of the impetus for positron innovations has been the pursuit of fundamental or academic questions. [36] Some of this type of work should continue as an investment in the future of materials science.

NEW POSITRON EXPERIMENTS

There are basically two ways to make an intense slow positron beam.

1) Electron bremmstrahlung sources (e.g. the 0.5 MW CEBAF FEL beam dump or neutron capture gamma rays from Cd in a reactor) are able to produce [37] on the order of 10^{10} e^+ s^{-1} in a 0.2 mm diameter source spot.

2) There are several practical radioactive positron sources that could be produced in a nuclear reactor and would yield more than 10^{11} e^+ s^{-1} from a 0.2 mm diameter source spot. Beta decay positrons are spin polarized and would therefore find a number of applications in the study of magnetic materials. The possibilities include a) Cu^{64} (in use at Brookhaven National Laboratory), b) Co^{58} (being considered for INEL), and c) Kr^{79} (being considered for ANS).

1. Super ACAR

The case for obtaining a pair of high efficiency position-sensitive cameras for doing 2D ACAR with an intense positron micro-beam is obvious. [38] One would be able to study samples that are only available in the form of thin films or tiny single crystals and one would be able to study fine details of the electron momentum distribution in a reasonable time. Figs 5a and 5b show cuts through the full reconstructed 3D pair momentum distribution in diamond and copper. [39,40] It is easy to see the discontinuity in the momentum distribution at the Fermi radius in Cu, broadened by the 0.1 a.u. = 0.7 mrad resolution of the apparatus. The distribution at the zone face in diamond, the Jones zone being filled, is broadened due to the large energy gap. However, there are many subtle effects that one would like to look for in electron momentum distributions. Take for example the theoretical predictions for the momentum density in YBCO in Fig 6. [41] There are many Fermi breaks on the order of 1% of the total distribution that would have to be resolved with 0.1 mrad = 0.014 a.u. to distinguish them from possible band-structure effects and convincingly identify them as discontinuities. Given the momentum of a thermalized positron in a solid at say 10 K, a momentum resolution of 0.01 a.u. is possible. We would need about 10^5 counts per pixel, about 10^{10} total counts per spectrum and 100 spectra at different angles of rotation of a single crystal sample to fully extract the information hinted in Fig 6. With a 10^{11} e^+ s^{-1} beam and an efficient high resolution detector we would be able to obtain the needed 10^{12} total counts in less than one day.

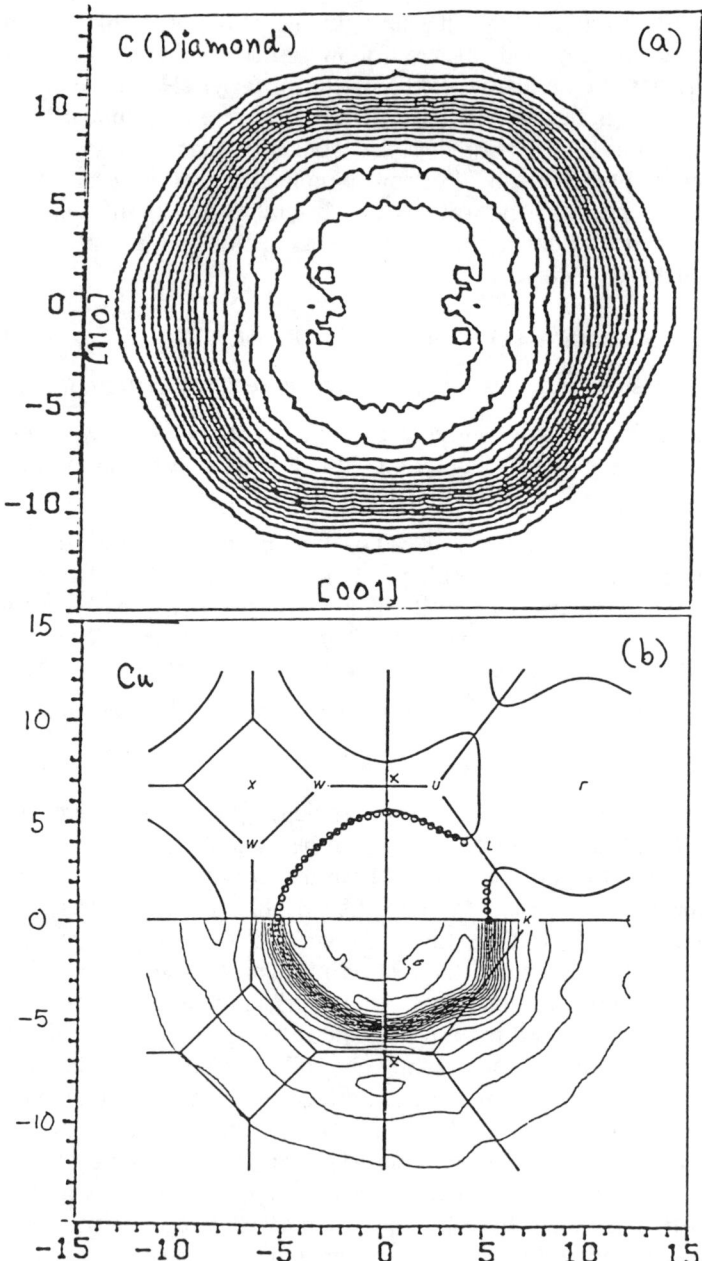

Figure 5. 3D-reconstructed ACAR momentum densities on a plane through $k=0$. (a) diamond, showing the full Jones zone and a broadening of the sharp cut-off of the momentum density at the zone face due to the electron energy gap. (b) Cu, data and theory for two orientations of the plane of the measurements. [From Refs 39 and 40]

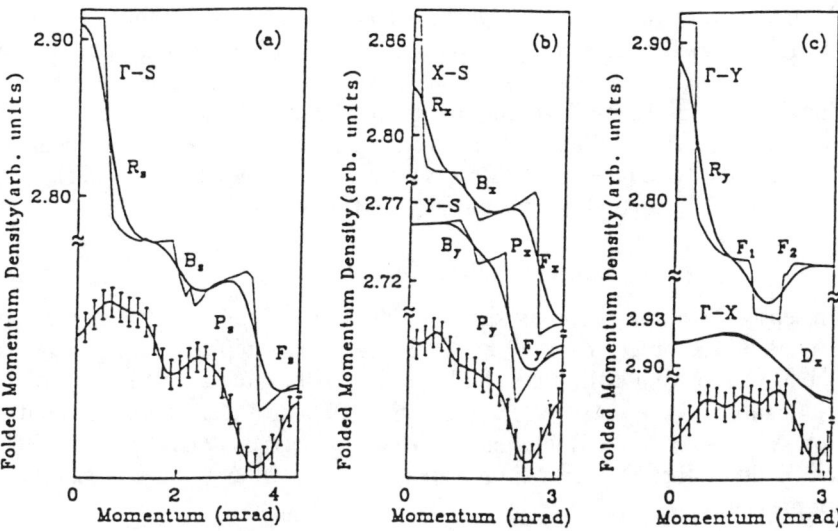

Figure 6. Calculated and measured projections on the ab plane of the Lock-Crisp-West (LCW) transformed electron-positron momentum densities in $YBa_2Cu_3O_7$. [From Ref 41]

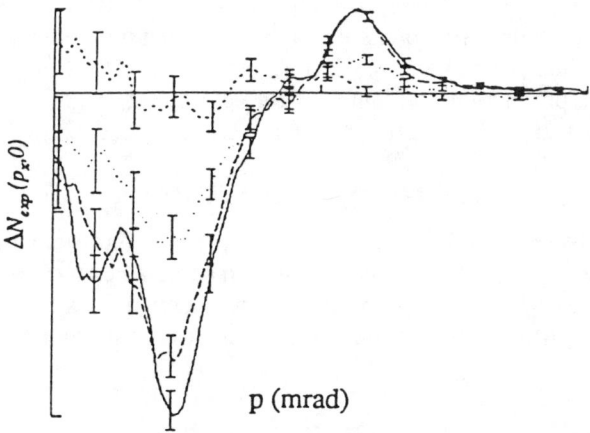

Figure 7. Relative spin density of Ni in the (100) plane along the [110] direction at different temperatures. Solid line, 4.2 K; long-dashed line, 300K; dotted line, 600K; short-dashed line, 660 K. [From Ref 42]

An example of the ability of the 2D ACAR method to see spin densities is shown in Fig 7. [42] It is clear that much could be learned about magnetic systems with better resolution and statistics. A further example of the usefulness of positrons in studying the electron momentum distribution in alloys is the study of LiMg alloys shown in Fig 8. [43] The addition of Mg to Li causes the Fermi radius to expand and to break through the zone face at a 28% concentration. The Fermi surface necks, seen for the first time in a bcc metal, are visible with text-book clarity.

2. Scanning defect microprobe

Presently positron lifetimes are measured using μCi Na^{22} sources and counting rates on the order of 10^2 s^{-1}. As Schrader has emphasized [44] one may actually obtain a spectrum of lifetimes by Laplace-transforming a time spectrum. Fig 9 shows a beautiful set of data by Jean [45] on a polymer under various amounts of applied pressure. The three lifetimes are easily resolved. The longest lifetime component is due to triplet positronium trapped in open volume defects that change size under pressure.

It is also possible to obtain spatially-resolved lifetime data by simply moving a small source about on the surface of a slice of crystal. [46] The positron lifetime in the CdHgTe samples of Fig 10 is sensitive to the presence of a vacancy concentration of around 10^{16} cm^{-3} or more. The ≈10-pixel image of Fig 10 was probably obtained by counting for about 1 d. It is not hard to see that increasing the resolution to the order of 1 μm and decreasing the counting time for a complete picture of a 100 mm wafer in a few minutes would be enormously useful. Let us suppose that the hypothetical 10^{11} e^+ s^{-1} ANS positron beam were brightness-enhanced and bunched to a 2 μm spot with pulses of 10^4 e^+ at a repetition rate of 10^6 s^{-1}. We could use an efficient detector to count 10^3 counts per pulse and obtain a complete million count lifetime spectrum in 1 ms and an image of 1 cm^2 in 6 h. This is too slow to look at every sample on a production line, but it could be a useful diagnostic. It is clear that one would always want more positrons for this type of microprobe.

3. Low energy positron diffraction

As currently practiced, a LEPD study of a compound semiconductor surface (Fig 3) requires several months of data acquisition using a beam of about 10^5 e^+ s^{-1}. [31] A meer factor of 10^3 increase in intensity would enable us to do much better in less than an hour. More intensity might make diffuse scattering and time-resolved experiments possible.

4. Other surface spectroscopies

Low energy Positronium (Ps) diffraction could be useful in studying surface contaminants and H in particular. [47] The full capabilities of a high intensity positron beam would be needed to make this technique useful.

Single photon annihilation in which a single photon is emitted and a nearby electron recoils to conserve energy and momentum [48] should have a branching ratio compared to 2γ annihilation of $R_{1\gamma} \approx 10^{-10}$. It would be most interesting to measure

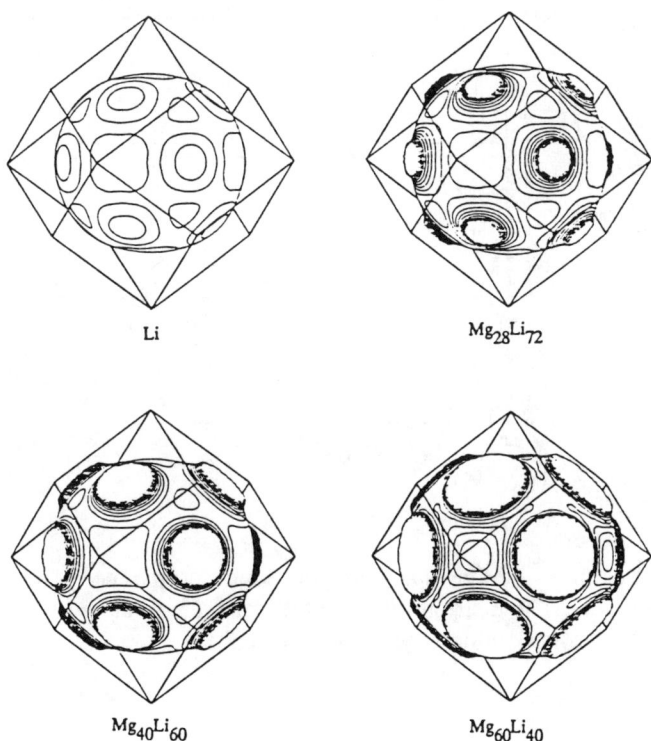

Figure 8. Fermi surfaces of pure Li and various LiMg alloys reconstructed in 3D from 2D ACAR measurements of the electron-positron momentum density. The first Brillouin zone is indicated. For all the alloys the Fermi surface is in contact with the Brillouin zone along the [110] direction. [From Ref 43]

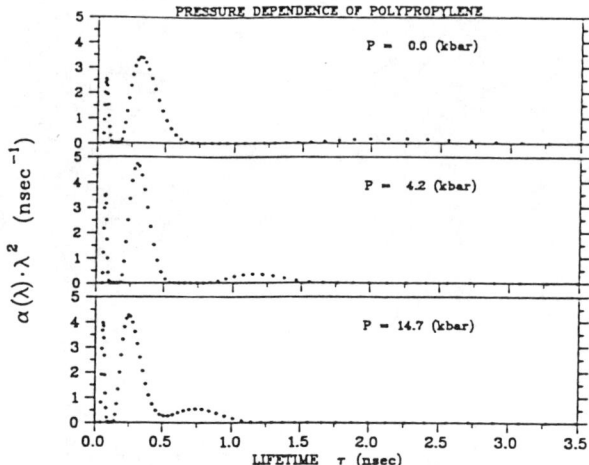

Figure 9. Positron lifetime *spectrum* of a polypropylene polymer under quasi-isotropic pressure. The distributions of lifetimes were obtained by Laplace-inversion of the decay time distribution spectra. [From Ref 45]

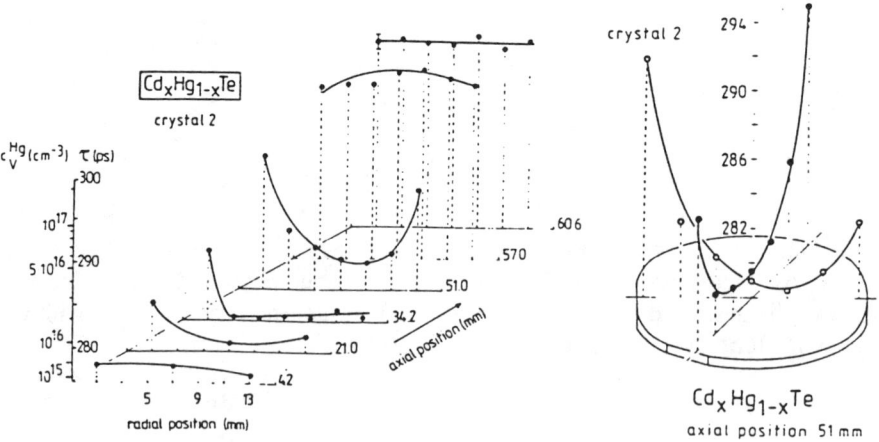

Figure 10. Spatially-resolved mean positron lifetime at various points in a HgCdTe single crystal. On the left, the lifetime is plotted versus radius for six wafers cut from the crystal. The right part of the figure shows two scans across the wafer with the position x=51 mm. [From Ref 46]

$R_{1\gamma}$ in various solids to determine something about the two-particle density matrix.

Annihilation-in-flight spectroscopy of high energy monoenergetic positrons with core electrons would be a probe similar to X-ray fluorescence for telling what elements are present.

The implantation of spin-polarized positrons and the measurement of their polarization after reemission or of the polarization of positronium formed at the surface could tell something about spin relaxation rates.

Positron induced Auger spectroscopy [49] uses the annihilation of core electrons by positrons rather than electron impact to Auger electrons for analyzing the constituents of surfaces. There is no background from secondary electrons and the positron seeks out the last layer of atoms, thus making the information complementary to that obtained using the electron-induced Auger effect. The extension of the method to time-dependent studies and polarization effects again requires an enormous increase in positron intensity.

Certain samples have a negative affinity for positrons and can be imaged with high resolution using the positron remission microscope. [50-52] The first experiments showed a sensitivity to defects in a thin Ni foil but the exposure times were many hours. With a high intensity beam we could have 1 nm resolution in a reasonable exposure time. An exciting possibility is that a Ni foil could be used as a microscope slide to examine biological specimens such as viruses. It is expected that the 1 eV reemitted positrons will cause much less radiation damage per unit of contrast that an electron microscope and will not require shadowing with metal.

5. Fundamental positron experiments

There are numerous experiments on positron plasmas, positron-induced ionization, positron-atom or molecule scattering cross sections, and positronium itself that would benefit from a high flux and good spatial resolution. At very high fluxes many-positron effects will occur in a solid when the e^+ Fermi energy becomes comparable to kT. A high intensity beam of positrons could also be useful in exploring such effects as the positronic superconductor and the intensity-correlation positron reemission microscope. Effects external to a solid such as Ps jets, Ps liquids, the Ps_2 molecule and its liquid would be possible.

RECOMMENDATIONS

Positrons have qualities that make them an excellent probe of momentum densities and defects. The present state of theory and experiment is ripe for an expansion of the positron facilities for materials studies. Recommendations that represent the opinions of the positron community are as follows.
1) As a first step we need an immediate upgrading of instruments and facilities. Newly developed instruments could be moved to better facilities when they become available. Specifically, a high resolution 2D camera pair is needed for momentum density measurements at the soon-to-be-commissioned Brookhaven HFBR positron microbeam. A high data rate scanning defect microprobe should be built at the

Livermore and/or Oak Ridge LINAC positron beams.

2) New methods of producing the highest fluxes of positrons should be tested and evaluated. Specifically, the rare gas Kr^{79} loop, favored by ANS designers because of its possible engineering simplicity, should be tested at the HIFR at Oak Ridge. An alternative reactor beam that uses fast neutrons to make Co^{58} could be tested at INEL. If the tests were favorable, the test beams would be excellent interim sources of positrons.

3) If the ANS is built, then it would definitely be a distinct advantage to incorporate an intense slow positron beam in the design. It is felt that specific recommendations about the exact design and whether for example there should be more than one facility will require a study that would include experts on costs and engineering as well as physicists.

REFERENCES

[1] S. DeBenedetti, C. E. Cowan, W. R. Konneker and H. Primakoff, *Phys. Rev.* **77**, 205 (1950).

[2] A. T. Stewart, *Phys. Rev.* **99**, 594 (1955).

[3] S. Berko and J. S. Plaskett, *Phys. Rev.* **109**, 399 (1958).

[4] K. Fujiwara and O. Sueoka, *J. Phys. Soc. Japan* **21**, 1947 (1966).

[5] S. Berko and J. Mader, *Appl. Phys.* **5**, 287 (1975).

[6] J. Mayers, J. D. McGervey, P. A. Walters and R. N. West, in *Positron Annihilation*, edited by R. R. Hasiguti and K. Fujiwara (Sendai, 1979) p. 417.

[7] H. Haghighi, et al., *Phys. Rev. Lett.* **67**, 382 (1991).

[8] I. K. MacKenzie, T. L. Khoo, A. B. MacDonald and B. T. A. McKee, *Phys. Rev. Lett.* **19**, 946 (1967).

[9] S. Berko and J. C. Erskine, *Phys. Rev. Lett.* **19**, 307 (1967).

[10] *Positron Annihilation*, edited by A. T. Stewart and L. O. Roellig (Academic, New York, 1967).

[11] *Positron Annihilation*, edited by G. Trumpy (unpublished, Helsingor, Denmark, 1976). p. 417.

[12] *Positron Annihilation*, edited by R. R. Hasiguti and K. Fujiwara (Sendai, 1979) p. 417.

[13] *Positron Annihilation*, edited by P. G. Coleman, S. C. Sharma and L. M. Diana (North-Holland, Amsterdam, 1982).

[14] *Positron Annihilation*, edited by P. C. Jain, R. M. Singru and K. P. Gopinathan (World Scientific, Singapore, 1985).

[15] *Positron Annihilation*, edited by L. Dorikens-Vanpraet, M. Dorikens and D. Segers (World Scientific, Singapore, 1989).

[16] *Positron Annihilation*, edited by Zs. Kajcsos and Cs. Szeles (Trans Tech, Switzerland, 1992).

[17] B. Y. Tong, *Phys. Rev.* **B5**, 1436 (1972).

[18] A. P. Mills, Jr., P. M. Platzman and B. L. Brown, *Phys. Rev. Lett.* **41**, 1076 (1978).

[19] A. P. Mills, Jr., Appl. Phys. **23**, 189 (1980).

[20] P. J. Schultz and K. G. Lynn, *Rev. Mod. Phys.* **60**, 701 (1988).

[21] *Momentum Distributions*, edited by R. N. Silver and P. E. Sokol (Plenum, New York, 1989).

[22] P. C. Hohenberg and P. M. Platzman, *Phys. Rev.* **152**, 198 (1966).

[23] H. A. Mook, R. Scherm, and M. K. Wilkinson, *Phys. Rev. A* **6**, 2268 (1972).

[24] See H. A. Mook in *Momentum Distributions*, Ref 21, p 159; P. E. Sokol, T. R. Sosnick, and W. M. Snow, *ibid*, p 139.

[25] P. M. Platzman in *Momentum Distributions*, Ref 21, p. 249.

[26] P. Hautojärvi, *Solid State Commun.* **11**, 1049 (1972).

[27] P. Pattison, S. Manninen, J. Felsteiner and M. Cooper, *Philos. Mag.* **30**, 973 (1974).

[28] L. C. Smedskjaer et al., *Phys. Rev. Lett.* **59**, 2479 (1987).

[29] A. P. Mills, Jr., L. Pfeiffer and P. M. Platzman, *Phys. Rev. Lett.* **51**, 1085 (1983).

[30] I. J. Rosenberg, A. H. Weiss and K. F. Canter, *Phys. Rev. Lett.* **44**, 1139 (1980).

[31] T. N. Horsky et al., *Phys. Rev. Lett.* **62**, 1876 (1989).

[32] R. Golub, D. J. Richardson and S. K. Lamoreaux, *Ultra-Cold Neutrons* (Adam Hilger, Bristol, 1991).

[33] C. G. Darwin, *Philos. Mag.* **27**, 675 (1914).

[34] E. M. Gullikson, A. P. Mills, Jr. and E. G. McRae, *Phys. Rev. B***37**, 588 (1988).

[35] M. J. Puska and R. M. Nieminen, *Phys. Rev. B***46**, 1278 (1992).

[36] K. F. Canter, A. P. Mills, Jr. and S. Berko, *Phys. Rev. Lett.* **34**, 177 (1975).

[37] R. H. Howell, R. A. Alverez and M. Stanek, *Appl. Phys. Lett.* **40**, 751 (1982).

[38] A. P. Mills, Jr., *J. Phys. Chem. Solids* **52**, 1589 (1991).

[39] W. Liu, S. Berko and A. P. Mills, Jr., in Ref 16, p. 743.

[40] S. Berko, in *Positron Solid State Physics*, edited by W. Brandt and A. Dupasquier (North-Holland, Amsterdam, 1983) p. 64.

[41] A. Bansil, P. E. Mijnarends and L. C. Smedskjaer, *Phys. Rev. B* **43**, 3667 (1991).

[42] P. Genoud, A. A. Manuel, E. Walker and M. Peter, in Ref 16, p. 639.

[43] W. Triftshäuser, A. Eckert, G. Kögel and P. Sperr, in Ref 16, p. 501.

[44] D. M. Schrader, in Ref 13, p. 912.

[45] Y. C. Jean, in Ref 16, p. 309.

[46] R. Krause et al., in Ref 16, p. 333.

[47] M. H. Weber et al., *Phys. Rev. Lett.* **61**, 2542 (1988).

[48] K. G. Lynn, D. N. Lowy and I. K MacKenzie, *J. Phys. C: Solid State Phys.* **13**, 919 (1980).

[49] A. Weiss, in Ref 16, p. 511.

[50] L. D. Hulett, J. M. Dale and S. Pendyala, *Mater. Sci. Forum* **2**, 133 (1984).

[51] J. Van House and A. Rich, *Phys. Rev. Lett.* **61**, 488 (1988).

[52] G. R. Brandes, K. F. Canter and A. P. Mills, Jr., *Phys. Rev. Lett.* **61**, 492 (1988).

THE DESIGN OF A NUCLEAR-REACTOR-BASED POSITRON BEAM FOR MATERIALS ANALYSIS

A. van Veen, H. Schut, and P.E. Mijnarends
Interfaculty Reactor Institute, Delft University of Technology,
Mekelweg 15, NL-2629 JB Delft, The Netherlands

L. Seijbel and P. Kruit
Department of Applied Physics, Delft University of Technology,
Lorentzweg 1, NL-2628 CJ Delft, The Netherlands

ABSTRACT

At the research reactor of the Delft University of Technology a positron beam facility is being developed which is designed to deliver an intense beam of mono-energetic positrons ($> 10^8$ e$^+$ s^{-1}) and will be used for defect studies in materials by positron microbeam analysis and by angular correlation measurements. The design of the facility, due to be in operation in 1993, is presented and discussed.

1. INTRODUCTION

Slow positron beams form a central tool in many studies in atomic physics[1] and materials science.[2] In recent years the intensities of the beams could be drastically improved by the application of better techniques for the moderation of fast positrons emerging from radioactive sources or generated by means of linear accelerators. Beams based on ^{22}Na and ^{58}Co sources and tungsten moderators are able to achieve intensities of the order of 10^5 - 10^6 e$^+$ s^{-1}. Positron beams with two orders of magnitude higher intensity can be created by LINAC-systems.[3] However, LINAC's produce pulsed positron beams, which is a disadvantage in applications requiring a continuous beam. Therefore, methods are being explored to stretch the pulses in time so that a pseudo-DC beam is obtained. At Brookhaven National Laboratory a beam exists which is based on the production of ^{64}Cu isotopes in the High Flux Beam Reactor. Intense beams of 10^7 e$^+$ s^{-1} and higher have been obtained on a regular basis.

Recently, several new plans have been developed to create intense positron beams. The impetus has been provided by the need for intense beams for positron microscopical research and for the application of angular correlation (ACAR) techniques to thin films and small crystals from e.g. high-T$_c$ superconductors.[4]

A few years ago plans were initiated at the Delft University of Technology to use the research reactor for generating an intense positron beam. Continuously activated copper foils positioned close to the reactor core would act as the primary positron source.[5] The basic idea was to neutron activate ^{64}Cu in copper foils and to use self-moderation in Cu or moderation in tungsten to generate slow positrons. Due to the moderate flux of the reactor at Delft a source had to be constructed with a large slow-positron emitting surface area. Along similar lines, but using pair formation by high-energy gamma quanta created by neutron capture in cadmium as the basic positron generating principle, Triftshäuser et al. drafted a proposal for a positron beam for the ILL high flux reactor at Grenoble.[6,7]

In the following the status of the Delft design will be described. In Sec. 2 an overview of the apparatus will be given. Section 3 discusses the source and presents estimates of the expected positron yields. The positron beam and its auxiliary equipment is discussed in Sec. 4, while Sec. 5 gives a brief description of the facilities to be added to the beam. The problem of radiation damage in the source is addressed in Sec. 6, and Sec. 7 lists the conclusions.

2. SETUP OF THE POSITRON BEAM FACILITY

The positron facility, called POSH (POSitrons at the HOR-reactor), is shown schematically in Fig. 1. Figure 2 shows two photographs of the part of the beam line outside the biological shield of the reactor. The facility consists of the following parts:

A: The vacuum tube which is introduced into an existing experimental channel of the reactor. The tube, made of aluminum with a diameter of 16 cm, houses the source configuration (at the front end, closest to the reactor core, for details see Fig. 3) and the positron-beam magnetic guiding system. The slow positrons emerging from the source surface are focused and accelerated to an energy of 5 keV. The source and the internal beam guiding tube are at a floating potential (maximum 40 kV) so that at the end the positrons can be post-accelerated or decelerated to the desired energy before hitting a target at ground potential.

B: Deflection part of the positron beam. In this section the positrons traveling with a transport energy of 5 keV are bent over an angle of 19° to allow the placement of a neutron beam stop in the line of sight from the reactor core.

C: A section made of stainless steel for further guiding and bending of the beam through the shielding material on its way to the experimental area. This section provides room for beam monitoring via a microchannel plate and a phosphorus screen device equipped with a camera.

D: A small blockhouse for biological shielding of neutron and gamma radiation leaking through the primary beam channel.

E: Experimental area where the primary positron beam is available for positron experiments, e.g., positron micro-analysis and two-dimensional angular correlation.

The total beam system is constructed according to Ultra High Vacuum standards and will be pumped by internal gettering material (SAES gettering

356 Nuclear-Reactor-Based Positron Beam

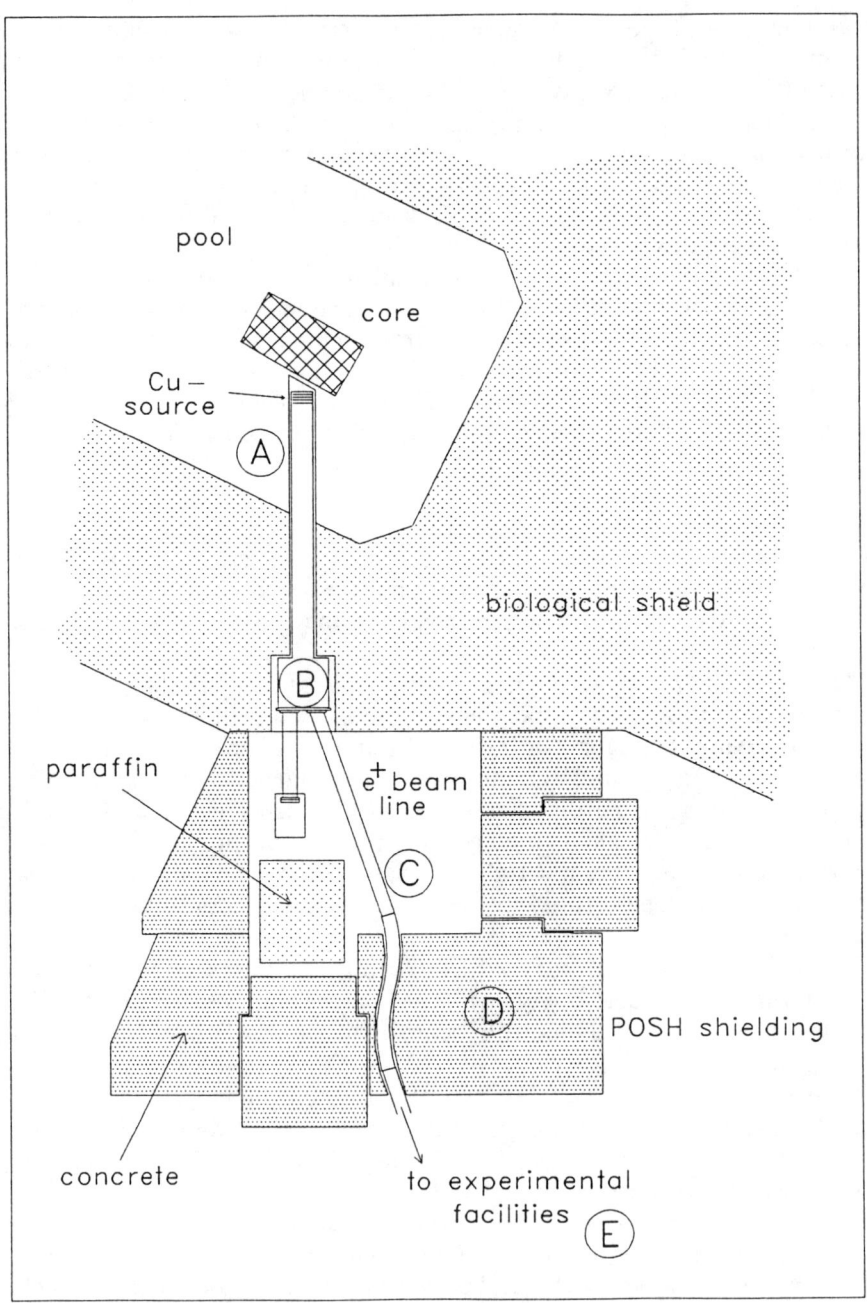

Figure 1. Schematic diagram of the Delft positron beam facility at the HOR-reactor. For explanation of the sections A through E see text.

ribbons mounted inside section A), ion gettering pumps, and titanium sublimation pumps. At the end of section C a turbo-molecular pump will be installed.

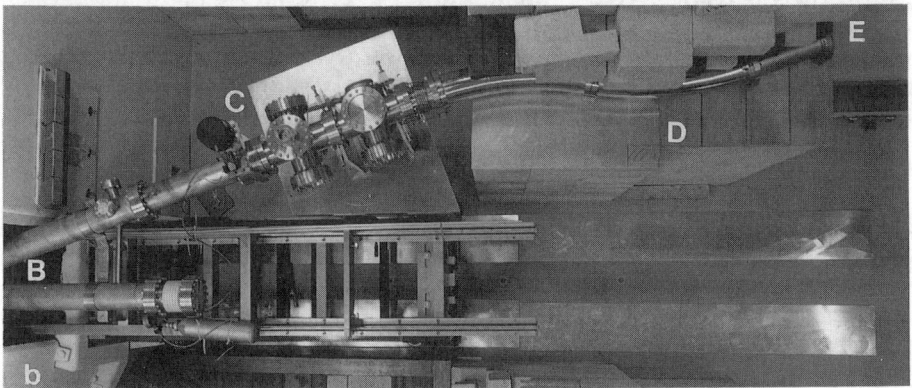

Fig. 2 a) Side view, b) top view of the positron beam line. B, C, and D label the different sections.

3. THE POSITRON SOURCE

3.1. Activation of copper

When copper is chosen as the positron generating material via the reaction $^{63}Cu(n,\gamma)^{64}Cu$ (halflife 12.8 h), a maximum positron activity per unit of volume $I = 1.8 \times 10^{11}$ e^+ cm^{-3}s^{-1} is attained after approximately 48 h. This figure is based on the use of natural Cu (69% ^{63}Cu) and a thermal neutron flux of 8×10^{12} n cm^{-2}s^{-1}. The yield of slow positrons from a copper surface is derived with the aid of a yield factor Y, which represents the number of re-emitted slow positrons per positron arriving at the Cu surface by diffusion, and the diffusion length $L=(D\tau)^{1/2}$, where D is the diffusion coefficient and τ

the positron lifetime in Cu. With Y=0.55 and L=104 nm (for defect-free Cu) one finds

$$e^+ \text{ yield} = Y \times I \times L = 1.03 \times 10^6 \text{ cm}^{-2}\text{s}^{-1}. \quad (1)$$

It is clear that for a high e$^+$ yield a large emitting surface area is required. Assuming an area of 1000 cm^3, a total yield of 1.03×10^9 should be feasible.

The above calculation assumes that the concentration of positrons is uniform throughout the source and that only positrons generated within a distance L from the surface contribute to the yield. Comparison with more exact yield calculations[8] shows that yields are predicted with 20% accuracy. A design requirement following from maximizing the yield is that the foils should be thicker than the mean penetration range (30 μm) of the fast positrons emitted by ^{64}Cu, or else a part of the generated positrons will leak from the source region without contributing to the positron concentration in the foils.

The copper surfaces can be covered with thin foils of tungsten so that one profits from the better moderating properties of tungsten in case the Cu surfaces become covered with impurities as a result of detoriation of the vacuum. If 3 μm tungsten foils are used the yield will only drop to 80 % of the value for uncovered copper. Furthermore, the extraction geometry should be optimized to extract the positrons with high efficiency from the source area.

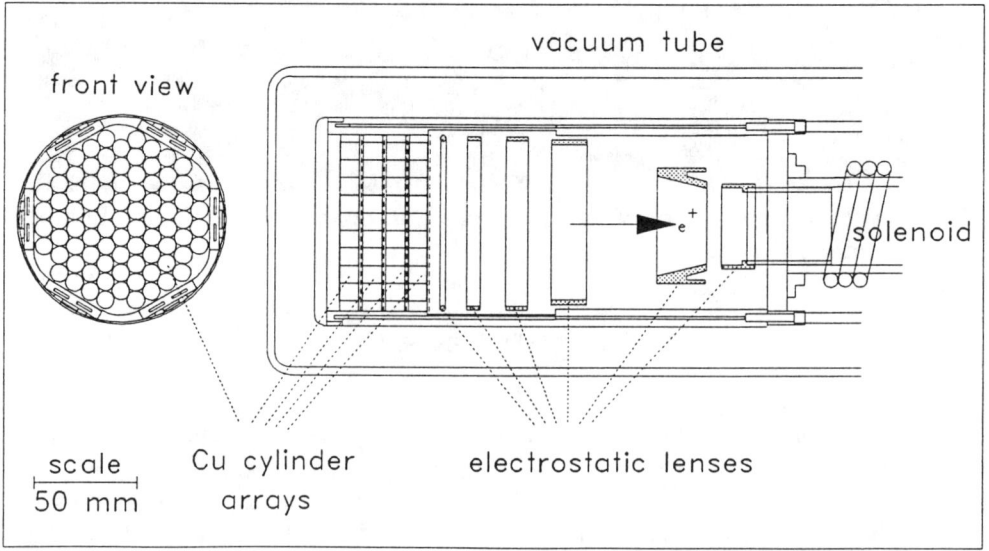

Figure 3. Details of the POSH source configuration (front and side view)

3.2. Pair formation

The flux of energetic γ radiation in the vicinity of the reactor core of the HOR is estimated to be of the order of 3.4×10^{12} cm^{-2}s^{-1} for γ energies above 2 MeV. Taking into account the attenuation of the flux in the source region, the average positron activity in the source material is calculated to be 2×10^{11} e$^+$ cm^{-3}s^{-1} when tungsten is used and ten times less when copper is used as source material. If Cd foil is positioned in front of the source a total absorption of the thermal neutrons can be achieved, yielding γ fluxes a factor three higher than without the cadmium. The resulting positron activities in tungsten will then be 6×10^{11} cm^{-3}s^{-1}. Thus, Cu activation and pair formation yield activities of a similar magnitude.

3.3. Extraction efficiency

In Fig. 4 results of trajectory calculations are shown for positrons emitted by the surfaces of the small cylinders which form the elements of the source array. On the basis of these calculations estimates have been made of the transmission efficiencies of positrons emitted from the back surface and from the cylinder surfaces 1 through 4. These estimates have been listed in Table I. It follows that the overall efficiency amounts to 67%. Thus, from a total positron emitting surface area of 1500 cm^2, 1000 cm^2 is effectively delivering positrons at the focusing part of the source. The focusing part is shown in Fig. 5. Positrons emerging from the last cylindrical section can be brought together in a beam with a waist of 2.2 cm diameter. Note that the positrons consist of different energy groups, since the four discs with cylinders are at different potentials. Subsequently, the positrons are injected into the magnetic guiding system consisting of a long solenoid generating a 0.02 T axial field. During

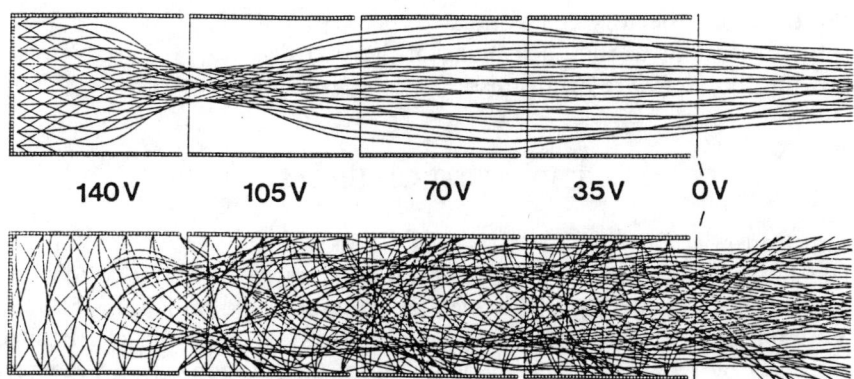

Figure 4. Calculated trajectories for positrons emitted by the surfaces of the small copper cylinders. Top: trajectories of positrons emitted by the back surface; bottom: trajectories of positrons emitted by the cylindrical surfaces. The potentials of the four disks and the end grid are shown.

focusing some additional 20% of the beam is lost. The transmission of the beam through the solenoid will be close to 100%.

The positron current expected at the exit of the solenoid will be of the order of 1×10^9 e$^+$ s^{-1}. The low brightness of this beam must be enhanced by remoderation in a tungsten or nickel foil. After remoderation we expect to have a parallel beam of 1 cm diameter with a five times reduced intensity, i.e., 2×10^8 e$^+$s^{-1}.

Table I Extraction efficiency of the positron source

transmission[a] through disk j of positrons originating in disk i[b]						effective surface area[c]	
						3	4 disks
i=	0	1	2	3	4		
j= 0	1.	0.91	0.91	0.91	0.91	71 (52)	71 (46)
1		0.96	0.56	0.56	0.56	198 (144)	198 (130)
2			0.96	0.56	0.56	198 (160)	198 (144)
3				0.96	0.56	338 (304)	198 (160)
4					0.96		338 (304)
		effective area				805 (660)	1003 (784) cm^2
		area				1134	1486 cm^2
		transmission				0.71 (0.58)	0.67 (0.53)
		intensity (10^8 e$^+$s^{-1})				8.8 (7.2)	11. (8.7)

a) on the basis of trajectory calculations
b) i, j =0 corresponds to the back plane of the first disk
c) figures for 100 % (90 %) transmission of the grids between the disks

4. THE POSITRON BEAM

4.1. Beam transport

The beam will be transported at a kinetic energy of about 5 keV in an electrically floating internal tube which is kept at a potential of -5kV with respect to the source. The source potential will be at most 45 kV. Positrons will be guided by axial magnetic fields of about 0.02 T generated by external solenoids and Helmholtz coils. The first remoderation to reduce the beam size will take place at the point at which the beam tube leaves the concrete blockhouse. After this remoderation the maximum beam energy will be 40 keV. Remoderation occurs in a zone where the magnetic field has been reduced to nearly zero. Further transport to the micro-analysis setup will be performed electrostatically. For transport over longer distances, e.g., to serve the 2D ACAR equipment, the beam will be re-introduced into a magnetic guiding system. At two locations a beam monitor will be installed to observe the beam size and beam quality. A chevron assembly of microchannel plates and a phosphor screen can be moved into a beam section. Via a mirror and a video camera the cross section of the beam can then be recorded.

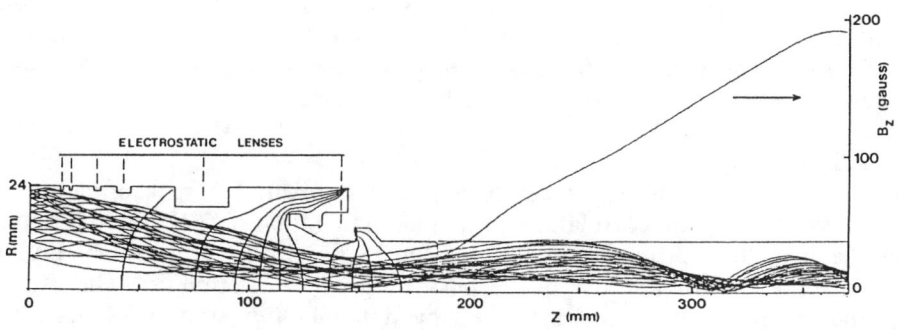

Figure 5. Calculated trajectories for positrons entering the focusing part of the positron source. The positrons move from the left to the right where they are injected into the magnetic guiding system.

4.2. Radiation shielding aspects

One of the problems encountered in the design of the POSH facility concerns the leakage of neutron and gamma radiation through the straight cylindrical duct in the shielding wall of the nuclear reactor. Without any measures taken, the primary neutron and γ flux densities just outside the biological shield would reach levels of the order of 10^8 n cm^{-2}s^{-1} and 10^9 γ cm^{-2}s^{-1}, respectively, at a reactor power of 2 MW thermal. In addition, one is faced with the production of secondary gammas due to the capture of thermal neutrons in the POSH construction materials such as flanges and other vacuum components. Capture γ rays emerge also from boron-doped paraffin used for thermalization and absorption of neutrons. Several computer codes such as neutron transport codes, point-kernel gamma shielding codes, and neutron activation codes have been used to design a shielding geometry capable of reducing the neutron and gamma dose rates below a level of 5 μSv/h. This has resulted in a concrete bunker (thickness of walls and ceiling approx. 1 m) adjacent to the biological shield of the reactor. Inside this bunker the positron beam is bent horizontally over 19° before it penetrates the wall of the bunker. In order to attenuate the primary neutron beam, a boron-doped paraffin beam stop (length 80 cm, diameter 60 cm) is provided.

5. EQUIPMENT FOR MATERIALS ANALYSIS

5.1. Positron micro-analysis

Brightness enhancement by remoderation in tungsten foils will be applied repeatedly to convert the large-diameter primary beam into a narrow parallel microbeam with a cross section less than 1 μm.[9-12] This beam will be introduced into a standard Philips Scanning Electron Microscope SEM 535 and guided toward the sample along the same path as the electrons. The apparatus will be equipped with detection facilities for γ radiation to allow perfoming Doppler-broadening measurements on pre-selected areas of the sample. By variation of the positron energy, depth profiles of defects can be measured or interface depths in layered structures be determined.

5.2. 2D ACAR

The equipment to be connected to the beam for measuring the two-dimensional angular correlation of annihilation radiation (2D ACAR) consists of the two 30 × 30 cm^2 detectors of the high-density proportional multiwire chamber type earlier in use in the positron group at Petten.[13,14] The efficiency of the chambers has been increased by doubling the amount of lead in the high-density converters. They will be placed at a relatively small distance from the sample in order to maximize the coincidence rate. Figure 6 shows the layout of the experimental hall to be built north of the reactor hall to accommodate the positron micro-analysis facility, the 2D ACAR setup, and guided neutron beams with equipment. The positron beam, which in the provisional arrangement shown in Fig. 1 leaves the reactor through the south wall, will then be transferred to a channel on the north side of the reactor.

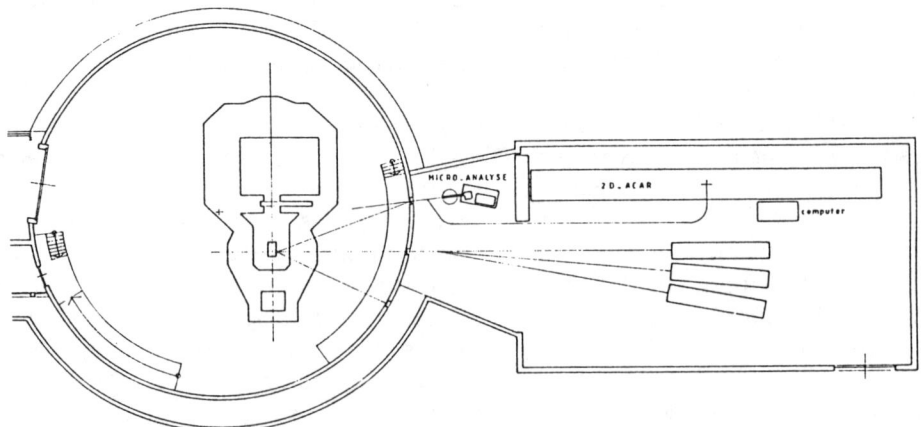

Figure 6. Layout of the experimental hall to be built north of the reactor hall.

6. DISCUSSION

The success of an in-core positron source depends largely on the vacuum and material conditions that can be achieved in the positron source tube. Radiation safety requires that the vacuum tube and other parts of the system consist mainly of low-activation materials, in our case aluminum and alumina. Aluminum UHV systems have proven to operate satisfactorily, but here we have to face the complication of an environment with rather intense neutron and γ radiation. Thus far, it is unknown how much gas will be released by the high-vacuum construction materials under irradiation, and how this will affect the positron re-emission properties of the copper or tungsten surfaces.

Although the reactor at Delft has a rather moderate neutron flux, radiation damage will eventually accumulate to a level where the diffusion length of the positrons is shortened considerably. The positron yield is proportional to the diffusion length and will therefore also be reduced. A reduction by a factor of two is expected when the defect concentration for tungsten accumulates to a level of 10 appm. Trapping rates at simple defects (vacancies) in Cu are a factor of 50 higher than in Mo (W).[15] Therefore, in Cu reduction of the diffusion length should occur much earlier. However, at the working temperature of 100 °C monovacancies anneal in Cu but not in tungsten. The epithermal neutrons (energy up to 1 MeV; flux 10^{12} cm^{-2}s^{-1}) will create about 10^3 displaced atoms per neutron. During one year of operation (250 days) the neutron fluence will be 2×10^{19} n cm^{-2}, giving rise to about 2×10^{-2} displacements per atom. Frenkel pair recombination processes taking place at the ambient temperature will cause damage levels to be a factor of 10-100 lower. Thus, concentrations of surviving defects are expected to be of the order of 200-2000 appm. We hope to reduce this concentration by periodic in-situ annealing.

7. CONCLUSIONS

A prototype of an in-core slow positron beam is under construction which is expected to reach the test phase in early 1993. The positrons are produced in a source with a large surface area which also may have application in other methods of positron generation such as pair formation. Information will have to be gained with respect to the behavior of the materials used in the source during one year of operation in a reactor environment. Positron micro-analysis equipment for use with the slow positron beam is under development, and we plan to add a 2D ACAR camera for defect analysis with slow positrons.

ACKNOWLEDGMENTS

P.E.M. wishes to thank the Netherlands Organization for Scientific Research (NWO) for a travel grant. These investigations in the program of the Foundation for Fundamental Research on Matter (FOM) have been supported in part by the Netherlands Technology Foundation (STW).

REFERENCES

1. *Atomic Physics with Positrons*, eds. J.W. Humberston and E.A.G. Armour, NATO ASI Series B: Physics 169 (Plenum Press, New York 1987).
2. *Positron Beams for Solids and Surfaces*, eds. P.J. Schultz, G.R. Massoumi, and P.J. Simpson, AIP Conference Proceedings Vol. 218, 1990.
3. R.H. Howell, M.J. Fluss, I.J. Rosenberg, and P. Meyer, Nucl. Instrum. and Methods B **10**, 373 (1985).
4. M. Peter and A.A. Manuel, Helv. Phys. Acta **63**, 458 (1990).
5. A. van Veen, POSH, Internal Report IRI-131-89-01, (1989) IRI-TUD Delft (unpublished).
6. W. Triftshäuser, G. Kögel, K. Schreckenbach, and B. Krusche, Helv. Phys. Acta **63**, 378 (1990).
7. B. Krusche and K. Schreckenbach, Nucl. Instrum. and Methods A **295**, 155 (1990).
8. K.G. Lynn, M. Weber, L.O. Roellig, A.P. Mills, Jr., and A.R. Moodenbaugh, in Ref. 1, p. 161.
9. L. Seijbel, P. Kruit, A. van Veen, and H. Schut, in: *Positron Annihilation*, eds. Zs. Kajcsos and Cs. Szeles [Mat. Sci. Forum **105-110**, 1977 (1992)].
10. K.F. Canter, G.R. Brandes, T.N. Horsky, P.H. Lippel, and A.P. Mills, Jr., in Ref. 1, p. 153.
11. W.E. Frieze, D.W. Gidley and K.G. Lynn, Phys. Rev. B **31**, 5628 (1985).
12. G.R. Brandes, K.F. Canter, T.N. Horsky, and A.P. Mills, Jr., in: *Positron Annihilation*, eds. L. Dorikens-VanPraet, M. Dorikens, and D. Segers (World Scientific, Singapore 1989), p. 600.
13. P. Zwart, L.P.L.M. Rabou, G.J. Langedijk, A.P. Jeavons, A.P. Kaan, H.J.M. Akkermans, and P.E. Mijnarends, in: *Positron Annihilation*, eds. P.C. Jain, R.M. Singru, and K.P. Gopinathan (World Scientific, Singapore 1985), p. 297.
14. L.P.L.M. Rabou, P. Zwart, G.J. Langedijk, and P.E. Mijnarends: PANDA II, An apparatus for positron annihilation angular correlation measurements in two dimensions. Report ECN-211 (Netherlands Energy Research Foundation ECN, Petten, The Netherlands 1988).
15. R.M. Nieminen and M.J. Manninen, in: *Positrons in Solids*, ed. P. Hautojärvi (Springer Verlag, Berlin 1979), p. 145.

Development of a High Intensity Low Energy Positron Beam

W.B. Waeber, M. Shi, D. Taqqu, U. Zimmermann, D. Gerola,
F. Hegedüs and L.O. Roellig

Paul Scherrer Institute, CH–5235 Villigen PSI, Switzerland

The purpose of our project is to greatly increase the moderation efficiency of high energy positrons (several hundred keV) to low energy positrons (a few eV). We will attempt to do this by first efficiently premoderating almost all of the positrons emitted from the radioactive source to energies less than 10 keV. These positrons will then be injected in a solid neon moderator which has been shown to have a moderation efficiency of $\sim 50\%$ at these energies.

The immediate objective is to show that confining and guiding electromagnetic fields together with repeated slowing down traversals through a thin solid foil leads to a high efficiency in the premoderation of much of the entire β^+ spectrum. We have performed computational simulation of the trajectories of positrons in the premoderation stage, and the point has now been reached where test experiments can be carried out. These will be done in the spring of 1993.

We are presently evaluating different positron emitting radioactive sources, and simultaneously considering various extraction mechanisms for removing positrons from a high magnetic field region to a field free space in addition to the originally proposed phase space shifter. Besides a status report of the project, which emphasises specific research and development work, a discussion will be given of the future steps and their time schedules, and of the planned initial beam experiments.

1 Milestones

The *motivation* and the search *for high intensity positron beams* has grown steadily over the last decade with the discovery and the development of new beam techniques [1]. Positron interactions with solid surfaces, for example, are understood well enough that information about the electronic and crystalline state of a sample can be obtained by positron spectroscopies. Surface positron spectroscopy techniques that are available today include positron reemission spectroscopy of work functions, low energy positron diffraction, positron measurement of defect profiles, positron induced Auger electron spectroscopy, positronium velocity spectroscopy of the electronic density of states, positron tunneling spectroscopy of surface layers and positron energy loss spectroscopy. One may focus the positrons to a microscopic spot to obtain spatially–resolved information [2,3], and use polarized positrons to selectively annihilate with electrons of a particular spin direction [4]. Specific combinations, for example of microscopy with annihilation techniques lead to novel kinds of information [5,6]. The first experiments in all these domains have indicated the potentials of the investigations with positrons. They are expected to provide information that is not accessible by any other method, and their systematic application requires a high intensity facility [7]. From the limited experience available up to now based on small scale (low intensity) laboratory experiments, it can be extrapolated that the use of a user–friendly (high intensity) facility will have a very strong impact on the development of these novel fields of research, and high intensity positron facilities should *make positron spectroscopy more generally available.* An intense positron beam may also find applications in atomic physics such as the production of antihydrogen and *positronium beams* [8]. Beams of such pseudo–atoms may for example be used for the study of the reflection and diffraction of Ps on surfaces, surface magnetism via Ps spin polarization, the melting of crystalline surfaces and Ps–Ps interactions.

In addition to *improved statistics* and shorter counting times a tunable high intensity beam with small phase space, small energy spread that can be polarized would have many applications because of these properties. Currently, the angular correlation ACAR technique suffers from drastic limits due to the low intensity of positron sources. This may disappear thanks to intense positron beams and the development of better detectors. In addition to the high spatial resolution the detectors should be fast enough to accommodate high counting rates made available by intense beams [9]. For this purpose the *development of fast detectors* with high spatial resolution should be promoted in parallel to the development of high intensity positron beams.

A milestone in the approach to intense positron beam production schemes was certainly the launching in 1989/90 of *a new idea and proposal for the high efficiency production of slow positrons* at PSI [10]. Some large projects providing 10^{10}e+/s are being studied in Germany, the United States and Japan. These projects are all based on very low efficiency positron moderation schemes while at PSI a new and very promising way to produce slow positrons with high efficiency has been proposed [11] and is now under investigation and development. Based on a cyclotron to produce the positron emitter isotopes, the PSI moderation scheme is about 50 times more efficient than other conventional schemes. The commitment of the PSI positron group to the realization of a high intensity beam dates from the 1989 international workshop at PSI [7]. Scientists from all over Europe, North America and Japan have acknowledged their interest in the Swiss project of an intense positron beam. They encouraged an investigation of the PSI concept without delay.

Joint funding by PSI and the Swiss Science Research Nationalfonds (NF) has made it possible to start 1991 a project of explorative nature with the aim of simulating, designing and setting–up an *intense lab–beam* [12], such that by 1993 first experimental tests and beam experiments can be realized and will offer and open up a new area of usable high intensity slow positron beams to the benefit of the whole positron community and of a wider scientific and industrial community in general.

If ever funding could be found for a last major step, to meet the general philosophy of PSI functioning as a user–lab for different scientific communities, a dedicated compact cyclotron could operate as base instrument and on–line production unit for the necessary radioisotopes combined with the high efficiency moderation concept in a *large–scale positron beam facility*.

2 The Source Concept

In contrast to standard laboratory beams where conventional radioactive source–moderator schemes (with moderation efficiencies of the order of $\sim 10^{-3}$) for the slow positron production are used, the presently planned high intensity beam is conceptually different with respect to its mode of production and with respect to the beam properties, as for example leading to intrinsically unpolarized slow positrons although being based

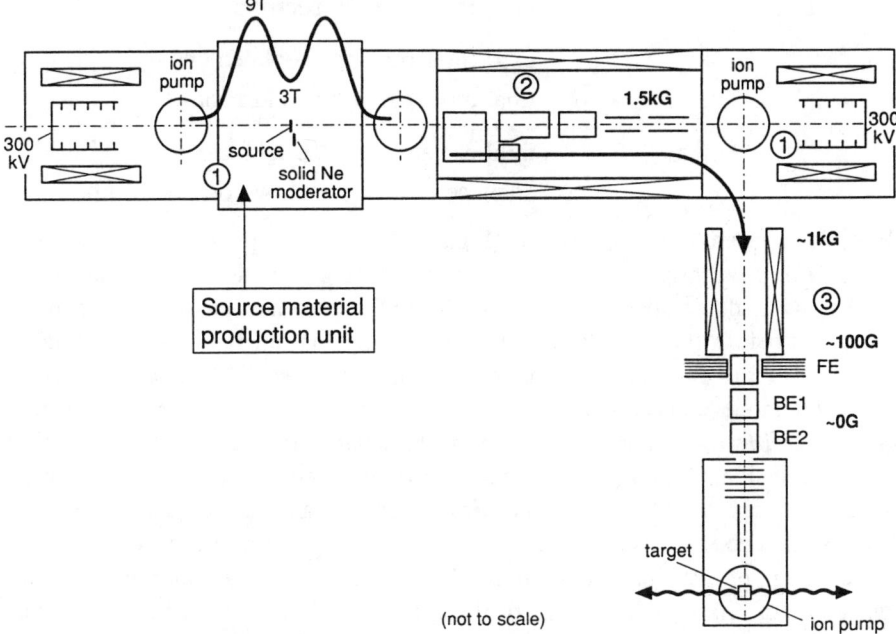

Figure 1: *Layout of the beam production stages (300kV – 300kV distance ~ 10 m, 300kV – target distance ~ 6 m). 1 = premoderation stage ($\sim 5-10$ keV) and 2 = extraction/final moderation stage ($\sim$ eV). They produce the high intensity slow positrons. 3 = field extraction/µbeam formation stage ($\sim \mu/eV$). It is forming the slow positrons into a usable µbeam in field free space. FE = field extraction, BE = brightness enhancement.*

on radioactive sources. The two major beam production units as shown in Figure 1 can be summarized into

- the efficient production of high intensity slow positrons, and

- the formation into a usable microbeam in field free space.

In both units rather unconventional slow positron production and beam formation methods will be used, which will call for unconventional technical solutions as will be shown below.

The outstanding character of this slow positron production concept is based on slowing down most of the high energy positrons emitted by a radioactive source to an energy of 5–10 keV, then extracting these positrons from the beam and having them strike a solid neon moderator which has an efficiency of $\sim 50\%$ at these energies [10].

Hence, the beam production concept which is adopted for the PSI high efficiency slow positron lab–beam can be structurized into the following operations (Figure 1):

slow positron microbeam	=	confinement and slowing–down of most of the β^+ spectrum
	+	extraction from the confinement and final moderation by solid neon
	+	field extraction and brightness enhancement of the slow positron beam.

Indeed, this concept uses three stages (Figure 1). The principle of the *premoderation* stage is to trap the high energy positrons within magnetic and electrostatic fields so that all slowing–down takes place in a thin solid foil placed at the center of a magnetic bottle. Just before they would stop in the foil the low energy positrons finally are shifted out of the slowing–down path and redirected onto the neon *moderator* in a reflection mode configuration. Extraction of the slow positrons and subsequent *phase space improvement operations* on the beam ultimately lead to an estimated overall conversion efficiency of the order of 50 times larger than the highest values of conventional schemes [13]. Together with a primary radioactive source strength of a few Ci β^+ this concept would lead to slow positron intensities in the 10^{10} e$^+$/s range.

The design of the premoderation, the intermediate extraction and beam formation stages calls for a thorough analysis of the charged particle motion in the electromagnetic confinement and guiding fields, with initial energies reaching values of several hundreds of keV up to the order of 1MeV. The objective of such simulations is the description, evaluation and, together with the planned measurements of the efficiencies, the optimization of these stages.

One of the inherent characteristics of this high efficiency positron conversion scheme is its loss of the polarization of the β^+ by having the acceptance of the positrons independent on the direction of emission from the radioactive source. From all possible kinds of beam features, which the present concept is able to provide, one of the most important properties – the polarization – would be doomed to remain inaccessible, if it were not for positron beam *repolarization* to be the only resource that could rescue the scheme from such a possible drawback [14]. This kind of a drawback is naturally the case too, when the positrons are produced via bremsstrahlung of high energy γ's as it is done in Livermore and planned in Japan, where an electron linac is the basic machine, or as proposed at the Grenoble reactor, where the γ's are produced via neutron capture.

3 Present Developments

For the realization of the beam outlined above there are intense planning, theoretical simulation, design and development activities going on. Within the framework of the technical baseline (section 2), various possibilities for resolving these problems are being studied in parallel. In the following, some of the related work is being highlighted.

3.1 Premoderation Efficiency Measurements

The premoderation efficiency is the most relevant quantity affecting the achievable slow positron intensity in the beam. Figure 2 shows the schematic situation of the premoderation stage with the relevant experimental parameters and the various efficiencies. The *confinement* efficiency is given by

$$\eta_{conf} = (N - n'_1)/N = n_1/N, \qquad (1)$$

where N is the total number of positrons emitted from the source, n'_1 is the number of positrons lost due to leaks (i.e. $\theta < \theta_c, E > eV_r$) in the electromagnetic confinement. The *extraction* efficiency (extraction from the slowing–down path of intermediate energy positrons) is obtained by setting

$$\eta_{extr} = (n_1 - n'_2)/n_1 = n_2/n_1, \qquad (2)$$

where n'_2 is the number of positrons stopped inside the source or the solid foil before they are available for intermediate extraction [13]. The *premoderation* efficiency and the *total conversion* efficiency are as given in the caption of Figure 2. A series of measurements is necessary in order to verify the theoretical values for the confinement and the extraction efficiencies as a function of various field configurations with the aim to optimize the slow positron production for maximum intensity. The confinement efficiency $\eta_{conf} = \eta_{conf}(V_r, B_0, t)$ is the fraction of emitted positrons that slow down to the lowest energies by multiple passages through the foil. It is obtained in absence of extraction by monitoring the rate of stopping in the source foil.

Information on the efficiency of extraction of the premoderated beam can be obtained in the initial stage (stage 1 of Figure 1, i.e. before setting up the extraction scheme stage 2 of Figure 1) by making use of a two foil configuration. The relative stopping rate of the two foils will depend on the potential difference ΔV between them (the foil with greater positive voltage should stop more positrons). The dependence can be analysed in terms of the energy distribution of the positrons outcoming the foil and compared to the predictions of the theoretical simulation on which the computation of the extraction efficiency is based (see ref. [13]).

Most of these experiments can be done with low activities of ^{18}F or ^{58}Co ($\leq$1mCi) deposited on the carbon foil. Some tests have already shown that sources of this kind should be feasible [15]. The measured quantities are the annihilation γ count rates

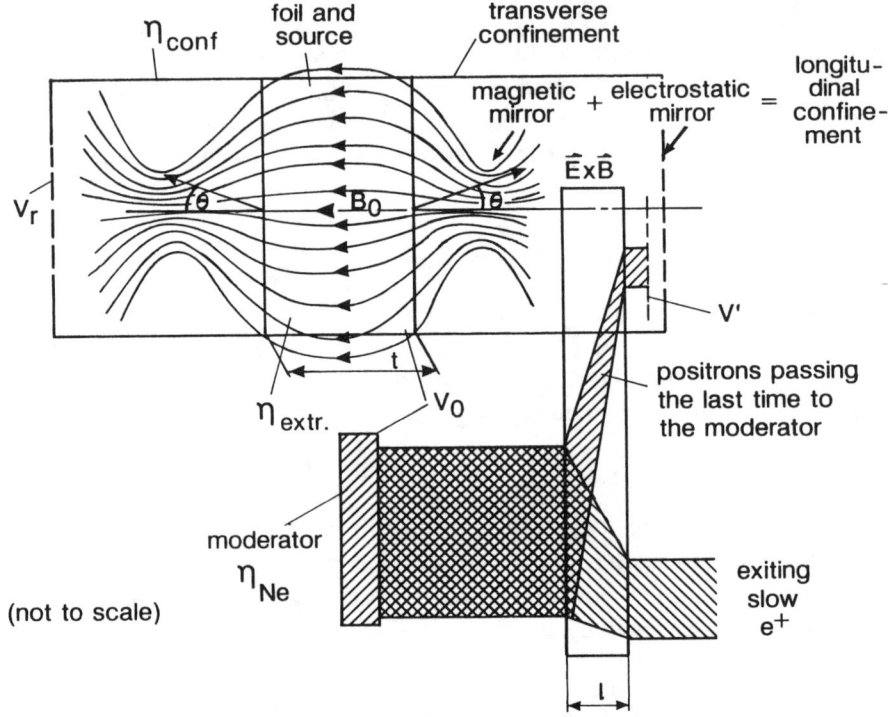

Figure 2: Premoderation stage (schematic) with indicated efficiencies and the relevant experimental parameters. The losses may be due to 'leaks' in the confinement and due to the stopping of the positrons in the foil. The definition of the efficiencies is given in ref. [13]. For the premoderation we have $\eta_{prem} = \eta_{conf}\eta_{extr}$, and the overall conversion efficiency is given by $\eta_{conv} = \eta_{prem}\eta_{Ne}$. V_r = high voltage mirror reflector potential, B_0 = magnetic field strength in the center of the magnetic bottle, V_0 = potential on the source/moderator, V' = max. potential on the intermediate reflector, t = source foil thickness.

originating from positrons stopping in the source–foil and in additional absorbers placed in or near the beam axis at various positions along the beam line. The additional absorbers may be thick (stopping all positrons) or thin carbon foils of the same thickness as the source foil. The annihilation radiation is detected with scintillation counters or Ge– or Si–detectors placed behind lead collimators outside the vacuum vessel. In some cases 2γ–coincidence measurements may also be useful. As a simple example we mention the measurement of the contributions of the electrostatic and magnetic mirrors to the confinement efficiency by switching the respective fields on and off and monitoring the 511keV intensity emerging from the source foil. Application of ^{58}Co has the advantage that calibration of the observed count rates is easier because the 810 keV γ–radiation can be used to normalize the rates. For ^{18}F the source foil has to be sandwiched between two absorbers in order to obtain the calibration constants.

3.2 Intermediate Extraction – Final Moderation

The separation of the $E_c \sim 5-10$ keV positrons from the high energy positrons is established by a trapping–drifting mechanism. After the slowing–down of the high energy positrons to energies less than the cutoff value $E_c = V' - V_0$ the lowest energy positrons are extracted from the slowing–down path by trapping them between the

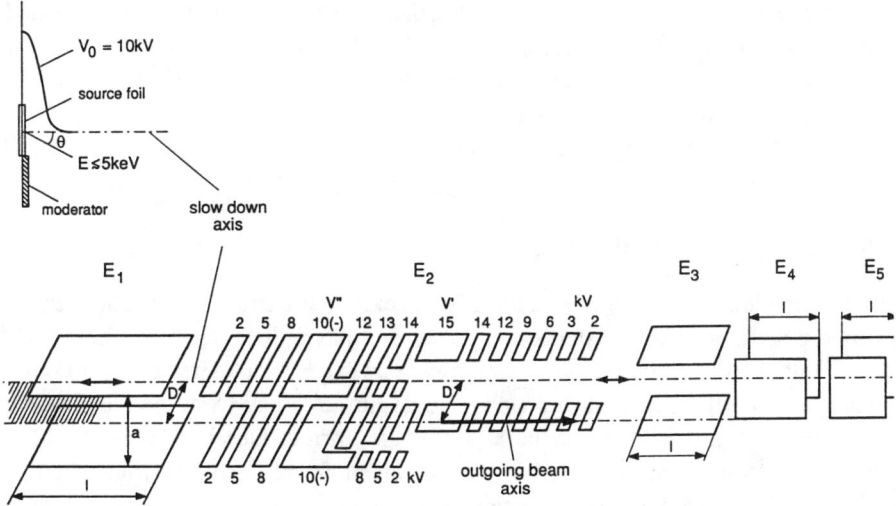

Figure 3: *Stage 2 of Figure 1: Extraction of the intermediate energy positrons from the confinement by trapping them between the moderator (at V_0) and the intermediate reflector E_2 (at V') and selective sidewards drift D of the lowest energy positrons. E_1, E_3, E_4, E_5 = E×B plates; The dimensions of the source foil and the moderator extension can be found in ref. [10]. Some typical parameter values are: $D \sim 8cm$, $a \sim 4cm$, $l \sim 70cm$, voltage$\sim \pm 5kV$.*

moderator at V_0 and the intermediate reflector E_2 at V' and by selectively sidewards drifting them by action of E_1 onto the outgoing beam axis (see Figure 3). The theoretical simulation of the positron trajectories through the set of $E \times B$ fields and the intermediate reflector E_2 (in plane plate geometry as shown in Figure 3, or in ring electrode geometry) is presently carried out and will lead to an optimal design of the whole extraction section.

The basic idea underlying the formation of a premoderated beam is to optimize the final moderation process by insuring a low maximum energy E_c of the positrons impinging the moderator. Positrons emitted from the source in the direction of the extraction electrodes with $E < E_c$ will not return back to the source but will be drifted away from the slowing-down path and hit the moderator. A low E_c maximizes the final moderation [16] but increases the fraction of positrons that are lost because they stop in the source-foil [13]. A high E_c has opposite effects so that there will be an optimal E_c associated to the given source thickness. Monte Carlo calculations have shown that for the proposed source of Li^{18}F on the carbon foil of together 22 μg/cm^2 thickness the optimal E_c is at about 7 keV [13]. At this energy the extraction efficiency of the premoderated beam is close to 90%. In case the optimal specific activity will not be reached the same source intensity will require a thicker source and lead to a higher E_c with less overall efficiency. This situation is amplified in the case of an intense ^{79}Kr source (section 3.5) which theoretically has significantly less specific activity and requires also tight vacuum windows. For that source E_c is expected to be significantly higher than 10 keV and lead to an appreciably reduced overall conversion efficiency. Further Monte Carlo calculations for the case of a ^{79}Kr source cell are in progress.

3.3 Slow–Positron Beam Bending

After final moderation the slow positron beam exits the extraction section parallel to the slowing down axis at a distance of $\sim$ 8cm and has a diameter of $\sim$ 3.5cm. Since the SF$_6$ tanks of the electrostatic mirror reflectors have a diameter of $\sim$ 50 cm it is necessary to bend the exiting beam (lab–layout) further away from the slowing–down path at a position just in front of the high voltage mirrors (see Figure 1).

The basic problem to be solved is the establishment of a field configuration, such that the bending or separation operation of the two beams takes place without perturbing the cross–sections of the slowing–down and exiting beam and by keeping the required adiabaticity of the particle movement in both beams. Careful analysis of various configurations have to be optimized as for coil geometry, the number of coil types and the overall power consumption.

Two configurations are presently being studied. A small diameter solenoid on the main axis with the slowing–down beam to pass inside, while the exiting beam is to follow the field lines outside it. Additional B–fields perpendicular to the slowing–down axis then establish a 90^0 bend on the slow positron beam however with some moderate distortions on the slow–down beam. This distortion is due to the broken cylindrical symmetry. It can be reduced considerably if 4 slanted coils (in a quadrupole–like configuration) are used. This is closest to the perfectly cylindrical symmetry within a certain frame of efforts, in which case the drift of the guiding center arising from any B–field gradient is in the azimuthal direction. Thus the result is only a rotation of the circular beam cross section about its axis without any modification of beam shape or dimension. – A second possible solution introduces an additional $E \times B$ field for drifting the low energy positrons a further 20 cm away from the main axis just in front of the high voltage mirror tube. At that distance from the slow–down axis the exiting beam can bypass the electrostatic mirror SF_6 tank and easily be bent out with further coils, not perturbing the slowing–down beam in any way owing to the larger distances of the bending coils from the latter. This last configuration requires an increased diameter of the B–field coils and leads to higher power consumption.

3.4 Beam Phase Space Operations

In order to provide microbeam formation via remoderator stages the positron beam has to be *extracted into the field free space*. If the present slow positron beam is produced in a high magnetic field we have to care about the very high field angular

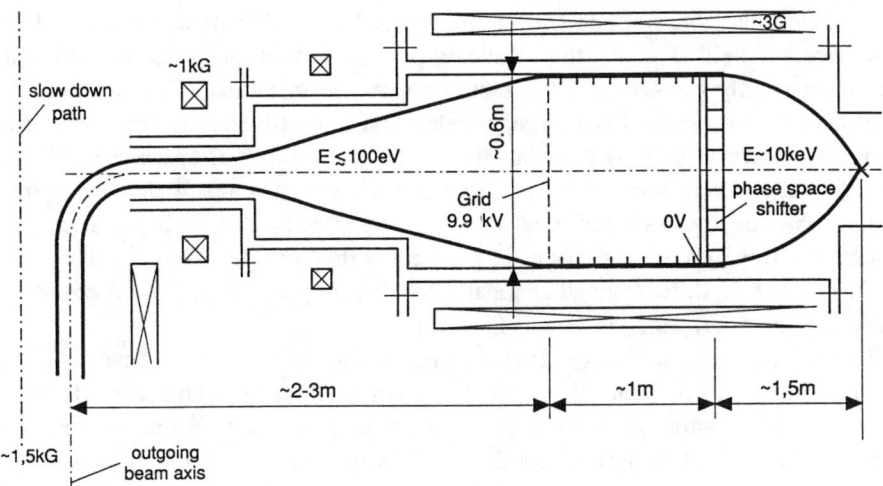

Figure 4: *Phase space shifter [17] showing the approximate measures of such an apparatus (for details we refer to the text)*

momentum $\vec{r} \times e\vec{A}$ in which the particles are embedded. As $e\vec{A} = erB/2$ reaches order of magnitudes higher values than the intrinsic particle transverse momentum p_θ special methods have to be used in order to hinder the field momentum to induce a blow up of the beam phase space. The problem that we have to overcome is buried in the fact that due to $\vec{r} \times \vec{P}_\perp = const$ and $P_\theta = mv_\theta + eA_\theta$, particle extraction into the $B = 0$ region is very difficult, because, owing to the presence of $B_r \neq 0$, considerably increasing beam diameter and beam divergency might be introduced, depending on the adiabatic or nonadiabatic character of the particle motion [1].

The approach for resolving the field extraction problem can be stated in the following short form: 'Field–extracting the beam = *adiabatically stretching* the phase space of the beam + *nonadiabatically shifting* the slanted phase space of the beam.' There are two solutions that are presently being analysed:

1. Beam extraction by the phase space shifter [17].

2. Beam extraction by field termination through a magnetic shield with beam aperture

    a) by introducing an azimuthal field E_θ for the compensation of A_θ changes in the shield aperture, or

    b) by introducing a magnetically conductive internal structure in the aperture, thus spreading out the B_r component in the shield aperture.

In a first step the *phase space shifter* brings the beam adiabatically from ~ 1.5 kG to the very low field $B \sim 3$G, thus reducing $p_\perp$ as $\sqrt{B}$ with the result that P_θ will be determined mainly by $A_\theta(r)$, at the same time the beam radius r increases as $1/\sqrt{B}$ (see Figure 4). In a second step, after acceleration to ~ 10keV, the beam would pass transverse deflection cells with an azimuthal electric field E_θ imparting a change in $p_\perp$ such that the eA_θ term in P_θ is compensated, with the result that the spiraling radius of the positron trajectories is $\rho = r/2$, and after half of their pitch length, all particles converge on a focal point on the axis with $P_\theta \approx p_\theta$. Thus, in this concept the phase space is shifted nonadiabatically by changing $p_\perp$ in $p_\perp^2/B$ ($\neq const$), and by keeping the homogeneous $B \neq 0$ field constant.

For the beam *extraction by field termination* again the exiting beam is brought adiabatically to a lower field $B \sim 100$ G, the beam diameter becomes ~ 10 cm, and a magnetic shield with an aperture of ~ 10 cm is introduced. Again $p_\perp$ reduces as $\sqrt{B}$, and the B–field is terminated $B_z \to 0$, while a radial field $B_r = -\partial A_\theta/\partial z$ in the region of the aperture is introduced. An azimuthal force $\vec{v} \times \vec{B}$ is effective, which a) can be compensated for particles far from the beam axis by introducing an azimuthal E–field with strength $v_\parallel B_r$ with the result that $p_\perp$ remains unchanged but eA_θ is compensated (see Figure 5), or b) the B_r field in the beam aperture can be spread out by introducing some internal structure to the aperture making B_r small even at larger radial distances from the axis. The result in this case is that the effect on the divergency remains limited because the azimuthal force effect of B_r on the particles remains small. Thus in this conceptual solution of field termination the phase space is shifted nonadiabatically by changing the B–field and by keeping $p_\perp$ almost constant in $p_\perp^2/B$ ($\neq const$).

---

[1] For a discussion of adiabaticity of the positron motion in the confining electromagnetic fields we refer to ref. [13].

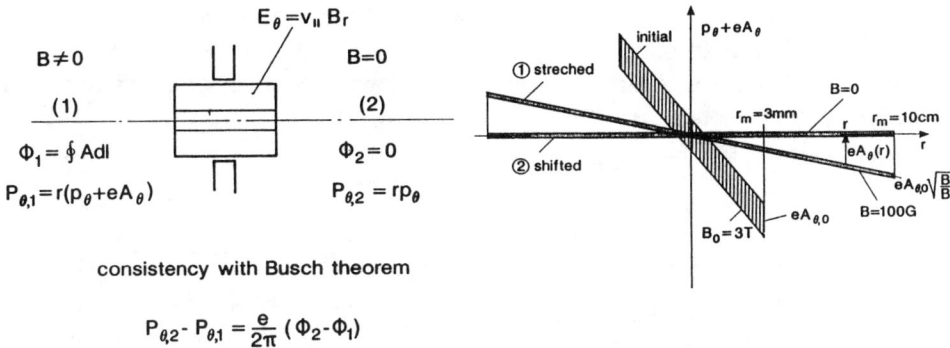

Figure 5: *Phase space situations before and after the field extraction (not to scale). For the meaning of the symbols and the operations on the beam phase space we refer to the text.*

3.5 Primary Source Arrangements

The problem to be resolved is to achieve an effective thickness of the slow–down foil + source spot to be smaller than the penetration range of ~ 5keV positrons, to produce a high surface specific activity and a small source spot diameter. We are presently evaluating various positron emitting radioactive sources. In addition to the originally considered ^{18}F [10,11] with a half–life of 110 min we plan to investigate the recently proposed ^{79}Kr isotope ($T_{1/2}$ = 35h). Although the production technology ^{18}O(p,n) for ^{18}F is standard at PSI (as used for medical applications), the development work for a precise and quick deposition of a solid Li^{18}F spot onto the carbon foil would be considerable, and beam operation would be close to a batch mode in ~ 2h cycles. The maximum energy of the β^+ spectrum is about the same, 600 keV, for both isotopes. The β^+ yield of ^{18}F is 100%, while only 8% for ^{79}Kr, however, due to the long half life and the better chemical characteristics (inert gas) ^{79}Kr seems to be more suitable for this purpose. While the initial proposal [20] suggested a reactor production scheme ^{78}Kr(n,γ)^{79}Kr, the poor specific activity achievable this way led us to look at the much more attractive production mode of

$$^{79}\text{Br}(p,n)^{79}\text{Kr}$$

which in principle allows the ^{79}Kr to be extracted from the target material (for example NaBr) in carrier free form. In addition, this way of producing ^{79}Kr can rely on existing technology developed and routinely used at PSI for the production of the isotope ^{123}I. The objective is to use the ^{79}Kr in the form of a gas (easily transported through piping systems), and to inject it continuously into a source–cell located in the confinement region of the premoderation stage of the beam producing facility.

In the following the needed production units for a continuous mode operation of the intense positron beam facility are briefly described, and the corresponding technical and operational implications will be indicated with emphasis on a dedicated cyclotron as base facility [11]. Figure 6 shows a diagram of the production line. The best way to produce ^{79}Kr is to bombard bromine targets with 60–70MeV protons. In a thick target (e.g. 8g/cm² of NaBr) the proton slows down to zero energy, therefore ^{79}Kr is produced by two reactions: ^{79}Br(p,n) and ^{81}Br(p,3n). There are very few data about these reactions [18,19]. The preliminary yield estimation is promising: 2.1×10^8

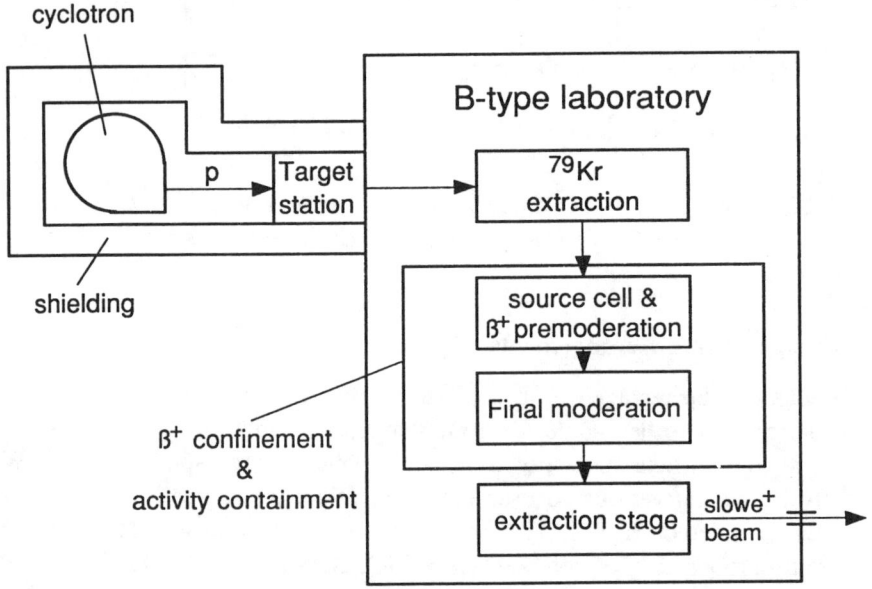

Figure 6: ^{79}Kr *production scheme and its 'continuous' injection into the source cell. The cyclotron would represent the source material production unit of Figure 1.*

Bq/µAh with 71.8MeV protons and 0.75×10^8 Bq/µAh with 20MeV protons. The corresponding accumulated yield for an irradiation of 35 hours with 50µA proton current is 2.7×10^{11} Bq and 0.94×10^{11} Bq respectively. After irradiation the target will be dissolved, the ^{79}Kr with helium gas extracted and carrier free, on a liquid nitrogen trap, collected (Figure 7). A series of experiments is planned with sandwich targets to measure the yield as a function of the proton energy. In a more refined version the chemical dissolution of the target could be eliminated by letting the dry carrier gas (He) flow directly into the hot target cell during the irradiation. In such a scheme the extraction (via a cold trap) and release of the ^{79}Kr activity takes place without any mechanical displacement of the target allowing long *continuous operation* time with negligible radioactive waste.

Figure 7 also shows the loading stage and the basic design requirements for the source–cell. The source cell will be a 1cm diameter tube of about 4cm length placed on the axis of the confining solenoid at its center. Ultra–thin vacuum tight foils (made of graphite or diamond) placed on both ends of this tube closes the cell. When ~ 4 Ci e+ source strength is needed in the cell for a final ~ 5 × 10¹⁰ slow e+/s beam (by assuming an overall conversion efficiency of ~ 35% [13]) the required pressure then is ~ 4 Torr. An 8% positron decay of ⁷⁹Kr and a source strength of 4 Ci e+ is equivalent

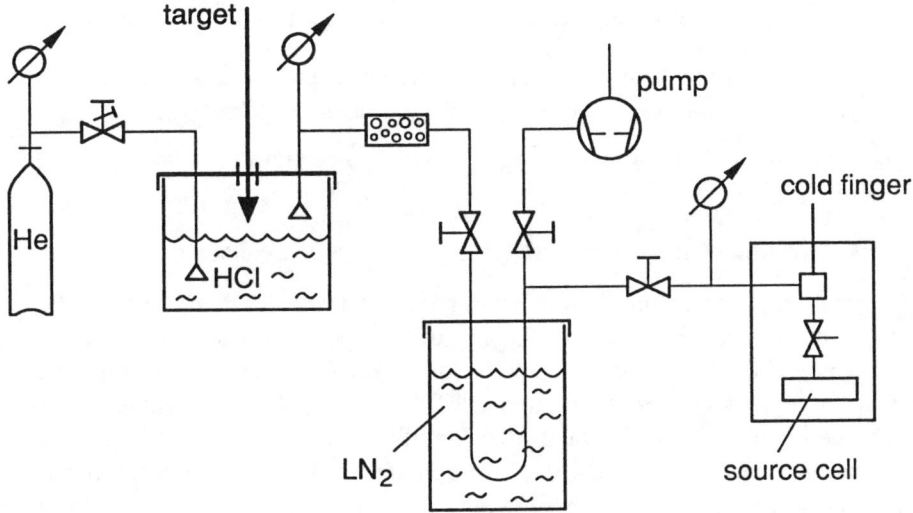

Figure 7: ⁷⁹Kr extraction either by chemical dissolution of the target and expelling the ⁷⁹Kr with dry helium, or extracting and release of the ⁷⁹Kr by direct flow of helium through the target cell during the irradiation.

to ~ 50 Ci ⁷⁹Kr activity. With a toxicity class of 9, it calls for a B–type laboratory for handling the activity [2], and operating the positron converter (premoderation and final moderation). It also means that biological shielding of most of the units indicated in Figure 6 is necessary. Moreover, the transfer of ⁷⁹Kr gas in piping systems may be risky due to the possibility of leaking. Thus a safety containment may be needed. However, the on–line system has the great (may be the all decisive) advantage of being 'closed', that is to say, no pipe–decoupling procedures, i.e. no escape possibilities of active gas rests have to be considered.

In conclusion, the requirements for the implementation of a Kr–technology can be summarized in the following main points of development: types of foils, ⁷⁹Kr yield determination, cell design, cold trap efficiency, trap–source–cell–transfer efficiency, cell test in operational conditions, safety considerations.

---
[2]regulations set by the Swiss governmental safety inspectorate

4 Future Steps and Perspectives

The research work that has been accomplished in the past year can be summarized into two parts. The first part of development work on the high intensity positron source consisted in the basic computer–simulation of the positron trajectories in an electromagnetic confinement which had to be optimized with respect to minimizing particle loss [13]. The specific development work done in this period was:

- Configuring the superconducting magnet (magnetic bottle) and the room temperature coils as well as the high voltage mirror reflectors

- Calculation of the 9T/3T – 1.5kG magnetic and 300 kV electrostatic field distributions

- Calculation of the positron trajectories in these confinements.

The results of these computer–simulations have defined the final *configuration*, the *layout*, the *design* and the *dimensions* of the components, which have already been fabricated and delivered for the complete assembly of the premoderation facility. In this budget period an overall investment of 0.5 million dollars for equipment has been established, and a further amount of 0.3 million dollars is needed to complete the microbeam. The second part of the positron source development work consisted in the Monte Carlo simulation of the slowing–down of positrons [13] in a thin carbon foil placed at the center of the magnetic bottle. Together with the results of the trajectory calculations they delivered

- the theoretical values for the confinement and extraction efficiencies [13].

These quantities are of central importance to the project, and they define, together with the final moderation, the overall conversion efficiency of the new beam production concept. We expect to compare these efficiencies with the experimental tests at the beginning of 1993.

The *planned* first high intensity *beam experiments* are mainly determined by the current research program on high–T_c materials. This beam is to be used for defect studies and also for surface and bulk electronic structure studies in HTSC single crystals. In particular in the case of

– the study of the irradiation induced defect structure and critical current enhancement with intense monoenergetic positron beams, and

– the investigation of the irradiation enhancement of critical currents in Cuprates through a study of the electronic structure with positrons.

Doppler broadening shape parameter depth profiling, PAES–spectroscopy [21] and 2D–ACAR measurements are planned. Application of these techniques envisage the correlation of experimental results with J_c–measurements and checks of electron–momentum distributions against theoretical models of bulk and surface electronic structure calculations (Fermi surface studies) respectively. In order to achieve these objectives in some cases a positron beam tunable in energy is needed. In the present concept it is planned to install a 75kV acceleration device and to have the target at high voltage, which asks for special care (sufficient resistance against dielectric break down and good thermal conductivity at ambient and at low temperatures). In addition, since a real *microbeam* will be used on the sample the target chamber has to be furnished with a high precision UHV manipulator, fully motorized, for the sample may turn into a strong γ–source of the order of Ci, owing to the high intensity beam annihilating in the sample.

Besides having a monoenergetic positron beam with a small energy spread, tunable in energy, a very small diameter and beam divergency, which may even be deviced to traverse a beam splitter or to be turned into a scanning microbeam, – further reaching beam qualities like a high degree of *polarization* is most important for a wide class of experiments [22]. A method to overcome the limitation of the present concept (and of positron factories in general) and to achieve a polarized intense slow positron beam has been presented at PSI in 1990 [17]. The basic idea is to repolarize the slow positrons by letting them interact with spin–polarized atomic hydrogen trapped in a high magnetic field B$\sim$10T at low temperature T$\sim$0.3K in a window–free gas cell ($\sim 10^{16}$ at/cm^3). Conservative estimates for the resulting beam polarization are around 80–98% with respective yields close to 25–40%. In comparison with the polarization and the yields that have been achieved with standard radioactive sources, the figures given above represent a significant improvement. This indicates that available laboratory beams would also benefit from such a repolarization scheme. However, the realization of the window–free spin–polarized gas cell involves a technical effort that is best invested within the frame of a positron factory. Furthermore, the outcoming polarized positron beam, with a minimal diameter of a few mm at 9T, and its extraction into the free field region requires the complex compensation procedures outlined before (see section 3.4). In our scheme such procedures need only to be done once by placing the repolarization stage before the phase shifting operation.

The future project and research plan anticipates the following items and time schedules:

1992/93 – taking into operation the premoderation stage of the high intensity lab–scale positron beam and measurement of the confinement and extraction efficiencies.

- simulation, design and set–up of the beam extraction section and of the microbeam formation line, beam testing.
- set–up of the PAES spectrometer and 2D–ACAR spectrometer including detectors.

1993/94 – Positron beam experiments (defect profiling and PAES) on irradiated HTSC single crystals to study defect structures.

- 2D–ACAR experiments on highest quality single crystals for Fermi surface studies.

> 1995 – Project up–grade to a large–scale positron factory with dedicated cyclotron with the aim to establish a user–lab for positron beam applications in various domains of science.

Acknowledgement

In many of the above indicated activities the positron group of the University of Geneva is collaborating with PSI, and without the substantial support of the Swiss Science Research Nationalfonds this explorative project could not have been started. We wish to acknowledge M. Peter, A.A. Manuel, A. Shukla, M. Sedlacek and F. Mulligan for continuous support and helpful discussions. We are indepted to W. Fischer and H.R. Ott for active interest, P. Gross for constructive design work, M. Werner for the help in magnetic coil design, and V. Vrancovic for help in the use of advanced simulation programs.

References

[1] P.J. Schultz and K.G. Lynn, *Rev. Mod. Phys.* **60** (1988) 701.

[2] K.F. Canter, T.M. Roach and A Bacshi, this conference.

[3] W.B. Waeber, *Helvetica Physica Acta* **63** (1990) 448, and references therein.

[4] J. Van House and P.W. Zitzewitz, *Phys. Rev.* **A29** (1984) 96.

[5] W.B. Waeber, U. Zimmermann and G. Solt in ref. [11], p.43–48.

[6] D.W. Gidley et al, this conference.

[7] W.B. Waeber, Ed., *Helv. Phys. Acta* **63** (1990) 377–470.

[8] M. Weber, S. Tang, R. Katri, S. Berko, K.F. Canter, K.G. Lynn, A.P. Mills, Jr., L.O. Roellig and A.J. Viescas, *Proceedings of the Workshop on Annihilation in Gases and Galaxies*, edited by Richard J. Drachman, NASA Conference Publication 3058 (1990) 137.

[9] A.A. Manuel and M. Peter, *Helv. Phys. Acta* **63** (1990) 397.

[10] D. Taqqu, *Helv. Phys. Acta* **63** (1990) 442.

[11] W.B. Waeber, D. Taqqu, U. Zimmermann and G. Solt, Eds., *PSI Report* **68**, Mai 1990.

[12] W.B. Waeber, D. Taqqu, U. Zimmermann and L.O. Roellig, *PSI Progress Report* **III** (1991) 80.

[13] Ming Shi, W.B. Waeber, D. Taqqu, F. Foroughi and J. Arkuszewski, this conference.

[14] D. Taqqu, *PSI Progress Report* **III** (1990) 101.

[15] U. Zimmermann et al, *PSI Progress Report* **III** (1991) 83.

[16] A. P. Mills, Jr. and E.M. Gullikson, *Appl. Phys. Lett.* **49** (1986) 1121.

[17] D. Taqqu, see ref. [11] p.75.

[18] D. Gandarias–Cruz and K. Okamoto, Status on the compilation of nuclear data for medical radioisotopes produced by accelerators, *IAEA Report* INDC–209–GZ, 1988, Vienna.

[19] J Lange und H. Münzel, Abschätzung unbekannter Anregungsfunktionen für (p,xn)–Reaktionen, Kernforschungszentrum Karlsruhe Report No KFK–767, 1968.

[20] A.P. Mills, Jr., *Nuclear Science and Engineering* **110** (1992) 165.

[21] A. Weiss in *Positron Beams for Solids and Surfaces*, Eds.: P.J. Schultz, G.R. Massoumi and P.J. Simpson, AIP Conference Proceedings **218** (1990) 61.

[22] A. Seeger and F. Banhart, *Helv. Phys. Acta* **63** (1990) 403.

SECTION 5

POSITRON MICROSCOPES

BRANDEIS SECOND GENERATION POSITRON REEMISSION MICROSCOPE

K.F. Canter, V. Dharmavaram, A.G. Smirnov,
S.A. Wesley, K.H. Wong, and R. Xie
Physics Department, Brandeis University, Waltham, MA 02254

G.R. Brandes[1] and A.P. Mills, Jr.
AT&T Bell Laboratories, Murray Hill, NJ 07974

The positron reemission microscope (PRM) is based on the proposal of Hulett et al.[1] to image thermalized positrons that are spontaneously reemitted from materials having negative positron affinities. In particular, a positron beam is focussed to a small spot on one side of a thin foil and a magnified image is then formed by those positrons diffusing to the other side of the foil and being reemitted into the vacuum. Since positrons are easily trapped by defects as they diffuse through the foil,[2] one should be able to observe defect structures. Also, the $\approx$1eV positron emission energy should allow the contours of overlying molecular structures on the surface of a relatively defect-free foil to be viewed without damage because of the extremely low energy of the reemitted positrons which back-illuminate the overlying structure. The low energy also guarantees strong scattering, and hence contrast, to structures as fragile as a patch of adsorbed hydrogen atoms, for example. Unlike the transmission positron microscope (TPM),[3] which is the direct analog of the conventional transmission electron microscope,[4] the principle of operation of the PRM is unique to positrons. While the PRM has its antecedent in the thermionic emission microscope,[5] and while there are also photo-excited negative affinity electron emitting surfaces,[6] there is presently no direct analogue of the PRM.

Positron reemission microscopy was first demonstrated by two different groups utilizing two different forms of the PRM: 1) a reflection mode PRM at the University of Michigan (maximum magnification 56× and 2.3μm resolution),[7] and 2) a transmission mode PRM at Brandeis University (maximum magnification 1150× and 0.2μm resolution).[8] The Brandeis PRM subsequently was able to reach a magnification of 4400× with a 0.08μm resolution in a 24 hour exposure.[9] The Michigan PRM was limited to a $\approx$ 50× magnification because it did not employ brightness enhancement[10] and because of inherent limitations in its off-axis incident positron beam geometry.[11] In order to maximize the magnification, and hence resolution, of a PRM it is necessary at the same time to maximize the number of incident positrons into the area from which reemitted positrons are to be imaged. This means maximizing the incident beam flux as well as focussing the incident positrons down to a small as possible spot on the sample.

The resolution of a PRM is limited by the statistical fluctuations due to finite counts/pixel, detector resolution, diffraction limits, and the spherical aberration of the immersion objective.[8] The chromatic aberration is negligible because positron reemission is highly monoenergetic with an energy spread of only $\approx$kT.[12,13] With a commercially available channel electron multiplier and resistive anode encoder detector having a spatial resolution of 60 μm,[14] a 50,000× magnification, for example, translates into a detector resolution of 1nm which can be effectively reduced to 0.5 nm by obtaining

---

[1]Present address: Advanced Technology Materials, 7 Commerce Drive, Danbury, CT 06810

slightly shifted image pairs. A more fundamental resolution limit is the ≈1nm de Broglie wavelength limit of the reemitted positrons. The small angle emission of positrons from a well ordered surface and a 75kV/cm field at the cathode (sample), which is typical in a transmission mode objective, will keep the spherical aberration below 1nm.[15] Thus, given a sufficiently intense positron beam, the resolution capability of a PRM with presently available detectors is 2nm with an ultimate potential for 1nm spatial resolution in the direct imaging mode.

In order to achieve 1nm resolution, $10^{17}e^+/cm^2$ incident time integrated flux density on the sample is required, assuming a 10% reemission efficiency of the foil and a 50% contrast in the feature being imaged.[8] This translates to $10^{10}e^+$ focussed into a 2 micron (FWHM) diameter spot, the smallest beam spot that can be produced with electrostatic focussing of a doubly remoderated beam.[16] Taking into account the 90% loss in primary beam flux resulting from the brightness enhancement needed to obtain a 2 micron beam at 5 kev, one would thus need the equivalent of $10^{11}e^+$ emitted from a 10 mm diameter primary moderator with a transverse energy of 0.2eV or better for each PRM exposure. With a $10^6e^+/s$ upper limit on the primary beam flux for our laboratory beam, we anticipate that it will not be practical to carry out experiments beyond a 10nm resolution at Brandeis without a higher flux beam. Figure 1 is a schematic of the second generation transmission mode PRM now in operation at Brandeis.

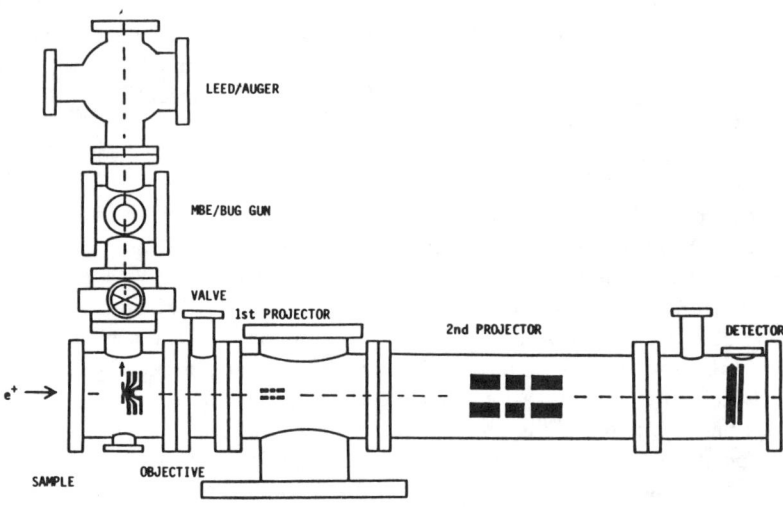

Figure 1: Brandeis second generation PRM

This PRM has a 50,000× magnification capability by virtue of the addition of a second projector lens. A new 4-element immersion objective allows a 75 kV/cm electric field at the sample.[15] Additional improvements over the first generation PRM also include a 4-fold higher resolution image detector, a new brightness enhancement chamber, and a sample translation system a with separate sample preparation chamber enabling electron and ion bombardment, molecular beam epitaxy, and electrostatic deposition of large molecules on the sample. The new brightness enhancement chamber is a larger version of the chamber which provided 500× brightness enhancement at Brandeis.[17] The significant change is that the larger chamber now allows the substitution of a W(110) second remoderator with a Ni(110) moderator that can be cooled to 100°K and can be treated with in situ glow discharge sputtering, electron bombardment annealing, and chlorine passivation. This should provide an ordered surface yielding positron remission with only thermal diffuse scattering[12,13] and consequently an overall brightness enhancement capability of 10,000×. In terms of the number of positrons per pixel in a PRM image, this upgrade in brightness enhancement is equivalent to a factor of 20 higher positron beam flux than was available for the 4400× PRM experiments at Brandeis. At present we are operating with the second Ni(100) remoderator only being heat treated in situ. The transverse energy ET for this remoderator has not been directly measured, but we have been able to achieve a 5 ± 1 μm FWHM spot diameter at 4 keV incident energy on the PRM sample foil. The PRM image of the positrons reemitted from a defect free portion of the film is shown in Fig. 2.

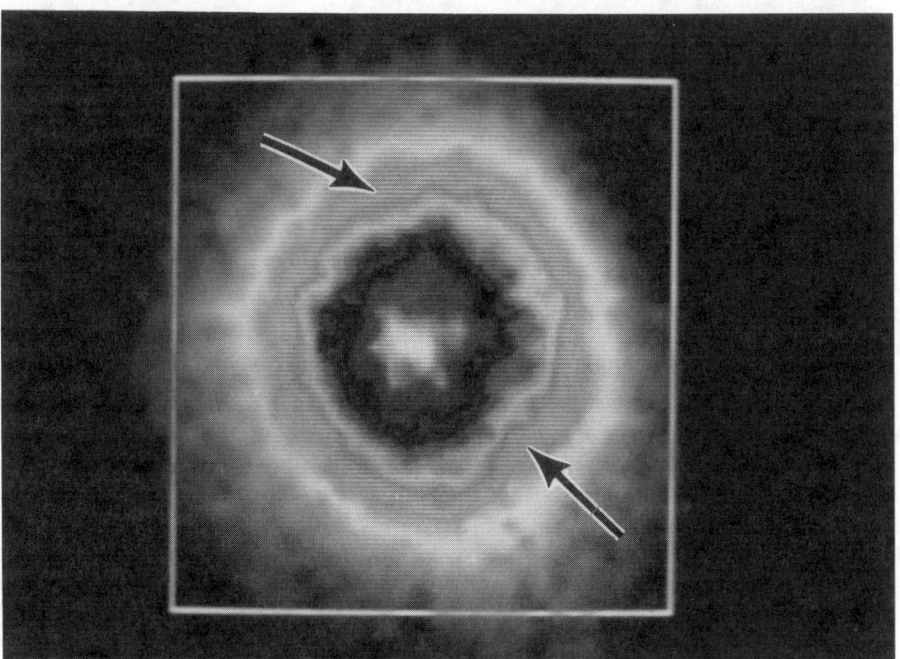

Figure 2: Black and white reproduction of a false color PRM image of a 5 keV microbeam. The central white area corresponds to 450-483 counts/pixel and a 170-190 counts/pixel contour is indicated by the arrows. The white box is 10μm × 10μm (26 × 26 pixels).

Based on the phase space requirements and the aberrations incurred in focussing the beam onto the sample,[16] we infer that ET $\leq$ 0.2 eV which would suggest that in situ heating of the Ni(100) is not sufficient alone for producing a well ordered surface. Figure 3 illustrates typical defect structures observed in our Ni(100) films when the

PRM is operated at 3900 magnification.

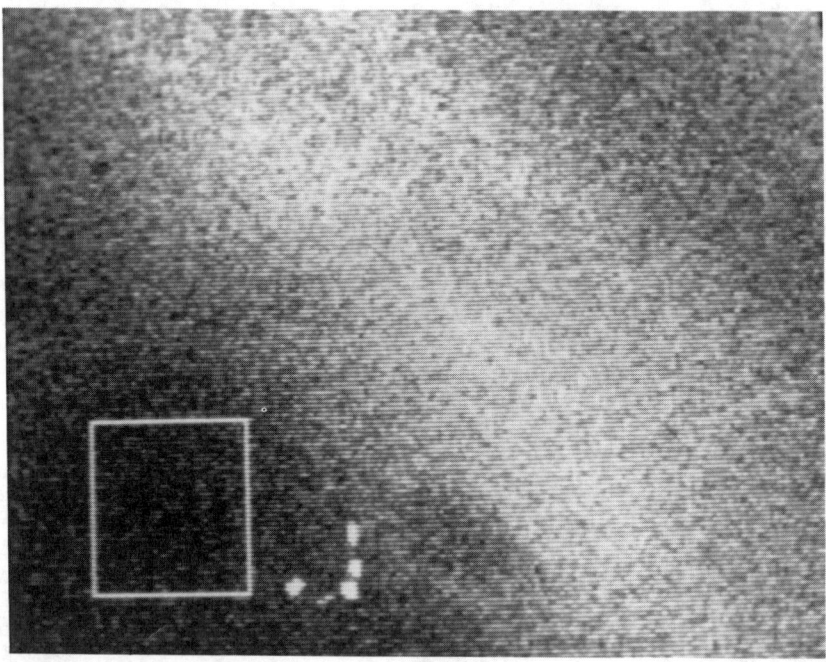

Figure 3: An 8 hour exposure PRM image of subsurface defects. White areas correspond to 81-90 counts/pixel. The dark grey area encompassed by the white box corresponds to 30 counts/pixel. The bright spots to the lower right of the box are due to detector "hot spots". The white box is 1μm × 1μm (40 × 40 pixels).

We have yet to observe a feature as a single or closely packed array of dislocation lines or low angle tilt boundaries near the outer surface so as to provide a measure of the PRM resolution as was possible by chance with our earlier PRM.[9] With the present primary beam flux at Brandeis we will principally limit the PRM experiments to studying large island growth of metals on Ni(100) at 10nm resolution. Overlaying structures should provide sharp contrast and boundaries which are easy to find as opposed to the hit and miss approach of looking for subsurface defects that happen to be close to the surface. After completing a program of experiments at 10nm resolution, our PRM will be moved to the $10^9 e^+$/s beam at Brookhaven for 1nm resolution investigations.

Our principal motivation for pushing the PRM to 1nm resolution is to be able to observe monovacancies which play a fundamental role in the onset of damage in metals and in several other mechanical properties of materials.[18] Since thermal[19] and near-thermal[20] positrons are readily trapped by monovacancies, a monovacancy at 1nm below the surface would produce a fuzzy spot with more than 50% contrast when viewed with a 1nm diffraction limited PRM at 50,000× magnification, for example. The contrast would decrease in a nearly linear fashion as the distance between the monovacancy and the surface increases, with an estimated viewing depth ranging to roughly 10 nm below the surface. By obtaining PRM images of a few monovacancies, it will be possible to track their thermally activated diffusion near the surface or within an interface. Most important will be the possible observation of monovacancy aggregation into divacancies and larger complexes which are most prevalent in the initial stages of metal fatigue. Atomic resolution PRM images of interfaces should also provide unique information about the nature of point defects in buried interfaces for overlying metal films of a few atomic layer thickness.

Another interesting area to investigate with the PRM's unique ability to image subsurface monovacanies in metals is to look at the initial stages of radiation damage. When ionizing radiation causes observable damage in most solids, it can be characterized by three stages. First the individual ionizing particle causes a low density distribution of monovacancies which mimics the incident particle's track, although there can also be a tendency for focussing the damage along close-packed directions in the lattice.[21] This is followed by a clustering of vacancies and microvoid formation as a result of many such ionizing particles and thermally activated migration.[22] The last stage can result in macroscopic damage produced by large scale thermally activated aggregation of the defects including possible interstitial impurities associated with the irradiating particles. The last two stages can be observed with a transmission electron microscope (TEM) and other conventional microscopy. However because of the low vacancy concentration the PRM provides the only way to directly observe this most important initial stage of radiation damage.

This work was supported in part by the National Science Foundation Grant DMR–9123386 and the New Energy and Industrial Technology Development Organization International Joint Research Grant.

REFERENCES

1. L.D. Hulett, J.M. Dale, and S. Pendyala, Mater. Sci. Forum.
2. C.H. Hodges, Phys. Rev. Lett. **25**, 284 (1970); see also J. Makinen, A. Vehanen, P. Hautojarvi, H. Huomo, J. Lahtinen, R.M. Nieminen, and S. Valkealahti, Surf. Sci. **175**, 385 (1986).
3. J. Van House and A. Rich, Phys. Rev. Lett. **60**, 169 (1988).
4. Ernst Ruska, Rev. Mod. Phys. **59**, 627 (1987), and references therein.
5. R.D. Heidenreich, J. Appl. Phys. **26**, 757 (1965).
6. H.-J. Drouhin, C. Hermann, and G. Lampel, Phys. Rev. **B31**, 3859 (1985), and references therein.
7. James Van House and Arthur Rich, Phys. Rev. Lett. **61**, 488 (1988).
8. G.R. Brandes, K.F. Canter, and A.P.Mills, Jr., Phys. Rev. Lett. **61**, 492 (1988).
9. G.R. Brandes, K.F. Canter, and A.P. Mills, Jr., Phys. Rev. **B43**, 10103 (1991).
10. A.P. Mills, Jr., Appl. Phys. **23**, 189 (1980).
11. A new reflection mode PRM is being developed at the University of Michigan which will incorporate normal incidence, SLOPOS 5 proceedings.
12. E.M. Gullikson, A.P. Mills, Jr., W.S. Crane, and B.L. Brown, Phys. Rev. **B32**, 5484-5486 (1985).

13. D.A. Fischer, K.G. Lynn, D.W. Gidley, Phys. Rev. **B33**, 4479 (1986).
14. Quantar Technology Incorporated, 3004 Mission Street, Santa Cruz, CA 95060.
15. For a review on photoelectron microscopy and immersion objective lenses, see O. Hayes Griffith and Gertrude F. Rempfer, in Advances in Optical and Electron Microscopy, vol. 10, edited by R. Barer and V.E. Cosslett (Academic Press, London, 1987), pp. 269-337.
16. K.F. Canter, G.R. Brandes, T.M. Roach, and A.P. Mills, Jr., in Positrons at Metallic Surfaces (Edited by A. Ishii, Trans Tech Publications, 1992), sec. 6.2.
17. K.F. Canter, G.R. Brandes, T.N. Horsky, P.H. Lippel, A.P. Mills, Jr., in Atomic Physics with Positrons, edited by J.W. Humberston, E.A.G. Armour (Plenum Press, New York 1987) pp. 153-160.
18. Point Defects in Solids - Vol.3, edited by J.H. Crawford and L.M. Slifkin (Plenum Press, New York, 1972).
19. C.H. Hodges, Phys. Rev. Lett. **25**, 284 (1970).
20. B. Nielsen, K.G. Lynn, and Y.-C. Chen, Phys. Rev. Lett. **57**, 1789 (1986); T. McMullen and M.J. Stott, Phys. Rev. **B15**, 8985 (1986).
21. A. Sosin and W. Bauer, "Atomic Displacement Mechanisms in Metals and Semi-Conductors," in Studies in Radiation Effects in Solids III, G.J. Dienes, ed. (Gordon and Breach, New York, 1969), p. 153.
22. A. Seeger, "The Nature of Radiation Damage in Metals," in Radiation Damage in Solids Vol. II, International Atomic Energy Agency, Vienna, 1962, p.101.

AN OVERVIEW OF
THE MICHIGAN POSITRON MICROSCOPE PROGRAM

D. W. Gidley, W. E. Frieze, T. L. Dull,
G. B. DeMaggio, E. Y. Yu, H. C. Griffin,*
M. Skalsey, R. S. Vallery, and B. D. Wissman

Department of Physics, *Department of Chemistry
University of Michigan
Ann Arbor, Michigan 48109

ABSTRACT

An overview of the Michigan Positron Microscope Program is presented with particular emphasis on the second generation microscope that is presently near completion. The design and intended applications of this microscope will be summarized.

INTRODUCTION

The invention of a variety of positron microscopes was reported in 1988. A transmission positron microscope (TPM), similar in operation to its electron counterpart, was demonstrated at the University of Michigan.[1] A scanning positron microprobe (SPM) was operated at Brandeis.[2] Two different versions of a positron reemission microscope (PRM) were invented later in 1988 at Michigan[3] and at Brandeis.[4] The PRM, SPM, and to a lesser extent the TPM, display totally unique image contrast mechanisms and thereby provide the materials researcher with a new view of phenomena that are not easily visible with existing microscopies. Positrons exhibit such behavior as trapping in open-volume defects as small as monovacancies, reemission from a surface due to a *negative* work function, electron capture to form positronium (similar to hydrogen formation), and annihilation with electrons yielding readily detectable gamma rays that reveal the annihilating pair's momentum. All of these unique positron properties lead to new forms of contrast in positron micrographs.

For the last three years the University of Michigan positron group has been funded by DoE's Advanced Energy Projects to design, construct, and begin testing a second generation microscope with greatly expanded capabilities as a PRM, SPM, and as an electron microscope. In the next section we will present a more detailed progress report on this new microscope–its design, modes of operation, and contrast mechanisms. At this point we will only present an overview of our microscope program which consists of four highly interrelated projects:

1. *The Microscope.* Construction of the second generation UHV microscope is the central element in the program. Designed to take advantage of thick, robust target samples, it is built as a *reflection*-style PRM compared to the Brandeis *transmission*-style PRM[4] which is based on thin film targets. As an SPM it is designed to operate with many existing positron techniques to generate the scanned signal. In addition, electron techniques of secondary electron microscopy, scanning Auger spectroscopy, and even low energy electron microscopy (LEEM) have been incorporated.

© 1994 American Institute of Physics

2. ^{58}Co Sources. To provide positron sources for the microscope we have been working on a totally in-house process for fabrication of thin ^{58}Co radioactive sources. Ni targets are irradiated in the University's Ford Nuclear Reactor. Co is chemically separated and then electroplated on a 3 mm diameter spot prior to encapsulation. In August a 40 gram Ni plate was completely processed and tested in a positron beam as a mock test of what would have been a 1 Ci ^{58}Co source if the Ni had been left in the reactor for several months. It was successful and we are confident that we will have such sources by summer 1993. By 1994 we hope to extend the process by roughly an order of magnitude.

3. INEL Collaboration. Realizing the longer-term need for even higher intensity sources we are actively collaborating with the Idaho National Engineering Laboratory (INEL) to scale up this ^{58}Co program using their EBR II reactor to the level of 10,000 to 100,000 Ci. The intense fast neutron flux of EBR II accounts for over a factor of 100 in the scaling. The goal of this collaboration is to construct (within 6 years) a third generation microscope facility capable of real-time imaging with ultimate positron microscope resolution. The design and intended applications of this facility are to be determined by our experience with the present microscope.

4. Applications. We have active non-microscopic surface physics and materials research programs utilizing positrons that are generating applications for the microscope effort. We have used positron beams to explore thin films with reemitted positron spectroscopy. In particular, we have carried out studies of pseudomorphic growth,[5] interdiffusion alloying,[6] near surface defects,[7] positron tunneling,[5,8] positron mean free paths,[6] and silicide thermal reaction dynamics.[9] All of these processes affect contrast in a PRM. We are also studying fatigue initiation in glassy polymers using positron lifetime techniques[10] and we are in the process of extending our slow positron beam techniques to include lifetime and Doppler broadening analysis.

With basic construction of the microscope completed and with 1 Ci ^{58}Co sources expected to be ready this summer, we will begin exploring a broad-based program in materials research and engineering, condensed matter physics, and biological science. The eventual need for more intense beams can be satisfied in the short term by improving our in-house source technique to 10 Ci and in the long run through our collaboration with INEL. In the meantime it is essential to explore those applications where positron microscopy can uniquely yield new information. It is precisely this goal that the present microscope is aimed at, with particular emphasis on materials research.

DESIGN AND FEATURES OF THE MICROSCOPE

A schematic representation of our second generation microscope is shown in Figure 1. The microscope incorporates both positrons and electrons because it can initially be difficult to interpret positron microscope images taken alone. Thus it is very important to be able to obtain well-characterized electron images of a sample *in situ*. In both cases, there are two possible imaging modes– magnifying mode and scanning mode. We summarize the charactistics of these various operational configurations below, and Table 1 summarizes some of the primary advantages and features the new microscope will have.

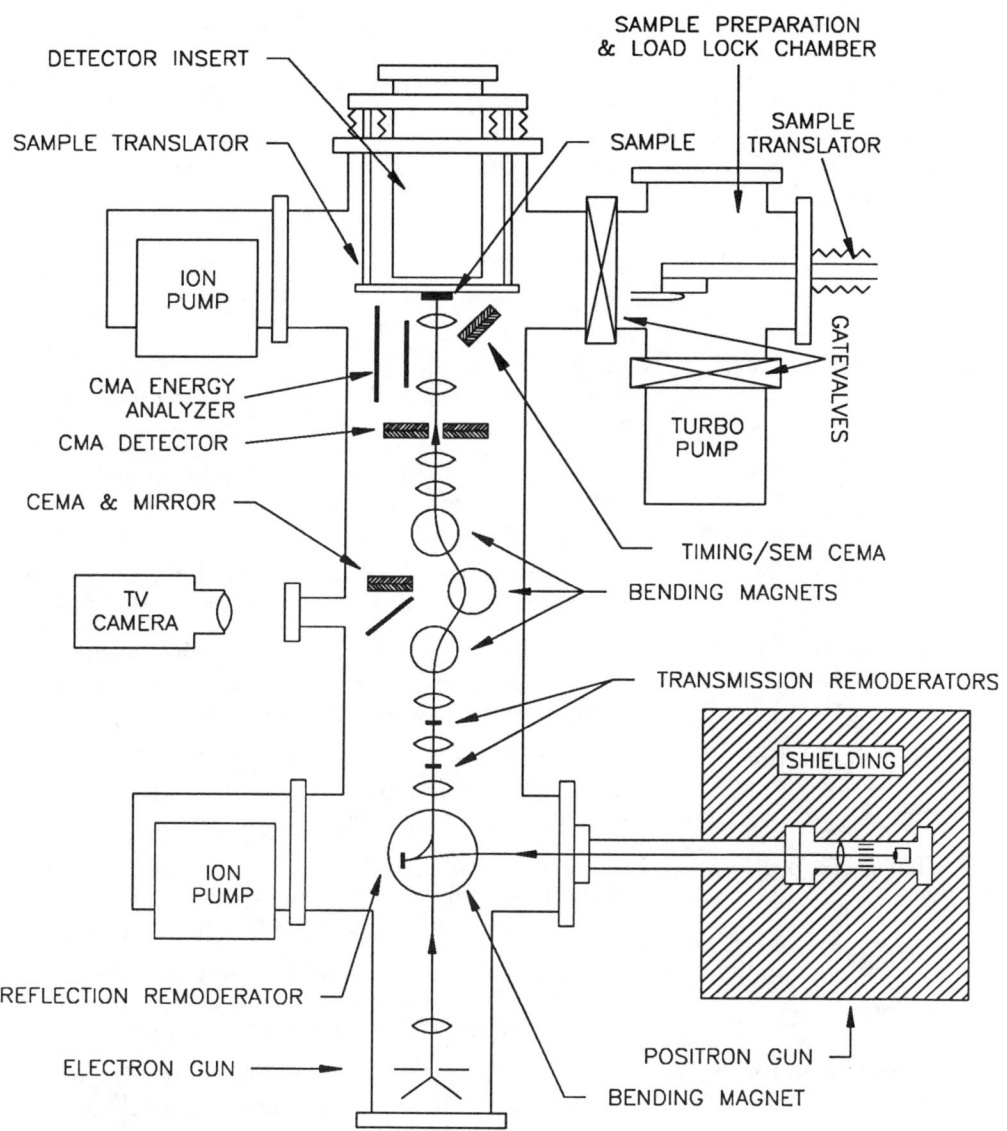

Fig. 1. Schematic Representation of the Microscope
This figure illustrates the basic design features of our positron microscope (not to scale). It shows the positron and electron guns, the three remoderators, the optics and image detector necessary for magnification mode, and some of the other detectors necessary for the various signals of scanning mode.

Table 1: Advantages and Features of the Microscope

Summary of some of the features our new positron microscope will have. In some cases, different features depend on the operating mode chosen. (For example, depth profiling and surface sensitivity appear contradictory, but in fact just depend on the choice of scanning mode vs. magnifying mode.)

- Multiprobe Capability — Multiple modes of microscopy will be possible: magnifying and scanning positron microscopy, LEEM, SEM, and SAM.

- Defect Sensitivity — Positrons offer the most sensitive measure of defect concentrations available; they are directly sensitive to the presence of lattice disorder.

- Defect Specificity — Microscope signals can differentiate among various defect types, yielding simultaneous information on several concentrations.

- Depth Profiling — Different implantation depths for different incident beam energies permit studies at specified depths.

- Surface Sensitivity — Some operating modes (such as positron tunneling microscopy) are highly surface specific, permitting study of purely surface phenomena.

- High Resolution — Ultimate lateral resolution in magnifying mode should be near 10 Å. Submonolayer depth resolution is possible in tunneling mode.

- Thick Samples — Samples can be used directly with no thinning; standard thick-sample preparation techniques and electron spectroscopies can be employed.

- In Situ Capability — Positron microscopy may be used simultaneously with sample treatments such as heating, evaporative coating, or application of electrical signals.

- Non-Destructivity — The small rates of positron beams require instrumentation sensitive to currents in the 10^{-15} A range or below, thus reducing sample damage to a minimum.

- High Vacuum — Base pressure in the microscope should be in the 10^{-9} torr range.

1. *Positron Magnifying Mode.* In this mode, positrons implanted in a sample are reemitted because of the negative work function. The reemitted positrons are then accelerated, magnified, and focused on an image detector. Any physical process which affects the probability of reemission, the energy, or the angular distribution can produce contrast in such an image. Magnifying mode should permit ultimate lateral resolutions near 10 Å, and can be performed in a depth profiled version (by varying the implantation energy of the incident positrons) up to a maximum depth of several thousand Å. This mode is often referred to as a positron reemission microscope, PRM.

2. *Positron Scanning Mode.* This method, analogous to scanning electron microscopy, involves a small diameter beam rastered across the sample while a signal of some sort is collected in synchronization. Positrons permit a wide variety of signals, including Doppler shift in the energy of the annihilation gamma rays, lifetime of the positrons in the sample prior to annihilation, and number of emitted positronium atoms. Lateral resolutions in scanning positron microscopy are limited by diffusion-induced spreading of the beam spot to roughly 1000 Å, but depth profiling is possible up to a maximum depth of several μm.

3. *Electron Microscopies.* Both magnifying and scanning mode will also be possible using electrons. Scanning electron microscopy and scanning Auger microscopy will be our primary traditional techniques for sample characterization. Fortunately, the optical system required for magnifying mode positron microscopy is identical to that required to perform the new technique of low energy electron microscopy (LEEM).[11] In this technique, an image is formed using the electrons from a single diffraction spot at an incident energy typically well below 100 eV. In effect, LEEM is imaged LEED with lateral resolution better than 100 Å. It has generated considerable excitement in the surface physics community as a surface probe with sub-monolayer depth resolution combined with the capability of producing real-time images of dynamic processes. Our instrument will be able to perform LEEM with only minimal changes in the voltage supplies of the lenses.

APPLICATIONS IN SURFACE STUDIES

There are many possible applications of positron microscopy that we would like to explore over the next several years. Some of the most promising cases are mentioned below. They are ordered according to the operational mode of the microscope: either positron reemission microscopy (magnifying mode) or defect imaging (scanning mode).

In reemission mode we expect the microscope to have unique contrast for thin films and overlayers, as well as sensitivity to surface structure, impurity coverage, and near-surface defect distribution. One very interesting case occurs when an overlayer has a higher positron energy level than the substrate, thus forming a barrier for emission from below. If the overlayer is thin enough, emission can occur by tunneling. Positron tunneling microscopy[8] should thus permit studies of overlayer growth and structure in suitable systems with sub-monolayer depth resolution. Some particular examples of studies we would like to initiate using magnifying mode are:

1. *Growth Mechanisms and Pseudomorphism.* Many of the phenomena associated with heteroepitaxial growth should be visible using positrons. Islanding, pin-holes, thickness variations and the like should be readily visible in

reemitted positron images. Even in cases where there may be no topographical contrast, physical processes that change the positron energy level of the overlayer such as pseudomorphic growth and interdiffusion alloying can produce contrast. For example in the Ni on Cu system, we have observed[5] that strained growth can eliminate the 0.5 eV energy barrier to positron emission normally presented by a Ni overlayer. We would like to acquire micrographs at various Ni film thicknesses to study the commensurent-to-incommensurent transition that occurs near the critical thickness (bright areas would be under strain, dark areas would be relaxed with strain localized at the misfit dislocations).

2. Surface Catalysis. Positron microscopy may be very useful in studying model systems in chemical surface catalysis where traditional electron microscopies are known to be inadequate, namely low Z adsorbates on high Z substrates. Carbon and sulfur for example, do not have sufficiently high electron scattering cross sections to be observed on highly scattering, high Z substrates like Ru and Pt. However, the positron emission process is independent of the thickness of the substrate, depending only on conditions near the surface. We should be able to observe the presence of adsorbed species by such mechanisms as trapping or scattering of the reemitted positrons, or changes in the positronium formation probability. LEEM may also provide complementary new information on these systems.

3. Thin Film and Interface Reaction Dynamics. The material specificity of positron reemission should be useful in studying the growth dynamics of systems such as metal silicide films produced by thermal reaction of metal films grown on silicon substrates. We have shown that all three major stoichiometries of the Co and Ni silicides have different positron work functions, and hence can be clearly distinguished in reemitted positron spectra.[9] With depth profiling we should be able to observe the 3D growth profile of the reaction.

APPLICATIONS IN DEFECT MAPPING

As a scanning defect microprobe the microscope will find application in a broad range of defect nucleation/damage initiation studies. Because positrons directly detect *disorder* in a crystal (as opposed to traditional techniques such as diffraction measurements which look for small departures from perfect order), they are the most sensitive probes of defects available. They are able to detect monovacancies, for example, at concentrations as small as 1 in 10^7. The wealth of signals available in scanning positron microscopy should permit detailed maps of the defects in a sample, maps which can be extended to three dimensions by depth profiling. Some particular examples of studies we would like to initiate using scanning mode are:

1. Electromigration Induced Voids. The *initiation* stages in the growth of electromigration induced voids in the conducting gateways of microelectronic devices are invisible with present electron techniques. High energy SEM can resolve such voids at the 0.1-1 μm level, but there is no hope of observing the nucleation phase at the 10 Å level. Defects of such a size are efficient positron traps, and hence will be visible in a positron micrograph. Thus we should be able to detect the defect nucleation phase of these voids well before they threaten the operation of the device, and at a size that is small even compared to the next several generations of downsized microelectronic devices.

2. Corrosion Initiation. A phenomenon similar to the growth of electromigration induced voids in device interconnects is the pitting of passivating oxide layers during corrosion of metals. In this case, impurity ions pair with vacancies to form Frenkel type defects. These provide the charged species that migrate under the influence of internal fields to agglomerate as voids that eventually become corrosion pits. Again, positrons should be able to map defects in the early initiation phase when current microscopies see nothing and before complete failure of the passivation. Here the ability to image thick samples is a distinct advantage. LEEM may also offer exciting new possibilities in this field.

3. Fatigue Initiation in Polymers, Ceramics, and Composites. For several years, we have been participating in a program (funded by the Division of Materials Science of DoE) of non-microscopic studies of polymer fatigue, using (among other techniques) positron lifetimes. Our measurements[10] have indicated that the physical changes induced by fatigue in glassy polymers such as polycarbonate are probably localized within a few μm of a very few nucleation sites. These sites eventually become visible as crazes and cracks, but at this point the sample is within 20% of the end of its useable life. The initiation phase, which is nominally 80% of the sample's fatigue lifetime, is invisible to conventional microscopies. Again, positron lifetime microscopy provides us with a technique for observing the nucleation of smaller defects as they evolve or coalesce into larger voids that become cracks. No other technique is sensitive to such small voids. A corollary study is planned for crack propagation in polymers, whereby the positron microprobe would be scanned in front of the crack to search for void coalescence and its spatial correspondence with the visible crack. This may provide the first firm evidence for microstructural changes at the 10 Å level induced by the propagating crack.

We suspect the positron microscope will find similar uses in some composite, ceramic, and ceramic-fiber-composite materials where crack formation is the dominant fatigue failure mechanism. These high performance materials in terms of strength, rigidity, or durability are of particular interest in the transportation and utility industries.

4. Radiation Damage. Crack formation is not limited to polymers of course, nor is the instigating agent limited to fatigue. Another interesting application is to radiation damage in metals. The abilities of a scanning positron microscope to map defects, to differentiate among different types of defects, and to nondestructively examine samples as the damage evolves, should be invaluable in the study of the complex defect dynamics that precede failure in materials damaged by radiation.

CONCLUSION

We anticipate completion of the basic microscope early in 1993. Additional capabilities will continue to be added to the instrument thereafter, but the ability to acquire images should begin at that time. The thrust of our research over the next several years is to explore the widest range of applications possible and to develop the most promising of them. Our main construction projects will be a sample manipulator/load lock system and a separate sample preparation chamber, with outfitting dictated by the needs of the particular applications under study. In the next several years we should have a much clearer assesment of the capability of positron microscopy to provide unique information in material systems of scientific and technological importance.

The construction of the second generation microscope was funded by DoE Grant DE-FG02-90ER12103 and by the Office of the Vice President for Research of the University of Michigan. Most of the ^{58}Co source development project has been funded by NSF Grant PHY-9119899, with some support from INEL. The application work has been mainly funded by NSF Grant DMR-9003987, with polymer studies funded by DoE Grant DE-FG02-88ER45366. This support is gratefully acknowledged.

REFERENCES

1. James Van House and Arthur Rich, Phys. Rev. Lett. 60, 169(1988).
2. G. R. Brandes, K. F. Canter, T. N. Horsky, and P. H. Lippel, Rev. Sci. Instrum. 59, 228(1988).
3. James Van House and Arthur Rich, Phys. Rev. Lett. 61, 488(1988).
4. G. R. Brandes, K. F. Canter, and A. P. Mills, Jr., Phys. Rev. Lett. 61, 492(1988).
5. D. W. Gidley, Phys. Rev. Lett. 62, 811(1989).
6. D. W. Gidley, W. E. Frieze, Phys. Rev. Lett. 60, 1193(1988).
7. D. W. Gidley, A. R. Köymen, T. W. Capehart, Phys. Rev. B, 37, 2465(1988).
8. W. E. Frieze, D. W. Gidley, B. D. Wissman, Solid State Comm., 74, 1079(1990).
9. B. D. Wissman, W. E. Frieze, D. W. Gidley, to be published Phys. Rev. B, see paper this proceedings.
10. L. Liu, A. Yee, D. W. Gidley, Journal of Polymer Science: Part B, Polymer Physics 30, 231(1991).
11. W. Telieps and E. Bauer, Ultramicroscopy, 17, 57(1985) and W. Telieps, Appl. Phys. A, 44, 55(1987).

SECTION 6

FUTURE USES OF POSITRONS

Why Antihydrogen - and not just Bare Antiprotons?[1]

B. W. Augenstein
RAND Corporation
1700 Main Street
Santa Monica, CA 90407

ABSTRACT

A next step of great interest and importance in the antimatter RDT&E field will be the merging of goals of both positron and antiproton experimenters, to produce neutral antihydrogen. This neutral form of antimatter - in the first instance in the atomic form - will have numerous critical implications, for both basic research and for applications of antimatter where the amounts needed can raise very difficult storage problems, if we were to store bare antiprotons.

This paper introduces basic issues of producing and storing antiprotons, and then discusses some uses where antihydrogen becomes vital. Antihydrogen production is also a basic step in producing cluster anti-ions, wherein the charge/mass ratio is much smaller than that of bare antiprotons.

The technologies in the positron and antiproton field appear ripe for active collaborative efforts, a collaboration necessary if the synthesis of antihydrogen is to become a routine reality in the very near future.

Chart 1 highlights the main topics of this paper. Three basic references are introduced.[1-3] These references will be called upon in what follows. The cluster ion production technique for producing macroscopic amounts of antimatter are described by Stwalley.[4] The scheme considers producing H and a catalyst H_N^+; the individual reaction steps potentially leading to the H_N "seed crystal" are reviewed in detail, as are the processes for prior normal matter simulations. The issue of accelerator production of heavier antimatter is discussed by Forward.[5] It is of some interest that, compared with antiproton delivery levels, delivery of antideuterons is down by "only" a factor, roughly, of about $d/\bar{P} \sim 10^{-4}$, at energies of ~ 200 GeV, validating the considerable interest in producing and collecting antideuterons for special experimental purposes. Alternative schemes for producing heavy anti-elements are discussed by Takahashi.[6] This scheme (suggested also by Forward) proposes using muon-catalyzed fusion from P using the μ^+ to make D or T, as a start. The processes are susceptible to preliminary tests using normal matter μ^- mesons and ordinary nuclei. Finally, the considerable and growing European interest in antihydrogen is reflected in a series of early papers.[7]

---

[1]This paper was actually given at the Workshop by Dr. Calvin Shipbaugh of RAND. For this and other active contributions, Dr. Shipbaugh is gratefully acknowledged.

Chart 2 develops some of the background on antiproton technologies needed to lay the basis for production of antihydrogen.

For production of antiprotons by interactions produced from the collision of a high energy proton beam on a metal target at rest in the laboratory system, the fundamental parameter of interest is the number of antiprotons produced per incident proton of energy E. This fixed target production is the standard technique used at CERN and FNAL. The simple production reaction is $P+P \rightarrow P+P+n(P+\bar{P})$, so that n nucleon-antinucleon pairs are produced. While a simple relativistic calculation shows that the threshold kinetic energy E_T to produce n nucleon-antinucleon pairs, with M the particle mass, is $E_T=2Mn(n+2)$, so that n antinucleons could result, the actual number resulting is very much smaller, because a great many competing reactions arise. Thus, a 100 GeV proton (FNAL usually operates at ~ 120 GeV), if all its energy could be devoted to producing appropriate antinucleons, could produce ~ 6 antiprotons. Competing processes in today's techniques lower this number to ~ 5×10^{-2} antiprotons, a factor of about 100 reduction. The value of ~ 5×10^{-2} is essentially supported by both theory and experiment.

These results say that FNAL should get about 5×10^{-2} antiprotons per proton. But actually FNAL gets still less - about 2×10^{-5} antiprotons per proton. Where does this added shortfall factor of ~ 2.5×10^3 originate? (The precise factor requires untangling of several effects.) The sources of the shortfall lie almost wholly in collection of the produced antiprotons, and are reasonably well understood. Detailed discussion would get into a number of technicalities of the FNAL antiproton production system, but major factors contributing to the shortfall include target characteristics (absorption, etc.), the momentum bite accepted by the collection system, the optics, depth of focus, etc. of the "lens" used, the ability to capture antiprotons over the whole conical angle of the emerging antiproton spray, and other loss parameters. In principle, given enough space, power, and money, most of this shortfall - between two and three orders of magnitude, roughly - could probably be avoided in a de novo design. Specific upgrading issues will be developed later.

Transportable storage options come in a number of possible implementations. The Penning trap technology is well-known and is described by Howe, et al.[8] The Penning trap typically uses electric fields to confine charged particles in the axial direction and magnetic fields for radial confinement (thus skirting around Earnshaw's Theorem). Radially, a force balance between Lorentz forces and centrifugal forces must be achieved; an exact force balance is called the Brillouin Limit. It is also possible to use a Paul trap to hold and cool both positive and negative ions. A Paul trap has the same electrode configuration as the Penning trap, but employs rf electric fields instead of static electric and magnetic fields to form the trapping region. A Paul trap could hold antiprotons and positrons and cool them to sufficiently low relative velocities so that conversion to H could take place. Complex arrangements of Penning traps could accomplish similar ends. Small rings are discussed in great detail by Cline.[9] A proposed design considers a ring about 4 meters by 2.5 meters; weighing less than 10 tons; storing up to 10^{12} antiprotons for $\geq$ 3500 hours with cooling; and capable of an energy range of about 200 MeV down to about 100 KeV. The main utility of rings for storage is their containment of relatively high energy antiprotons (a few hundred MeV, for example). Such energies are suitable for certain biomedical purposes - e.g., for antiprotons to penetrate several tens of centimeters of body tissue - without going through the hassle of re-accelerating antiprotons from the very low storage energies compatible with Penning trap technologies.

Production of antihydrogen from antiprotons and positrons, using a variety of techniques, has been extensively studied; see the review article by Mitchell.[10] Additional early detailed studies are by Poth, et al.,[11] Deutch, et al,[12] and Gabrielse, et al.[13] Although apparently antihydrogen has not yet been produced in the laboratory (as of August 1992), extensive plans persist to do so.[14-16] There seems little doubt of the feasibility of doing this, and on a scale adequate to satisfy foreseeable needs. Recall that early normal matter experiments at the Soviet NAP-m device and at CERN, about 15 years ago, where merging beams of protons and electrons (for proton cooling) produced relatively impressive amounts of normal hydrogen, lend a considerable plausibility to at least one of such possibilities. We believe we know processes capable of very high rates of antihydrogen production (see the referenced papers). A near-term goal would be to create antihydrogen at rates compatible with antiproton production rates; we believe this to be a distinct possibility, based on the arguments cited in the references. Finally, schemes for storing (neutral) antihydrogen are also fairly well formulated.

Chart 3 discusses issues of increasing antiproton production. Starting from the current embodiment of antiproton production machines, as reflected in the current phase by FNAL (which would however need add-on facilities to produce antiprotons at the appropriate low energies) and CERN, what are some of the next steps which could be taken towards increased antiproton production, particularly by advancing to the fixed target, enveloping plasma type of facility? A next phase could take one major path:

> Exploit embodiments of proposed Hadron/Kaon machines. These designs are typically proton synchrotons at ~ 30-60 GeV energy, stretchable to ~ 100 GeV, and with beam currents of perhaps 100-200 μA. Such machines would allow full testing of scaleup RDT&E - the pacing issue is cooling of the antiprotons - and could with little difficulty raise production and collection levels to the tens to hundreds of micrograms annual level. Collection could be very significantly enhanced using the enveloping collector. A machine of comparable production used only for antiproton creation would be expected to be less costly than the broad capability Hadron/Kaon machines. Even more extreme scaleups are considered by Mills[17] and Augenstein.[18]

The practical problems of producing and accumulating one milligram of antiprotons per year have been subjected to rather careful analysis by Fred Mills of FNAL in Ref. 1. The very nice Mills paper has a reasonably complete discussion of a number of topics: production cross sections, collector and accelerator types (including luminosity requirements to make colliding beam systems work), candidate accelerator types, antiproton cooling methods, and the essential Research and Development areas. The latter contains proposals for scaleup R&D in some seventeen critical areas. The Mills paper is highly recommended to anyone wishing to become acquainted with the manifold technical aspects of scaleup issues and performance requirements for antiproton production, from the point of view of a preeminent machine expert. From this paper one can also derive a sense that in certain ways antiproton production scaleup to the vicinity of one milligram per year reflects about the *maximal* step one could take using today's and near future information base on basic machine parameters and embodiments. However, as the Mills paper also stresses, we have a good grasp of the basic RDT&E we can undertake to have some sensible prospects for moving significantly above the milligram per year level.

Variations of the scheme of using a fixed target for antiproton production have of course been proposed. Antiproton yields in relativistic nucleus-nucleus collisions have been measured[19] to be significantly increased (e.g., about a 60-fold increase when carbon nuclei are collided with copper nuclei at energies of 3.65 GeV per nucleon). Effective routine use of such schemes is still somewhat problematical and in the future. Takahashi and Powell[20] propose possible enhancement of antiproton production by multiple collisions in a thick target.

Chart 4 summarizes the future possibilities for antiproton production if we very carefully plan for and optimize the engineering arrangement and resolve some of the outstanding R&D problems, particularly in cooling of the antiprotons. Note that the power levels on the chart hold only for *future optimized systems*. The power level for *current* FNAL operations is typically in the several MW range, much more than necessary for optimized systems. While the power investments at the high end of the production may seem high, there are extensive possibilities to arrange for "self-powering" of such systems. The schemes for this are considered in Augenstein.[21] Those who are frightened of the use of Gigawatts of power may recall that at one time the U.S. gaseous diffusion plant for producing special nuclear materials consumed over 6 Gigawatts of input electrical power.

Chart 5 reviews some of the antiproton trap storage issues, and estimates (probably conservatively) the weights relevant to traps storing a respectable number of antiprotons. We assume in the "Go for broke" category, perhaps possible in a few years, that a weight of the order of one ton (or perhaps even somewhat less) to store about 10^{12} antiprotons may become realized.

Chart 6 discusses some options for increasing trap storage capabilities. A fundamental review paper of some of these areas is by Campbell.[22] This paper discusses in a comprehensive way some of the more "exotic" possibilities for antiproton storage. The difficulties one may encounter in this set of R&D paths emphasized one of the key practical reasons for creating antihydrogen (or cluster anti-ions) - circumventing basic space charge constraints one encounters in storing many particles of like charge.

Chart 7 suggests in summary some of the basic reasons for emphasizing the production of antihydrogen. The rest of this paper develops some of the themes indicated on Chart 7.

Chart 8 gives a few examples of antihydrogen use in basic physics experiments.

The use of antiprotons to test the gravitational interactions of antimatter has been extensively developed, primarily by Nieto and Goldman.[23] This paper gives a wealth of several hundred references, dealing with all aspects of the relevant issues. While there are now cosmological and astrophysical bounds on how much difference one might expect for the gravitational accelerations of antiprotons vice protons, of course even the minutest of differences would be of tremendous interest. Actual experiments using antiprotons have been carefully planned. After extensive cooling of the antiprotons to low temperatures in a "release trap," at time t=0 about 100 particles at a time are released (to minimize Coulomb force effects), with some distribution of energies, to head upward in a vertical drift tube with signal detection at the top of the drift tube. Particles with relatively large energies will race up the tube, while those with smaller and smaller kinetic energies will arrive later and later. Finally, there will be the last antiproton with just enough energy to make it up the drift tube against the force of gravity, with an arrival time of t=τ. From τ the gravitational acceleration of the antiproton can be determined. Over many experiment repetitions, a time-of-flight spectrum can be built up, so that good statistics for τ can be developed.

Although the larger mass of the antiproton compared with that of the electron or positron greatly reduces the error sources of the classic Fairbanks electron gravitational experiment[24] (see additional references in Ref. 23)), one still has relatively large errors which can be minimized by using neutral antihydrogen. The subject is also extensively discussed by Gabrielse[25] and by Beverini, et al.[26] Note that either a null experiment (antiprotons fall exactly as does normal matter), or an experimental result suggesting different gravitational accelerations for antiprotons, is very important.

The history of the study of the gravitational acceleration of elementary particles and elementary anti-particles is long and complex; Ref. 23 discusses the salient issues comprehensively. Reference 23 also emphasizes that the antiproton experiment is of importance not only in testing the existence of non-Newtonian gravitational forces, but also in studying the possibility of new non-inverse square and/or composition-dependent components of gravity.

Quantum Electrodynamics (QED) and its associated normal matter tests today give us some of the most precise insights on fundamental physics. QED-like tests using antimatter are discussed in many places; see, e.g., Bonner and Nieto.[27] Other commentators include Neumann,[28] and Hughes and Deutch.[29] The last reference is compelling, and points out for example the details of why spectroscopic measurements on antihydrogen could yield an improved test of the antiproton's charge, to about one part in 10^{11}. The measurement would combine precision cyclotron frequency measurements with spectroscopic measurements on antiprotonic atoms. Indeed, both the proton-antiproton mass and charge ratios would be testable to a precision of one part in 10^{11}.

Chart 9 summarizes what may be one of the most compelling uses for antiprotons - biomedical uses. The topic is treated at length by Kalogeropoulos, et al.[30] The chart should be fairly self-explanatory.

Imaging appears to be perhaps one of the most promising single near-term biomedical application for antiprotons. As an example of the potential of antiprotons, 10^7 antiprotons could give the same quality image as a computer tomography scan, with 1/15 the dose and with none of the artifacts that can cluster in a CT image. An entire image requires only 10^9 antiprotons, which is also well within the portable storage capabilities envisioned.

For therapy the doses must be increased one or two orders of magnitude, and at those levels more information is needed about the local energy deposition in biological targets. One potential application for antiprotons in therapy is as a tool for testing, monitoring, simulating, and improving proton and heavy ion therapies.

406 Why Antihydrogen-and Not Just Bare Antiprotons?

Because antiprotons annihilate at the end of their range and send out products that can be traced back to the annihilation point, they are unique among portable particle beams in their ability to determine accurately where the therapeutic effects are taking place.

The third interesting area for medical experimentation with antiprotons, using X-ray emissions or nuclear gammas, is in the general area of "mesic chemistry" or imaging elemental atoms in-vivo or in-vitro. Antiprotons have several advantages over muons used for the same purpose and, with portable storage devices, promise the ability to monitor all elements in the living body. Oxygen, carbon, hydrogen, nitrogen, calcium, and phosphorus - in fact, all elements at once - can be imaged by events with 10^9 antiprotons (i.e., ~ 1 rad), with images of constituents up to phosphorus made with millions of events.

Finally, there are a number of special uses, including the in situ transmutation of elements in the body into natural positron emitters, thus permitting the easy, direct use of PET technologies for special diagnostic purposes. This pleasant marriage of two antimatter technologies is one example of a theme emphasized in this paper - it is timely to bring together antiproton and positron experimenters, in a synergistic way producing highly productive results.

It should be noted that the same properties of antiprotons which make them useful for biomedical purposes also hold very substantial promise for industrial inspection, general NDE purposes, and processing purposes; see Greszczuk.[31] In particular, inspection speeds may be raised by about three orders of magnitude, compared with, say, conventional CT techniques.

Chart 10 should again be self-explanatory. The argument is that very large scale uses of antiprotons for biomedical purposes - uses still compatible with the anticipated upgrading of antiproton production levels in a relatively few years - would quickly run into logistic and sheer convenience problems for storage. Thus a very high payoff would accrue to improved storage, and storage in antihydrogen or cluster anti-ion form would have great interest.

Chart 11 introduces one of the very interesting potential engineering applications of antiprotons - their use for various kinds of rocket propulsion. The basics of this use have been reviewed by, among others, Morgan,[32] Forward,[33] and Augenstein.[34] The basic problem that arises is that large numbers of antiprotons are needed - numbers which are in excess of realistic near-term scaling plans. Thus alternative plans are reviewed which could reduce the numbers of antiprotons needed. A start on such considerations was undertaken in another context by Polikanov.[35] This work was quickly elaborated on in a fundamental paper by Solem,[36] who discussed the general theoretical basis for using antiprotons for opacity and equation-of-state measurements otherwise accessible only through nuclear weapons tests. The work was then carried forward by, e.g., Smith, et al.,[37] who undertook a comprehensive series of studies on using antiprotons to boost microfission processes whose goal is to initiate self-propagating fission reactions in the smallest amount of material. Calculations have now been carried out using a number of the most competent codes available. The intent finally is to combine properly this fission burn to aid fusion burning also - hence *antiproton-catalyzed micro-fission and fusion burn*. Using such techniques, it is possible that we can achieve ICF propulsion using many orders of magnitude fewer antiprotons than pure antiproton propulsion would require. As a matter of fact, enthusiasts in this field will suggest that antiproton catalysis may be a means for generally realizing ICF results earlier and more surely. These possibilities are, for example, discussed in Augenstein,[38] and in Chiang, Lewis, Smith, et al.[39] Both these papers contain extensive references.

A series of experiments on micro-fission are now planned for the SHIVA Star facility at the USAF Phillips Laboratory.

SHIVA Star is a solid liner imploder operated at the Phillips Laboratory, Kirtland AFB, Albuquerque, New Mexico. Energy is provided from a 9.45MJ capacitor bank, which is inductively discharged through an anode-cathode structure. The resultant multi-megamp current which flows down the liner forces it to collapse inwardly. For the antiproton-boosted microfission application, the liner will be used to compress a working fluid, such as hydrogen, with a large speed of sound to avoid shock formation. This produces a pressure field around the target which is positioned at the center of the liner, leading to a fast (several hundred ns) uniform pressure up to ~ 40 Mbar on the spherical target.

It is important to find conditions under which the concept of antiproton-boosted microfission can be tested. The SHIVA Star testing is designed to provide such conditions. This can be done under relaxed conditions of compression work and pressure (typically 200-1000 KJ and 10-40 Mbar, respectively), provided the target mass is in the tens of gram range. Calculations show that, with one thousand antiprotons injected into such a target in a short burst, significant subcritical neutron multiplication (20-40%) can be expected. We assume a U(235) target, 250KJ compression work, a neutron detector with 5% net efficiency, and a 10 ns pulse of only 10^3 antiprotons. The test would be considered a success if neutron multiplication relative to uncompressed targets in the mass range 10-30 grams (1000-1200 versus 900-950 counts) were observed, and scaled in proportion to the number of antiprotons.

This is the standard description of the proposed SHIVA Star test series. It is however evident that many exciting variants of such tests exist, to provide broad experience with fission and fusion burns in small targets. The experiments as a whole could constitute a valuable precursor to the more general notion of a Laboratory Micro Fusion capability (LMF), for example.

In the meantime, it is useful to know that all the nuclear pulse propulsion techniques - ranging from the many variants of the ORION concept, which uses nuclear bombs to push or pull space vehicles, to the ICF, employing many very small explosion concepts - form a continuum of possibilities which can be characterized by a relatively small number of basic parameters. This is discussed in Augenstein.[40]

Chart 12 now discusses a specific ICF propulsion use, for a Mars mission, based on Reference 39. We see from this example that the number of antiprotons needed - about 10^{16} - is then much more compatible with current FNAL antiproton delivery levels, and is in fact well within the current actual production (but not collection) capabilities of FNAL. But now antiproton storage issues become paramount. To keep within the take-off mass limits assumed for this Mars mission (2000 metric tons), we would need to store about 10^{12} antiprotons in ~ 100 Kg storage containers (the containers would include all elements needed for storage - vacuum vessels, magnets, controls, diagnostics, etc.). While the earlier discussion - Charts 5 and 6 - suggests that one might not be able to rule out such technologies completely, it would clearly require tour-de-force developments. In addition, apart from weight issues, there would likely remain rather serious issues of complexity in such

408 Why Antihydrogen-and Not Just Bare Antiprotons?

arrangements. The payoff from more efficient storage would be enormous - for example, a levitated antihydrogen ice mass could in principle be stored at the container weight of one antiproton storage container. Remember also that the mass of the assumed heavy metal mixer in the propulsion system of the example is itself about 500 metric tons. Thus we get uncomfortably close to the limits of possibility, if we had no better options for antimatter storage than storing bare antiprotons.

There remain a great many detailed engineering challenges connected with the storage and use of antihydrogen for this propulsive purpose. For example, for the actual manipulation of the antihydrogen from storage to the ICF chamber, do we try to guide antihydrogen itself; or do we first remove a small amount of antihydrogen, strip the positrons from it by one of several techniques, and then guide the remaining antiprotons into the ICF chamber? Clearly very numerous intriguing issues remain to be explored.

Chart 13 reviews some aspects of using relatively small amounts of antiprotons for power generation. The Brillouin limit argument shows clearly that unless one has a means for multiplying the energy release from bare antiprotons by other subsidiary processes, there is in general no merit energetically for extracting power from bare antiprotons, since we already have in the trap confining fields at least as much energy as we can get from the antiprotons. Multiplying the antiproton energies, as we propose in the ICF-related uses discussed for propulsion, requires a considerable added mass of supplementary machinery, and that added mass would vitiate the desire for a compact, lightweight, all-environment-compatible means for power generation.

We then again have very great payoffs in finding a storage means in which the stored antiproton annihilation energy far exceeds any energy needed to confine the antiprotons. Therefore again storage in antihydrogen (neutral) form or in cluster anti-ion (charge/mass ratio much smaller than bare antiprotons) form is very attractive.

It should be noted that no careful, critical study of the relative applications merits (for appropriate uses) of antihydrogen vice cluster anti-ions appears to exist. Here is a nice opportunity for authoritative comparisons and assessments.

Aside from the storage issues per se, there are many very interesting questions on how one uses the antimatter for power generation. These questions are sketched in Augenstein.[41] A specific case from the sketch where the antimatter is used as a simple heater is noted in Chart 13.

Another kind of possibility is also given in Ref. 41 - an example possibility for direct electricity production. γ-rays are produced via several processes in annihilation and any subsequent electromagnetic cascade. Characteristic γ-ray energies are ~ 100 MeV (a broad spectrum is produced). Compton scattered electrons from 100 MeV γs are highly peaked in the forward direction, leading to the suggestion of a "battery" of several plates from which γs drive electrons, and produce large potentials between pairs of plates. The annihilation source is placed at one end of such an array. Important analytical issues are, for example, the solid angle subtended by the plates vis-a-vis the annihilation source, the capacitance of the whole assembly, leakage currents between plates, and total Compton cross sections of each plate. This speculative possibility and special integrated applications are raised in a 1988 RAND document by Solem, Mayer, and Augenstein. The possibility would require concerted analysis and RDT&E to assess practicality; but the point again is that the necessary program is fully definable, and its payoffs could be large in several fields.

Reference 41 discusses a number of applications for power generation in specific environmental circumstances. Many very interesting applications emerge, when we begin to have available amounts of antimatter in the range of tens to hundreds of micrograms, or more - at which levels storage in forms other than bare antiprotons becomes mandatory, of course. For example, one can consider small unmanned vehicles periodically flying in the atmosphere of Mars and landing - an exploration vehicle of unmatched mobility. Such propeller driven vehicles have been considered by Clarke, et al.,[42] and by Augenstein.[43] One can go through detailed performance analyses (see, e.g., Reference 43), but rather direct calculations are also possible when one has a very compact power source such as may be possible using antimatter.

The extremely low mass of antimatter needed leads to a number of very simple performance relations. Equating the energy in a mass m of antimatter, used at an efficiency ε, with C = light velocity, to the energy needed to fly a mass M a distance R, over a body of radius R_B with a gravitational acceleration of G_B, one gets: $m = M/2\varepsilon C^2 \cdot G_B R_B \cdot D/L \cdot R/R_B$. D/L is the vehicle drag to lift ratio. Complete circumnavigation of Mars in a great circle route ($R = 2\pi R_B$), with a vehicle of 400 kg, D/L = 1/30, ε = 1/3, would require $m \approx 18$ micrograms of antimatter, as one performance example. This example assumes an antimatter heated engine, employing a closed cycle Stirling engine to drive a propeller.

As an aside, since propulsion and power generation research work have a number of commonalities, it would not seem to be out of the question to *collocate* propulsion and power generation RDT&E facilities and programs. Further, much of the propulsion and power work with antimatter has, of necessity, large intersections with basic physics investigations with antimatter, and with the disciplines of atomic and nuclear physics. Thus the number of laboratories capable of conducting power and/or propulsion RDT&E is large, and spans university, industry, and national laboratory facilities, leading to many siting possibilities.

Chart 14 summarizes some of the themes which are ubiquitous in current antimatter research of both a basic science and an applications nature. It is useful to reemphasize, as Chart 14 does, that many means for antihydrogen production remain under active consideration. The bibliography of relevant articles is now quite large. In recent literature, for example, three routes to antihydrogen synthesis and capture at low temperatures again stress reactions between laser excited positronium and cooled, trapped antiprotons - Deutch[44] - positrons and antiprotons at a few degrees Kelvin temperature in nested ion traps - G. Gabrielse, et al.[45] - and positronium and antiprotons that are bound in exotic helium-like atoms - Yamazaki.[46] Once the basic antihydrogen synthesis is well in hand, there will be a plethora of interesting and important problems to consider - storage, and in what form; appropriate extraction, manipulation, and control; and so on. The *collaboration of positron and antiproton experimenters will become increasingly important and critical for rapid progress.*

An example of such collaborations already exists implicitly in the work done at some accelerator laboratories. Here are cases where electrons, positrons, protons, and antiprotons are available, to run a variety of tests fundamental to the handling of intense particle beams, to the production of antihydrogen, and to other issues basic to the antimatter field. Some of this work is to be described in this Workshop - Derbenev, et al.[47] Other particulars are found in past publications - Derbenev,[48] and Artamonov, et al.[49] In general, one expects the field of routine antihydrogen production - and extensions possibly into the fields of cluster anti-ion production, and towards production of heavier anti-elements - to expand rapidly via such collaborations; and so should also the manifold applications of antihydrogen and its further elaborations.

Finally, Chart 15 summarizes the main points of this paper. Antihydrogen is *needed* for many compelling applications; the production technologies are *ripe* for *exploitation*; and the joining of forces of *positron and antiproton experimenters* will benefit both parties in very rewarding ways.

Why Antihydrogen?

- → • Basic physics
    - Some experiments easier; more interesting; more fundamental
- • Applied physics
    - Study cluster anti-ions (Stwalley)
    - Bridge between bare antiprotons, neutral antihydrogen
- → • Applications
    - Where large amounts of antimatter are needed
    - Where space charge constraints become dominant
- • Step along the way to heavier antimatter
    - Accelerator production minimal
        - E.g., $\overline{He}/\overline{P} \sim 10^{-12} - 10^{-14}$
    - Experiment with other production routes (Takahashi)
- • Note: Extensive overseas interest in antihydrogen

──▶ = Emphasized in this talk

Chart 1

Some Background on Antiprotons

- • Production of antiprotons — example:
    - Fermilab system <u>produces</u> about 1 microgram/year
    - Fermilab system <u>collects</u> about 1 nanogram/year
- →   • We know how to upgrade (Mills, Augenstein)
- • Once antiprotons are produced and collected — then what?
    - We want them stored in <u>transportable</u> containers
    - Almost all <u>future uses remote</u> from production site
        • Use: in <u>any</u> lab, facility, area, environment
- • Transportable storage options (some experiments; extensive studies)
- →   - Ion (Penning) traps — store at low energies (≤50 KeV)
    - Small rings — store at few hundred MeV (Cline)
    - Definition of <u>transportable</u> — truck, aircraft movement
- • With copious supplies of <u>positrons, many</u> antihydrogen routes
    - E.g., 3-body recombination in traps (Mitchell, Rich, Gabrielse)
        • Theoretically, very high production at very high rates
- →   - Assume: Problems of antihydrogen formation, capture, storage <u>solvable</u>
        • But there will be many non-trivial problems

──▶ = Emphasized in this talk

Chart 2

Upgrading Antiproton Production/Collection

- We understand how to improve <u>collection</u>
  - –Increase collection efficiencies from $<10^{-3}$ to about 10^{-1}
- We understand how to improve <u>production</u>
  - –Accelerators: can design, build
  - –Targetry: scalable, with R&D
  - –Cooling: Innovations will have high payoff
- If we put it all together:
  - –<u>Deliver</u> ~1–100 μg antiprotons/year
  - –At <u>acceptable</u> power investments, R&D risk
  - –<u>Early</u> in next century
- ➔ But: 1–100 μg antiprotons extraordinarily difficult, cumbersome
  - –To store, transport, manipulate efficiently
  - –One motivation for seeking neutral antimatter forms
    - Or, cluster anti-ions (e/m very small)

➔ = Emphasized in this talk

P717–## 7/92 R

Chart 3

Antiproton Production

GGB30m-3 9/90 RAND

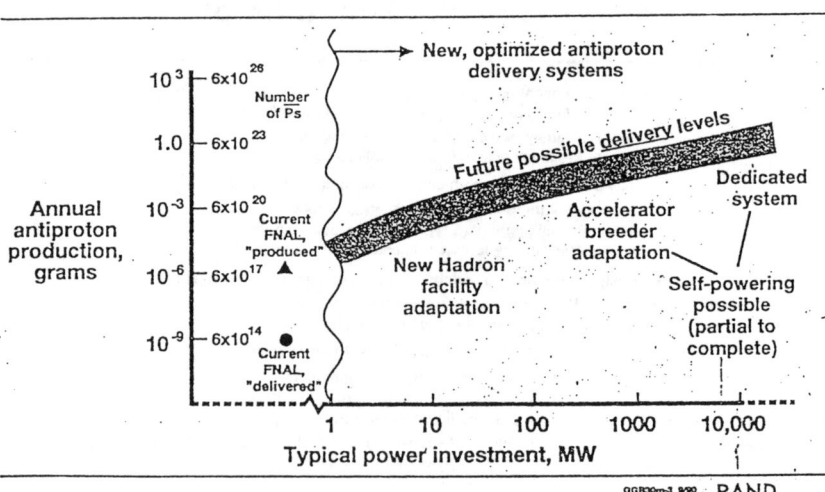

Chart 4

Antiproton Trap Storage Issues

- Basic constraint (Brillouin limit) for confinement of $\bar{P}$
  - Energy density of confining field ≥ rest mass energy density of $\bar{P}$
  - ∴ storing $\bar{P}$ solely for <u>direct energy storage</u> is nonsense
- $\bar{P}$ density ultimately limited by space charge
  - $\bar{P}$ density of $2.5 \times 10^{3}/cm^3$ ⇌ 100 atmospheres pressure
  - That pressure must be handled by confining field
- For <u>current</u> macroscopic trap technologies (Hynes, Howe):
  - Operate about factor of 10^2–10^3 <u>below</u> Brillouin limit
  - "Practical" storage ~10^{12}–10^{13} $\bar{P}$
  - Complete system (vacuum, magnets, controls, power, extraction)
    - Weight: order of 10 tons
    - Size: several meters in each dimension
    - Many technology issues
      - Vacuums
      - Plasma physics
      - Safety
- What are we <u>confident</u> of realizing in single macroscopic traps?
  - Achieved storage > 10^6 $\bar{P}$
  - "Easy" upgrades — 10^8 (10^{10} ?) $\bar{P}$
  - → "Go for broke" in a few years — ~10^{12} $\bar{P}$?
- ~10^{12} $\bar{P}$ about current practical limit for <u>small ring storage</u> also

Chart 5

What Are Current $\bar{P}$ Trap Enhancement Options?

- Operate much closer to Brillouin limit
  - Sharpens all technology demands
- Just accept need for multiple traps
  - If you need 10^{16} $\bar{P}$, use 10^3–10^4 traps
  - Possible, but not attractive in many cases
- <u>Down scale</u> traps — microtraps (Campbell)
  - If mean trap dimension is L
  - Effective charge density ~$1/L$, field needed ~$1/L^{1/2}$
  - At few micron size, <u>might</u> get ~10^{17}–10^{18} $\bar{P}/meter^3$
  - Very difficult fabrication; energy density of field enormous
  - Perhaps possible, but not attractive generally
- Quantum traps (Campbell)
  - Some schemes for gases might be possible (Stark saddles, etc.)

THE MORAL
→
- For ≥10^{12} $\bar{P}$/trap, begin to look for other storage solutions
- Keep up current kinds of trap research
- But: add other storage RDT&E paths
  - Antihydrogen
  - Cluster anti-ions

→ = Emphasized in this talk

Chart 6

Why Antihydrogen-and Not Just Bare Antiprotons?

A Few Examples of Where Antihydrogen Is Desirable/Vital

Class A: — Compatible with current $\bar{P}$ production, collection, and storage
— But antihydrogen is wanted anyway

- Examples in basic physics
  - Antimatter gravitational interactions
  - QED tests
  - Antihydrogen formation, storage, manipulation tests

Class B: — Would need all current or increased $\bar{P}$ production, collection
— Storage as $\bar{P}$ ranges from very awkward to wholly impractical

- Examples in applications
  - Biomedical uses
  - Propulsion uses
  - Power generation uses

Note: Defining experiments in Class B are generally doable with current $\bar{P}$ production, collection, storage technologies

Chart 7

Basic Physics Examples — Antihydrogen Use

- Antimatter gravitational interactions
  - Test weak equivalence principle
  - Several "time-of-flight" experiments formulated with $\bar{P}$
  - For good statistics want $\sim 10^{10}$ (to 10^{12}) $\bar{P}$
  → — Advantage of antihydrogen
    - External perturbations (electric forces, etc.) neutralized
    - Orders of magnitude greater measurement precision
- QED tests, etc.
  - Consider: magnetic moment, transition amplitudes, decay rates, energy shifts, etc.
  - CPT theorem states analogous quantities exactly predictable by CPT for antihydrogen
  - Classical tests (Lamb shift/vacuum polarization, etc.) repeatable using antihydrogen
  - For good statistics might need $\sim 10^{10}$–10^{12} antiparticles
  → — Advantage of antihydrogen
    - Very sensitive tests
    - Repeat all measurements which form basis of QED
- → Test fundamentals of antihydrogen or cluster anti-ion formation, collection, storage
  - Formation rates; $\bar{H}, \bar{H}_2$, etc. — complex issues
  - Type of storage; form of stored antihydrogen
  - Manipulation of antihydrogen (or cluster anti-ion)
    - E.g., for actual use, remove positron for extraction?
  - Tests of individual reaction steps for cluster anti-ions (Stwalley)
  - Numbers of initial $\bar{P}$ needed — a few, to perhaps $\sim 10^{10}$

→ = Emphasized in this talk

Chart 8

$\bar{P}$ Biomedical Uses

- Findings (Kalageropoulos, et al):
  - Imaging
    - Annihilation products give very good dE/dX imaging
    - 3D; no CT deconvolution artifacts; very fast; simple detectors
    - Pbar dose <1/10 CT scan dose for comparable quality
    - Micro imaging possible
    - 10^9 Pbars image human head (1500 cc) on 2 mm grid
  - Therapy
    - Precision beam control allows treatment of very small tumors (mm size)
    - Treatment of malignancies requires ~10^9–10^{10} Pbar/cc
    - Extraordinary control: can image and treat at same time
    - Very small numbers of Pbars could guide normal matter treatments
    - Overall advantage 2–4 times better than any normal matter beams
  - Diagnostics
    - Problem: image biomedical atoms *in vivo*, *in vitro*
    - Far superior to muons (no portable sources; low mass ⇒ large stopping region)
    - Pbar mesic chemistry
      - Portable sources
      - All elements provide signatures via X-ray emissions or nuclear gammas
      - Whole head dose of 1 rad (~10^9 Pbar) images all elements at once
  - Special
    - In situ transmutation of O-16 into O-15 for PET scan

Chart 9

$\bar{P}$ Biomedical Uses on Large Scale

⟶ • Why antihydrogen?
  - Large scale uses would swamp $\bar{P}$ storage
- Needs estimates for U.S.:
  - One <u>image</u>/year per person
    - 10^9 $\bar{P}$/image
    - Use efficiency ~1/2
    - Total $\bar{P}$ ~5 x 10^{17}
  - <u>Therapy</u> uses
    - 5 x 10^5 patients/year
    - Average 20 cc tumor
    - ~3 treatments per patient, use efficiency ~1/2
    - Total $\bar{P}$ ~6 x 10^{17}
- Even if 100 use sites were available:
  - Total usage could approach ~10^{16} $\bar{P}$/site
  - Approach ~10^4 $\bar{P}$ storage containers/site/year
  - Would pose extremely demanding logistics, cost issues

⟶ • High payoff for much better storage capabilities
  - Role for antihydrogen, cluster anti-ion storage

⟶ = Emphasized in this talk

Chart 10

Propulsion Uses of $\bar{P}$

- Pure $\bar{P}$ propulsion possible, but requires enormous numbers of $\bar{P}$
  - E.g., to put 1 ton in low earth orbit ⟶ ~/mg (6×10^{20}) $\bar{P}$
- One recent alternative proposes ICF (inertial confinement fusion) approach
  - <u>But</u>, uses $\bar{P}$ to ignite ICF micro-fission/fusion reaction
  - Thus, $\bar{P}$ <u>catalyzed</u> ICF propulsion (Smith, Augenstein)
- Why $\bar{P}$ catalysis mechanism?
  - $\bar{P}$-induced fission in uranium at CERN (Smith)
  - About 16 neutrons produced per $\bar{P}$
  - Phenomenology (Smith, Wienke)
    - Speed up reactions prior to material disassembly
    - $\bar{P}$ ⟶ neutrons ⟶ early <u>fission</u> reactions
    - Bypass early stages of pure fusion burn
    - Use <u>combined</u> fission/fusion burn
    - Calculations suggest high total burn
  - <u>Typical</u> calculation
    - 10 ns burst of ~10^7–10^9 $\bar{P}$ into compressed U/DT target
    - ~10 GJ energy (plasma + radiative) from 1 gram target
    - Mix with heavy metal to capture some radiative energy
    - Jet energy ~.4–.5 GJ, I_{sp} ~13,000, thrust ~30,000 N
    - 150 day round trip to Mars, 100 MT payload, 2,000 MT mass
- ⟶ Series of <u>experiments</u> planned (1993–97) to <u>test $\bar{P}$ micro-fission</u> concepts
  - SHIVA Star facility at Albuquerque (Phillips Laboratory)

Chart 11

Propulsion Uses — $\bar{P}$ Needs

- Mars mission characteristics
  - 5 Hz ICF pulses
  - About 50 days burn (~4×10^6 seconds)
  - ~10^7–10^9 $\bar{P}$/pulse (use ~2×10^8)
  - $\bar{P}$ use efficiency ~ 40%
  - Total mission $\bar{P}$ use ~10^{16}
- Number of $\bar{P}$ storage containers needed
  - In range ~1,000–10,000
  - Not wholly impossible, perhaps — but:
    - High weights associated with $\bar{P}$ storage
    - High volumes associated with $\bar{P}$ storage
    - High complexity: switching, manipulation
- ⟶ Advantage of equivalent antihydrogen (or cluster anti-ion) storage
  - Might store in weight, volume of <u>one</u> $\bar{P}$ storage container
  - Very large <u>complexity</u> reduction
- ⟶ Therefore: high possible payoff for antihydrogen use

⟶ = Emphasized in this talk

Chart 12

Power Generation Uses for $\bar{P}$

- <u>No</u> role for stored bare $\bar{P}$ (Brillouin limit)
  - ∴ Require alternate storage (where holding energy « rest energy)
- Features <u>favoring</u> $\bar{P}$ (in form of <u>antihydrogen</u> or <u>cluster anti-ions</u>)
  - Infinitely scalable energy yield
  - Some relief from problems of reactors
    - Actual radioactivity can be significantly less
    - Antimatter not subject to legal constraints of U, Pu
  - Great range of implementation possibilities
    - From special means of direct electricity production
    - To replacements for other internal heat sources
- One design example for power production — Stirling engine
  - Cycle efficiency intrinsically high
  - Extensive design base; well documented design principles
  - Cycle adaptable to electricity production or direct motive power
  - Very flexible design alternatives
  - Well suited to antimatter use:
    - Central feature: large heater to heat retained working gas
    - Heater can be region where $\bar{P}$ annihilates
    - Net efficiencies (electricity/annihilation) might be 10–20%
- High outputs using small amounts
  - At 20% efficiency, 10μg $\bar{P}$ → ~100 Kw hrs

⟹ = Emphasized in this talk

Chart 13

Some Common Themes and Reminders

- We need positron, antiproton <u>experimenter</u> collaboration
- We need <u>trap designer</u> collaboration
  - Much work revolves around <u>low temperature</u> traps
  - Nested traps storing positrons, antiprotons simultaneously
  - Test 3-body recombination theory exhaustively
- Many technology issues can be <u>pretested</u> using normal matter
- There are many <u>common/analogous</u> steps in:
  - Forming antihydrogen, cluster anti-ions (Mitchell, Stwalley)
  - Manipulating antihydrogen, cluster anti-ions
- There are <u>many proposals</u> for antihydrogen production
  - Radiative recombination; positronium charge exchange; 3-body recombination
  - There <u>may be</u> alternative options preferred
    - For lab scale experimental amounts
    - For larger routine production amounts
- <u>Many issues</u> left to explore in:
  - Antihydrogen storage (∴ role for cluster anti-ions?)
  - Antihydrogen manipulation (∴ role for cluster anti-ions?)
- <u>Very broad</u> experimental fields in <u>producing</u> such forms of antimatter
  - As broad as fields for <u>using</u> such forms of antimatter

Chart 14

Summary Conclusions — Antihydrogen, Cluster Anti-ion Roles

- High <u>motivations</u> to develop technologies
  - Basic physics
    - Where experiments with bare $\bar{P}$ not best possible
    - Although we could store enough bare $\bar{P}$
  - Applications
    - In principle could use bare $\bar{P}$
    - But storage in bare $\bar{P}$ form impractical
- Basic <u>tests</u> of antihydrogen, cluster anti-ion technologies
  - Doable well within <u>today's</u> $\bar{P}$ delivery levels
  - Range of formation, collection, manipulation techniques
  - Major <u>intersections</u> with today's trap technologies
    - Blend positron, antiproton technologies
    - 3-body recombination route to antihydrogen
- Next steps
  - Time for positron, antiproton experimenters to <u>join forces</u>
  - In experiments compatible with <u>today's and tomorrow's</u> $\bar{P}$ levels
- <u>Very rich experiment spectrum</u> to develop antihydrogen, cluster anti-ion technologies
- These antihydrogen, cluster anti-ion technologies <u>are needed</u>
- Let's <u>formulate</u> program, <u>get on</u> with it!
- Joining of e^+, $\bar{P}$ communities can achieve <u>critical mass</u>
  - For <u>future</u> adventurous, compelling antimatter program

Chart 15

P717-## 7/92   R

REFERENCES

1. *Proceedings of the RAND Workshop on Antiproton Science and Technology.* Editors: Augenstein (RAND), Bonner (Rice University, Mills (FNAL), and Nieto (LANL), Published by World Scientific, Singapore, New Jersey, and Hong Kong, 1988.

2. RAND Workshop on Antiproton Science and Technology, October 6-9, 1987: *Annotated Executive Summary*, RAND Note N-2763-AF, Bruno W. Augenstein, October 1988.

3. *Production and Investigation of Atomic Antimatter.* Editors: H. Poth, A. Wolf, Proceedings of a Symposium held at Kernforschungszentrum, Karlsruhe (KFK), November 30-December 2, 1987, published by J. C. Baltzer AG, Basel, Switzerland, 1988.

4. W. C. Stwalley (Univ. of Iowa), "Synthesis of Large Cluster Ions from Elementary Constituents - Possible Route to Antimatter"; and "Bibliography of Hydrogen Cluster Ions" - in Reference 1.

5. R. L. Forward (Hughes Research Labs), "Production of Heavy Antinuclei: Review of Experimental Results" - in Reference 1.

6. H. Takahashi (BNL), "Some thoughts on the muon-catalyzed process for antimatter propulsion and for the production of high A mass numbers nuclei" - in Reference 1.

7. See for example the series of 15 papers on Reference 3, pages 257-411.

8. S. D. Howe, M. Hynes, and A. Pickleshimer (LANL), "Portable Pbars, Traps that Travel" - in Reference 1.

9. D. Cline (UCLA), "A Storage Ring for Antimatter Transport" - in Reference 1.

10. J. B. A. Mitchell (University of Western Ontario), "Antihydrogen Production Schemes" - in Reference 1.

11. H. Poth, et al., "Antihydrogen production in a merged beam arrangement" - in Reference 3.

12. B. I. Deutch, et al., "Antihydrogen by positronium-antiproton collisions" - in Reference 3.

13. G. Gabrielse, et al., "Possible antihydrogen production using trapped plasmas" - in Reference 3.

14. B. Deutch, Hyperfine Interactions 73, 175 (1992).

15. R. J. Hughes, in Nature 353, 700 (1991) and references therein.

16. T. W. Hansch, in *The Hydrogen Atom*, edited by G. Bassani, et al., Springer, 1989, p. 93.

17. F. E. Mills (FNAL), "Scaleup of Antiproton Production Facilities to 1 mg/year" - in Reference 1.

18. B. W. Augenstein, "Introduction and context for the four ICENES '91 Antimatter Annihilation Papers," *Fusion Technology*, Vol. 20, Dec. 1991, pp. 1035-1039.

19. Baldin, et al., "Antiproton yield in the collision of carbon nuclei with copper nuclei at energy of 3.65 GeV per nucleon," JETP LETT. Vol., 48, No. 3, 1988, p. 137.

20. H. Takahashi, and J. Powell (BNL), "Multiple Collision Effects on Antiproton Production by High-Energy Protons (100 GeV-1000 GeV)" - in Reference 1.

21. B. W. Augenstein, Concepts, *Problems, and Opportunities for Use of Annihilation Energy: An Annotated Briefing on Near Term RDT&E to Assess Feasibility*, N-2302-AF/RC, RAND Corporation, June 1985.

22. L. J. Campbell (LANL), "Normal Matter Storage of Antiprotons" - in Reference 1.

23. M. M. Nieto and T. Goldman, "The arguments against 'antigravity' and the gravitational acceleration of antimatter," *Physics Reports,* Vol. 205, No. 5, July 1991.

24. F Witteborn, W. Fairbank, "Experimental comparison of the gravitational force on freely falling electrons and metallic electrons," *Phys. Rev. Lett.*, 19 (1967) pp. 1049-1052.

25. G. Gabrielse, "Trapped antihydrogen for spectroscopy and gravitation studies: is it possible?" - in Reference 3.

26. N. Beverini, et al., "Possible measurements of the gravitational acceleration with neutral antimatter" - in Reference 3.

27. B. E. Bonner, M. M. Nieto, "Basic Physics Program for a Low Energy Antiproton Source in North America" - in Reference 1.

28. R. Neumann, "Fast Antihydrogen Beam Spectroscopy" - in Reference 3.

29. R. J. Hughes, B. I. Deutch, "Electric charges of positrons and antiprotons," *Phys. Rev. Lett.*, Vol. 69, No. 5 (1992), pp. 578-581.

30. T. Kalogeropoulos, et al., "Biomedical Potential of Antiprotons" - in Reference 1.

31. L. B. Greszczuk, "Potential Applications of Antiprotons for Inspection and Processing of Composites" - in Reference 1.

32. D. Morgan (LLL), "Propulsion Test Facility - Antiproton Stopping and Annihilation in Various Antimatter Engine Types - Needed Experimental Information" - in Reference 1.

33. R. L. Forward, *Antiproton Annihilation Propulsion,* Report AFRPL TR-85-034, prepared for the Air Force Rocket Propulsion Laboratory, September 1985.

34. B. W. Augenstein, "On Antiproton Science and Technology - Implications for Antimatter Propulsion RDT&E," AIAA Presentation, Session 4, February 14, 1989.

35. S. Polikanov, "Could Antiprotons be Used to Get a Hot, Dense Plasma?" in *Physics at LEAR with Low Energy Cooled Antiprotons,* edited by Gastaldi and Klepisch, 1982.

36. J. C. Solem (LANL), "Extreme States of Matter - Could Antiprotons be Used to Power Table-top Equation-of-State or Opacity Measurements?" - in Reference 1.

37. R. A. Lewis, G. A. Smith, et al., "Antiproton Boosted Microfission," *Fusion Technology,* Vol. 20, December 1991.

38. B. W. Augenstein, "Antiproton-initiated fusion propulsion - the big intermediate step," IAF-90-234, paper presented at the 41st Congress of the International Astronautical Federation, Oct. 6-12, 1990, Dresden, GDR.

39. P.-R. Chiang, R. A. Lewis, G. A. Smith, et al., "An Antiproton driver for inertial confinement fusion propulsion," Paper PSU LEPS 92/05, given at the meeting on Nuclear Technologies for Space Exploration, at Jackson, Wyoming, August 16-19, 1992.

40. B. W. Augenstein, "Some aspects of interstellar space exploration - New ORION systems, early precursor missions," Paper IAA-91-716, given at the 42nd Congress of the International Astronautical Federation, October 5-11, 1991, Montreal, PQ, Canada.

41. B. W. Augenstein, "Antiprotons for Power Generation," RAND informal memorandum, March 1989.

42. V. C. Clarke, et al., "A Mars Airplane?" A79-17875, AIAA, 1979.

43. B. W. Augenstein, "The Mars Airplane Revived - Global Mars Surface Surveys," AAS 87-270, in *The Case for Mars III,* Vol. 75, 1989, published for the American Astronautical Society by Univelt, Inc.

44. B. I. Deutch, Reference 14.

45. G. Gabrielse, et al., *Phys. Lett.* A 129, 38 (1988).

46. T. Yamazaki, *Z. Phys.* A 341, 223 (1992).

47. Ya. Derbenev, A. J. Artamonov, "Magnetization effects in electron cooling of positrons and generation of intense antihydrogen beams," this Workshop.

48. Ya. Derbenev, "On possibilities of fast cooling of heavy particle beams," *Conference Proceedings No. 253,* Editors: W. Destler, S. Gaharay, AIP, 1992.

49. A. Artamonov, et al., "Electron Cooling Storage Ring for Positrons to Produce Antihydrogen," *Particle Accelerators,* 1988, Vol. 23, pp. 79-92.

FORMATION OF ELECTRON - POSITRON PLASMAS IN THE LABORATORY

Heinrich Boehmer
University of California, Department of Physics
Irvine, CA 92717

ABSTRACT

Electron - positron plasmas have properties distinctly different from those of conventional electron - ion plasmas. Particularly, electromagnetic wave excitation and propagation in a magnetized plasma, Alfven and whistler waves as well as non-linear effects are modified or non-existent. It is demonstrated that it should be feasible to generate an electron - positron plasma by trapping externally injected, low energy electrons and positrons by electron cyclotron resonance heating in a magnetic mirror. By balancing the positron flux from a commercial Na-22 source, moderated by a tungsten foil, with the potential loss mechanisms, it is expected that an equilibrium density of 10^6 - 10^8 cm^{-3} can be achieved.

INTRODUCTION

Positrons have become an important research tool both in the colliders of high energy physics[1] as well as in solid state physics[2]. Depending on the applications, positrons are generated either by pair production during the interaction of energetic primary electrons with high Z targets or collected from radioactive sources. In either case, the technology has advanced to a degree that positrons are routinely generated and manipulated in copious quantities. Particularly because of the development of radioactive source - moderator combinations[2,3], it seems now feasible to attempt the formation of an electron - positron plasma, which, as will be shown, has unique properties not found in the usual electron - ion plasma and which can serve as a model for astrophysical and relativistic plasmas.

Unlike the basic electron - ion plasma the electron - positron plasma can not be produced <u>in situ</u>. Both species have to be formed separately, transported to and, because of the limited flux of conventional positron sources, have to be trapped and accumulated in the experimental chamber.

Single species electrons and positrons have been trapped in Penning type configurations where the particles are confined radially by a magnetic field and axially by an electrostatic potential[4,5]. The pioneering work for trapping single species particles in a Penning geometry was performed by Malmberg[4] using electrons. Most of the positron traps are intended for accumulation of positrons to form high density bunches for various purposes: injection into accelerators[6], formation of antihydrogen[7,8,9], plasma diagnostics purpose[10,11], and to improve the signal-to-noise ratio in positron scattering and solids interaction studies[2].

Since the electrostatic confinement operates on a single charge, only, different methods have to be found to confine electrons and positrons simultaneously. We propose to use the well researched magnetic mirror as a confinement scheme. While the mirror was also considered by Tsytovich and Wharton[12], their positron formation and trapping method is different from the one considered here.

ELECTRON - POSITRON PLASMA PROPERTIES AND RELATION TO ASTROPHYSICAL PLASMAS

The difference between a normal electron - ion and an electron - positron plasma is, at the first glance, trivial: The ion mass is replaced by the electron mass. This has the primary consequence that the plasma frequency ω_p is replaced by $\sqrt{2}\omega_p$ and the Debye length becomes $v_{th}/\sqrt{2}\omega_p$. There are no low frequency ion waves and their dispersion branches merge into those of the electron branches.

The more interesting situation arises with the inclusion of a magnetic field since now for equal electron and positron densities and temperatures only the diagonal terms of the dielectric tensor[13] survive. For example, for the simplest case of electromagnetic wave propagation parallel to the magnetic field, the dispersion relations for the right (R) and left (L) handed circular polarized waves are, assuming equal species densities, for
electron - ion plasmas:

$$\frac{k_{R,L}^2 c^2}{\omega^2} = 1 - \frac{\omega_p^2}{\omega(\omega \mp \omega_c)} \tag{1}$$

electron - positron plasmas:

$$\frac{k_{R,L}^2 c^2}{\omega^2} = 1 - \frac{\omega_p^2}{(\omega^2 - \omega_c^2)} \tag{2}$$

where $k_{R,L}$ are the wave vectors and ω_c the electron (positron) cyclotron frequency. In the latter case, the dispersion relations for right and left handed polarization are identical. Consequently, the plane of polarization of a linearly polarized electromagnetic wave does not change, Faraday rotation does not occur. This may have important consequences for astrophysical observations, since the magnitude of stellar magnetic fields are frequently deduced from observed Faraday rotations.

Among other plasma phenomena, the Alfven speed is greatly modified since it is inversely proportional to the plasma <u>mass</u> density, and the whistler wave ceases to exist. Also, Tsytovich and Wharton[12] find that the non-linear plasma behavior is greatly modified, since the sum of electron and positron non-linear responses vanishes. This changes non-linear wave - wave interactions and, therefore, non-linear Landau damping.

Of particular interest is the active interaction of electromagnetic waves with plasmas and their emission by the plasmas since these processes are an important link between astrophysical processes and their observation. It is generally agreed that the electromagnetic radiation emitted from stellar objects is due to synchrotron radiation. While the frequency of these waves agrees with the assumed, or form observation deduced, magnetic fields, the amplitude of the waves is usually much larger than that attributable to incoherent emission.

Enhanced, coherent emission can occur either because of the interaction of energetic electron (positron) components with electrostatic waves in the two stream instability[14,15] or with electromagnetic waves in the cyclotron maser instability[16]. While these interactions can occur in pure electron plasmas, Yoon and Chang[17] have shown that the inclusion of positrons in the formalism of the cyclotron maser interaction generates a new dispersion branch which is linearly polarized. This can explain the observation that the cyclotron radiation received from extragalactic nuclei, quasi stellar objects and the so-called core jets associated with these objects is frequently almost linearly polarized and not circular polarized as expected for a pure electron plasma. The electron - positron plasmas in these objects as well in the pulsar magnetosphere[14] is assumed to be generated by a cascading shower of pair production and bremsstrahlung radiation.

In summary, electron - positron plasmas differ in substantial aspects from their conventional counterparts and will be important to investigate, especially with respect to astrophysical phenomena.

POSITRON SOURCES

Most of the advancement in positron source development for small scale experiments is the result of intense research in positron - gas scattering and positron-solid state interactions. Among the long life β^+ emitters, Na-22 is widely used which has a half life of 2.6 years, a β^+ fraction of 90% and a maximum β^+ energy of 0.54 MeV. These sources are commercially available[18] with a maximum activity of 150 mC.

For many applications, the high β^+ energy, the wide energy distribution and the low brightness limit seriously the usefulness of the radioactive sources. To overcome this disadvantage, the primary β^+ beam is moderated by letting the particles enter thin solid targets, like tungsten foils, where they undergo inelastic scattering and capture in lattice defects[2,3,19]. A small number diffuses to the surface where they undergo desorption because of their negative work function. Typically, the energy of the moderated positrons is 2 eV with a yield as high as 10^{-3}. Such moderated Na-22 sources have yielded fluxes of 10^6 moderated positrons using a 100 mC source[20]. Recently, solid Neon moderators have been developed which improve the moderated positron flux by a factor 100[21].

It will be shown below that the low energy positrons are much easier to trap than the primary energetic particles, although this means a sacrifice in the available positron flux.

Using Cu-64, which has a half life of 2.8 hours, as a β^+ source, Lynn, et al.,[22] report achieving a self moderated activity of 10^8 self moderated positrons per second. Large diameter Co-58 positron sources using solid neon moderators were proposed to generate fluxes of 10^{12} moderated positrons per second[23]. Reactor based positron sources are being considered to yield 1.1×10^{11} e$^+$/sec[24] in one scheme and 10^8 in another[25].

Positrons are also routinely generated by successive bremsstrahlung and pair production precesses in high Z targets of energetic primary electron beams from RF linear accelerators. For research with low energy positrons, the energetic positrons can again be moderated to form high brightness beams. In this way, moderated positron fluxes of 10^9 e$^+$/sec have been achieved[26].

ELECTRON - POSITRON PLASMA TRAPPING AND CONFINEMENT

The simplest configuration to confine a two charge sign species plasma is the magnetic mirror, although means have to be found to trap the externally generated species and to eliminate instabilities associated with this configuration.

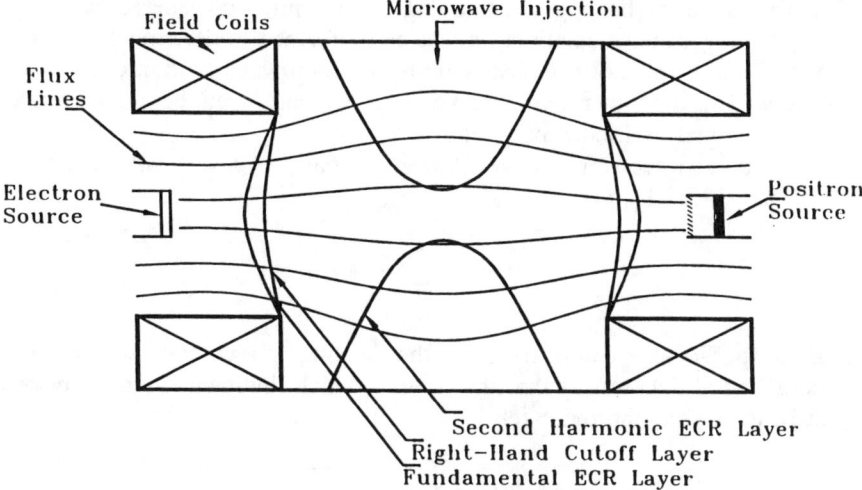

Figure 1. Magnetic mirror with positron and electron sources, lines of constant flux, fundamental and second harmonic cyclotron resonance layer, and right-hand cutoff layer.

Figure 1 depicts schematically a magnetic mirror with the electron and positron sources located in the mirror throats. Particles entering the mirror will be lost at the opposite end unless their pitch angle θ is sufficiently increased while inside the mirror to position them outside the loss-cone angle θ_{lc} which is given by

$$\sin^2\theta_{lc} = \frac{V_{0\perp}^2}{V_0^2} = \frac{B_0}{B_m} = \frac{1}{R_m} \qquad (3)$$

where the mirror ratio R_m is the ratio of the magnetic fields in the mirror throat (B_m) and the center (B_o). $v_{o\perp}/v_o$ is the ratio of perpendicular to total particle velocity at the mirror center. The necessary increase of the perpendicular momentum during transit can be accomplished be electron (positron) cyclotron resonance heating (ECRH). For $R_m = 2$, the fundamental resonance layer, given by $\omega = \omega_{ce} = eB/m$, is positioned approximately as shown in Figure 1. Numerically $\omega_{ce}/2\pi(GHz) = 2.8*B$ (kG). This means that, for example, a 900 Gauss field can be addressed by the output of low cost magnetrons of conventional microwave ovens.

ECRH has been studied extensively in magnetic fusion related experiments and is very well understood. In both mirror based fusion concepts ECRH is used, in the Tandem Mirror to form the proper potential distribution of the thermal barrier[27] and in the EBT concept to form mirror trapped transverse current distributions for the stabilization of flute instabilities[28,29]. In either case, electrons from an untrapped "stream" plasma are heated to form the trapped distributions. Particularly a single cell of EBT is similar to the heating and trapping configuration considered here.

Ray tracing code calculations[29] must be performed to determine properly the energy deposition of the, not necessarily uniform, microwave field into the electrons and positrons. On the other hand, reasonable estimates can be made using the formalism developed by Lichtenberg, et al.[30].

A particle traversing the cyclotron resonance layer will increase its perpendicular velocity by

$$\Delta v_\perp = \frac{e}{m} E \, t_R \tag{4}$$

where E is the local electric field of the heating wave. For $\Delta v_\perp \ll v_z, v_\perp$, Lichtenberg[30], et al., show that the time t_R the particle interacts with the resonance layer can be expressed approximately as

$$t_R = 1.13 \left\{ \frac{2}{|\alpha v_{zR}|\omega} \right\}^{1/2} \tag{5}$$

Here, v_{zR} is the axial velocity at resonance and $\alpha = (1/B)(dB/dz)$.

Using for v_{zR} the velocity corresponding to the average slow positron energy of 2 eV, assuming further a microwave frequency of 10 GHz and a field gradient of $\alpha = 0.03$, which is typical for mirrors, one finds from equation (5) $t_R = 4 \times 10^{-9}$ sec. For a mirror ratio of 2, the perpendicular velocity has to increase by $\Delta v_\perp = \sqrt{2} \, v_z$ to effect trapping. Using Equation (4), one finds then for the necessary heating field E = 15 V/cm. If the vacuum chamber can be described by a cavity of 10 cm diameter and a Q of 1, one finds that microwave power of 380 Watts is necessary for trapping which is well within the range of available X-band microwave sources. It is also evident from these considerations that it would be impractical to trap the primary β^+ particles since the required heating field would be impossible to achieve.

Although highly simplified, estimates like these have given reasonable answers for actual experimental systems[30].

Once trapped in the mirror, the particles can perform multiple bounces and can be heated further during each transit through the resonance layer. Heating is stopped when, at large energies, the phases of bounce and cyclotron motion become correlated[31] (superadiabatic orbits). For typical magnetic mirrors, the final temperature is in the multi-keV range. Further heating can be achieved by introducing an element of stochaticity, for example multi-frequencies and by second harmonic heating[32,33]. In an axisymmetric mirror, hot electron densities of 10^{11} cm^{-3} at temperatures of 0.5 MeV have been obtained this way with heating powers of less than 1kW[34].

COMPUTER SIMULATIONS OF ELECTRON AND POSITRON HEATING AND TRAPPING IN A MAGNETIC MIRROR

Because of the complexity of full codes like those discussed in Reference 26, a 1d, 3v bounded code, developed be C.K. Birdsall and his group[35], was used to demonstrate ECR heating and trapping of electrons and positrons in a simple mirror. The code was modified to include spatially variable magnetic fields.

The cyclotron resonance heating field was simulated by prescribing a constant electric field perpendicular to the system axis with an axial extend of one cell, located at x=1/4 L and x=3/4 L, where L, containing 200 grid points, is the system length between mirror throats. Cell size and magnetic field strength were chosen such, that the particles underwent less than one cyclotron orbit during the transit through the resonance layer. Electron and positron streams were injected from opposite ends using various longitudinal and transverse velocity distributions with an average energy of 2 eV.

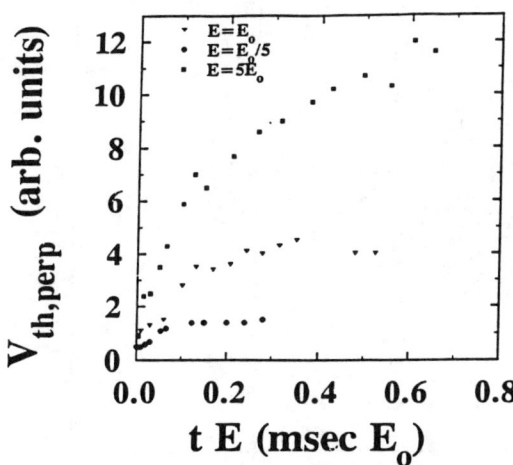

Figure 2. Temporal variation of transverse particle thermal velocity for 3 heating fields, E_o = 2 V/cm.

The width of the particle distribution function, $f(v_\perp)$, measured at the midplane, was monitored as a function of time. Examples for three heating fields are shown in Figure 2. As expected, the heating saturates after a certain time. The final temperature of 10 keV for the highest heating field agrees with past experiments[34] and the theories[30,31] mentioned above. Figure 3 shows the phase space distribution of both species after a certain time, clearly demonstrating the trapping in the mirror volume.

PLASMA LOSSES AND FINAL DENSITY ESTIMATES

The electron - positron plasma density that can be achieved depends on the particle source and loss rates. The flux of the moderated positron source can be taken as the source term, assuming that most positrons entering the mirror volume will be trapped. Both particle number and energy loss can occur. Energy losses due to free-free bremsstahlung, synchrotron radiation and Coulomb drag can be estimated from the expressions given in Reference 34 and are small for the parameters considered here.

Several particle loss mechanisms can occur, all of which have been investigated in great detail for mirror fusion systems: 1. 90° collisions on background ions, 90° collisions on primary plasma constituents, 3. Classical cross-field diffusion, 4. Neoclassical transport associated with quadrupole components of the confining field, 5. Anomalous transport due to turbulence, 6. Loss due to the ECR heating field, and 7. Losses due to annihilation and positronium formation.

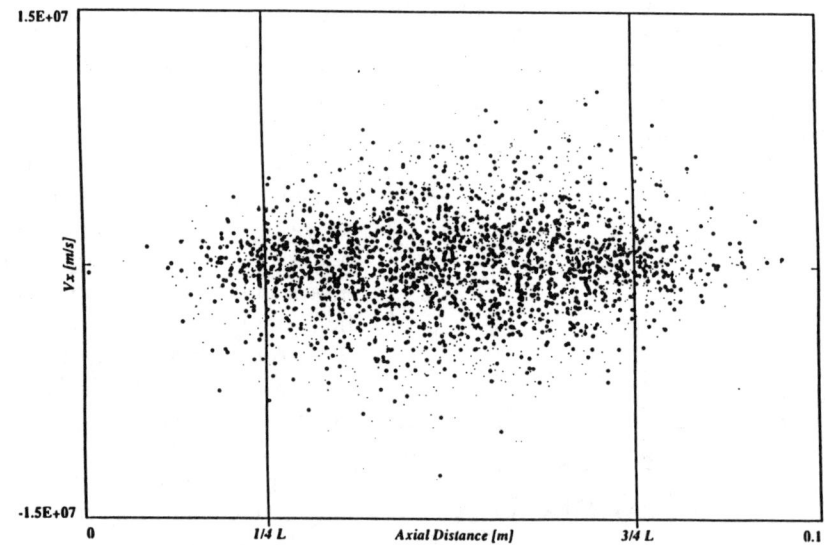

Figure 3. Example of phase space distribution of electrons [dots] and positrons [circles] in a computer simulation of ECRH heating and trapping in a magnetic mirror.

<u>Scattering into Loss Cone:</u> The 90° collision time for electrons on plasma positrons (e^+) and background ions (i^+) can be written as[36]

$$\tau_{e^+,i^+} = \alpha_{e^+,i^+} \frac{T_e^{3/2}}{n_{e^+,i^+}} \ln \Lambda \tag{6}$$

where $\alpha_{e+} = 8 \times 10^5$ for e^- - e^+ collisions and $\alpha_{i+} = 4 \times 10^5$ for e^{-+} - i^+ collisions. T_e is in eV, n in cm^{-3} and lnΛ, the Coulomb logarithm, is of the order of 10. The 90° collision time is interpreted as a confinement time since a 90° collision will scatter the confined, large pitch angle particles into the loss cone.

Scattering on ions, formed by ionization of background gas, becomes negligible at background pressure below 10^{-8} Torr.

The equilibrium density can be estimated from the balance of source and loss terms:

$$\frac{dn_e}{dt} = S - <\sigma v> n_e^2 = 0 \qquad (7)$$

$<\sigma v>$ is related to the 90° collision time via

$$\tau = 1/n_e <\sigma v> \qquad (8)$$

For a plasma volume of 30 cm^3, a primary flux of 10^6/sec and an electron (positron) temperature of 1 keV, one finds for the equilibrium density n = 10^7 cm^{-3}. The confinement time (Eq. 6) is then 240 sec. In comparison, a confinement time of 1 sec was reported in Reference 34 for a 100 keV electron component, heated and trapped in a mirror by ECRH in the presence of a background plasma density of 10^{12} cm^{-3}, in reasonable agreement with Equation (6).

Cross Field Diffusion: The classical diffusion coefficient for cross - field transport, $D_\perp$, has been derived in numerous publications[36]. If length is expressed in cm, electron temperature T_e in eV, and the confining field B in Gauss, one finds

$$D_\perp = 9 \times 10^{-5} \frac{n_e \ln \Lambda}{B^2 T_e^{3/2}} \quad (cm/sec). \qquad (9)$$

$D_\perp$ is related to the flux Γ via

$$\Gamma = D_\perp dn/dr \; . \qquad (10)$$

Obviously, the electron temperature and, especially, the confining field should be high. Assuming T_e = 1 keV, B = 3 kG, a diffusion scale length of 1 cm, a plasma surface area of 60 cm^2 and balancing the radial flux by a positron source flux of 10^6 e$^+$/sec, one arrives at an equilibrium density of n = 7×10^7 cm^{-3}, which is significantly higher than that derived from scattering loss.

Neoclassical losses due to imperfections of the solenoidal magnetic field are difficult to estimate, even if the actual field distribution would be known. For axisymmetric systems, they play a significant role for very large confinement times, only[4].

Anomalous losses due to turbulence occur in ordinary plasmas due to low frequency waves which, in the case of electron - positron plasmas, does not occur because of the absence of ions.

Annihilation and positronium formation losses are small compared to scattering losses because their cross section is small, of the order of πr_o^2, where r_o is the classical electron radius.

PLASMA STABILITY

Because of the strong anisotropy of the trapped particles, the plasma can be unstable to high frequency instabilities like electrostatic electron cyclotron waves or the cyclotron maser interaction. From the plasma physics point of view, they are the main reasons for investigating the electron - positron plasma. They will contribute to transport in velocity space, not in real space.

A serious instability which can cause catastrophic loss of a mirror confined plasma is the flute instability, the magnetized plasma equivalent of the Rayleigh-Taylor instability. The source of free energy of the flute instability is the charge sign dependent azimuthal drift of confined particles, caused by the transverse field gradient ∇B of the magnetic mirror. Because both charge components drift in opposite directions, a spontaneous density, and therefore potential, fluctuation will grow if a radial density gradient exists. Such flute modes can cause plasma loss on a millisecond time scale. The flute modes can simply be stabilized by short circuiting the potential fluctuations via a highly conducting electron component to a conducting and electron emitting end plate[37]. This will be the additional function of both the electron emitter and the moderator of the positron source.

DIAGNOSTICS

The most commonly used plasma diagnostic techniques, namely diamagnetic loops to measure the product of total particle number and temperature, microwave interferometers to measure the line density, and microwave cavities to measure the plasma density are marginal at best for the expected plasma parameters. Probes are inconvenient because the intercepted particle flux is too large.

Non-disturbing diagnostic techniques of the electron - positron plasma rely on the emitted radiation. Three radiation mechanism will be present: synchrotron radiation, free-free bremsstahlung and annihilation radiation. While the radiated power due to these processes will not contribute measurably to the plasma energy loss, it is still large enough for diagnostic purposes.

<u>Synchrotron Radiation:</u> The power emitted from the plasma by synchrotron radiation from uncorrelated electrons (positrons) can be written as[38]:

$$P_s = 6.2 \times 10^{-20} B^2 n_e V T_e \left(1 + \frac{T_e}{2 \times 10^5}\right) \quad (Watts) \qquad (11)$$

where T_e is in eV and B in Tesla. Assuming B = 3 kG, $n_e = 10^7$ cm^{-3}, V = 100 cm^3, and T_e = 5 keV, one finds from Equation 11 for the total radiated power $P_s = 3 \times 10^{-8}$ Watts. Modern microwave receiver in the X-band range have a noise figure of 6 dB or better. Since this is equivalent to a noise temperature of 900° K, the receiver noise power is kTΔf ≈ 10^{-15} Watts, assuming a receiver bandwidth of Δf = 100 MHz. The incoherently emitted synchrotron radiation, therefore, should be easily detectable.

The frequency spectrum of the synchrotron radiation is a sensitive function of the velocity distribution of the radiating particles. It is, therefore, frequently employed as a diagnostic tool in fusion devices. In the presence of electrostatic instabilities, the observed synchrotron radiation will deviate drastically, both in amplitude and frequency spectrum, from an incoherently emitted spectrum. Measuring the frequency spectrum of the synchrotron radiation will, therefore, give important information about the state of the trapped electron - positron plasma.

Free-Free Bremsstrahlung: The x-ray analysis of radiation emitted due to free-free bremsstrahlung of energetic electrons on ions is also a well developed diagnostic technique. To apply this technique to the electron - positron plasma, a low density heavy ion beam could be transmitted transversely through the system, allowing the measurement of electron (positron) energy spectra as well as radial density profiles.

Annihilation Radiation: In the energy range of interest here, electrons and positrons will interact via two photon annihilation, $e^- + e^+ \rightarrow 2\gamma$, with an approximate cross section $\sigma = \pi r_o^2$. Assuming a Maxwellian energy distribution for both species, one finds for the interaction

$$<\sigma v> = \frac{4\sigma}{(2\pi m)^{1/2}} \sqrt{kT} \quad . \tag{12}$$

Remembering that for each interaction two photons are emitted, the total number of annihilation photons emitted per second into a fractional solid angle $\Omega/4\pi$ from a volume V is given by

$$N = 2n_{+,-}^2 <\sigma v> v \, \Omega/4\pi \quad . \tag{13}$$

For an electron - positron density of 10^8 cm^{-3} and V = 100 cm^3, one finds from Equations (12) and (13) that the total number of photons emitted into Ω = 1 sr ranges between 160/sec for 1 keV and 370/sec for 5 keV particle temperatures.

Experimentally, the annihilation photons can be detected in a standard fashion with a NaI(Tl*) scintillator - photomultiplier combination or HPGe detectors.

EXPERIMENTAL SYSTEM

The experimental system will have the following components (see Figure 4): Solenoidal coils to generate the confining magnetic field with a minimum field of 3.5 kG in the mirror throats, 2. the vacuum chamber with appropriate windows for microwave injection and diagnostics, 3. positron and electron sources, 4. the ECRH microwave source at about 10 GHz and 1 kW power level, 5. microwave receivers and 6. γ-ray detectors for diagnostic purposes.

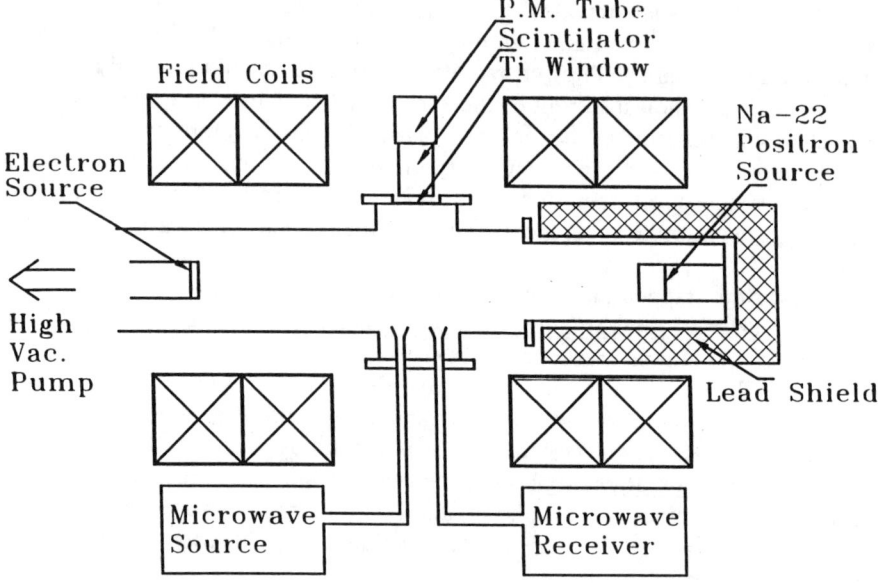

Figure 4. Experimental setup.

CONCLUSION

It is shown that the development of high flux positron sources together with the application of the well researched electron (positron) cyclotron resonance heating to trap the particles makes the formation of a magnetic mirror confined electron - positron plasma feasible. For positron fluxes of standard Na-22 sources with moderators, plasma densities in the 10^7 - 10^8 cm^{-3} range are expected. The velocity distributions are expected to be highly anisotropic so that phenomena of high interest from the plasma physics point of view, like high frequency instabilities and radiation processes, can be studied. The positron trapping scheme discussed here could also be used to effect positron accumulation for other research purposes.

ACKNOWLEDGEMENTS

The author would like to thank Prof. N. Rynn, who first suggested the idea of generating an electron - positron plasma confined in a magnetic mirror to the author, for valuable discussions and encouragement. The contribution by T. Ross for performing the computer simulations is gratefully acknowledged.

This work was supported by the National Science Foundation, Grants No. PHY-9014129 and No. PHY-9024667.

REFERENCES

1. S. Ecklund, in Intense Positron Beams, E.H. Ottowitte and W. Kells, eds., World Scientific. N.J., 1988, p. 42.
2. A.P. Mills, Jr, in Positron Solid State Physics, H.E. Ottowitte and W.Kells, eds., North-Holland Publishing Co., N.Y., 1983, p. 432.
3. A large number of publications discuss positron moderators, see, e.g., Ref. 2 and: C.D. Beling, et al., in Atomic Physics with Positrons, J.W. Humberston and E.P.G. Amour, eds., Plenum Press, N.Y., 1987, p. 175.
   W. Raith, in Fundamantal Processes of Atomic Dynamics, J. Briggs, H. Kleinpoppen, and H. Lutz, eds., Plenum Press, N.Y.
4. J.H. Malmberg and C.F. Driscoll, Phys. Rev. Lett. 44, 654 (1980).
   J.H. Malmberg, in Low Energy Antimatter, D.B. Cline, ed., World Scientific. N.J., 1986, p. 184.
5. C. Surko, et al., in Positron Studies of Solids, Surfaces, and Atoms, A.D. Mills, Jr., W.S. Crane and K.F. Canters, eds., World Scientific, N.J., 1986, p. 221.
6. W.P. Kells, in Intense Positron Beams, H.E. Ottowitte and W. Kells, eds., World Scientific, N.J., 1988, p. 135.
7. W.P. Kells, ibidem, p. 74.
8. W.P. Kells, ibidem, p. 223.
9. H. Poth, in Atomic Physics with Positrons, J.W. Humberston and E.A.G. Armour, eds., Plenum Press, N.Y., 1987, p. 307.
10. C.M. Surco, et at., Rev. Sci. Instr., 57, 1862 (1986).
11. T.J. Murphy, Plasma Physics and Controlled Fusion, 29, 549 (1987).
12. V. Tsytovich and C.B. Wharton, Comments on Plasma Physics and Controlled Fusion, 4, 91 (1978).
13. T.H. Stix, The Theory of Plasma Waves, McGraw Hill, N.Y., 1962.
14. P.A. Sturrock, Astrophys. J., 164, 229 (1971).
    M.A. Ruderman and P.G. Sutherland, Astrophys. J., 196, 51 (1975).
15. A. Sagion, Astr. Astrophys., 44, 285 (1975).
16. See for example: K.R. Chu and J.C. Hirshfield, Phys. Fluids, 21, 461 (1978).
17. P.H. Yoon and T. Chang, Astrophys. J., 343, 31 (1989).
18. G.P. Tercho, in Intense Positron Beams, World Scientific, N.Y., 1988, p. 3.
19. G. Sinapius and H.L. Ravn, in Atomic Physics with Positrons, J.W. Humberston and E.A.G. Armour, eds., Plenum Press, N.Y., 1987, p. 185.
20. G. Sinapius and M. Weber, University of Bielefeld, private communication.
21. A.P. Mills, Jr. and E.M. Gullikson, Appl. Phys. Let., 49, 1121 (1086).
    R. Khartiri, et al., Appl. Phys. Lett., 57, 2374 (1990).
22. K.G. Lynn, et al., in Atomic Physics with Positrons, J.W. Humberston and E.A.G. Armour, eds., Plenum Press, 1987, p. 161.
23. B. Brown and H. Makowitz, this proceedings.
24. W. Triftshäuser, et al., this proceedings.

25. A. Van Veen, et al., this proceedings.
26. R,H. Howell, et al., in Intense Positron Beams, E.H. Ottowitte and W. Kells, eds., World Scientific, N.J., 1988, p. 27.
27. D.E. Baldwin, et al., Plasma Physics and Controlled Fusion Research, IAEA, Vienna, 1981, vol.1, p. 133.
28. R.A. Dandl, et al., Plasma Physics and Controlled Fusion Research, IAEA, Vienna, 1975, Vol 2, p. 141.
29. D.B. Batchelor and R.C. Goldfinger, Nuclear Fusion, 20, 403 (1980).
    D.B. Batchelor, et al., ORNL Report # TM-8770 (1983).
30. A.J. Lichtenberg, et al., Plasma Physics, 11, 1073 (1969).
31. N.C. Wyeth, et al., Plasma Physics, 17, 679 (1975).
32. B.H. Quon, et al., in Hot Electron Ring Physics, Vol. 2, ORNL Rep. Conf. 911203, 1982, p.441.
33. T.D. Rognlien, Nuclear Fusion, 23, 163 (1983).
34. H. Boehmer, et al., Phys. Fluids, 28, 3099 (1985).
35. C.K. Birdsall and A.B. Langdon, in Plasma Physics via Computer Simulations, McGraw Hill, N.Y., 1985.
    W.S. Lawson, J. Comput. Physics, 80, 253 (1989).
36. See for example: J.D. Rose and M Clark, Plasma Physics and Controlled Fusion, MIT Press, Cambridge, MA, 1961.
37. D. Segal, et al., Phys. Fluids, 25, 1485 (1982).
38. G.H. Haste, in Proceedings of the 2nd Workshop of Hot Electron Rings, ORNL Report TN, 1982, p. 339.

SECTION 7

SLOW POSITRON BEAMS—ARTS AND TECHNIQUES

EXTRACTION OF SLOW POSITRONS FROM THE MAGNETIC FIELD

Takashi Akahane
National Institute for Research in Inorganic Materials
1-1 Namiki, Tsukuba-shi, Ibaraki 305, Japan

ABSTRACT

Extraction of slow positrons from a guiding magnetic field was investigated experimentally. With moderate acceleration, more than 40% of slow positrons could be collected in the field free region. Experimental results indicated that minimization of the magnetic field around a moderator and precise alignments of an accelerating electric field with the guiding magnetic field were important in order to attain high extraction efficiency.

INTRODUCTION

Magnetic guiding method has been the predominant choice to transport slow positrons from a primary moderator to a remote target because of its ease of implementation and high transport efficiency. For example, magnetic guiding systems have been commonly used for transport of slow positrons generated with electron LINACs, in which case positrons should travel long distance to avoid intense radiation at a converter-moderator position. Magnetic field is also indispensable for a Penning trap which is used to stretch slow positron pulses produced with an electron LINAC[1]. But some applications of slow positrons such as high precision LEPD and positron microscopy require electrostatic focusing in zero magnetic field. Feasibility of a hybridized electrostatic/magnetic system is worth to be examined. Extraction of slow positrons from the guiding magnetic field is a key to a success of the hybrid system. It has been already discussed[2] and tested experimentally[3,4]. In particular, Ito et al.[4] reported more than 90 % transport efficiency to the field-free region with acceleration of slow positrons up to 1 keV. However, more experiments are necessary because extract efficiencies and resultant beam sizes are of critical importance for the hybrid beam system and they strongly depend on characteristics of the primary beams and

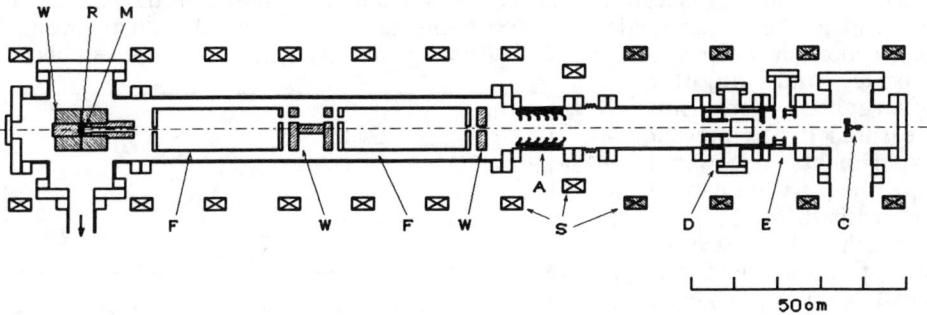

Fig. 1. Experimental setup. R: ^{22}Na positron source, W: tungsten shield, M: moderator, F: E x B filter, A: accelerator tube, D: deflector, E: einzel lens, C: channel electron multiplier, S: solenoid coils. Solenoid coils shown with hatching were turned off for extraction experiments.

© 1994 American Institute of Physics

details of the experimental setups, though they are limited by phase space relations in principle.

EXPERIMENTS

Extraction of slow positrons was tested with an apparatus shown schematically in fig. 1. Positrons emitted from a ^{22}Na radioisotope were moderated in a well-annealed polycrystalline tungsten foil. Slow positrons were accelerated in the magnetic field after they passed E x B filters. They were detected by a Ceratron, a kind of channel electron multiplier with an effective aperture of 10 mm diameter. A deflector and an einzel lens were placed in front of the detector to adjust beam path and to improve the collection efficiency. Solenoid coils with hatching in fig. 1 were turned off for extraction experiments. The magnetic field at the detector position was less than 1 Gauss as shown in fig. 2. Extraction efficiencies were defined as the ratio of numbers of detected positrons to those obtained with all the solenoids turned on. Complementary experiments were also performed with electrons. In these experiments, the radioisotope-moderator assembly and the channel electron multiplier were replaced with a tungsten filament and a Faraday cup, respectively. A fluorescent screen was also used to trace the position of electron beams.

RESULTS

An example of results of positron extraction experiments is shown in fig. 3. Open circles represent the efficiencies obtained using an einzel lens. In case of acceleration voltage of 5 kV, an increase of extraction efficiency by the use of einzel lens was very small. More than 40 % of slow positrons were collected with 3kV acceleration when the einzel lens was used. A result of electron experiments is shown in fig. 4. Extraction efficiencies were several times smaller than those for positrons. One of the reasons for this difference was the field strength where beams were produced. Slow positrons were emitted in a magnetic field less than 50 Gauss (see fig. 2), while electron experiments were carried out in the two times stronger field. Theoretically, beam radius in the field free region is proportional to the field strength where the beam is produced, if transverse momentum components of the particles in the primary beam are neglected. Experimental results agreed with theoretical expectation qualitatively, though it was difficult to discuss quantitatively because beam profile and angular divergence were not measured in the present experiments. Several experiments were carried out with different field strength. Decrease of slow positron yield was small when the magnetic field was changed from 100 to 50 Gauss. But considerable change of the yield was observed when field was decreased to 25 Gauss. The other difference between positron and electron

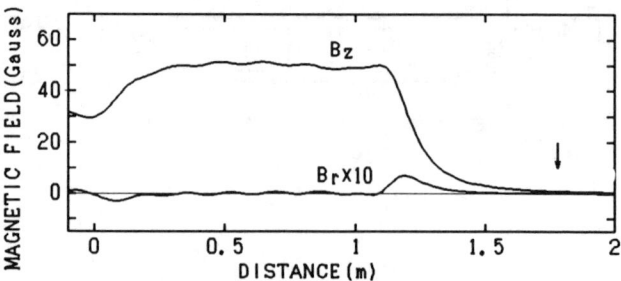

Fig. 2. Magnetic field when solenoids with hatching in fig.1 were turned off.
Bz : axial field on the symmetry axis, Br: radial field at the point 5 mm away from the axis. An arrow in the figure indicates the detector position in the experimental setup.

experiments was characteristics of the primary beams. Electron beams probably had broader energy and geometrical distributions because they were produced with thermal emission. The energy spread of positron beams was less than several electron volts in FWHM.

The decreases of the efficiency with increasing acceleration potential more than 3 keV in both positron and electron experiments contradicted with theoretical expectations and results of computer simulations. The reason is not clear at present. Misalignments of the magnetic and the electric field were suspected to be responsible. Figure 5 shows typical behaviors of the electron beams observed in the fluorescent screen when the guiding magnetic field was decreased stepwise in the experimental setup similar to one shown in fig. 1. In these experiments, diameter of the primary electron beams was limited by an aperture. If the alignment of the electric and the magnetic field was perfect and beams were produced exactly on the axis of the field, beams should stay at the same position with decreasing the guiding field. In experiments, however, beam drifts were always observed together with increase of the beam size. It was necessary to adjust beam positions using the deflector and the small steering coils(not shown in fig.1) to get optimum results for each acceleration voltage in the experiments shown in fig. 3 and fig. 4. The scatter of experimental results was at least partly due to the necessity of adjustment for each experimental condition. These adjustments became more difficult with increasing acceleration voltage, which caused the unexpected results for higher voltages. The other point to be noted on fig. 5 is that extraction was easier when beams were produced in weaker magnetic field. The fields around an electron gun were 100 Gauss in the upper case, while 50 Gauss

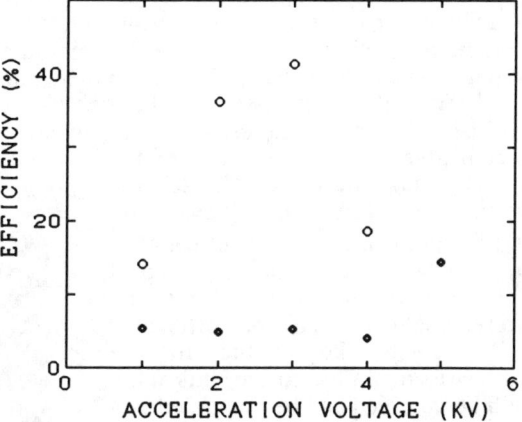

Fig. 3. Extraction efficiency of slow positrons from the magnetic field of 50 Gauss. Open circles represent efficiencies when the einzel lens was used. Black circles show efficiencis obtained without using the einzel lens.

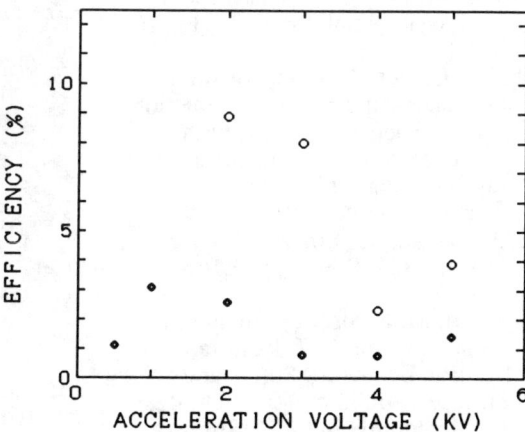

Fig. 4. Extraction efficiency of electrons from the magnetic field of 100 Gauss. Open circles represent efficiencies when the einzel lens was used. Black circles show the efficiencies obtained without using the einzel lens.

in the lower case. It is clear that beam sizes after extraction were small in the lower case.

CONCLUSION

Slow positrons were extracted from the guiding magnetic field with an efficiency of more than 40%. The feasibility of a hybrid beam system was confirmed. In order to attain high extraction efficiency, alignment of the accelerating electric fields with the guiding magnetic field was practically important. The resultant beam sizes depended strongly on the magnetic field around a moderator.

Several improvements can be tried to get higher extraction efficiency. Effects will be examined on the usage of a magnetic shield made of soft iron and permalloy which gives steeper decrease of the guiding magnetic field. The scheme where one start with electrostatic acceleration in zero magnetic field[2] and inject into a magnetic transport system should be compared with the present method. An over all efficiency will be the key issue in the comparison. In the future, this setup will be modified and be integrated as a primary beam of an electrostatic beam line with a brightness enhancement.

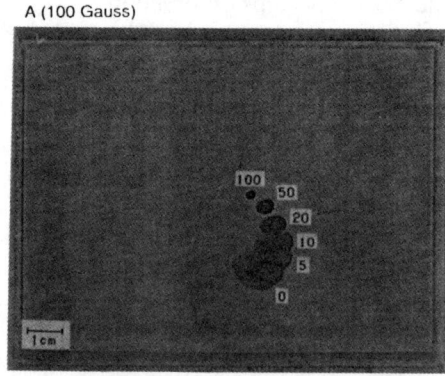

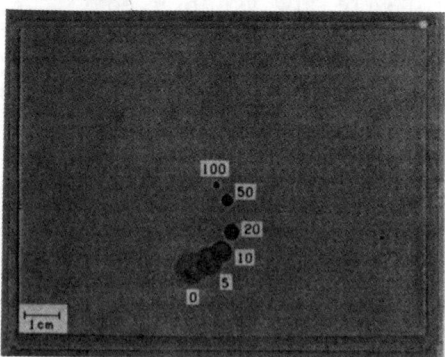

Fig. 5. An example of behaviors of electron beams when the guiding magnetic field was decreased stepwise. A: electron beam was produced in the 100 Gauss field. B: electron beam was produced in the 50 Gauss field. Numbers in the figure represent the field strength around the detector. The acceleration potential was 3 KeV.

REFERENCES

1. T. Akahane, T. Chiba, N. Shiotani, S.Tanigawa, T. Mikado, R. Suzuki, M. Chiwaki, T. Yamazaki, and T. Tomimasu, Appl. Phys. A 51, 146 (1990)
2. K. F. Canter, Abstracts of Inter national Symposium on Production of Low-Energy Positrons with Accelerators and Applications (Justus-Liebig-University of Giessen, Giessen, 1986), p.20.
3. O. Sueoka, M. Ymazaki, and Y.Ito, Jpn. J. Appl. Phys., 28, L1663 (1989)
4. Y. Ito, M. Hirose, O. Sueoka, I. Kanazawa, and S. Takamura, Positron Beams for Solids and Surfaces, edited by P. J. Schultz, G. Massoumi, and P. J. Simpson, (American Institute of Physics, New York, 1991), p.252.
5. D. T. Pierce, C. F. Kuyatt, and R. J. Celotta, Rev. Sci. Instrum. 50, 1467(1979)

Electrostatic lenses and beam optics calculations

G. Amarendra and B. Viswanathan
Materials Science Division
Indira Gandhi Centre for Atomic Research
Kalpakkam - 603102, INDIA

ABSTRACT

The basics of electrostatic lens optics in terms of the details of convergence properties and important lens parameters are discussed. An outline of beam optics calculation based on the transfer matrix method is described for standard lenses. The above scheme has been employed to calculate the positron trajectories towards optimising the beam transmission in our positron beam set up and these are illustrated.

INTRODUCTION

Electrostatic lenses are required in a positron beam set up for extraction, acceleration and focusing of slow positrons emanating from the surface of a moderator. Towards having an optimal beam transmission from the moderator to the target, it is quite necessary to calculate the beam trajectories so as to decide the critical lens parameters. The present paper discusses the details of commonly used electrostatic lenses and outlines the various steps involved in calculating the beam trajectories.

ELECTROSTATIC LENSES

A basic electrostatic lens arrangement [1-3], which is a combination of two tubes is shown in Fig. 1(a). The first and second tubes are at voltages V_1 and V_2 respectively, where V_1 is more positive with respect to V_2. The potential $\phi(z)$ felt by the particle across the gap can be obtained by solving the Laplace equation. The variations of potential $\phi(z)$ and its derivatives $\phi'(z)$ and $\phi''(z)$ across the lens gap are shown schematically in Fig. 1(b). As the positron goes through the lens gap, it experiences a radial force given by

$$F_r = e\phi''(z)r/2 \qquad (1)$$

The force acting on the particle is proportional to the height of the particle trajectory with respect to the Z-axis. Further, the direction of the force will depend on the curvature of the equipotential surface. In the regions, where the curvature $\phi''(z)$

is negative, the radial force acting on the positron will be pointing towards the axis, leading to convergence action. On the other hand, for regions where $\phi''(z)$ is positive, the resultant action will be divergent in nature. Fig. 1(c) shows the equipotential surfaces and the resulting particle trajectories across the lens gap. The lens gap comprises a combination of convergent and divergent regions, as shown in Fig. 1(d). However, as is the case with the light optics, the net result of these two opposite actions is still convergent. Thus, the combination of two tubes separated by a small gap acts as a convergent lens. In the above example, where $V_1 > V_2$, the lens acts in accelerating mode for positrons. However, even for deceleration mode with $V_1 < V_2$, the lens will also act as a convergent lens. This is so because in both the cases, the positron tends to spend relatively longer time in the region of convergence across the lens gap.

The various commonly used electrostatic lenses are shown in Fig.2. They are two tube lens (TTL), two aperture lens (TAL) and three tube lens. For all these lenses, the focal properties critically depend on three parameters viz., the diameter D of the lens, the separation of the tubes or the lens gap g and the voltage ratio of the respective elements. Basically, these three parameters affect the variation of the potential $\phi(z)$ across the lens gap and in turn affect the first and second derivatives of the potential. Thus, by varying any of these parameters one can alter the focal properties of the lens. If D or g is increased, the lens becomes weaker, while an increase of the voltage ratio of the elements results in strong lensing action. Normally, the lens diameter is so chosen that it is at least a factor of two larger than the possible beam diameter across the gap, so as to minimize the aberrations. The lens gap is chosen to be a fraction of the lens diameter, typically 0.1D or 0.5D for TTL or TAL and 0.1D in the case of three tube lens. Having chosen D and g, the voltage ratio can be varied so as to optimise the beam transmission and beam characteristics.

The three tube lens is commonly known as einzel lens and there are two types of operation of this lens viz., symmetric and asymmetric. In the case of symmetric einzel lens (SEL), the first and third tubes have the same voltage applied to them i.e., $V_1 = V_3$. The lens can be optimised by varying the voltage ratio V_2/V_1. For asymmetric einzel lens (AEL), the first and the third tubes are at different voltages and the AEL can be tuned by varying the voltage ratios V_2/V_1 and V_3/V_1. A feature of SEL is that the beam size and beam angle can be varied without affecting the beam energy. On the other hand, the beam energy gets changed upon going through the TTL and AEL. As compared to TTL, three tube lenses have better maneuverability in the sense that the focal properties can be altered over a significant range by a small change of the parameters V_2/V_1 and V_3/V_1. Towards transporting the beam from the object plane to the image plane, any combination of TTL, TAL, SEL and AEL can be used depending on the specific needs. However, it is to be ensured that the successive lenses are separated by at least three times the lens diameter so that the lenses are non-interacting.

EVALUATION OF BEAM TRAJECTORY

Once the potential $\phi(z)$ and its derivatives are known from the solution of the Laplace equation, the positron trajectory can be calculated from the following paraxial ray equation [1-3],

$$d^2r/dz^2 + (\phi'(z)/2\phi(z)) \cdot dr/dz + (\phi''(z)/4\phi(z)) \cdot r = 0 \qquad (2)$$

The particle trajectory is characterized by two parameters viz., its height r from the Z-axis and its slope dr/dz or angle θ with respect to the Z-axis. By solving the above equation, one could obtain (r,θ) of the particle at various points along the Z-axis. For commonly used electrostatic lenses with standard geometry, the above equation has already been solved and the focal properties documented in terms of focal and mid-focal lengths [4]. These documented data can be used for the optimisation of beam transport, using transfer matrix method.

TRANSFER MATRIX METHOD

If the lensing action of a given electrostatic lens can be expressed as a matrix, then one could obtain the coordinates of the particle (r', θ') past the lens gap, by simply operating the lens matrix on the coordinates of the particle (r,θ) before the gap. The matrix representing the lensing action is known as transfer matrix and the elements of this (2x2) matrix are expressed in terms of the mid-focal and focal lengths of the lens [3,5]. It is customary to define the transfer matrix elements in such a way that the distance over which the particle is transferred across the lens gap is taken to be the diameter D of the lens [3,4] i.e., from -D/2 to +D/2 with respect to the center of the lens.

The transfer matrix method is a convenient representation of the lensing action for the trajectory calculation and Fig.3 shows schematically the calculation steps. At the object plane, normally taken to be the location of the moderator surface in the context of slow positron set up, the particle coordinates are (r_o, θ_o). From the object plane upto the first lens L_1 (D/2 to the left of the lens gap), the particle travels in field-free region, wherein the angle θ remains constant while its height r will get changed. The transfer matrix corresponding to field-free transport is given by,

$$TM^{FF}(z) = \begin{pmatrix} 1 & z \\ 0 & 1 \end{pmatrix} \qquad (3)$$

where z is the distance travelled in the field-free region. Then corresponding to the lens L_1, the transfer matrix TM(1) is operated on the particle coordinates, resulting in modified (r,θ) at a distance of D/2 to the right of lens gap. Then, the particle is again field-free transported further till the second lens gap, corresponding to L_2. Accordingly, in the present case, the transport of the particle from the object plane to the image plane is given by,

$$\begin{pmatrix} r_i \\ \theta_i \end{pmatrix} = TM^{FF}(z_3).TM(2).TM^{FF}(z_2).TM(1).TM^{FF}(z_1) \begin{pmatrix} r_o \\ \theta_o \end{pmatrix} \quad (4)$$

where (r_i, θ_i) are the coordinates of the particle at the image plane. Here TM(1) and TM(2) are the transfer matrices of lens L_1 and L_2, respectively. The ray tracing between the object and image planes is generally carried out using two characteristic rays. The rest of particle trajectories can be expressed by a linear combination of these two characteristic ray trajectories [3].

The transfer matrix subroutines corresponding to TTL, SEL and AEL have already been documented [5]. The transfer matrix of any other standard lens can be constructed along similar lines [4,5].

PRESENT BEAM OPTICS PROGRAM

Based on the transfer matrix scheme discussed above, we have developed beam optics program to calculate the beam trajectories and optimise the beam transmission for (a) a combination of AEL and TTL and (b) two aperture lenses. The former has been carried out for extraction of positrons emanating from the surface of a moderator and injection into a defining aperture located at the entrance of the magnetic transport line of our slow positron set up [6]. The calculations of aperture lens optics have been made for envisaged use in the target side optics. Fig. 4 shows, by way of illustration, the calculated positron trajectories corresponding to case (a) above. Here, the lens diameter D is 22 mm and the lens gaps have been chosen to be 0.1D for both AEL and TTL. The optimised separation between AEL and TTL is 164 mm. The beam trajectories have been calculated corresponding to a moderator size of 5 mm and positron launch angle of $\pm 10°$. Two characteristic rays of (+2.5 mm, 0.175 rad) and (+2.5 mm, -0.175 rad) and their symmetric counterparts have been traced for beam transmission. Fig. 5 (a) and (b) show the phase space (r, θ) plots at the moderator and aperture respectively. In the optimised condition of positron extraction energy of 225 eV, the beam size is about 2.5 mm with a divergence of $\pm 1°$ at the aperture located at a distance of 369.5 mm from the moderator.

Distortions in the beam characteristics due to spherical aberrations can also be computed by using the tabulated spherical aberration coefficients of various lenses [4]. We plan to incorporate this in our beam optics calculations.

CONCLUSIONS

The transfer matrix scheme is a simple and convenient method for calculating beam trajectories, as demonstrated originally by Canter and coworkers [5] in the context of slow positron beam transport. This scheme has been adopted by us towards the design and optimization of electrostatic transport in our beam set up.

REFERENCES

[1] "Electron beams, lenses and optics" Ed. by A.B. El-Kareh and J.C.J. El-Kareh (Academic Press, NY, 1970) Vol.1

[2] "Electron Optics" Ed. by P. Grivet (Pergamon Press, NY, 1965)

[3] "Introduction to Electron and Ion Optics" Ed. by P. Dahl (Academic Press, NY, 1973)

[4] "Electrostatic Lenses" Ed. by E. Harting and F.H. Read (Elsevier, Amsterdam, 1976)

[5] K.F. Canter, P.H. Lippel and D.T. Nguyen, in "Positron Studies of Solids, Surfaces and Atoms", Ed. by A.P. Mills Jr., W.S. Crane and K.F. Canter (World Scientific, Singapore, 1986) p. 207

[6] G. Amarendra, B. Viswanathan, G. Venugopal Rao, K. V. Thomas Kutty, B. Purniah and K. P. Gopinathan, in these proceedings.

Figure captions

Fig.1   (a) A basic electrostatic lens comprising of two tubes of diameter D, separated by a gap g (b) Schematic variations of the potential $\phi(z)$ and its derivatives $\phi'(z)$ and $\phi''(z)$ felt by positron across the lens gap (c) Equipotential surfaces and the resulting particle trajectories (d) The convergent and divergent regions across the lens gap pictorially shown as convex and concave lenses.

Fig.2   Commonly used electrostatic lenses (a) Two tube lens with a lens diameter D and lens gap g (b) Two aperture lens with a lens diameter D and lens gap g (c) Three tube lens with a lens diameter D and lens gap g=0.1D. The separation between the two centers of gaps is 1.0D.

Fig.3   Schematic representation of various steps involved in beam trajectory calculation using transfer matrix method. Z_1, Z_2 and Z_3 indicate the field-free transport regions while L_1 and L_2 are the lenses.

Fig.4   Calculated positron beam trajectories from the moderator to the aperture in our positron beam set up, using a combination of AEL and TTL.

Fig.5   Phase space (r,θ) plots at the location of (a) moderator and (b) aperture.

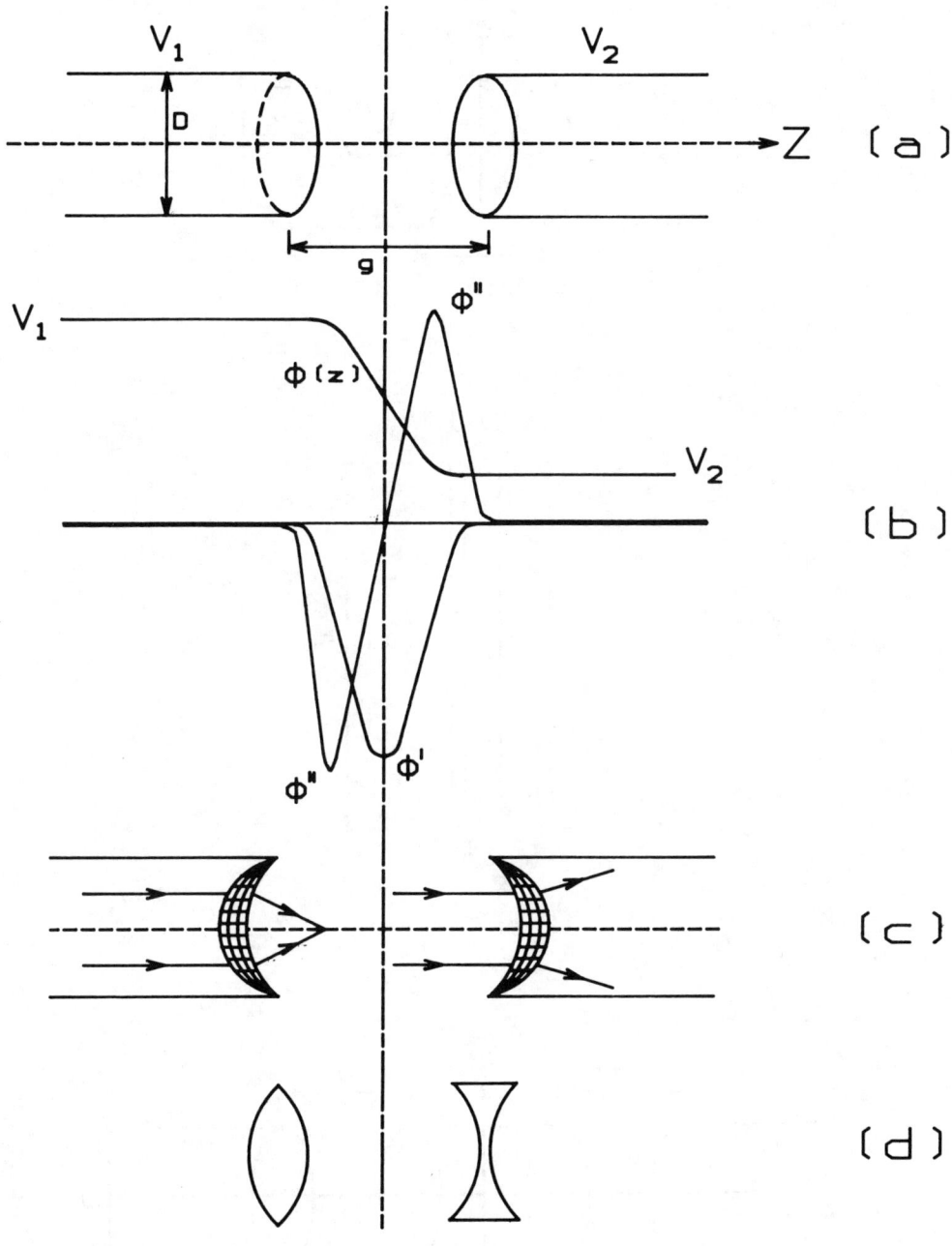

Fig. 1

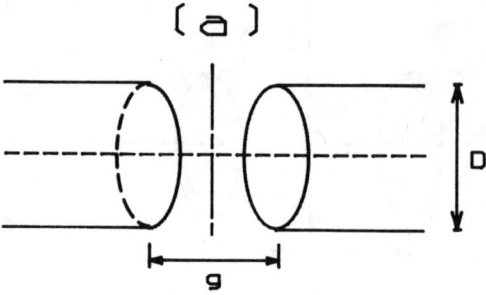
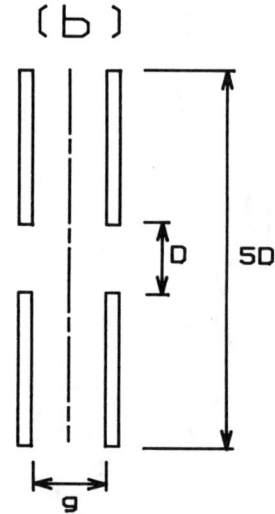
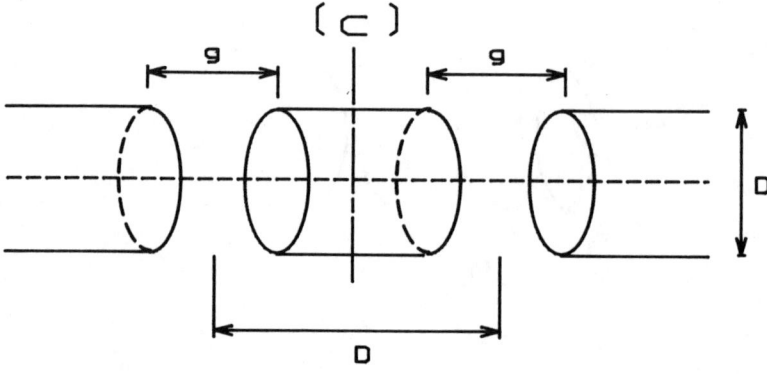

Fig. 2

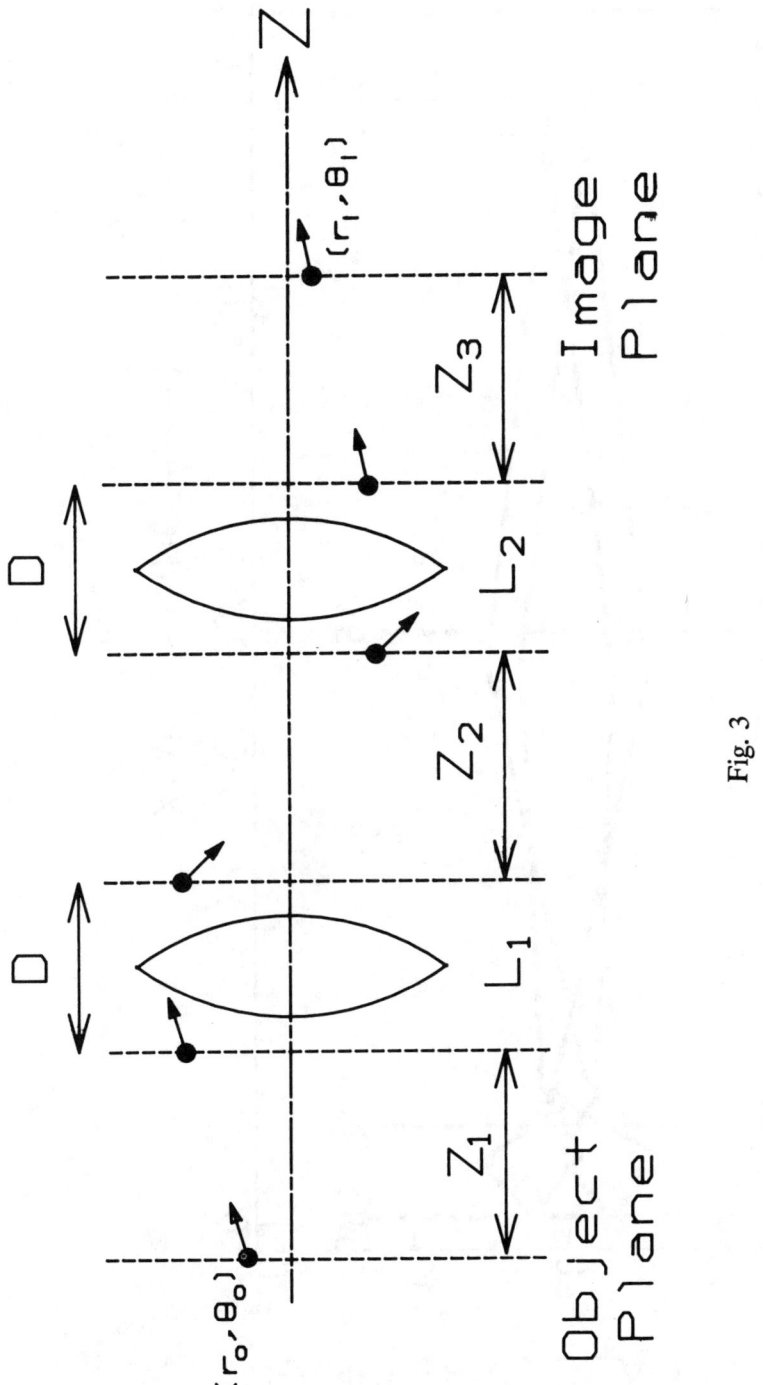

Fig. 3

450 Electrostatic Lenses and Beam Optics Calculations

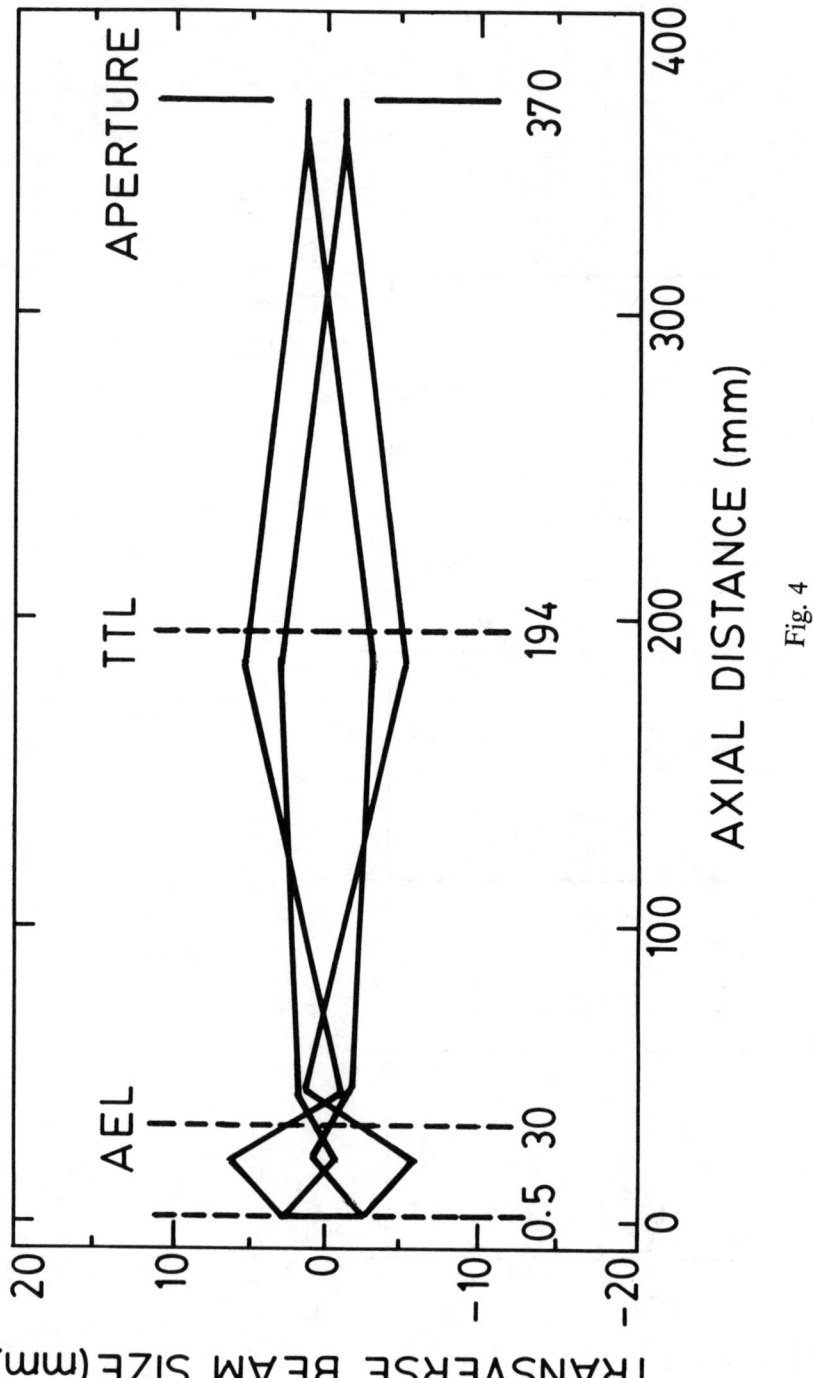

Fig. 4

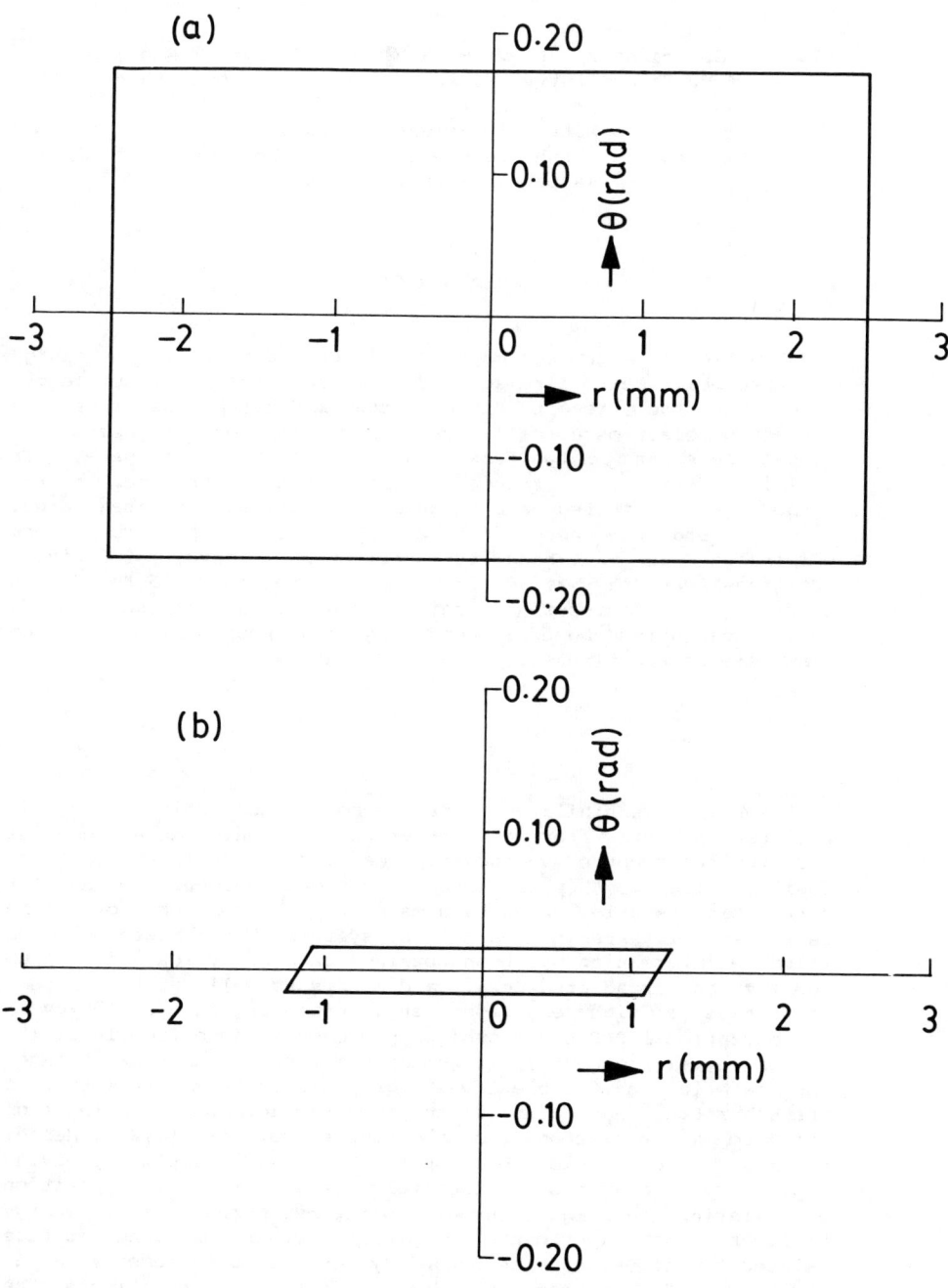

Fig. 5

Slow positron beam set up at Kalpakkam - A progress report

G. Amarendra, B. Viswanathan, G. Venugopal Rao,
K.V. Thomas Kutty, B. Purniah and K.P. Gopinathan

Materials Science Division
Indira Gandhi Centre for Atomic Research
Kalpakkam - 603102, INDIA.

ABSTRACT

Construction of a compact magnetically guided variable low energy positron beam set up is discussed and the present status report given. In the present design, the moderated positrons are extracted electrostatically, using a combination of asymmetric einzel lens and a two tube lens whose parameters have been optimised using a ray tracing program based on transfer matrix methods. The extracted beam is guided magnetically to the target chamber and the non-moderated high energy positrons are eliminated in a U-bend transport tube. Energy of incident positrons can be varied from 100 eV to about 35 KeV. Depth profiling of near-surface defects and study of solid-solid interfaces using Doppler broadening and positronium fraction measurements are planned.

INTRODUCTION

Positron annihilation spectroscopy (PAS) using energetic positrons directly from radioactive sources, has been a powerful tool for the study of vacancy-type defects in metals and alloys. Positron lifetime, Doppler broadening and one-dimensional angular correlation studies have been made in our laboratory over the last several years in a variety of systems. From these studies valuable information has been obtained on micro-voids and helium bubbles in irradiated metals and alloys as well as on defect properties in high temperature superconducting oxides. However, in conventional PAS techniques sample depths range from 10 to 100 microns, yielding depth averaged information in the bulk state of the material. A variable low energy positron beam facility, on the other hand, has potential to probe sample depths ranging from zero to a few microns from the sample surface [1,2]. Hence, setting up of a slow positron facility with depth resolving capability augments effectively the existing positron annihilation techniques in our laboratory. Moreover, low energy positron beam experiments complement other existing surface related techniques such as Secondary Ion Mass Spectrometry (SIMS) and Thermal Helium Desorption Spectrometry. These constitute the motivation for the present project whose details are discussed in this paper.

In constructing a variable low energy positron beam facility, one is faced with questions concerning the choice of the primary source, type of moderator to be used, nature of transport of moderated positrons from the source region to the target area viz., electrostatic or magnetic and the acceleration to the desired energy [2,3]. It is without doubt that the final design depends on the type of measurements to be carried out. The design and construction of the present set up under discussion is based on the need to build a compact and relatively simple system mostly from indigenously available materials for the generation of laboratory beams of moderate intensity and a few mm in diameter. In the present set up, slow positron extraction from the moderator has been designed based on the calculations of electrostatic lens optics. Magnetic beam transport and acceleration at the target have been incorporated following design features reported earlier [4,5].

CONSTRUCTION DETAILS

The system has been fabricated out of ultra high vacuum compatible SS 304 material and consists of three main parts : source side beam line with moderator and extraction optics, beam transport line with velocity selection and target chamber with accessories and acceleration stage. The primary source is a 50 mCi ^{22}Na of active diameter 3 mm deposited on to a titanium capsule and sealed with an electron beam welded 5 µm thick Ti foil window. The moderator is a W(100) single crystal foil of thickness in the range of 7000 - 10,000 Å obtained from the University of Aarhus. A cross-sectional view of the set up is shown in Fig.1 with constituent parts. The primary source disc is held on to a bellowed linear motion arm (A) with micrometer attachment. The source can be positioned as close as possible ($\leq$ 1 mm) with respect to the moderator foil (B) used in the transmission geometry. The moderator foil will be given high temperature annealing treatment in a separate UHV chamber. After heat treatment, the foil is transferred in air quickly to the beam line. This procedure has been shown [6] to give satisfactory yields stable over long periods of time. The source side vacuum tube of 48 mm OD has UHV ports for evacuation, vacuum monitoring and electrical feedthrough for lenses. The source side beam line is evacuated by a 170 l/s Turbo molecular pump.

A beam transport program to calculate the beam trajectories has been developed based on the transfer matrix methods coupled with the documented data on the focal properties of commonly used electrostatic lenses [7,8]. It is found from our beam optics calculations [9] that a combination of asymmetric einzel lens (AEL) and a two tube lens (TTL) provides satisfactory slow positron extraction from the moderator and focusing into an aperture with initial acceleration of a few hundred volts. Figure 2 shows the optical arrangement designed for slow positron extraction. The source disc on a linear drive is positioned close to the moderator foil. The moderator foil, seated between two annular rings of 25 μm thick W foils and held between two SS discs, is mounted in position. The elements of AEL and TTL are shown in Fig. 2. The diameter of lens tubes is fixed as 22 mm and the lens gaps are chosen to be 0.1 times the lens diameter. The optimum separation between AEL and TTL is 164 mm. The voltages corresponding to the moderator and the lenses have been optimised as indicated in Fig.2. The lens arrangement focuses the beam onto the SS aperture of 8 mm diameter, located at a axial distance of 370 mm from the moderator. The extraction assembly is mounted inside an optics tube of 33 mm OD. Alignment of the optics tube with the magnetic transport line, and easy mounting and de-mounting of the tube in the beam line have been ensured. The calculated beam trajectories from the moderator to the focusing aperture are shown in Fig.3. Slow positrons emanate from the moderator surface of diameter 5 mm with a divergence of $\pm 10°$, typical of positron emission from W moderators. The beam focused onto the aperture has a diameter of 2.5 mm with a divergence of $\pm 1°$, corresponding to an extraction energy of 225 eV. Beam transmission at lower energies is also satisfactory according to our beam trajectory simulation.

The details of the rest of the parts of the set up shown in Fig.1 are as follows. Between the extraction stage and the target chamber, magnetic guidance in the form of $180°$ bent U-tube column [5] has been installed. This transport section is made of an SS vacuum tube of 60 mm OD with a radius of curvature of 300 mm. The linear sections of the transport tube at the entrance and exit of the U-bend are of length 150 mm and 300 mm, respectively. The $180°$ bent section has been incorporated for the following reasons : (1) Firstly, it acts as an energy filter by which high energy (unmoderated) positrons are eliminated in the beam line. (2) The curved section prevents direct line of sight between the source and the target positioned at the end of the transport line. (3) Energy selection in the present case is simple and effective, without the use of a rather complex system like ExB filter. The longitudinal homogeneous guiding magnetic field is generated by double layers of solenoidal coils (H) made of 1.2 mm super-enameled Cu wires wound closely on the U-tube. Guiding fields upto a strength of 120 G can be achieved in the beam line. The strength of the guiding field is chosen to be around 70 G (7 mT), corresponding to which positrons injected with an energy of 225 eV spiral along the guide tube with a radius of gyration of < 1 mm. Higher energy positrons with larger Larmor gyration pitch, compared to the length scale of the curved magnetic field lines, are not able to follow the guiding field and hence annihilate at the walls of the vacuum vessel beyond the first SS aperture. The resulting high energy gamma background is eliminated by effective lead shielding (G) of the linear section of the transport tube starting from the source. As the positrons traverse the bend, they experience a gradient drift velocity v_D proportional to R x B [10] where R is the radius of the bend and B the magnetic field. Corresponding to positron transport energy of 225 eV, R = 300 mm and B = 70 G, the strength of the correction field for drift compensation is estimated to be of the order of 2 G (0.2 mT). For the correction of beam drifts from the axis around the bend, a pair of coils have been positioned along the inner and outer curved sections of the U-tube respectively. In addition, another pair of correction coils rotated by 90° (which are perpendicular to the plane of curvature) have been located along the top and bottom sections of the transport tube. The above pairs of orthogonal correction coils, denoted by J in Fig.1 serve to optimise the drift compensation. In addition, two sets of steering coils rotated by 90° have been positioned, one each at the entrance and exit sections of the U-transport tube as denoted by I and K, respectively in Fig.1. These help to steer the beam axially in the transport tube and counter drifts due to the earth's field and other extraneous magnetic effects. The various coils are energised by independent current regulated power supply units. For the purpose of beam diagnostics, a second SS aperture is positioned in the transport tube before entry into the target chamber.

The target chamber (M) mounted at the end of the transport tube, has an ID of 200 mm and of length 220 mm. The chamber is evacuated by a 500 l/s Turbo pump which is connected to the bottom flange of the chamber through an I section (not shown in Fig.1). The U-transport tube is mounted with a 45° inclination so that the target position in the chamber is at a height with respect to the source side beam line for convenience of operation. The target chamber has UHV ports for beam entry, vacuum monitoring, high voltage and low voltage feedthroughs and load lock for sample transfer. The magnetic guiding field is extended into the target chamber by Helmholtz coils (L), having the same nominal field strength as the solenoid. The use of Helmholtz coils in the target chamber is compelled by geometric reasons. By extending the guiding field right upto the target, the target side optics is of simple magnetic focusing type. Final acceleration to the desired positron energy is achieved with the target biased to negative high voltage with respect to the beam line. Provision exists to incorporate HV aperture rings connected via a resistor chain for gradual increase of the acceleration potential on the target, if found necessary. These potential rings, which can be viewed as two-aperture lenses, have been optimised by our beam optics calculation. The presently adopted practice of floating the target at high voltage is in contrast to the more prevalent method [2] of floating the source side beam line at high voltage and keeping the target at ground potential. The reason for this is to allow a future provision in our set up for pulsing the beam for slow positron lifetime experiments [5] without major modification in the design. The sample holder is attached to the conducting tip of the HV feedthrough denoted by N in Fig.1. The HV feedthrough has capability for use upto 50 KV. A HPGe detector of 35% efficiency is positioned at a distance of 5 cm behind the target for Doppler broadening and Ps fraction measurements as a function of incident positron energy. Between the target and the detector, a tungsten collimator (as shown in Fig.1) is fixed to eliminate the annihilation events due to back-scattered positrons. The set up has provision for surface characterization of the sample under study in a separate analysis chamber denoted by Q in Fig.1. The analysis chamber is equipped with dedicated pumping system, sample manipulator, ion gun sputtering and load lock for sample transfer to the positron beam chamber. The surface analysis chamber, which has been installed and tested, is kept ready for coupling to the positron beam chamber, whenever required. The two chambers are isolated by a gate valve P as shown in Fig.1.

PRESENT STATUS

The slow positron beam set up fabricated mostly from indigenously available materials, has been assembled on a 2 x 1.5 m SS table, with control and power supply units mounted in a separate panel. The beam line with in situ optics assembly and the target chamber have been baked at 150°C and vacuum tested. After baking the vacuum obtained in the beam line is 2×10^{-8} Torr. Improvement in the ultimate vacuum of the beam line is envisaged with the addition of Ti-sublimation pump. Insulations in the various coils have been satisfactorily tested at the baking temperature. The guiding field homogeneity in the solenoidal column varies by less than 3% along the beam line. The long term stability of both longitudinal and transverse magnetic fields is quite satisfactory. For the acceleration of the beam, a DC power supply with switch mode type of regulation and variable output from 0 to 50 KV has been locally fabricated. The HV power supply tested for stability and regulation is ready for use. With the expected arrival of the primary source in sealed capsule, which is under procurement from M/s. Dupont, beam tuning at the target and preliminary slow positron measurements will be commenced soon. Testing of the beam line with electrons has not been attempted, since it is not considered a faithful reproduction of beam transmission characteristics, as emanating from the positron source-moderator configuration.

The complementary infrastructure at our centre include a 400 KeV heavy ion accelerator, SIMS and thin film sputtering facility. Our planned areas of study with slow positrons are depth profiling of near-surface defects in ion implanted metals, interfaces, superconducting thin films and practical problems in surface diagnostics.

ACKNOWLEDGMENTS

We thank Prof. W. Triftshauser for useful discussions and suggestions. We would like to thank Dr. Kanwar Krishan for his support and Drs. C.S. Sundar and A. Bharathi for their keen interest.

REFERENCES

[1] A.P. Mills Jr., in "Positron Solid State Physics" Ed. by W. Brandt and A. Dupasquier (North Holland, Amsterdam, 1983) p. 432

[2] P.J. Schultz and K.G. Lynn, Rev. Mod. Phys. $\underline{60}$ (3) 701 (1988)

[3] K.F. Canter, G.R. Brandes, T.N. Horsky, P.H. Lippel and A.P. Mills Jr., in "Positron Annihilation" Ed. by L. Dorikens-Vanpraet, M. Dorikens and D. Segers (World Scientific, Singapore, 1989) p.18

[4] W. Triftshauser and G. Koegel, in "Positron Annihilation" Ed. by P.G. Coleman, S.C. Sharma and L.M. Diana (North Holland, Amsterdam, 1982) p. 142

[5] D. Schoedlbauer, P. Sperr, G. Koegel and W. Triftshauser, Nucl. Inst. and Methods in Phys. Research B $\underline{34}$ 258 (1988)

[6] N. Zafar, T. Chevallier, F.M. Jacobsen, M. Charlton and G. Laricchia, Appl. Phys. A$\underline{47}$ 409 (1988)

[7] K.F. Canter, P.H. Lippel and D.T. Nguyen, in "Positron Studies of Solids, Surfaces and Atoms" Ed. by A.P. Mills Jr., W.S. Crane and K.F. Canter (World Scientific, Singapore, 1986) p. 207

[8] "Electrostatic Lenses" Ed. by E. Harting and F.H. Read (Elsevier, Amsterdam, 1976)

[9] G. Amarendra and B. Viswanathan, these proceedings

[10] W.E. Kauppila, T.S. Stein, G. Jesion, M.S. Dababneh and V. Pol, Rev. Sci. Instrum. $\underline{48}$ (7) 822 (1977)

Figure Captions

Fig. 1. : A cross-sectional view of the slow positron beam set up shown with constituent parts. A - source mounted on a linear drive; B - moderator foil; C - electrostatic lens elements; D - port for vacuum monitoring; E - port for pumping; F - ports for electrical feedthrough; G - lead shield; H - solenoidal guiding magnetic field; I - steering coils at the entrance; J - orthogonal coils for drift correction; K - steering coils at the exit; L - Helmholtz coils; M - target chamber; N - HV feedthrough cum sample holder; O - HPGe detector; P - isolation valve; Q - surface analysis chamber.

Fig. 2. : Arrangement of the positron extraction optics. The optimised voltages for the elements are : moderator at +225 V; AEL elements at +224.5, +217 and +165 V; TTL elements at +165, 0 V. Beyond the aperture, positrons are confined by the magnetic field.

Fig. 3. : Calculated positron beam trajectories with rays traced from the moderator to the focusing aperture. The positions of the moderator, AEL, TTL and aperture are shown marked in the figure.

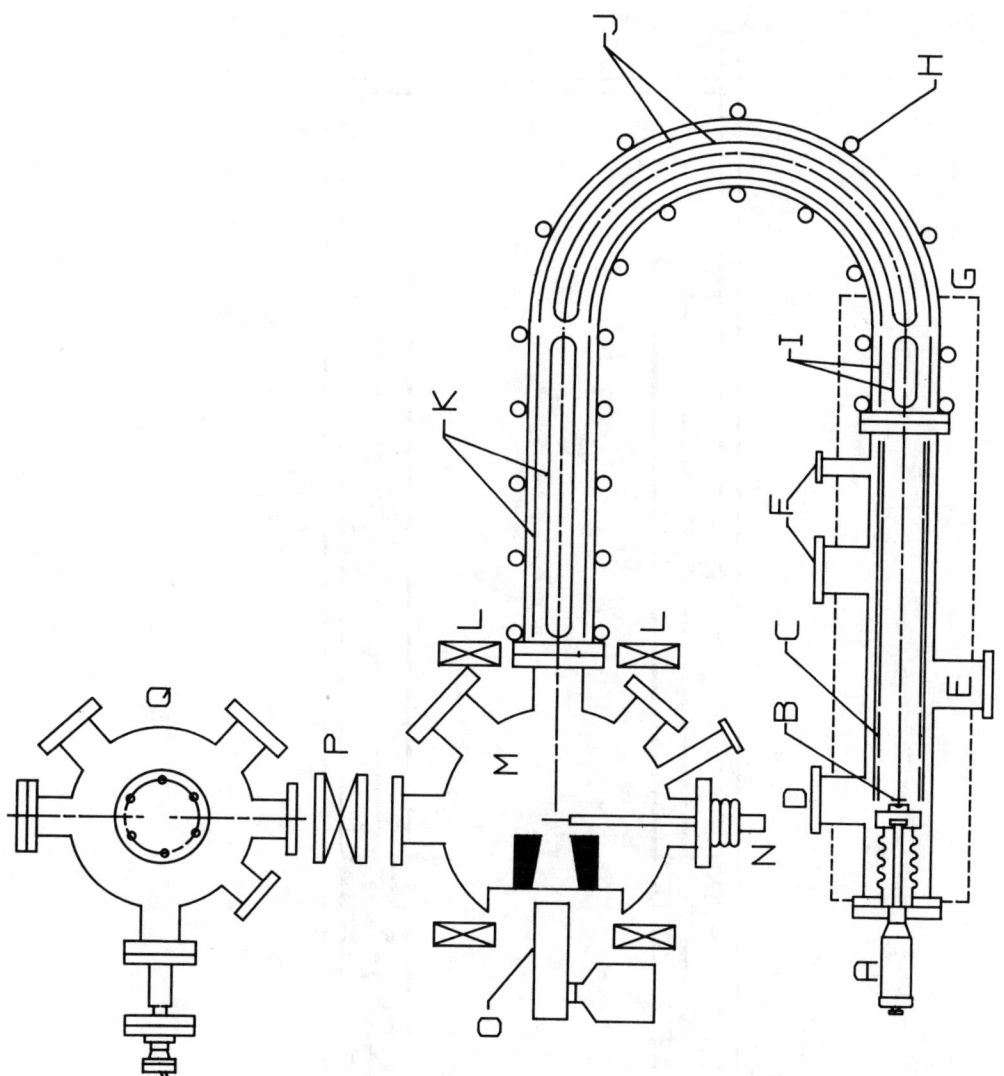

Fig. 1.

460 Slow Positron Beam Set Up at Kalpakkam

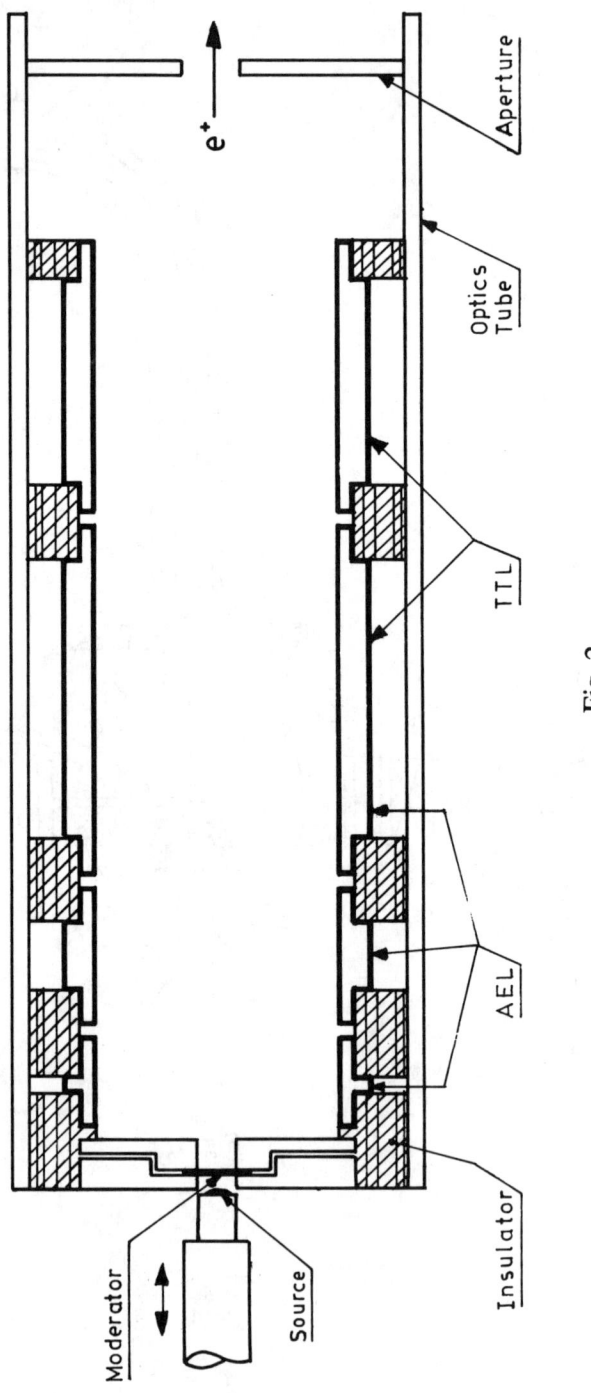

Fig. 2.

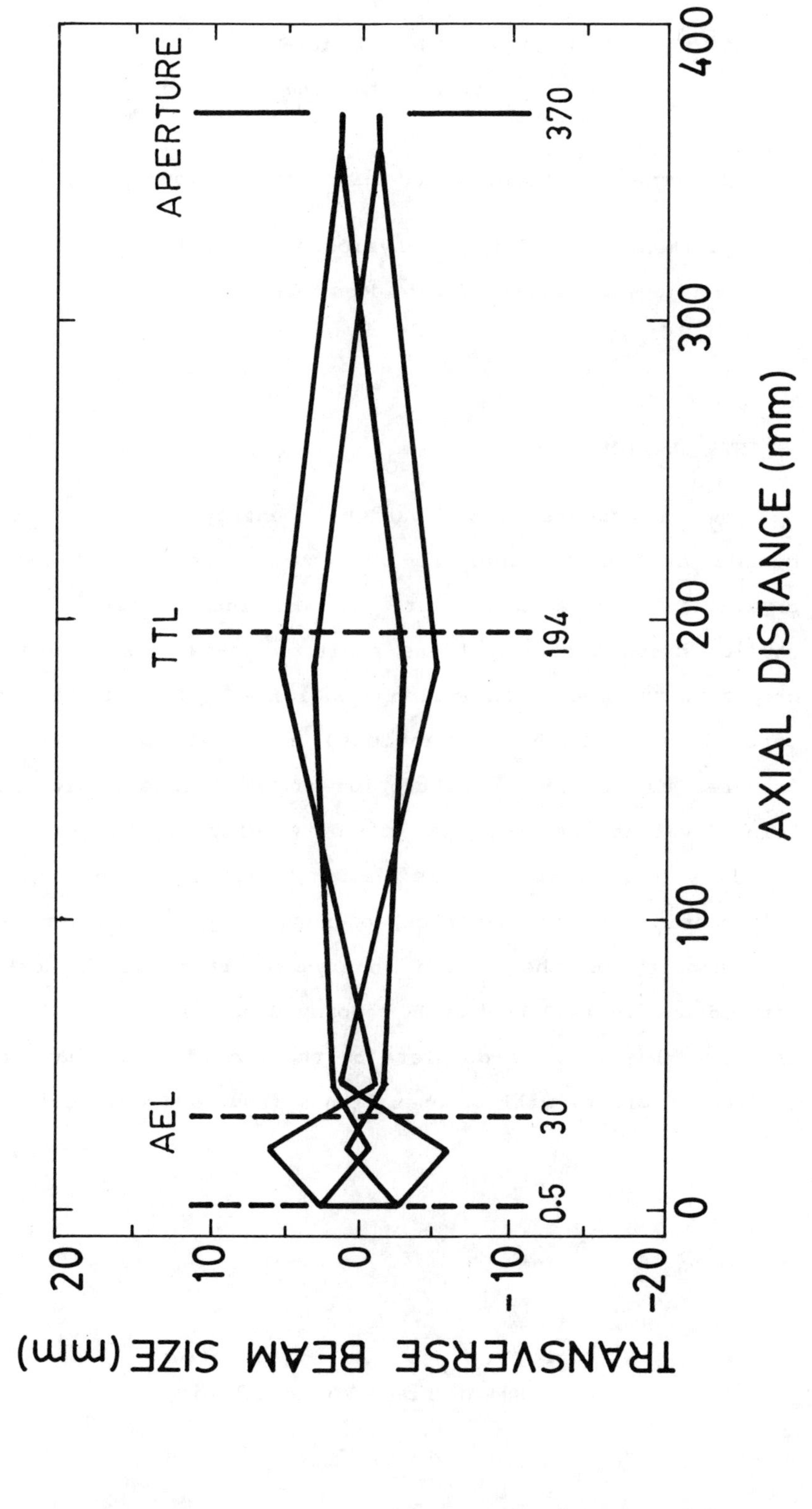
APERTURE
TTL
AEL
370
194
30
0.5
0
100
200
300
400
20
10
0
−10
−20
AXIAL DISTANCE (mm)
TRANSVERSE BEAM SIZE (mm)

Fig. 3.

THE SLOW-POSITRON BEAM FACILITY AT THE UNIVERSITY OF HONG KONG

C.D.Beling, S.Fung, H.M.Weng[1], C.V.Reddy,
S.W.Fan, Y.Y.Shan and C.C.Ling

Department of Physics, The University of Hong Kong, Hong Kong.

[1]Department of Modern Physics, University of Science and Technology of China, Hefei, PR of China.

1. INTRODUCTION :

Slow-positron beams of variable energy have over the last decade become a powerful probe for the investigation of crystalline defects at surfaces and interfaces[1]. The new physics brought out by these positron beams owes much of the credit to the depth resolution capability. This depth profiling capability that the conventional positron bulk studies lack inspired us in the construction of an intense, slow-positron beam of variable energy at the University of Hong Kong. This facility was first commissioned by the UPGC (University and Polytechnics Grants Committee) of Hong Kong in October of 1988. The majority of the design and vacuum fabrication works were carried out in 1989 by the UK company Caburn-MDC Ltd. It is hoped that assembly will be complete by the end of this year and that the beam facility will be operational from 1993 onwards[2].

2. PHILOSOPHY OF THE HKU BEAM :

i) The beam is to be easily replenishable and inexpensive to maintain:

The design seeks to overcome a major problem encountered by many other designs in that the amount of ^{22}Na activity can be maintained at a suitable level without too much difficulty. The mechanism by which this is achieved is that of forming an annular source out of eight independent sources. Each source may be extracted from the system when it has decayed significantly, and more ^{22}Na directly deposited onto the W backed source tips. The eight sources may be withdrawn through vacuum interlocks and replenished without the need to break the UHV and loose beam time. The exact positions of the eight sources relative to the moderator may be varied slightly to either optimise the beam intensity or to keep the beam intensity at some desired value.

ii) The beam is to be intense for fast data collection:

We have aimed at producing a highly efficient source-moderator geometry so as to allow the possibility of achieving beam intensities of around 10^6 e^+s^{-1}, and thus fast data collection. To obtain the necessary efficiency, we use a single crystal of W(110) as a moderator in the backscattering geometry. If the crystal were point source irradiated the efficiency would be 3 X 10^{-3}[3]; but, since the solid angle is approximately three times worse for our annular geometry, we expect an efficiency of 10^{-3}[4]. Taking a total activity of 100 mCi (absolute maximum) we obtain an expected beam intensity of :

$$I = 100 \times 10^{-3} \times 3.7 \times 10^{10} \times 0.9 \times 10^{-3} = 3.3 \times 10^6 \, e^+ \, s^{-1}$$

Of course the beam can be operated at much lower intensities, if required. In particular, we point to the fact that 8 X 1.25 mCi (10 mCi total) of activity would yield quite a useable beam intensity of ~6 X 10^5 e^+s^{-1}. [Assuming an efficiency of 2 X 10^{-3} is attainable]

iii) **The beam is to be focussed for semiconductor applications:**

High integrity semiconductor contacts can only be made in small dimensions easily - typically 1 X 1 mm^2. The obvious advantage of having a millimetre diameter beam need not be stated. What we have done by careful computer simulation using the SIMION software[5] is to produce a hybrid system using electrostatic focussing into a magnetic solenoidal transport to obtain the required small beam diameters, while retaining the convenience of magnetic transport[6].

3. DESCRIPTION OF THE BEAM APPARATUS:

The schematic diagram of the slow-positron beam of variable energy in the range 0 to 60 KeV at the University of Hong Kong is shown in Fig.1. With this energy range available, the facility is useful not only for near-surface work but also for depth profiling of defects, investigating properties of multilayered structures such as quantum well lasers, diodes etc., and positron channelling.

The beam apparatus essentially consists of three interconnected parts, namely (i) Source tank, (ii) Filter tank and (iii) Target chamber. The source tank houses the primary radioactive sources, moderator, the beam extraction optics and the acceleration stage. The filter tank consists mainly of the E x B velocity filter. The target chamber is equipped with all necessary sample preparation and characterisation equipment. The entire apparatus is maintained in UHV (Ultra High Vacuum), which is necessary for all kinds of experiments with positrons at surfaces. The nominal vacuum throughout the system without baking is around 10^{-8} torr and about 10^{-10} torr following bake-out of the entire system at 200° C for 48 hours, as measured with a nude ionisation gauge.

(i) Source tank:

This chamber is of 258 mm OD, 80 cm length, is fitted with two 340 mm OD flanges on either sides, and has 16 ports located at different convenient positions. Besides this a turbo pump and a sputter-ion pump are attached to this tank through suitable ports. Of the 16 ports, eight are dedicated for mounting the linear drives that hold the radioactive source sticks which are arranged orderly in a convenient fashion, so that all the eight sticks meet at the centre of the axis of the cylinder thus forming the annular source. Out of the remainder of the ports, a few serve as view ports, some as HT feedthrough ports and two are dedicated for mounting an ionisation guage to measure the inside pressure and a residual gas analyser. The moderator linear drive is fixed to the rear-end flange whereas the accelerated beam enters the filter tank through a port fixed at the centre of the front-end flange.

Source-Moderator Configuration :

A new source-moderator configuration has been employed in our beam design, as shown in Fig. 2, for the efficient extraction of slow positrons so as to obtain an intense beam as well as to maintain safety standards while the beam is in operation.

^{22}Na in the form of NaCl is used as a primary radioactive source and a single crystalline W(110) is employed as the moderator in our beam. As stated earlier, the efficiency of the moderator for our source-moderator configuration is 10^{-3}. However, with cone enhancement to the source-moderator geometry[7], we have estimated that the efficiency can be improved by a factor of approximately two. The gold coated stainless steel cone will backscatter not only the fast positrons (directly coming from the source) but will also reflect energetic backscattered positrons emitted from the moderator back to the moderator for a second attempt at moderation.

The radioactive source is deposited on eight different sticks with W backing which may be moved independently to the positron moderator position or taken away by means of the linear transporters attached to each stick. Together they form an annulus of ^{22}Na and the total surface area is large enough to prevent significant source absorption from occurring. The important advantage of this kind of source arrangement is the ability to maintain the beam intensity with time. Essentially the weak sources can be removed independently through vacuum interlocks and replaced with more active ones without closing down the beam. In this way the beam intensity can be built up to the desired strength over longer periods of time. Another advantage with this kind of arrangement can easily be explained by considering a simple example as follows : If 100 mCi of ^{22}Na is required to get a flux of > 3 X 10^6 e^+s^{-1}, it would be dangerous to handle this amount of radioactive material as a single source, whereas with our design, the source can be divided into eight equal parts (12.5 mCi each) which is less hazardous to handle.

A high precision machined component, the "locator", made of molybdenum, allows the accurate convergence of the eight radioactive sources and the moderator at the position "M2" as shown in Fig.3. The moderator is transported to the position "M1" through the moderator linear drive and annealing is carried out at an elevated temperature of 2000°C by e-beam heating for approximately 10 minutes in order to remove the defects in the moderator crystal in order to maximise the slow positron yield. As shown in Fig.3, necessary care has been taken in the design so as to prevent the other parts of the beam apparatus being exposed to this high temperature radiation. After annealing, the moderator is driven back to the position "M2" for normal application.

As may also be seen from Fig.3., the locator, the moderator, and sources when converged at position "M2", are surrounded by a large cylindrical block of antimonial lead (4% Sb). This block which we call the "radiation shield", shields gamma rays (1.28 M eV) emanating from "M2", by an attenuation factor of 10^3. It also has the important function of supporting the "locator", and itself stands on insulating ceramics so that it may be raised to the accelerating potential.

Focussing and Acceleration:

A new focussing and accelerating system has been employed in our beam design. The beam will start with an essentially electrostatic extraction from the moderator, followed by magnetic transport. As stated earlier, the beam is magnetically transported with a background field of 100 gauss. This field is reduced to approximately 10 gauss or even less at the moderator position and in the region around the mouth of a long focal length two-element electrostatic lens (labelled as the "extractor"). This is achieved by means of a solenoid which produces an opposing magnetic field to the background field. The net result of this is to achieve a funnel-shaped magnetic field gradient, which will focus the beam from 8 mm to less than 2 mm diameter. The beam can be further focussed down to < 1 mm diameter by using an increasing magnetic field along the beam axis towards the target chamber.

The function of the Extractor assembly is two-fold. This is essentially an electrostatic element that sits inside the locator, as shown in Fig.3. Firstly, it allows a small extraction potential to be applied close to the region "M2" which, due to the increasing magnetic field along the axis of this element, is necessary to prevent magnetic reflection of the positrons. A small positive potential applied to this element will also switch off the beam. Secondly, apart from the small extraction potential this electrode is at the acceleration potential and thus in conjunction with the funnel action of the magnetic field, allows some focussing action of the positrons by the funnel shaped end section at the elements extreme right.

The moderated positrons may also, if desired, be accelerated to the desired potential by a conventional accelerator. This consists of ten stainless steel annular plates stacked-up with ceramic breaks between them. The plates are electrically connected with 150 MΩ UHV resistors, thus allowing a linear potential drop across the accelerator stack. The conventional accelerator is simply switched into action by raising the first of the annular plates to extractor potential. With this plate earthed the conventional accelerator is rendered inactive and previously described acceleration-focussing action occurs. Thus the switching from one mode to the other does not require breaking the vacuum.

(ii) Filter tank:

This chamber is a 258 mm OD and 130 cm long cylinder containing only the E X B velocity filter stage.

In the earliest magnetic transport systems, the slow positrons were separated from the high energy gamma rays and source positrons, by guiding them round a gently curving magnetic field[8]. More recently, because of the compactness and geometrical convenience, there has been an increasing use of the E X B velocity filter[9-11]. The basic principle of this is well known. An electric field is applied in the direction perpendicular to that of the axial guiding magnetic field "B" causing the gyration centers of incoming charged particles to drift with a velocity v, where v is given by

$$v = \frac{E \times B}{B^2}$$

The filter, located at the centre of the beam axis, consists of a set of parallel planar plates, producing a transverse electric field. The positrons are deflected 10 mm downwards and pass through an 8 mm aperture. The small deflection over the plates, which are 1m in length, is necessary to prevent a significant increase in tangential velocities and consequent beam diameter broadening. Since the voltage V_D applied to the plates is a function of beam energy, a computer is used to set V_D at different positron energies. An advantage of this arrangement is the ability to monitor the energies of positrons and Auger electrons emanating from the target, by placing a channeltron detector and retarding grids at the position shown.

(iii) **Target chamber:**

This is a single stage vertical chamber with five 200mm OD flanges and sixteen 70 mm OD ports, and has a capacity of around 40 liters. It is connected to the beam through a UHV standard stainless steel pipe of 2 meter length. It consists mainly of a sample manipulator, germanium detector, channeltron, vacuum gauge and other equipment relating to sample preparation.

The sample manipulator, which sits on the top of the 200 mm OD flange, is designed in such a way that it can handle two samples at a time. It has a vertical motion of 100 mm and horizontal motions of ± 50 mm with rotational motion about the vertical axis of the chamber. The temperature of the sample can be varied from 100 K to 1500 K and it is measured with chromel-alumel thermocouple. An intrinsic germanium detector for the measurement of annihilation gamma radiation is mounted on another 200 mm OD flange at a distance of 50mm from the sample. A channel electron multiplier for secondary electron or re-emitted positron detection is also fixed to this chamber.

The target chamber can be isolated from the rest of the beam apparatus by means of a gate valve while loading new samples. The advantage of this isolation is two-fold. Firstly, the closing down of the entire beam apparatus while loading new samples is thus eliminated. The second advantage is to prevent the moderator from being exposed to atmospheric contamination, each time the beam is closed. However, once the target chamber is pumped down to the beam pressure with the help of its dedicated diffusion pump, it is then connected back to the rest of the system by simply opening the valve.

iv) The UHV System:

The total capacity of the entire beam apparatus is around 250 liters. It can be pumped down to a base pressure of 10^{-10} torr by means of two sputter-ion pumps each combined with a TSP. A turbomolecular pump is used initially to evacuate the system to 10^{-8} torr and is used throughout bake-out, but is switched off later as the Ion pump-TSP combinations hold the system in UHV. A liquid nitrogen trapped 4 inch diffusion pump is fixed at the bottom of the target chamber, and together with another TSP inside the chamber again produces base pressures of 10^{-10} torr.

4. ADVANTAGES AND DISADVANTAGES:

There are a number of advantages and disadvantages associated with our beam when compared to other beams[3,10-12] and these may be briefly listed as follows:

(a) Although a conventional accelerator may be employed, the advantage of using the focussing accelerator exists when smaller beam diameters (< 2 mm) are desirable, and, can be switched into action with out breaking the vacuum.

(b) The filtering stage comes after the accelerator. The advantage in doing this is that the filter will block any backscattered positrons from the target and lift them away from the beam axis, whereas in the opposite case, a fraction of reflected positrons may give a possible source of error in interpreting the results.

(c) The annular positron source being a composite of 8 smaller sources, allows for more operator control and maintenance of the beam intensity, and moreover makes for safety in handling smaller sources.

(d) The high accelerating voltage is kept inside the vacuum and the need to raise pumps and the vacuum chamber to high potential is thus eliminated.

(e) One of the disadvantages associated with our beam is that it is necessary to change the potentials on the filter plates each time the beam energy is changed, but this is expected to be relatively simple to automate with microcomputer control.

(f) Another disadvantage that may be considered is the extra cost and complexity of our source-moderator geometry.

5. CHARACTERISTICS OF THE BEAM:

a) Energy of the beam : 0 - 60 KeV
b) Intensity of the beam : ~ 10^6 e^+s^{-1}
c) Radioactive source : ^{22}NaCl on W backing
d) Moderator : W(110) single crystal annealed at 2000°C
e) Size of the beam : < 2 mm diameter
f) Type of Filter : E X B velocity filter
g) Transport : Magnetic, with some electrostatic focussing.
h) Attainable vacuum pressure: Around 10^{-10} torr
i) Special features : Easily replenishable and intensity controlled.
j) Applications : -Metal-semiconductor interfaces
-Depth profiling of defects
-DLPTS

6. APPLICATIONS OF THE BEAM:

In the above sections, we have described a variable energy positron beam facility at HKU. The beam has been designed for a variety of experiments that have been carried out elsewhere or planned for the future. Since a number of reviews have already described the various techniques employed with slow-positron annihilation spectroscopy (SPAS), we restrict our discussion only to the novel study, which we have proposed, that is, Deep Level Positron Transient Spectroscopy (DLPTS), which is a result of combining the salient features of deep level transient spectroscopy (DLTS)[13,14] and slow-positron beam annihilation techniques[15-17]. In a conventional DLTS system, a capacitance transient is sampled at times t_1 and t_2 and the difference is plotted as a function of temperature. The capacitance transient results due to the repetitive pulsing of the junction device (which is kept under constant reverse bias) into either forward or zero bias in order to fill the trap levels with charge carriers. Depending upon the thermal environment, the trapped charge carriers will make a transition to the respective band and this results in a capacitance transient. In the case of our proposed DLPTS technique, the S-parameter is sampled at two different times and the difference is plotted as a function of temperature. Such a technique should be able to give information on whether deep levels are vacancy related or not. This is certainly one of our aims at the University of Hong Kong.

ACKNOWLEDGEMENTS:

The authors would like to thank the UPGC of Hong Kong for providing the funds required to build the positron beam facility at the University of Hong Kong. Prof. Weng would like to thank the University of Hong Kong for awarding him a Visiting Scholarship.

REFERENCES:

[1] Peter J.Schultz and K.G.Lynn, Rev. Mod. Phys., **60**, 701 (1988).

[2] C.D.Beling and S.Fung, Proceedings of the International Workshop on the Industrial Applications of Surface Analysis Techniques, The Commercial Press (Hong Kong) Ltd., 1991.

[3] A.P.Mills, Jr., Appl. Phys. Lett., **37**, 667 (1980).

[4] R.S.Brusa, R.Grisenti, S.Oss, A.Zecca and A.Dupasquier., Rev. Sci. Instrum., **56**, 1531 (1985).

[5] Dahl, David A., MS 2208, EG & G Idaho Inc., Idaho National Engineering Laboratory, P.O.Box 1625, Idaho Falls, ID 83415.

[6] W.K.Wong, R.Unwin, C.D.Beling and S.Fung., "The Focussing of a positron beam by an electrostatic and an increasing magnetic field" presented in the 9th International Workshop on Positron Annihilation, Hungary (1991).

[7] Xiao-ye Wu, Peter Dull and K.G.Lynn., Appl. Phys. Lett., **57**, 998 (1990).

[8] K.G.Lynn and H.Lutz., Rev. Sci. Instrum., **51**, 977 (1980).

[9] S.M.Hutchins, P.G.Coleman, R.J.Stone and R.N.West., J.Phys.E: Sci.Instrum., **19**, 282 (1986).

[10] Peter J. Schultz., Nucl. Instr. Meth., **B30**, 94 (1988).

[11] J.Lahtinen, A.Vehanen, H.Huomo, J.Makinen, P.Huttunen, K.Rytsola, M.Bentzon and P.Hautojarvi., Nucl. Instr. Meth., **B17**, 73 (1986).

[12] W.E.Frieze, D.W.Gidley and K.G.Lynn, Phys.Rev.B., **31**, 5628 (1985).

[13] D.V.Lang, J.Appl.Phys., **45**, 3014, 1974.

[14] F.W.Sexton and William D. Brown, IEEE Trans. Instrum. Meas., **IM-30**, 186 (1981).

[15] A.Vehanen, Hyperfine Interactions, **45**, 179 (1989).

[16] K.G.Lynn, Helvetica Physica Acta, **63**, 389 (1990).

[17] To Chi Leung, Ph.D. Dissertation submitted to the State University of New York at Stony Brook, December 1991.

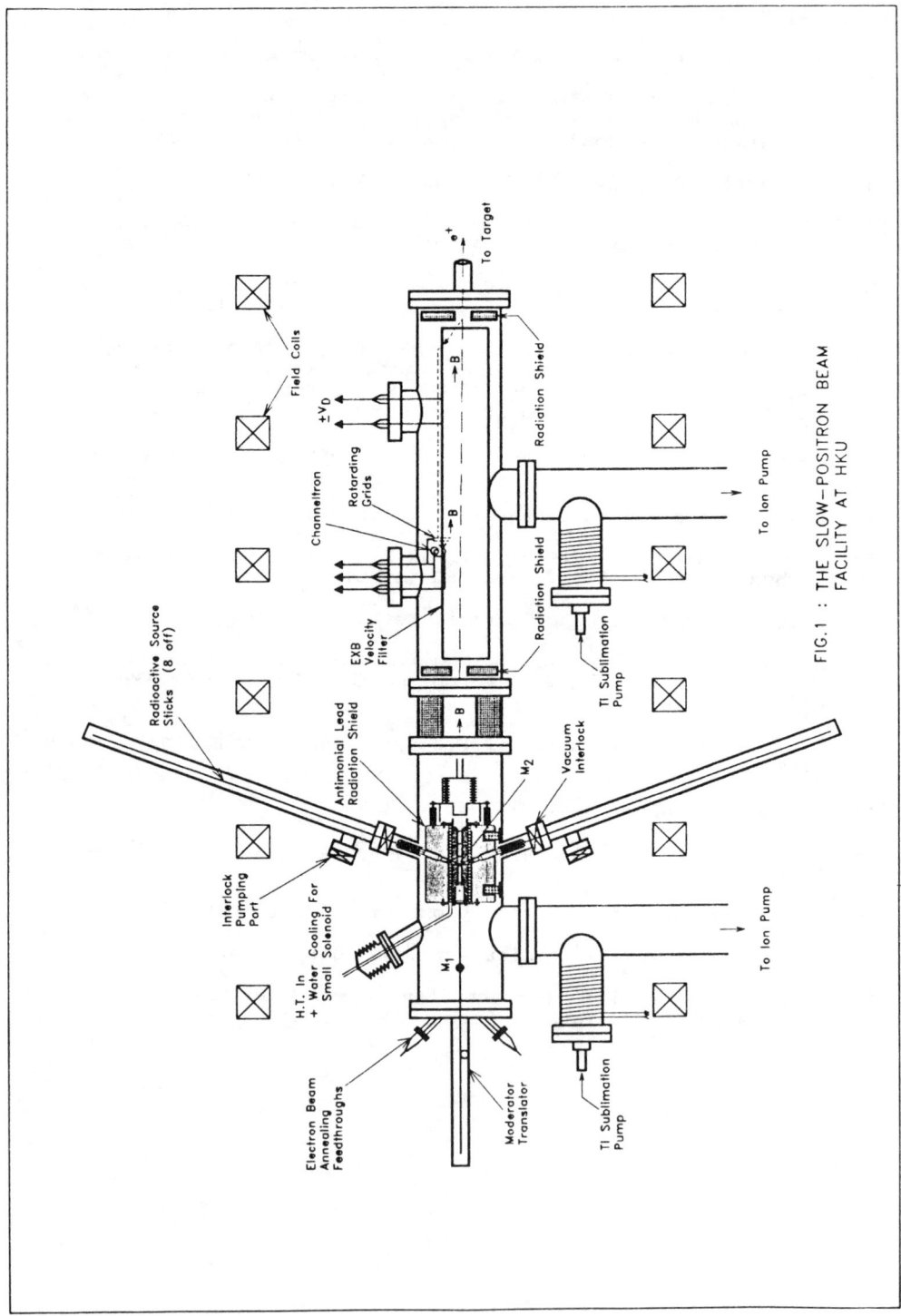

FIG.1 : THE SLOW-POSITRON BEAM FACILITY AT HKU

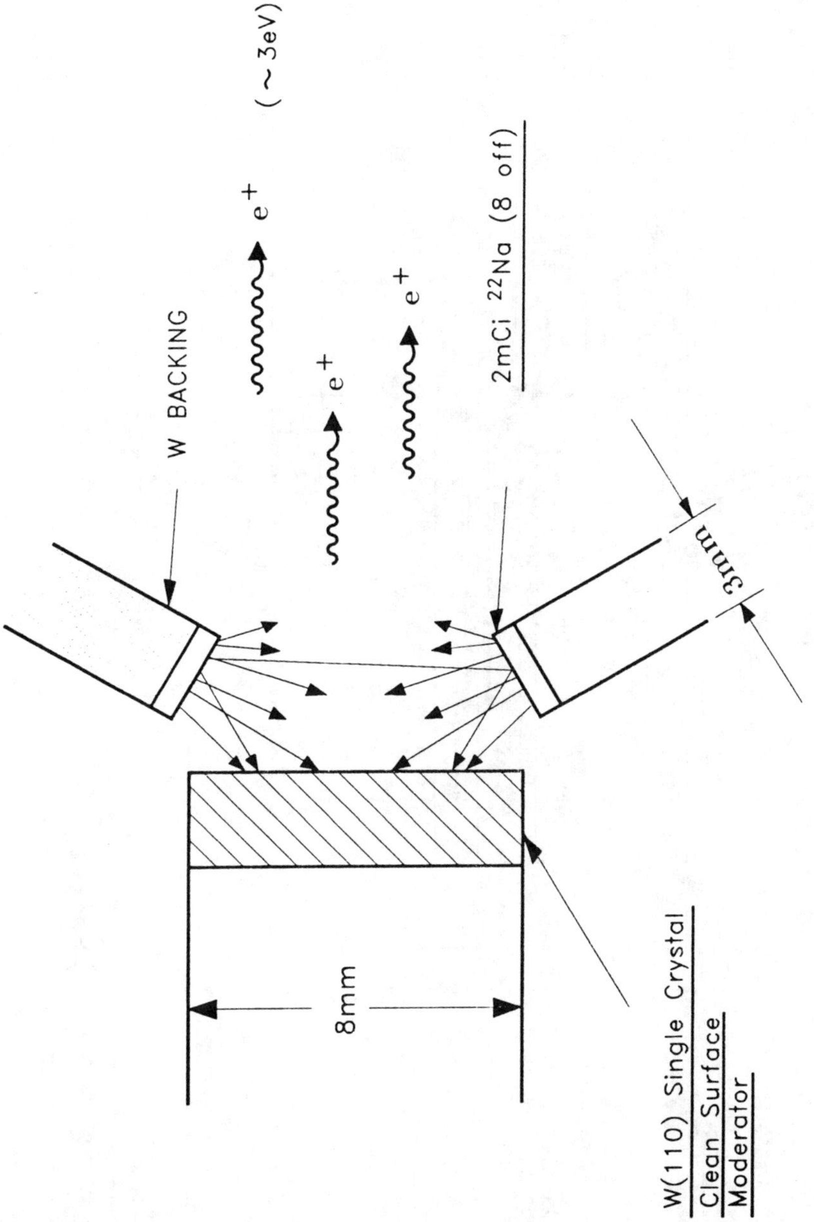

FIG.2 : DETAIL OF THE POSITRON MODERATOR GEOMETRY BEING USED IN HKU

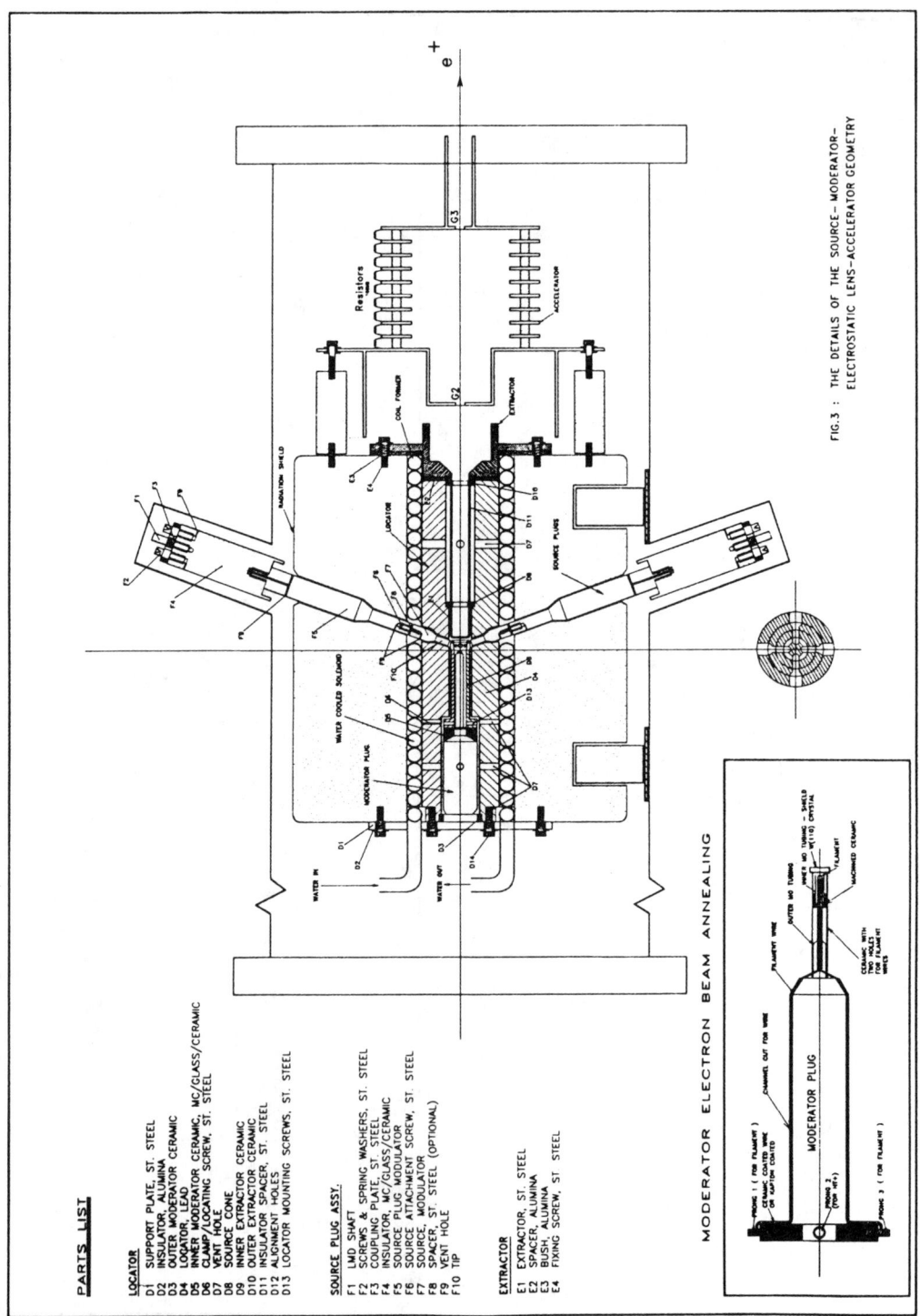

FIG.3: THE DETAILS OF THE SOURCE- MODERATOR- ELECTROSTATIC LENS-ACCELERATOR GEOMETRY

CAPTIONS FOR FIGURES

Fig.1. The Slow-Positron Beam Facility at HKU.

Fig.2. Detail of the Positron Moderator Geometry being used in HKU.

Fig.3. The Details of the Source-Moderator-Electrostatic Lens-Accelerator geometry.

Heating of a Thin Tungsten Foil for Efficient Positron Moderation in a 4.2 K Environment

Benjamin L. Brown, Tamara S. Andrew,
Margaret S. Clarkson, Sarah K. Makoski,
Sonal Parikh, Sujata Vemuri

Physics Department
Mount Holyoke College
South Hadley, MA, USA

Abstract

We have explored the feasibility of heating a Tungsten (W) foil in a 4.2 K environment. Motivation for this work is to increase the moderation efficiency of positrons (e^+) by raising the W foil temperature to at least 20 K. This may make storage of large numbers of positrons (10^5) possible for anti-hydrogen production. This work is part of a continuing collaboration with G. Gabrielse, L. Haarsma, and K. Abdulah of Harvard University.

Trapping positrons in a Penning trap at 4.2 K in a 6 T field is desirable for the production of antihydrogen at low energies. The present configuration of the Harvard positron trap, designed to trap ~ 10^5 -10^6 positrons at 4.2 K, is shown in Fig. 1.[1,2] This scheme is a variation on an unmoderated Penning trap used by Dehmelt.[3] Slow positron emission in this configuration has yet to be demonstrated. It is not yet clear if there is a physical reason for non-emission of positrons at 4.2 K, but one possibility is the quantum reflection of positrons.[4] It has been established that slow positrons are emitted relatively efficiently at 20 K.[5] Thus, heating the foil may provide for more efficient slow positron emission in the experimental configuration described. It is also conceivable that the intense magnetic field (6T) might inhibit the free diffusion of positrons in the foil. With a heated foil, this possibility can be explored.

Moderator heating can be accomplished by electron bombardment heating or ohmic heating. The most promising method appears to be electron bombardment of a thermally isolated foil, using a field emission point (FEP). Various parameters of both methods are calculated for comparison, including the power transferred by radiation, thermal conduction, and direct heating using different thicknesses, lengths, temperatures of the foil, and currents in the foil. These comparisons are useful in estimating the power required to raise a W foil from 4.2 K to 20 K and 77 K, and the power which is deposited in the He dewar.

The ohmic heating method is detailed in Table I. Estimates are presented for the resistance of the foil, R, the radiated power, P_{rad} , the

Table 1 Ohmic Heating

Temperature (K)	4.2	20	77
$R = \rho L/wt$, (Ω)	9×10^{-4}	2×10^{-3}	2×10^{-2}
Radiation $P_{rad} = \sigma(T_2^4 - T_1^4)$, (W)		4×10^{-4}	0.2
Conduction P_{con} , (W)		6×10^{-3}	7×10^{-2}
Current $I = (P/R)^{1/2}$, (A)		2	5

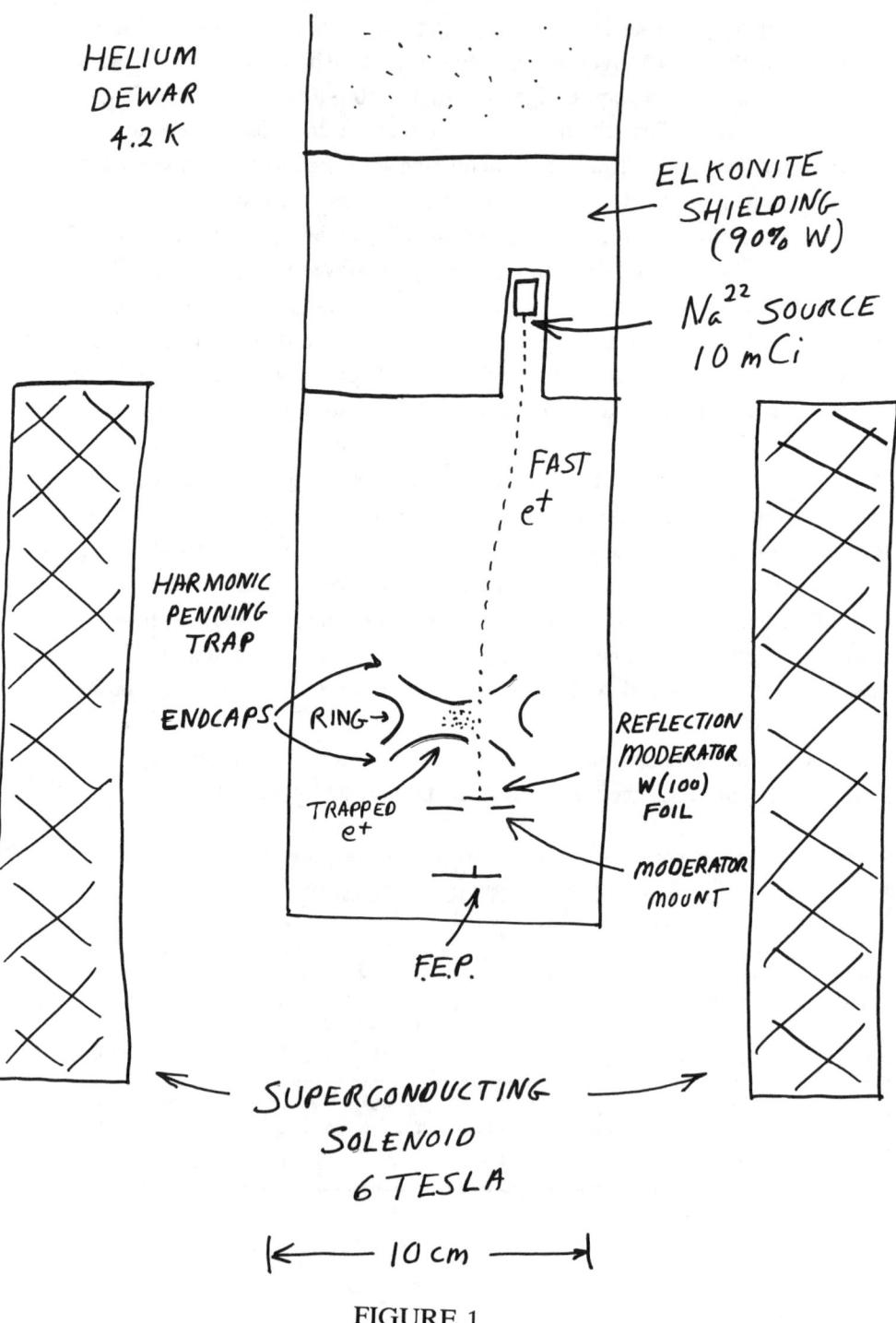

FIGURE 1

conduction power, P_{con}, and the ohmic heating current, I, at temperatures of 4.2 K, 20 K, and 77 K. Advantages to ohmic heating of the foil include the possibility of cleaning and annealing of the foil at room temperature before immersion in the helium dewar. Figure 2 illustrates the one possible arrangement for ohmic heating. The currents are rather high and this might lead to foil failure due to the Lorentz force, **(BxI)**L, in a 6 T field. The foil has been found to survive in a 200 G field conducting 3 A of current at 2400 K (20,000 Å thick foil), but the durability with similar currents in a larger field has not been demonstrated. Certainly, heating reliably to this temperature range would be useful for cleaning and annealing of the foil in situ before placing the apparatus in the Helium dewar and magnet.[6]

Future tests will be needed if ohmic heating to elevate the foil in the helium dewar to at least 20 K is to be seriously considered. Transferring the 2-5 amp current required (refer to Table 1) down to the dewar will be difficult without substantially increasing the helium loss rate due to heat conduction in the wires. The power requirement to heat the foil is in the tens of milliwatt range which itself represents a significant heat load. A superconducting transformer (e.g. a turns ratio of 1000:1) could be used to couple the power efficiency from the outside to the foil. This would allow a high voltage and low current input to be carried by small "constantan" wires which would create negligible heat loss to the dewar system. The frequency of such a system could be in the audio range. This would be easier to filter and thus less detrimental to the sensitive Penning trap system than radio frequency.

Calculations for FEP heating are summarized in Table 2. The typical heat provided by the FEP can be 10^{-4} W for 100 nA current at 1 KV. The heating to beyond 73 K appears to be possible with a minimal impact on the helium evaporation rate. One possible FEP configuration is illustrated in Fig. 3, where the positrons are incident on the foil from below. The moderator could also be mounted on thin wires instead of the teflon foil, with a corresponding rise in thermal conduction. The moderator cleaning and annealing before insertion in the helium dewar and magnet would have to be done as a separate step, by electron bombardment heating[6] or by laser heating[7]

Heating of a Thin Tungsten Foil

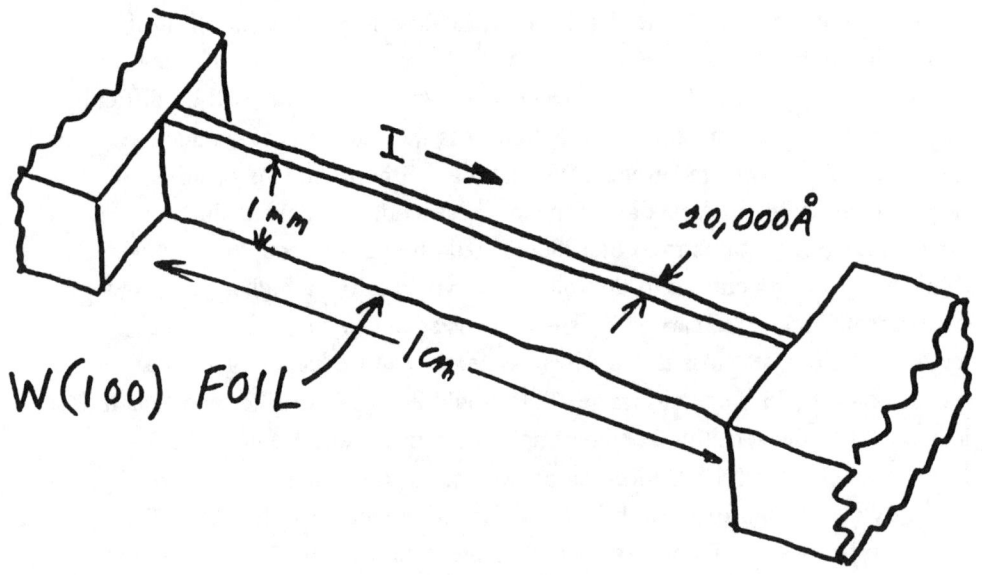

FIGURE 2

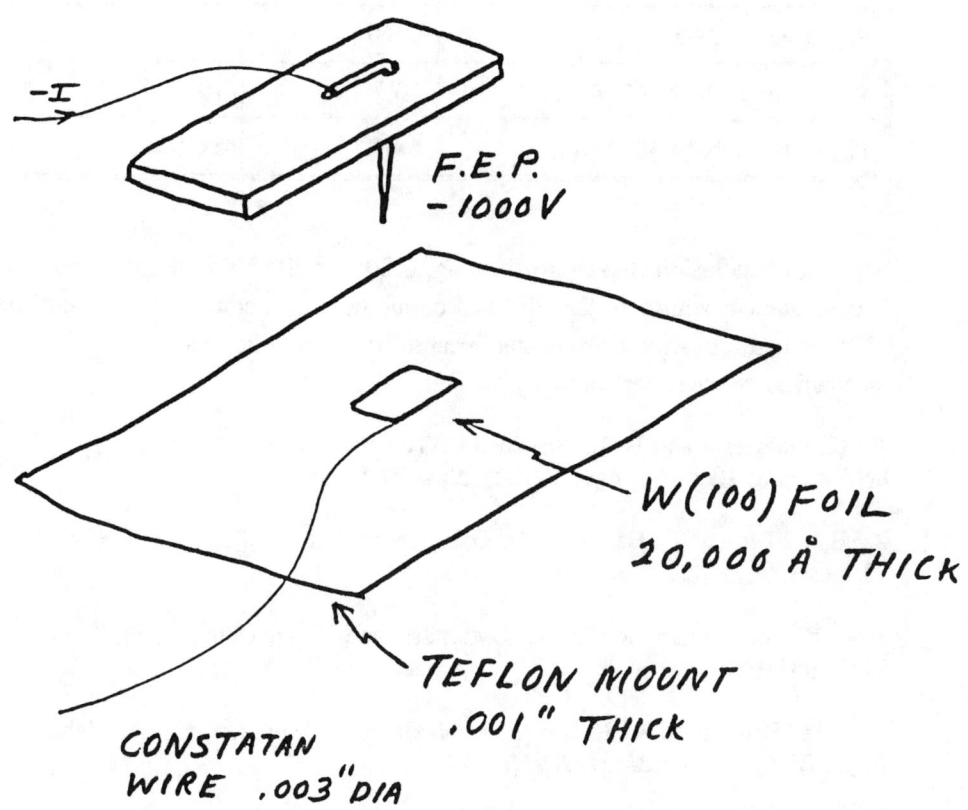

FIGURE 3

Table 2 FEP Heating		
	Teflon	Constantan
g (W/K-m)	3×10^{-2}	1
P_{con} at 16 K (W)	2×10^{-5}	5×10^{-6}
P_{con} at 73 K (W)	9×10^{-5}	2×10^{-5}
P_{rad} from 16 K to 4.2 K (W)	10^{-9}	10^{-10}
P_{rad} from 73 K to 4.2 K (W)	6×10^{-7}	5×10^{-8}

In conclusion, moderator heating to 20 K and 77 K is feasible in a 4.2 K environment with both the FEP and ohmic heating methods. At present the FEP method looks most promising because of the simplicity and the lower mechanical stress experienced by the foil.

1. G. Gabrielse and B. L. Brown, in : The Hydrogen Atom, Proc. Symp. held in Pisa, Italy (Springer Verlag, New York, 1989)

2. B. L. Brown, L. Haarsma, G. Gabrielse and K. Abdulah, Hyperfine Int. **73**,193(1992).

3. P. B. Schwinberg, R. S. Van Dyck, Jr., and H. G. Dehmelt, Phys. Lett, **81**A(1981)119.

4. D. T. Britton, P. A. Huttunen, J. Mäkinen, E. Soininen, and A. Vehaanen, Phys. Rev. Lett. **62**,2413(1989).

5. B. L. Brown, W. S. Crane, and A. P. Mills, Jr., Appl. Phys. Lett. **41**, 1076(1987).

6. N. Zafar, J. Chevallier, F. M. Jacobsen, M. Charlton, and G. Laricchia, Appl. Phys. A **47**, 409(1988).

7. F. M. Jacobsen, M Charlton, J. Chevallier, B. I. Deutch, G. Laricchia, J. Appl. Phys. **67**575(1990).

ACCUMULATION AND BUNCHING OF POSITRONS

B. Ghaffari, R.S. Conti, and T.D. Steiger
Randall Laboratory of Physics, University of Michigan, Ann Arbor, MI 48109

ABSTRACT

Results from a positron accumulator that operates efficiently over a range of repetition rates from 100 to 1000 Hz are presented. Moderated β-decay positrons from a radioactive source are accumulated in a Penning-style trap. At a repetition rate of 250 Hz an accumulation efficiency of $\sim 25\%$ has been achieved. Two techniques for reducing the time spread of the positron pulses have been investigated. The most successful method reduces the pulse width from 120 ns to 20 ns.

INTRODUCTION

In many of the experiments that involve positrons it is advantageous to use a pulsed positron beam. Such a beam can provide slow, time-tagged positrons for measuring decay rates, for example of positronium (Ps),[1] or positrons in matter.[2] A source of pulsed positrons could furnish high positron densities for antihydrogen-formation experiments.[3] A pulsed positron beam is also invaluable in experiments that employ other pulsed particle or laser beams. In an experiment in progress at the University of Michigan[4] a pulsed laser is required to measure the fine structure intervals $2^3S_1 \rightarrow 2^3P_J$ (J = 0,1,2) in Ps. The necessity of using an intense pulsed positron beam which is synchronous with this laser motivated the work presented in this paper.

A common source of pulsed positrons is bremsstrahlung pair production from pulsed relativistic electrons. This method produces intense microsecond-long positron pulses that need to be further time-compressed to be useful for many experiments.[2,5] The cost of constructing and operating these pulsed positron sources is quite high.

A more economical beam of pulsed positrons can be provided by accumulating positrons from a radioactive source. Several mechanisms for trapping moderated positrons have been investigated. An rf cavity tuned to the positron cyclotron resonance frequency has been used to provide accumulation by increasing the transverse energy spread of the positrons.[6] A harmonic bunching technique subsequently decreases the time spread of the positron pulses. The minimum repetition rate for efficient accumulation in this apparatus is greater than 1 kHz, which is high for most of the experiments mentioned above. Another technique to accumulate positrons is based upon the use of inelastic collisions with neutral gas molecules[7] to cool the positrons. This design requires extensive differential pumping and the optimum repetition rate is much lower than 100 Hz. The design for an accumulator which operates efficiently at moderate repetition rates (100–1000 Hz) has been tested at the University of Michigan. The work presented here marks a significant improvement over the preliminary results reported previously.[8]

ACCUMULATION

The design of the accumulator is based on a cylindrical open-endcap Penning trap.[10] The electrostatic lensing elements and the applied electric potentials are shown schematically in fig. 1. Beta-decay positrons from a 3-mm diameter, 15-mCi ^{22}Na source (S) are moderated using a tungsten-vane moderator (M) in a Venetian-blind geometry.[9] The trap is divided into 13 cylindrical sections (T_1–T_{13}) which are ~ 5.1 cm in diameter and ~ 5.5 cm long. The gate (G) is a high-transmission grid placed at the opposite end of the trap. Radial confinement is provided by a solenoidal magnetic field of 90 G and the effects of the Earth's magnetic field are cancelled using a pair of Helmholtz coils.

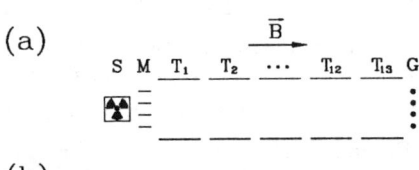

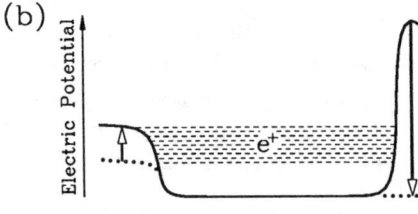

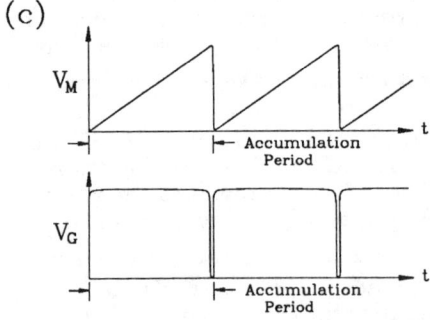

Fig. 1. The electrostatic lensing elements of the accumulator are schematically shown in (a). The electric potential applied to the radioactive source, moderator, sections of the trap, and the gate are denoted by V_S, V_M, V_{T_i}, and V_G respectively. These potentials are shown directly below in (b). The arrows indicate voltages which are changing in time. The time dependence of V_M and V_G are explicitly shown in (c). V_S is always ~ 3 V higher than V_M.

Slow positrons are emitted into the trap with a moderation efficiency of approximately 2×10^{-4}. The kinetic energy of these positrons in the trap is given by

$$E_K = e\left(V_M(t) - V_{T_i}\right) - \phi_+ \quad (1)$$

where e is the positron charge, $\phi_+ \simeq -2.8$ eV is the work function of the tungsten moderator, and V_M and V_{T_i} are defined in fig. 1.

A high positive potential is applied to the gate during accumulation. Slow positrons are reflected from this potential back toward the moderator. The time required for positrons to travel from the moderator to the gate and back is called the trap period and is a function of E_K. During the accumulation period (see fig. 1c) V_S and V_M are continuously rising at a rate denoted by R_{inc}. If these potentials increase by more than $|\phi_+/e| \simeq 2.8$ V during one trap period, the returning positrons will not be energetic enough to reach the moderator. Thus, these positrons become axially trapped between the moderator and the gate.

However, it is possible to accumulate positrons efficiently (10 − 45%) even when V_M increases by < 100 mV during a trap period rather than 2.8 V. Several possible explanations for this were investigated. Processes such as neutral gas scattering, positive ion scattering, and electron cooling were ruled out. The observed high accumulation efficiency is the result of magnetron motion as discussed below.

MAGNETRON MOTION

In order to accumulate efficiently, it is necessary to prevent the positrons returning toward the moderator from striking the vanes. If moderated positrons strike the tungsten vanes, they may be remoderated and reenter the trap. However, even with a remoderation efficiency of $\sim 20\%$[11] less than 1% of these positrons will survive after three trap periods. Magnetron motion aids accumulation by preventing positrons from hitting the moderator vanes. Positrons undergo an $\vec{E} \times \vec{B}$ drift—the magnetron motion—while reflecting from the electric potential of the gate. The drift velocity is given by

$$\vec{v}_d = \frac{\vec{E} \times \vec{B}}{B^2}. \qquad (2)$$

The details of the magnetron motion depend upon the ultimate proximity of positrons to the gate compared to its grid-wire spacing. If the field near the gate is strong enough that positrons only approach to within a few times the grid-wire spacing, then the gate's grid structure is of no consequence. Hence, the positrons remain within a region in which the electric field has only axial and radial components and, as a result, the drift velocity has only an azimuthal component. The projection of the resultant azimuthal displacement on a plane perpendicular to the magnetic field is a section of a circle centered on the cylindrical symmetry axis of the electric field lines. In this case positrons which are farther away from this symmetry axis are displaced more than those near the center because they sample a larger radial electric field.

On the other hand, if the field near the gate is weak enough for positrons to approach to within a grid-wire spacing, then the positrons will sample a region of the electric field that is strongly influenced by the grid structure. In this case, though $|\vec{E}|$ is relatively small, the positrons sample an electric field which has components in all directions. Therefore, magnetron motion occurs in both the azimuthal *and* radial directions. This displacement does not depend upon the gross cylindrical symmetry of the field and hence may strongly effect positrons irregardless of their distance from the symmetry axis.

In either case the displacement caused by the magnetron motion can aid accumulation. Due to the magnetron motion positrons returning to the vicinity of the moderator may fail to strike a vane. Such positrons are reflected by V_S and reenter the trap (see fig. 2). Thus, the increase in V_S and V_M required for accumulation ($|\phi_+/e|$) may be applied over many trap periods.

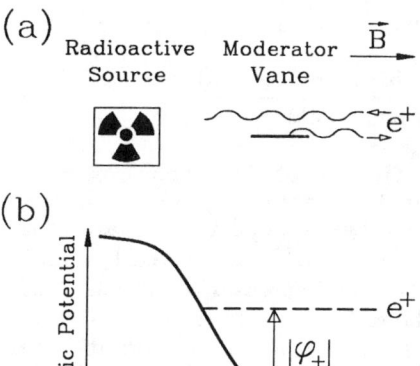

Fig. 2. A schematic illustrating the effect of magnetron motion is shown in (a). The cyclotron orbit of a positron is shown both before and after magnetron motion occurs. The returning positron does not strike the vane from which it was emitted and is therefore reflected back into the trap by V_S. The relevant electric potentials and the corresponding positron energy are directly below in (b).

ANALYSIS AND RESULTS

A channel electron multiplier array (CEMA) assembly is placed after the gate. This assembly includes a phosphor screen to observe the beam profile. A 5 cm × 5 cm plastic scintillator is placed outside the vacuum chamber to detect γ rays from the annihilation of positrons on the channel plate. In order to calibrate these detectors and optimize the positron optics, the accumulator was operated in a continuous mode in which V_G was held below V_M. Coincidences between the signals from the CEMA and the γ-ray detector can be used to calculate the slow positron beam rate (B) as well as the efficiencies of the CEMA (η_β) and the γ-ray detector (η_γ) as follows:

$$B = \frac{R_\beta R_\gamma}{R_{coinc}}, \quad \eta_\beta = \frac{R_{coinc}}{R_\gamma}, \quad \eta_\gamma = \frac{R_{coinc}}{R_\beta}. \tag{4}$$

In the above equations R_β, R_γ, and R_{coinc} denote the background-corrected CEMA, γ and coincidence rates respectively.

The signal from the CEMA cannot be used to detect pulsed positrons because of noise induced by the high-voltage gate pulse. The pulsed positrons are therefore only detected using the γ-ray detector. A time spectrum of the positron pulse is obtained by collecting coincidences between the gate pulse and γ-detector signal. It is important to note that the use of a γ-ray detector guarantees that the observed signal is unambiguously due to positrons.

The accumulation efficiency is defined as

$$\eta_{acc} = \frac{\text{Number of pulsed positrons per second}}{\text{Slow positron beam rate}}. \tag{3}$$

This efficiency strongly depends on the effectiveness of the magnetron motion in initially trapping the positrons. The magnetron motion, in turn, has a complicated dependence on V_M, $V_{T_{13}}$ and the voltage applied to the gate during accumulation (V_G^{max}). The accumulation efficiency can be improved by increasing the difference between V_G^{max} and $V_{T_{13}}$ since this increases the radial electric field that causes the magnetron motion. The difference between V_M and $V_{T_{13}}$ determines the speed of positrons near the gate and hence the period of time during which magnetron motion occurs (see eq. 1). On the other hand, the difference between V_G^{max} and V_M determines the ultimate proximity of positrons to the gate which in turn determines the nature of the magnetron motion. Furthermore, V_M is not held constant during the accumulation period and, as a result, the trapping efficiency is a function of time. It is the combined effect of this time-dependent trapping probability and the ability of the rising moderator voltage to accumulate the positrons after several trap periods that determines the total accumulation efficiency.

The accumulation efficiency is plotted versus $V_G^{max} - V_M^{min}$ and $V_{T_{13}} - V_M^{min}$ for three different values R_{inc} in fig. 3. For the sets of data plotted in fig. 3a, both R_{inc} and $V_{T_{13}}$ are held constant while V_G^{max} is varied. Therefore it is V_G^{max} that determines the nature of the magnetron motion. When $V_G^{max} \sim V_M$ the effect of the gate's grid structure on the electric field is the dominant factor determining the trapping efficiency. The smaller the difference between

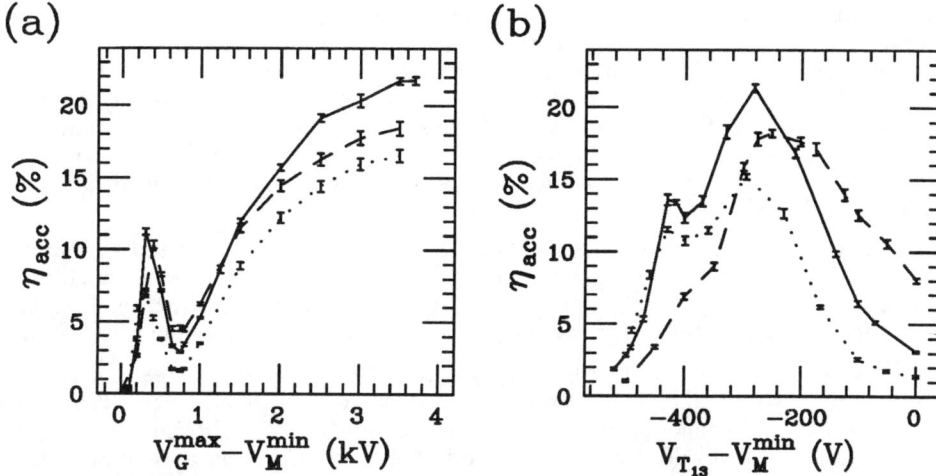

Fig. 3. The accumulation efficiency at a repetition rate of 250 Hz was measured for three different rates of moderator voltage increase. Trap sections $T_1 - T_{12}$ are held at 0 V and $V_M^{min} = 0$ V. The data points are connected with straight lines to be distinguished from one another. Dotted lines are used for the moderator voltage increasing at a rate of $R_{inc} = 37.5$ V/ms, solid lines for the optimum rate of $R_{inc} = 60$ V/ms and dashed lines for $R_{inc} = 90$ V/ms. The effect of V_G^{max} on the accumulation efficiency is shown in (a) where $V_{T_{13}} = -280$ V. The effect of $V_{T_{13}}$ for $V_G^{max} = 3.5$ kV is shown in (b).

V_G^{max} and V_M, the closer positrons come to the gate. These positrons sample an electric field which has greater azimuthal and radial components and therefore higher trapping efficiencies are possible. Consequently for the cases in which positrons are allowed to approach the gate to within a grid-wire spacing the accumulation efficiency is optimized for $V_G^{max} \sim V_M^{max}$ (see fig. 3a). As V_G^{max} is raised above V_M^{max}, positrons do not approach the gate as closely and, thus, the effect of the magnetron motion is reduced and the accumulation efficiency decreases sharply. However, when the voltage on the gate is raised even further, the overall radial component of the electric field becomes large enough to cause a substantial azimuthal drift velocity (see eq. 2) and the trapping efficiency increases. The data in fig. 3b are taken with the same rates of moderator voltage increase as in fig. 3a, but in this case V_G^{max} is held constant while $V_{T_{13}}$ is varied. Increasing $|V_{T_{13}}|$ increases the radial electric field component that causes the magnetron motion. However, this also increases the speed of positrons reflecting from the gate potential and thus decreases the amount of time positrons undergo magnetron motion. These two factors affect the trapping efficiency in opposite directions and lead to the broad maxima seen in fig. 3b. The optimum values for R_{inc} and $V_{T_{13}}$ can be experimentally determined for different repetition rates as demonstrated in fig. 3 for 250 Hz.

The accumulation efficiency could potentially be improved by increasing the pressure of the system. Inelastic collisions with neutral gas molecules can lead to accumulation by cooling the positrons.[7] Elastic scattering from positive ions or neutral molecules can also help accumulation if the scattering angle is sufficiently large. Such large-angle scatterings transfer part of the axial momentum of positrons into cyclotron motion. Positrons become trapped when their axial momentum is insufficient to reach the moderator. However, increasing the pressure also increases the probability of forming Ps. It was found experimentally that an increase in pressure leads to a decrease in the accumulation efficiency which indicates that Ps formation has a larger effect than the gain mechanisms described above.

The positron accumulator described in this paper was optimized to match a pulsed laser with a maximum repetition rate of 250 Hz. The highest accumulation efficiency achieved at 250 Hz is $\sim 25\%$. The parameters which yielded this efficiency and the resultant positron pulse time spectrum are shown in fig. 4.

The accumulator has also been tested at other repetition rates and reasonable accumulation efficiencies were achieved even though all parameters were not optimized. An accumulation efficiency of $\sim 45\%$ was obtained at 670 Hz. This result and tests done at 1000 Hz indicate that achieving an accumulation efficiency of $\sim 50\%$ at 1000 Hz is quite feasible. The best efficiency achieved at 100 Hz is $\sim 10\%$. The accumulation efficiency is generally lower at low repetition rates because positrons have a finite lifetime in the trap. Positronium formation is the most important factor limiting the efficiency for long accumulation periods.

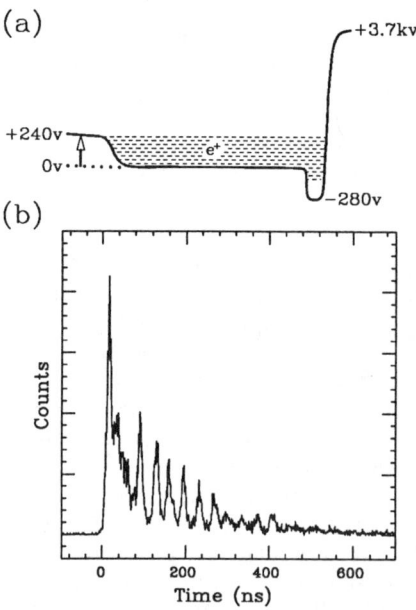

Fig. 4. The electric potential of the lensing elements leading to the highest accumulation efficiency at 250 Hz is schematically shown in (a). The positron pulse time spectrum is shown in (b). The pressure is $\sim 2 \times 10^{-8}$ torr. The moderator voltage is raised from 0 V to 240 V during the 4-ms period. The accumulation efficiency is $\sim 23\%$ resulting in ~ 8 e^+/pulse per mCi of radioactivity.

The initial sharp peak in the time spectrum (see fig. 4b) is due to positrons localized in the small potential well next to the gate (see fig. 4a). The well is generated by applying a higher voltage to T_1 through T_{12} than to T_{13}. This potential configuration increases the trap period (see eq. 1) without changing the trapping efficiency (which depends on $V_{T_{13}}$ as illustrated by fig. 3). The result is a higher accumulation efficiency because positrons return fewer times to the moderator and therefore are less likely to strike a vane and be lost.

The other peaks in the time spectrum are not due to an effect inherent in the accumulation process. Noise induced on the trap sections by the high-voltage gate pulse modulates the positron beam leading to the peaks observed in figs. 4, 5, and 6. This noise has since been greatly reduced and the peaks mentioned above have smoothed out. These peaks have since been eliminated by placing a ground braid around the gate pulse line and diminishing ground loops.

During accumulation the moderator voltage is raised by 240 V and therefore the energy spread of the pulsed positrons is ~ 240 eV (see eq. 1). This energy spread is small enough that a remoderation efficiency as high as 20% can be achieved[11] if thermal positrons are required. However, for the present application remoderation will not be necessary. The formation probability for n=2 Ps decreases with increasing positron energy but is reasonably constant from 10 to 150 eV.[12] Therefore, the energy spread is not a concern in the fine structure measurement cited above.

COMPRESSION AND BUNCHING

The typical width of the positron pulses (see fig. 4b) is ~ 120 ns (FWHM) which is too large for many applications. Two techniques of time compressing these pulses were investigated. One of these techniques relies on decreasing the physical volume in which the positrons are trapped after accumulation has been completed. This is done by raising the voltage of the trap sections sequentially starting with V_{T_1} (fig. 5). The resulting potential gradient compresses the positrons into an ever smaller volume next to the gate before it is lowered.

During this process positrons within the section which is rising in potential may gain a considerable amount of energy. This not only increases the energy spread of the positrons (which is undesirable for many experiments) but it can also result in detrapping. An effort was made to minimize this energy gain by raising the potentials of the sections slowly and by raising the voltage of one section at a time. Even when the trap sections were raised with a 10-90% rise time of 100 ns and an 80 ns delay between consecutive sections, the accumulated positrons were able to gain enough energy to leave the trap before the gate was lowered. Some of these positrons were even observed to have gained enough energy to exit the trap over the 3.7 kV barrier of the gate. As a result, this method of time compression continuously decreases the total number of positrons in the trap. The best results are therefore obtained if the gate potential is lowered before the compression process has been completed.

The most successful time compression technique is similar to the method of harmonic bunching.[5,6] At the end of the accumulation period a linearly-sloped electric potential is suddenly applied to the trap (fig. 6). This produces a constant axial electric field pointing toward the gate. Positrons which are closer to the moderator are accelerated by the resultant force for a longer period of time and hence leave the trap with a greater speed. The accelerated positrons therefore reach their intended target in less time which can reduce the positron pulse time spread. The farther away from the gate the target is, the smaller the required slope to achieve a given pulse width. Bunching the positron pulse in fig. 4 with a linear slope of 6.6 V/cm and a rise time of 30 ns ($10-90\%$) results in a pulse with a ~ 20 ns FWHM as shown in fig. 6.

Using the linear buncher to time compress the pulse also increases the energy spread of the positrons. The energy spread of the bunched positrons was measured by varying the amplitude of the pulse on the gate. Plotting the number of accumulated positrons which were allowed to exit the trap vs. the minimum voltage on the gate results in an S-shaped curve. The 10–90% rise of this curve is a measure of the axial momentum of the accumulated positrons. For the pulse in fig. 6 the energy spread was measured to be ~ 400 eV.

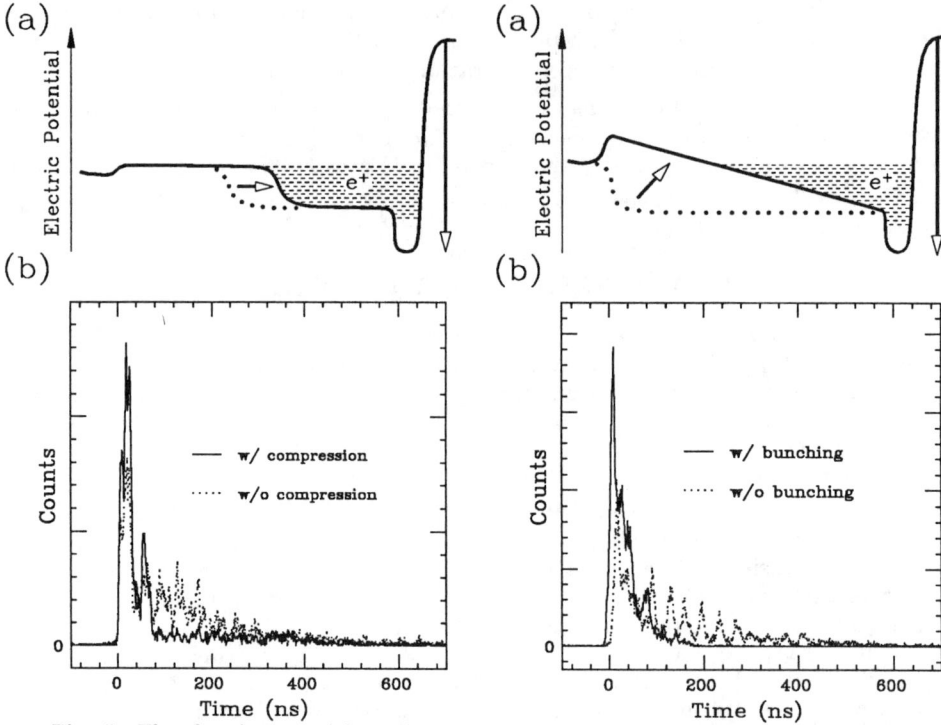

Fig. 5. The electric potential applied to the lensing elements during compression are schematically shown in (a). The time compression achieved on the positron pulse in fig. 4b by raising the sections 300 V in a 10-90% rise time of 100 ns is shown in (b). Consecutive trap sections were raised with an 80-ns delay and the gate was lowered when 9 of the sections were raised.

Fig. 6. The electric potentials applied to the lensing elements before and after bunching are schematically shown in (a). The time-compression achieved on the positron pulse in fig. 4b with a potential slope of 6.6 V/cm over the 71-cm trap is shown in (b). The electric potential on the gate was lowered 70 ns after the sloped potential was applied.

Many of the experiments that would benefit from having a pulsed positron beam must be done in a region which is free of magnetic fields. In the present system extraction of the pulsed positrons from the magnetic field is complicated by their relatively large energy spread. The details of extraction depend on the particular system used and consequently will not be discussed here. It is worth mentioning, however, that the bunched positrons shown in fig. 6b have been extracted to a region where the magnetic field is $\sim$ 1-2 G. An extraction efficiency of $\sim$ 60% has been achieved and the time spread of the pulse was increased by only $\sim$ 15%.

ACKNOWLEDGEMENTS

The authors would like to acknowledge useful discussions with the members of the Michigan positron group. This research has been supported by NSF grant PHY91-19899.

REFERENCES

1. J.S. Nico, D.W. Gidley, A. Rich, and P.W. Zitzewitz, *Phys. Rev. Lett.* **65**, 1344 (1990).
2. R. Suzuki, Y. Kobayashi, T. Mikado, H. Ohgaki, M. Chiwaki, T. Yamazaki, and T. Tomimasu, in *Proc. Int. Conf. on Evolution in Beam Applications* (Takasaki, 1991) B1-07.
3. R.S. Conti, B. Ghaffari, and T.D. Steiger, presented at the Antihydrogen Workshop (Munich, 1992), proceedings to be published in *Hyp. Int.*
4. T.D. Steiger, B. Ghaffari, A. Rich, and R.S. Conti, *Abstracts of Contributed Papers* 12th Int. Conf. on Atomic Physics, Ann Arbor, 1990, eds. W.E. Baylis, G.W.F. Drake, and J.M. McConkey (University of Windsor, Windsor, 1990) p. I-22.
5. A.P. Mills, Jr., E.D. Shaw, R.J. Chichester, and D.M. Zuckerman, *Rev. Sci. Instr.* **60**, 825 (1989).
6. A.P. Mills, Jr., in *Positron Scattering in Gases*, eds. J.W. Humberston and M.R.C. McDowell (Plenum, New York, 1984) p. 121.
7. C.M. Surko, M. Leventhal, and A. Passner, *Phys. Rev. Lett.* **62**, 901 (1989).
8. R.S. Conti, B. Ghaffari, and T.D. Steiger, *Nucl. Instr. and Meth.* **A299**, 420 (1990).
9. J. Van House and P.W. Zitzewitz, *Phys. Rev.* **A29**, 96 (1984).
10. G. Gabrielse, L. Haarsma, and S.L. Rolston, *Int. Jour. Mass Spec.* **88**, (1989) 319 and errata **93**, 121 (1989).
11. P.J. Shultz, E.M. Gullikson, and A.P. Mills, Jr., *Phys. Rev.* **B34**, 442 (1986).
12. T.D. Steiger and R.S. Conti, *Phys. Rev.* **A45**, 2744 (1992).

MONTE-CARLO SIMULATION OF THE POSITRON PRODUCTION AT A LINAC-BASED SLOW POSITRON BEAM

D.Segers, M.Dorikens, J.Paridaens and L.Dorikens-Vanpraet

PositronCentre, Laboratory for Nuclear Physics, Proeftuinstraat 86,
B-9000 Gent, Belgium

ABSTRACT

At accelerator-based slow positron beams the slow positron yield depends on the efficiency of the primary production of positrons by the pair-production process in the converter. This efficiency depends on the electron energy and the converter thickness. No simple relation exists which correlates the converter thickness to the energy of the impinging electrons. We performed Monte-Carlo simulations as a function of the electron energy and the thickness of the converter to obtain the optimum thickness of the converter. The angular spread and the energy distribution of the outcoming positrons was also studied.

INTRODUCTION

It is known from literature that for accelerator-based slow positron beams the slow positron yield is a function of the energy of the primary electron beam and also depends on the chosen thickness of the electron-positron converter. The fast positrons coming out of the converter have a certain energy distribution. The angular spread of those positrons also depends on the electron energy and the thickness of the converter. A moderator has to be installed behind the converter. In order to intercept as much fast positrons as possible, it is important to know the angular spread of the positrons. The moderation of the positrons is a complicated process. The moderator geometry depends greatly on the energy distribution of the incoming fast positrons.

From the early beginning that slow positron beams were installed at electron accelerators it was noticed that the slow positron yield depends very critically on the moderator thickness[1] and on the electron energy[1,2].

In the work of the Livermore group[1] it was noticed that for 100MeV electrons impinging on a tantalum converter the maximum positron yield was obtained at a thickness of three radiation lengths. It was also seen that the positron yield steadily increased with increasing electron energy in the range from 60 to 120MeV. The same conclusions on the energy dependence of the slow positron production were obtained by the Mainz group[2] for the electron energy range between 80 and 240MeV.

The Giessen group[3] studied the positron yield in the electron energy range between 10 to 40MeV and observed a nearly linear increase in slow positron yield.

Japanese groups[4,5] also observed a dependence of the slow positron intensity as a function of the Ta converter thickness. An optimum in the converter thickness was found for the specific electron energy used.

No systematic study of the positron yield as a function of the converter thickness and as a function of the electron energy exists. To do this experimentally is nearly impossible. A possible way to study the dependence of the slow positron yield on the different parameters (converter thickness and electron energy) is to perform Monte-Carlo simulations of the fast positron production.

RESULTS

The EGS4 code[6] was used to perform the Monte-Carlo simulations. As a material we used Ta. The geometry of the converter is represented in figure 1. It was considered being an infinite slab (represented in figure 1 by medium 2). The slab is surrounded by vacuum (this is medium 1 and medium 3 in figure 1). The thickness of this slab (region 2) was varied between 1 and 10mm (in steps of 1mm) for impinging electron energies ranging between 20 and 70MeV (in steps of 10MeV). The electrons were considered at normal incidence and a total of 100,000 events were followed. The Monte-Carlo process was followed for electrons and positrons down to energies of 100keV and for photons down to an energy of 1MeV (no pair production occurs below this energy). The simulations were carried out on a MicroVAX 3500 computer. Some of the simulations (for a particular electron energy and tantalum thickness) lasted up to 40h of CPU time.

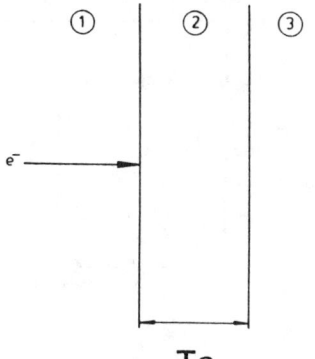

Figure 1: Representation of the considered converter geometry for the Monte-Carlo simulations. Medium 1 and medium 3 is vacuum. Medium 2 is tantalum. The thickness of the tantalum slab can be varied. The electrons impinge on the slab at normal incidence. Their energy was also varied.

Results of the Monte-Carlo simulations are represented in figures 2 to 5. Figure 2 illustrates the positron yield as a function of the electron energy and the Ta converter thickness. This is represented in a 3D plot with projections on the different planes, illustrating for example the dependence of the positron yield on the electron energy and converter thickness. Figures 3 and 4 give the angular dependence of the outcoming positrons. For this purpose the halfspace was divided into 128 concentric rings subtending the same solid angle ($2\pi/128$). Figure 3 illustrates the results obtained by irradiating a 5mm thick Ta target with 40MeV electrons and figure 4 gives the results for a 6mm thick Ta target irradiated with 70MeV electrons. The statistical fluctuation in the 70MeV electron irradiated sample is much less in comparison with the 40MeV irradiated Ta converter. This is due to the fact that the 70MeV electron beam creates much more positrons in comparison to the 40MeV electron beam (see the horizontal scale in both the figures 3 and 4).

Figure 5 displays the positron energy spectra as a function of the incoming electron energy. The thickness for the tantalum converter was kept constant and is 5mm.

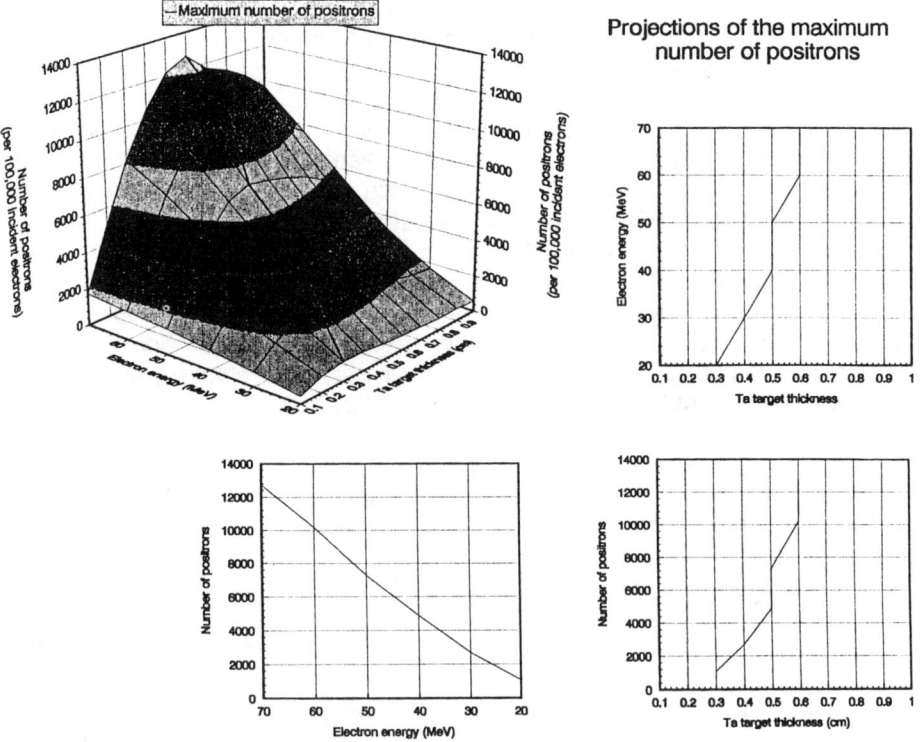

Figure 2: Three dimensional representation of the positron yield as a function of the electron energy (from 20MeV to 70MeV) and the Ta converter thickness.

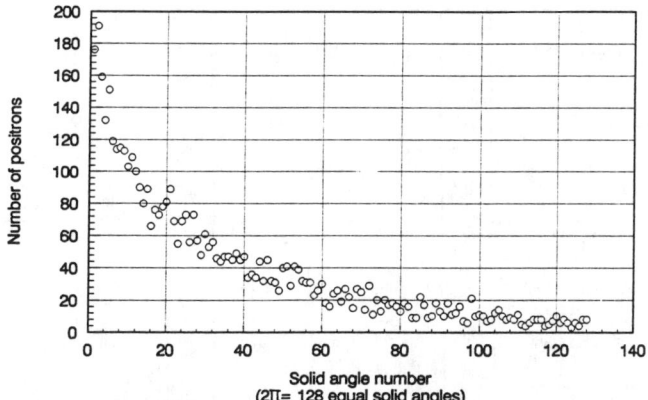

Figure 3: Angular distribution of the number of positrons as a function of the solid angle for 100,00 electrons with an energy of 40MeV impinging upon a 5mm thick Ta converter.

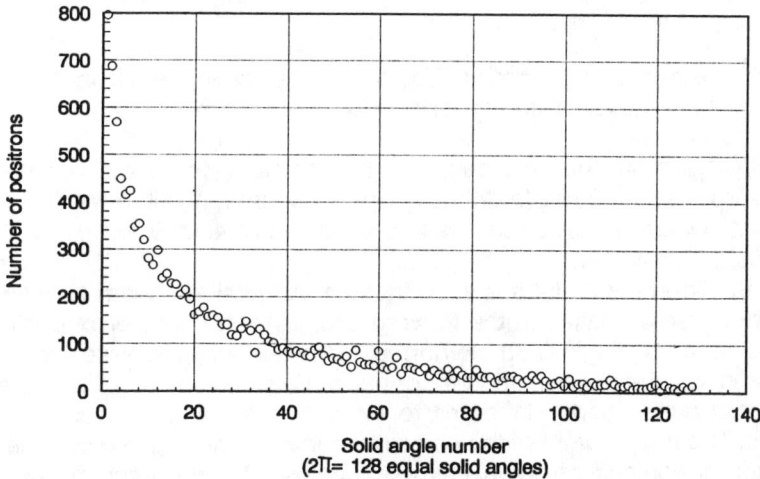

Figure 4: Angular distribution of the number of positrons as a function of the solid angle for 100,00 electrons with an energy of 70MeV impinging upon a 6mm thick Ta converter.

DISCUSSION

The Monte-Carlo results display some interesting features. Figure 2 clearly illustrates that the maximum total number of positrons increases very fast with the electron energy (from about 1,000 up to about 13,000 positrons when the electron energy is changed from 20 up to 70MeV). To obtain the maximum

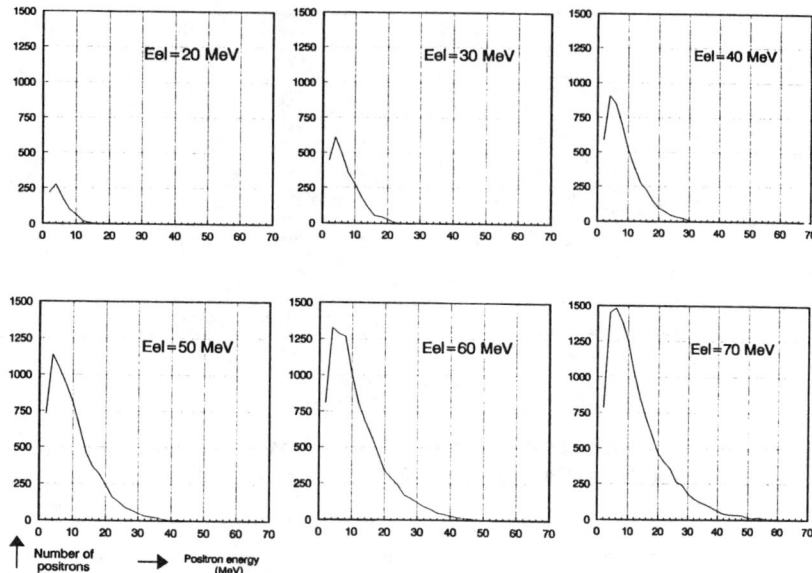

Figure 5: Positron energy spectra for different incoming electron energies. The thickness of the tantalum converter is constant: 5mm.

positron production the thickness of the tantalum converter has to be changed from about 3mm (at 20MeV) up to about 6mm (at 70MeV). In our equipment[7] where the electron beam is operated at 40MeV we use a 5mm thick tantalum converter.

From the figures 3 and 4 it is seen that the fast positron beam coming out of the converter is peaked in the forward direction. The number of positrons as a function of the angle (with the normal to the surface) decreases very fast. This means that the positron moderator has to be placed as close as possible to the positron converter in order to intercept as many as possible fast positrons. The temperature of the converter can rise quite high so that the first moderator foil can suffer from this temperature rise. The moderator is also put at a high tension. If the space between the moderator and the converter is too small electric discharges can occur when the accelerator beam is on.

From figure 5 it follows that the positron energy spectra (for a 5mm thick tantalum converter) are all peaked at energies below 10MeV (in the electron energy range investigated here). The increase in the total number of positrons as a function of the electron energy is clearly seen. In accelerator based beams the moderator is mostly of a venetian blind geometry and has a restricted total thickness. This means that a lot of fast positrons are not absorbed in the moderator and are lost. A better converter-moderter geometry could certainly increase the slow positron yield.

In a recent publication by Akahane[5], a simple model to estimate the optimal converter thickness was explained. It was assumed that the electron energy E is shared between the electron-positron pair production process and the Bremsstrahlung process. Their model predicts that the optimum thickness of

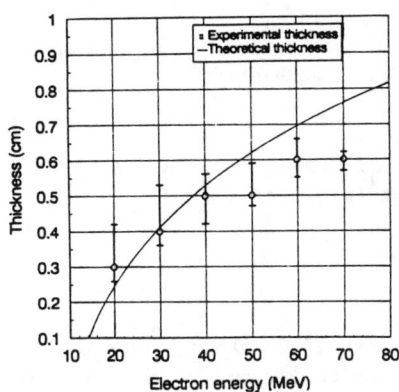

Figure 6: Predictions for the converter thickness made by the Akahane[5] model as compared with the results obtained from our Monte-Carlo simulations.

the converter is obtained by: $x = X_0 \cdot \ln(E/E_c)$. with X_0 and E_c constants depending on the material. For tantalum $X_0 = 4.1$mm and $E_c = 11.0$MeV. In figure 6 we plotted the expected optimum converter thickness (as predicted by the model) and compared it with the converter thicknesses (producing the maximim of fast positrons) as obtained from our Monte-Carlo simulations. We see that there is a discrepancy between the two results. The model used by Akahane[5] is very simple and only serves as a first estimate.

CONCLUSION

It is shown that Monte-Carlo simulations of the fast positron-production at accelerator based slow positron beams give a lot of useful information for the construction of slow positron beams. The slow positron beam production at those accelerator based beams is a complicated process of positron production in the converter (with a certain energy distribution) and the moderation process. A lot of fast positrons are lost in most of the moderator geometries.

REFERENCES

1. R.H.Howell, R.A.Alvarez, K.A.Woodle, S.Dhawan, P.O.Egan, V.W.Hughes, M.W.Ritter, IEEE Trans Nucl. Sc. 30, 1438 (1983)
2. G.Gräff, R.Ley, A.Osipowicz, G.Werth, J.Ahrens, Appl.Phys. A33, 59 (1984)
3. F.Ebel, W.Faust, C.Hahn, S.Langer, H.Schneider, Appl.Phys. A44, 119 (1987)
4. O.Sueoko, Y.Ito, T.Azuma, S.Mori, Y.Katsumura, H.Kobayashi, Y.Tabata, Jap. Jour. Appl. Phys. 24, 222 (1985)
5. T.Akahane, T.Chiba, N.Shiotani, S.Tanigawa, T.Mikado, R.Suzuki, M.Chiwaki, T.Yamazaki, T.Tomimasu, Appl. Phys. A51, 146 (1990)
6. W.R.Nelson, H.Hirayama, D.W.O.Roger, SLAC-265 (1985)
7. J.Paridaens, D.Segers, M.Dorikens, L.Dorikens-Vanpraet, Nucl. Instr. & Meth. A287, 359 (1990)

OPTICAL DESIGN OF A REMODERATION SECTION

L.J. Seijbel, P. Kruit, J.E. Barth
Particle Optics group, Department of Applied Physics, Delft University of Technology,
Lorentzweg 1, NL-2628 CJ Delft, The Netherlands

A. van Veen, H. Schut
Reactor Physics group, Interfaculty Reactor Institute, Delft University of Technology,
Mekelweg 15, NL-2629 JB Delft, The Netherlands

ABSTRACT

This paper discusses the design of a remoderation section for the Delft positron microprobe. The positron beam that emerges from the Delft nuclear reactor has a very low brightness. In order to have the maximum current in a microprobe this brightness must be enhanced using four thin remoderation foils. The paper shows the design for the last remoderation step in which the beam is demagnified from 26 µm to 0.8 µm and thus the brightness is enhanced by about a factor 300 depending on the re-emission yield.

INTRODUCTION

At the Delft nuclear reactor a positron beam facility is being built. In these proceedings the design of this facility is described by van Veen et al.[1]. The goal of the project is to get a positron beam in the energy range of 0.2-20 keV with 10^6 positrons per second. The smallest diameter of this beam will be 100 nm. To get the maximum number of positrons in this spot the beam must be remoderated[2] four times after which it will be guided into a Philips SEM 535-M. In the SEM focusing lenses and a scanning system that can be used for the positron microanalysis are already present. New in the SEM are a 90°-deflector to deflect the positron beam onto the optical axis of the microscope and a Ge-detector to measure the energy of the annihilation photons. In Fig. 1 the whole beam setup is given with nuclear reactor, transport section, remoderation part, and SEM.

In this paper a general description of the last remoderation section is given. For this reason first a particle optics description of the beam characteristics is given, after which some calculations on lenses are shown. Since aberrations reduce the effective brightness, they must be kept as small as possible. A more detailed description of the total remoderation section will be given later[3].

PARTICLE OPTICS

When the beam emerges from the reactor the brightness of the beam is very low. When focusing this beam into a spot of 100 nm only 1 positron every 100 seconds would enter the specimen. Therefore we have to enhance the brightness of the beam. It can be calculated that we need four remoderation steps to get the optimum brightness[4]. The brightness of the beam is

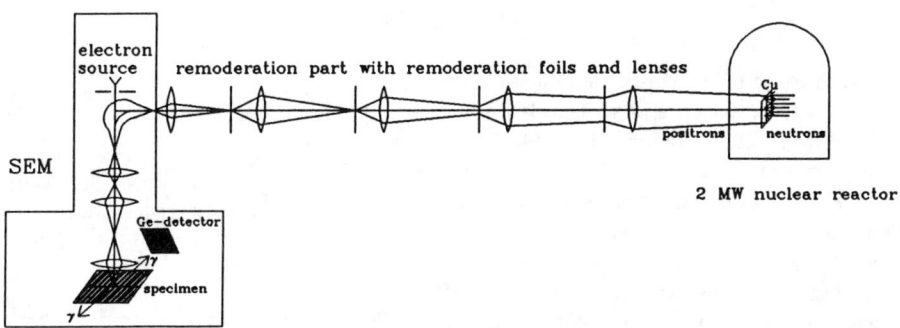

Fig. 1 Setup of the Delft positron microbeam. The figure is purely schematic since the remoderation section is only few tens of centimeters long and the transport section from the reactor to the first foil is a few meters long.

defined as the current(I) per area(A), solid angle(Ω), energy(E), or in formula:

$$B = \frac{I}{A\Omega E}. \qquad (1)$$

Brightness enhancement is the reduction of the product of solid angle and energy. Inside a thin single crystal the positrons are slowed down to thermal energies and are re-emitted with an energy as large as the absolute value of the negative workfunction for positrons of that material. Therefore the beam must be focused into the smallest diameter onto the foil using some lenses. During the remoderation process the beam diameter is not enlarged, where the number of positrons is reduced. If this reduction is not too big another remoderation step can be performed.

When designing a lens system for remoderation not only the first order properties like magnification and strength are important, the 3rd-order aberrations are also of importance. The most important aberration is spherical aberration. The contribution of spherical aberration to the spot size can be determined as $C_s \alpha^3$. To keep this contribution low C_s and α must be kept small. To keep α, the opening angle small is a matter of taking the magnification not too great. C_s, the coefficient for spherical aberration can be kept small by reducing the size of the lens and placing the image in a region with a strong electric or magnetic field. Another aspect on spherical aberration is that the disk of least confusion is not in the image plane. The place where 100% of the beam has the smallest diameter is at a position $0.75\ C_s \alpha^2$ in front of the image plane. This is the best place to put the remoderation foil.

Another important aberration when working with larger beams is coma. The contribution of this aberration to the spot size is not only a function of the opening angle but also a function of the initial spot size($\propto \alpha^2 d_0$). In contrast to spherical aberration coma can be compensated by using an extra lens. This extra lens must have its back focal plane exactly in the center of the demagnifying lens. Then the positrons will go through the center of the demagnifying lens. Knowing this an optimized lens system for the remoderation section can be built.

OPTICAL DESIGN

In this part of the paper the design of the last remoderation section is described. It can be calculated from the aberrations of the SEM that the spot size on the last remoderator must be 0.8 μm. From this point the maximum demagnification must be found by calculating different lens systems. So the strategy is to calculate back from the smallest spot in the SEM to the diameter of the beam at the positron source. With estimates for C_s it can be calculated that 4 remoderation steps are necessary,[4] where the optimum acceleration voltage is 4500 V.

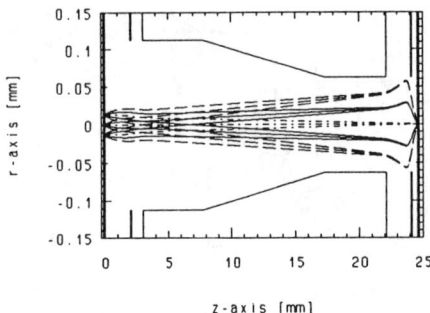

Fig. 2 Design of the two lens remoderation section containing calculated positron trajectories.

The design for the remoderation section needs two lenses between the two remoderation foils. To build the lenses three other electrodes are necessary. In Fig. 2 the two foils with the three electrodes are drawn. The middle (big) electrode is at the same potential as the last foil. The most important lens is the demagnifying lens. This lens consists of the middle electrode, the last electrode and the last remoderation foil. Since we need to compensate coma a first lens for the coma compensation is used. This lens consists of the first foil, the first electrode and the middle electrode.

Through this system several positron trajectories are calculated. First the electrostatic fields are calculated using the finite element method program package ELD[5]. When the field is calculated the positron trajectories through this field can be calculated using a ray trace program[6]. It can not be seen in this figure that the foil is placed at the disk of least confusion. Therefore a closer look to the end of the trajectories is necessary.

In Fig. 3 five of the calculated trajectories are shown. These rays all come from the rim of the spot of the one but last remoderator i.e. a distance to the optical axis of 13 μm. The angles are 0.10, 0.05, 0.00, -0.05, and -0.10 rad. With these rays the total spot size can be calculated. The smallest spot with all positrons in it, is there where the -0.10 rad ray intersects the 0.05 rad ray as a consequence of the spherical aberration. In the figure it can be seen that these rays do intersect just in front of the last remoderator. The coefficient for spherical aberration is calculated to be 4.68 mm.

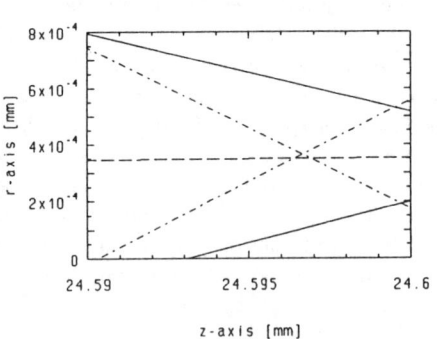

Fig. 3 End of some of the trajectories of Fig. 2.

Furthermore it can be seen that coma is almost totally compensated by letting the zero angle rays go through the center of the demagnifying lens(Fig. 2). In case of no coma the rays which have an angle are symmetrically distributed around the zero angle ray as can be seen in Fig. 3.

In Fig. 3 it can also be seen that the size of the spot is larger than the 0.8 μm that was aimed for. The reason for this is that the first rough calculation was an estimate based on quadratic summation of all the components. This summation already holds that not all the particles are in this spot size. In the figure the most outer rays of the beam are shown. To decide how important these rays are for the intensity of the beam, 10000 trajectories of positrons with randomly calculated begin position and angles are calculated. The result of these calculations is shown in Fig. 4. In this figure it can be seen that over 90% of the positrons end inside a spot with a diameter of 0.8 μm.

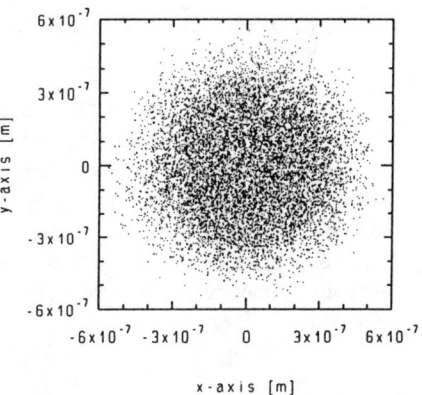

Fig. 4 End positions of 10000 randomly calculated trajectories.

CONCLUSION

From these calculations it can be concluded that a spot size of 800 nm can be obtained. The brightness of the beam is at that position high enough to do positron analysis with a probe diameter of 100 nm, the diffusion length of positrons inside many materials. The exact brightness depends on the re-emission yield, it is expected to be 10^{-3} Acm^{-2}sr^{-1}eV^{-1}.

The design is now under construction together with similar designs for the other stages.

ACKNOWLEDGEMENTS

These investigations in the program of the Foundation for Fundamental Research on Matter (FOM) have been supported by the Netherlands Technology Foundation (STW).

REFERENCES

1. A. van Veen, H. Schut, P.E. Mijnarends, L.J. Seijbel, P. Kruit, these proceedings
2. A.P. Mills Jr., Appl. Phys. 23, 189 (1980).
3. L.J. Seijbel, J.E. Barth, P. Kruit, A. van Veen, H. Schut, to be published.
4. L.J. Seijbel, P. Kruit, A. van Veen, H. Schut, Mat. Sci. Forum 105-110, 1977 (1992).
5. B. Lencova, Manual for the ELD-program package, Information attainable at the Particle Optics Foundation, Lorentzweg 1, NL-2628 CJ Delft, The Netherlands
6. G. Wisselink, Manual for the ray trace program TRASYS, not published.

Theoretical Simulation of Positron Premoderation

M. Shi, W.B. Waeber, D. Taqqu, F. Foroughi and J. Arkuszewski

Paul Scherrer Institute, CH–5235 Villigen PSI, Switzerland

A discussion will be given of the magnetic and electrostatic confinement which will be used in the premoderation stage in the PSI positron beam project [1]. An analysis of the positron trajectories based upon various computer codes and Monte Carlo calculations has given us the angular and energy distribution of positrons slowing down in a carbon foil at the center of the confinement. These analyses will be compared with the planned experimental measurements to optimize the efficiency of our system to premoderate positrons from several hundreds of keV to 5–10 keV.

1 Introduction

Conventional positron source–moderation schemes as used in standard lab–beams have moderation efficiencies of the order of 10^{-4} to several times 10^{-3} [2,3,4]. It is however possible to minimize moderation losses by letting the positrons enter the moderator only after going through certain controlled slowing–down processes allowing them to reach energy ranges that are low enough for the penetration depth in the moderator to be minimized. This is the basic idea of a new slow positron beam production scheme [5]. It uses two steps, a *premoderation* step and a final moderation step. The principle of the premoderation step is to trap the high energy positrons within magnetic and electrostatic fields so that all slowing down takes place in a thin solid foil placed at the center of a magnetic bottle. Just before they would stop in the foil the low energy positrons finally are shifted out of the slowing down path and redirected onto the moderator in a reflection mode configuration. Extraction of the slow positrons and subsequent phase space improvement operations on the beam [1] ultimately lead to an estimated overall conversion efficiency of the order of 50 times larger than the highest values of conventional schemes. Together with a primary radioactive source strength of a few Ci β^+ this concept would lead to slow positron *microbeam* intensities in the $10^{10} e^+/s$ range.

The design of the premoderation stage calls for a thorough analysis of the charged particle motion in the electromagnetic confinement fields, with initial energies reaching values of several hundreds of keV up to the order of 1MeV. The objective of such simulations and the scope of this paper is the description, evaluation (and later optimization) of the premoderation processes – the confinement and slowing down – and the calculation of their efficiencies. Simultaneously this yields the prescriptions for the specific experimental set–up of the proposed scheme [6,1].

2 Trapped Positron Motion

In the present concept the positrons are forced by the confining fields to perform multiple passages, forward and backward, through a thin solid foil at the center of a magnetic bottle. Thus the question of central importance concerning the electromagnetic confinement is: 'Under what conditions would positrons return to the source and slowing down foil placed at the center of the magnetic bottle (longitudinal confinement = magnetic bottle + electrostatic mirror reflectors), and under what conditions would a positron's trajectory preserve the distance of its guiding center from the axis and its Larmor radius unchanged (transverse confinement) when it returns to the slowing down foil due to reflections?'

2.1 Adiabatic Motion and Confinement

The confining fields have to be configured in such a way that the trapped positron motion is *adiabatic*, i.e. well behaved and controlled such that losses of particles due to nonadiabatic motion are avoided. In this way the confinement of positrons by the fields can be optimized.

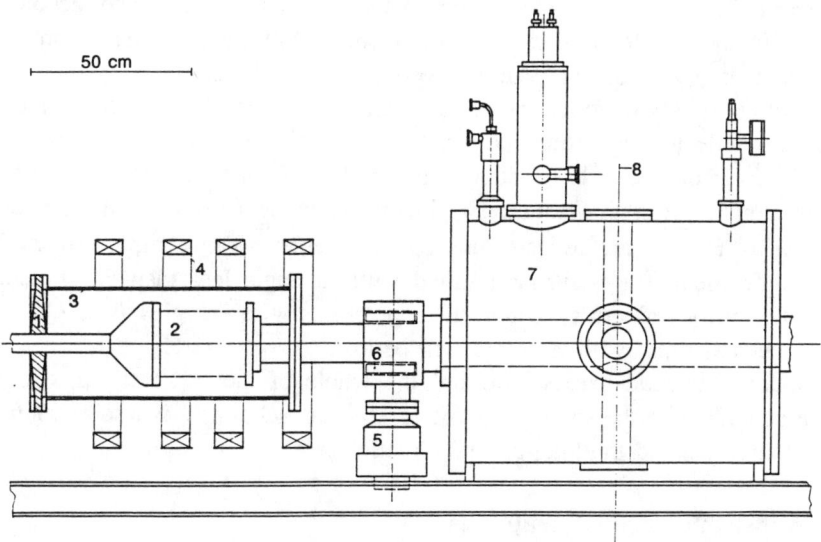

Figure 1: *Design of the premoderation test experiment: 1=insulating flange, 2=300 kV acceleration tube, 3=SF$_6$ tank, 4=6=1.5 kG magnetic coils, 5=ion pump, 7=9T/3T superconducting magnet (cryostat), 8=symmetry axis with respect to the* **B** *and* **E** *field configuration on both sides of the cryostat.*

Our field configuration shown in Figure 1 has cylindrical symmetry. The development

$$\frac{v_{c\perp}^2}{B} + A_0 + A_1\epsilon + A_2\epsilon^2 + o(\epsilon^3) = const \qquad (1)$$

shows 'adiabatic invariance' of $v_{c\perp}^2/B$, oscillating around a mean value with negligible amplitude whenever $\dot{B} = 0$, if the smallness parameter ϵ is very small (the meaning of ϵ is discussed in Appendix 1). Here

$$A_0 = \pm \frac{r_0 v_{o\perp} \dot{B}}{B^2} \sqrt{v^2 - \frac{2M}{m} B \cos \lambda}, \tag{2}$$

where the positron with mass m and velocity v has the velocity component $v_{c\perp}$ perpendicular to the cylindrical axis, r_o is the radial coordinate of the guiding center and $v_{o\perp}$ is $v_{c\perp}$ at $z = 0$, M is the equivalent magnetic moment of the current loop of the orbiting positron, λ is the phase of the positron motion on its orbit of gyration and $\dot{B} = \partial B(0, z)/\partial z$. In cases where **B** is parallel to the axis of cylindrical symmetry and $\mathbf{E}_\perp$ has the order of ϵ, the drift velocity of the guiding center is described by

$$\dot{\mathbf{R}}_\perp = \frac{\epsilon}{B^2}(v^2 + \frac{1}{2}v_\parallel^2)\mathbf{B} \times \nabla B - \frac{\mathbf{B} \times \mathbf{E}}{B^2}, \tag{3}$$

which means that viewed along the flux line, at any $z = const$ plane, due to the cylindrical symmetry, and in 1st order, the movement of the guiding center is only in the *azimuthal* direction where B does not change its value. Hence, according to equations (1) and (3), for positrons returning to the foil the *transverse* confinement is guaranteed at any z position where $\dot{B} = 0$ or sufficiently small.

The critical points of reflection (*longitudinal* confinement) between the electrostatic mirror reflectors are found at positions at which $v_{c\perp}^2 = v^2$. From equation (1), the first mirror action is introduced by the 9T/3T magnetic bottle such that all positrons emitted from the centered foil with an angle θ to the axis greater than $\theta_c = arcsin\sqrt{B_0/B_m}$ are reflected back to the center. The complementary longitudinal confinement, for positrons emitted with an angle less than θ_c, is added by electrostatic mirror reflectors, so that these positrons with energies $E < eV_r$ will be reflected whereas those with $E > eV_r$ will be lost.

Thus, for 'adiabatic particle motion' the whole of the β^+ spectrum, except for emission conditions at the source foil with $\theta < \theta_c$ and $E > eV_r$, is totally confined in both the transverse and the longitudinal directions.

2.2 Field Distributions and Trajectories

The important factor which makes the adiabatic approximation to become more accurate is to keep $\rho|\nabla B|/B < \frac{mc}{e}|\nabla B|/B^2$ much less than 1 in the region of the positron motion (ρ = Larmor radius) so that ϵ is a very small quantity. By making

$$\frac{mc}{e}|\nabla B|/B^2 \ll 1 \tag{4}$$

the geometry (Figure 1), current density, the potential on the electrodes and the position of the coils can be optimized and they lead to the magnetic field and electrostatic

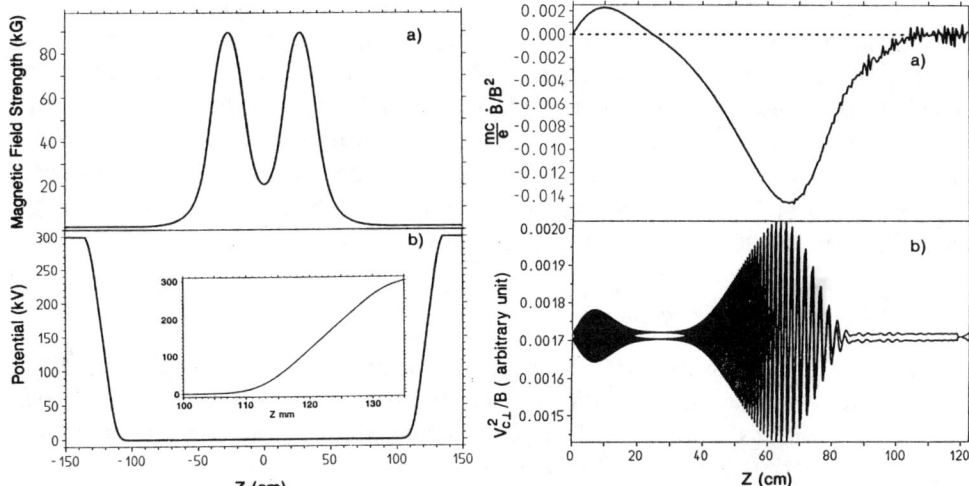

Figure 2: (a) Magnetic field distribution produced by the 9T/3T superconducting coils and the normal conducting 1.5 kG coils on both sides of the cryostat as shown in Figure 1. (b) Potential distribution on the axis produced by the electrostatic mirror reflector (max. potential $= V_r$), the inset shows the potential rise with enlarged scale.

Figure 3: (a) The characteristics of the magnetic field given by the leading factor of A_0 along the axis of symmetry. (b) 'Adiabatic constant' corresponding to the field distribution of Figure 2. The difference between the forward motion (upper curve) and backward motion (lower curve) is due to numerical errors, which has been checked by the scaling law and the SLAC program. The initial conditions are identical to those indicated in Figure 4.

potential distributions as shown in Figure 2. The corresponding values of $\frac{mc}{e}|\nabla B|/B^2$ are plotted as a function of z in Figure 3a. [1]

The trajectory calculations have confirmed that, for the particular magnetic field and electrostatic potential distributions, the predictions of section 2.1 hold true. An example of a positron trajectory in these field distributions is shown in Figure 4 and the corresponding 'adiabatic invariance' $v_{c\perp}^2/B$ is plotted in Figure 3b.

3 Simulation of the Premoderation Efficiency

The most powerful theoretical technique to study the slowing down of positrons in solids is Monte Carlo simulation of particle trajectories. The purpose of this simulation is to answer the question: 'How many of the emitted positrons from a source on a thin solid foil placed at the center of the electromagnetic confinement can be slowed

---

[1] The programs employed in this section are SLAC, POISSON and BPOSI. The latter has been developed by F. Foroughi at PSI, in which the Runge–Kutta method is used to solve three coupled differential equations of motion derived from the Lorentz equation, and which does not assume a prefixed mesh structure for the calculation of the magnetic field strength.

down by the source foil to low energies below which they would be available for intermediate extraction and for redirection to the final moderator?' The code employed in this simulation work is MCNP version 4 [7]. Energy losses and scattering angles are sampled from tables derived from multiple–scattering theory and the theory of energy loss due to ionization according to the prescriptions in refs. [8,9]. In further studies and for comparison reasons we shall also use the programs developed by Nieminen and others [10,11,12]. The positrons with energies less than a certain cutoff energy are considered lost if they are found inside the source foil.

3.1 The Definition of Efficiencies

The definition of the relevant efficiencies characterizing the premoderation process can be given in different ways. We adopted the following path, where the various numbers of positrons involved can be represented in terms of integrals over appropriate angular and energy distribution functions (see Appendix 2):

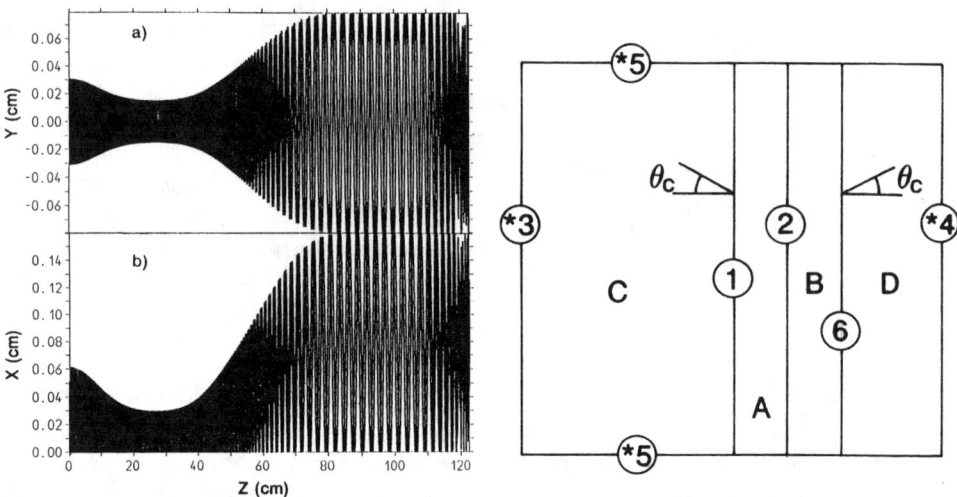

Figure 4: (a) Projection onto the y–z plane of a positron trajectory in the magnetic field and electrostatic potential as shown in Figure 2. (b) The same as for (a), but projected onto the x–z plane. Initial conditions of the positron: point of emission at the origin, with the momentum on the y–z plane and its angle to the z-axis $\theta = 28^0$, $P = 0.4$Mev/c.

Figure 5: Model of the confinement and slowing–down: Surfaces 1, 2, 3, 4, 6 are plane and surface 5 is cylindrical. The stars indicate that the corresponding surfaces are reflecting surfaces. Cell A is the carbon foil with thickness 5 $\mu g/cm^2$, cell B is the $Li^{18}F$ source material with thickness 17 $\mu g/cm^2$, cell C and D are empty. In cell D extraction takes place. θ_c is the escape angle (critical angle), it is 30^0 for the calculation of the confinement efficiency and 0^0 for the calculation of the extraction efficiency respectively.

The *confinement* efficiency is given by

$$\eta_{conf} = (N - n'_1)/N = n_1/N, \qquad (5)$$

where N is the total number of positrons emitted from the source, n'_1 is the number of positrons lost due to leaks (i.e. $\theta < \theta_c$, $E > eV_r$) in the electromagnetic confinement, and otherwise we assumed 100% efficiency of the transverse confinement.

The *extraction* efficiency (extraction from the slowing down path of intermediate energy positrons) is obtained by setting

$$\eta_{extr} = (n_1 - n'_2)/n_1 = n_2/n_1, \qquad (6)$$

where n'_2 is the number of positrons stopped inside the source or the solid foil before they are available for intermediate extraction. It follows, that the *premoderation* efficiency then becomes the simple product of the two previous quantities

$$\eta_{prem} = n_2/N = \eta_{conf}\eta_{extr}. \qquad (7)$$

The *total conversion* efficiency is given by

$$\eta_{conv} = n_3/N = \eta_{Ne}(\bar{E}, 0)\eta_{prem}, \qquad (8)$$

which in this work will conservatively be approximated by $\eta_{conv} \approx \eta_{Ne}(E_{c2})\eta_{prem}(E_{c2})$. A model description for the calculation of these quantities with the Monte Carlo technique is given below.

3.2 Model Description for the Simulation

The model situation is shown in Figure 5. The transverse confinement by the magnetic field is simulated by the reflecting surface *5. The longitudinal confinement (the 9T/3T magnetic bottle and electrostatic mirror reflectors) is simulated by the reflecting surfaces *3 and *4 and the critical angle θ_c. If the positrons emitted from the source foil with energy larger than eV_r the θ_c is set to be $arcsin\sqrt{B_0/B_m}$ and for other energies $\theta_c = 0$. Only those positrons emitted with an angle θ larger than θ_c are reflected by surfaces *3 and *4, otherwise they pass through the surfaces, can not return and are counted lost. Thus, all positrons with emitting angle $\theta < \theta_c$ cannot be confined. We simulate the termination of the slowing down process by setting different cutoff energies in different cells. In cells A, B and C the cutoff energy is E_{c1} below which we consider the positrons to be stopped in the foil and counted lost. In cell D the cutoff energy is set to be E_{c2}, the upper limit of intermediate extraction [13]. By counting the number of positrons which end up in cell D and which have energies $E_{c1} < E < E_{c2}$ (and which are therefore available for extraction), finally gives the extraction efficiency.

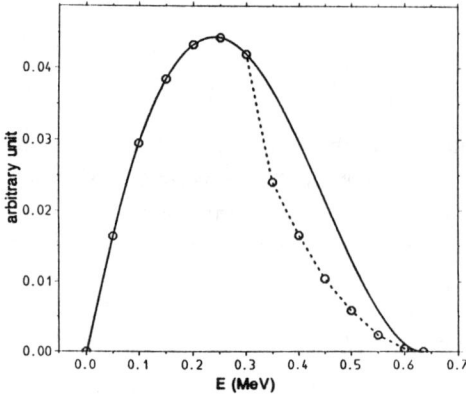
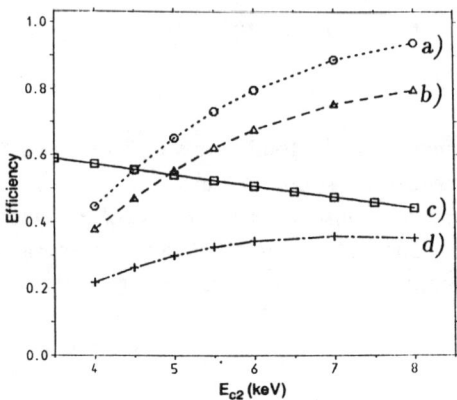

Figure 6: The spectrum of ^{18}F (bold line) and the confined spectrum (dotted line). The confinement efficiency in this case is $\eta_{conf} = 85\%$.

Figure 7: As a function of the cutoff energy E_{c2} are shown (a) the extraction efficiency, (b) the premoderation efficiency, (c) the Neon moderation efficiency taken from ref [14] and (d) the total conversion efficiency.

4 Results and Conclusion

Since the motion of positrons, as emitted from a β^+ source in the center of a magnetic bottle B_m/B_o, is required to be adiabatic and totally confined in both the transverse and the longitudinal direction with respect to the slowing down axis, the corresponding computer–simulation has been used to optimize the adiabaticity of the positron trajectories. This has led to a self–consistent field configuration with minimal losses of positrons that are exclusively due to the inherent limitation of the maximum value V_r of the electrostatic potential of the mirror reflectors in the proposed concept.

The field configuration found for the confinement and slowing down totally confines the entire β^+ spectrum except for positron emission conditions at the source foil surface during the slowing down process with emission angle $\theta < arcsin\sqrt{B_0/B_m}$ and initial energy $E > eV_r$.

For one specific set of parameters, $arcsin\sqrt{B_0/B_m} = 30°$, $E_{c1} = 1$keV $V_r = 300$kV, source foil thickness $t = 22\mu g/cm^2$, the results for the confinement efficiency and for the extraction efficiency have been obtained and are presented in Figures 6 and 7. It can be concluded that, after parameter optimization, the confinement efficiency can be higher than 85%, and by taking Mills' measured efficiencies of a solid Neon moderator at $0°$ implantation angle [14] an overall conversion efficiency of more than 35% can be expected. Based on realistic simulation studies, this is the first theoretical confirmation of estimates that have been stated in the original proposal for a high efficiency moderation scheme [5].

The data for the magnetic coils, the superconducting magnet and the acceleration sections as well as their relative positions along the magnetic axis have been taken as the basic input for the design, the technical specifications, the fabrication of these components and their assembly in the experimental set-up. The results of this work will be compared, for purposes of optimization, with the planned measurements of the efficiencies as a function of experimental parameters, like B_o, V_r, t, etc.

Acknowledgement

This work is partially supported by the Swiss Science Research Nationalfonds. We wish to acknowledge M. Peter and A.A. Manuel for continuous support and active interest in this work, and M. Sedlacek for providing his experience. We are also indebted to L.O. Roellig and U. Zimmermann for many helpful discussions.

Appendix 1

The motion of a charged particle in a magnetostatic and electrostatic field is described by the equation of motion

$$\frac{m}{e}\ddot{\mathbf{r}} = \dot{\mathbf{r}} \times \mathbf{B}(\mathbf{r}) + \mathbf{E}(\mathbf{r}). \tag{9}$$

Let $\tilde{\mathbf{B}} = \mathbf{B}(\mathbf{r})/B_o$, $\tilde{\mathbf{E}} = (1/v_o B_o)\mathbf{E}(\mathbf{r})$, $\tilde{T} = (v_o t/L)$ and $\tilde{\mathbf{R}} = \mathbf{r}/L$, where v_o is the initial particle velocity, B_o is the magnetic field at a typical point, and L is a characteristic dimension or length over which the fields change. We get the equation of motion in the dimensionless form [15]

$$\frac{mv_o}{eB_o L}\frac{d^2\tilde{\mathbf{R}}}{d\tilde{T}^2} = \frac{d\tilde{\mathbf{R}}}{d\tilde{T}} \times \tilde{\mathbf{B}}(\tilde{\mathbf{R}}) + \tilde{\mathbf{E}}(\tilde{\mathbf{R}}), \tag{10}$$

where $\frac{mv_o}{eB_o L}$ is just the ratio of the radius of gyration to the characteristic length over which fields change and therefore is the quantity which should be made small in order for the adiabatic approximation to become valid.

Considering $\frac{mv_o}{eB_o L}$ as small, in some cases, we can get an approximate solution of equation (10) up to some orders. This solution can be translated into the solution of equation (9) by back substituting $\tilde{\mathbf{R}} \to \mathbf{r}$ and so on. On the other hand, because equation (9) is formally identical with equation (10) for any field configurations, we can directly deal with equation (9) by using the dimensional quantity $\frac{m}{e} \equiv \epsilon$ as the smallness parameter and get the solution up to some orders. However, in order to ensure that the approximation is correct, the corresponding parameter in the dimensionless equation (10), $\frac{mv_o}{eB_o L}$, really has to be a small number. So, if we assume ϵ to be very small it always implies that $\frac{mv_o}{eB_o L}$ is very small.

Appendix 2

In the premoderation step there are two possibilities for losing intensity (see also section 3.2). One is due to the inherently nonperfect confinement, and it can be described by

$$n'_1 = \int_0^{\theta_c} \int_{eV_r}^{\infty} f_1(E,\theta) dE d\theta, \qquad (11)$$

where $f_1(E,\theta)$ is the number of positrons with energy E and angle θ to the axis when they exit the source foil on any side (to the left of cell A or to the right of cell B, see Figure 5) during the slowing down process. The other comes from the positrons stopping inside the source foil: It has the form

$$n'_2 = \int_0^{\pi/2} \int_{E_{c2}}^{eV_r} f_2(E,\theta) P_{E_{c1}}(E,\theta) dE d\theta + \int_{\theta_c}^{\pi/2} \int_{eV_r}^{\infty} f_2(E,\theta) P_{E_{c1}}(E,\theta) dE d\theta, \qquad (12)$$

where $f_2(E,\theta)$ has the same meaning as $f_1(E,\theta)$ except that only the positrons on one source side to the right of cell B in Figure 5 are being considered, $P_{E_{c1}}(E,\theta)$ is the probability with which a positron with energy E and angle θ to the axis stops inside the source foil with cutoff energy E_{c1} when it *passes through the foil twice*.

The number of positrons that will be moderated is given by

$$n_3 = \int\int g_1(E,\theta) \eta_{Ne}(E,\theta) dE d\theta, \qquad (13)$$

where $\int\int g_1(E,\theta) dE d\theta = n_2$ is the number of positrons that are available for extraction after slowing down, and $\eta_{Ne}(E,\theta)$ is the solid Neon *moderator* efficiency as a function of implantation energy at implantation angle θ. Since the angular dependence in $\eta(E,\theta)$ is not known we conservatively neglect the angular dependence of the positrons impinging onto the Neon moderator. Equation (13) can then be written in the form $n_3 = \eta_{Ne}(\bar{E},0) n_2$.

References

[1] W.B. Waeber et al, this proceedings.

[2] P.J. Schultz and K.G. Lynn, Rev. Mod. Phys. **60** (1988) 701.

[3] K.F. Canter and A.P. Mills, Jr., Can. J. Phys. **60** (1982) 551.

[4] A. Dupasquier and A. Zecca, Riv. Nuovo Cimento **8** (1985) 1.

[5] D. Taqqu, Helv. Phys. Acta **63** (1990) 442.

[6] W.B. Waeber, D. Taqqu, U. Zimmermann and G. Solt, PSI Report **68**, Mai 1990.

[7] J. Briesmeister ed. LANL Report LA-7396-M, Rev. 2 (1986)

[8] J.A. Halbleib, T.A. Mehlkorn, SANDIA Report, SAND84-0573 (1984).

[9] J. Briesmeister, J. Hendricks, *LANL MCNP4 Newsletter X–6:JFB–90–368* (1990).

[10] R. Nieminen, talk on 'Positron–Surface Interactions' at *Fifth International Workshop on Slow–Positron Beam Techniques for Solids and Surfaces*, August 6–10, 1992 in Jackson Hole, Wyoming, USA.

[11] Kjeld O. Jensen, Alison B. Walker, and Nadir Bouarissa in *Positron Beams for Solids and Surfaces*, P.J. Schultz, G.R. Massoumi and P.J. Simpson, Eds., AIP Conference Proceedings 218, American Institute of Physics, New York, 1990, p19.

[12] K.A. Ritley, M. McKeown, and K.G. Lynn in *Positron Beams for Solids and Surfaces*, P.J. Schultz, G.R. Massoumi and P.J. Simpson, Eds., AIP Conference Proceedings 218, American Institute of Physics, New York, 1990, p3.

[13] E_{c2} is identical to E_c in ref. [5]

[14] A. P. Mills, Jr. and E.M. Gullikson, *Appl. Phys. Lett.* **49** (1986) 1121.

[15] T.G. Northrop, *The Adiabatic Motion of Charged Particles*, John Wiley & Sons, New York, 1963.

A Low Energy Positron Flux Generator for Microstructural Characterization of Thin Polymer Films

J. J. Singh, A. Eftekhari and T. L. St. Clair
NASA Langley Research Center
Hampton, VA 23681

ABSTRACT

A low energy positron flux generator using well-annealed polycrystalline moderators and a Na^{22} positron source has been developed for microstructural characterization of thin polymer films. A 200 μc Na^{22} source, deposited on a thin (2.54 μm) aluminized mylar film, is sandwiched between two 0.0127cm x 2.54cm x 2.54cm tungsten strips. Two identical test polymer films, whose thicknesses may range from 0.001 to 0.01 cm, insulate the two tungsten moderator strips from the aluminized source film. A potential difference of 10 - 100 volts, depending on the test film thickness, is applied between the tungsten strips and the aluminized source film. Thermalized positrons diffusing out of the moderator strips are attracted to, or repelled from, the source foil depending on the polarity of the potential difference between the moderator strips and the source foil. Thus, more positrons are expected to stop/annihilate in the test films when the source is at a negative potential than when it is at a positive potential. The difference in positron lifetime spectra with the source at $\mp$(V) volts with respect to the moderator is entirely due to the positrons annihilating in the test films. Two examples showing the usefulness of the new device for monitoring the morphology of thin polymer films are presented.

INTRODUCTION

Thin, high performance polymer films are finding widespread applications in industry, communication networks and aerospace research. Since the microstructure of these films determines how well they meet their theoretical promise, it is necessary to develop efficient techniques for providing reasonable fluxes of low energy positrons needed for investigating thin polymer films. We have developed[1] an efficient scheme for generating high fluxes of slow positrons. Figure-1 shows a schematic diagram of the slow positron generator. It takes advantage of the negative work function of positrons in tungsten. After entering the tungsten moderator strips, the Na^{22} positrons are quickly thermalized. Some of the surviving positrons diffuse back to the entrance surface from which they are emitted with an energy of ~3ev. These positrons can be injected into the test films, where they will eventually annihilate, by applying an appropriate negative potential difference between the moderator strips and the source film. Monte Carlo calculations indicate that majority of the back-diffusing positrons are ejected within 30 picoseconds of their time of injection into the moderator strips. Thus, the lifetime of the positrons injected into the test films after their emergence from the moderator strips can be referenced to their time of birth at the Na^{22} nucleus.

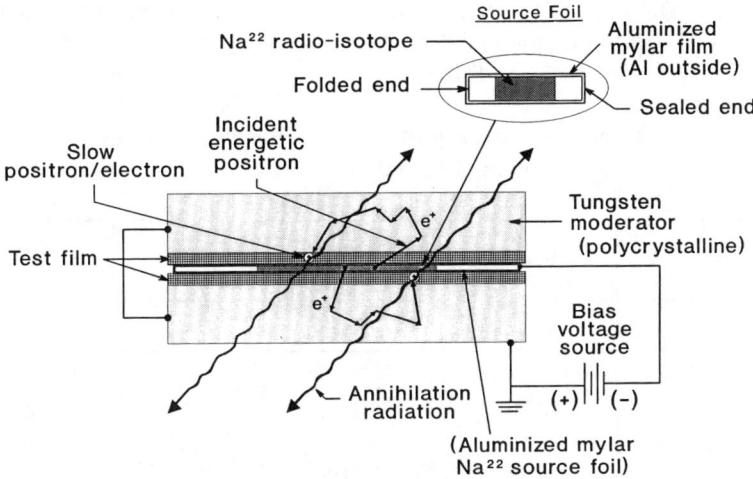

Figure 1. Schematic Diagram of the Moderator-Source Assembly.

Lifetime measurements have been made in several selected linear polymer films using the generator described above. The lifetime data have been used to calculate free volume cell sizes[2] and free volume fractions[3] in them. It has been noted that the free volume cell sizes in linear polymers depend on their molecular weights. The free volume fraction, on the other hand, correlates well with the dieletric constants of the test films. These results are described in the following sections.

EXPERIMENTAL PROCEDURE

Two types of poly(arylene ether ketone) films of various molecular weights and a series of special-purpose polyimide films developed for aerospace communication networks were selected for positron lifetime measurements in them. Details of their synthesis and characterization are described elsewhere[3,4].

Sample Preparation:
Thin(30 μm) films of Meta- and Para- poly(arylene ether ketone)s were prepared by casting them on glass plates from m-cresol solution. These films were air-dried for two days in a low humidity enclosure, then were gradually heated in an air-circulating oven to 250 °C, which is well above their glass

transition temperature(T_g). Finally, they were cooled rapidly to room temperature by removing the plates from the oven and placing them on a cool surface. Wide angle x-ray scattering showed the films to be free from any measurable crystallinity. The molecular weights(M_n) of various fractions, determined by gel permeation chromatography and low angle laser light scattering(GPC/LALLS) techniques are summarized in Table-I.

Table I. Summary of Physical Properties of the Poly(arylene ether ketone) Films

Sample	M_n (*) (g/mole)	Density (g/cm^3)	η_{inh} (dl/g)	T_g (°C)
PARA				
Whole Polymer	19.0x10^3	1.2058	0.50	-
Fraction A	47.0	1.2060	0.78	165
Fraction B1	44.5	1.2041	0.70	163
Fraction C	17.6	1.2052	0.37	159
Fraction D	16.3	1.2061	0.38	160
META				
Whole Polymer	20.8x10^3	1.2072	0.39	148
Fraction AB1	60.8	1.2077	0.65	153
Fraction C	19.8	1.2074	0.35	148

(*) Number-averaged molecular weights determined by GPC/LALLS techniques. These values have an error of ±10 percent.

The physical and electrical properties of the aromatic polyimide films used in this study are summarized in Table-II. The polyamic acid precursor solutions were prepared in closed vessels at ambient temperature by reacting stoichiometric amounts of diamine and dianhydride in dimethylacetamide at a concentration of 15-20 percent solids by weight. The resulting high molecular weight polyamic acid solutions were cast onto glass plates in a dust-free chamber at a relative humidity of 10 percent. The polyamic acid films were thermally converted into corresponding polyimide films by heating in a forced-air oven for one hour each at 100 °C, 200 °C and 300 °C.

Table II. Physical and Electrical Properties of Polyimide Films Tested

Sample	Density (g/cm^3)	Sat. Moist. Content, (Vol. %)	Dielectric Constant ε (At 10 GHz)
KAPTON (Reference)	1.431	2.02	3.20±0.03
BFDA+ODA	1.384	0.74	2.63±0.03
BFDA+4-BDAF	1.400	1.38	2.44±0.03
6FDA+DDSO$_2$	1.486	0.74	2.86±0.03
BFDA+DABTF	1.440	0.49	2.55±0.03
6FDA+APB	1.434	0.53	2.71±0.03
BTDA+ODA	1.380	1.21	3.15±0.03

<u>Positron Lifetime Measurements:</u>

Positron lifetimes in the test films were measured using the low energy positron flux generator described above. The lifetime data were acquired using a 200 μc Na^{22} positron source and a standard fast-fast coincidence measurement system with a time resolution of about 225 picoseconds. The lifetime spectra were analyzed using PAPLS[5] and POSFIT EXTENDED[6] programs. All measurements were made at room temperature and atmospheric pressure.

EXPERIMENTAL RESULTS

(a) Results in poly(arylene ether ketone) samples.

The positron lifetime results in Para- and Meta- PAEK films are summarized in Table-III. τ_1, I_1 and τ_2, I_2 refer to the lifetime and intensity of the first and second lifetime components, respectively. τ_3 and I_3 represent the lifetime and intensity of orthopositronium annihilations. The orthopositronium lifetime(τ_3) in nanoseconds is related to the free volume cell radius(R) in nanometers as follows[2]:

$$\frac{1}{2\tau_3} = (1 - \frac{R}{R_0} + \frac{1}{2\pi} sin2\pi \frac{R}{R_0}) \qquad (1)$$

where $R_0 = R + \Delta R$, $\Delta R = 0.1659$ nanometers

The free cell volume(V_f) is given by $(4/3)\pi R^3$. It is expected[7] to be a function of the molecular weight(M) of the test polymer. Figures 2 and 3 show V_f as a function of M for the Para-PAEK and Meta-PAEK samples, respectively. It is apparent from the data in these figures that V_f varies with M according to an equation of the form $V_f = AM^B$, where A and B are structural constants.

Table III. Summary of Positron Lifetime Parameters in Poly(arylene ether ketone) Films

Sample	$\tau_1(ps)/I_1(\%)$	$\tau_2(ps)/I_2(\%)$	$\tau_3(ps)/I_3(\%)$
PARA			
Whole Polymer	187±12 / 37±3	409±16 / 59±3	905±60 / 4±1
Fraction A	171±9 / 44±3	407±16 / 53±2	1069±70 / 3±1
Fraction B1	139±16 / 34±4	371±24 / 60±2	1037±70 / 6±1
Fraction C	196±8 / 41±2	416±20 / 56±3	869±75 / 3±1
Fraction D	165±17 / 33±5	390±24 / 63±3	850±65 / 4±1
META			
Whole Polymer	183±4 / 51±2	442±8 / 43±1	1698±40 / 6±1
Fraction AB1	211±3 / 59±1	516±23 / 36±1	2273±77 / 5±1
Fraction C	181±7 / 46±3	433±12 / 47±2	1651±52 / 7±1

The comparison between the molecular weights of the test samples calculated from $V_f = AM^B$ and measured by other techniques is summarized in Table-IV.

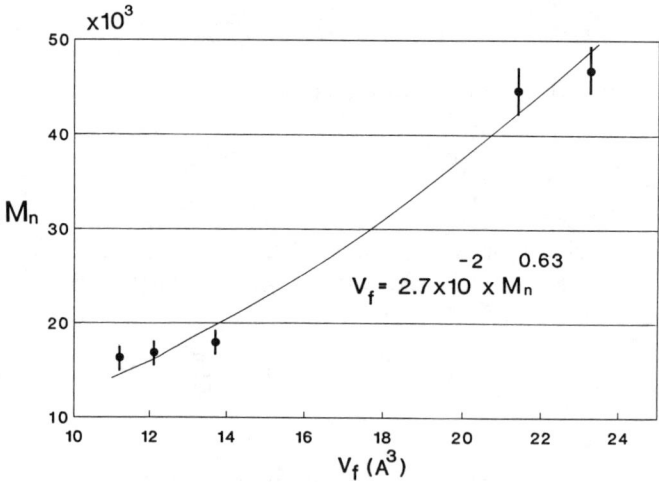

Figure-2. Free Volume Cell Size as a Function of Molecular Weight in Para-PAEK Samples.

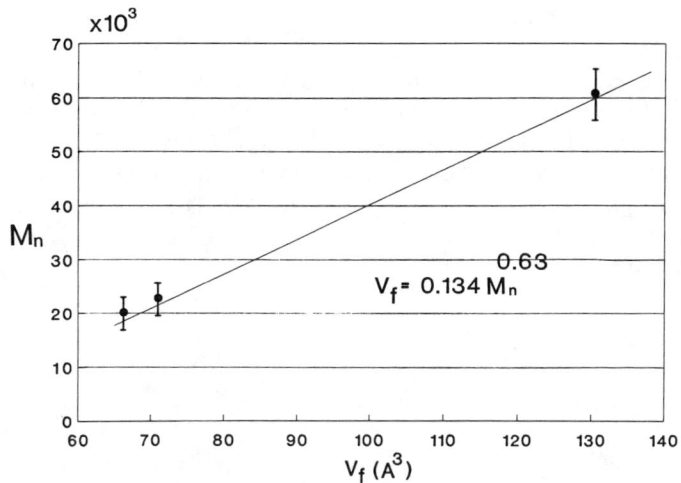

Figure-3. Free Volume Cell Size as a Function of Molecular Weight in Meta-PAEK Samples.

Table IV. Comparison Between the Measured and Computed Molecular Weights of Poly(arylene ether ketone) Films

Sample	$V_f(\text{Å}^3)$	$M_n^{(*)}$	$M_n^{(**)}$
PARA			
Whole Polymer	13.7±0.6	19.0 x 10^3	(21.3±1.5)x10^3
Fraction A	23.2±0.7	47.0	51.3±2.4
Fraction B1	21.3±0.6	44.5	43.2±1.9
Fraction C	12.1±0.8	17.6	17.5±1.9
Fraction D	11.2±0.8	16.3	15.5±2.0
META			
Whole Polymer	70.8±0.4	20.8 x 10^3	(22.8±0.2)x10^3
Fraction AB1	130.2±0.4	60.8	60.3±0.3
Fraction C	66.2±0.6	19.8	20.4±0.3

(*) Number-averaged molecular weights measured by GPC/LALLS
(**) Calculated from $V_f = AM^B$. Para: $A=2.7\times10^{-2}$, $B=0.63$; Meta: $A=1.34\times10^{-1}$, $B=0.63$.

(b) Results in Polyimide Films.

The positron lifetime values and the free volume cell sizes in the test polyimide films are summarized in Table-V. Also included in this table are the values of the free volume fraction(f) in the test films. Only two lifetime components were observed in these films. The longer lifetime component(τ_2) corresponds to the positrons trapped in the potential defects(free volume cells). The free volume cell radius(R) in nanometers and the trapped positron lifetime(τ_2) in nanoseconds are related as follows[3]:

$$\frac{1}{2.5\tau_2} = (1 - \frac{R}{R_0} + \frac{1}{2\pi}Sin2\pi\frac{R}{R_0}) \qquad (2)$$

Table V. Summary of Positron Lifetimes, Free Volume Cell Sizes and Free Volume Fractions in the Test Polyimide Films

Sample	Positron Lifetime Parameters		Free Volume Cell Size	Free Volume Fraction f(%)
	τ_1/I_1 (ps/%)	τ_2/I_2 (ps/%)	$V_f(Å^3)$	
KAPTON (Ref)	114±3/65.3±1.0	471±6/34.7±1.0	1.54±0.19	2.02±0.27
BFDA+ODA	223±4/73.2±1.0	699±17/26.8±1.0	12.26±1.10	12.43±1.09
BFDA+4-BDAF	131±2/70.8±1.0	790±12/28.1±1.0	18.36±0.85	19.56±0.91
6FDA+DDSO$_2$	170±3/72.5±1.0	623±14/27.5±1.0	7.94±0.75	8.26±0.79
BFDA+DABTF	135±4/68.6±1.0	653±15/31.4±1.0	9.57±0.96	11.41±1.14
6FDA+APB	254±4/86.2±1.0	867±39/13.8±1.0	23.93±2.83	12.5±1.48
BTDA+ODA	124±3/60.9±1.0	531±7/39.1±1.0	3.7±0.25	5.48±0.37

Equation-2 differs slightly from equation-1. This difference has been dictated by the observation that 400 picoseconds is the minimum observed trapped positron lifetime in all the polyimide films studied in this laboratory. The free volume fraction(f) in the test films were calculated as follows[3]:

$$f = CI_2 V_f \qquad (3)$$

where C is a structural constant and I_2 is the intensity of the trapped positron lifetime component.

The structural constant(C) was calculated by equating free volume fraction in Kapton(reference) with its saturation moisture fraction by volume. It has been assumed that the structural constant C has the same value for all the test polyimide films. We have calculated the effect of free volume fraction(f) on the dielectric constant(ε) of the test films by an expression of the following form[3]:

$$\frac{1}{\varepsilon} = \frac{1-f}{\varepsilon_R} + \frac{f}{\varepsilon_{vac}} \tag{4}$$

The value of ε_R, which corresponds to the value of ε for zero free volume fraction, was obtained from ε vs f data illustrated in figure-4. It has been found to be 3.55. The measured and calculated values of the dielectric constants for various test films are summarized in Table-VI.

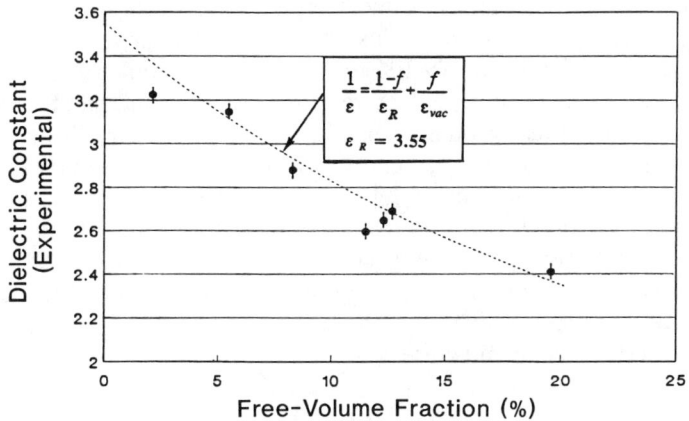

Figure-4. Experimental Dielectric Constant VS. Free-Volume Fraction in Thin Test Films.

Table VI. Comparison Between the Experimental and
Calculated Values of the Dielectric Constants

Sample	f(%)	ε (expt)	ε (calc)
KAPTON (Ref)	2.02±0.27	3.20±0.03	3.31±0.03
BFDA+ODA	12.43±1.09	2.63±0.03	2.69±0.05
BFDA+4-BDAF	19.56±0.91	2.44±0.03	2.37±0.04
6FDA+DDSO$_2$	8.26±0.79	2.86±0.03	2.93±0.05
BFDA+DABTF	11.41±1.14	2.55±0.03	2.74±0.06
6FDA+APB	12.5±1.48	2.71±0.03	2.70±0.07
BTDA+ODA	5.48±0.37	3.15±0.03	3.12±0.03

DISCUSSION

From the data summarized in Table-IV, and illustrated in figures 2 and 3, it appears that V_f values track the M_n values of the test polymers quite well. As expected, a unique set of A and B values predict the test polymer M_n values as long as their molecular structure remains unchanged. A change in the molecular structure(as one goes from Para-PAEK to Meta-PAEK geometry) necessitates a different set of constants. A notable feature of the results is the fact that V_f values are much larger in the Meta geometry than in the Para geometry. This, however, is quite consistent with different packing behavior of the two geometries.

From the data summarized in Table-VI and illustrated in figure-4, it is obvious that the free volume fractions of the test films strongly influence their dielectric constants. The test films with the largest free volume fractions have the lowest values of the dielectric constants as suggested by equation-4. The large free volumes in these samples are due to the presence of Meta-linkages and bulky CF_3 groups in their molecular architecture.

CONCLUSIONS

The Langley slow positron flux generator has been successfully applied to measure positron lifetime values in thin polymer films. The lifetime data indicate that positron lifetime spectroscopy can provide a reasonably accurate technique for measuring molecular weights of linear polymers. It also provides a sensitive technique for characterizing thin polyimide films in terms of their dielectric constants. The presence of bulky CF_3 groups and Meta linkages in their structure increase their inter-molecular spacing resulting in high free volume fraction and, consequently, lower dielectric constant.

REFERENCES

1- J.J. Singh, A. Eftekhari and T.L. St. Clair., Nucl. Instr. and Meth. B53, 342, 1991.
2- H. Nakanishi and Y.C. Jean., in: Positron and Positronium Chemistry, eds. D.M. Schrader and Y.C. Jean(Elsevier, Amsterdam, 1988), p.159.
3- A. Eftekhari, A.K. St. Clair, D.M. Stoakley and J.J. Singh., Polym. Matl. Scin. and Engg. 66, 279, 1992.
4- J.A. Hinkley, R.A. Crook and J.R.J. Davis., High Perf. Polymers. 1, 61, 1989.
5- J.J. Singh, G.H. Mall and D.R. Sprinkle., NASA Technical Paper TP-2853, 1988.
6- P. Kirkegaard, M. Eldrup, O.E. Mogensen and N.J. Pedersen., Comput. Phys. Commun. 23, 307, 1981.
7- J.J. Singh and A. Eftekhari., Nucl. Instr. and Meth. B63, 477, 1992.

AN INTENSE PULSED POSITRON BEAM AND ITS APPLICATIONS

R. Suzuki, T. Mikado, H. Ohgaki, M. Chiwaki and T. Yamazaki
Electrotechnical Laboratory, 1-1-4 Umezono, Tsukuba, Ibaraki 305, Japan

Y. Kobayashi
National Chemical Laboratory for Industry, 1-1 Higashi, Tsukuba,
Ibaraki 305, Japan

ABSTRACT

A positron pulsing system for variable-energy positron lifetime spectroscopy and time-of-flight measurements is reported. The system generates high-intensity pulsed positron beam of variable energy from an intense positron beam produced by an electron linac. The pulsed beam enables us to perform positron lifetime measurements on surfaces or near surface regions with a high resolution, high count rate, high peak-to-background ratio, and wide time range. Preliminary experimental results of age-momentum correlation spectroscopy, time-of-flight energy analyses of positronium atoms and Auger electrons emitted from surfaces is also presented.

§1. INTRODUCTION

The possibility of using slow positron beams to study surface and near surface phenomena has been demonstrated by a number of researches.[1-3] For some measurements with slow positrons, such as positron lifetime spectroscopy, time-of-flight (TOF) measurements of positronium or secondary emitted particles, *etc*, it has been necessary to obtain the precise time at which a positron has entered a sample as a start timing signal. Several techniques to obtain the start signal have been developed (see reviews in ref.1 and 2) and many interesting experiments with these techniques have been carried out.[2,4] In particular, some results of positron lifetime measurements with pulsed positron beams have shown the great potential of variable-energy pulsed positron beams for surface and near-surface studies.[5-10] For the positron lifetime spectroscopy, high resolution, high peak-to-background ratio and wide measurable time range have been desired. However, because the lifetime spectroscopy with a pulsed positron beam is not a coincidence technique, it is difficult to distinguish between annihilation γ-ray signal and background signal whose pulse height is close to that of annihilation γ-rays. Thus, to obtain high quality positron lifetime spectra, an intense pulsed positron beam has been required.

In 1987, a facility to produce an intense slow positron beam was constructed at the Electrotechnical Laboratory (ETL).[11] We have developed a pulsing system for the intense positron beam to perform variable-energy positron lifetime spectroscopy (VEPLS) and TOF experiments with high peak-to-background ratio, high count rate, and wide measurable time range.[12]

In this paper, we report the pulsing system, applications of the pulsed positron beam—variable-energy positron lifetime spectroscopy, positronium TOF spectroscopy, and positron annihilation induced Auger electron spectroscopy.

§2. POSITRON PULSING SYSTEM

An intense positron beam of about 10^7 e$^+$/s is produced with a high energy (~70 MeV) electron beam of the ETL linac.[11] The positron beam is initially a pulsed beam corresponding to the linac pulse. However, the repetition rate of the linac is too low (50 pps - 100 pps) and the pulse width is too wide (1 μs - 4 μs) for high count-rate positron lifetime and TOF experiments. Thus, the positron beam is stored temporarily in the linear storage section,[13] and the pulse-stretched beam is used for the pulsing system. The pulsing system consists of a chopper, a sub-harmonic pre-buncher (SHPB) and a double harmonic buncher as shown in Fig. 1. An axial magnetic field of ~0.01 T is applied to the pulsing system by Helmholtz coils for beam guiding. Details of the system are reported in ref. 14.

The chopper generates a pulsed beam (~5 ns, ~250 eV) from the pulse-stretched beam. The chopper consists of three grids, and utilizes a pulse electric field longitudinal to the beam direction. The main feature of this chopper is that the performance is not affected by the strength of the axial magnetic field; thus, it can be operated at any frequency by applying appropriate electric pulses. The SHPB compresses the pulse width of the chopped beam to ~1 ns (FWHM). This beam can be used for time-of-flight experiments. For the positron lifetime experiments, the positron pulse is further compressed to ~150 ps at the sample position by the buncher. The fundamental frequency of the buncher, f_B, is 150 MHz; the operating frequency of the SHPB is a quarter of the buncher frequency, and the chopper frequency is $f_B/(4 \cdot n)$, ($n=1$, 2, ...). There are accelerating rings and a drift tube between the buncher and the sample. By adjusting the applied voltages to the drift tube and the accelerating rings, the incident positron energy at the sample can be varied from 0.5 keV to 30 keV. As a result, the positron pulsing system generates a pulsed beam of high intensity, variable energy, and variable pulse period.

Figure 2 shows a block diagram of the electronic system. A signal generator (SG; Panasonic VP8300A) generates a fundamental rf signal which matches the first resonant frequency of the buncher cavity. The signal is split to two signals: one is fed to the buncher control unit; the other is fed to the timing pulse generator and the SHPB control unit through a $f/4$ frequency divider. The buncher control unit consists of a signal divider, a harmonic generator, phase shifters, variable attenuators, and a combiner. The output signal of the buncher control unit is amplified and coupled into the buncher. The amplitude and phase of the buncher pick-up signal are monitored by a vector volt meter (VVM; HP 8508A). The SHPB control unit, which consists of a filter, a phase shifter, a variable attenuator and amplifiers, generates a sinusoidal wave of $f_B/4$. The outputs of the timing pulse generator are connected to a time-to-amplitude converter (TAC; Ortec model 567) and a chopper control unit. The outputs of the SHPB control unit and the chopper pulse generator are fed to the SHPB and the chopper respectively. The timing signal of the linac is used for the gate signal of the TAC in order to cut initial burst and to reduce background radiation noises.

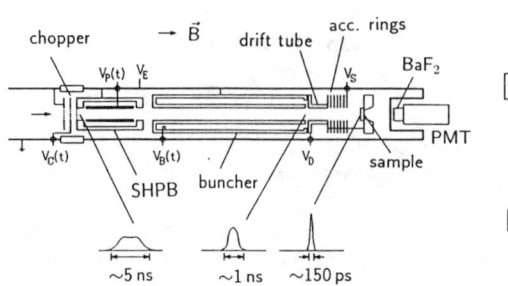
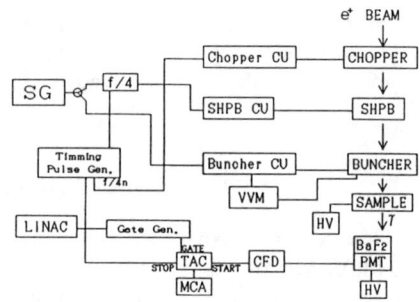

Fig. 1. Schematics of the positron pulsing system.

Fig. 2. Block diagram of the electronic system.

§3. APPLICATIONS OF THE PULSED POSITRON BEAM

3.1 POSITRON LIFETIME MEASUREMENT

A positron lifetime spectrum is obtained by measuring the time interval between the timing signal of the pulsing system and timing signal of an annihilation γ-ray detected with a BaF$_2$ scintillation detector. The γ-ray signal is fed into the start input of the TAC and the signal of the pulsing system is fed into the stop input in order to reduce the dead time. At present time, we can measure positron lifetime spectra with a count rate of 100-500 cps and with the time resolution of ~250 ps. A peak-to-background ratio of 2000-6000 has been attained by adjusting the storage time of the linear storage section and the gate timing for a time-to-amplitude converter. The high peak-to background ratio, wide measurable time range and high time resolution make it possible to measure both long- and short-lived components in near surface regions, surfaces, interfaces, and thin films. Several experimental results using the pulsed positron beam have proved usefulness of the variable-energy positron lifetime spectroscopy.[7-10] Some positron lifetime studies on thin films are presented in another contribution.[10]

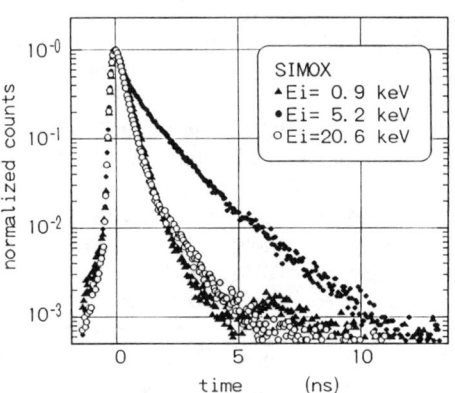

Fig. 3. Positron lifetime spectra of SIMOX.

Figure 3 shows an example of positron lifetime experiments. The specimen is a SIMOX (separation by implantation of oxygen) sample obtained by oxygen ion implantation (200 keV, 2.2×10^{18}/cm^2) and annealing. The lifetime spectrum is drastically changed as a function of incident positron energy. The spectra of 0.9, 5.2, 20.6 keV, corresponding to the mean implantation depths of 15 nm, 250 nm, 2.3 μm respectively, are shown in Fig. 3. The second peak of the 0.9 keV spectrum at around 7 ns is due to the backscattered positrons. At 5.2 keV, we found long-lived (~ 1.5 ns) component which can be attributed to o-Ps formed at free volume in SiO_2 network. This spectrum is quite similar to that of thermally grown SiO_2 film.[8] At 0.9 keV and 20.6 keV, the intensity of the long-lived component is low because most of positrons would be annihilate in the Si layer or the Si substrate. The lifetime spectrum is also depend on both the implantation and annealing conditions. Detailed results and discussion of the SIMOX samples is presented in another contribution.[15]

3.2 AGE-MOMENTUM CORRELATION SPECTROSCOPY

Age-momentum correlation spectroscopy (AMOCS) is known to provide information which is unavailable from individual measurement of age (positron lifetime spectroscopy) and/or that of momentum (measurement of Doppler broadening or angular correlation of annihilation radiation). Several results proved that the AMOCS is especially useful to study annihilation process of positronium.[16-19] Conventional AMOCS system requires triple coincidence technique for the start β or γ ray signal and two annihilation γ-ray signals. Thus, the maximum count rate is limited to about 100 cps. On the contrary, AMOCS system with a pulsed positron beam requires only double coincidence technique, because the start signal can be obtained electronically from the pulsing system. Thus, it is possible to perform high count rate AMOCS. Furthermore, we can obtain AMOCS data of a desired depth because we can control the incident positron energy. Therefore, we have tried age-momentum correlation measurement using the variable-energy pulsed positron beam.

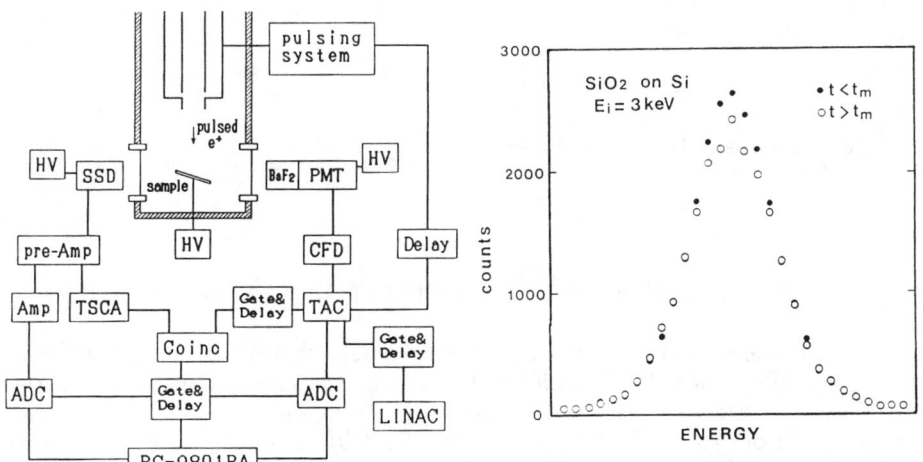

Fig. 4. Experimental set up of the AMOCS system.

Fig. 5. Doppler broadening spectra of short age region and long age region.

Figure 4 shows the experimental setup of the AMOCS system. The annihilation γ-ray energy and annihilation timing are measured by a solid state detector (SSD) and BaF$_2$ scintillation detector respectively. The distance between the SSD and scintillation detector is 20 cm. The spectrum is taken by a two-dimensional multichannel analyser which consists of two analogue-to-digital converters (ADC) and a personal computer (NEC:PC9801RA). Figure 5 shows a preliminary result of AMOCS—Doppler broadening spectra of short age region and of long age region. These spectra indicate that there are low momentum and short lifetime component, and relatively high momentum and long-lived component. This suggests that the intensity of positronium components is high, and that the o-Ps atoms annihilated by pick-off process. The count rate of ~10 cps was achieved with the pulsed beam of the ETL. The present count rate is, however, insufficient for actual experiments because the total coincidence counts should be more than 10^6 counts for AMOCS but the beam time of the linac is limited to a few hours. If a higher-intensity positron beam is available, the count rate can be easily increased.

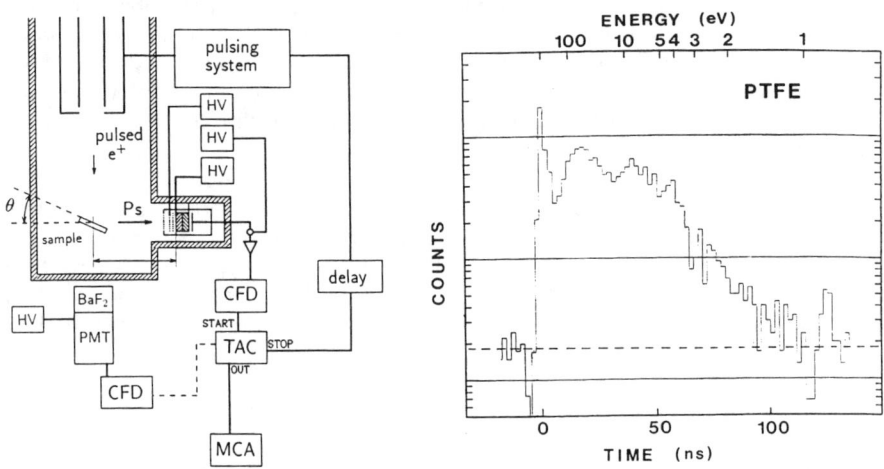

Fig. 6. Experimental set up of the Ps-TOF system.

Fig. 7. Ps-TOF spectrum of PTFE.

3.3 POSITRONIUM TIME-OF-FLIGHT SPECTROSCOPY

Several experimental and theoretical results have shown that energy analysis of positronium (Ps) emitted from solid surfaces gives valuable information on surface properties.[20–23] We can measure the energy spectrum with a TOF technique using a pulsed positron beam. Thus, we have tried Ps-TOF spectroscopy with an experimental setup as shown in Fig. 6. Positronium atoms are detected by a single anode micro-channel-plate (MCP) detector. TOF spectra are obtained by measuring the time interval between the signal of MCP detector and that of the pulsing system. Figure 7 shows a Ps-TOF spectrum of polytetrafluoroethylene (PTFE) at an incident positron energy of 200 eV. A peak at around 0 ns is due to γ-rays of positron annihilation at the sample. In Fig. 7, the counts of Ps atoms are suddenly decreased from around 50 ns (~5 eV) and no significant Ps signal was observed at more than 90 ns (less than 1.5 eV). The variable-energy positron lifetime experiment on PTFE revealed that significant energy dependence of incident positrons was not observed in the energy range of 0.5 keV and 25 keV,[12] in contrast to the case of other polymers, e.g., polyethylene[8] and EPOXY[24]. These results suggest that there is no or very small probability of emmision of thermalized positronium from the PTFE surface. Meanwhile, positronium at the high energy region, which would be non-thermalized positronium, was clearly observed. This indicates a realization possibility of the positronium formation spectroscopy proposed by Ishii.[25]

3.4 POSITRON ANNIHILATION INDUCED AUGER ELECTRON SPECTROSCOPY

Positron annihilation-induced Auger electron spectroscopy (PAES) is known to have significant advantages over conventional Auger electron spectroscopy (AES).[26] In PAES, core electrons are removed by matter-antimatter annihilation. Therefore, the incident beam energy in PAES need not be larger than the core level ionization energy, permitting the elimination of the large secondary electron background. Furthermore, the surface sensitivity of this technique is extremely high[27] since the positrons are trapped in an image correlation well just outside the surface. Several works using a position sensitive or single slit energy analyzer have demonstrated the possibility of PAES.[26-31] However, the energy resolution of these systems is lower than conventional AES since the resolution is mainly limited by the beam diameter.

Thus, we have tried PAES experiment with a TOF technique to increase energy resolution of the spectrum. Figure 8 shows the experimental set up of TOF-PAES. An axial magnetic field of 0.007 T is applied to the system for beam guiding. A Nd-Fe-B permanent magnet placed behind the sample is used to minimize their angular divergence.[29] The electron detector and $\vec{E} \times \vec{B}$ deflection plate are placed between the sample and pulsing devices. The distance between the sample and detector is 70 cm. Electrons are detected by a single anode MCP detector which has three mesh grids in front of the MCP to cut both positive particles and low energy negative particles. TOF spectrum was obtained by measuring the time interval between the signal of pulsing system and signal of MCP detector with a pulsed positron beam. There is another TOF technique which uses a continuous positron beam and which detects annihilation γ-rays as a start signal. We tested both techniques and found that the TOF technique with a pulsed positron beam provides lower background noise in the case of our system.

Figure 9 shows a PAES spectrum of highly oriented pyrolytic graphite (HOPG), obtained with the pulsed positron beam. A significant peak at around 250 eV corresponding to Auger peak of carbon was observed. This measurement is preliminary, nevertheless, the resolution of carbon Auger peak is significantly higher than that of previously reported Auger peak of the PAES experiment.[30]

The present count rate of the Ps-TOF system and the TOF-PAES system is insufficient for high statistic measurement. One reason of this is that the pulsing efficiency of the present system, which is optimized for lifetime measurements but not for TOF measurement, is low (a few %). Therefore, if we construct an optimized pulsing system for TOF experiments, the pulsing efficiency will increase about one order of magnitude.

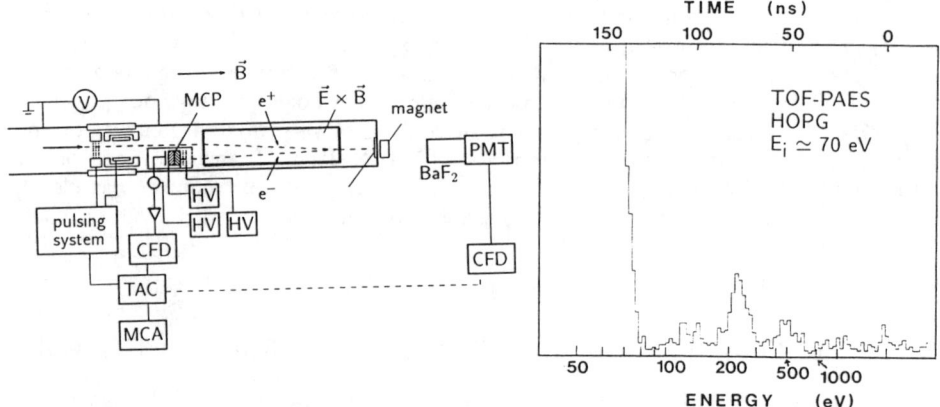

Fig. 8. Experimental set up of the PAES system.

Fig. 9. PAES spectrum of HOPG.

§4. CONCLUSION

The slow positron pulsing system at the ETL positron facility and application examples of the pulsed positron beam were described. We demonstrated that it is possible to carry out variable-energy positron lifetime spectroscopy, age-momentum correlation spectroscopy, Ps-TOF and TOF-PAES measurements.

The main feature of measurements using a pulsed positron beam is that there is almost no limitation on the count rate since the start timing signal can be obtained electronically from the pulsing system. Therefore, we can obtain data of these measurements with high statistics in a short-time scale when higher intensity positron beams, e.g., 10^9-10^{12} e$^+$/s, become available. Then, these measurements will be a powerfull tool to study dynamical processes at surfaces or in near surface regions.

The authors would like to thank Drs. T. Noguchi, S. Sugiyama, K. Yamada, A. Uedono, S. Tanigawa, Y.C. Jean, T. Azuma, H. Saito, K. Kawashima and T. Hyodo for discussions and collaboration in some of the experiments.

REFERENCES

1. A.P. Mills Jr, *Proc. Int. School of Physics 'Enrico Fermi' Course LXXXIII* (North-Holland, Amsterdam, 1983), p.432.

2. P. J. Schultz and K. G. Lynn, Rev. Mod. Phys. 60, 701 (1988).

3. A. Vehanen, Positron Annihilation, eds L. Dorikens-Vanpraet, M. Dorikens and D. Segers, (World Scientific, Singapore, 1989) p. 39.

4. for example, Positron Annihilation, Zs. Kajcsos and Cs. Szeles ed., Materials Science Forum, 105-110 (1992).

5. G. Kögel, D. Schödlbauer, W. Triftshäuser and J. Winter, Phys. Rev. Lett. 60 (1988) 1550; J. Nucl. Mater. 162-164, 876 (1989).

6. R. Steindl, G. Kögel, P. Sperr, P. Willutzki, D. T. Britton, W. Triftshäuser, Materials Science Forum, 105-110, 1455 (1992).

7. R. Suzuki, Y. Kobayashi, T. Mikado, A. Matsuda, P. J. McElheny, S. Mashima, H. Ohgaki, M. Chiwaki, T. Yamazaki and T. Tomimasu,Jpn. J. Appl. Phys., 30, 2438 (1991).

8. R. Suzuki, Y. Kobayashi, T. Mikado, H. Ohgaki, M. Chiwaki, T. Yamazaki and T. Tomimasu, Materials Science Forum, 105-110, 1159 (1992).

9. R. Suzuki, Y. Kobayashi, T. Mikado, H. Ohgaki, M. Chiwaki, T. Yamazaki, A. Uedono, S. Tanigawa and H. Funamoto, Jpn. J. Appl. Phys., 31, 2237 (1992).

10. R. Suzuki, T. Mikado, H. Ohgaki, M. Chiwaki, T. Yamazaki, K. Awazu, A. Matsuda, Y. Kobayashi, A. Uedono and S. Tanigawa, in this conference.

11. T. Mikado, R. Suzuki, M. Chiwaki, T. Yamazaki, N. Hayashi, T. Tomimasu, S. Tanigawa, T. Chiba, T. Akahane and N. Shiotani, Proc. 3rd Japan-China Joint Symp. on Accelerators for Nuclear Science and Their Applications (Riken, Wako, 1987) p. 163.

12. R. Suzuki, Y. Kobayashi, T. Mikado, H. Ohgaki, M. Chiwaki, T. Yamazaki and T. Tomimasu, Jpn. J. Appl. Phys. 30, L532 (1991).

13. T. Akahane, T. Chiba, N. Shiotani, S. Tanigawa, T. Mikado, R. Suzuki, M. Chiwaki, T. Yamazaki, and T. Tomimasu, Appl. Phys. A 51, 146 (1990).

14. R. Suzuki, Y. Kobayashi, T. Mikado, H. Ohgaki, M. Chiwaki, T. Yamazaki and T. Tomimasu, Proc. Int. Conf. Evolution in Beam Applications, (Radiation Application Development Association, Tokyo) p.357, 1992.

15. A. Uedono, S. Watauchi, Y. Ujihira, L. Wei, S. Tanigawa, R. Suzuki, H. Ohgaki, T. Mikado, H. Kametani, H. Akiyama, Y. Yamaguchi and M. Koumaru, in this conference.

16. F.H.H. Hsu and C.S. Wu, Phys. Rev. Lett., 18, 889 (1967).

17. I.K. MacKenzie and P. Sen, Phys. Rev. Lett, 37, 1296 (1976).

18. Y. Kishimoto and S. Tanigawa, Positron Annihilation, eds P.G. Coleman, S.C. Sharma, L.M. Diana, (North-Holland, Amsterdam, 1982) p. 790.

19. T. Chang, M. Xu and X. Zeng, Phys. Rev. Lett. A, 126, 189 (1987).

20. A. Ishii, J.B. Pendry and D.K. Saldin, Positron Annihilation, eds L. Dorikens-Vanpraet, M. Dorikens and D. Segers, (World Scientific, Singapore, 1989) p. 326.

21. A.P. Mills, Jr., L. Pfeiffer, Phys. Rev. B **32**, 53 (1985).
22. R.H. Howell, P. Meyer, I.J. Rosenberg and M.J. Fluss, Phys. Rev. Lett. **54**, 1698 (1985).
23. P. Sferlazzo, S. Berko, and K.F. Canter, Phys. Rev. B **35**, 5315 (1987).
24. Y.C. Jean, G.H Dai, H. Shi, R. Suzuki, Y. Kobayashi, in this conference.
25. A. Ishii, in this conference.
26. A. Weiss, R. Mayer, M. Jibaly, C. Lei, D. Mehl, and K.G. Lynn, Phys. Rev. Lett. **61**, 2245 (1988).
27. A. Weiss, N. Fazleev, J.H. Kim, K.O. Jensen, A.R. Koymen, K.H. Lee, L.W. Tyan, G. Yang, S. Yang and H.Q. Zhou, in this conference.
28. R. Mayer, A. Schwab, A. Weiss, Phys. Rev. B **42**, 1881 (1990).
29. C. Lei, D. Mehl, A. R. Koymen, F. Gotwald, M. Jibaly and A. Weiss, Rev. Sci. Instrum. **60**, 3656 (1989).
30. E. Soininen, A. Schwab, and K.G. Lynn, Phys. Rev. B **43**, 10051 (1991).
31. A. Weiss, Materials Science Forum, **105-110**, 511 (1992).

DEVELOPMENT OF THE SLOW POSITRON BEAM SYSTEM AS A COMMERCIAL PROTOTYPE

Hikaru Ueno, Osamu Azuma, Takafumi Ogawa, Takashi Sato, Tatsumi Kawaratani and Okitada Hara

Mechatronics Development Center,
Ishikawajima-Harima Heavy Industries Co.,Ltd.
1-15, Toyosu 3-Chome, Koto-ku, Tokyo 135, JAPAN

ABSTRACT

In this paper, the slow positron beam system which is developed in IHI(Ishikawajima-Harima Heavy Industries Co., Ltd.) for measuring S-parameter will be introduced. This system has magnetic coils and the ExB energy filter. The positron source is ^{22}Na, and the activity is 3GBq at the maximum. The accelerating voltage is 50kV at the maximum. The radiation shield is composed of lead tightly covered by the stainless steel container around the vacuum chamber. And this container can be easily moved by external pressurized slider gas bearings.

INTRODUCTION

Recently in the field of the material science, the use of the slow positron beam is paid attention to, thanks to many workers in the world who work about lattice defects[1], surfaces or interfaces[2] and microprobe[3] using the slow positron[4]. The availability of the slow positron beam begins to be recognized by degrees. Also in Japan the analysis of material surface using the positron beam recently begins to attract many semiconductor companies in accordance with the increment of integrated circuit density. IHI has been developing the slow positron beam system as a commercial prototype and as a machine for our own use from April, 1991.

DESCRIPTION OF THE SYSTEM

The sketch of this system is shown in Fig. 1. The dimensions of the system are 3.8m long, 1.43m wide and 1.63m high. The weight is about 4tons, including the weight of shields.

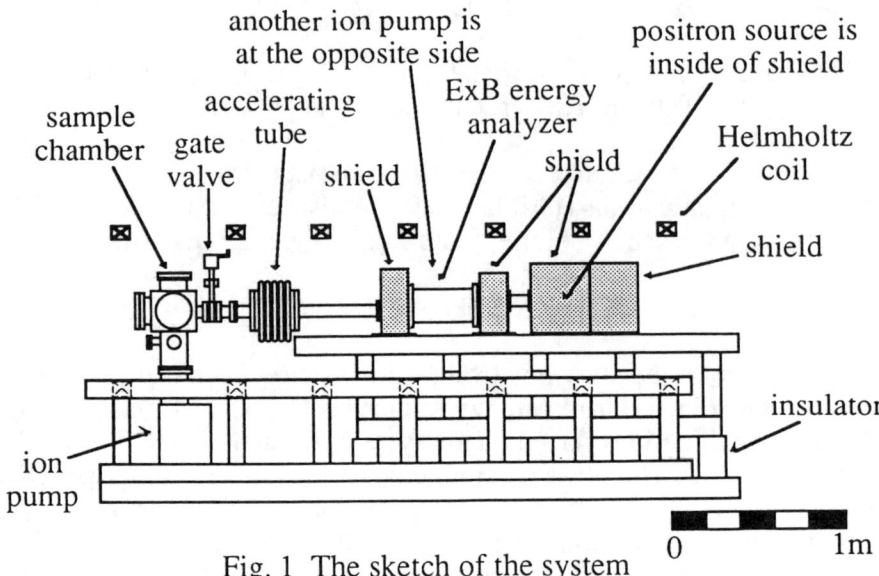

Fig. 1 The sketch of the system

In the following, main parts are described.

SOURCE PART

The source part consists of the source, moderator[5], grid, shields and vacuum chamber. The moderator is made of tungsten ribbons. The width of these ribbons is 5mm, and thickness is 25.4μm. These are annealed at 2200°C for 2 hours in vacuum(2.7x10^{-5}Pa). Using the 3.07MBq ^{22}Na source which is gamma reference source, the count of the positrons are 10 e+/s(the count of the γ-rays are 0.12cps.) at the sample. When the efficiency of the moderator is defined by Eq. (1), the efficiency is about 3.3x10^{-6}.

$$\text{Efficiency} = \frac{e^+ \text{ count at the sample}}{\text{source activity}} \quad (1)$$

This efficiency is quite low, comparing to other systems. The reasons of the low efficiency are that the distance between the source and the moderator was long, about 10mm, and that the only 30% positrons went out from this source because of 0.1mm aluminum window. The 3.07MBq ^{22}Na source is just for system adjustment and the source will be replaced on the next stage.

The grid is made of tungsten mesh which has 100 meshs per one square inch. Slow positrons are extracted from the moderator by 200V between the grid and the moderator. The moderator is insulated by kapton film between the grid and source. The Fig. 2 shows source, moderator and grid.

The source is ^{22}Na. The 3GBq ^{22}Na can be used at the maximum, which is limited by the shield thickness. At the present we use the 3.07MBq ^{22}Na for the system adjustment.

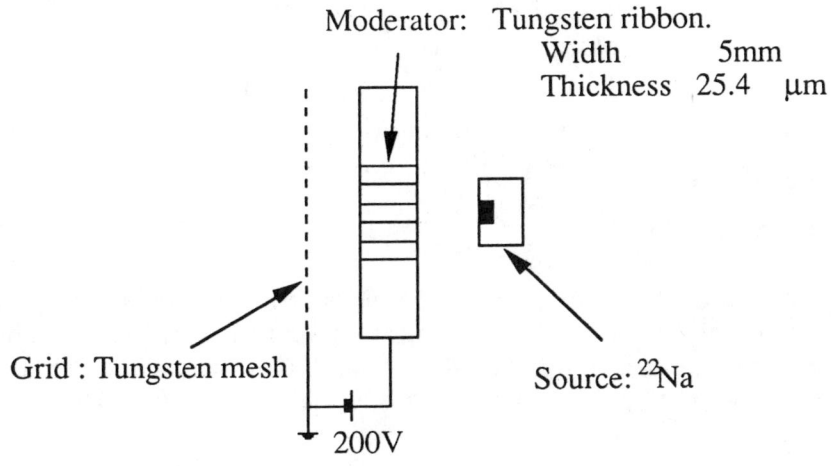

Fig. 2. Source and moderator

ExB ENERGY FILTER

The electrode of the ExB is the coaxial cylindrical structure[6]. The electric field is 14kV/m between the electrode when the positron beam energy is 200eV, and the magnetic field is 0.01T, which is generated by seven Helmholtz coils.

The positron beam is shifted downward by 5cm when the beam passes through this part(Fig. 3).

There are shields on both side of the ExB, which shut out the γ-rays from the source.

ACCELERATING TUBE

The monoenergetic positrons, which is 200eV, are accelerated to 50keV at the maximum by this accelerating tube. The method of the acceleration is that the upstream of the accelerating tube is raised up to

50kV at the maximum, and the downstream of the accelerating tube is always at ground level.

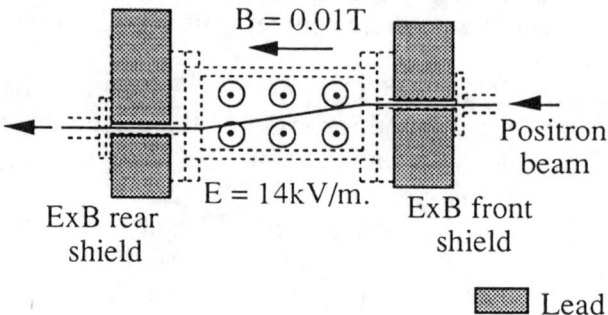

Fig. 3 ExB energy filter.
Beam shift width is 5cm.

VACUUM SYSTEM

There are 140 l/s ion pump and 150 l/s turbo molecular pump in the sample part, and there is 60 l/s ion pump in the ExB part(Fig. 4). The vacuum pressures are about 4×10^{-5}Pa at the ExB part and about 5×10^{-6}Pa at the sample part by these pumps, without baking.

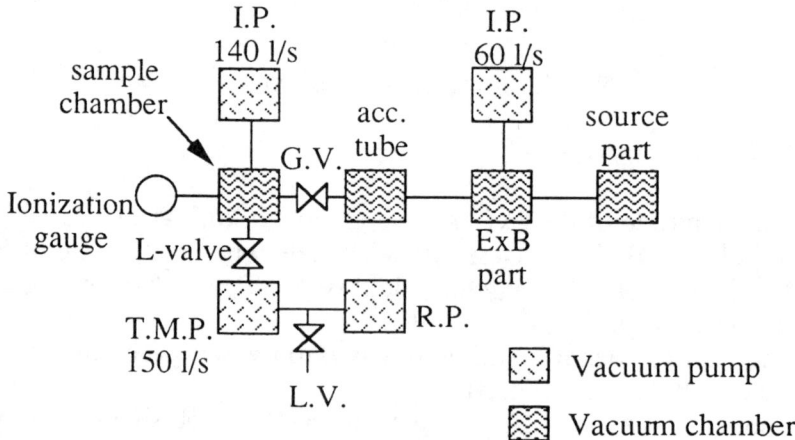

Fig. 4 Vacuum system

MAGNET

There are seven magnetic coils which diameter is 1m. The distance of the each coils is 67cm at the sample chamber and, 50cm at the other parts. The magnetic field is 0.01T at the center of the coils. The magnetomotive force of the coil is 4.5kA•turn.

SHIELDS

These are composed of lead tightly covered by the stainless steel container around the vacuum chamber. This can be easily moved by external pressurized slider gas bearings. So the shield behind the source part can be easily moved backward when the source or the moderator will be exchanged. Fig. 5 shows the shields for the γ-rays from the source and annihilation γ-rays. The dose equivalent is 0.03μSv/h at the coil using the 3.07MBq ^{22}Na. This value is same as background before carrying the source into the room.

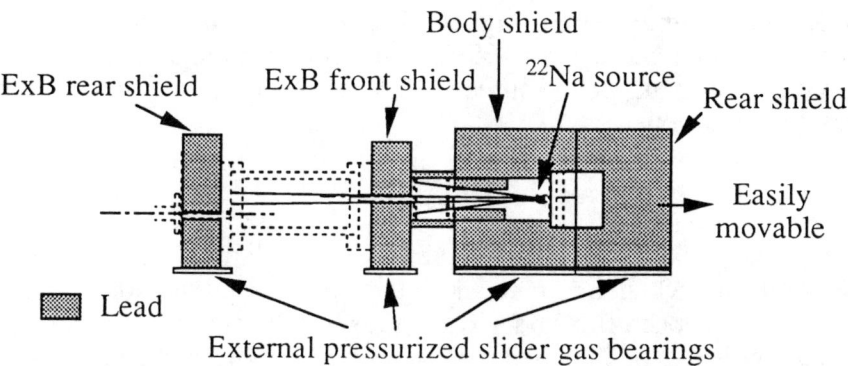

Fig. 5 γ-ray shield

GAS BEARINGS

These gas bearings are equiped beneath all four shields(Fig. 6). The four air intakes are located on the upper face of the bottom plate, there are four air rooms on the lower face of the bottom plate, which diameters are about φ50mm~φ100mm centering around the air intake, and the depth is 0.5mm. The flatness of the lower face of the bottom plate is 0.03mm, and the flatness of the table, which the shields slide on, is also about 0.03mm. The pressure of the air is about 3.92×10^5Pa.

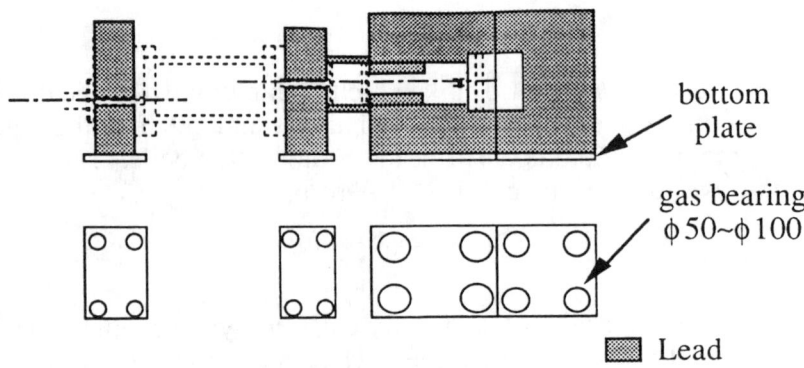

Fig. 6 External pressurized slider gas bearings

BEAM TRANSPORTING OPERATION OF THE SYSTEM

At the beam transporting operation, the electron gun was used instead of the RI(positron source). That is the reason why it is safer than RI, and beam current is higher. The result was that the transmission rate was almost 100%. After that, we are adjusting whole the system using the weak source, 3.07MBq ^{22}Na, in order to avoid the excessive radiation.

SUMMARY

At present, we have detected the annihilation γ-rays of the slow positrons, which are injected on the sample, and the 50kV high voltage test has been finished. At the present, the automatic controlling system is going to be installed for the power supplies and data acquisition system. After this automatic system, the source will be exchanged to 185MBq ^{22}Na, to which the moderator is matched(the distance is short, about 1.5mm, between the source and moderator.).

The addition of the life time measurement system, the improvement of the slow positron current, the down sizing of the system are scheduled in order.

ACKNOWLEDGMENT

The authors are greatly indebted to Dr.S.Tanigawa of University of Tsukuba for stimulating the development of the slow positron beam system.

REFERENCES

1. Shoichiro Tanigawa, Jpn. J. Appl. Phys. 50, 237 (1981)
2. B. Nielsen, K. G. Lynn, D. O. Welch and T. C. Leung, Phys. Rev. B 40, 1434.(1989)
3. I. J. Rosenberg, A. H. Weiss and K. F. Canter, Phys. Rev. Lett. 44, 1139 (1980)
4. P. J. Schultz and K. G. Lynn, Rev. Mod. Phys. 60, 701 (1988)
5. Akira Uedono and Shoichiro Tanigawa, RADIOISOTOPES 37, 217 (1988)
6. S. M. Hutchins, P. G. Coleman, A. Alam and R. N. West, The UEA Positron Beam Surface Spectrometer on Positron Annihilation (World Scientific, Singapore, 1985), p. 983.

RETARDING FIELD AND TIMING MEASUREMENTS OF POSITRONS FROM A W[100] FOIL

P. Willutzki, J. Störmer, D.T. Britton, G. Kögel, P. Sperr, R. Steindl and W. Triftshäuser

Institut für Nukleare Festkörperphysik, Universität der Bundeswehr München, D-8014 Neubiberg, Werner-Heisenberg-Weg 39, Germany

ABSTRACT

Measurements with the new prebuncher for the existing pulsed positron beam (presented at ICPA-9) show the possibility of bunching a dc-beam with a time structure of 2.1ns (FWHM). The pulse width is very sensitive to the longitudinal energy distribution of the moderated positrons. A pulse width of 2.1ns corresponds to a longitudinal energy distribution of 0.35(5)eV.
With a retarding grid configuration we measured the positron yield. Measurements with inhomogenic magnetic guiding fields clearly show the influence of elastically scattered events. Their fraction is very sensitive to the surface condition of the moderator. The nonscattered positrons are very well represented by a gaussian distribution function. Its FWHM is in good overall agreement with the timing measurements.

INTRODUCTION

We are planning to add a new prebuncher [1] to our existing Pulsed Positron Beam [2]. It compresses the incoming dc-beam to pulses of 2ns FWHM (full width at half maximum). The existing chopper then cuts the background between these pulses and the following main buncher provides a further compression of the pulse width to a FWHM of less than 120ps at the target position. The schematic overview is given in Fig.1:

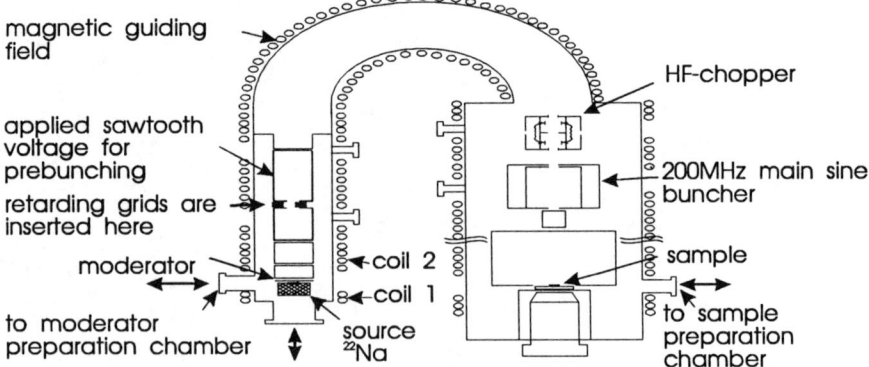

Fig1: Schematic overview of the system (not to scale)

The advantage of the sequence "dc-beam / prebunching / chopping / main bunching" is a gain in count rate compared to the existing "dc-beam / chopping / main bunching", where only 10 to 20% of the incoming positrons are used. The attainable pulse width of the new prebuncher is restricted by the longitudinal energy spread of the incoming (moderated) positrons. Fig.2 shows the achieved results and Fig.3 the dependence on the longitudinal energy distribution.

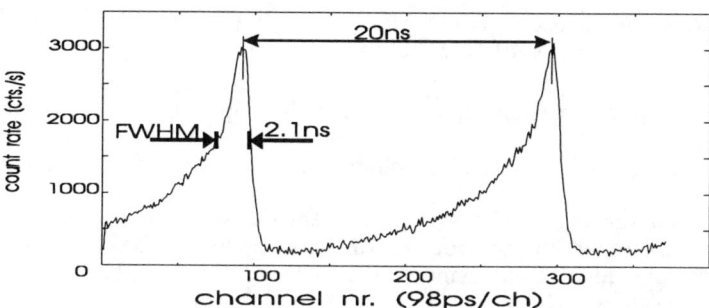

Fig.2: Achieved results for the time structure of the new prebuncher after correction for background.

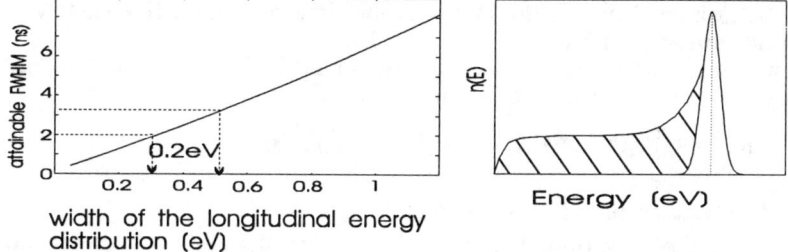

Fig.3: Attainable FWHM of the pulse width

Fig.4: Schematic energy distribution of moderated positrons; hatched part: energy loss tail

QUESTIONS

The FWHM of 2.1 ns corresponds to a longitudinal energy spread of 0.38eV. We have placed retarding grids into the first bunching gap (Fig.1) to see if our results are in agreement with the longitudinal energy distribution and the asymmetry of the pulse structure visible in Fig.2 is due to the low energy tail of the moderated positrons (hatched part in Fig4). The arrangement of the system is now similar to that of [3].

Further more general physical questions were the following:

- How does the longitudinal energy distribution of the moderated positrons depend on the treatment of the moderator foil?

- Is the energy loss of the positrons mainly elastic or inelastic? (Measurements with various combinations of the magnetic guiding field Bz at the position of the moderator and the retarding grids)
- What role does the surface contamination play on the moderator efficiency and the fraction of scattered positrons? (Measurements depending on the time after cleaning)

GENERAL CONDITIONS AND SETTINGS FOR THE RETARDING FIELD MESUREMENTS

- The retarding grids are tungsten meshes with a spacing of 0.6mm and a transmission of 90%.
- The positrons are accelerated to an energy E_{bias} = 6eV before entering the retarding grids.
- To get a clean surface for the time dependent measurements the moderator was flashed to a temperature of about 2500K with an electron beam (1min).
- In the time dependent measurements the pressure shortly after flashing (1. measuring cycle) was $3 \cdot 10^{-8}$ mbar.
- The time for measuring one cycle for time dependent and time independent measurements was 90s and 480s respectively.
- The resolution of the retarding grid was approximately 0.2 eV (lower limit). This can be estimated due to the fact, that the time resolution (FWHM) with inserted grids drops from 2.1ns to 3.5ns (Fig.3).
- The diameter of the source (^{22}Na, approx. 3mCi) is 3.2(2)mm whereas the retarding grids sit in an aperture of 5mm.

NUMERICAL TREATMENT OF THE DATA (Fig. 5)

- All the data were corrected for background.
- The part of the spectrum, which deviates from the distribution function (indictated by a * in Fig.4), was fitted with a cubic spline.
- The fraction of positrons in the energy loss tail was chosen to be the difference F in Fig. 5 (hatched part in Fig.4).

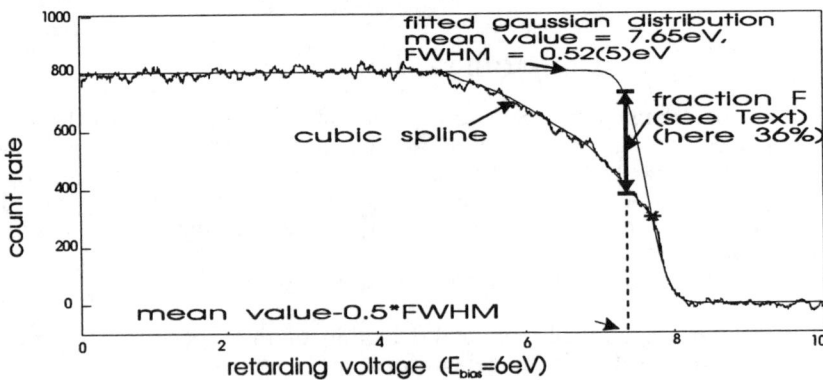

Fig.5: Retarding energy spectrum 16h after flashing the moderator, measuring time 480s.

CONDITIONING OF THE MODERATOR

The moderator (W [100] 1μm selfsupporting [4]) was only rinsed in alcohol before inserting in the system. In this condition it shows no moderation effect. It was then heated to 2500K with an electron beam for 30min. Afterwards it was flashed several times for 1min at 2500K. After the third cycle no more changes in the moderation properties occured (Fig.6). After exposure to air for 10min we found a decrease in count rate of 35% (Fig. 7).

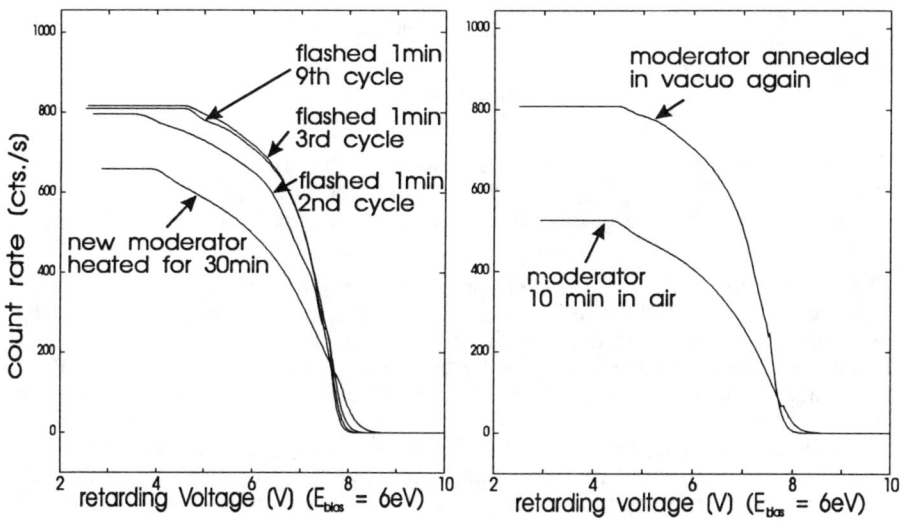

Fig. 6 and 7: Conditioning of the moderator and behaviour after exposure to air for 10min.

IS THE LONGITUDINAL ENERGY BROADENING OF THE EMITTED POSITRONS MAINLY ELASTIC OR INELASTIC?

There have been many speculations as to whether the energy loss process of the positrons emitted from a metal surface is mainly elastic or inelastic (for a review see [5]). Because of the invariance of flux [6,7], a decrease of the magnetic guiding field Bz can shift transversal to longitudinal energy.
By applying different currents to the parts of the coil before and behind the moderator (coils 1 and 2 in Fig.1) we were able to vary the ratio $R=B_{g(rid)}/B_{m(oderator)}$ from 0.4 to 1.9.
The magnetic fields at the different positions were calculated with the magnetic lens design program MLD [8] within an accuracy of 10 Gauss.
Fig.8 shows the data obtained with various ratios R. Fig.9 shows the same data after shifting the energy loss part by a factor of 1/R.

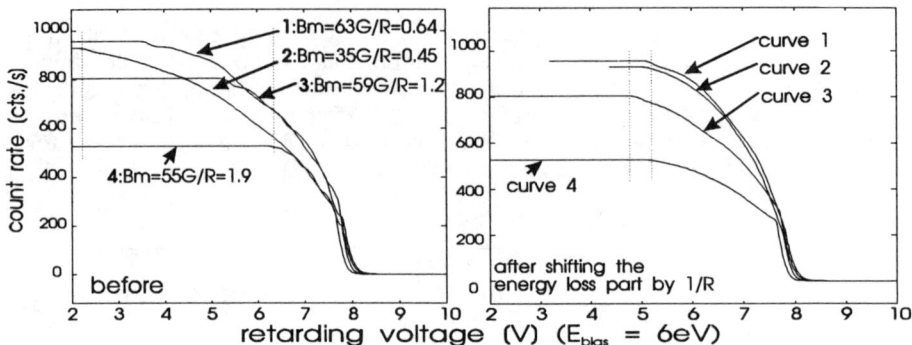

Fig. 8 and 9: Spectra with various values of the magnetic field B_m and different ratios $R = B_g/B_m$. The dashed vertical lines show the compression of the endpoints of the energy loss parts, when shifting the spectra by the factor of 1/R (see text).

The width and the form of the energy loss parts are now very similar. This indicates that the total energy [$E_{l(ongitudinal)} + E_{t(ransversal)}$] is conserved and the positrons therefore are scattered under high angles.
If the positrons at the beginning of the decrease of the curve have lost all their longitudinal energy, the work function would be 2.7(3)eV (in agreement with [5]).
The difference of this starting point to E_{bias} is attributed to contact potentials.
The different total count rates are due to the fact that the positrons with high transverse energy fractions annihilate at the grid aperture. From calculating the beam radii at the aperture (depending on E_t and B) we can assume that for curve 1 (Fig.9) nearly all the positrons pass the retarding grids. When all the losses in E_l in this spectrum are shifted to the transverse direction and by summation over all the positrons, starting from different moderator positions, we can calculate the total transmission rate J_{tot} through the aperture of the retarding grids (Fig.10). The good agreement of the calculated and measured transition rates are a further indication for the mainly elastic scattering process.

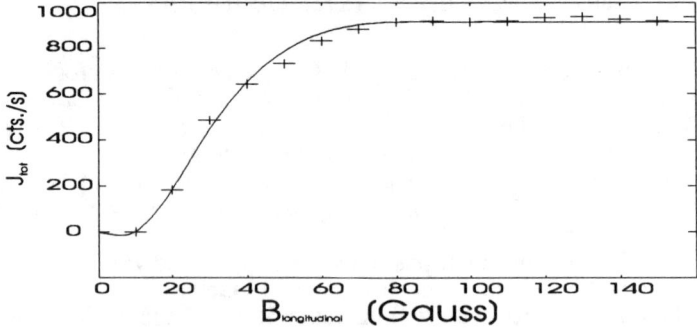

Fig.10: Calculated and measured total transmission rates J_{tot} through the aperture of the retarding grid system.

WHAT ROLE DOES THE SURFACE CONTAMINATION PLAY?

When flashing the moderator, a quadrupole mass spectrum analysis shows that C and O are the main impurities. This is consistent with [7]. The sticking coefficient of pure W[100] for oxygen is 1 and drops to 0 for 1.2 monolayer contamination [9]. At a pressure of $3·10^{-8}$ mbar the first monolayer will grow within the first minutes. The measurements show the following results (Fig.11):

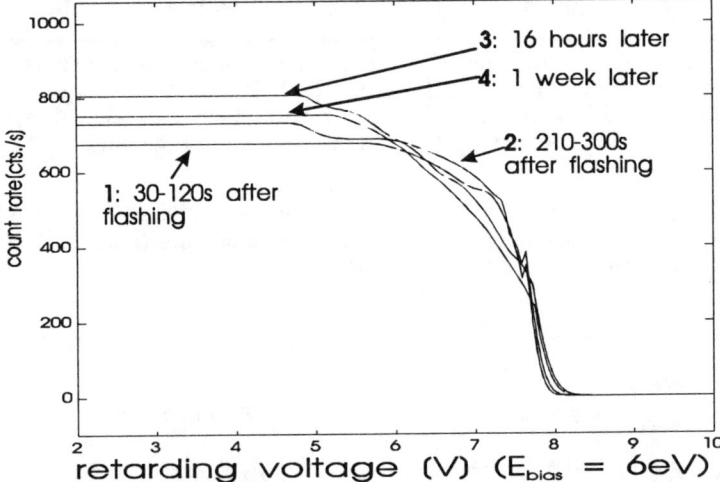

Fig. 11: Measurements depending on the time after annealing the moderator. An increase of the count rate up to 19% was observed. After one week due to the thicker surface contamination, especially the wide angle scattered positrons are reduced. The parameters of the magnetic guiding field were: Bm=59G, R=1.2; The spectra have been shifted by 1/R.

Several things are clearly visible:
- The surface contamination increases the moderation efficiency of the moderator.
- At the same time the fraction of scattered positrons is substantially increased (Fig.11).
- The mean value of the gaussian distribution function does not shift within the measurement limits. The surface contamination only changes the dipole term in the positron and electron workfunction. The difference measured here remains unchanged [5].
- The FWHM of the energy distribution for the nonscattered positrons drops from 0.34(4) to 0.49(4)eV as an influence of the broadened energy loss part (Tab.1). The value of 0.44(4)eV (energy resolution of 0.2eV assumed) is in good agreement with the timing measurements (Fig.3).
- It is possible to get moderated positrons with an energy spread less than 0.5eV from a tungsten transition moderator, even when using a sodium source. The literature reports on comparable measurements give 0.76eV and were obtained with monoenergetic positrons [10].

curve	time after annealing the moderator	FWHM (eV)	FWHM single (eV)	fraction F of positrons in the enegy loss tail (%)
1	30-120s	0.34(4)	0.27(4)	26
2	210-300s	0.38(4)	0.32(4)	30
3	16h	0.52(4)	0.48(4)	36
4	1week	0.49(4)	0.44(4)	41

Tab.1: FWHM of the fitted gaussian distribution function for the spectra, taken at different times after annealing. For the "single FWHM" an energy resolution for the grids of 0.2eV is assumed. The fraction F in the last column is the fraction of positrons in the energy loss tail as defined in Fig.4 and Fig.5.

The best agreement with the gaussian distribution function is shown in the measurement with Bm= 55G/R=0.45 short after annealing (Fig.12). Here there are nearly no more scattered positrons visible.

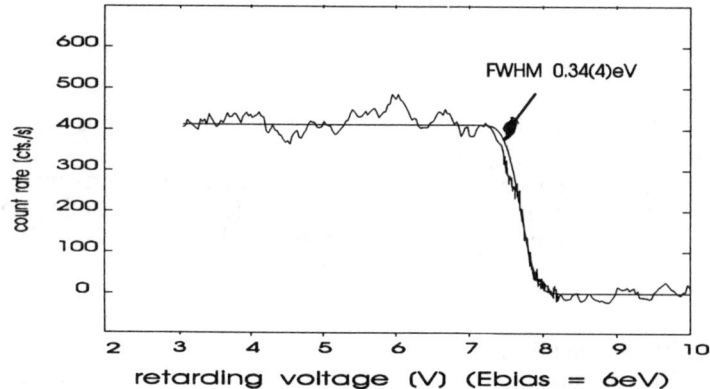

Fig.12: Spectrum taken 30-120s after annealing, Bm=55G/R=1.9. The spectrum is not corrected. To gain statistics the measuring time for the

CONCLUSION

We have performed timing and retarding field measurements with our new prebuncher. It is possible to form pulses of 2.1ns FWHM from a dc-beam. This is consistent with the FWHM of the longitudinal energy distribution, obtained from the retarding field mesurements. The tails of the pulses are due to scattered positrons which occur with the growing surface contamination on the moderator. Although there are reports for some distinct energy losses on different materials (for example CO on Ni [1]), we can conclude from our spectra, that the energy loss process for positrons emitted from W[100] foils is mainly elastic.

The influence of the surface contamination is a remarkable increase in count rate (efficiency) and a higher probability for high angle scattering. In conclusion we can say, that if it is possible to prepare a clean moderator surface (for example by active methods such as electron- or photo-desorption) it should be possible to form a positron beam with a low angular spread. Also designs with a decrease of the magnetic guiding field in combination with apertures can reduce the higher transverse positron energies.

REFERENCES

1. Proceedings of the **ICPA-9**
2. D. Schödlbauer, P. Sperr, G. Kögel, and W. Triftshäuser, Nucl. Instrum. Methods, **B34**, 258, (1988)
3. E. Gramsch, J. Throwe, and K.G. Lynn, Appl. Phys. Lett. **51**(22),1862, (1987)
4. moderator foils provided by J. Chevalier, Arhus University
5. P.J. Schultz, K.G. Lynn, Rev. Mod. Phys., Vol.**60**, No.3, July 1988
6. J.D. Jackson, Klassische Elektrodynamik, de Gruyter, Berlin-NewYork
7. E.M. Gullikson, A.P. Mills Jr., W.S. Crane, and B.L. Brown, Phys.Rev. **B32**, 5484, (1985), and reference therein
8. MLD: Magnetic Lens Design Program, Delft Particle Optics Foundation
9. B.A. Chuikov, V.D. Osovskii, Yu.G. Ptushinskii, and V.G.Sukretnyi, Surf.Sci. **213**, 359, (1989)
10. D.M Chen, K.G. Lynn, R. Pareja, and B. Nielsen, Rev.Phys. **B31**,4123, (1985)
11. D.A. Fischer, K.G. Lynn, and W.F. Frieze, Phys.Rev. Lett., 50, 1149, (1983)

THE ORNL SLOW POSITRON FACILITY AND QUADRATIC-POTENTIAL TIME-OF-FLIGHT MASS SPECTROMETER

Jun Xu, Lester D. Hulett, Jr., and T. A. Lewis
Oak Ridge National Laboratory
P. O. Box 2008, MS 6142
Oak Ridge, TN 37831-6142

ABSTRACT

This paper presents the production and characterization of the slow positrons (0.5 - 100 eV) produced at the Oak Ridge Electron Linear Accelerator and discusses the advantages and characteristics of the quadratic-potential time-of-flight mass spectrometer used in studies of the ionization of large organic molecules.

INTRODUCTION

Low-energy positrons described in this work refer to the positrons having energies in the range of 0.5 eV to 100 eV. These low-energy positrons have been demonstrated to be important in surface studies [1,2] and in studies of the positron ionization/fragmentation of organic molecules [3,4]. In this work, we present the production and characterization of the slow positrons produced at the Oak Ridge Electron Linear Accelerator (ORELA). In addition, we describe a specially designed quadratic-potential time-of-flight spectrometer for measurements of ions produced by low energy positrons interacting with organic molecules.

FAST POSITRON GENERATION

The experimental setup for fast positron generation is shown in Figure 1 [5]. The ORELA provides a pulsed electron beam (pulse width: 10-20 ns, beam energy: 150 MeV, repetition rate: 800 Hz, power: 10 kW). The electron beam bombards a tantalum target and produces intense gamma rays. The forward scattered gamma photons are captured by an array of tungsten plates that converts them to electron-positron pairs. The tungsten plates have been annealed in such a manner that they also serve as efficient moderators of the fast positrons. Positrons of two energy ranges are produced: 1) high-energy positrons having energies from 10 to 120 keV and 2) low-energy positrons whose energies (2-3 eV) are determined by the negative positron work function of the annealed tungsten plates.

An extraction tube is placed immediately in front of the moderator. The opposite end of the tube runs through a pair of ferrite toroids to ground. The high inductive reactance of the ferrite makes the tube AC isolated and DC grounded. A negative 3-kV pulse is applied to the extraction tube for accelerating the 2-3 eV positrons to 3 keV. The 3 keV positrons and the high energy positrons are guided out of the target room by solenoidal magnetic fields.

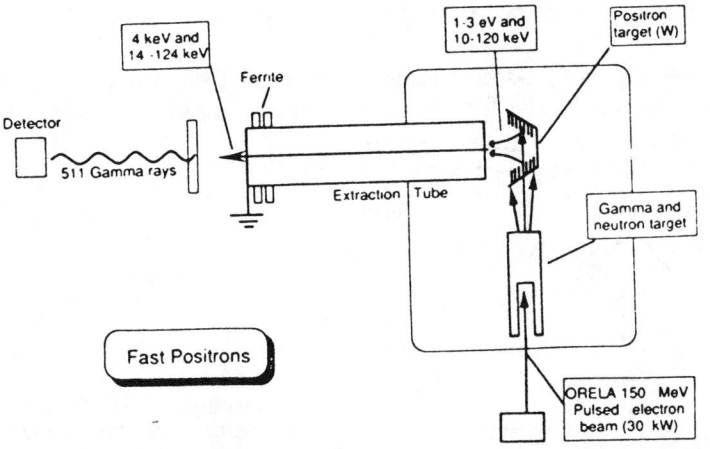

Figure 1. Experimental setup for producing fast positrons.

To obtain a quantitative measure of positron energy and intensity, time-of-flight spectra of the fast positrons were measured, as shown in Figure 2. The ORELA pulse was used for starting signals. The 511 keV gamma rays, produced by annihilation of the positron on a microchannel plate, were used for stop signals. The upper spectrum shows a measurement with the extraction tube pulsed at -3 kV. Three groups of gamma rays are observed under these conditions: 1) the initial gamma flash from the accelerator, 2) the 511 gamma rays produced by the high-energy (10-120 keV) positron annihilation, and 3) the 511 gamma rays generated by the 3-keV positron annihilation. The lower spectrum shows a measurement without the HV pulse on. As expected, the gamma rays associated with the 3 keV positrons disappear with absence of the -3 kV pulse. The current of the 3 keV positrons was measured at 1.0×10^8 positrons per second for an ORELA power of 33 kW.

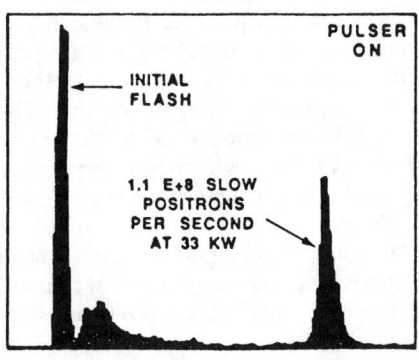

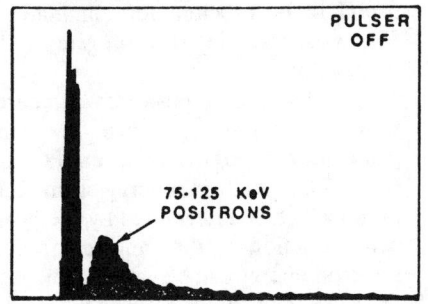

Figure 2. TOF spectra of fast positrons.

SLOW POSITRON BEAM PRODUCTION

The experimental setup for producing a slow positron beam is illustrated in Figure 3. As described above, positrons leaving from the target room consist of two components

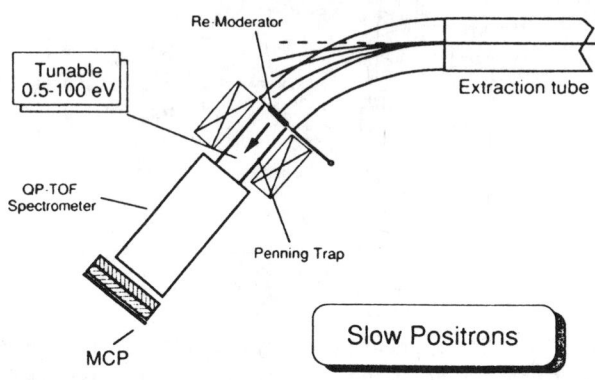

Figure 3. Experimental setup for producing a slow positron beam

in energy: 1) the accelerated 3-keV positrons, which will produce tunable low-energy positrons after bombardment of a tungsten-foil re-moderator, and 2) high-energy positrons with energy between 10 - 120 keV. If these high-energy positrons pass through a tungsten-foil re-moderator, they will produce positrons with energies continuously distributed from 5 eV to 100 keV. To remove the high energy component, we used a 50-gauss magnetic bending solenoid (200 cm in bending diameter). The bending coil guides the low-energy positrons to the experimental chamber, but not the high-energy positrons since the field is too weak. This method has been found to be efficient for eliminating the fast positrons in the tungsten-foil re-moderator.

The 3 keV positrons bombard the tungsten foil to produce low energy positrons emitted from the rear face of the foil. The positron energy is in 2-3 eV range which is determined by the positron work function of the tungsten foil. The positron energy can be varied by negatively or positively biasing the tungsten re-moderator foil.

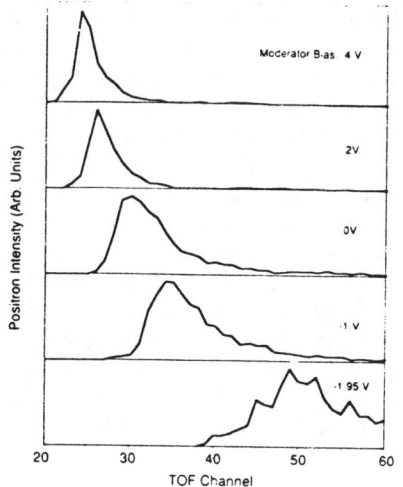

fig. 4. TOF spectra of slow positrons

To characterize these low-energy positrons, we measured the time-of-flight spectra of the re-emitted positrons as a function of the re-moderator voltages, as shown in Figure 4. The moderator was biased from -2 to 10 eV; the TOF distribution of the positrons was found to vary accordingly. We converted the TOF spectra to kinetic energy distributions. At the zero-moderator bias, positron energy peaks at 2.25 eV. The positron energy is calibrated by measuring the positronium formation thresholds of the gases.

QUADRATIC-POTENTIAL TIME-OF-FLIGHT MASS SPECTROMETER

Hulett and co-workers [6] designed an ionization Penning trap jointly equipped with a quadratic-potential time-of-flight spectrometer (QP-TOF) for measuring the masses of ions produced by low-energy positrons interacting with molecules and atoms. The following discussions will demonstrate that this spectrometer can enhance the sensitivity and produce a high resolution.

The Penning trap consists of input grid and output grid, spaced apart by 10 cylindrically-symmetric lenses, as shown in Figure 5. two coils produce 1000 Gauss of magnetic field in the center region of the trap. The trap is operated at a base pressure about 5×10^{-8} Torr. Vapors are admitted to the trap through a leak valve. Pressures range from the base pressure to 1.0×10^{-6} Torr. As the remoderated positrons emerge from the backside of the W film, the input grid is held at ground potential. After they pass, the input grid is returned to a positive potential. The positively biased input and output grids prevent longitudinal escape of the positrons and ions from the trap. The high magnetic field prevents radial escape.

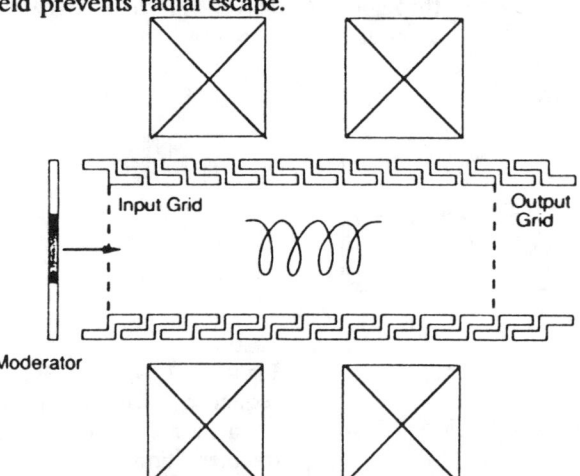

Figure 5. The Penning trap including input grid, output grid, ten lenses, and two solenoids.

The Penning trap increases the collision path between positrons and molecules, thereby enhancing the ionization counting rate. However, this apparatus makes the ionization volume large. Consequently, the large ion-containing volume would result in bad resolution in a conventional TOF spectrometer. To overcome this large volume effect, a varying electrostatic field is used in which the potential varies as a square function of the distance of the ion from the detector. The idea is illustrated in Figure 6. With this quadratic potential applied to the lenses, any ion with the same mass will reach the detector at the same time, insensitive to the starting position of the ion.

The experiments are operated by a pulse sequence, as shown in Figure 7: (1) A start pulse from the ORELA (V_{sync}) initiates the sequence. (2) As positrons emerge from the tungsten plates, a negative high voltage pulse (V_{extr}) is applied to the extraction tube to accelerate the positrons out of the target room. (3) About 250 nsec is required for the positrons to travel the 10-meter distance to the foil moderator. As the positrons emerge from the backside of the foil, the input grid of the Penning trap is pulsed to zero potential ($V_{I/G}$) to allow the slow positrons to enter.

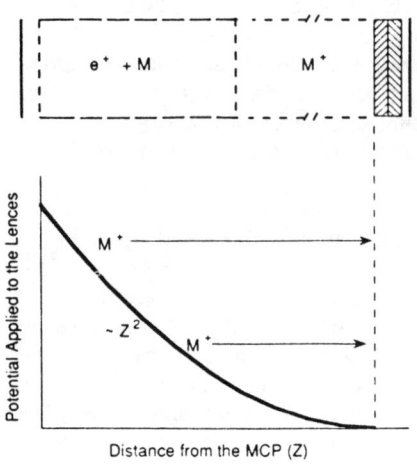

Figure 6. Illustration of a quadratic-potential time-of-flight spectrometer.

The positive potential of the input grid is restored for retaining the positrons. (4) We can retain the positrons in the trap for periods of 20-1800 microseconds. The ions are produced during this period. Then the input grid opens a zero-potential window again to allow residual positrons to escape from the trap. (5) Immediately following the positron dump, a quadratic potential pulse is applied to the lenses of the spectrometer to start the record of flight times of the ions. (6) Ion signals (V_{sig}) received by the detector serve as the stop of the pulse sequence. The time-of-flight spectra are recorded by a time-to-digital convertor (TDC), interfaced through a CAMAC unit to a personal computer.

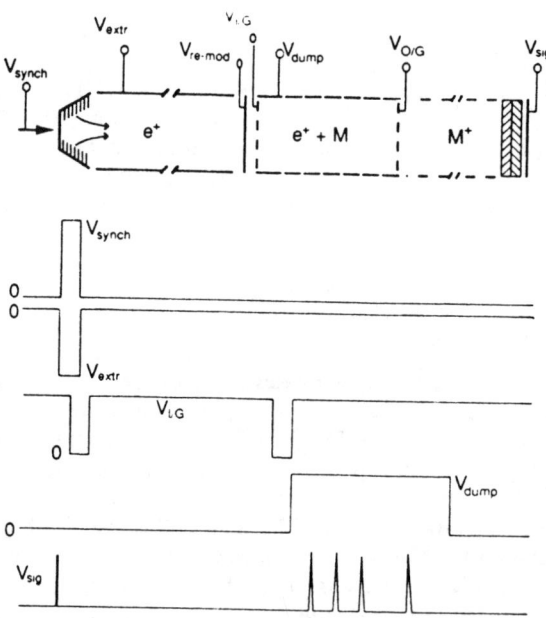

Figure 7. a Pulse sequence for positron ionization and fragmentation

POSITRON-INDUCED DISSOCIATIVE IONIZATION

As an application of the tunable, low-energy positron beam, we studied positron-induced dissociative ionization of organic molecules. Figure 8(a) shows a time-of-fight spectrum of decane under 7-eV positron impact. The largest mass peak is attributed to the parent ion, or molecular ion (C_{10}). The other peaks are respectively attributed to C_8, C_7, C_6, C_5, C_4, C_3 fragment ions. As seen in the figure, the spectrum has adequate resolution for studies of positron ionization mass spectrometry.

Since the positrons in the Penning trap are energy-tunable, we studied the energy dependence of the parent ion and the fragment ions. Figure 8(b) shows the positron energy dependence of the C_3, C_5, and C_7 fragment ions, and the C_{10} molecular ion. The C_4, C_6 and C_8 fragment ions showed similar patterns. The decane parent ions were produced at about the positronium formation threshold, 2.85 eV (IP-6.8 eV). The fragment ions are produced at energies above the Ps threshold.

The data clearly show different energy thresholds for different fragment ions. Detailed examination reveals that the onset energies (the energy above the Ps threshold) of the fragment ions are linearly correlated to the dissociation energy of the parent ion [4].

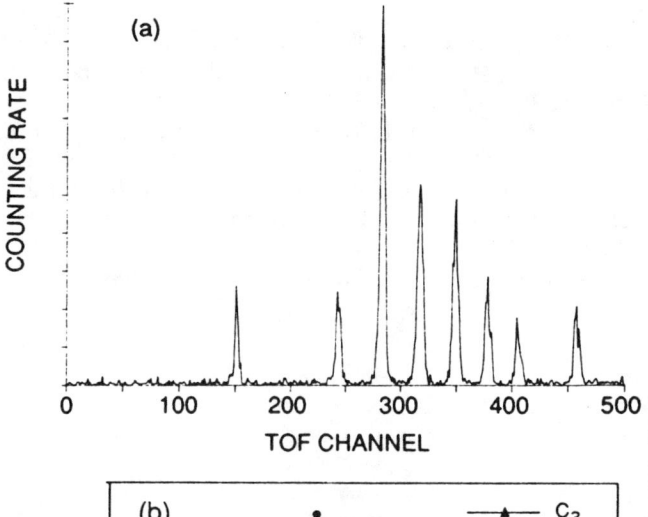

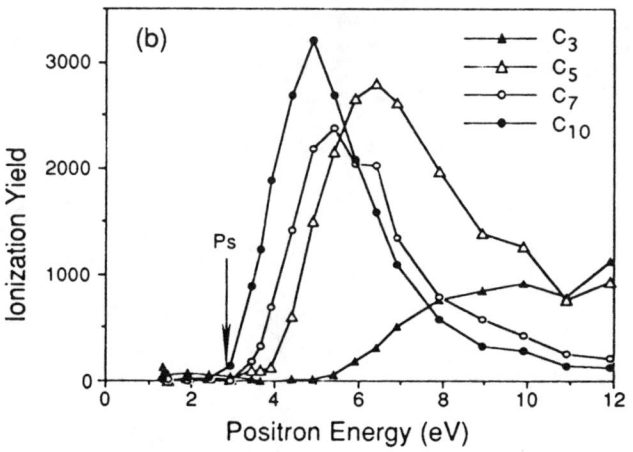

Figure 8. (a) Time-of-flight mass spectrum of decane, (b) energy dependences of the parent ion and the fragment ions.

In summary, we demonstrated the production, characterization, and energy-tunability of the low-energy positrons by measuring time-of-flight spectra of the positrons produced in ORELA. We presented a quadratic time-of-flight mass spectrometer which provides the high counting rate and the best resolution in positron ionization studies. As an application of the facility, we studied positron-induced dissociative ionization of organic molecules.

ACKNOWLEDGEMENT

We thank the staff of the Oak Ridge Electron Linear Accelerator for their helpful support. This work is supported by the U.S. Department of Energy, Office of Basic Energy Sciences, under contract DE-AC05-84OR21400 with Martin Marietta Energy Systems, Inc.

REFERENCES

1. Alex Weiss, R. Mayer, M. Jibaly, C. Lei, D. Mehl, and K. G. Lynn, *Phys. Rev. Lett.* **61**, 2245 (1988).
2. R. Suzuki, Y. Kobayashi, T. Mikado, H. Ohgaki, these proceedings.
3. D. L. Donohue, L. D. Hulett Jr., B. A. Eckenrode, S. A. McLuckey, and G. L. Glish, *Chem. Phys. Lett.* **168**, 37 (1990).
4. Jun Xu, L. D. Hulett, Jr., T. A. Lewis, D. L. Donohue, S. A. McLuckey, G. L. Glish, Submitted to *Phys. Rev. A*, (1992).
5. L. D. Hulett, Jr., T. A. Lewis, D. L. Donohue, and S. Pendyala, In: Positron Annihilation, eds. L. Dorikens-Vanpraet, M. Dorikens, and D. Seegers (World Scientific, Singapore, 1989) p. 586.
6. L. D. Hulett Jr., D. L. Donohue, and T. A. Lewis, *Rev. Sci. Instrum.* **62**, 2131 (1991).

SECTION 8

FUNDAMENTAL PHYSICS STUDIES

OBSERVATION OF RESONANCE-LIKE STRUCTURES IN POSITRON-KRYPTON ELASTIC SCATTERING*

L. Dou, W.E. Kauppila, C.K. Kwan, and T.S. Stein
Department of Physics and Astronomy, Wayne State University
Detroit, Michigan 48202

ABSTRACT

Absolute measurements of differential cross sections (DCSs) for positrons elastically scattered by krypton atoms reveal unexpected resonance-like structures at 25 and 200 eV when the DCSs at the fixed scattering angles of 30°, 60°, 90°, and 120° are plotted versus energy. The resonance-like structures may represent a new type of scattering resonance arising from coupling effects between the elastic and inelastic scattering channels.

INTRODUCTION

Some initial direct experimental evidence for the possible existence of resonances [1] in positron-atom scattering has previously been reported only for positron differential elastic scattering by argon atoms. Prior to the above measurements [1] it was expected that the most likely place to make the first direct observations of positron scattering resonances would be in the vicinities of inelastic scattering thresholds, as discussed by Ho [2]. The resonance-like structure observed [1] for positron-argon differential elastic scattering was puzzling because it occurred at an intermediate energy of between 55-60 eV, which is well above any known inelastic thresholds for argon. A theoretical prediction by Higgins and Burke [3], using an R-matrix method, of a "coupled-channel shape resonance" in both the elastic scattering and the positronium (Ps) formation cross sections for positron-hydrogen scattering may be relevant and possibly similar to the observations for argon [1] because the predicted positron-hydrogen resonance occurs at 35.6 eV, which is also well-above the inelastic thresholds of atomic hydrogen, and has an energy width of 4.22 eV. This predicted resonance occurs only when both the elastic scattering and Ps formation channels are considered together. A more recent and somewhat more elaborate calculation by Hewitt et al. [4] confirms the existence of the resonance predicted by Higgins and Burke [3] for positron-hydrogen scattering, although at a lower energy of 20.37 eV (with a width of 3.675 eV).

EXPERIMENTAL METHOD

The present absolute DCS measurements are obtained with the same experimental setup and technique as used previously by our group [1] where variable energy positron and electron beams (obtained from an annealed tungsten moderator placed in front of a ^{22}Na radioactive source) are crossed with a constant intensity Kr atom beam effusing from a multichannel capillary array. Channeltron electron multipliers (CEMs) are used to detect the primary beam (CEM1) and the elastically scattered positrons and electrons (CEM2 and CEM3). The range of detectable scattering angles is from 30° to 135°.

© 1994 American Institute of Physics

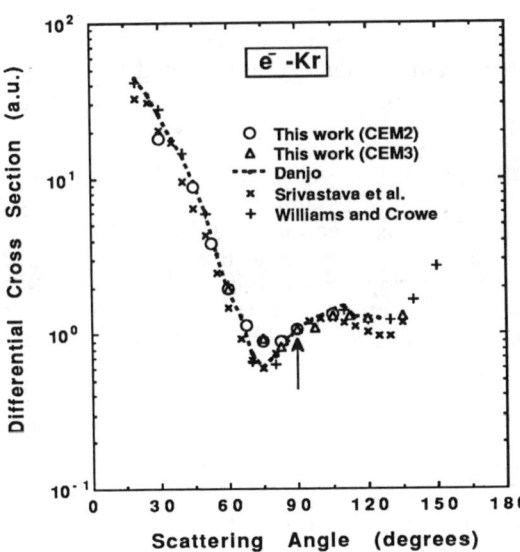

Fig. 1 Differential cross section for e⁻-Kr scattering at 20 eV.

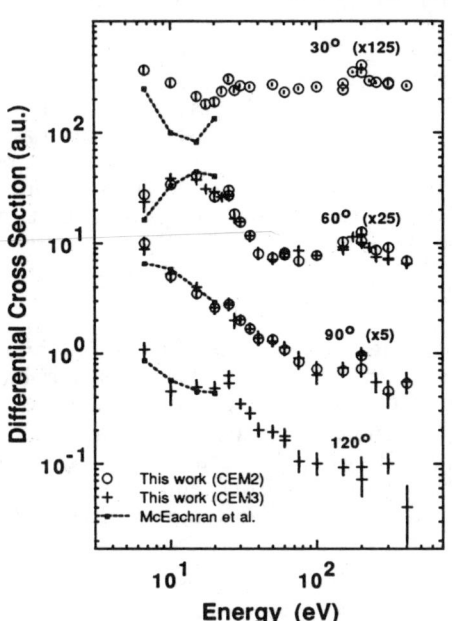

Fig. 2 DCSs versus energy for e⁺-Kr elastic scattering.

With this experimental setup direct comparisons can be made between relative DCSs obtained for positron and electron scattering. To obtain absolute DCSs for positrons, we first normalized our 20 eV e⁻-Kr relative DCS measurements to the experimental values of Danjo [5] at 90°, where there is good agreement with the measurements of Williams and Crowe [6] and Srivastava et al. [7], as shown in Fig. 1. An angle of 90° is also desirable for the normalization because it minimizes the effect of our finite angular acceptance (±8°) on our measured DCSs. We then compared our relative positron DCS measurements at any desired positron energy with the normalized e⁻-Kr results at 20 eV to obtain absolute values.

RESULTS AND DISCUSSION

The most striking features of the present results shown in Fig. 2 are the resonance-like peaks appearing at positron energies of 25 and 200 eV for 30°, 60° and 90° and 25 eV for 120°. Each of these peaks is represented by two or more data points taken during different data runs over a period of several months to help ensure that they were reproducible. This along with the statistical uncertainties with each data point clearly indicate that the peaks are statistically significant. It is to be noted that the shape of the DCS versus energy curves in Fig. 2 should be little affected by the uncertainty in the determination of the absolute values due to the normalization procedure. A comparison of the present results with a polarized orbital calculation of McEachran et al. [8]

shows best agreement for a scattering angle of 120°.

The existence of these resonance-like peaks at 25 and 200 eV is similar to the earlier observation [1] of structure for positron-Ar elastic scattering between 55-60 eV in that the present peaks also occur well above any of the familiar inelastic thresholds for Kr (e.g., Ps formation at 7.2 eV, 9.9 eV for excitation, and 14.0 eV for ionization). It would seem likely that these "resonances" may be two additional examples of "coupled-channel shape resonances", as discussed by Higgins and Burke [3] for e^+-H elastic scattering and Ps formation. In the case of e^+-Ar scattering it was noticed [1] that the observed resonance-like structure occurs near where the ionization cross section (Q_{ion}) reaches a maximum (60 eV) and accounts for more than half of the total cross section (Q_t), while Ps formation (Q_{Ps}) accounts for only about 25%, suggesting that the ionization channel may be more likely to play a role in the existence of the "resonance". It may be relevant in trying to understand the origin of the peaks for Kr that at a positron energy of 25 eV Q_{Ps} [9] represents close to 50% of the Q_t [10], while at 200 eV Q_{Ps} represents only about 25% of Q_t. While we are not aware of any measurements of Q_{ion} for e^+-Kr scattering, it would seem likely that at 200 eV Q_{ion} would be larger than Q_{Ps}, as is the case for He [11] and Ar [1]. This might suggest that the 25 eV peak for Kr may relate to the Ps formation channel, while the 200 eV peak may relate to the ionization channel. We are not aware of any published calculations which consider the possible cross-channel coupling effect of ionization on the elastic scattering channel for e^+-atom scattering. It would seem that there is much interesting work remaining to be done, both theoretically and experimentally, in the intermediate energy range for e^+-atom scattering where elastic scattering, Ps formation and ionization are known to be significant contributors to Q_t and cross-channel coupling effects are likely to play a very important role.

* This work was supported in part by NSF Grant No. PHY90-21044.

REFERENCES

[1] L. Dou, W.E. Kauppila, C.K. Kwan and T.S. Stein, Phys. Rev. Lett. <u>68</u>, 2913 (1992).
[2] Y.K. Ho, in proceedings of the "Workshop on Positrons in Gases" held at Macquarie University, Sydney, Australia, 3-5 July 1991, Hyperfine Interactions <u>73</u>, 109 (1992).
[3] K. Higgins and P.G. Burke, J. Phys. B <u>24</u>, L343 (1991).
[4] R.N. Hewitt, C.J. Noble and B.H. Bransden, J. Phys. B <u>24</u>, L635 (1992).
[5] A. Danjo, J. Phys. B <u>21</u>, 3759 (1988).
[6] J.F. Williams and A. Crowe, J. Phys. B <u>8</u>, 2233 (1975).
[7] S.K. Srivastava, H. Tanaka, A. Chutjian and S. Trajmar, Phys. Rev. A <u>23</u>, 2156 (1981).
[8] R.P. McEachran, A.D. Stauffer and L.E.M. Campbell, J. Phys. B <u>13</u>, 1281 (1980).
[9] L.M. Diana, P.G. Coleman, D.L. Brooks and R.L. Chaplin, in *Atomic Physics with Positrons*, edited by J.W. Humberston and E.A.G. Armour, NATO ASI Series B, Vol. 169, (Plenum, New York, 1987), pp. 55-69.
[10] M.S. Dababneh, W.E. Kauppila, J.P. Downing, F. Laperriere, V. Pol, J.H. Smart and T.S. Stein, Phys. Rev. A <u>22</u>, 1872 (1980).
[11] W.E. Kauppila and T.S. Stein, Adv. At. Mol. Opt. Phys. <u>26</u>, 1 (1990).

MEASUREMENTS OF POSITRONIUM FORMATION CROSS SECTIONS FOR POSITRON - POTASSIUM ATOM SCATTERING*

L. Jiang, W.E. Kauppila, C.K. Kwan, S. P. Parikh, T.S. Stein, and S. Zhou
Department of Physics and Astronomy, Wayne State University,
Detroit, Michigan 48202

ABSTRACT

We report preliminary measurements of positronium (Ps) formation cross sections for 1 - 20 eV positron - potassium atom collisions. Our results are compared with prior theoretical results.

INTRODUCTION

Up to the present time, there has been only one report (also by our group) of the measurements of Ps formation cross sections for positron-alkali atom collisions [1]. One of the important differences between this collision system and that involving room-temperature gases is that positrons of arbitrarily small energy can form Ps in collisions with alkali atoms, whereas for the room-temperature gases, the threshold for forming Ps is at least several eV. Since the Ps channel is open at all positron energies for the alkali atoms, there is no direct indication provided by Q_T measurements (as there is in the case of the room-temperature gases) of the extent of the role played by Ps formation. In this paper, we present our preliminary measurements of Ps formation cross sections for K.

EXPERIMENT

The experimental apparatus and procedure are the same as used by Zhou et al. [2] except for the scattering cell shown in Fig. 1, which is the same as that used for our Q_T measurements [3] for the alkali atoms. Additional information on the experimental apparatus and procedure are provided in Ref. [1].

RESULTS

Using the same technique described by Zhou et al. [2], we have made preliminary determinations of upper limits (zero retard) and lower limits on Q_{Ps} for positron-K collisions and these results are shown in Fig. 2, together with the results of prior first Born approximation (FBAa and FBAb) and distorted wave approximation (DWAa and DWAb) calculations [4] of Q_{Ps}. The present results supersede our initial measurements [1] due to improved determinations of detector efficiencies. The FBA results are close to our lower limit results at 10 and 20 eV, but become significantly larger than our upper limit results below 10 eV, whereas the DWA results are less than one third of our lower limit results at energies above 5 eV.

* Research supported by NSF Grant PHY90-21044.

REFERENCES

[1] T.S. Stein, W.E. Kauppila, C.K. Kwan, S.P. Parikh and S. Zhou, Hyperfine Interactions, 73, 53 (1992).

[2] S. Zhou, L. Jiang, W.E. Kauppila, C.K. Kwan, S.P. Parikh, and T.S. Stein, contribution to this conference.
[3] C.K. Kwan, W.E. Kauppila, R.A. Lukaszew, S.P. Parikh, T.S. Stein, Y.J. Wan and M.S. Dababneh, Phys. Rev. A 44, 1620 (1991).
[4] S. Guha and P. Mandal, J. Phys., B 13, 1919 (1980).

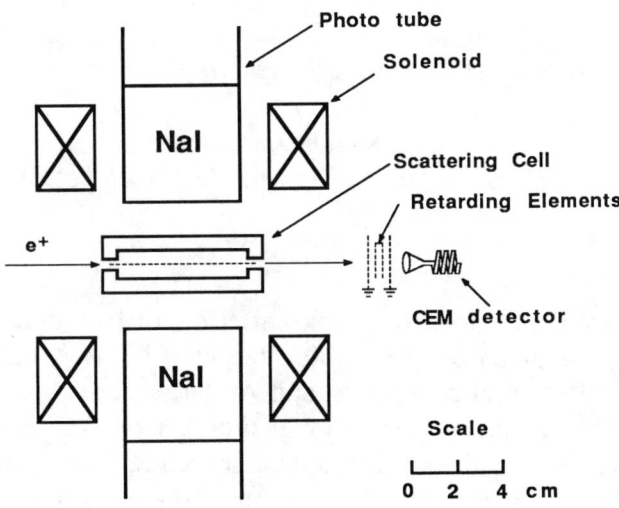

Fig. 1. Apparatus for measuring Ps formation cross sections for e^+ - K collisions.

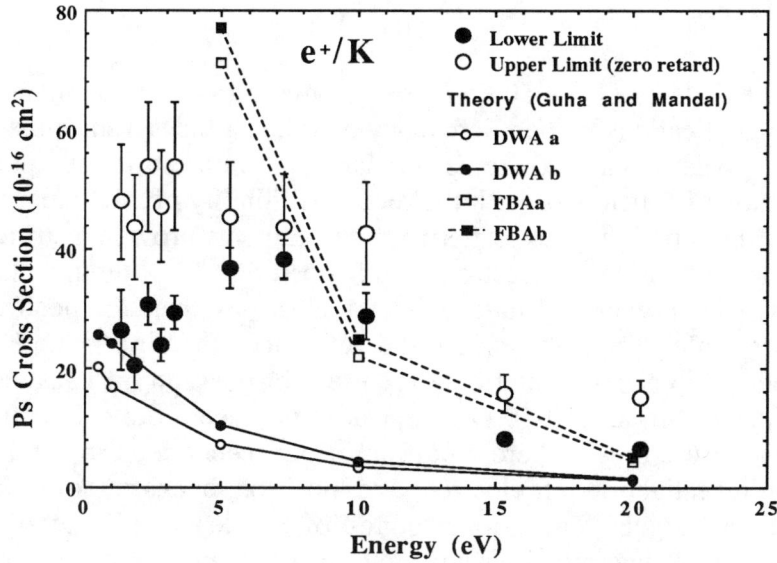

Fig. 2. Ps formation cross sections for positron-K collisions.

EXPERIMENTAL AND MONTE-CARLO STUDIES OF ELECTRON AND POSITRON BACKSCATTERING

G.R. Massoumi, W.N. Lennard, and Peter J. Schultz
Department of Physics, The University of Western Ontario, London, Ontario, Canada N6A 3K7

A.B. Walker
School of Physics, University of East Anglia, Norwich NR4 7TJ, U.K.

Kjeld O. Jensen
Department of Physics, University of Essex, Colchester CO4 3SQ, U.K

ABSTRACT

Electron and positron backscattering distributions for projectiles normally incident on different thick targets ($4 \leq Z \leq 82$) at 35 keV are reported. Differential data for Al and Au targets have also been measured for various incidence angles. For e^+ incident on Au, and both e^- and e^+ incident on Al, evidence for specular scattering is observed in the backscattering distribution. In contrast, the e^- scattering from Au is nearly symmetric about the surface normal. For positrons we find excellent agreement between experimental data and Monte Carlo simulations.

For keV e^- and e^+ (10-50 keV), backscattering is a useful analytical probe with applications to the microscopy of solid surfaces, thin films and interfaces. Additionally, there is fundamental interest in comparing particle/antiparticle behavior with the recent availibility of monoenergetic positron beams. In fact, backscattering measurements provide a stringent test of theory for both electrons and positrons. Although more experimental backscattering data exist for electrons than for positrons, [see, for example, Neidrig's review and references therein][1-3], there has been in total little experimental data reported for positron backscattering at intermediate energies (10-100 keV) prior to the recent flurry of activity in this area.[4-7] We present here a portion of our recent experimental and theoretical investigation of electron and positron backscattering from different thick targets for various angles of incidence. Some of the experimental and theoretical results presented in this paper have been published elsewhere in preliminary or abbreviated form.[4,7]

A detailed description of the experimental facility can be found in references 4 and 8. Energetic positrons from an isotopic ^{58}Co or ^{22}Na source were moderated in solid Ar, from which they are emitted with ~14.2±1.7 eV energy.[9] The positron beam was formed in a modified Soa gun[10] and electrostatically guided through a Heddle 5-element afocal lens,[11] a home-built electrostatic accelerator, and a final focusing Einzel lens. The electron beam was produced using a standard electron gun in place of the positron source and moderator. A rotatable energy dispersive surface barrier detector (SBD) recorded signals arising from backscattered particles, see Fig. 1. The target could be tilted by an angle α, defined as the angle of the surface normal (n̂) relative to the incident beam direction (-k̂). The SBD was rotatable in the same plane (defined by k̂ and n̂), where γ is the angle of the SBD relative to n̂. Periodically, the target was retracted and the detector moved to θ=0° (where θ is the scattering angle) to measure the incident flux, I_o.

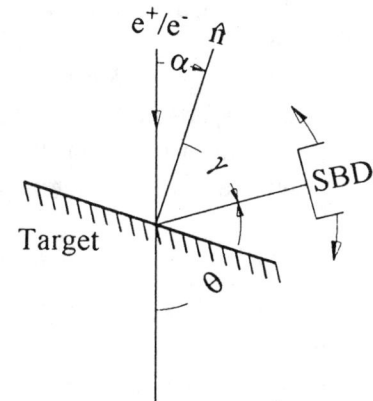

Fig. 1 - Schematic of the experimental geometry.

Doubly differential backscattering data, $d^2\eta/d\Omega dE$, were measured

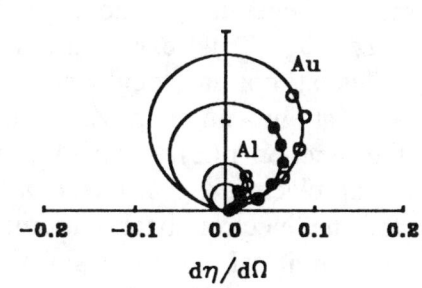

Fig. 2- Angular backscattering distributions, $d\eta^{\pm}/d\Omega$, for 35 keV positrons and electrons normally incident on planar Au and Al targets. The smooth curve results from fitting the data using Legendre polynomials.

for both positrons and electrons at normal and tilted angles of incidence. Total intensities for the differential backscattered yields (integrated over energy), $d\eta(\gamma)/d\Omega$, were determined by measuring the ratio $I(\gamma)/I_o\Omega_d$.

Due to detector noise limitations, the energy distributions could not be directly measured below E_{min} ~15 keV, which was approximately twice the detector electronic noise level. In order to derive the energy-integrated yield at a given angle, we assumed the intensity varied linearly with energy from E=0 up to E_{min}.

In order to obtain the total backscattering coefficients, we fitted the data with a sum of Legendre polynomials and then integrated over the

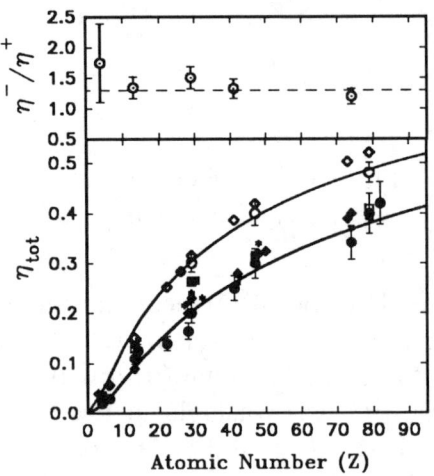

Fig. 3- Positron and electron backscattering coefficients, $\eta^{\pm}_{tot}$, as a function of target atomic number, Z, for normal incidence ($\alpha=0°$). ○ - present electron data; ◇ - electron data from ref. 28; ● - present positron data; ◆ - positron data from refs. 29; ■ - positron data from ref. 5; □ - positron data from ref. 6; ★ - positron results (35 keV) from Monte Carlo calculations.

backward hemisphere (2π sr). Fig. 2 shows the Legendre polynomial fits to both the electron and positron data for $\alpha = 0°$. Total backscattering coefficients obtained for 35 keV projectiles at normal incidence are shown in Fig. 3 as functions of the target atomic number, Z.[4] The smooth curves are fits to our data using the function $\eta(Z) = \exp(-AZ^{-\frac{1}{2}})$, where A=8.59 for positrons and A=6.40 for electrons. Arnal et al.[12] derived the equation $\eta^- = 1/(1+\cos\alpha)^p$ for the electron backscattering coefficient from bulk specimens from an empirical equation for the transmission coefficient of electrons, where $p = C/Z^{1/2}$ for normal incidence ($\alpha=0°$). Using $\eta^- = 2^{-9.4/\sqrt{Z}}$, they obtained a good fit to electron experimental data in the energy region from 10 to 50 keV for a large range of target atomic numbers.[1] Our electron data are consistent, being well fit with the function $\eta^- = 2^{-9.2/\sqrt{Z}}$.

For positrons, our data can be fit using C=12.4. Mäkinen et al. fit their 30 keV positron data with C=11.5.[6] The uncertainty estimated for each datum is ~10% for Z>20, arising mainly from the integrations over angle and energy as discussed earlier. For Z<20, the uncertainties are larger on account of (*i*) the yields are small, and (*ii*) a significant fraction of the e^+ or e^- yield is contained in the low energy region which falls in the electronic noise region (<15 keV) of the SBD response.

In addition to our own experimental results, all other relevant data are shown in Fig. 3. The criterion for comparison is that the experiment employed a beam of e^+ or e^- at normal incidence, as opposed to a 'diffuse' beam or one of undefined geometry.[4] Our electron data points are compared with the mean of data for 20, 40, and 60 keV electrons, taken from Neubert and Rogaschewski.[3] Our positron data are compared with data taken from refs. 13 (200 keV positrons), 5 (35 keV positrons), and 6 (30 keV positrons). The results of the Monte Carlo calculations for 35 keV positrons are also shown. The reader is referred to ref. 7 for a description of the procedure used for the MC simulations. The agreement among all the data shown in Fig. 3 is quite reasonable, given the experimental uncertainties. Slight deviations from the monotonic increase with Z around Z=30 have been reported by both Heinrich[14] and Mäkinen et al.[6]. Our description for the Z-dependence of η is based on a simple structureless analytical expression, as has been done by other experimenters.[15-17] While the quality of the experimental data do not justify a more detailed description at this time, there is in principle no *a priori* reason to expect a featureless dependence of η on Z, and the apparent experimental discrepancy near Z=30 may well be an indication of interesting new phenomena that have yet to be explored. Indeed, structure in this region of Z has been theoretically predicted.[18]

In the upper portion of Fig. 3, we show the ratio η^-/η^+ as a function of Z. It has long been thought that this ratio has a value ~1.3 over the entire periodic table, initially suggested by experiment.[15,16] This constant value is indicated by the horizontal dashed line. Our measured ratios suggest that η^-/η^+ may be a slightly decreasing function of Z. Further, at least one theoretical calculation predicts a ratio $\eta^-/\eta^+ = 1.16$ for Z=80.[19] In general, experimental uncertainties are still too large to preclude a constant ratio, but the agreement now emerging in the data are convincing evidence that the relationship is not so simple, and provoke interest in the obvious need for more precise measurements of the η^-/η^+-ratio, especially for low-Z materials.

Fig. 4 shows experimental results of the angular backscattering distributions, $d\eta/d\Omega$, for both 35 keV positrons and electrons from Au for different values of α (10°≤α≤60°). The solid curves are results of MC calculations for 35 keV positrons only. The agreement between

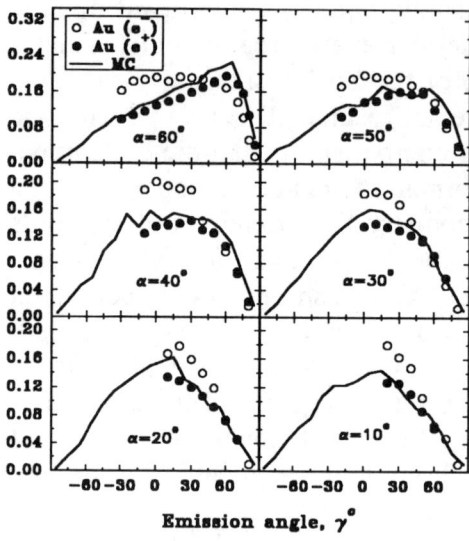

Fig. 4- Angular backscattering distributions, dη/dΩ, for Z=79 (Au) as a function of emission angle, γ, for the φ=0° plane for 35 keV positrons and electrons incident at different angles, 10°≤α≤60°. The solid line shows the MC calculation only for 35 keV positrons.

computations and experiment is excellent - a result that we find particularly encouraging in view of the fact that there are *no adjustable parameters*. It is noticeable that at lower incident angles, the distribution has the characteristic (approximate) cosine distribution that is observed for normal incidence.[1] As α increases, specular reflection becomes more evident for e^+ than for e^- for both Al and Au. This difference is evenmore pronounced for Au. In general, at oblique incident angles, the backscattered intensity peaks near the reflection angle (α≈γ). In the case of Al (low-Z bulk material), e^+ and e^- behave qualitatively the same. For Au at α ≥ 40°, both the electron and positron angular backscattering distributions behave neither qualitatively nor quantitatively the same. The specular distribution observed for positrons at oblique angles of incidence contrasts sharply with the angular distribution of backscattered electrons. We suggest that this observation arises directly from the fundamental difference in the elastic and inelastic cross sections for electrons and positrons - differences which are expected to increase with increasing Z. This observation is explained fully in reference 20.

Fig. 5 illustrates two examples (γ=60° and γ=0°) of the full doubly differential positron scattering distributions, $d^2η/dEdΩ$, for each of Al and

Fig. 5- Measured and MC calculated energy spectra of backscattered positrons, corresponding to the doubly differential backscattering probabilities, $d^2\eta/dEd\Omega$, for an incident angle $\alpha=60°$ on Al and for emission angles $\gamma=0°$ and $60°$. The smooth curves showing the MC results have been convoluted with the energy resolution of the detector. The histogram shows the Monte Carlo results for $\gamma=60°$ before convolution.

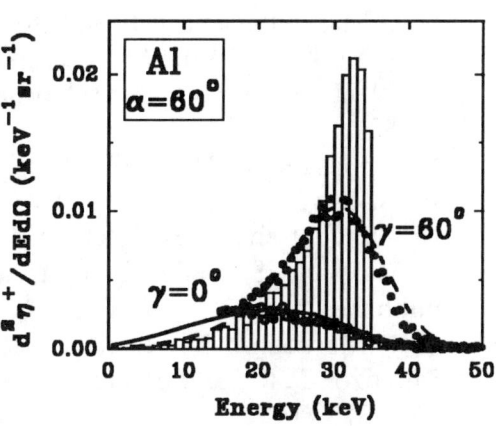

Au at $\alpha=60°$. The solid and dashed curves are the MC results convoluted with the surface barrier detector resolution function (FWHM ~11 keV). The excellent agreement lends confidence to the MC histogram, shown for the $\gamma=60°$ case only. This result illustrates the sharply peaked (particularly for Au) elastic-like scattering which dominates in all cases.

In summary, we report measurements of doubly differential backscattering yields for both electrons and positrons for a variety of thick elemental targets ($4 \leq Z \leq 82$) as a function of the target tilt angle, α. We observe quasi-specular scattering for both e^+ and e^- incident on Al. For Au, the lack of a specular feature for e^- is understood in terms of elastic scattering coupled with electron indistinguishability.[20] Monte Carlo calculations for positrons are in excellent agreement with the experimental results.

The authors are grateful for the technical assistance of I. Schmidt, B. Campbell and P. Perquin. Funding for this research has been provided by the Natural Sciences and Engineering Research Council (NSERC) of Canada (WNL, PJS) and by the Network of Centres of Excellence in Molecular and Interfacial Dynamics (PJS).

REFERENCES

[1] H. Niedrig, J. Appl. Phys. **53**, R15 (1982).

[2] G.R. Massoumi, W.N. Lennard, H.H. Jorch and Peter J. Schultz, *Positron Beams for Solids and Surfaces*, edited by Peter J. Schultz, G.R. Massoumi and P.J. Simpson, AIP Conf. Proc. No. 218 (AIP, New York, 1990) p.39.

[3] G. Neubert and S. Rogaschewski, Phys. Stat. Sol. (a) **59**, 35 (1980).

[4] G.R. Massoumi, N. Hozhabri, W.N. Lennard and Peter J. Schultz, Phys. Rev. **B44**, 3486 (1991).

[5] L. Albrecht and P.G. Coleman, to be published.

[6] J. Mäkinen, S. Palko, J. Martikainen and P. Hautojärvi, (to be published).

[7] G.R. Massoumi, N. Hozhabri, K.O. Jensen, W.N. Lennard, M.S. Lorenzo, Peter J. Schultz and A.B. Walker, Phys. Rev. Lett. **68**, 3873 (1992).

[8] G.R. Massoumi, N. Hozhabri, W.N. Lennard, Peter J. Schultz, S.F. Baert, H.H. Jorch and A.H. Weiss, Rev. Sci. Instr. **62**, 1460 (1991).

[9] E.M. Gullikson and A.P. Mills Jr., Phys. Rev. Lett. **57**, 376 (1986).

[10] K.F. Canter, *Positron Studies of Solids, Surfaces, and Atoms*, edited by A.P. Mills Jr., W.S. Crane, and K.F. Canter (World Scientific, Singapore, 1986).

[11] D.W.O. Heddle, J. Phys. E **4**, 981 (1971); D.W.O. Heddle and N. Papadovassilakis, *ibid* **17**, 599 (1984).

[12] F. Arnal, P. Verdier, and P.D. Vincensini, C. R. Acad. Sci. **268**, 1526 (1969).

[13] P.U. Arifov, A.R. Grupper and H. Alimkulov, *Positron Annihilation*, edited by P.G. Coleman, S.C. Sharma, and L.M. Diana (North-Holland, Amsterdam, 1982) p. 699.

[14] K.F.J. Heinrich, Appl. Spectrosc. **22**, 395 (1968).

[15] A. Bisi and L. Braicovich, Nucl. Phys. **58**, 171 (1964).

[16] H.H. Seliger, Phys. Rev. **88**, 408 (1952).

[17] I.K. MacKenzie, C.W. Schulte, T.E. Jackman and J.L. Campbell, Phys. Rev. **A7**, 135 (1973).

[18] L.R. Logan, M.G. Cottam, Peter J. Schultz, and H.H. Jorch, *Positron Annihilation*, edited by L.Dorikens-Vanpraet, M. Dorikens and D. Segers (World Scientific, 1988) p. 300.

[19] W. Miller, Phys. Rev. **82**, 452 (1951).

[20] G.R. Massoumi, W.N. Lennard, P.J. Schultz, A.B. Walker, and K.O. Jensen, (to be submitted in PRB).

SEARCH FOR RESONANCE-LIKE STRUCTURE IN THE TOTAL SCATTERING OF POSITRONS BY ARGON AND KRYPTON

D. Przybyla, C. K. Kwan, R. A. Lukaszew, W. E. Kauppila, and T. S. Stein
Department of Physics and Astronomy, Wayne State University,
Detroit, Michigan 48202

ABSTRACT

Total scattering cross sections (Q_T's) have been measured for positrons scattered by argon and krypton using a beam transmission technique. We are specifically searching for resonance-like structure in the Q_T's.

INTRODUCTION

Our initial motivation was to investigate whether a resonance-like structure recently seen[1] in the elastic differential cross section (DCS) (when plotted versus energy at fixed scattering angles) for positron-argon scattering between 55 and 60 eV would give rise to some corresponding structure in Q_T's measured over the same energy range. This structure is notable in that it does not occur in the vicinity of any known inelastic scattering thresholds. Additional measurements were then made of Q_T's for positron-krypton scattering at and near 25 and 200 eV to investigate these regions where resonance-like structures have been observed[2] in the elastic DCS (again plotted versus energy at fixed scattering angles).

EXPERIMENTAL TECHNIQUE

Our original beam transmission experiment[3] has been modified by replacing the original ^{11}C positron source with a ^{22}Na source and an annealed tungsten moderator, and by the partial automation of our data acquisition. The full-width at half-maximum of our present positron beam is about 1.5 eV. Previously, for each energy at which we measured a Q_T value, we would maximize the intensity of the positron beam at the channeltron detector (i.e. 'tune' the beam). We have found that for a particular tuning, we can also make measurements of Q_T at nearby energies and get reasonable results. Thus, automating allows us to program a 'sweep' through an energy range and accumulate a large amount of data without tedious retuning for each energy. It should be noted, however, that we expect the cross section to be most reliable at the tuning energy, and that cross sections measured far from that tuning energy would be less reliable than they would be had the positron beam been tuned at those energies. A local maximum may be observed at or near the tuning energy in a sweep over a large energy range.

RESULTS

The present positron-argon Q_T measurements for several tunings at different energies are shown in Fig. 1 along with prior measurements[4,5] made by our group. The present results are in good agreement with the prior results except below 9 eV, where our current measurements become lower due primarily to a broader positron beam energy width (which results in poorer angular discrimination[5]). We have measured Q_T's from 52 to 65 eV in 0.5 eV steps. The Q_T results show no definite structural features between 55 and 60 eV greater than a 1% deviation from a smooth curve.

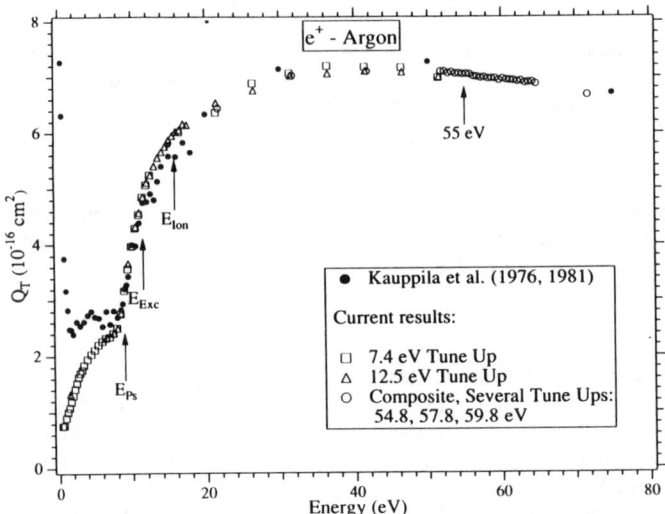

Fig. 1. Positron-argon total cross sections.

Our present measured positron-krypton Q_T's are shown in Fig. 2 along with prior measurements[6,7] made by our group. We agree with Dababneh et al.[6,7] in the shape of the Q_T curve, although our Q_T's are from about 1% to 10% lower. In the vicinity of 200 eV, there appears to be a very slight bump in the Q_T's for tuning energies of 178 and 204 eV. In Fig. 3 we have attempted to scale our Q_T measurements to enhance these features. It is not clear that there is any meaningful structure centered around 200 eV. Near 25 eV we see a maximum in Q_T, but such a structure has been seen in earlier measurements, and there is no reason to believe this is related to the resonance-like structure observed in the elastic DCS.

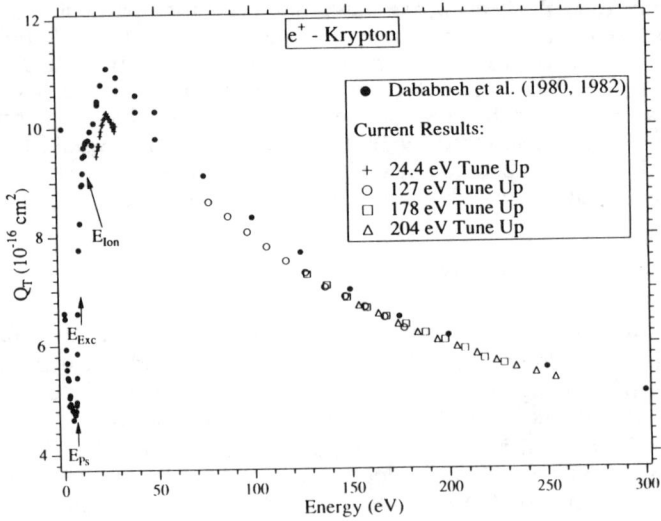

Fig. 2. Positron-krypton total cross sections.

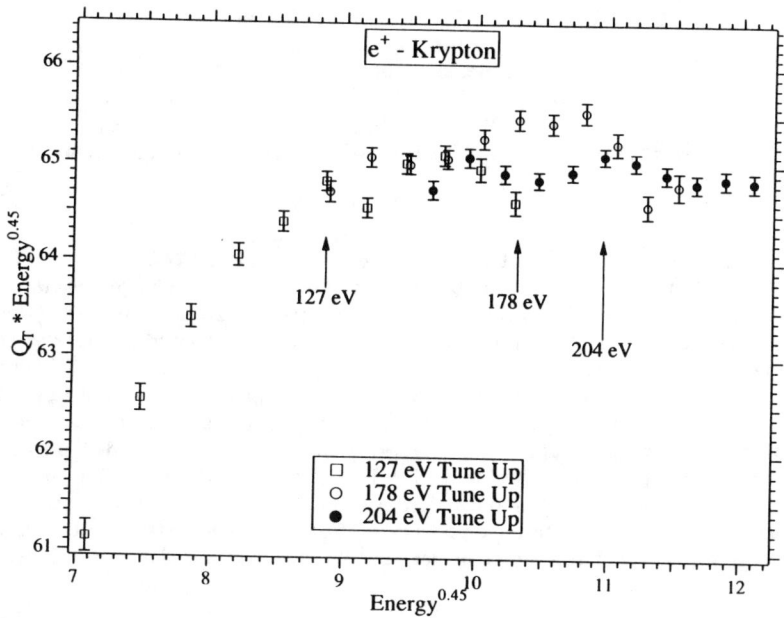

Fig. 3. Positron-krypton scaled total cross sections.

ACKNOWLEDGEMENTS

This work was supported in part by NSF grant PHY90-21044.

REFERENCES

1. L. Dou, W. E. Kauppila, C. K. Kwan, and T. S. Stein, Phys. Rev. Lett. **68**, 2913 (1992).
2. L. Dou, W. E. Kauppila, C. K. Kwan, D. Przybyla, Steven J. Smith, and T. S. Stein, (published in these proceedings).
3. W. E. Kauppila, T. S. Stein, G. Jesion, M. S. Dababneh, and V. Pol, Rev. Sci. Instrum. **48**, 822 (1977).
4. W. E. Kauppila, T. S. Stein, and G. Jesion, Phys. Rev. Lett. **36**, 580 (1976).
5. W. E. Kauppila, T. S. Stein, J. H. Smart, M. S. Dababneh, Y. K. Ho, J. P. Downing, and V. Pol, Phys. Rev. A **24**, 725 (1981).
6. M. S. Dababneh, W. E. Kauppila, J. P. Downing, F. Laperriere, V. Pol, J. H. Smart, and T. S. Stein, Phys. Rev A **22**, 1872 (1980).
7. M. S. Dababneh, Y.-F. Hsieh, W. E. Kauppila, V. Pol, and T. S. Stein, Phys. Rev. A **26**, 1252 (1982).

EXPERIMENTS WITH LOW ENERGY POSITRONS

H. Schneider*, I. Tobehn*, M. Rückert*,
U. Brinkmann**, R. Hippler**
*Strahlenzentrum der Universität Giessen,
Leihgesterner Weg 217, D-6300 Giessen, Germany
**Fakultät für Physik, Universität Bielefeld,
Universitätsstrasse 25, D-4800 Bielefeld 1, Germany

ABSTRACT

The positron source TEPOS at the Giessen LINAC of the Strahlenzentrum delivers intense beams in the energy range between few 10 eV and 5 keV; with postacceleration up to about 80 keV. Results for remoderation of the positrons will be discussed. Energy-losses of positrons in thin aluminium and gold foils at incident energies of 6-20 keV were measured. The results are presented shortly. Relative cross sections for K- and L-shell ionization of silver and gold targets by positron and electron impact were determined at projectile energies of 30-70 keV. The cross sections for positrons decrease more rapidly towards threshold than for electrons. Calculations in plane wave Born approximation (PWBA) which take into account the Coulomb interaction with the target nucleus will also be discussed.

INTRODUCTION

An intense source for low-energy positrons (TEPOS) with a magnetic transporting system has been provided at the Giessen 65 MeV pulsed electron linear accelerator (LINAC) at the Strahlenzentrum of the University. This facility is described elsewhere in detail [1,2].

EXPERIMENTAL RESULTS

To improve the beam quality of our positron source TEPOS (smaller beam diameter and lower energy spread) we performed experiments with various remoderation arrangements in transmission mode (1000 Å W-foil) with different electrostatic positron extraction systems. At the end of the set-up a deflecting spherical condensor (with an adequate diaphragm) to analyse the energy of the positrons, and channel plates are installed. The incoming positrons penetrate into the W-foil, where some of them obtain thermal energies. Then, they diffuse in the foil and a part arrive at the surface without annihilation. There, they may be reemitted (remoderated positrons) with kinetic energies corresponding to the work function of tungsten and are subsequently submitted to a postaccelerating system. The maximum yield of remoderated positrons is achieved at an energy of the incoming positrons of about 5 keV[2,3].

An arrangement for remoderation in a 1000 Å thick W-foil or for measurements of the energy-loss of positrons in thin aluminium foils at incident energies of 6-20 keV, respectively, is shown in fig. 1.

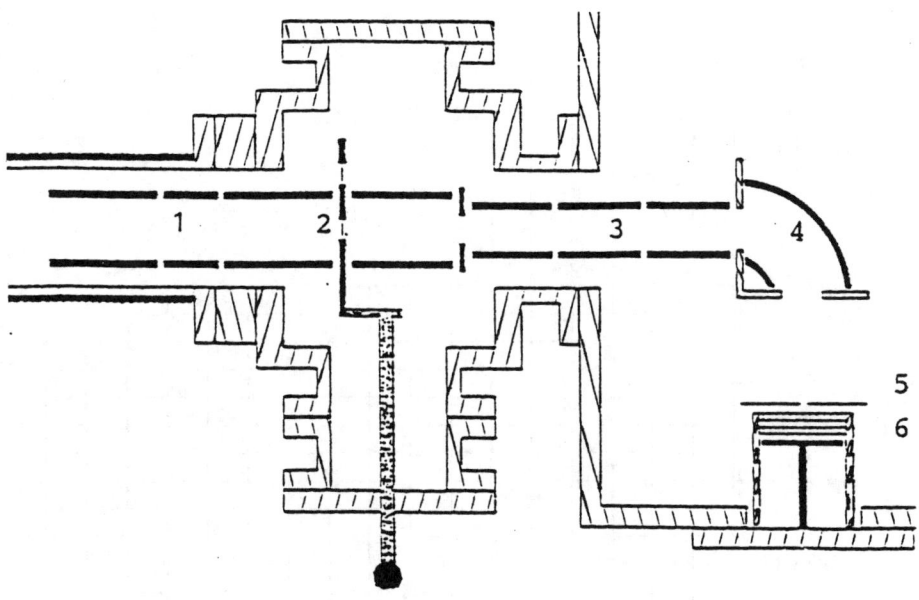

Fig. 1. Arrangement for remoderation or measurement of energy-losses alternatively. 1,3: Electrostatic lenses, 2: Target, 4: Spherical capacitor or tandem spectrometer, 5: Slit, 6: Channel plates or channeltron, respectively.

An electrostatic tandem spectrometer with a channeltron is used alternatively for analysing the mean energy-losses instead of the spherical condensor. First experimental results for forward-scattered positrons are presented in fig. 2.

The ionization of inner atomic shells, for example, is one of the processes which are of fundamental importance for our understanding of collision dynamics [4-8]. For a quantitative analysis, reliable cross sections for impact ionization are required.

Here we present experimental results for K- and L-shell ionization of atoms by electron and positron impact, respectively, at incident energies close to ionization threshold. At the present experiments the positrons or electrons enter a vacuum chamber, containing the target support with the target foil [9,10]. The kinetic energy E_0 of the impinging positrons or electrons, respectively, can be varied by a suitably chosen electrical potential at which the target is held. The target foil is hit under 45°. The x-rays emitted from the target had to pass through a 50 μm mylar window, and are registered perpendicular to the incident beam by a Si(Li) detector.

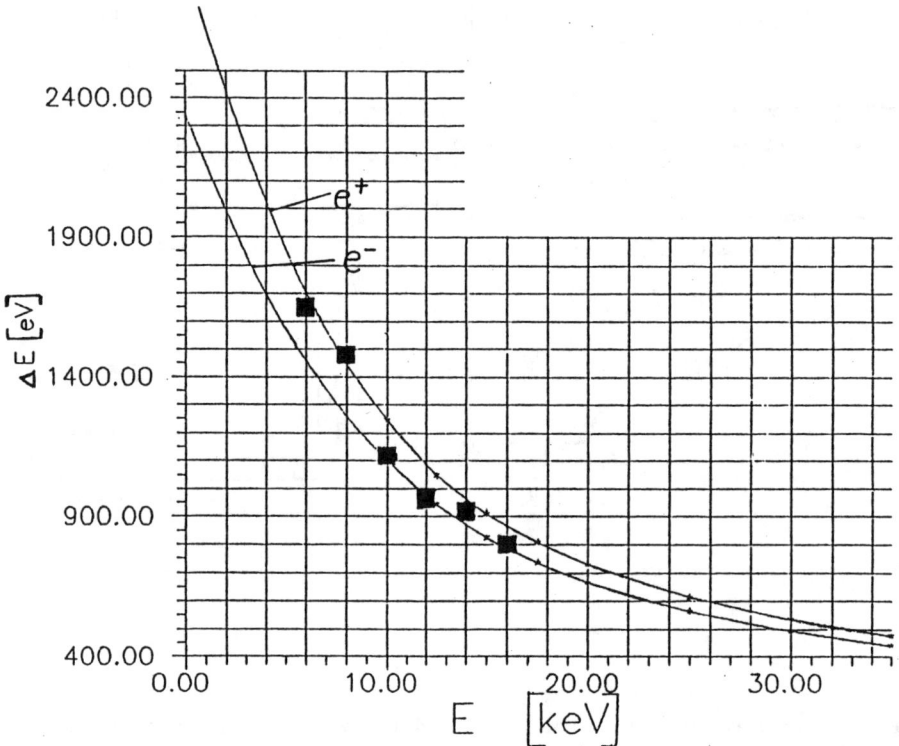

Fig. 2. Measured energy-losses for positrons in a 2500 Å thick Al-foil (■). The curves represent tabulated stopping powers (M.J. Berger and S.M. Seltzer, National Bureau of Standards, Vahington, D.C., USA).

Former experiments on *K-shell ionization* were performed for silver and copper targets [9]; the thickness of these foils ranged from 100 to 520 μg/cm². In the silver case the x-rays resulting from K-shell ionization (I= 25.52 keV; E_0/I= 1,7...2.7) were normalized to the simultaneously recorded x-rays from the silver L_{III}-shell (I= 3.35 keV; corresponding E_0/I= 12...21). Such a normalization procedure is useful as long as the cross section ratio σ^-_L/σ^+_L for L-shell ionization by electron and positron impact equals unity, which holds for $E_0/I > 9$.

In this communication we present and discuss also new experimental results for *L-shell ionization* of gold by lepton (electron, positron) impact at incident energies between 30 and 70 keV [10]. We performed the experiments with *Au/Ag-multilayer-targets*; several targets were used, e.g. Au 400 Å/Ag 125 Å, Au 600 Å/Ag 1000 Å, Au 800 Å/Ag 1800 Å on a carbon-backing and Au 1450 Å/Ag 3500 Å. The purpose of these multilayer targets was to use also the Ag-L x-rays as reference data to which the Au-L x-rays data were normalized. The obtained cross section ratios σ^-_K/σ^+_K and σ^-_L/σ^+_L for positron and electron impact are displayed in fig. 3 (cp. [9,10]).

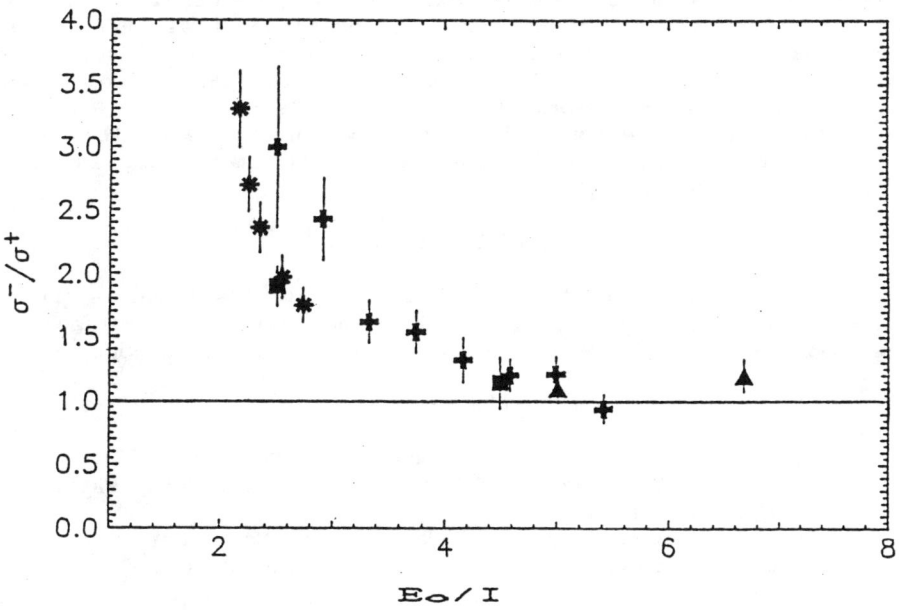

Fig. 3. Cross section ratios σ^-/σ^+. (✻): Ag-K-shell [9], (▲): Cu-K-shell [9], (✚): Au-L-shell [10], (■): Cu-K-shell [6].

As is shown by our calculations, the cross section ratio $\sigma_{e^-}/\sigma_{e^+}$ as a function of the reduced incident energy E_0/I displays a pronounced energy dependence below $E_0/I \lesssim 5$, whereas due to the near cancellation of the combined Coulomb and exchange effects this ratio is close to unity already for $E_0/I \gtrsim 10$. In the energy region of interest here corresponding to $E_0/I = 2.5...5.5$ for Au-L_{III} (I≃ 11.92 keV) but 9...19 for Ag-L our calculations give $\sigma(Ag - \Sigma L)_{e^+}/\sigma(Ag - \Sigma L)_{e^-} = 1.047\pm0.011$. This implies that the silver L-shell data can serve as a reference to which the Au-L_{III} data can be normalized. With the above result we can form the double ratio10

$$\sigma^-/\sigma^+ \equiv$$
$$[\sigma(Au - L_\alpha)/\sigma(Ag - \Sigma L)]_{e^-} / [\sigma(Au - L_\alpha)/\sigma(Ag - \Sigma L)]_{e^+} =$$
$$\sigma(Au - L_\alpha)_{e^-}/\sigma(Au - L_\alpha)_{e^+} \cdot (1.047\pm0.011).$$

CONCLUSION

The measurements of energy-losses are due to further investigation. Of interest are the differences between positrons and electrons.

The departure from the prediction of PWBA observed for K- and L- shell ionization by electron and positron impact [7,8] is mainly due to two effects: (i) the projectile - target nucleus interaction (Coulomb effect, C), and (ii) electron exchange between the incident and the bound electron (exchange effect, Ex). The measured ratio increases with decreasing projectile energy; this result is in fair agreement with our present calculations and can be explained by the Coulomb effect which is most important at small E_0/I. This means that the cross section ratio σ^-/σ^+ increases with decreasing energy of the projectiles, because electrons are accelerated, whereas positrons are slowed down in the nuclear field of the target. This effect overcompensates the exchange effect. The cross section ratio would otherwise increase with increasing incident projectile energy (cp. [11]).

ACKNOWLEDGEMENTS

The technical support by the operating staff of the linac is gratefully acknowledged. We thank the Deutsche Forschungsgemeinschaft, Bonn-Bad Godesberg (Germany), for generous financial support.

REFERENCES

1. F. Ebel, W. Faust, C. Hahn, S. Langer, M. Rückert, H. Schneider, A. Singe, I. Tobehn, Nucl. Instr. Meth. Phys. Res. A272, 626 (1988).
2. F. Ebel, W. Faust, C. Hahn, M. Rückert, H. Schneider, A. Singe, I. Tobehn, Nucl. Instr. Meth. Phys. Res. B50, 328 (1990).
3. W. Faust, C. Hahn, M. Rückert, H. Schneider, A. Singe, I. Tobehn, Nucl. Instr. Meth. Phys. Res. B56/57, 575 (1991).
4. C.J. Powell, Rev. Mod. Phys. 48, 33 (1976); and in: Electron Impact Ionization (T.D. Märk, G.H. Dunn, Eds.), Chap. 6, Springer, Berlin (1986).
5. S. Ito, S. Shimizu, T. Kawaratani, K. Kubota, Phys. Rev. A22, 407 (1980).
6. P.J. Schultz, J.L. Campbell, Phys. Lett. A112, 316 (1985).
7. R. Hippler, Phys. Letters A144 81 (1990).
8. H. Schneider, I. Tobehn, R. Hippler, Hyperfine Interactions 73, 17 (1992).
9. F. Ebel, W. Faust, C. Hahn, M. Rückert, H. Schneider, A. Singe, I. Tobehn, Phys. Letters A140, 114 (1989).
10. H. Schneider, I. Tobehn, R. Hippler, Phys. Letters A156, 303 (1991), Nucl. Instr. Phys. Res. B68, 491 (1992).
11. V.N. Lennard, P.J. Schultz, G.R. Massoumi, L.R. Logan, Phys. Rev. Lett. 61, 2428 (1988).

FORMATION OF POSITRONIUM COMPOUNDS IN SLOW POSITRON BEAMS

D. M. Schrader
Chemistry Department, Marquette University, Milwaukee, WI 53233

ABSTRACT

The formation of stable positronium compounds in rearrangement collisions of positrons and molecules is considered. The discussion focuses on energetics and quantum mechanics, with emphasis being given to the role of the intermediate resonance state. The calculation of the cross section and the role of spin is considered. Some conjectures concerning possible future work are given.

1. INTRODUCTION

We consider the reaction:

$$e^+ + AB \rightleftharpoons [e^+,AB] \rightarrow PsA + B^+ \qquad (1)$$

A and B are two parts of any molecule. This is a familiar type of chemical reaction in that two reactants come together and form an intermediate which can dissociate either to the original reactants or to two new entities, which we call products. Its only novelty as a chemical reaction is that one of the reactants is antimatter. In physics, reaction (1) is also familiar as an example of dissociative attachment or rearrangement, or a rearrangement collision.[1]

The intermediate is called, in chemistry, an activated complex, or a transition state; in physics, it is called a resonance. This short-lived structure plays a crucial role in the reaction (1).

The product of interest, PsA, is considered to be stable if its energy is below that of any separated parts, such as Ps + A. PsA annihilates, of course, and may be said to be unstable in that sense. This discussion is about chemical stability, not annihilative stability.

One specific example of the generic reaction (1) has A = H and B = CH_3; this was the subject of a recent experiment.[2] Other interesting examples include the formation of positronium chloride, PsCl, (A = Cl and B = CH_3) and sodium positride, NaPs, (A = Na and B = Na_{n-1}^{2+}), in which the targets are, respectively, methyl chloride, CH_3Cl, and a singly charged cluster of n sodium atoms.

2. ENERGETICS

2.1 The Threshold

In order for the reaction (1) to take place, energy is required to break the A-B bond and to ionize B. Energy is gained in the formation of positronium out of the incident positron and the electron from B, and in the formation of the Ps-A bond. Thermodynamically,

$$\Delta E = BE(A-B) + IP(B) - 6.8eV - BE(Ps-A). \qquad (2)$$

This is the threshold for the reaction, which is the minimum energy requirement in order for the reaction to be possible. Normally, AB is in its lowest quantum state and possesses energy of the order kT, which is contained in translational, rotational, and possible low frequency vibrational modes. Thus the energy ΔE must be supplied by the positron.

© 1994 American Institute of Physics

2.2 The Appearance Potential of B⁺

In practice, one proves that the formation of the compound PsA occurs by detecting the fragment ion B⁺ for a positron energy in a region in which (1) is the only possible source of B⁺. It is important to realize that the first step in the reaction, the formation of the resonance [e⁺,AB], takes place within an interval of time which is short compared to the period of a molecular vibration; and the second step, the dissociation of the resonance, is much slower because massive molecular fragments must recede from each other. They do this by proceeding along the potential energy curve of the resonance.

The situation is illustrated in fig. 1 in which H_2 is the target AB. The formation of the resonance is a vertical process starting from the ground state of the target. The potential energy curves for the resonance are not known. Two hypothetical curves labelled (a) and (b) are shown for the purposes of this discussion.

If (a) is the correct curve, then the signature ion H⁺ begins to form at a positron energy of 10.22eV, which is ΔE for this case. The next threshold for proton production is that for dissociative positronium formation, which is 11.28eV. The portion of the Franck-Condon region which lies between these two thresholds is doubly

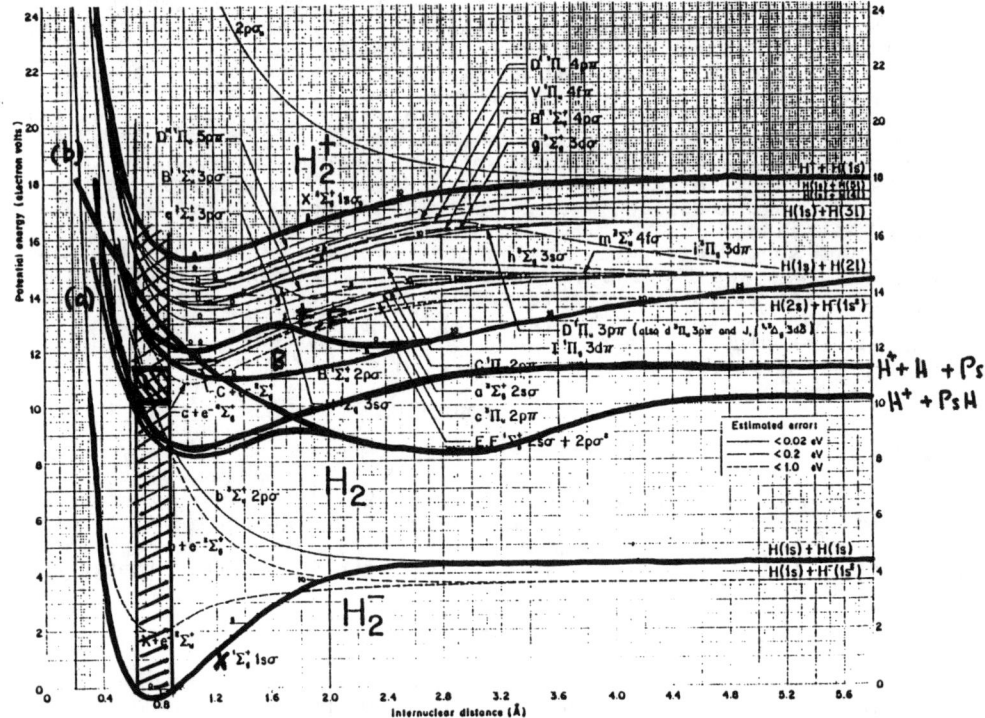

Fig. 1. Potential energy curves for hydrogen, H_2.[3] States important to the discussion are highlighted. The Franck-Condon region, in which the scattering takes place, is shaded. Curves (a) and (b) are two hypothetical potential energy curves for the resonance [e⁺,H_2]. The curve labelled "H⁺ + H + Ps" in the right margin is the H_2^+ potential curve drawn 6.8eV below its proper place.

shaded in fig. 1. We call this region the "window" for PsH formation. In the window, reaction (1) is the only possible process for forming a proton. Therefore, the detection of a proton for a positron beam energy in this region is proof of the formation of PsH.

If, on the other hand, (b) is the correct curve, then the minimum energy requirement for reaction (1) is the energy of the lowest crossing of the curve (b) with the Franck-Condon region, which is about 12.7eV. This is greater than the threshold for the formation of positronium in a dissociative reaction, so protons will be formed above 12.7eV from both processes. The two can be distinguished by the energy of the receding molecular fragments. If PsH is formed at a positron energy of 12.7eV, then 2.5eV will be shared between it and H^+; but if dissociative positronium formation takes place instead, 1.4eV will be shared between the fragments H^+, H, and Ps. In the former case, the two fragments are of virtually the same mass, so the energy will be equally shared and the proton will recede with a kinetic energy of 1.2eV; in the latter, the kinematics is more complicated, so a simple quantitative deduction cannot be made. One can envision a fast process in which the incoming positron extracts an electron from the target and leaves as Ps, followed by a slow process in which the proton and the hydrogen atom recede from each other. In this case, that part of the 1.4eV left behind by the departed positronium will be equally shared by H^+ and H, so the maximum proton energy is about 0.7eV. An experimental determination of the energy of the signature proton which has an uncertainty of less than half a volt at an energy of one electron volt will distinguish between the two processes.

In case (a), the appearance potential of the proton is the same as the threshold ΔE; in case (b), the appearance potential of the proton exceeds ΔE by 2.5eV, the energy difference between the bottom of the window and the minimum-energy crossing of curve (b) with the Franck-Condon region. In general, we may write

$$AP(B^+) = \Delta E + \varepsilon, \qquad (3)$$

where ε is defined by the immediately preceding text.

2.3 The Binding Energy of PsA

The binding energies of only a handful of positronium compounds are known. In an experiment, the appearance potential of B^+ is measured directly. If the kinetic energy of B^+ is also measured, then ε and hence ΔE can be determined. If the bond energy and ionization potential in eq. (2) are known, the binding energy of PsA can then be deduced. On the other hand, if the appearance potential of B^+ but not its kinetic energy is measured, then only a lower bound on the PsA binding energy is obtained. If this quantity is positive, stability is proven; if several measurements on different compounds yield the same lower bound, then one would hope that this lower bound is also the lowest possible value, the PsA binding energy.

3. QUANTUM MECHANICS

It takes 10^{-16} sec for a 10eV positron to traverse a hydrogen molecule; this is to be identified as the time required for the resonance to form in competition with elastic scattering. The period of one vibration of H_2 is 10^{-14} sec; this is about how long it takes for the resonance to dissociate into the products H^+ and PsH. At any time during this dissociation, the resonance can autoionize; i.e., rearrange itself into a positron and molecular hydrogen in its ground electronic state. This is the reverse of resonance formation; presumably it requires 10^{-16} sec also. Autoionization is represented in (1) by the left-pointing arrow. Other autoionizing channels may also be open, depending on whether the potential energy curve for the resonance crosses that for an electronically excited state of H_2. The lowest such state is the b $^3\Sigma_u^+$ state, but since the Hamiltonian contains no spin terms, spin quantum numbers are constants of motion. Therefore, autoionization involving the formation of this state, or any other triplet, is not allowed.

Positronium formation, accompanied by the formation of H_2^+, also requires 10^{-16} sec. This process is another form of autoionization, and it competes with dissociation to PsH until the protons separate farther than about 2Å, if we take fig. 1 literally.

Processes which, like resonance formation, autoionization, and positronium formation, take place in times substantially shorter than the period of a molecular vibration are vertical processes; i.e., the Franck-Condon principle applies.

Annihilation requires 10^{-9} sec, and so it takes place long after the chemistry of reaction (1) is over. We do not consider it further in this discussion.

3.1 The Calculation of the Cross Section

The cross section for the formation of PsA, σ_{PsA}, is the product of two factors: the cross section for the formation of the resonance in the Franck-Condon region, and probability that the resonance will dissociate to PsA + B^+ before it autoionizes. Each of these three processes, formation of the resonance, dissociation to products, and autoionization to reactants, is represented by an appropriate arrow in the reaction (1).

The cross section for the formation of the resonance, σ_{res}, may be calculated by any of several means. It is shown elsewhere[4] that, in the case of AB = H_2, this cross section is probably close to that for the excitation of the B $^1\Sigma_u^+$ state by electron impact.

The time required for dissociation to products is calculated with reasonable accuracy by treating the motion classically. The relative velocity v of the products is given by

$$\tfrac{1}{2}\mu v^2 = E_{beam} - V_{res}(R), \qquad (4)$$

where μ is the reduced mass of the receding products, E_{beam} is the energy of the positron beam, and V_{res} is the potential energy curve of the resonance. Since the latter depends upon the internuclear distance, as indicated, so does v.

The autoionization rate γ is a strongly varying function of the proton-proton distance R. This is plausible if one considers the structure of the resonance. In the Franck-Condon region all the leptons are close to both protons, so rearrangement to a free positron and the X state of the target molecule, which contributes to elastic scattering, must be an important process. But for large R, all the leptons are on one proton, and autoionization requires the long-range transfer of an electron to the other proton. We expect γ to be large in the Franck-Condon region and to decrease strongly as R increases. This is true for all targets, not just H_2.

The probability for the resonance to have survived autoionization at time t (measured from the formation of the resonance) is $e^{-\rho}$, where ρ is

$$\rho = \int_0^t \gamma(R)\, dt' = \int_{R_0}^{R_{asy}} \frac{\gamma(R)}{v(R)}\, dR. \qquad (5)$$

R_0 is the internuclear distance in the Franck-Condon region, and R_{asy}

is a value of R large enough so that autoionization is negligible.

3.2 The Structure of the Resonance

In order to calculate ρ and hence σ_{PsA}, we need first to calculate γ and V_{res}. These tasks requires the wave function of the resonance. The case of AB = H_2, considered elsewhere recently,[4] is instructive as well as simple. As noted above, spin quantum numbers are conserved throughout reaction (1), and for a two-electron target, that means that the spin wave function is unchanged during the reaction. Hence, spin plays no part in the dynamics; we suppress the spin wave functions in the remainder of this discussion.

If ψ_a is the wave function for PsH with the leptons on proton "a," then the wave functions for the product PsH + H^+ are, for large R:

$$\Psi_g = \frac{1}{\sqrt{2}}(\psi_a + \psi_b); \quad {}^{2,1}\Sigma_g^+$$
$$\Psi_u = \frac{1}{\sqrt{2}}(\psi_a - \psi_b); \quad {}^{2,1}\Sigma_u^+ \quad (6)$$

From this we see that, since scattering takes place from the H_2 ${}^1\Sigma_g^+$ ground state, only σ_g- and σ_u-wave positron scattering in the incident channel contributes to PsH formation in the exit channel. The notation "2,1" in the term symbols above are the values of 2S+1 for all three leptons and for only the electrons, respectively.[5] The energy for each of these states is 10.22eV on the right side of fig. 1. They are the resonance in the asymptotic region, and they correlate with two states of the resonance for all other R. Curves (a) and (b) in fig. 1 are hypothetical potential curves for these two states. In addition to not knowing where these curves really are in fig. 1 (except as R→∞), we also do not know whether the g- or the u-state has the lower energy in the Franck-Condon region.

Other states in the asymptotic region are those for which PsH is not involved. Those significant to the present discussion are:

$$ {}^{2,1}\Sigma_g^+ {}^{2,1}\Sigma_u^+$$

$$e^+ + H_2(X) \quad f_g(ab+ba) \quad f_u(ab+ba) \quad (7)$$
$$e^+ + H_2(B) \quad f_u(\chi_a - \chi_b) \quad f_g(\chi_a - \chi_b) \quad (8)$$
$$e^+ + H_2(E,F) \quad f_g(\chi_a + \chi_b) \quad f_u(\chi_a + \chi_b) \quad (9)$$
$$Ps + H_2^+ \quad F_g\varphi + \varphi F_g \quad F_u\varphi + \varphi F_u \quad (10)$$

f and F are scattering waves for the positron and positronium, resp.; a and b in (7) stand for a 1s orbital on proton "a" and "b", resp.; χ_a is the wave function for the hydride ion, H^-, on proton "a"; and φ is the ground state wave function for H_2^+. The order of arguments for the electrons is meant to be "1" first, then "2" in all expressions above; spin conservation requires that all the wave functions considered here be symmetric under interchange of electrons. Hence the appearance of line (10).

By writing $f_g \approx ks_a + ks_b$ and similarly for f_u, one can expand the four wave functions in lines (8) and (9) and show that each is comprised equally of two parts: one of these parts is one of the PsH + H^+ wave functions of (6), and the other part is:

$$\frac{1}{\sqrt{2}}(ks_a \chi_b \pm ks_b \chi_a) \qquad (11)$$

Conversely, each of the wave functions in (6) can be decomposed equally into two parts, one of which has the B state of H_2 as its electronic parent, and the other, the E,F state. This is true only in the asymptotic region, but it has interesting implications for the Franck-Condon region, where the resonance forms. There, the wave function is still a mixture of B- and E,F-like parental parts, but those two states are about one electron volt apart in energy. It seems that one of the resonance states will be mostly B-like in its electronic parent, and the other, mostly E,F-like. These two parental states are no longer strongly ionic at that internuclear distance,[6,7] so perhaps the resonance energy will lie above the parent states. Only a good quantum mechanical calculation of the energy will tell.

In (11), the positron is on the "wrong" proton; this state correlates to the electronic parent $H^+ + H^-$, which (interestingly) is not shown in Sharp's figure (fig. 1). Its energy is 17.33eV.

3.3 Angular Dependence of PsA Formation

Both the incident positron and the molecular target in reaction (1) define the symmetry of the system. The conservation of this symmetry gives rise to a dependence of the cross section for the formation of PsA upon the angle defined by the internuclear axis of the target molecule and the direction of approach of the positron. This angle is the outgoing direction of PsA and B^+ since the dissociation proceeds along the internuclear axis and the rotational period of the target is long compared to the time required for dissociation. It follows that the angle of detection of the signature ion with respect to the positron beam direction should be selected with the angular dependence of the cross section in mind.

This angular dependence has been considered by Dunn,[8] who has worked out selection rules for dissociative electron attachment to a diatomic molecule for two important special cases: the axis of the ion detector parallel to, and perpendicular to the positron beam axis. His results apply equally well to reaction (1) for a diatomic target, and his ideas can be extended to more complicated targets as well.[9] For the target H_2, Dunn's results show that only σ_g- and σ_u-wave positrons are capable of forming $PsH + H^+$ for either orientation, and that for the perpendicular orientation, which is the one used in the only experiment performed to date,[2] the σ_g-wave gives rise to the $^{2,1}\Sigma_g^+$ resonance, and the σ_u-wave, to the $^{2,1}\Sigma_u^+$. Thus, both possible potential curves are accessible for the perpendicular orientation for this target.

In spite of the lack of specificity for this case, one should make these considerations for each target in order to avoid falling into the trap of trying to do an impossible experiment.

3.4 Spin Conservation

We have already mentioned that spin is conserved throughout reaction (1). This gives rise to additional selection rules. For example, if the target AB is a singlet and PsA has a total electron spin of zero, then the spin state of the signature ion B^+ must also be a singlet. One should not, therefore, try to form PsH by scattering positrons off HCl and detecting Cl^+, for the ground state of that ion is ^3P. The lowest singlet state of Cl^+ is 1.44eV above the ground state, but 0.36eV below the opening of the channel for dissociative positronium formation. Therefore, before any PsH forms, there will be a copious production of Cl^*, and the only way to distinguish between the two channels is to distinguish between triplet and singlet Cl^*.

4. CONCLUSIONS AND CONJECTURES

Although high quality information about binding energies is available for only a handful of positronium compounds, we nevertheless state categorically that many of them are stable. We believe that positronium can substitute for the hydrogen atom in many compounds. For example, we know that PsH,[10] PsCl,[11] and the other positronium halides[12] are stable; we believe that PsOH, $PsNO_2$, CH_3Ps, C_6H_5Ps, NaPs, etc., are also stable. It seems important to have a quantitative understanding of these exotic molecules in order to carry out meaningful work in using positrons as probes of surfaces, voids, etc.

Predicting whether these compounds will yield their binding energies in a simple experiment such as the one already carried out[2] requires further developments in the theory. We do not know, for example, why H_2 did not give a proton signal in the PsH-formation energy window. Is it because the resonance potential crosses the Franck-Condon region above the window? And is this because all the electrons are in the K-shell? Or because H_2 is homonuclear instead of just nonpolar?

It would seem, paradoxically, that nonpolar targets may more easily produce positronium compounds than polar. This is plausible because compound formation requires an ion-pair electronic parent in the dissociation limit, and this correlates with an ionic parent target molecule. Ground states wave functions have to be orthogonal with those of excited states, and one way for this to be is for the excited state wave function to be ionic. So low-lying ionic excited states might be available for most nonpolar targets. Polar ground state targets, on the other hand, will have nonpolar excited states, which are poor candidates for attachment by a positron.

Similarly, polar targets, like HCl, might have low lying excited states which are "reverse ionic," i.e., H^-Cl^+, and might more easily form the unexpected product (PsH rather than PsCl).

Clearly, both the qualitative and quantitative quantum mechanics of the formation process needs further development; it is desirable to measure the energy of the signature ion; and it is important to think about spin and about the symmetry restrictions of the detector geometry.

Besides the immediate spin-off of supplying basic energetic information to materials science, radiation chemistry of positrons, surface science, and so forth, the study of stable positronium compounds will shed light on questions having to do with the nature of Surko's hypothesized vibrational resonances,[13] the mystery of the giant Z_{eff} values[14] of simple organic molecules, and the surprising fragmentation patterns in positronic mass spectrometry recently observed.[15]

In addition, the study of stable positronium compounds is important for its own sake.

REFERENCES

1. H. S. W. Massey, Electronic and Ionic Impact Phenomena, 2nd ed., vol. II (Clarendon Press, Oxford, 1969).
2. D. M. Schrader, F. M. Jacobsen, N.-P. Frandsen, and U. Mikkelsen, Phys. Rev. Lett. 57, 6 (1992).
3. T. E. Sharp, Atomic Data 2, 119 (1971).
4. D. M. Schrader, Theoret. Chim. Acta 82, 425 (1992).

5. C. F. Lebeda and D. M. Schrader, Phys. Rev. 178, 24 (1969).
6. W. Kołos and L. Wolniewicz, J. Chem. Phys. 45, 509 (1966).
7. E. R. Davidson, J. Chem. Phys. 33, 1577 (1960).
8. G. H. Dunn, Phys. Rev. Lett. 8, 62 (1962).
9. J. N. Murrell, S. Carter, S. C. Farantos, P. Huxley, and A. J. C. Varandas, Molecular Potential Energy Functions (Wiley, 1984).
10. Y. K. Ho, Phys. Rev. A 34, 609 (1986).
11. D. M. Schrader, T. Yoshida, and K. Iguchi, Phys. Rev. Lett. 68, 3281 (1992).
12. D. M. Schrader, T. Yoshida, and K. Iguchi, unpublished.
13. A. Passner, C. M. Surko, M. Leventhal, and A. P. Mills, Jr., Phys. Rev. A 39, 3706 (1989).
14. J. D. McNutt, V. B. Summerour, A. D. Ray, and P. H. Huang, J. Chem. Phys. 62, 1777 (1975).
15. D. L. Donohue, G. L. Glish, S. A. McLuckey, L. D. Hulett, Jr., H. S. McKown, and S. Pendyala, in Positron Annihilation, edited by L. Dorikens-Vanpraet, M. Dorikens, and D. Segers (World Scientific Publ. Co., Singapore, 1989), p. 612.

Energy Loss of Subionizing Charged Particles in Polar Media

Sergey V. Stepanov
ITEP, Moscow, 117259, Russia

ABSTRACT

After the well-known paper of Froehlich and Platzman [1] the problem of energy loss of subionizing particles (which does not excite electron transitions in molecules) in dielectrics has been discussed many times in radiation-chemical literature.

Nevertheless the available formulas for dielectric loss rate $-dW/dt$ are distinguished in some cases even parametrically. It is enough for example to compare eqs.(35,43) from [2] in the limit $\frac{\varepsilon_\infty}{\varepsilon_0} \cdot \frac{v\tau}{a} \ll 1$ and eq.(23) in [3], which describe dielectric energy loss of a charged particle of radius a moving with the velocity v in a polar media. Here ε_0 and ε_∞ are the static and high frequency dielectric constants and τ is the relaxation time.

The aim of the present paper is to recalculate analytically a dielectric energy loss rate of the particle moving straightforwardly or diffusively in polar media in the frame of excluded sphere approximation. The obtained formula reproduce as the limiting cases the known particular results [1,3]. We have also presented a semiphenomenological approximation for taking into account molecular vibrations. It is shown in accordance with [4] that the energy losses of subionizing electrons in dielectrics are caused mainly by excitations of intra- and intermolecular vibrations. The contribution of the temperature dependent Debye losses is small.

1. Introduction

After the well-known paper of Froehlich and Platzman [1] the problem of energy loss of subionizing particles (which does not excite electron transitions in molecules) in dielectrics has been discussed many times in radiation-chemical literature.

Subionizing particles (electrons, positrons, muons, protons, ions) can loose their energy in direct collisions exciting intra- and intermolecular vibrations. In this case a particle transfers to the medium a definite quanta of energy. In polar media there is an additional way of slowing down of the particle. It is in essence purely classical because the particle loses its energy quasi continuously slightly orienting a large number of molecules around its trajectory.

For quantitative consideration of the first possibility (direct energy losses) it is necessary to know the wave functions of the particle in initial and final states and the type of interaction. Indirect losses may be discussed in the frame of classical electrodynamics.

The problem of energy loss in the first place attracts an attention of radiochemists and radiobiologists. High energy radiation produces a great number of secondary electrons with the relatively low energies (about several eV). Having no possibility to ionize molecules they spent energy exciting molecular vibrations. In polar media there is an additional way: energy loss to orientation of the dipole molecules. In water and aqueous solutions of complicated organic compounds such a process leads to the rupture of hydrogen bonds and denaturation of biomolecules i.e. to radiobiological effects. For chemistry of exotic atoms (positronium, muonium, tritium) this problem is important in view of diffusion recombination model of the formation of these particles.

An attempt of general consideration of the indirect losses for a straight-line and random motions of the finite size particle was made by Tachiya and Sano [2]. In spite of the agreement of their results and that from [1] it is puzzled an appreciate discrepancy between their formula and the Zwanzig expression for dielectric friction force in the case of slow moving particle [3]. Therefore we have decided to reexamine that papers which are considered as classical since their publications.

The aim of the present paper is to recalculate analytically a dielectric energy loss rate of the charged particle moving straightforwardly or diffusively in polar media. The obtained formula reproduce as the limiting cases the known particular results [1,3]. We have also presented a semiphenomenological approximation for taking into account molecular vibrations. It is shown in accordance with [4] that the energy losses of subionizing electrons in dielectrics are caused mainly by excitations of intra- and intermolecular vibrations. The contribution of the temperature dependent Debye losses is small.

2. General Formulation

The rate $-\dot{W}$ of energy loss of a charged particle moving in a dielectric medium is given by [5]

$$-\dot{W} = \frac{1}{4\pi} \int d^3r \cdot \vec{E}(\vec{r},t) \frac{\partial}{\partial t} \vec{D}(\vec{r},t) , \qquad (1)$$

where $\vec{E}(\vec{r},t)$ and $\vec{D}(\vec{r},t)$ are the electric field and displacement at point $\vec{r}$ and time t produced by the particle placed at $\vec{x}(t)$. Here an integration range is the whole space except a spherical volume of the radius a around the particle (so called "excluded sphere approximation" (ESA)).

The nature of the dielectric losses is that the polar medium can not rapidly adjust to the instantaneous particle' position. The delayed reorientation of the dipoles results in an electrical drag force on a moving particle. Such a retarded response of polarization $\vec{P}(\vec{r},t)$ may be described in terms of memory (aftereffect) function $\gamma(t)$

$$\vec{P}(\vec{r},t) = \frac{1}{4\pi} \int_0^\infty dt_1 \gamma(t_1) \vec{D}(\vec{r}, t-t_1) . \qquad (2)$$

On Fourier inversion in frequency this leads to the relationship between $\gamma(t)$ and reciprocal dielectric susceptibility

$$\int_0^\infty dt\, e^{i\omega t} \gamma(t) = 1 - \frac{1}{\varepsilon(\omega)} . \qquad (3)$$

Eqs. (2,3) assume that

$$\vec{D}(\vec{r},t) = \vec{E}(\vec{r},t) + 4\pi \vec{P}(\vec{r},t) \qquad (4)$$

in the time representation and

$$\vec{D}(\vec{r},\omega) = \varepsilon(\omega) \cdot \vec{E}(\vec{r},\omega) = \vec{E}(\vec{r},\omega) + 4\pi \vec{P}(\vec{r},\omega) \qquad (5)$$

in the Fourier transform. Using eq. (4) we can rewrite eq. (1) as

$$-\dot{W} = - \int d^3r \cdot \vec{P}(\vec{r},t) \frac{\partial}{\partial t} \vec{D}(\vec{r},t) , \qquad (6)$$

because the term $\int d^3r \cdot \vec{D}(\vec{r},t) \frac{\partial}{\partial t} \vec{D}(\vec{r},t)$ is equal zero (see eq. (16) $\psi(0)=0$). Due to this reason we can omit a singular term ($\sim \delta(t)$) in the memory function $\gamma(t)$ having the following structure

$$\gamma(t) = \gamma_{sing} \cdot \delta(t) + \gamma_{reg}(t). \qquad (7)$$

This term gives no contribution to the retardation of the particle. Therefore further we will hold only regular part $\gamma_{reg}(t)$ of the memory function.

Substituting eq. (2) into eq. (6) one obtains

$$-\dot{W} = - \frac{1}{4\pi} \int_0^\infty dt_1 \gamma_{reg}(t_1) \cdot \psi(t, t_1) , \qquad (8)$$

$$\psi(t,t_1) = \int d^3r \cdot \vec{D}(\vec{r}, t-t_1) \frac{\partial}{\partial t} \vec{D}(\vec{r}, t) . \qquad (9)$$

Dashed region shows an integration range. We assume that ψ is invariant to the shifting over t i.e. the particle moves uniformly. This means that $\psi(t,t_1) = \psi(t+\Delta t, t_1) = \psi(t_1)$. Therefore

$$\psi(t_1) = \int d^3r \cdot \vec{D}(\vec{r}, t) \frac{\partial}{\partial t_1} \vec{D}(\vec{r}, t+t_1) . \qquad (10)$$

Calculating the space integral in eqs. (8,9) when the displacement $\vec{D}$ is created by a particle moving with a constant velocity, Zwanzig [3] have expressed the friction force $F_{drag} = W/v$ through $\gamma(t)$. However it seems more convenient to obtain an expression for $-\dot{W}$ through an experimentally measurable function $\varepsilon(\omega)$. For that it is enough to integrate eq. (10) by parts:

$$-\dot{W} = \frac{1}{4\pi} \int_0^\infty dt \cdot \gamma'_{reg}(t) \int_0^t \psi(t_1) dt_1 , \qquad \gamma'_{reg}(t) = \frac{d\gamma_{reg}(t)}{dt} \qquad (11)$$

It is important to note that for convenience of integration by parts it is impossible to change an order of a differentiation over t_1 and integration over r in eq. (10) because of the integration range dependence on t_1. Therefore the relationship

$$\int d^3r \cdot \vec{D}(\vec{r},t) \frac{\partial}{\partial t_1} \vec{D}(\vec{r},t+t_1) = \frac{\partial}{\partial t_1} \int d^3r \cdot \vec{D}(\vec{r},t) \vec{D}(\vec{r},t+t_1)$$

WRONG !   (12)

is wrong. Unfortunately Tachiya and Sano have made use of this equation. And this is the first reason which has led to their disagreement with Zwanzig's result and to erroneous formulas in [2].

Approach of Tachiya and Sano [2] outlined in frequency decomposition (here we carry out calculations in the time representation) is valid either for point particles [5] or in the "excluded cylinder approximation" (ECA), used in [1], when the energy losses in the cylinder of radius a along the straight-line trajectory of the particle are not taken into account. In these cases integration ranges over $\vec{r}$ in $-\dot{W}$ do not depend on t_1 (or $\vec{x}(t_1)$). That is why Zwanzig had abandoned that approach considering the dielectric energy loss of the particle of finite size in ESA. The incorrect operation in [2] is the substitution of the order of integration over t and R in eq.(5,[2]) which is used for obtaining the final eq.(10,[2]).

In reality the dependence ψ vs t_1 is implicit. The changing of $\vec{D}(\vec{r})$ is due to the variation of $\vec{x}(t_1)$. Therefore it is more correctly to consider ψ as a function of a displacement $\vec{\Delta}$ of the particle which happens during t_1.

For straight-line uniform motion $\vec{\Delta} = \vec{v} t_1$. In the case of diffusion motion the loss rate becomes a time-dependent or more correctly a function of Δ. It is interesting to calculate the rate $\langle -\dot{W} \rangle$ averaged over the time essentially greater then the characteristic time of the considered type of motion. It is possible to do that using the propagator $P(\vec{\Delta},t)$ of the motion. $P(\vec{\Delta},t)$ presents itself the probability of finding the particle at time t at point $\vec{\Delta}$ if at t=0 it was in the origin of coordinates. The average of the eq.(8) is

$$\langle -\dot{W} \rangle = \frac{1}{4\pi} \int_0^\infty \gamma'_{reg}(t) \tilde{\psi}(t) dt , \qquad (13)$$

$$\tilde{\psi}(t) = \int d^3\Delta \cdot P(\vec{\Delta},t) \cdot \int_0^\Delta d\Delta_1 \psi(\Delta_1) , \qquad (14)$$

where $\psi(\Delta_1)$ in accordance with eq.(10) has the form

$$\psi(\Delta) = \int_{\vec{\Delta},a} d^3r \cdot \vec{D}(\vec{r},0) \frac{\partial}{\partial \vec{\Delta}} \vec{D}(\vec{r},\vec{\Delta}) , \qquad (15)$$

$$\vec{D}(\vec{r},\vec{\Delta}) = e \cdot \frac{\vec{r}-\vec{\Delta}}{|\vec{r}-\vec{\Delta}|^3} .$$

It is convenient to calculate $\psi(\Delta)$ in bipolar coordinates. The result is

$$\psi(\Delta) = -\frac{8\pi e^2}{3a^2} \cdot \begin{cases} \Delta/a &, \Delta < a \\ (a/\Delta)^2 &, \Delta > a . \end{cases} \qquad (16)$$

For a straight-line motion

$$P(\vec{\Delta},t) = \delta^3(\vec{\Delta}-\vec{v}t) , \qquad (17)$$

and for a diffusion motion

$$P(\vec{\Delta},t) = \frac{1}{(4\pi Dt)^{3/2}} \cdot \exp\left(-\frac{\Delta^2}{4Dt}\right) , \qquad (18)$$

where D is the diffusion coefficient of the particle. Inserting eq. (16) into eq. (14) we have calculated the function $\tilde{\psi}(t)$ for a straight-line ($\tilde{\psi}_{sl}(t)$) and diffusion ($\tilde{\psi}_{diff}(t)$) motions, respectively:

$$\tilde{\psi}_{sl}(t) = -\frac{8\pi e^2}{3a^2} \cdot \begin{cases} \delta^2/2 & , \ 0 \le \delta \le 1 \\ 1.5 - 1/\delta & , \ \delta \ge 1 \end{cases} \quad , \ \delta = \frac{vt}{a}, \qquad (19)$$

$$\tilde{\psi}_{diff}(t) = -\frac{4\pi e^2}{a}\left[1 - \mathrm{erf}(y^{-1/2}) + \frac{y}{\pi^{1/2}} \cdot \int_0^{1/y} \sqrt{x}\, e^{-x} dx\right], \qquad (20)$$

$$y = \frac{4Dt}{a^2}, \qquad \mathrm{erf}(y) = \frac{2}{\sqrt{\pi}} \int_0^y \exp(-x^2) dx \ .$$

Now we have to express the time derivative γ'_{reg} through $\varepsilon(\omega)$. After some transformations of the eq. (3) we arrive at

$$\frac{d\gamma_{reg}(t)}{dt} = \frac{1}{\pi} \int_{-\infty}^{+\infty} d\omega \cdot \cos\omega t \cdot \frac{\omega \varepsilon''(\omega)}{|\varepsilon(\omega)|^2}, \quad t > 0. \qquad (21)$$

Here $\varepsilon''(\omega)$ is an imaginary part of $\varepsilon(\omega)$. Taking into account eq. (21) we can rewrite eq. (10) in the following form

$$\langle -\dot{W}\rangle = \frac{1}{2\pi} \int_0^\infty \frac{\omega \varepsilon''(\omega)}{|\varepsilon(\omega)|^2} \cdot J(\omega) d\omega, \qquad (22)$$

$$J(\omega) = \frac{1}{\pi} \int_0^\infty \cos\omega t \cdot \tilde{\psi}(t) dt \ . \qquad (23)$$

Calculation of spectral function $J_{sl}(\omega)$ is elementary for a straight-line motion and a bit more complicated for a diffusion one ($J_{diff}(\omega)$). The results are $\quad (Ciw = \int_w^\infty \frac{\cos x}{x} dx)$

$$J_{sl}(\omega) = \frac{8e^2}{3v}\left\{\frac{\sin w}{w} - \frac{\cos w}{w^2} + \frac{\sin w}{w^3} - Ciw\right\} =$$

$$= \frac{e^2}{v} \cdot \begin{cases} 2.67 \cdot \ln 1/w + 2.017, & w \ll 1 \\ \frac{8\sin w}{w^3}, & w \gg 1 \end{cases} \qquad w = \frac{\omega a}{v}. \qquad (24)$$

$$J_{diff}(\omega) = \frac{e^2 a}{D}\cdot\left[\frac{\sin\sqrt{2\Omega}}{\Omega} \cdot e^{-\sqrt{2\Omega}} - \frac{1}{\sqrt{\pi}}\int_0^\infty y\cdot\cos\Omega y\cdot dy \int_0^{1/y} \sqrt{x}\, e^{-x} dx\right] =$$

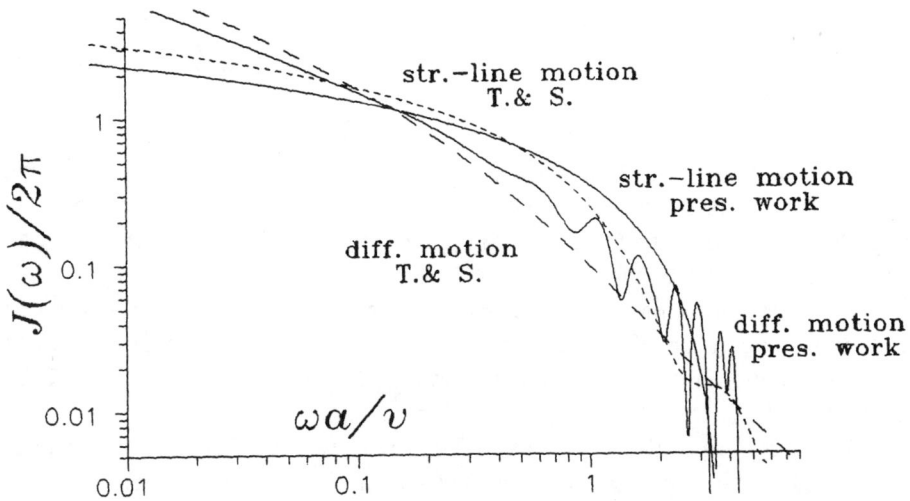

Fig.1 Spectral functions for straight-line and diffusion motions. Comparison between results of present work (solid lines, eqs.(24, 25)) and that of Tachiya and Sano [2] (dashed lines). The ordinate $J(\omega)/2\pi$ is measured in units e^2/v. For uniform representation of J_{sl} and J_{diff} we have set $D = av/6$.

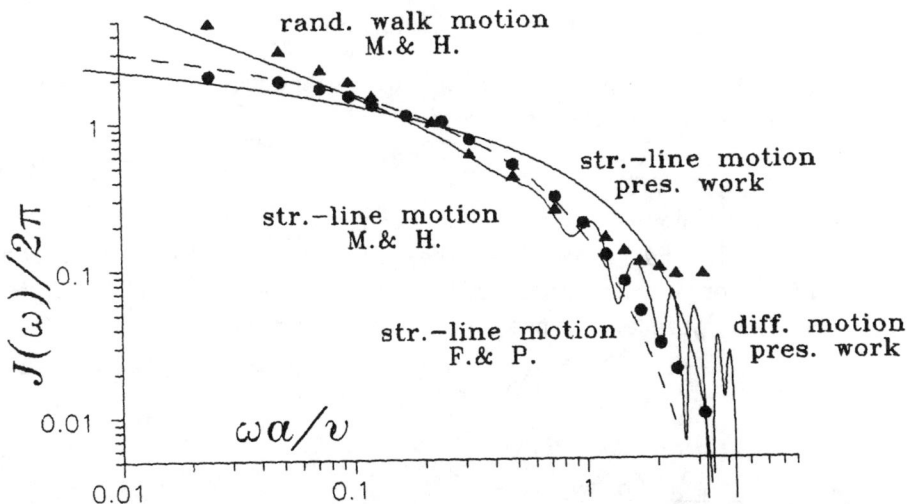

Fig.2 Spectral functions for straight-line and diffusion motions. Comparison between the present work (solid lines, eqs.(24, 25)), Froehlich and Platzman [1] (dashed line) and Magee and Helman [6] (● show $J_{sl}(\omega)$ for a straight-line track, ▲ present $J(\omega)$ for a random-walk track). See also the legend to fig.1.

$$= \frac{e^2 a}{D} \begin{cases} \frac{2}{3}\sqrt{\frac{2}{\Omega}} - 1, & \Omega \ll 1, \\ \frac{1}{2\Omega^2}, & \Omega \gg 1, \end{cases} \qquad \Omega = \frac{\omega a^2}{4D}. \qquad (25)$$

Behavior of the $J_{sl}(\omega)$ and $J_{diff}(\omega)$ is shown on figs. 1,2. We must mention here that comparison of our $J_{diff}(\omega)$ with the corresponding function obtained numerically by Magee and Helman [6] is rather questionable because

1) Magee and Helman treated the random walk of the electron discretely both in time and in space (see a discussion in [2]) while we have used the continuous diffusion propagator eq. (18) averaging $-\dot{W}$;

2) Magee and Helman considered a point particle and how do they treat with a divergence on k (see a discussion at the end of this paper) is not clear.

Additionally for the sake of comparison we were needed (following Tachiya and Sano) to put the mean free path l from [6] equal to our radius a of the excluded sphere.

3. Energy loss to dipolar relaxation

Evaluation of the integral in eq. (22) requires the knowledge of $\varepsilon(\omega)$. We use Debye dispersion low

$$\varepsilon(\omega) = \varepsilon_\infty + \frac{\varepsilon_0 - \varepsilon_\infty}{1 - i\omega\tau}. \qquad (26)$$

Here ε_0 and ε_∞ are the static and high frequency dielectric constants and τ is the relaxation time. Now we are able to calculate dielectric energy loss rate taking into account eqs. (24,25) for $J(\omega)$ for two different types of motions, respectively

$$\langle -\dot{W}_{sl} \rangle = \frac{2e^2}{3a\tau} \cdot \frac{\varepsilon_0 - \varepsilon_\infty}{\varepsilon_\infty^2} \cdot \left\{ \mathit{æ}_s^2 - \frac{1}{\mathit{æ}_s} E_1(\frac{1}{\mathit{æ}_s}) - \exp(-\frac{1}{\mathit{æ}_s}) \cdot (\mathit{æ}_s^2 + \mathit{æ}_s - 1) \right\} =$$

$$= \frac{e^2}{a\tau} \cdot \frac{\varepsilon_0 - \varepsilon_\infty}{\varepsilon_\infty^2} \cdot \begin{cases} 1 - \frac{2\ln\mathit{æ}_s}{3\mathit{æ}_s}, & \mathit{æ}_s \gg 1, & (27a) \\ \frac{2}{3}\mathit{æ}_s^2[1 - 3\exp(-\frac{1}{\mathit{æ}_s})], & \mathit{æ}_s \ll 1, & (27b) \end{cases}$$

$$\langle -\dot{W}_{diff} \rangle = \frac{e^2}{a\tau} \cdot \frac{\varepsilon_0 - \varepsilon_\infty}{\varepsilon_\infty^2} \cdot \left\{ \frac{4}{\pi} \int_0^\infty \frac{z^3 \sin z \cdot e^{-z}}{(2/\mathit{æ}_d)^2 + z^4} dz + \mathit{æ}_d - \right.$$

$$-\frac{1}{\sqrt{\pi}} \cdot \int_0^\infty \frac{dz}{\sqrt{z}} (1+æ_d z) \cdot \exp(-z - \frac{1}{zæ_d})\} =$$

$$= \frac{e^2}{a\tau} \cdot \frac{\varepsilon_0 - \varepsilon_\infty}{\varepsilon_\infty^2} \cdot \begin{cases} 1, & æ_d \gg 1, \\ æ_d/2, & æ_d \ll 1, \end{cases} \quad (28)$$

where $æ_s = \frac{\varepsilon_\infty}{\varepsilon_0} \cdot \frac{v\tau}{a}$, $æ_d = \frac{\varepsilon_\infty}{\varepsilon_0} \cdot \frac{4D\tau}{a^2}$ and $E_1(y) = \int_y^\infty e^{-x} \frac{dx}{x}$.

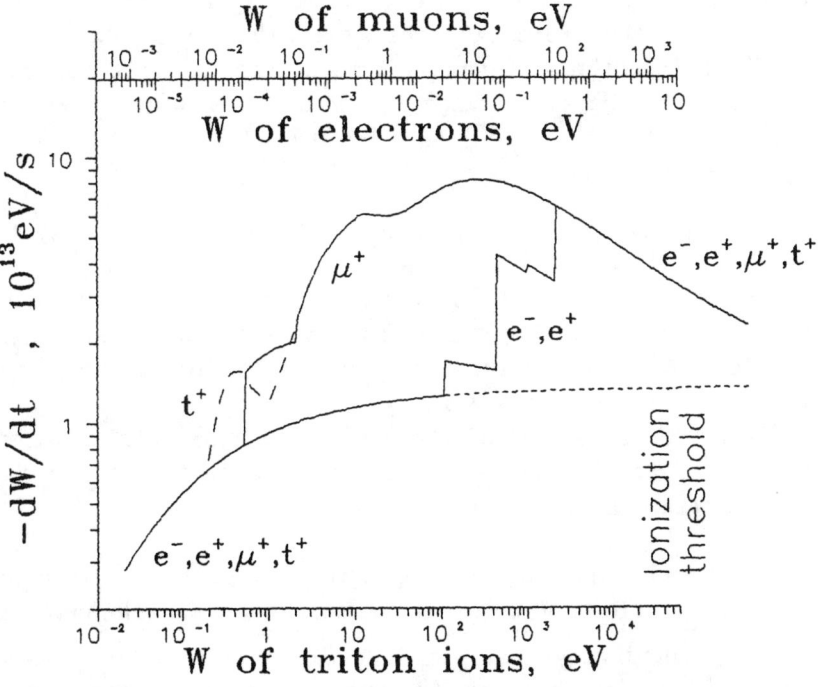

Fig. 3. Energy loss rate vs kinetic energy W of the e^-, e^+, μ^+ and t^+ (straight-line motion is assumed). The lower smoothly raised curve presents a contribution to $-\dot W$ of Debye dielectric energy losses (indirect losses). Upper curves show the contribution of direct processes (excitations of intra- and intermolecular vibrations).

In the case of rapid motion ($æ_s \gg 1$ and $æ_d \gg 1$) energy loss rate does not depend on the type of motion

$$\langle -\dot{W}_{sl}\rangle = \langle -\dot{W}_{diff}\rangle = \frac{e^2}{a\tau} \cdot \frac{\varepsilon_0 - \varepsilon_\infty}{\varepsilon_\infty^2} . \qquad (29)$$

This result agrees parametrically with Froehlich and Platzman [1]. Small deviation in multiplier in their formula from our one ($\pi/4$ instead unity in eq.(27a)) is connected with that in [1] the authors did not take into account the losses in the cylinder around particle trajectory (they worked within ECA). Eq.(27a) was obtained in [2] but in essence by chance: at first Tachiya and Sano used incorrect relationship (12), secondly aiming to correct drawbacks coming from ECA the authors declared that they have accounted energy losses in the whole space outside the particle. However when they calculate the integral $\int d^3r \cdot \vec{D}(t+t_1) \cdot \vec{D}(t)$ (see (19,23) in [2]) the forbidden volume of the particle was excluded twice. Their integration range in our notation was

Therefore in [2] was really used "excluded two sphere approximation".

For slow moving particles ($\varkappa_s \ll 1$ and $\varkappa_d \ll 1$) energy loss rate has quite different expressions for the straight-line $\langle -\dot{W}_{sl}\rangle \sim \varkappa_s^2$ (eq.(27b) exactly reproduces the Zwanzig's formula [3]) and the diffusion motion $\langle -\dot{W}_{diff}\rangle \sim \varkappa_{diff}$.

Energy dependencies of the dielectric loss rate in water ($\tau = 0.85 \cdot 10^{-11}$ s at room temperatures) for some particles are shown as an illustration on fig.3.

4. Semiphenomenological approximation of taking into account excitations of molecular vibrations ("direct" processes).

It is obviously that eqs.(27,28) do not include the energy loss to molecular vibrational excitations of a slowing down particle. To estimate a contribution of these direct processes into the loss rate it is possible to make use of the energy absorption owing to these excitations is described by the resonance (δ-function) contributions to $\varepsilon''(\omega)$. Semiphenomenological approach accounting a contribution of such a quantum processes in the frame of puarely classical formula (22) implies that the energy is transferred to the medium vibrational excitations quasicontinuously i.e. by small portions, but with one "quantum-mechanical" restriction: the particle with the energy W could not excite vibrational quanta $\hbar\omega_i$ if $W < \hbar\omega_i$. Therefore with these assumptions eq.(22) attains the form [4]

$$\langle -\dot{W} \rangle = \frac{1}{2\pi} \left[\sum_{\omega_i} I(\omega_i) J(\omega_i) \vartheta(W) \hbar\omega_i) + \int_0^\infty \frac{\omega\varepsilon''(\omega)}{|\varepsilon(\omega)|^2} \cdot J(\omega) d\omega \right]. \quad (30)$$

$$\vartheta(W > \hbar\omega_i) = \begin{cases} 1, & W > \hbar\omega_i \\ 0, & W < \hbar\omega_i \end{cases}$$

Here ω_i is the resonance frequency of molecular vibration of the i-th type, $I(\omega_i) = \frac{1}{|\varepsilon(\omega_i)|^2} \cdot \int \omega\varepsilon_i''(\omega) d\omega$ is a characteristics of an integral absorption in the i-th resonance (here $\varepsilon(\omega)$ represents a continuous part of the dispersion low).

Let us consider as an example the slowing down of e^-, μ^+ and triton ions in water. Ionization potential I (or energy of the lowest electronic excitation) of liquid water is about 7.5-8 eV [7]. Therefore ionization threshold for e^-, μ^+ and t^+ are I, $\frac{m_M}{m_e}I$ and $\frac{m_t}{m_e}I$, respectively (m_M and m_t are the masses of muon and triton). Particles having smaller energies are subionizing.

There are four resonance vibrating maxima of absorption in water [4]:

$\hbar\omega_1 = 0.4$ eV, $I_1 = 1.4 \cdot 10^{28}$ s^{-2},
$\hbar\omega_2 = 0.2$ eV, $I_2 = 7.0 \cdot 10^{26}$ s^{-2},
$\hbar\omega_3 = 0.08$ eV, $I_3 = 3.5 \cdot 10^{27}$ s^{-2},
$\hbar\omega_4 = 0.02$ eV, $I_4 = 2.3 \cdot 10^{26}$ s^{-2}.

For straight-line motion of the particles energy loss rate vs W is shown on the fig.3 according to eq.(30) and accounting these data. It is seen that the direct processes contribute several times greater than the Debye relaxation mechanism. Owing to that and because the values I_i and ω_i do not differ significantly in liquid water and ice the loss rate is also practically the same in these media. Temperature dependent contribution of continuous component in $\varepsilon(\omega)$ to the $\langle -\dot{W} \rangle$ is small.

It is followed from the fig.3 that for e^- and e^+ with the initial kinetic energy about several eV a thermalization time τ_{th} is about 10^{-13} s. Strictly speaking this is an estimation from below because we did not include quasi-elastic processes, rotational excitations etc. It seems, however, that their contribution is negligible. Our τ_{th} is in a reasonable agreement with the electron solvation time τ_s in water ($3 \cdot 10^{-13}$ s [8]) other assessments [9]. τ_s may be considered as an upper limit for τ_{th}.

5. Conclusions

In the present paper we have obtained the formulas for dielectric energy loss rate of charged subionizing particle moving (straightforwardly or diffusively) in the polar media. Two mistakes in the paper [2] devoted to the same subject were pointed out.

The further development is made to account semiphenomenologically energy losses to intra- and intermolecular vibrations (direct losses).

At last let us note the following advantage of ESA or in other words the finite size particle approach. It is not difficult to obtain the expression

$$-\dot{W} = \frac{e^2}{2\pi^2} \cdot \int d^3k \, \frac{\vec{k}\vec{v}}{k^2} \, \text{Im} \, \frac{1}{\varepsilon(\vec{k},\vec{k}\vec{v})} \quad (31)$$

presented the dielectric loss rate of a straightforwardly moving particle [5,10]. Nevertheless the integral over k in eq.(31) diverges if the typical $\varepsilon(\vec{k},\vec{k}\vec{v})$ are used [10]. Therefore it is necessary to introduce "by hands" the cuttings on the upper limit (to set for example $k < k_{max}$). It is chosen usually $k_{max} \sim 1/d$, where d is an intermolecular distance. It is difficult to propose more accurate definition for k_{max}. That is why Hubburd and Stiles [10] had obtained an expression for $-\dot{W}_{sl}$ at $æ_s \ll 1$ which agrees with Zwanzig's result only up a to numerical multiplier. It seems to be better to choose k_{max} getting a numerical accordance with exactly-solvable models like [3 and present paper] which has no adjustable parameters.

6. Acknowledgments

The author thanks Professors V.M.Byakov and F.S.Dzheparov for valuable discussions of the problem.

REFERENCES

1. H. Froehlich, R. Platzman, Phys. Rev., V.92(5), P.1152, 1953
2. M. Tachiya, H. Sano, J. Chem. Phys., V.67(11), P.5111, 1978
3. R. Zwanzig, Phys. Rev., V.38(7), P.1603, 1963
4. B.M. Garin, V.M. Byakov, High Energy Chem., V.22(3), P.195, 1988
5. L.D. Landau, E.M. Lifshitz, Electrodynamics of Continues Media, Pergamon, New York, 1960
6. J.L. Magee, W.P. Helman, J. Chem. Phys., V.66(1), P.310, 1977
7. V.M. Byakov, F.G. Nichiporov Intratrack Chemical Processes (in Russian), Moscow, Energoatomizdat, 1985
8. A.K. Pikaev, High Energy Chemistry, V.25(1), P.4, 1991
9. A. Mozumder, Radiat. Phys. Chem., V.32(2), P.287, 1988 (Int. J. Radiat. Appl. Instr. Part C)
10. J.B. Hubbard, P.J. Stiles, Chem. Phys. Lett., V.114(1), P.121, 1985

MEASUREMENTS OF POSITRONIUM FORMATION CROSS SECTIONS FOR POSITRON - ARGON SCATTERING*

S. Zhou, L. Jiang, W.E. Kauppila, C.K. Kwan, S. P. Parikh, and T.S. Stein
Department of Physics and Astronomy, Wayne State University,
Detroit, Michigan 48202

ABSTRACT

We report preliminary measurements of positronium (Ps) formation cross sections (Q_{Ps}) for 9 - 452 eV positron - argon collisions. The present Q_{Ps} lower and upper limits are used to check for consistency with prior measurements.

INTRODUCTION

Positronium (Ps) formation has been found to play a major role in the scattering of positrons from room-temperature gases when the positron's kinetic energy is within several tens of eV above the Ps formation threshold (the ionization threshold energy of the target atom minus the binding energy [6.8 eV] of Ps in its ground state). As an indication of the importance of Ps formation in the positron - atom scattering process, the total scattering cross section for positrons colliding with the room-temperature gases shows an abrupt increase just above the predicted Ps formation thresholds [1]. The clear indications of the importance of this process in positron-atom (molecule) collisions have stimulated a number of different efforts to measure and to calculate Ps formation cross sections for positrons colliding with a variety of atoms and molecules in recent years. In this paper, we present our preliminary measurements of Ps formation cross sections for Ar.

EXPERIMENT

A brief description of the apparatus and procedure is given below. Additional details are provided in Ref. [2]. Our positron beam is derived from a ^{22}Na source which has an activity of approximately 7 mCi. The positrons emanating from the source pass through a 0.004 mm thick, annealed tungsten moderator and a beam is formed using an electrostatic lens system. The beam is transported in an axial magnetic field. Two different scattering cells [2] are used for Ar as shown in Fig. 1. When the beam leaves the scattering cell, it passes through a retarding grid assembly which is used to determine the energy distribution of the positron beam. The primary beam intensity is monitored by a Channeltron electron multiplier (CEM) which is located just beyond the retarding grid assembly. Typical detected beam intensities are about 5000 positrons/sec at 200 eV and 900 positrons/sec at 1 eV, and the measured energy width of the beam (full-width at half-maximum) is of the order of 1 eV.

Our experimental approach involves setting lower and upper limits on Ps formation cross sections using a combination of (1) the detection of the coincidences of 511 keV annihilation gamma rays produced by the decay of para-Ps and by the interaction of ortho-Ps with the walls of the scattering cell in which the Ps is formed, and (2) the determination of scattering cross sections associated with the measured transmission of the positron beam through the gas in our scattering cell with the angular discrimination of our apparatus deliberately made as poor as possible.

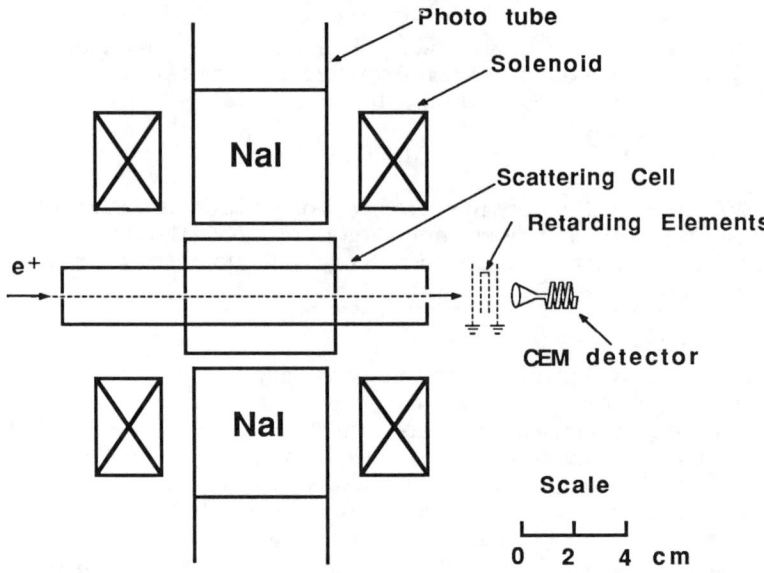

Fig. 1. Apparatus for measuring Ps formation cross sections for e⁺ - Ar collisions.

We obtain an upper limit on Q_{Ps} by determining the scattering cross sections associated with the measured transmission of the positron beam through the gas in our scattering cell using no retarding potential on the retarding elements, a sufficiently high axial magnetic field and a sufficiently large exit aperture size that, for the most part, positrons which have been scattered into the forward hemisphere by any process other than Ps formation will pass through the cell's exit aperture and be detected by the CEM. The cross sections obtained in this way are considered to be upper limits to Q_{Ps} because these cross sections include not only the contributions from positrons which form positronium, but also the contributions from positrons scattered into the backward hemisphere. Because this upper limit is obtained by setting the potential on the retarding element to zero, we refer to it as the "zero-retard" upper limit on Q_{Ps}.

Another approach that we use to obtain information on Q_{Ps} is to use the photomultiplier tubes with their attached NaI scintillators shown in Fig. 1 to detect coincidences of 511 keV annihilation gamma rays which are produced when para-Ps decays. If there were no contribution from ortho-Ps to our detected signal, then the value of our resulting measured cross section, which we call $Q_{para-Ps}$ could be multiplied by four to obtain the total Ps formation cross section Q_{Ps}, since para-Ps constitutes one-fourth of the total amount of Ps which is produced, while ortho-Ps constitutes the remaining three-fourths. However, if we consider the possibility of the interactions of the ortho-Ps with the scattering cell's inner surface, which could give rise to the production of two 511 keV annihilation gamma rays, then the measured value of $Q_{para-Ps}$ times four can still serve as another upper limit on Q_{Ps}. In addition, the measured value of $Q_{para-Ps}$ can serve as a lower limit on Q_{Ps}.

RESULTS

Our best upper and lower limits (as determined by the methods described in Ref. [2]) are shown in Fig. 2. The present results supersede our earlier results [2] due to improved determinations of detector efficiencies. It is quite intriguing to us that our best lower limits are close to our best upper limits over a significant part of the energy range that we have investigated, and are very close to the measurements by Fornari et al.[3] over most of the energy range of overlap. This suggests that a large fraction of the ortho-Ps formed in our scattering cell is being converted into two 511 keV gamma rays, presumably via the interaction of the ortho-Ps with the scattering cell's inner surface. Our limits on Q_{Ps} exclude most of the measurements of Diana et al.[4] and we see no evidence of the structure indicated by their measurements.

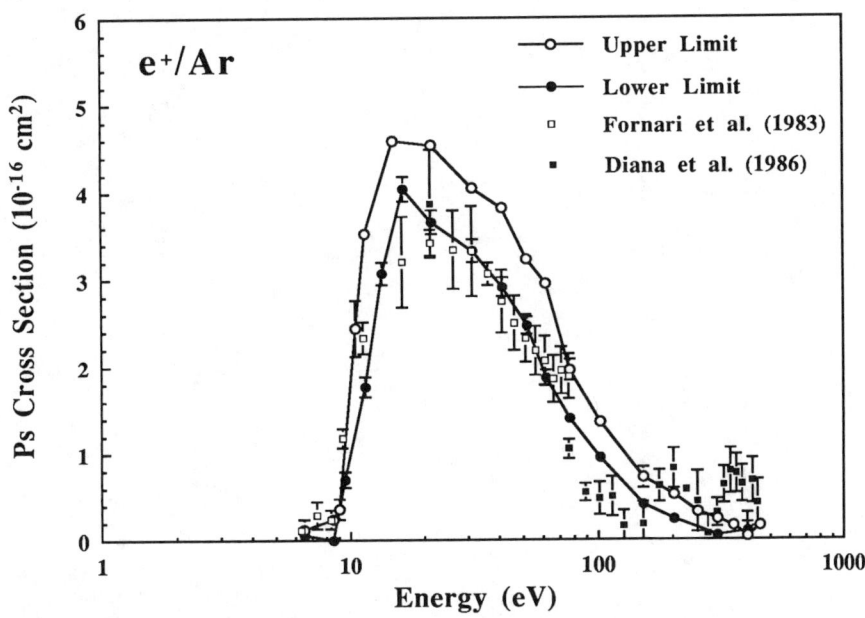

Fig. 2. Positronium formation cross sections for positron-Ar collisions.

* Research supported by NSF Grant PHY90-21044.

REFERENCES

[1] W.E. Kauppila and T.S. Stein, Adv. At., Mol., and Opt. Phys. 26, 1 (1990).
[2] T.S. Stein, W.E. Kauppila, C.K. Kwan, S.P. Parikh and S. Zhou, Hyperfine Interactions. 73, 53 (1992).
[3] L.S. Fornari, L.M. Diana and P.G. Coleman, Phys. Rev. Lett. 51, 2276 (1983).
[4] L.M. Diana, P.G. Coleman, D.L. Brooks, P.K. Pendleton, D.M. Norman, B.E. Seay and S.C. Sharma, in Positron (Electron) - Gas Scattering, W.E. Kauppila, T.S. Stein and J.M. Wadehra, eds. (World Scientific, Singapore, 1986) pp. 296-8.

LIST OF PARTICIPANTS
(AND THEIR INTERESTS)

G. C. Aers
National Research Council
Microstructural Science Lab
Ottawa, Ontario, Canada K1A OR6
613-993-9403
(Defect Profiling Calculations)

Takashi Akahane
National Institute for Research
in Inorganic Materials
1-1 Namiki, Tsukuba, Ibaraki Japan 305
81-298-51-3351
(Production of Slow Positron Beams;
Characterization of Solids and Surfaces
Using Positron Beams)

G. Amarendra
Indira Gandhi Centre for Atomic
Research
Materials Science Division
Kalpakkam Tamilnadu, India 603102
91-4117-385
(Defect Studies using Positron Beams)

Geoffrey Anderson
The University of Western Ontario
Chemistry Bldg.
London, Ontario, Canada N6A 5B7
519-661-2166
(Positron Re-Emission; Ultra Thin Film
Growth; Surface and Interface Defects)

P. Asoka-Kumar
Brookhaven National Lab
510B Dept. of Physics
Upton, NY 11973
516-282-4574
(Positron Studies in Silicon Base
Systems; Positron Fundamental Studies
in the MeV Regions, etc.)

Bruno W. Augenstein
Rand Corporation
1700 Main St. P. O. Box 2138, MS 1E
Santa Monica, CA 90407-2138
310-393-0411, ext. 6605

Heinz Boehmer
University Cal/Irvine
Dept. of Physics
Irvine, CA 92717
714-725-2710
(Plasma Physics)

Gerhard Brauer
Forschungzentrum Rossendorf e. V.
Institut für Ionenstrahlphysik und
Materialforschung, Postfach 19
Dresden, Germany D-O-8051
49-351-591-3167

D. T. Britton
Universität der Bundeswehr München
Inst. für Nükleare Fest-
korperphys.,Werner Heisenberg Weg 59
Neubiberg, Germany D-8014
49-89-6004-3513
(Positron Diffusion, Surface States,
Layered Semiconductors, Magnetic
Optics.)

Ben Brown
Mount Holyoke College
Dept. of Physics
S. Hadley, MA 01075-1461
413-538-2238

Karl F. Canter
Brandeis University
Department of Physics, 415 South Street
Waltham, MA 02254
617-736-2915
(Surface Studies)

Florin Chifu
University of Cincinnati, Coll. of Engin.
Dept. of Mat. Science and Engin.
498 Rhodes Hall (ML 12)
Cincinnati, OH 45221-0012
513-556-5473; 961-4661
(Materials Science, Surfaces)

Neil B. Chilton
Nippon Steel Corporation
Advanced Mat. & Technol. Research
Lab., 1618 Ida, Nakahara-ku
Kawasaki Japan 211
81-44-777-4111

P. G. Coleman
University of East Anglia
School of Physics
Norfolk Norwich, United Kingdom
NR4 7TJ
44-603-56161

Paul Csonka
University of Oregon
Institute of Theoretical Science
Eugene, OR 97403
503-346-5205
(Accelerators; Particle Beams)

G. H. Dai
University of Missouri-Kansas City,
Dept. of Chemistry, 5100 Rockhill Rd.
Kansas City, MO 64110
816-276-2296

Qin Deng
University of Missouri-Kansas City,
Dept. of Chemistry, 5100 Rockhill Rd.
Kansas City, MO 64110
816-276-2296

Art Denison
Idaho National Engineering Laboratory
P.O. Box 1625
Idaho Falls, ID 83415
208-526-1294

Greg DeMaggio
University of Michigan
Dept. of Physics
Ann Arbor, MI 48109-1120
313-936-1134
(Positron Re-Emission Microscopy)

Yaroslav Derbenev
University of Michigan
Randall Labs
Ann Arbor, MI 48109-1120
313-936-1027; 764-2534
(Slow Positron Beams Novel Appl.
Fundamental Physics Studies)

Vasumathi Dharmavaram
Brandeis University
Physics Dept.
Waltham, MA 02254
617-736-2885

M. Dorikens
Rijksuniversiteit Gent
Lab. Nucl. Physics, Proeftuinstraat 86
Gent, Belgium B-9000
32-91-64-65-38

L. Dorikens-Vanpraet
Rijksuniversiteit Gent
Lab. Nuclear Physics, Proeftuinstraat 86
Gent, Belgium B-9000
32-91-64-65-38

Lie Dou
Wayne State University
Dept. of Physics & Astronomy
Detroit, MI 48202
313-577-2780
(Positron Scattering by Atoms and
Molecules, Experimental)

Jerzy Dryzek
Inst. of Nuclear Physics, Krakow
Ul.Radzikowiego 152
Krakow, Poland PL-31-342

Terry Dull
University of Michigan
Dept. of Physics
Ann Arbor, MI 48109-1120
313-936-1134
(Positron Microbeams and Positron Microscopy)

Abe Eftekhari
Hampton University Physics Dept.
Hampton, Virginia 23668
804-864-4764
(Polymers and Microstructures)

Pedro A. Encarnacion
University of Michigan
Dept. of Physics
Ann Arbor, MI 48109-1120
313-936-1134

Na'il G. Fazleev
University of Texas at Arlington
Dept. of Physics, Box 19059
Arlington, Texas 76013-0059
817-273-2512
(PAES)

Masanori Fujinami
Nippon Steel Corporation
Materials Characterization Labs, R. & D. I., 1618 Ida, Nakahara-Ku
Kawasaki, Japan 211
81-44-777-4411

Bita Ghaffari
University of Michigan
Dept. of Physics
Ann Arbor, MI 48109-1120
313-936-1134
(Pulsed Positron Beams w/o Using Accelerators)

Vinita J. Ghosh
Brookhaven National Laboratories
Department of Applied Science,
Bldg. 480
Upton, NY 11973
516-282-3527

David Gidley
University of Michigan
Department of Physics
Ann Arbor, MI 48109-1120
313-936-1134

Alec Goodyear
University of East Anglia
School of Physics
Norwich, UK NR4 7TJ
44-603-592-592

Henry Griffin
University of Michigan
Dept. of Chemistry
Ann Arbor, MI 48109-1120
313-764-1438

R. A. Hakvoort
Delft University of Technology,
Interfac. Reactor Inst., Mekelweg 15
JB Delft, The Netherlands NL-2629
31-15-783155

A. Halec
The University of Western Ontario
Department of Physics
London, Ontario, Canada N6A 3K7
519-661-3390

Richard H. Howell
Lawrence Livermore National Lab.
P.O. Box 808 L-280
Livermore, CA 94550
510-422-1977

Nader Hozhabri
University Texas at Arlington
P.O. Box 194513
Arlington, TX 76019
817-273-2874

List of Participants

Lester Hulett
Oak Ridge Nat. Lab
P.O. Box 2008, MS 6142
Oak Ridge, TN 37831-6142
615-574-4879
(Intense Positron Facilities Sources; Gas Phase Spectroscopy; Materials Sciences)

A. Ishii
Fritz-Haber-Institut der Max-Planck-Gesellschaft
Abteilung Theorie, Faradayweg 4-6,
Berlin 33, Fed. Rep. of Germany D-1000
49-30-8305-1
(Surface Studies)

Finn M. Jacobsen
University of Aarhus
Instit. of Physics
Aarhus C, Denmark DK-8000
45-86-128899

Y. C. Jean
University of MO-Kansas City,
Dept. of Chemistry, 5100 Rockhill Rd.
Kansas City, MO 64110
816-276-2295
(e+; Ps)

Licai Jiang
Wayne State University
Dept. of Physics & Astronomy
Detroit, MI 48202
313-577-2780
(Positron Scattering and Positronium)

Geraint O. Jones
University College London
Dept. of Physics & Astronomy, Gower Street
London, UK WCIE 6BT
44-71-387-7050 x 3468
(Atomic Physics; Slow Positron Beams; Novel Applications)

W. E. Kauppila
Wayne State University
Dept. of Physics and Astronomy
Detroit, Michigan 48202
313-577-2780
(Positron Scattering by Atoms and Molecules, Including Total Differential Elastic and Positronium Formation Cross Section Measurements)

Jae Hong Kim
University of Texas at Arlington
502 Yates Str., P.O. Box 19059
Arlington, TX 76019
817-273-2480

Andrew Knights
University of East Anglia
School of Math and Physics
Norwich, UK NR4 7TJ
44-603-592-592

Jack Kossler
College of William & Mary
496 Burnham Rd
Williamsburg, VA 23185
804 221-3519

Katarina Kristiakova
Slovak Academy of Sciences
Dubravska Cesta 9, Inst. of Physics
Bratislava, Czecho-Slovakia 842 28
42-7-378-2135
(Apparatus for Ps Lifetime Measurements; Measurements of Doppler Broadening)

Toshikazu Kurihara
Photon Factory
Natl. Lab. for High Energy Physics
1-1 Oho, Tsukuba-shi Ibaraki-ken,
Japan 305
81-298-64-1171

Ching-Kwan Kwan
Wayne State University
Dept. of Physics & Astronomy
Detroit, MI 48202
313-577-2780
(Positron Scattering by Atoms and Molecules, Experimental)

Derek W. Lawther
Dalhousie University
Dept. of Physics
Halifax, Nova Scotia, Canada B3H 3J5
902-494-2337

Keunho Lee
University at Arlington
Physics Dept. P.O. Box 19059
Arlington, Texas 76019
817-273-2480

Del Lessor
Pacific Northwest Lab.
P.O. Box 999, K7-15
Richland, Washington 99352
509-375-2382
(Low Energy Positron Diffraction)

T. C. Leung
The University of Western Ontario
Department of Physics
London, Ontario, Canada N6A 3K7
519-661-3390

Kelvin Lynn
Brookhaven National Laboratory
Physics Dept. 510B
Upton, NY 11973-5000
516-282-3501
(Positron Beams)

Henry Makowitz
EG&G Idaho, Inc.
MS 2211, P.O. Box 1625
Idaho Falls, ID 83418
208-526-9301

Alfred A. Manuel
University of Geneva
Physics Dept.,24 Quai Ansermet
Geneva 4, Switzerland CH-1211
41-22-702-6264
(ACAR Studies; Slow e+ Beams)

Guiti R. Massoumi
The University of Western Ontario
Dept. of Physics
London, Ontario, Canada N6A 3K7
519-661-2111 ext 6422
(Slow e^+ Beams; Fundamental Physics Studies; Tutorials on Optics; Defect Profiling and Moderators)

Alexander Melker
St. Petersburg State Technical University
Physics and Mech. Faculty
Leningrad, Russia SU-171187

J. P. Merrison
University College London
Dept. of Physics & Astron. Gower Street
London, England WC1E 6BT
44-71-387-7050
(Fundamental Physics; Slow e+ Beams)

P. E. Mijnarends
Delft University of Technology
Interfaculty Reactor Institute,
Mekelweg 15
JB Delft, The Netherlands 2629
31-2246-4528
(2D-ACAR; Electronic Structure)

Allen Mills
AT&T Bell Labs
Rm 1D 338, 600 Mountain Ave.
Murray Hill, NJ 07974
908-582-4162

Bent Nielsen
Brookhaven National Lab
Mat. Science Div., Dept. of Applied Science
Upton, NY 11973
516-282-3525

R. Nieminen
Helsinki University of Technology
Laboratory of Physics
Espoo 15, Finland SF-02150
358-0-4574344

Eric H. Ottewitte
575 Murdock Ln.
Idaho Falls, ID   83402
208-522-9534
(Development of new nuclear and
subnuclear technologies, e+ beams,
antimatter engineering)

Mogens Rysholt Poulsen
Aarhus University
Institute of Physics and Astronomy
Ny Munkegade
Aarhus C, Denmark DK-8000
45-86128899
(Surface Studies; Antihydrogen)

David Przybyla
Wayne State University
Dept. of Physics & Astronomy
Detroit, MI 48202
313-577-2780
(Techniques for Making Moderators)

C. Venkataramana Reddy
The University of Hong Kong
Dept. of Physics, Pok Fu Lam Road
Hong Kong
852-859-2358
(Slow-Positron Beam Applications)

Peter Rice-Evans
Dept. of Phys., Royal Holloway
College, University of London
Egham, Surrey,
United Kingdom TW20 0EX
44-784-443446

Kenneth A. Ritley
University of Ill., Urbana-Champaign
Mat. Resrch. Lab., 1110 W. Green,
Urbana, IL 61801
217-344-0209
(Simulation of Positron Energy Loss
Processes)

Anna Rubaszek
Polish Academy of Sciences
W. Trzebiatowski Institute of Low
Temperature and Structure Research
P.O. Box 937
Wroclaw 2, Poland PL-50-950
48-71-35021

D. M. Schrader
Marquette University
Chemistry Dept.
Milwaukee, WI   53233
414-288-3332
(Dissociative Attachment Collisions -
leading to the formation of positronium
compounds; Theory and Calculations of
Various Cross Sections for Low Energy
Processes; Calculations of Bound States
(energies, lifetimes, angular correlation
curves and surfaces) of Molecular
Materials)

P. J. Schultz
The University of Western Ontario
Dept. of Physics
London, Ontario, Canada   N6A 3K7
519-661-3390

H. Schut
Interfacultair Reactor Instituut
Mekelweg 15
JB Delft, The Netherlands 2629
31-15-781961

D. Segers
Rijksuniversiteit Gent
Lab. Nucl. Physics, Proeftuinstraat 86
Gent, Belgium B-9000
32-91-64-38

Leon Seijbel
TU Delft
Dept. of Applied Physics, Lorentzweg 1
CJ Delft, The Netherlands NL-2628
31-15-786-108

S. C. Sharma
University of Texas at Arlington,
P.O. Box 19059, Dept. of Physics
Arlington, TX 76019
817-273-2470

Hao Shi
University of MO-Kansas City,
Dept. of Chemistry, 5100 Rockhill Rd.
Kansas City, MO 64110
816-276-2296

Ming Shi
Paul Scherrer Inst.
Villigen PSI, Switzerland CH-5232
41-56-99-2463
(Beam Production; Defect Profiling;
Optics; Moderator)

Calvin Shipbaugh
Rand Corporation
1700 Main Street, P.O. Box 2138
Santa Monica, CA 90406-2138
310-393-0411

Peter J. Simpson
Brookhaven National Laboratory
Dept. of Physics Bldg. 510B
Upton, NY 11973
516-282-7579
(Defects; Electronic Materials)

J. J. Singh
NASA - Langley Res. Center
Instr. Resrch. Div., Mail Stop 235
Hampton, VA 23655-5225
804-864-4760
(High Intensity Low Energy Positron
Beams; Microstructural Characterization
of Polymers.)

S. V. Stepanov
ITEP, Moscow
Moscow, Russia SU-117259

Hermann Stoll
Max-Planck Inst. für Metallforsch.
Heisenbergstrasse 1, Postfach 800665
Stuttgart 80, Germany D-7000
49-7116860-551
(Doppler, Positron Lifetime; Age-
Momentum-Correlation Measurements in
Solids and Liquids with an MeV Beam)

Ryoichi Suzuki
Electrotechnical Lab.
High Energy Radiation Sect.,1-1-4
Umezono Tsukuba-shi
Ibaraki, Japan 305
81-298-58-5681
(Slow Positron Beam Techniques;
Application of Slow Positron Beams)

Csoba Szeles
Brookhaven National Laboratory
Dept. of Physics, Bldg. 510B
Upton, NY 11973
617-736-2883

Shoichiro Tanigawa
University of Tsukuba, Tsukuba-shi
Institute of Materials Sci.
Ibaraki, Japan 305
298-53-5135
(Defects in Semiconductors)

W. Triftshäuser
Universitaet der Bundeswehr München
Inst. für Nuklear Festkörperphysik,
Heisenberg Weg 39
Neubiberg, Germany D-8014
49-89-60043505
(Positron Beams; Positron Scanning
Microscope; High Intense Positron
Source)

List of Participants

J. F. Tsai
University of MO-Kansas City,
Dept. of Chemistry, 5100 Rockhill Rd.
Kansas City, MO 64110
816-276-2296

Akira Uedono
University of Tokyo, 7-3-1 Hongo,
Bunkyo-ku
Dept. of Industrial Chem, Fac. of Eng.
Tokyo, Japan 113
81-3-3812-2111 ext 7243

Hikaru Ueno
Ishikawajima-Harima Heavy Industries
Co., Ltd.
Mechatron. Devel. Ctr. 1-15
Toyosu 3-Chome, Koto-ku
Tokyo, Japan 135
81-3-3534-3425
(Slow Positron Beam System as
Commerce)

Rich Vallery
University of Michigan
Dept. of Physics
Ann Arbor, MI 48109-1120
313-936-1134
(Intense Positron Sources)

A. Van Veen
Delft University of Technology
Interfac. React. Institute, Mekelweg 15
JB Delft, The Netherlands 2629
31-15-782801
(Defect Profiling Applications)

W. B. Waeber
Paul Scherrer Institute
Villigen PSI, Switzerland CH-5232
41-56-99-2395
(Beam Production; Bulk and Surface
Electronic Structure; 2D ACAR; PAES;
Defect Profiling)

Shao Jie Wang
Wuhan University
Dept. of Physics
Wuhan, Peoples Republic of China
430072
86-27-812712-969
(Slow Positron Beams; Defect Profiiling;
Surface Studies)

Marc Weber
Universität Bielefeld
Fakultät für Physik, D1
Bielefeld 1, Germany W-4800
49-521-106-5401
(Positron Beams; Crossed Beam
Scattering; Polarized Beams; Reactor
Produced Sources)

Long Wei
University of Tsukuba
Inst. of Materials Sci.
Tsukuba, Ibaraki, Japan 305
81-298-535357
(Defects in Semiconductors)

Alex Weiss
University of Texas at Arlington
Physics Dept., Box 19059
Arlington, TX 76019
817-272-2459

Paul Willutzki
University der Bundeswehr, München
Fakultät LRT / Physik, Heisenberg Weg
39
Neubiberg, Germany D-8014
49-896004-3253

Barry Wissman
University of Michigan
Dept. of Physics
Ann Arbor, MI 48109-1120
313-936-1134
(Surface Physics; Depth Profiling;
Doppler Broadening)

Rong Xie
Brookhaven National Lab
Physics Dept. Bldg. 510B
Upton, NY 11973
617-736-2883

Jun Xu
Oak Ridge National Lab.
P.O. Box 2008, MS 6142
Oak Ridge, TN 37831-6142
615-574-4879
(Intense Positron Facilities; Gas Phase Spectroscopy; Materials Science)

Gi-Mo Yang
University of Texas at Arl.
501 Yates Street, P.O. Box 19059
Arlington, TX 76019
817-273-2480

Shu Yang
University of Texas at Arlington
Dept. of Physics, Box 19059
Arlington, TX 76013-0059
817-273-2480

Eileen Yu
University of Michigan
Dept. of Physics
Ann Arbor, MI 48109-1120
313-936-1134
(Biophysical Applications of the Positron Microscope)

Hong Zhang
University of MO-Kansas City,
Dept. of Chemistry, 5100 Rockhill Rd.
Kansas City, MO 64110
816-276-2296

Q. W. Zhang
University of Missouri-Kansas City,
Dept. of Chemistry, 5100 Rockhill Rd.
Kansas City, MO 64110
816-276-2296

Haiqing (Amy) Zhou
University of Texas at Arlington
Dept. of Physics, Box 19059
Arlington, TX 76019
817-273-2480

Shangjing Zhou
Wayne State University
Dept. of Physics & Astronomy
Detroit, MI 48202
313-577-2780
(Positron Scattering; Positronium)

SLOPOS-5 PROGRAM

Wednesday 8/5/92 Events
0900   Meeting with workers by Room 196
1300-1900   Registration, information by Room 196
1900-2200   Reception, registration, information in Tag's (restaurant) back room

Thursday 8/6/92 Events
0730   Registration, Information, Breakfast
0830   Welcome and Introductions (Eric, INEL, Alex)

8:50am **T1. Tutorials on Defect Profiling** (30min + 10min)@. Peter Mijnarends, Chair.
0850   Kelvin Lynn
0930   Peter Schultz
1010   BREAK

10:30am **A. Defects** (14 min talks + 5 min disc.)@.

Tom Van Veen, Chair
1030   G. C. Aers (National Research Council of Canada), K. Jensen, A. Walker, *Calculation of Positron Diffusion in Layered Systems*
1049   D.T. Britton (U. Bundeswehr München), *Diffusion and Annihilation of Positrons in Solids*
1108   M. Fujinami (Nippon Steel Corp.), N.B. Chilton, T. Fujita and I. Hamaguchi, *A Slow Positron Study of Oxygen Related Defects in Si*
1127   BREAK

1145   Lunch in SLOPOS5 Lounge. Talks on *CRISES AND OPPORTUNITIES IN EASTERN EUROPE PHYSICS*

Y.C. Jean, Chair
1315   V.J. Ghosh (BNL), D.O. Welch, and K.G. Lynn, *Monte Carlo Studies of Positron Implantation in Elemental Metallic and Bimetallic Systems*
1334   R.A. Hakvoort (T. U. Delft), A. van Veen, M.J. van den Boogaard, A. Berntsen, S. Roorda, P.A. Stolk, A.H. Reader, *Characterization of Amorphous Silicon*
1353   B.Nielsen (BNL), and S.M. Heald, *Defects and Density of Thin Metal Films*
1412   BREAK

Richard Howell, Chair
1432   C. Smith (Royal Holloway College), P.C. Rice-Evans and N. Shaw, *Positron Experiments with Cadmium Mercury Telluride*
1450   S. Tanagawa (Institute of Materials Science, Tsukuba), L. Wei, and Y. Tabuki, *Microscopic Observation of the Kirkendall Effect in Metal/Si Systems by Means of a Variable-Energy Positron Beam*
1509   A. Uedono (U. Tokyo) S. Fujii, L. Wei, and S. Tanigawa, *Point Defects in As-Grown and Ion Implanted GaAs Probed by a Monoenergetic Positron Beam*
1528   SHORT BREAK

3:40pm **T2. Tutorial on Creation and Evolution of Vacancy Clusters in Solids Under Irradiation.** Alexander Melker (St. Petersburg State Technical Univ.) <u>M. Dorikens, Chair</u>
1610   SHORT BREAK

4:20pm **B. Lifetime Techniques** (14 min. talks + 5 min. disc.)@. <u>S. Tanigawa, Chair</u>

1620   S.C. Sharma (U. Texas Arlington), R.C. Hyer, M. Green, J.M. Perez, A.R. Chourasia, and D.R. Chopra, *Variable Energy Positron Beam and Positron Lifetime Measurements on Diamond Films Correlations Between Results Obtained from Positron Annihilation, Raman, Scanning Tunneling Microscopy, Scanning Electron Microscopy, and X-Ray Photoelectron Spectroscopy*
1633   Jag J. Singh (NASA), Abe Eftekhari, Terry L. St. Clair, *A Low-Energy Positron Flux Generator for Microstructural Characterization of Thin Polymer Films*
1657   H. Stoll (Max-Plank-Institute für Metallforschung), *Lifetime Measurements with the DC Positron Beam at the Stuttgart Pelletron Accelerator*
1716   Ryoichi Suzuki (Electrotechnical Laboratory), Yoshinori Kobiyashi, Tomohisa Mikado, Hideaki Ohgaki, Mitsukuni Chiwaki and Tesuo Yamazaki, Tsukuba, *An Intense Pulsed Positron Beam and its Applications*
1735   BREAK

1755   Leave for SLOPOS5 picture, paid Chuck Wagon dinner and show.

Friday 8/7/92 Events
0715 - 0810 Food service in SLOPOS5 lounge

8:15am **T3. Tutorials on Positron Interactions with Solids** (30 +10 min)@. <u>Kelvin Lynn, Chair</u>
0815   Allen Mills, Jr.
0855   Risto Niemenen
0935   BREAK

0945 Leave for 1015 free orchestra rehearsal or cable car ride ($14 adults) in Grand Teton Village, float trips or other options

1200 - 1315 Food service in SLOPOS5 lounge

1:45pm **C. Surface Studies** (15 min talks + 5 min disc.)@.

<u>Dave Gidley, Chair</u>
1345   G.W. Anderson (U. W. Ontario), K.O. Jensen, T.D. Pope, K. Griffiths, P.R.Norton, and P.J. Schultz, *Positron Studies of the Growth and Annealing Properties of Palladium Overlayers on Cu(100)*
1405   A. Goodyear (U. of East Anglia), A. Knights and P.G. Coleman, *Energy Spectrum of Work-Function and Epithermal Positrons from Solid Surfaces*
1425   Akira Ishii (Fritz-Haber Institut der Max-Plank-Gesellschaft), *Adsorbate-Sensitive Ps Formation and Its Application to Determine Atomic Structure of Randomly Adsorbed Adatom*
1445   BREAK

<u>Karl Canter, Chair</u>
1505   Delbert L. Lessor (Pacific Northwest Laboratory), Karl Canter, and C.B. Duke, *Low Energy Positron and Electron Diffraction from Surfaces: What You Learn and How They Differ*

1525 A. Rubaszek (W. Trzebiatowski Institute), A. Kiejna and S. Daniuk, *Annihilation Characterisics for Positrons Trapped at the Surfaces of Simple Metals*
1545 S.J. Wang (Wuhan U.), *Defects and Phase Transitions at Physisorbed Surface Studied by Positron Annihilation*
1605 BREAK

Walter Kauppila, Chair
1625 B.D. Wissman (U. Michigan), W.E. Frieze and D.W. Gidley, *Reemitted Positron Spectroscopy of Cobalt and Nickel Silicide Films*
1645 Alex Weiss (U. Texas Arlington), Ali Koymen, Shu Yang, K.H. Lee, G. Yang, H-Q. Zhou, L-W Chen, and J.H. Kim, *Positron Annihilation Induced Auger Electron Spectroscopy*
1703 N. G. Fazleev (U. Texas Arlington), J.L. Fry, J. H. Kaiser, A.R. Koymen, K. Kuttler, T.D. Niedzwiecki and Alex Weiss, *A Model for the PAES Cu M_{23v} Signal Versus Cs Coverage on the Cu(100) Surface*
1718 K.H. Lee (U. Texas Arlington), Gimo Yang, A.R. Koymen, and A.H.Weiss, *Surface Study of Submonolayer Film of Au on Cu(100) Using Positron Annihilation Induced Auger Electron Spectroscopy*
1733 BREAK

1735 Options including float trips

1800 - 1900 Food service in SLOPOS5 lounge

8:45pm Lecture by Dr. Mark Boyce (U. Wyoming) on *THE GREATER YELLOWSTONE ECOSYSTEM: REDEFINING AMERICA'S WILDERNESS HERITAGE*. Refreshments

Saturday 8/8/92 Events
0715 - 0840 Food service in SLOPOS5 lounge

9:00am **D. Round Table on Intense Positron Beam Facilities**. Chair: Arthur Denison
(~10 min. opening presentations, 2 hours total).

0900 Alexander Artamonov and Y.S. Derbenev (Institute of Complete Electric Drive and U. Michigan), *Magnetization Effect in Electron Cooling of Positrons and Generation of Intense Antihydrogen Beams*
0910 H.C. Griffin (U. of Michigan, Ann Arbor), W.E. Frieze, D.W. Gidley, and R.S. Vallery, *Prospects for Fabricating Very Intense ^{58}Co Sources for Positron Beams*
0920 Jack Kossler (Coll. William and Mary), *Feasibility Studies for Slow Positron and Surface Muon Beams at CEBAF*
0930 K.G. Lynn (BNL), *Progress of the BNL High Flux Positron Beam*
0940 Henry Makowitz (EG&G Idaho, Inc.), *The Intense Slow Positron Source Concept a Theoretical Perspective on a Proposed INEL Facility*
0950 Ben Brown (Mount Holyoke College), H. Makowitz, INEL , D.W. Gidley, H. Griffin, and W. Frieze, U. of Michigan, *A Proposed Intense Positron Source An Experimental Overview*
1000 BREAK

1015 W. Triftshauser (U. der Bundeswehr München), *High Intensity Positron Beam*
1025 A. van Veen (T. U. Delft), H. Schut, P.E. Mijnarends, L. Seijbel, P. Kruit, *The Design of a Nuclear Reactor Based Positron Beam for Materials Analysis*
1035 W. B. Waeber (Paul Scherrer Institute), *The PSI Intense Positron Beam - A Project Status Report*

1045 DISCUSSION
1115 BREAK

11:30am   E. Positron Microscopes - Round Table Discussion (15 min. talk + 5 min. opening statement, 1 hour 15 min. Total).   Lester Hulett, Jr., Chair

1130   K.F. Canter (Brandeis U.), T.M Roach and A. Bacshi, *Principals of Positron Microfocussing for the Positron Annihilation Microprobe and Positron Reemission Microscope*
1145   D.W. Gidley (U. of Michigan), W.E.Frieze, T.L. Dull, G.B. DeMaggio, E.Y. Yu, H.C. Griffin, M. Skalsey, R.S. Vallery, and B.D. Wissman, *An Overview of the Michigan Positron Microscope Program*
1200   A. Goodyear (U. of East Anglia), and P.G. Coleman, *The UEA Positron Reemission Microscope - A Progress Report*
1215   ROUND TABLE DISCUSSION

1245 Options, including lunch in SLOPOS5 lounge and float trips

3:30pm - 5:00pm **Attended-Poster Session I**. Refreshments

5:05pm **F. Future Uses of Positrons** (~20 min + 5 min)@. E. H. Ottewitte, Chair

1705   B. Augenstein (RAND Corp), *Prospects and Utility of Antihydrogen in Large Amounts*
1730   H. Boehmer (U.Cal. at Irvine), *Laboratory Formation of Electron-Positron Plasmas*

1800   Options
1800 - 1900   Food service in SLOPOS5 lounge

Sunday 8/9/92 Events
0715 - 0815 Food service in SLOPOS5 lounge

Options including float trips, Yellowstone day trips, church services

1130 - 1300 Food service in SLOPOS5 lounge
1715 - 1830 Food service in SLOPOS5 lounge

7:00pm **Attended-Poster Session II**. Refreshments

8:30pm **G. Collaboration and Small Group Discussions**

Monday 8/10/92 Events
0715 - 0810 Food service in SLOPOS5 lounge

T. Akahane, Chair
8:30 **T4. Tutorial on Positron Optics** (30min +10min) - Karl Canter

9:10am **H. Slow Positron Beams Arts and Techniques** (15 min talks + 5 min disc.)@

0910   G. Amarendra (Indira Ganhi Center for Atomic Research), and B. Viswanathan, K.V. Thomas Kutty, B. Purniah and A. V. Rao, *Slow Positron Beam Set Up at Kalpakkan - A Progress Report*

0930   C.D. Beling (U. of Hong Kong), S. Fung, H.M. Weng, C.V. Reddy, S. W. Fan, Y.Y. Shan, C. C. Ling, *The Slow-Positron Beam Facility at the University of Hong Kong*
0950   BREAK

S.J. Wang, Chair
1010   B. Ghaffari (U. of Michigan), R.S. Conti, T.D. Steiger, *Accumulation and Bunching of Positrons*
1030   Leon Seijbel (T. U. Delft), J. E. Barth, P. Kruit, A. van Veen, H. Schut, *Optical Design for a Remoderation System*
1050   Hikaru Ueno (Ishikawajima-Harima Heavy Industries Co., Ltd.), Osamu Azuma, Takafumi Ogawa, Takashi Sato, Tatsumi Kawartani and Okitada Hara, *The Construction of the Slow Positron Beam System as a Commercial Prototype*
1110   Jun Xu (Oak Ridge National Lab.), Lester Hulett, Jr. and T.A. Lewis, *The ORNL Slow Positron Facility and Quadratic-Potential Time-of-Flight Mass Spectrometer Instrumentation*
1130   BREAK

1200 -1315 Lunch

1:45pm **I. Fundamental Physics Studies** (15 min talks + 5 min disc.)@

Paul Coleman, Chair
1345   L. Albrecht (U. of East Anglia), and P.G. Coleman, *Total Positron Backscattering Coefficients*
1405   G.R. Massoumi (U.W.Ontario), N. Hozhabri, W.N. Lennard, M.S. Lorenz and P.J. Schultz, *Positron and Electron Backscattering from Solids*
1425   P. Asoka-Kumar (BNL 0, J. C. Palathingal, U. of Puerto Rico, and K.G. Lynn, BNL, *Single Quantum Annihilation of Positrons*
1445   BREAK

Walter Kauppila, Chair
1505   L. Dou (Wayne State U).W. E. Kauppila, C. K. Kwan and T. S. Stein, *Observation of Resonance-Like Structure in Positron Krypton Elastic Scattering*
1525   R.H. Howell (LLNL), T. Cowan, R. Rohatgi, B. Beck, J.L. McDonald, J.H. Hartley, *Present Status of the LLNL Positron-Electron Scattering Experiment*
1545   L.D. Hulett, Jr. (Oak Ridge Nat. Lab.), Jun Xu, T.A. Lewis, D.L. Donohue, G.L. Glish, S.A. McLuckey, *The Ionization of Large Organic Molecules by Slow Positrons at Energies Above and Below the Positronium Formation Threshold*
1605   BREAK

L. Dorikens, Chair
1625   Finn Jacobsen (Univ. of Aarhus), *On the Formation of PsH in Collisions Between e^+ and CH_4*
1645   D.M. Schrader (Marquette U.), *The Formation of Positronium Compounds in Positron Beams*
1705   Sergey V. Stepanov, ITEP, Moscow, *Energy Loss of Subionizing Electrons (Positrons) in Dielectric Liquids*
1725   M. Weber, U. Bielefeld, W. Sperber, A. Hofmann, W. Raith, and K.G. Lynn, *Positronium Formation in Positron Atomic Hydrogen Collisions*

1800 - 1900 Food service in SLOPOS5 lounge

7:30pm **Closing Get-Together.** Refreshments, collaboration and small group discussions, poster demounting ceremonies, etc.

Tuesday 8/11/92 Events
No food service or use of convention rooms.
Limited coordination of float and bus trips and airport transport.

Posters

(Authors should try to be present in the next one or two breaks following the announcement of their poster and in at least one of the Attended-Poster Sessions I or II)

A1P. Defects - Near Surface

1. Neil B. Chilton (Nippon Steel Corporation), and Masonori Fujinami, *A Slow Positron Investigation of Ion Implanted Silicon Wafers*
2. P. Maguire (Univ. of Western Ontario), A. Halec, P.J. Simpson, P.J. Schultz, T.E. Jackson, and P. Marshall, *Positron Stopping in Germanium*
3. K. A. Ritley (U. of Illinois at Urbana-Champaign), K.G. Lynn, V Ghosh, and D.O. Welch, *A New Calculation of Positron Implantation Profiles*
4. H. Schut (T.U.Delft), A. van Veen, R.A. Hakvoort, M.J.W. Greuter and L. Nielsen, *The Effect of Channeling on the Defect Depth Distribution in 110 keV Rb Implanted Poly-W*
5. P.J. Simpson (BNL), P.J. Schultz, S. Tong Lee, S. Chen and G. Braunstein, Kodak, *Void Formation in Si - Implanted GaAs*
6. Cs. Szeles (BNL), P. Asoka-Kumar, K.G. Lynn, E.H. Poindexter, *Variable Energy Positron Beam Study of Charge Density Effects at SiO_2/Si Interfaces*
7. A. Uedono (U. of Tokyo), S. Watauchi, Y. Ujihira, L. Wei, S. Tanigawa, R. Suzuki, H. Ohgaki, T. Mikado, H. Kametani, H. Akiyama, Y. Yamguchi, and M. Koumaru, *Characterization of SiO_2 by Monoenergetic Positron Beams*
8. L. Wei (Univ. of Tsukuba), S. Tanigawa, Y. Hiroyama, T. Motooka and T. Tokuyama, *Defect in Amorphous Silicon Prepared by Ion Implantation*
9. L. Wei (Univ. of Tsukuba), S. Tanigawa, Y. Jia, A. Yamada and M. Konagai, *Investigation of Vacancy-Related Defects in Heavily Phosphorus-Doped SiP Grown by Plasma Chemical Vapor Deposition*

A2P. Defects - Bulk

1. G. Brauer (Forschungszentrum Rossendorf), Th. Daniel, W. Faust, Z. Michno, H. Schneider, *Positron Lifetime Measurements in γ-Irradiated Polyethylene under Different Conditions*
2. Jerzy Dryzek (Institute for Nuclear Physics, Cracow), *Calculations of Positron Resonant Trapping*.
3. N. Hozhabri (U. Texas at Arlington), S.C.Sharma, R.N. Pathak, K. Alavi, J.Y. Ma, *Variable Energy Positron Beam and Infrared Transmission Measurements on GaAs Epilayers Grown at Low Temperatures by Molecular Beam Epitaxy*
4. K. Kristiakova (Intitute of Physics, Slovak Academy of Sciences), J. Kristiak, O. Sausa, M. Morhac, *The Correlation Between Lifetime and Momentum of e+ - e- Pairs*
5. G. Kontrym-Sznajd (W. Trzebiatowski Institute, Poland), and A. Rubaszek, *On the Interpretation of ACPAR Spectra with Respect to Electron Momentum Density in Real Metals*

CP. Surface Studies

1. C.E. Haynes and P.C. Rice-Evans (U. London), *Positronium at a Nitric Oxide Monolayer on Graphite*
2. Y.C. Jean (U. of Missouri), G.H. Dai, H. Shi, R. Suzuki, Y. Kobayashi, *A Slow Positron Study of an Epoxy Polymer*
3. J.H. Kim (U. Texas Arlington), G.Yang, S.Yang, K.H. Lee, A.R. Koymen and A.H. Weiss,

Temperature Dependence of the Surface Composition of an Au/Cu(100) System
4. Mogens R. Poulsen (Aarhus Univ.), *Ps Formation from Silver*
5. Ryoichi Suzuki (Electrotechnical Laboratory), Yoshinori Kobayashi, Tomohisa Mikado, Hideaki Ohgaki, Mitsukuni Chiwaki, and Tetsuo Yamazaki, Tsukuba, *Positron Lifetime in Thin Films Studied by a Pulsed Positron Beam*
6. G. Yang (U. Texas Arlington), S. Yang, J.H. Kim, K.H. Lee, A.R. Koymen, and A.H. Weiss, *Positron Annihilation Induced Auger Electron Spectroscopy (PAES) Study of the Structure of the Rh/Ag(100) Su*
7. H.Q. Zhou (U. Texas Arlington), Shu Yang, A.R. Koymen, and A.H. Weiss, *Electrostatic Positron Beam Design for Use in High Resolution Auger Line Shape Studies*

EP. Positron Microscopes

1. K.F. Canter (Brandeis U.), V. Dharmavaram, A.G. Simrnov, S.A. Wesley, K.H. Wong, and R. Xie, G.R. Brandes and A.P. Mills,Jr., AT&T Bell Labs, *Brandeis Second Generation Positron Reemission Microscope*
2. G.B. DeMaggio (U. of Michigan), T.L. Dull, E.Y. Yu, W.E. Frieze, and D.W. Gidley, *The Positron Reemission Microscope at Michigan*
3. T.L. Dull (U. of Michigan), G.B. DeMaggio, E.Y. Yu, W.E. Frieze and D.W. Gidley, U. of Michigan, *The Scanning Positron Microscope at Michigan*
4. A. Goodyear (U. of East Anglia), and P.G. Coleman, *The UEA Positron Reemission Microscope - A Progress Report*
5. E.Y. Yu (U. of Michigan), W.E. Frieze, D.W. Gidley, and M. Skalsey, *Positron Microscopy of Biological Samples*

HP. Slow Positron Beams Arts and Techniques

1. Takahi Akahane (National Institute for Research in Inorganic Materials), *Extraction of Slow Positrons from the Magnetic Field*
2. G. Amarendra (Indira Gandhi Center for Atomic Research), and B. Viswanathan, *Electrostatic Lenses and Beam Optics Calculations*
3. Benjamin L. Brown (Mount Holyoke College), Sonal Parikh, Sujata Vemuri, Tamara S. Andrew, Margaret S. Clarkson, Sarah K Makoski, *Heating of a Thin Tungsten Foil for Efficient Moderation of Positrons in a 4.2K Environment*
4. Abe Eftekhari (Hampton U., Virginia), Trina C. Veals, Anne St. Clair, Diane M. Stoackly, Danny R. Sprinkle, and Jag J. Singh, *Free Volume Model for Dielectric Constant of Polymer Films*
5. Pedro A. Encarnacion (INEL), Benjamin L. Brown, Mount Holyoke, Henry Makowitz, *INEL SIMION PC Calculations of Various ISPS First-Stage Optics Concepts*
6. A. Enomoto, H. Kobayashi, T. Kurihara, K. Nakahara, T. Shidara and A. Shirakawa, *Status of a Slow-Positron Source Using the Photon Factory Electron Linac*
7. Masonori Fujinami (Nippon Steel Corporation), and Neil B. Chilton, *Design and Construction of a Slow Positron Beam at Nippon Steel Corporation*
8. T.C. Leung (U. of Western Ontario), P.J. Simpson, I. Schmidt, P. Perguin, and P.J. Schultz, *A New Magnetic Positron-Beam Facility for Defect Profiling*
9. J.P. Merrison (U. College London), M. Charlton, B.I. Deutch, and I.V. Jurgensen, *Field Assisted Positron Extraction by Surface Charging of Rare Gas Solids*
10. D. Segers (Positron Centre, Lab. Nuclear Physics, Belgium), M. Dorikens, J. Paridaens and L. Dorikens-Vanpraet, *Monte Carlo Simulation of the Positron Production at the LINAC-Based Slow Positron Beam*
11. Ming Shi (Paul Scherrer Institute), *Positron Confinement and Slowing Down - The Premoderation Step in the PSI Beam Project*
12. R. S. Vallery (U. of Michigan), W.E. Frieze, D. W. Gidley, H.C. Griffin, W.A. Loinaz, R. Makidon, M. Skalsey, *Fabrication of 1-Curie ^{58}Co Positron Sources*

13. S.J. Wang (Wuhan U.), *Slow Positron Beam in China*
14. P. Willutzki (U. Der BundesWehr München), J. Stormer, D.T. Britton, G. Kogel, P. Sperr, R. Steindl and W. Triftshauser, *Retarding Field and Timing Measurements of Positrons from a W[100] Foil*

IP. Fundamental Physics Studies

1. L. Jiang (Wayne State U.), W. E. Kauppila, C.K. Kwan, S.P. Parikh, and S. Zhou, *Measurements of Positronium Formation Cross Sections for Positron-Potassium Atom Scattering*
2. G. O. Jones (U. College London), M. Charlton, J.A. Slevin, G. Laricchia, A. Kover, M.R. Poulsen, and S. Nic Chormaic, *Ionization of Atomic Hydrogen by Positrons*
3. D. Przybyla (Wayne State U.), C.K. Kwan, R.A. Lukaszew, W.E. Kauppila and T. S. Stein, *Search for Resonance-Like Structure in the Total Scattering of Positrons by Argon and Krypton*
4. H. Schneider (U. Giessen), I. Tobehn, M. Ruckert, U. Brinkmann, R. Hippler, *Experiments with Low-Energy Positrons*
5. S. Zhou (Wayne State U.), L. Jiang, W.E. Kauppila, C.K. Kwan , S.P. Parikh, and T.S. Stein, *Measurements of Positronium Formation Cross Sections for Positron-Argon Scattering*

We would like to acknowledge the support of EG&G Idaho, Inc., U. Texas at Arlington, Idaho State University, du Pont Radiopharmaceuticals, Kosciuszko Foundation, American Physical Society, U. Wyoming, Idaho National Engineering Laboratory and the U.S. Department of Energy in sponsoring this workshop.

Author Index

A

Abbé, J. Ch., 179
Abrashoff, J. D., 305
Aers, G. C., 13, 53
Akahane, T., 437
Akiyama, H., 101
Alavi, K., 58
Albano, R. K., 305
Amarendra, G., 441, 452
Anderson, G. W., 193
Andrew, T. S., 200, 480
Arkuszewski, J., 506
Augenstein, B. W., 1, 401
Awazu, K., 84
Azuma, O., 535

B

Bandžuch, P., 150
Barth, J. E., 502
Beling, C. D., 462
Berntsen, A. J. M., 48
Bharuth-Ram, K., 179
Billard, I., 179
Boehmer, H., 422
Brandes, G. R., 385
Brauer, G., 123
Brinkmann, U., 574
Britton, D. T., 20, 542
Brown, B. L., 200, 289, 480
Bundy, K., 200

C

Canter, K. F., 246, 385
Chilton, N. B., 25, 31
Chiwaki, M., 84, 526
Clarkson, M. S., 200, 480
Coleman, P. G., 218
Connell, S. H., 179
Conti, R. S., 487

D

Dai, G. H., 129
Daniel, Th., 123
Daniuk, S., 249
DeMaggio, G. B., 391
Denison, A. B., 1, 200, 289
Dharmavaram, V., 385
Dorikens, M., 496
Dorikens-Vanpraet, L., 496
Dou, L. 559
Duke, C. B., 246
Dull, T. L., 391
Duplâtre, G., 179

E

Eftekhari, A., 128, 516
Encarnación, P., 200, 289

F

Fan, S. W., 462
Faust, W., 123
Fazleev, N. G., 208
Foroughi, F., 506
Frieze, B., 289
Frieze, W. E., 264, 391
Fry, J. L., 208
Fujii, S., 92
Fujinami, M., 25, 31
Fung, S., 462

G

Gerola, D., 365
Ghaffari, B., 487
Ghosh, V. J., 1, 37, 64
Gidley, D. W., 264, 289, 391
Goodyear, A., 218
Gopinathan, K. P., 452
Greer, A. J., 296

Greuter, M. J. W., 73
Griffen, H. C., 289, 391
Griffiths, K., 193

H

Hakvoort, R. A., 48, 73
Halec, A., 53
Hara, O., 535
Haricharun, H., 179
Haynes, C. E., 223
Hegedüs, F., 365
Hippler, R., 574
Hiroyama, Y., 107
Hozhabri, N., 58
Hulett, L. D., Jr., 296

I

Ishii, A., 227

J

Jackman, T. E., 53
Jacobsen, F. M., 1
Jean, Y. C., 129
Jensen, K. O., 13, 193, 564
Jia, Y., 113
Jiang, L., 562, 599

K

Kaiser, J. H., 208
Kametani, H., 101
Kauppila, W. E., 559, 562, 571, 599
Kawaratani, T., 535
Kiejna, A., 249
Kim, J. H., 234, 274
Knights, A. P., 218
Kobayashi, Y., 84, 129, 526
Koch, M., 179
Kögel, G., 542
Konagai, M., 113
Kontrym-Sznajd, G., 140

Kossler, W. J., 296
Koumaru, M., 101
Koymen, A. R., 208, 234, 239, 274, 279
Krištiak, J., 150
Krištiaková, K., 1, 150
Kruit, P., 354, 502
Kutty, K. V. T., 452
Kwan, C. K., 559, 562, 571, 599

L

Landman, W. H., 305
Larson, J. D., 305
Lauff, U., 179
Lee, K. H., 234, 239, 274
Lennard, W. N., 564
Lessor, D. L., 246
Lewis, T. A., 550
Ling, C. C. 462
Lukaszew, R. A., 571
Lynn, K. G., 1, 37, 64

M

Maguire, P., 53
Maier, K., 179
Major, J., 179
Makoski, S. K., 480
Makowitz, H., 200, 289, 305
Marshall, P., 53
Massoumi, G. R., 564
Matsuda, A., 84
Melker, A. I., 156
Michno, Z., 123
Mijnarends, P. E., 354
Mikado, T., 84, 101, 526
Mills, A. P., Jr., 335, 385
Morháč, M., 150
Motooka, T., 107

N

Niedzwiecki, T. D., 208
Nieminen, R. M., 1
Niesen, L., 73

Norton, P. R., 193

O

Ogawa, T., 535
Ohgaki, H., 84, 101, 526
Ottewitte, E. H., 1

P

Paridaens, J., 496
Parikh, S. P., 480, 562, 599
Pathak, R. N., 58
Pope, T. D., 193
Przybyla, D., 571
Purniah, B., 452

R

Reader, A. H., 48
Reddy, C. V., 462
Rice-Evans, P. C., 78, 223
Ritley, K. A., 1, 64
Roellig, L. O., 365
Roorda, S., 48
Rubaszek, A., 140, 249
Rückert, M., 574

S

Sato, T., 535
Šauša, O., 150
Schneider, H., 123, 574
Schrader, D. M., 579
Schultz, P. J., 53, 193, 564
Schut, H., 48, 73, 354, 502
Seeger, A., 179
Segers, D., 496
Seijbel, L. J., 354, 502
Sellschop, J. P. F., 179
Shan, Y. Y., 462
Sharma, S. C., 58
Shaw, N., 78
Shi, H., 129

Shi, M., 365, 506
Sideras-Haddad, E., 179
Simpson, P. J., 1, 53
Singh, J. J., 128, 516
Skalsey, M., 391
Smirnov, A. G., 385
Smith, C., 78
Sperr, P., 542
Sprinkle, D. R., 128
St. Clair, A., 128
St Clair, T. L., 516
Steiger, T. D., 487
Stein, T. S., 559, 562, 571, 599
Steindl, R., 542
Stepanov, S. V., 587
Stockly, D. M., 128
Stolk, P. A., 48
Stoll, H., 179
Störmer, J., 542
Sutton, C. S., 200
Suzuki, R., 84, 101, 129, 526

T

Tajima, T., 305
Tanigawa, S., 84, 92, 101, 107, 113
Taqqu, D., 365, 506
Tobehn, I., 574
Tokuyama, T., 107
Triftshäuser, W., 542

U

Uedono, A., 84, 92, 101
Ueno, H., 535
Ujihira, Y., 101

V

van den Boogaard, M. J., 48
van Veen, A., 48, 73, 354, 502
Vallery, R. S., 391
Vemuri, S., 480
Venugopal Rao, G., 452
Viswanathan, B., 441, 452

W

Waeber, W. B., 365, 506
Walker, A. B., 13, 564
Wang, S. J., 259
Watauchi, S., 101
Wei, L., 92, 101, 107, 113
Weiss, A. H., 1, 208, 234, 239, 274, 279
Welch, D. O., 37, 64
Weng, H. M., 462
Wesley, S. A., 385
Wesolowski, P., 179
Willutzki, P., 542
Wissman, B. D., 264, 391
Wong, K. H., 385

X

Xie, R., 385

Xu, J., 550

Y

Yamada, A., 113
Yamaguchi, Y., 101
Yamazaki, T., 84, 526
Yang, G., 234, 239, 274
Yang, S., 234, 274, 279
Yang, S. J., 259
Yu, E. Y., 391

Z

Zhou, H. Q., 279
Zhou, S., 562, 599
Zimmerman, U., 365

AIP Conference Proceedings

		L.C. Number	ISBN
No. 276	Very High Energy Cosmic-Ray Interactions: VIIth International Symposium (Ann Arbor, MI, 1992)	93-71342	1-56396-038-9
No. 277	The World at Risk: Natural Hazards and Climate Change (Cambridge, MA, 1992)	93-71333	1-56396-066-4
No. 278	Back to the Galaxy (College Park, MD, 1992)	93-71543	1-56396-227-6
No. 279	Advanced Accelerator Concepts (Port Jefferson, NY, 1992)	93-71773	1-56396-191-1
No. 280	Compton Gamma-Ray Observatory (St. Louis, MO, 1992)	93-71830	1-56396-104-0
No. 281	Accelerator Instrumentation Fourth Annual Workshop (Berkeley, CA, 1992)	93-072110	1-56396-190-3
No. 282	Quantum 1/f Noise & Other Low Frequency Fluctuations in Electronic Devices (St. Louis, MO, 1992)	93-072366	1-56396-252-7
No. 283	Earth and Space Science Information Systems (Pasadena, CA, 1992)	93-072360	1-56396-094-X
No. 284	US-Japan Workshop on Ion Temperature Gradient-Driven Turbulent Transport (Austin, TX, 1993)	93-72460	1-56396-221-7
No. 285	Noise in Physical Systems and 1/f Fluctuations (St. Louis, MO, 1993)	93-72575	1-56396-270-5
No. 286	Ordering Disorder: Prospect and Retrospect in Condensed Matter Physics: Proceedings of the Indo-U.S. Workshop (Hyderabad, India, 1993)	93-072549	1-56396-255-1
No. 287	Production and Neutralization of Negative Ions and Beams: Sixth International Symposium (Upton, NY, 1992)	93-72821	1-56396-103-2
No. 288	Laser Ablation: Mechanismas and Applications-II: Second International Conference (Knoxville, TN, 1993)	93-73040	1-56396-226-8
No. 289	Radio Frequency Power in Plasmas: Tenth Topical Conference (Boston, MA, 1993)	93-72964	1-56396-264-0

No. 290	Laser Spectroscopy: XIth International Conference (Hot Springs, VA, 1993)	93-73050	1-56396-262-4
No. 291	Prairie View Summer Science Academy (Prairie View, TX, 1992)	93-73081	1-56396-133-4
No. 292	Stability of Particle Motion in Storage Rings (Upton, NY, 1992)	93-73534	1-56396-225-X
No. 293	Polarized Ion Sources and Polarized Gas Targets (Madison, WI, 1993)	93-74102	1-56396-220-9
No. 294	High-Energy Solar Phenomena A New Era of Spacecraft Measurements (Waterville Valley, NH, 1993)	93-74147	1-56396-291-8
No. 295	The Physics of Electronic and Atomic Collisions: XVIII International Conference (Aarhus, Denmark, 1993)	93-74103	1-56396-290-X
No. 296	The Chaos Paradigm: Developments an Applications in Engineering and Science (Mystic, CT, 1993)	93-74146	1-56396-254-3
No. 297	Computational Accelerator Physics (Los Alamos, NM, 1993)	93-74205	1-56396-222-5
No. 298	Ultrafast Reaction Dynamics and Solvent Effects (Royaumont, France, 1993)	93-074354	1-56396-280-2
No. 299	Dense Z-Pinches: Third International Conference (London, 1993)	93-074569	1-56396-297-7
No. 300	Discovery of Weak Neutral Currents: The Weak Interaction Before and After (Santa Monica, CA, 1993)	94-70515	1-56396-306-X
No. 301	Eleventh Symposium Space Nuclear Power and Propulsion (3 Vols.) (Albuquerque, NM, 1994)	92-75162	1-56396-305-1 (Set) 156396-301-9 (pbk. set)
No. 302	Lepton and Photon Interactions/ XVI International Symposium (Ithaca, NY, 1993)	94-70079	1-56396-106-7
No. 304	The Second Compton Symposium (College Park, MD, 1993)	94-70742	1-56396-261-6
No. 305	Stress-Induced Phenomena in Metallization Second International Workshop (Austin, TX, 1993)	94-70650	1-56396-251-9
No. 306	12th NREL Photovoltaic Program Review (Denver, CO, 1993)	94-70748	1-56396-315-9